GLOBAL INDUSTRY, LOCAL INNOVATION: THE HISTORY OF CANE SUGAR PRODUCTION IN AUSTRALIA, 1820-1995

GLOBAL INDUSTRY, LOCAL INNOVATION: THE HISTORY OF CANE SUGAR PRODUCTION IN AUSTRALIA, 1820–1995

PETER D. GRIGGS

PETER LANG
Bern · Berlin · Bruxelles · Frankfurt am Main · New York · Oxford · Wien

Bibliographic information published by Die Deutsche Nationalbibliothek
Die Deutsche Nationalbibliothek lists this publication in the Deutsche Nationalbibliografie; detailed bibliographic data is available on the Internet at ‹http://dnb.d-nb.de›.

British Library Cataloguing-in-Publication Data: A catalogue record for this book is available from The British Library, Great Britain

Library of Congress Cataloging-in-Publication Data

Griggs, P. D. (Peter D.)
Global industry, local innovation : the history of cane sugar production in Australia, 1820-1995 / Peter D. Griggs.
p. cm.
Includes bibliographical references.
ISBN 978-3-0343-0431-3
1. Sugar trade–Australia–History–19th century. 2. Sugar trade–Australia–History–20th century. 3. Sugar growing–Australia–History–19th century. 4. Sugar growing–Australia–History–20th century. 5. Sugar–Manufacture and refining–Australia–History–19th century. 6. Sugar–Manufacture and refining–Australia–History–20th century. 7. Sugarcane–Australia–History–19th century. 8. Sugarcane–Australia–History–20th century. 9. Australia–Economic conditions–19th century. 10. Australia–Economic conditions–20th century. I. Title.
HD9118.A8G75 2010
338.1'73610994–dc22

2010045491

Cover illustration: Cover Illustration of *The Australian cane sugar industry,* issued by the sugar industry organisations, 1938.
Cover design: Thomas Jaberg, Peter Lang AG

ISBN 978-3-0343-0431-3

Hochfeldstrasse 32, CH-3012 Bern, Switzerland
info@peterlang.com, www.peterlang.com, www.peterlang.net

For Tracey and in memory of Emma J. Fells

Contents

Acknowledgements

The research and writing for this volume has been conducted over the past fifteen years, being sandwiched between my normal teaching commitments. It has involved a journey to Scotland in 1995 and numerous journeys to libraries and archives in Canberra, Brisbane, Sydney and Melbourne and to the sugar-producing districts throughout eastern Australia. In a work of such duration and complexity, I have contracted far more debts than can be acknowledged in detail. Many people have contributed freely with material, information and advice and by discussion and criticism. Their encouragement and shared interest have assisted greatly, and I here record my warm thanks to all of them.

James Cook University has supported the work through financial assistance and the granting of four periods of study leave. The maps and diagrams were drawn by Adella Edwards in the Cartography Office, School of Earth and Environmental Sciences, James Cook University. Her skill in turning my hand-drawn draft maps into such appealing finished maps never ceases to amaze me. The librarians at the Cairns Campus have been so supportive, especially Vaurian Lewin and Beth Barnes who managed the inter-campus loans, and Lyn Clark and Janine Meakin who arranged for the numerous inter-library loans. Jean Main, Ann Bolton, Sandi Robb and Tara Day were very helpful research assistants during the project.

This book would not have been possible without access to the archival records of CSR Ltd. I am extremely indebted to CSR Ltd. for allowing me to read through the firm's pre-1980 historical records and for permission to reproduce illustrative material which appears as Figure 6.5 and Plates 7.3, 8.3, 8.6, 11.1, 12.1, 12.2, 12.3, 15.2 and 16.3. Several other sugar milling companies also granted access to their records and provided hospitality during research visits. I gratefully acknowledge the assistance provided by the Proserpine Cooperative Sugar Milling Association Ltd., Mossman Central Mill Company Ltd. and the Mackay Sugar Cooperative Association Ltd. In addition, the New South Wales Cane Growers' Association allowed access to their historical records, enabling a New South Wales perspective to be added to this book.

I owe thanks to the following institutions and their staffs for the use of material in their custody and for help over the years: in Victoria, the State Library of Victoria, the ANZ Group Archive, the National Australia Bank Group Archives, Melbourne University Archives and the Public Record

Office Victoria; in New South Wales, the State Records Authority of New South Wales, the State Library of New South Wales and Southern Cross University Library; in Queensland, the Fryer Memorial Library, The University of Queensland, Queensland State Archives, Queensland Department of Primary Industries Library and the Queensland Department of Natural Resources and Water Library; in Scotland, Glasgow University Archives; and in Canberra, the National Library, National Archives of Australia and the Noel Butlin Archives Centre, Australian National University. In particular, I am very indebted to the following current and former staff at the Noel Butlin Archives Centre: Penny Pemberton, Maureen Purtell, Emma Jolley, Tatiana Antsoupova, Mary Paton and Sarah Lethbridge. These archivists over the years retrieved countless boxes of CSR records from 'the tunnel' and provided warm welcomes and hot cups of tea to a visitor from Queensland who often ended up in Canberra during its coldest months. Deborah Martin, Librarian and Records Archivist, at BSES Limited, Brisbane, assisted greatly by providing biographical details on past scientists who worked for the Queensland Bureau of Sugar Experiment Stations.

Chapter Six contains part of my 1997 paper 'The origins and development of the small cane farming system in Queensland, 1870–1915' that originally appeared in the *Journal of Historical Geography* and my 2000 paper 'Sugar plantations in Queensland, 1864–1912' that originally appeared in *Agricultural History*. Sections of Chapter 7 appeared originally in my 2003 paper 'Australian scientists, sugar cane growers and the search for new gummosis-resistant and sucrose-rich varieties of sugar cane, 1890–1920' that was published in the *Historical Records of Australian Science* and my 2004 paper 'Improving agricultural practices: science and the Australian canegrowers, 1864–1915' that was published in *Agricultural History*. Chapter 9 includes part of my 2001 paper 'The decline of competition: the emergence of a duopoly in the Australian sugar refinery sector, 1841–1915' that was originally published in the *Australian Geographer*. The detail on soil erosion in Chapter 12 comes from my 2005 paper 'Soil erosion, scientists and the development of conservation tillage techniques in the Queensland sugar industry, 1935–1995', which originally appeared in *Environment and History*. Information about the attempts to control cane grubs in Chapter 14 originally appeared in my 2005 paper, 'Entomology in the service of the state: Queensland scientists and the campaign against cane beetles, 1895–1950' and the methods developed to control cane diseases discussed in Chapter 15 was published in my 2007 paper, 'Defeating cane diseases: plant pathologists and the development of disease control strategies in the Australian sugar industry, 1920–1950'. Both were published in the *Historical Records of Australian Science*. I am grateful to the editors of the following journals and copyright owners for permission to reproduce this material: Agricultural History Society (*Agricultural History*); The White Horse Press (*Environment and History*); Elsevier (*Journal of Historical Geography*); Geographical Society of New South Wales (*Australian Geographer*); and CSIRO Publishing (*Historical Records of Australian Science*).

I gratefully acknowledge the permission of the following organisations to reproduce their illustrative material as photographs in this book: Part 1 photograph; Mitchell Library, State Library of New South Wales; Plates 8.2 and 8.4; Mackay Regional Council Libraries; Plates 5.1, 6.2, 7.1, 8.1, 8.7, 16.1 and 18.1; John Oxley Library, State Library of Queensland; Part 2 photograph and Plate 6.1; Fryer Library, The University of Queensland; Plates 7.2, 9.1 and 9.2; State Library of Victoria; Part 3 photograph and Plates 12.4 and 16.5; National Library of Australia; Plate 11.4; National Archives of Australia; Plates 11. 7 and 18.4; CANEGROWERS, Brisbane; Plates 7.5, 11.5, 11.6, 12.5, 13.2, 13.3, 14.1, 14.3, 14.4, 15.1, 15.5 and 17.5; BSES Limited; Plate 16.2; Gold Coast City Council Local Studies Library; Plates 16.6, 17.1 and 17.2; Sunshine Coast Libraries; Plates 16.7, 16.9, 16.10, 16.11, 17.4, 17.6, 17.7, 18.2 and 18.5; Rural Press Queensland Pty. Ltd.; Plate 16.8; Burdekin Shire Library; Plate 18.3; Queensland Department of Transport and Main Roads; and Plate 17.3; Australian Cane Farmers Association Limited.

Several colleagues have read different parts of this book. I am particularly indebted to Professor John Weaver, McMaster University, for his encouragement over the years, his very helpful comments on the entire manuscript and his suggestion about the book's title. I am also very appreciative of the comments provided by the following academics and scientists over the years: Professor Emeritus Jock Galloway, University of Toronto; Professor Emeritus Phillip Courtenay, James Cook University; Dr Mac Hogarth, current Editor of the *Proceedings of the International Society of Sugar Cane Technologists* and former Group Manager for Cane Breeding and Improvement, Queensland Bureau of Sugar Experiment Stations; Mr Brian Egan, the former Coordinator of the Cane Pest and Disease Control Boards in Queensland and pathologist with the Queensland Bureau of Sugar Experiment Stations; and Dr Ralph Shlomowitz, Flinders University.

I am extremely grateful to Daniela Christen at Peter Lang AG, who initially showed such enthusiasm for the book. More recently, I have been ably guided by Katrin Forrer who has seen the project through to completion. Her advice and guidance has been much appreciated.

Families ultimately become involved in the research for a book. Mine have been no exception. My special thanks are extended to my parents, Jack and Veronica Griggs, and my father-in-law, John Rowan. They answered numerous questions about the historical development of the Australian sugar industry between 1930 and 1985, drawing upon their experiences as owners of a cane farm and as an employee of Pioneer Sugar Mills Pty. Ltd., respectively.

My greatest thanks, however, are to my partner Tracey, and our children Thomas, Llewellyn and Emma, who accepted the absence of their father and partner during his many trips away and who survived the ordeal of the last eighteen months of virtual non-stop writing with such generosity and patience.

P.D.G.
Cairns

A note on statistics and currency

Metric measures have been used in this book, except where there is significance in the Imperial units. No attempt, however, was made to convert predecimal Australian currency units into modern day equivalents. All currency units are expressed in Australian pounds (£), shillings (s), and pence (d), unless stated otherwise.

A note on geographical boundaries and place names

Throughout this book, reference is made to specific regions in Queensland where sugar cane is cultivated. Definitions of these regions are as follows: North Queensland is Mackay northwards; Far North Queensland is Tully northwards to Mossman; Central Queensland encompasses the Proserpine and Mackay districts; Southern Queensland is Bundaberg southwards; and the Wide Bay-Burnett region included the Bundaberg, Isis (or Childers) and Maryborough districts.

Modern spellings of place names and modern place names have been used throughout the book.

Abbreviations in footnotes and table and figure captions

ABARE	Australian Bureau of Agricultural and Resource Economics
AC	*Australian Canegrower*
ADB	*Australian Dictionary of Biography*
AEMCo.	Australian Estates and Mortgage Company Limited
AER	Australian Estates and Mortgage Company Limited Records
AONSW	Archives Office of New South Wales, Sydney (now State Records Authority of New South Wales)
ANZA	ANZ Group Archive, Melbourne
APCQ	Air Pollution Council of Queensland
AR	Annual Report
ASMC	Australian Sugar Milling Council
ASSCT	Australian Society of Sugar Cane Technologists
ASJ	*Australian Sugar Journal*
ATCJ	*Australian Town and Country Journal*
BC	*Brisbane Courier*
BCSM	Bureau of Central Sugar Mills, Queensland.
BOA	Bank of Australasia
BOAR	Bank of Australasia Records
BRC	Business Records Centre, Glasgow University Archives
BSES	Queensland Bureau of Sugar Experiment Stations
CBCS	Commercial Banking Company of Sydney
CBCSR	Commercial Banking Company of Sydney Records, National Australia Bank Group Archives, Melbourne
CGQB	*Cane Growers' Quarterly Bulletin*
CRE	*Clarence and Richmond Examiner*
CPP	*Commonwealth Parliamentary Papers*
CSR	Colonial Sugar Refining Company
CSRR	Colonial Sugar Refining Company Records
Doc.	Document
Ev.	Evidence
FISCC	Fruit Industry Sugar Concession Committee
HRA	*Historical Records of Australia*
HRNSW	*Historical Records of New South Wales*
HRV	*Historical Records of Victoria*
IAC	Industries Assistance Commission

Insp.	Inspector
ISSCT	International Society of Sugar Cane Technologists
IWSC	Irrigation and Water Supply Commission, Queensland.
J.	Journal
JCUA	James Cook University Archives, Townsville
JOL	John Oxley Library, Brisbane
LW	*Logan Witness*
Mang.	Manager
MC	*Maryborough Chronicle*
Melb.	Melbourne
ML	Mitchell Library, Sydney
MM	*Mackay Mercury*
MUA	Melbourne University Archives
NBAC	Noel Butlin Archives Centre, Australian National University, Canberra
NBAM	National Australia Bank Group Archives, Melbourne
NS	*Northern Star*
NSW	New South Wales
NSWCGA	New South Wales Cane Growers' Association
NSWVP	*New South Wales Votes and Proceedings*
PDT	*Port Denison Times*
PMR	Pioneer Mill Records
Proc.	Proceedings
Q	*Queenslander*
QAG	Queensland Auditor-General
QAJ	*Queensland Agricultural Journal*
QCGC	Queensland Cane Growers' Council
QGG	*Queensland Government Gazette*
Qld.	Queensland
QNB	Queensland National Bank
QNBR	Queensland National Bank Records
QPP	*Queensland Parliamentary Papers*
QPD	*Queensland Parliamentary Debates*
QSA	Queensland State Archives
QSB	Queensland Sugar Board
QSC	Queensland Sugar Corporation
QSSCT	Queensland Society of Sugar Cane Technologists
QVP	*Queensland Votes and Proceedings*
RRE	*Richmond River Express*

SAPP	*South Australian Parliamentary Papers*
Sec.	Secretary
SJTC	*Sugar Journal and Tropical Cultivator*
SRI	Sugar Research Institute, Queensland
UB	Union Bank of Australia
UBR	Union Bank of Australia Records
VPRO	Victorian Public Record Office, Melbourne
WQCQ	Water Quality Council of Queensland

Note: All archival material of the Colonial Sugar Refining Company was located at the Noel Butlin Archives Centre, Australian National University, Canberra, unless stated otherwise.

Illustrations

Plates

Figures

Tables

Glossary of Sugar Terms and Definitions

Affination: The first step in the refining of raw sugar. It involves blending the raw sugar with hot concentrated syrup and spinning off the softened outer syrup layer of the raw sugar crystal in a centrifugal.

Assignment: A right, given by the Central Sugar Cane Prices Board, to grow cane only on a defined section of a property. It is not illegal to grow sugar cane on unassigned land, but raw sugar produced from such cane was paid a penalty price. The Queensland sugar mill to which cane grown on the assignment must be supplied is also specified.

Bagasse: The fibrous residues after the harvested stalks of cane have gone through the sets of crushing rollers. Consists of fibre, water and a small quantity of sugar. It may be used as fuel in a sugar mill's boilers or as a soil enhancement agent.

Barracks: Residential accommodation provided for gangs of canecutters during the crushing season (see crushing season).

Bin: A tramway wagon enclosed with metal or strong wire mesh for carrying cane billets (see cane billets).

Brand 1: The primary Queensland export brand of raw sugar (see raw sugar).

Brand JA: A brand of raw sugar with lower polarisation than Brand 1. It is made for the Japanese market (see raw sugar).

Brix: A unit used to express the concentration of solids in aqueous sugar solutions. For example, 60° Brix sugar solution contains 60% by weight of sugar.

Calandria pan: In a calandria vacuum pan, the heating surface consists of a large number of vertical tubes inside the vessel. The juice flows through the tubes, while steam or vapour is directed onto the outside of the tubes.

---

* This glossary draws heavily on the glossaries contained in the Queensland Sugar Corporation's, *Sugar Notes* and G. Burrows and C. Morton, *The Canecutters* (Melbourne, 1986), pp. 262-267.

Cane billets: The common name for the chopped lengths of cane (approximately 250–400mm) produced by a mechanical chopper cane harvester during the harvesting operation.

Cane bin: see bin

Cane carrier: A conveyor that carries the harvested cane from the unloading point to the crushing mills (see crushing mill).

Cane inspector: A sugar mill employee who acts as a link between the mill and canegrowers, particularly with regards to estimating the size of the crop and harvesting arrangements.

Cane invert: A liquid sugar product produced at a sugar mill, containing a relatively high proportion mixture of glucose and fructose.

Cane juice: A liquid consisting of water with sugar and other substances dissolved in it and a proportion of insoluble particles suspended in it.

Carbonation: The clarification process of removing impurities from sugar liquor by the addition of lime and carbon dioxide to form calcium carbonate which traps the other impurities. This process is undertaken at a sugar refinery (see liquor).

ccs: Commercial cane sugar. An estimate of the weight of raw sugar that could be obtained from sugar cane, expressed as a percentage of a tonne of cane.

Central/cooperative mill: A term referring to sugar mills built in Queensland either with government grants or loans under Acts in 1886, 1893, 1911 and 1922. These sugar mills were owned by supplier canegrowers (usually between 150 and 300) with suppliers owning the shares in the central milling company.

Centrifugal (fugal): A perforated basket which spins inside a casing to separate sugar crystals from molasses.

Chopper harvester: A mechanical cane harvester which chops the cane stalks into short billets. These are discharged into bins hauled alongside the harvester by tractors (see bin and cane billets).

Clarification: The process of separating insoluble suspended matter and some soluble substances from cane juice, to produce a clear juice. The process generally involves adding lime and heating the limed cane juice. Carried out in special vessels known as clarifiers or subsiders (see clarifier).

Clarifier: Large, specially designed vessels where clarification of the cane juice is undertaken. Sometimes known as a subsider. The clear sugar juice is run off from the top of each clarifier. Muddy juice from the bottom of the clarifiers is filtered again using different processes and machinery (e.g. plate and frame presses; rotary vacuum filters) to recover more sugar (see filter mud).

Coefficient of work: A measure of the efficiency of a sugar mill's operation. It represents the recovery of raw sugar (tonnes 94 n.t.) per 100 tonnes ccs in cane.

Crushing mill: Three large rollers arranged in a triangular formation. The shredded cane is fed through these rollers. The process separates the sugar juice from the fibrous material known as bagasse. From 1880 onwards, most Australian sugar mills had two or more sets of crushing rollers known as double, treble, quadruple or quintuple crushing. This arrangement was called a mill tandem or mill train.

Crushing season: The period during which the sugar cane is harvested and crushed at the sugar mills. It generally lasts for 20 to 22 weeks from June to November, but in some years has extended through December and January.

Crystallisation: The process of "growing" sugar crystals by boiling them with syrup in a vessel under vacuum (see massecuite).

Decolourisation: The sugar refinery process whereby the colour of filtered sugar liquor is reduced by passing it over either bone char, granular carbon or ion exchange resin (see liquor).

Diffusion: The process of immersing bagasse or slices of cane in water to 'wash' out the sugar juices. Completed in special units known as diffusion plants or diffusers.

Drop planter: A labour-saving machine that mechanised the planting of setts; first developed in the late 1890s, but widely adopted during the 1920s (see sett).

Effet: Cylindrical steel vessel or evaporator in which a steam heated surface is used to concentrate clear juice from the clarifiers by boiling under a vacuum.

Multiple effet evaporation occurs in a series of connected effets known as double, triple, quadruple and quintuple effets. The concentrated juice is called syrup and is sent to a vacuum pan (see vacuum pan).

Excess cane: Cane delivered in excess of farm peak (see farm peak).

Excess sugar: Sugar produced by a sugar mill above its peak (see mill peak and Peak Year Scheme).

Extraction: The process of removing sugar and juice from stalks of cane, billets or bagasse, by milling or diffusion.

Extraneous matter: Any material other than harvested cane stalks (e.g. stones, soil, roots, tops and trash) sent to the mill (see trash and top).

Farm peak: A delivery quota which represents a farmer's right to supply cane to a sugar mill to produce a given quantity of sugar. This production control measure was introduced in Queensland in 1939.

Filterability: The filterability of a raw sugar during refining is measured by comparing the filtration rate of the sugar with that of a standard sucrose solution under specified conditions. The result is expressed as a percentage of the filtration rate of the standard sugar.

Filter mud: In clarifying cane juice, the insoluble matter extracted from the juice forms a mud which is removed from the clarifiers, filtered and washed to recover the sugar it contains. Filter mud is used as a fertiliser. Also known as subsider mud, filter cake or filter press (see clarifier).

Filtration: A sugar refinery process whereby cloudy carbonated sugar liquor is passed through filter cloth to produce clear liquor.

Fugal: See centrifugal.

Headland: Areas at the end or head of canefields deliberately left clear of cane. Mechanical harvesters and farm implements can be turned in this space without damaging the crop.

Imbibition: The process of spraying water or cane juice over the blanket of bagasse emerging from each set of crushing rollers to help wash out the sugar.

Juice mill: A primitive sugar mill in which only the cane stalks were crushed. The juice was then sent to another sugar mill equipped with machinery to convert the juice into raw sugar. Found only in the Maryborough, Isis and Bundaberg districts, Queensland.

Levellers: Rotating knives fixed radially on a shaft across the cane carrier. Used to level off thick spots on the carrier feed or to disintegrate the cane stalks before they enter the crushing rollers.

Liquor: Sugar crystals created after affination are dissolved in hot water to form a sugar liquor. This liquor is then carbonated or phosphated to remove remaining impurities (see affination, carbonation and phosphatation).

Local Award or Mill Award: The award made each year by the local board is a contract between the mill-owners and each assignment holder in a particular Queensland mill supply area. It contains an outline of all penalties that applied to diseased, frosted and burnt cane and covers all matters relating to the harvesting and delivery of cane by the canegrower and the transport, payment, handling and crushing of the cane by the mill-owner. Local awards were issued for the first time in 1915 (see mill supply area).

Lodged cane: A crop which has been flattened to ground level by floodwaters, high winds and/or rain.

Maceration: The process in which the bagasse is steeped in an excess of water or cane juice, generally at high temperatures, in order to dissolve more juice out of the baggase.

Massecuite: The mixture of sugar crystals and molasses produced by crystallisation in a vacuum pan. The term is French for 'cooked mass'. A, B and C are grades of purity. A and B are the purest and produce high-grade sugars. Low-grade sugar comes from C massecuite.

Mill Peak: Production quotas specified in tonnes of raw sugar allocated to Queensland sugar mills. It is the sum of farm peaks in that mill supply area. Raw sugar produced in excess of peak receives the No. 2 Pool price. First introduced in 1930 as a method to avoid over-production (see Pools).

Mill supply area: Localities from which an Australian sugar mill drew its cane supplies. In Queensland after 1915 these localities became fixed and were related to assignments. Mostly incorporated between 100 and 300 farms (see assignment).

Mill tandem: Two, three, four, five or six sets of crushing rollers. Also known as a mill train. Such an arrangement improved the juice extraction capabilities of sugar mills (see crushing mill).

Mill whites: A partially refined sugar produced in a sugar mill. It is not as pure as refined white sugar.

Molasses: The black syrup, commonly known as 'C' syrup, remaining after the sugar syrup has been boiled and passed through the centrifugals for the last time in a sugar mill or sugar refinery. The sugar it contains cannot be removed economically. This product is stored for later sale, sometimes as a fertiliser, stock feed or to distilleries (see syrup).

Mud: See filter mud.

Net titre (n.t.): An historical measure of the commercial value of raw sugar for refining purposes. Net titre provided a method for expressing different sugar at a standard value and was used for statistical and payment purposes.

Pans: see vacuum pans

Plant cane: The first growth from cane planted the previous season (see ratoon cane).

Peak Year Scheme: An arrangement introduced in 1930 whereby the highest output of sugar for each Queensland sugar mill in any one year since 1915 was taken as the limit of any future year's production. Essentially a measure designed to avoid over-production (see mill peak).

Phosphatation: A process conducted in a sugar refinery. It involves trapping the impurities in a precipitate formed by adding lime and phosphoric acid to the sugar liquor and removing them as a layer from the top of a flotation clarifier.

pol: Polarisation. The sucrose content of sugar, expressed as degrees pol, which is approximately a percentage. A sugar of 98° pol would contain about 98% sucrose.

Pools: In Queensland, payment for raw sugar production was made on a pooling basis. For payment purposes, two different pools were calculated between 1930 and 1995. The No. 1 Pool price was paid on all sugar produced within the Mill Peak and the No. 2 Pool price was paid for the remaining sugar produced from cane which had been grown on assigned land.

Ratoon cane: Cane is harvested by cutting the plant off at ground level. The stubble remaining puts out new shoots, and grows into what is known as a ratoon crop. One plant crop is usually allowed to grow two to four ratoon crops. After a final ratoon crop has been harvested the ground is usually ploughed and fallowed for a year. Legumes are sometimes grown on the fallow ground.

Raw sugar: The sugar crystals separated in a centrifugal in a sugar mill. Australian raw sugar is usually produced in two grades of about 98.8° pol (Brand 1) or 97.6 ° pol (Brand JA). Pol is varied to satisfy the requirements of customers (see Brand 1 and Brand JA).

Refined sugar: Sugar which has passed through the refining process that has involved the removal of impurities which impair colour, flavour and longevity, making it more suitable for direct human consumption. Also known as white sugar (see mill whites).

Sett: A piece of cane stalk used as planting material.

Shredder: Machinery used to reduce and shred the harvested stalks of cane into fibrous material and rupture the juice cells.

Stand-over cane: Mature cane, sometimes in excess of peak, left unharvested at the end of the crushing season. This cane can either be allowed to grow another year or may be destroyed.

Stool: The cluster of cane stalks arising from the germination of setts or the regrowth which comes from the stumps and roots of harvested cane (see sett).

Subsider: See Clarifier.

Syrup: A concentrated sugar solution produced by evaporation of water from sugar cane juice in the effets. In sugar refineries, syrup refers to the less pure solution which is spun off crystals in centrifugals (see effets).

Top: Thick stalk and leaf material above the juice-bearing part of the stalk. This part of the cane plant should be removed during harvesting (see extraneous matter).

Trash: Leaf matter attached to the cane stalk, which ideally should be removed before milling, by either burning or through mechanical harvester cleaning mechanisms.

Toll refining: Toll refining refers to the use of refining assets, for a fee, to process raw sugar into refined sugar. The refined product remains the property of the raw sugar supplier.

Two-year cane: Cane allowed to mature for two years before harvesting, rather than the normal twelve months. Common only in New South Wales.

Vacuum pan: Cylindrical steel vessel in which a steam heated surface is used to boil sugar syrups under partial vacuum at relatively low temperatures (see syrup and massecuite).

White sugar: The common name for refined sugar.

CHAPTER 1
An Introduction

Between 1860 and 1915, the river valleys of most coastal regions between Grafton in New South Wales and Mossman in North Queensland were transformed as the forests were cleared and replaced by fields of sugar cane. Several thousand Europeans eventually settled along these river valleys, making their living from the growing of sugar cane, either directly as farmers or as workers in the sugar mills erected to crush the crops of cane. Other settlers followed, providing the goods and services needed by the farmers and workers. At times during this period, thousands of indentured Pacific Islanders also called these river valleys home; historians have estimated that some 63 000 Pacific Islanders, mainly from the Solomon Islands and Vanuatu, laboured in the Australian sugar cane fields between 1863 and 1904.[1] They were also joined by Chinese, Japanese, Javanese and Indians in lesser numbers. The Aboriginal peoples – the traditional occupiers of these river valleys – sometimes resisted this invasion by European agriculturalists and their indentured workers, attacking the livestock on the sugar plantations or occasionally burning crops of cane.[2] Eventually, some Aborigines worked as labourers on the sugar plantations and/or small sugar cane farms. This great transformation in the tropical and sub-tropical littoral of north east Australia took place to satisfy those Europeans living in temperate Australia, who demanded ever increasing amounts of sugar.

Europeans during the eighteenth and nineteenth centuries, especially the British and French, developed a taste for sweet things. As a result, sugar had become an important commodity. Traditionally, sugar for European consumption had been manufactured in the Lesser Antilles (or West Indies), Cuba, Brazil and Louisiana. During the nineteenth century, sugar production in these traditional areas expanded to meet this demand. In addition, sugar production spread to new regions such as Java, Argentina, Natal, Peru, Hawaii, Fiji and Australia. The emergence of an Australian sugar industry was, therefore, part of a wider nineteenth century dispersal of commercial sugar production around the globe.

The Australian sugar industry that developed between 1864 and 1884 had many similarities with other cane sugar industries around the world. The sugar cane was grown on plantations or large estates where indentured labourers worked in both the fields and the estate's own sugar mill. These indentured workers, however, were mostly Pacific Islanders, instead of Indians, Chinese or Japanese who made up the bulk of indentured sugar workers elsewhere around the globe. The sugar produced in northern Australia was sent south to capital city-based sugar refineries, much like the movement of sugar from the West Indies to the sugar refineries in Bristol, Glasgow, Liverpool or London. The milling sector of the Australia sugar industry, like the sugar factories in other sugar-producing regions, had been modernised by the adoption of improvements in machinery (e.g. vacuum pans; multiple evaporators) and industrial processes (e.g. double or treble crushing). In contrast, the agricultural sector of the Australian sugar industry remained labour-intensive, just like the field practices of canegrowers in the Americas, Asia and rest of Oceania.

Despite these similarities, there were early signs that the Australian sugar industry would not evolve exactly as its overseas counterparts. During the 1870s, the central milling system emerged in New South Wales. Here, European small farmers, mostly without the aid of indentured workers, grew the sugar cane, and the Colonial Sugar Refining Company (CSR), a joint stock company, operated three, large central sugar mills equipped with the most up-to-date machinery. This arrangement spread to Queensland after 1884, although the Queensland government, not private enterprise, provided the funds to farmer co-operatives to erect their own central sugar mills. This transformation came about because of Queensland government policies that supported the creation of a class of European farmers for social and political reasons. Other Queensland sugar planters during the 1890s and 1900s became proprietary central mill-owners, creating tenanted estates and concentrating their resources on the manufacturing side of operations. In addition, most indentured workers brought to Queensland were not allowed to settle in Australia as workers or small farmers, unlike other sugar-producing regions (e.g. Fiji; South Africa; Mauritius; British Guiana). The newly formed Commonwealth government of Australia introduced legislation in 1901 which applied a protective duty to sugar imports, prohibited the recruitment of Pacific Islanders into Australia after 31 March 1904 and required the majority of Pacific Islanders to be repatriated to their home islands after 1906. Europeans who were paid higher wages took their places in the sugar mills and fields. Consequently, the Australian sugar industry in 1915 was characterised by central mills supplied with cane from family farms owned by Europeans and European field and mill workers (mostly of Anglo-Celtic descent). This system was unique amongst the world's cane sugar industries.[3]

The development trajectory of the Australian cane sugar industry over the period 1915–1995 continued making it different from other cane sugar industries. Sugar was produced universally and exclusively by Europeans paid high wages. This feature led to a high degree of mechanisation in both the fields and sugar mills. In addition, the Australian sugar industry until the mid-1980s was controlled comprehensively by government, being 'honeycombed with statutory rules and regulations'.[4] These regulations ensured Europeans could farm successfully in the tropical and sub-tropical parts of Australia, avoiding over-production and protected from cheaper imports of sugar produced in other countries where workers received considerably lower wages than Australians.[5]

As the Australian sugar industry was deregulated during the late 1980s and early 1990s, it began resembling other sugar industries with regards to its exposure to competition from foreign sugar producers and limited government intervention in marketing arrangements (e.g. Mexico).[6] Yet the Australian sugar industry stood apart from the cane sugar industries in Brazil and the United States, for example, where government policies protected domestic producers and regulated marketing operations. Thus, the marked differences which once existed between the Australian and other cane sugar industries may have narrowed, but the Australian sugar industry in the mid-1990s was still unique, being one of the world's most low cost, technically efficient and deregulated sugar industries, leading the world in the bulk storage, loading and shipping of the product. Moreover, no other major exporter, apart from Cuba, exported such a high proportion of its raw sugar production.[7]

Historical accounts on aspects of the Australian sugar industry have been written by journalists, historians, economic historians and geographers. Their output is voluminous. Published material on the topic, however, falls into four broad categories. Publications in the first category include those glossy booklets or the more substantial and well-researched books produced to celebrate some milestone in the history of a sugar mill.[8] The late John Kerr, statistician and part-time historian, made a speciality out of writing sugar mill histories, having completed accounts about six Queensland sugar mills and the history of the Bundaberg Sugar Co., which operated four sugar mills during the 1980s.[9] The second category of publications contains theses and books by both nineteenth and twentieth century authors on the history of sugar production in New South Wales, Queensland or the Northern Territory.[10] Material in the third category includes the numerous journal articles and books dealing with the labour supply for the Australian sugar industry, particularly the trade in Pacific Islanders during the nineteenth century,[11] the rise of trade unionism amongst European sugar workers and the origins and development of the cane cutting teams that took over the annual cane harvest during the 1900s,[12] and the mechanisation of cane harvesting in the 1950s and 1960s.[13] The fourth category contains studies that deal with the historical development of sugar cane growing as part of a local history or regional study.[14]

Given that so much has been written about different aspects of the Australian sugar industry, how is this book different from previous studies? Firstly, this work takes a national approach. The evolution of sugar production in Queensland, New South Wales, Western Australia and the Northern Territory are linked together, in order to show the broader, overall picture. Moreover, this research acknowledges that although the sugar cane was grown and processed into raw sugar in northern Australia, other parts of the industry – the refiners who further processed the raw sugar into refined sugar and the merchants who sold the raw and refined sugar to consumers – were located mostly in southern Australia. The inter-colonial and inter-state trade in sugar joined Australian regions of sugar production with Australian regions of sugar consumption. In addition, flows of capital from southern Australia during the nineteenth century, especially Victoria, financed the establishment and growth of sugar production in northern Australia.

Secondly, this account is a study in which geographical and environmental questions are paramount. Publications into aspects of the Australian sugar industry by historians and economic historians have often ignored the importance of place, regional differences and the environmental consequences of sugar production. The economic and political forces shaping the development of Australia's sugar industry did not create a homogenous economic or cultural landscape, and the environmental impact of sugar production varied between regions. Commercial sugar production emerged under different arrangements in the various sugar-producing districts throughout the country, the speed of change in the production system during the nineteenth century was not uniform and neither was the impact of post-1930 regulations, particularly controls over production. In addition, quite marked regional differences occurred in the varieties of sugar cane cultivated, the adoption rate of new machinery and farming techniques, the types of disease and pest outbreaks and the ethnicity of workers employed, especially during the late nineteenth century.

Thirdly, this work is an integrated study covering all sectors of the Australian sugar industry. Researchers often overlook the dual nature of sugar production. In any cane sugar industry there is an agricultural sector and an industrial sector that is split into milling and refining operations. Historians and economic historians in their studies of the Australian sugar industry have concentrated on the agricultural sector, particularly its labour supply and the transformation of the production system from plantations to central sugar mills supplied by small farmers (i.e. the period 1885–1915).[15] In this account, however, details have been systematically assembled for the first time on the changing cultivation methods and machinery employed by Australian canegrowers, their fight against pests and diseases, and the assistance they received from the country's agricultural scientists to breed new cane varieties that were suited to the different environments where the crop is cultivated in Australia. Chapters on the evolution of the milling and refining sectors and the marketing of sugar are also provided. These sectors have not been considered previously in any depth or not at all in the standard works on the Australian sugar industry.

The book begins with a brief account of the biology and environmental requirements of sugar cane and explains how sugar cane is converted into raw (or brown) and refined (or white) sugar (Chapter 2). The book is then structured in three sections, covering the main epochs in the historical development of the Australian sugar industry. Part 1 covers the beginnings of the Australian sugar industry from the arrival of Europeans in New South Wales in 1788 to 1863, the year before the commercial production of raw sugar was finally established permanently in Australia. The discussion documents the unsuccessful intermittent attempts to manufacture raw sugar in Australia (Chapter 3), and the growing demand for sugar by convicts and free settlers and its supply by merchants and then the first sugar refiners (Chapter 4). Part 2 comprises five chapters covering the years from the sustained commercial production of raw sugar in 1864 to the outbreak of World War I. This period is when the Australian sugar industry finally becomes established and the foundations for its post-1915 character emerged. Topics covered include the following: the rapid increase in the area cultivated with sugar cane in different parts of Australia (Chapter 5); the transformation in the production unit from plantations to large numbers of small, European-owned family farms supplying sugar cane to cooperative or proprietary central sugar mills (Chapter 6); earliest cultivation methods (Chapter 7); technological change in the milling sector (Chapter 8); and CSR achieving almost a monopoly over the refining and marketing of sugar in Australia by 1914 (Chapter 9).

The final section covers the years from 1915, when government regulations start controlling all aspects of the production and marketing of sugar in Australia, until 1995, when the Australian sugar industry operated in a mostly deregulated environment. Chapters cover the following topics: the regulation of the growth in the area cultivated with sugar cane to ensure supply matched the domestic and global demand for sugar (Chapter 10); increasing use of labour-saving machinery, fertilisers and herbicides (Chapter 11); the upsurge in drainage- and irrigation-related activities and measures to reduce soil erosion (Chapter 12); breeding new cane varieties (Chapter 13); efforts to combat the impact of pests and diseases upon yields (Chapters 14 and 15); the harvesting of cane and its transport to the sugar mills (Chapter 16); the trend towards fewer sugar mills with quicker and greater processing capacities (Chapter 17); and the inter-linking of the marketing and refining of sugar and its pricing until deregulation in the 1990s (Chapter 18).

Throughout this narrative, primary sources have been used extensively. They have been considered from a different perspective and some material has been examined in a more exhaustive manner than has hitherto been attempted. In particular, a considerable amount of time was devoted to reading through pre-1930 newspapers, magazines, sugar industry journals and mill records and the voluminous pre-1980 archival records of CSR, which revealed not only much about the operation of that company, but details about the agricultural and business practices of other Australian canegrowers and mill-owners. In addition, the following new material was examined: the historical records of CSR's now closed Goondi and Hambledon Mills and the Technical and Field Department of Macknade Mill which were transferred to the Noel Butlin Archives Centre, Canberra, in the mid-1990s; the financial records of the sugar mills and plantations operated by the Queensland National Bank, the Commercial Banking Company of Sydney and the Union Bank of Australia Ltd. (located at the archives of the National Australia and ANZ Banks, Melbourne); and the order books of the Scottish engineering firms involved in the manufacturing of the machinery for the sugar mills in Queensland during the nineteenth century (found at the Business Records Centre, Glasgow University Archives). To my knowledge, none of this material has been considered previously in any depth, if at all, for information about the Australian sugar industry. Therefore, the primary sources used as the basis for the narrative in this book will add a new dimension to understanding the historical development of the Australian sugar industry.

Notes

1 C. Price with E. Baker, 'Origins of Pacific Island labourers in Queensland, 1863–1904: a research note', *Journal of Pacific History*, Vol. 11 (1976), p. 116.

2 D. Jones, *Trinity Phoenix. A History of Cairns* (Cairns, 1976), p. 162; J. Kerr, *Sugar at Maryborough: 120 years of Challenge* (Maryborough, 1987), p. 15; *North Queensland Herald*, 6 April 1892, p. 13; S. May (ed.), 'The Diary of Arthur Neame, 1870–1897', pp. 33 & 36, JOL.

3 H. Blume, *Geography of Sugar Cane* (Berlin, 1985), p. 210.

4 F. Maxwell, *Economic Aspects of Cane Sugar Production* (London, 1927), pp. 9–10.

5 A. Graves, 'Crisis and change in the Australian sugar industry, 1914–1939', in W. Albert & A. Graves (eds.), *The World Sugar Economy in War and Depression* (London, 1988), pp. 142–156.

6 For these measures see G. Robinson, 'Deregulation and restructuring of the Australian cane sugar industry', *Australian Geographical Studies*, Vol. 32, no. 2 (1995), pp. 221–227.

7 Sugar Industry Review Working Party, *Sugar–winning globally. Report of the Sugar Industry Review Working Party*, pp. 1 & 7; for details about the regulation of different sugar industries see T. Sheales, S. Gordon, A. Hafi and C. Toyne, *Sugar: International Policies Affecting Market Expansion.* ABARE Research Report 99.14 (Canberra, 1999).

8 For these booklets see these references listed in the bibliography: Anon (1984); Landells (1937); & Pioneer Sugar Mills (1984); for more substantial publications on the same theme see the following references: Alcorn and Dunn (1997); Hudson (1995); Hudson (2000); Langford and Thomis (1979); Manning (1983); Morton (1995); & Rolleston (1987).

9 See these references in the bibliography: Kerr (1979); Kerr (1983); Kerr (1987); Kerr (1988); Kerr (1991); Kerr (1996); & Kerr (1997).

10 For nineteenth century authors see the following references listed in the bibliography: Roth (1880); Mackay (1883) & Fletcher (1886); for twentieth century studies on Queensland see the following references in the bibliography: Adamson (1953); Easterby (c. 1933); & Shogren (1980); for twentieth century studies on New South Wales see the following references in the bibliography: Higman (1968) & Jeans (1972, pp. 239–250); for the Northern Territory see Hillock (2000).

11 For nineteenth century commentators see the following references in the bibliography: Roth (1883) & Black (1894); for more recent discussions see the following references in the bibliography: Corris (1970); Corris (1973); Gistitin (1995); Graves (1993); McGrath (1976); Mercer (1995); Moore (1985); Saunders (1979); Saunders (1982); Shlomowitz (1981); Shlomowitz (1982); & Shlomowitz (1985).

12 See the following references in the bibliography: Armstrong (1983); Hunt (1978); Saunders (1978); & Shlomowitz (1979).

13 See the following references in the bibliography: Burrows and Shlomowitz (1992); Department of Labour and National Service, Australia (1970); Kerr and Blyth (1993); & Willis (1972).

14 For nineteenth century accounts see the following references in the bibliography: Goodwin (1883) & Munro (1895); for more recent publications on Queensland see the following references in the bibliography: Collinson (1945); Dignan (1964);

Gaylard (1967); Gregory (1991); Griggs (1990); Jones (1973); Jones (1976); Kerr (1980); Kerr (1994); May (1984); & Wegner (1984); for more recent publications on the New South Wales sugar industry see the following references in the bibliography: Daley (1966); Davies (1953); & Smith (1991).

15 For examples see A. Graves, 'Crisis and change in the Queensland sugar industry, 1862–1906', in W. Albert & A. Graves (eds.), *Crisis and Change in the International Sugar Economy 1860–1914* (Norwich, 1984), pp. 261–279 & R. Shlomowitz, 'The search for institutional equilibrium in Queensland's sugar industry, 1884–1913', *Australian Economic History Review*, Vol. 19 (1979), pp. 91–122.

Chapter 2

Sugar Cane and The Production of Sugar

Sugar cane is a member of the grass family and is classified in the genus *Saccharum*. Six species belong to the genus *Saccharum*. Only two species, S. *spontaneum* and *S. robustum*, occur in the wild state, growing mostly around the watercourses throughout Papua New Guinea and Southeast Asia. The cultivated species include *S. officinarum*, *S. barberi*, *S. sinese* and *S. edule*. They have less fibre and higher sucrose content than the wild species, but are susceptible to diseases. *S. officinarum* became known as the 'noble' cane because of its thick stalks containing high sugar content. *S. edule* is not grown for its sugar, but its edible tassel which resembles an elongated cauliflower. The commercial sugar cane varieties (clones) cultivated today are complex hybrids involving interbreeding between two or more species of *Saccharum*.[1]

Sugar cane differs from many common grasses in two main respects. Firstly, the plant has evolved to have a long period of vegetative growth (i.e. nine to twelve months), resulting in the crop's large size and unusually high biomass per unit area. Secondly, instead of storing starch in the seed heads like the major grain crops, sugar cane has evolved to store sugar (sucrose) in its stalk.[2] These stalks are the visible part of the plant above the ground and support the leaves and flowers, while below the ground is attached the root system (Plate 2.1). Each has an important role in crop growth, crop maturation and the production of new plants and new varieties, although the stalk is of greatest interest to the mill-owners and canegrowers, for the sucrose is stored in the stem.

Cane stalks form a cluster known as a stool, with most stools having between six and ten stalks depending upon the cane variety (Plate 2.1). Stalks are variously coloured from the purplish black of the Badila and Black Tanna cane varieties, the yellow-green or green of the Goru, Bourbon, China or P.O.J. 2878 cane varieties and the reds of the Rappoe or Rose Bamboo cane variety. Normally, stalks are 1.2–4 m in length, although ones of greater length are not uncommon. In diameter, stalks vary between 0.5 and 8 cm. Most commercial cane varieties have stalks with diameters ranging from 2–4 cm. The cane stalk is approximately cylindrical in cross-section with a hard outer rind enclosing softer internal flesh or pith and fibres. The sugar is stored in the cells making up the softer internal tissue of the stalk. Low fibre content and low hardness of

Plate 2.1. A cluster of cane stalks, known as a stool.
Source: Anon, *The Australian Sugar Industry* (Sydney, 1925), p. 2.

rind contribute to good millability, with cane varieties not having these characteristics requiring special treatment during the process of juice extraction.[3]

Sugar cane is propagated commercially by the vegetative method, which involves planting and covering sections of the stalk with moist soil. These sections are known as stem cuttings, setts or seed-pieces, although in Australia since the early 1970s they have also been referred to as cane billets – chopped lengths of cane stalk, approximately 200–300 mm in length. The sections of stalk contain all the plant food and water required for buds to germinate, but the germination process is sensitive to temperature. Thus, under the right conditions, a new shoot will grow from the bud. This process produces new plants in all respects like the canes from which the setts were taken. However, asexual propagation of cane enables particular diseases (e.g. gumming) to be spread very easily, and special control measures (e.g. quarantine zones; hot-water treatment) are often needed to ensure that canegrowers are provided with 'clean' or disease-free setts.[4]

Planting fields of cane was a labour-intensive task, although in parts of the world the process has now been mechanised. One planting, however, usually gives several crops. After the harvest of the first crop, known as the plant cane, the roots and lower part of the stalk with its buds remain in the ground. New shoots emerge from the buds to form another stool of cane. Each of the crops succeeding the plant cane is known as a ratoon crop – first ratoons, second ratoons and so forth. The number of ratoon crops allowed from a plant crop depends upon the cane variety, incidence of disease, yield and local environmental conditions. Sugar yields from ratoon crops are usually lower than from the plant cane, so a point is reached at which further ratoons are no longer economic. Some Australian canegrowers during the late nineteenth century anticipated up to six or eight ratoon crops. Today in Australia it is rare to grow more than four ratoon crops before ploughing out the crop and replanting the field.[5]

The leaves of the sugar cane plant have been described as their 'photosynthetic engines', as they produce the sucrose that is stored in the stalks.[6] The leaves of the cane plant occur alternatively on the opposite sides of the cane stalk and develop into a fan-like formation. The leaf edges are generally serrated and the leaf sheaths (i.e. parts of the leaf that are attached to the stalk) are frequently clothed with a growth of fine hairs that makes handling unpleasant in some cane varieties. As the stalk grows, new leaves form at the top and shade the older leaves lower down; sometimes these older leaves will turn yellow and eventually die. Leaf size and number increase from germination, reaching about ten per stalk at maturity. The two youngest leaves are usually vertical, while older leaves become progressively more horizontal. Leaf length depends upon cane variety and environmental conditions, but is usually between 0.9 and 2.0 m in mature leaves. Leaves are generally stripped or removed from the cane at harvest, for there is little point sending the leaves to the sugar mill for processing as they contain little sucrose. In some parts of the world, the cane crops are deliberately burnt, thereby clearing away the dead leaves in order to make the harvest easier.

Sugar cane and its environment

Sugar cane is a highly versatile plant that can be cultivated successfully under various environmental conditions, although it is considered a tropical crop. Therefore, the ideal climate for sugar cane would have high amounts of solar radiation and warm weather throughout the growing season. Satisfactory growth will take place at 21 °C, but higher mean day temperatures between 27 °C and 35 °C are desirable. The ripening and harvesting season should be cool and dry, with mean day temperatures between 10 °C and 20 °C, but frost free. Sugar cane requires an abundance of water during the growing season to keep the plant in continuous growth. Generations of experience throughout the world's sugar-producing regions have shown that sugar cane can survive the normal fluctuations in rainfall around an annual mean of 1 200 mm, but that lower rainfall regions are not suited to the crop. The successful cultivation of sugar cane in regions of low rainfall and/or marked dry seasons can only be achieved with irrigation. Currently, irrigation is used extensively in Australia to grow crops of sugar cane in the Atherton Tableland, Lower Burdekin, Proserpine, Mackay, Bundaberg, Isis and Maryborough districts.[7] The location of these sugar-producing districts is shown in Figure 2.1.

Temperature is the critical factor limiting sugar cane production. Sugar cane can survive extreme temperatures of above 40 °C, although some scorching of leaves may occur under these conditions, and the plant will eventually die if the temperatures exceed 60 °C. High temperature stress is not often encountered in the field if water is available to the plant. Low temperature stress frequently occurs in sugar-producing regions. Below 21 °C, growth rates for sugar cane are reduced and no growth at all occurs below 11 °C to 13 °C. Sugar cane is more tolerant of frost than some other tropical crops (e.g. the banana or coffee), with the actual impact of frost depending upon the cane variety, its health and on soil moisture. Light frost of less than an hour will cause only limited damage to the leaves or leaf scorch. More severe frosts with temperatures of minus 3 °C, however, kill the leaves of young plants, while even lower temperatures kill the entire above-ground portion of the stalk. Air temperatures of minus 11 °C for three hours will result in the death of the underground portion of the stools.[8] In Australia, therefore, much of the northern part of the continent is most suited climatically to the cultivation of sugar cane because it is frost-free. The sub-tropical coastal lands of eastern Australia from Rockhampton southwards to Grafton are capable of supporting crops of sugar cane, but they suffer from frosts

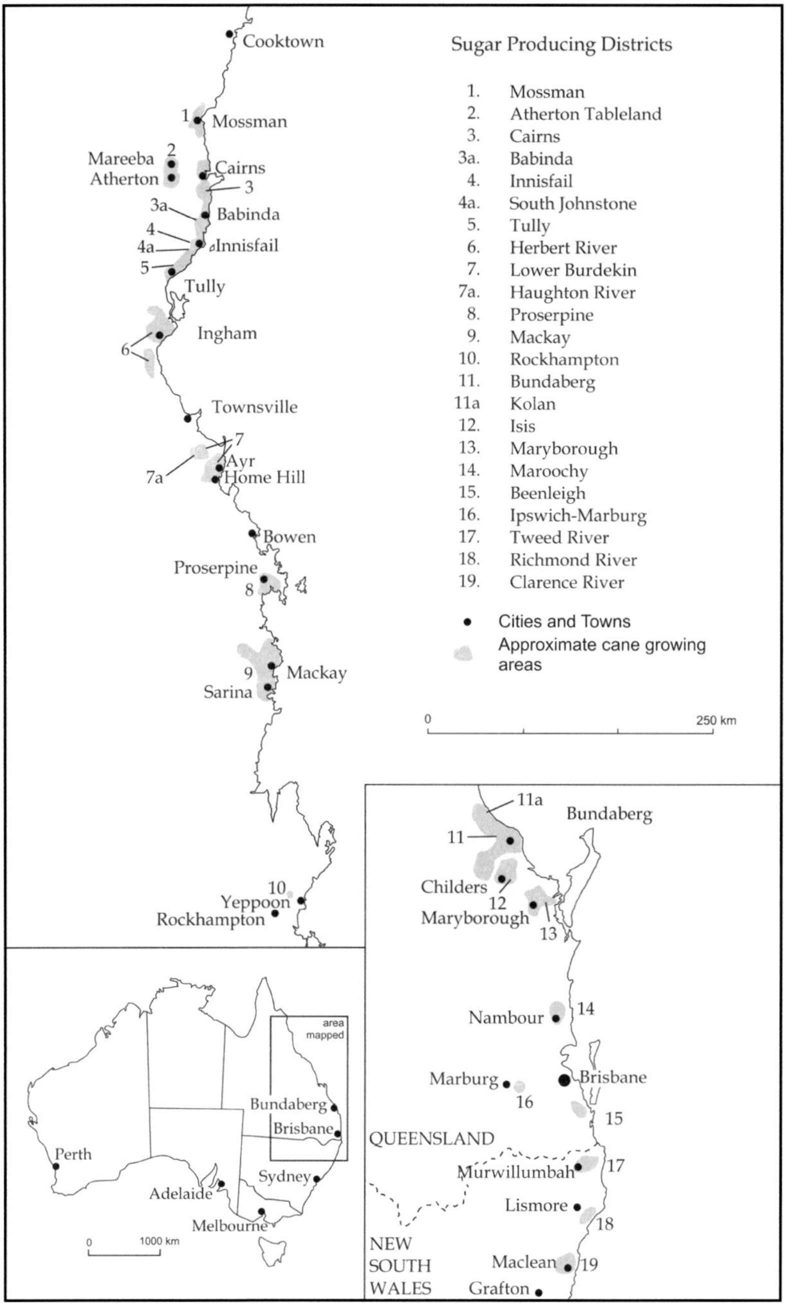

Figure 2.1. The main Australian sugar-producing districts (past and present).

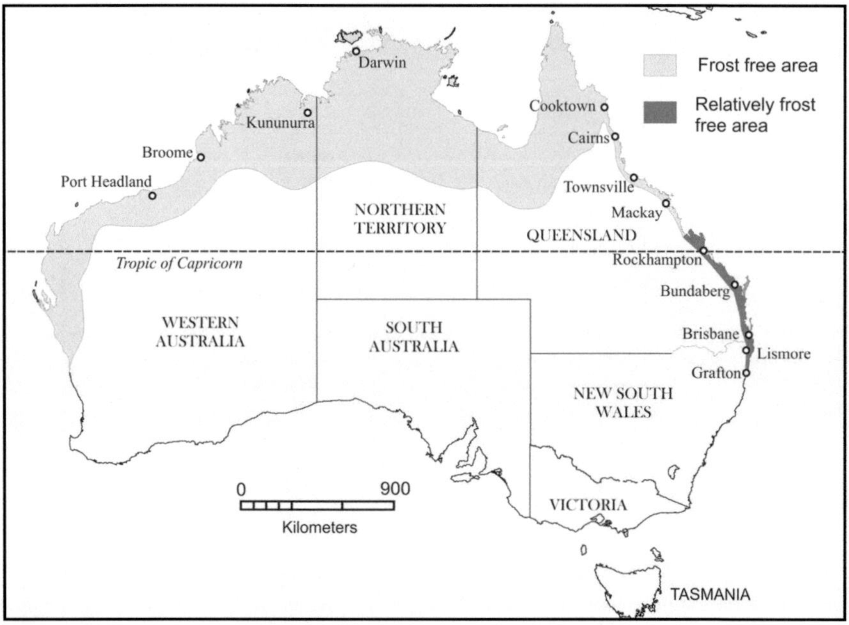

Figure 2.2. Frost free areas of Australia.

Source: Based upon a map in Australian Division of National Mapping, *Atlas of Australian Resources*. Third Series. Volume 4: Climate (Canberra, 1986), p. 32.

(Figure 2.2). In particular, frost damage occurs to sugar cane in Australia throughout the low-lying and inland sections of the Bundaberg, Isis, Maryborough, Richmond River and Clarence River districts.[9]

Sugar cane requires large supplies of water to support its growth, but cannot tolerate its roots being submerged in water for long periods of time. Waterlogged conditions damage the root system, with the roots dying from a lack of oxygen. The plants are killed or seriously stunted when the soil is waterlogged or covered by stagnant water for over a week. Therefore, while ample water is needed, a well-drained soil is essential. In addition, inundation of farms with floodwaters will cause retardation in growth at any time during the growing season, with the degree of damage being dependent on the depth and duration of inundation, and the time of season. Inundation kills young cane less than three months old, particularly when the growing point is submerged for longer than three to four days, especially if the water is silt-laden. More mature cane can be inundated for longer periods, although its growth will be arrested. Young cane sitting in ponded water can also be easily scalded as the water is heated during hot summer days. On the other hand, lack of soil moisture can retard growth and substantially reduce yields. Drought symptoms are wilting or rolling of leaves and a browning of the tips of older leaves, followed by the eventual death of all leaves. In some cane varieties, the plant may survive and regrow when moisture becomes available.[10]

Soils and landforms are less critical environmental factors than either temperature or soil moisture. Sugar cane thrives on soils that vary in texture from light sands to heavy clays, with no particular type being considered 'ideal'. In Australia, sugar cane flourishes in soils as diverse as the red-volcanic soils of the Bundaberg, Isis and South Johnstone districts and the alluvial loams throughout the Cairns, Herbert and Lower Burdekin districts. Successful sugar cane ventures in Australia have also been established on lands with slopes varying from flat to steeply undulating (i.e. slopes greater than 10%). Ideally, the land should have long, regular, smooth slopes of up to 1 ° to 3 °, which enables the widest choice of field layout design and the most economical cultivation and harvesting techniques. Completely flat land is less desirable because of the lack of natural surface run off and the necessity of establishing a slight slope for water-distribution purposes when irrigation is practised.[11]

The sugar making process

The first stage of manufacturing sugar, whether in cottage industries or in a large factory, is to extract the juice containing the sugar from the harvested cane, either by crushing or diffusion – a crushing and washing process. This stage is known as the extraction or crushing phase and has involved the use of various types of equipment throughout the long history of sugar production. At one end of the technological spectrum, juice extraction can be carried out by using a mortar and pestle or pressing two stones together. By 1 000 A.D., screw presses, like those used to express oil from olives, were employed in the extraction phase throughout the sugar-producing regions around the Mediterranean Sea. From the mid-fifteenth century onwards, vertical roller mills were used to crush pieces of cane. This equipment comprised three wooden rollers, encased in iron bands and geared together, with motion imparted to the middle roller. Motive power to move the rollers was generated by slaves, draught animals, windmills or waterwheels. During the late eighteenth and nineteenth centuries, the vertical mills were replaced with horizontal mills or solid rollers, with hydraulic feed attachments to give equal compression whatever the thickness of the feed. Steam engines superseded animals and waterwheels as the providers of motive power.[12]

During the nineteenth century, the extraction phase became increasing elaborate in more technologically advanced sugar mills. On its way to the crushing rollers, the cane received pre-treatment by passing under sets of rotating knives or hammers and/or through crushers and shredders. Such treatments smash the hard outer surface of the cane, so that the material reaching the first set of crushing rollers is a coarse fibrous blanket, enabling the crushing to proceed more effectively. By the late 1880s, the crushed sugar cane fibre began to be passed through two, three, four or even five sets of crushing rollers (known as a mill tandem or a mill train). The crushed fibre can be sprayed with fine jets of hot or cold water as it passes from one set of crushing rollers to the next (i.e. imbibition) or it is saturated entirely with either hot or cold water (i.e. maceration). Extraction by the diffusion process still involves crushing the cane in one or more three-roller mills, but then subjecting the crushed fibre known as bagasse to a continuous treatment with water or dilute cane juice in a tank or diffuser divided into seven or eight compartments. The process removes any residual juice from the bagasse. Dried bagasse can be used as a fuel in the mill's boiler furnaces or as a soil enhancement agent on paddocks where sugar cane is cultivated.[13]

Following the extraction phase, the juice is then ready to be made into sugar. At this stage, the juice is opaque, dark green-brown in colour, acidic and may contain many impurities, including suspended dirt particles and organic matter. The process of transforming the juice into raw sugar commences with the removal of any impurities from the juice (i.e. clarification) and concentrating the juice by heating it over one or more stages to produce a syrup containing sugar crystals (i.e. evaporation phase). These phases may be interlinked. Clarification involves removing any surface debris, passing the juice through screens or filter bags to exclude impurities and/or defecation, which is adding lime to the heated juice and removing the scum formed during the process. Traditionally clarification was done in a battery, copper wall or Jamaica train – a series of open, hemispherical copper or iron pans set over furnaces. By the late nineteenth century this phase was being undertaken in clarifiers or open, steam heated rectangular iron boxes or closed, steam-heated vessels known as subsiders.

Further heating of the juice occurs after it is cleared of impurities, either in additional open, hemispherical copper or iron pans or from 1832 onwards under vacuum in a series of connected vessels called evaporators or effets. Eventually, the heated juice becomes syrup, containing about 65–70% sugar crystals, although not of uniform size. In more primitive factories, this syrup was sent to shallow, wooden troughs called coolers and allowed to crystallise (i.e. crystallisation phase) before being placed in draining boxes where the crystals of sugar separated from the remaining liquor known as molasses. In more technologically advanced factories, this syrup was further concentrated by boiling in a vacuum pan to produce massecuite or a mixture of crystals of uniform size and syrup. The massecuite is then sent to the centrifugals, where the sugar crystals are separated from the molasses. Sugar from the centrifugals can be dried as it tumbles through a stream of warm air in a rotating drum.[14]

Some mill-owners proceed further and make white sugars for direct consumption, known as plantation or mill whites. In these factories the cane juice is treated with either sulfuric or phosphoric acid, before being filtered through iron cylinders filled with pulverised animal charcoal (or bone char) to remove most impurities. This clear juice is then put through the same process to make raw sugar, except whiter sugar is manufactured. This sugar competes directly with white sugars made in the refineries. The designation 'refined' implies freedom from all forms of contamination. The process to achieve refined or white sugar begins when the raw sugar or refining crystals are washed in hot water and mixed with a warm, heavy sugar syrup (i.e. affination). This liquor is purified by the addition of sulfuric or phosphoric acid or carbon dioxide and lime. It is then filtered through granulated carbon or animal charcoal. This decolourised liquid is processed into a variety of products: caster sugar; icing sugars; liquid sugars; sugar cubes; or white sugar for household use.

Notes

1 M. Julien, J. Irvine and G. Benda, 'Sugar cane anatomy, morphology and physiology', in C. Ricaud *et al.* (eds.), *Diseases of Sugar Cane. Major Diseases* (Amsterdam 1989), p. 1; T. Bull, 'The sugar cane plant', in M. Hogarth and P. Allsopp (eds.), *Manual of Cane Growing* (Brisbane, 2001) p. 71.

2 Julien, Irvine and Benda, 'Sugar cane anatomy, morphology and physiology', p. 2.

3 N. King, R. Mungomery, C. Hughes, *Manual of Cane-Growing* (Sydney, 1953), p. 8.

4 King, Mungomery, Hughes, *Manual of Cane-Growing*, 1953 edition, pp. 9–10; Bull, 'The Sugarcane Plant', pp. 75–76.

5 Galloway, *The Sugar Cane Industry*, p. 13; Bull, 'The sugar cane plant', pp. 79–80; for more than three ratoon crops see *Q*, 25 March 1876, p. 21 & 26 May 1883, p. 83 & W. Maxwell, *Report Upon an Investigation into the Condition of the Sugar Industry in Queensland* (Brisbane, 1900), pp. 10–11.

6 Bull, 'The sugar cane plant', p. 74.

7 F. Blackburn, *Sugar-cane* (London, 1984), pp. 65–71.

8 Julien, Irvine and Benda, 'Sugar cane anatomy, morphology and physiology', pp. 16–17.

9 G. Kingston, 'Climate and the management of sugarcane', in M. Hogarth and P. Allsopp (eds.), *Manual of Cane Growing* (Brisbane, 2001), p. 13.

10 Blackburn, *Sugar-cane*, p. 71; R. Ridge and J. Reghenzani, 'Drainage', in M. Hogarth and P. Allsopp (eds.), *Manual of Cane Growing* (Brisbane, 2001), pp. 229–230.

11 Blackburn, *Sugar-cane*, p. 43.

12 Galloway, *The Sugar Cane Industry*, pp. 37–40; Blackburn, *Sugar-cane*, pp. 1 & 4–5; A. Barnes, *The Sugar Cane* (New York, 1974), pp. 5–7.

13 Barnes, *The Sugar Cane*, pp. 423–427; Blackburn, *Sugar-cane*, pp. 316–317.

14 Blackburn, *Sugar-cane*, pp. 296–305; Barnes, *The Sugar Cane*, pp. 427–430.

Part One

1788 TO 1863: BEGINNINGS

The Australian Sugar Company's Refinery, Canterbury, c. 1843.
Source: Image reproduction courtesy of the Mitchell Library, State Library of New South Wales.

Chapter 3
Intermittent Attempts at Sugar Production

In 1783, an Englishman named James Matra drew up a proposal for using convicts to establish a settlement in New South Wales, enlisted the support of Sir Joseph Banks and formally submitted his plans to the British government. As part of his proposal, Matra suggested that New South Wales was well suited for the production of 'sugar cane, tea, coffee, silk, cotton, indigo, tobacco and the other articles of commerce that have been so advantageous to the maritime powers of Europe'. Matra's proposal was not acted upon immediately, but three years later the British government finally decided to send convicts to Botany Bay as part of a scheme containing some of the elements suggested by Matra. Lord Sydney, the Home Secretary, wrote to the Lords Commissioners of the Treasury in August 1876, ordering them to charter ships and buy all the provisions thought necessary to support an expedition to establish a new penal colony. The letter contained fifteen paragraphs; most were concerned with issues dealing with the transport of convicts. None, however, promoted the growing of sugar cane or other tropical products in New South Wales.[1]

Captain Arthur Phillip R.N. was appointed to command the expedition to Botany Bay. On the voyage during 1787, the First Fleet spent several weeks each at Rio de Janeiro and Cape Town. At both ports, the ships were loaded with seeds and growing specimens of most edible plants, both tropical and temperate. Phillip was recorded as collecting unidentified varieties of sugar cane from the Dutch Governor's gardens at Cape Town.[2] Details about the fate of these plants are unknown, although it is probable that some of the First Fleet sugar cane was taken to Norfolk Island when it was occupied in March 1788. The settlement's first commandant, Lieutenant Philip Gidley King, wrote in a diary entry of 24 March 1788 that he 'put some sugar-cane into ye ground.'[3] These cuttings most likely survived, for the settlement's second Commandant, Major Robert Ross, reported in December 1790 that the island's sugar cane was in a 'luxuriant state'.[4]

In 1791, Lieutenant King, back on Norfolk Island, decided to boil the juice from some of the island's sugar cane to test its suitability for producing sugar. He manufactured a small sample of sugar, which he described as 'very tolerable' and sent it to the Colonial Office in London.[5] King must have been optimistic about the potential of the island's sugar cane, for he requested in March 1793 that the settlement be provided with 'rollers and cogs for making a mill or two'.[6] The equipment was never supplied, but King reported in May 1793 that he had made approximately 30 kg of sugar from the island's crop, but 'the want of a sugar mill renders it very tedious to press the juice from the cane.'[7] King never managed to erect a sugar mill on Norfolk Island, and he eventually left the settlement in October 1796. As Governor of New South Wales, King received representations in 1804 from the Norfolk Island Commandant, Lieutenant Colonel Joseph Foveaux, that some Norfolk Island settlers were interested in making sugar. King announced a reward of one cow to the Norfolk Island settler who made the first 500 pounds (225 kg) of sugar from the sugar cane still growing on the island. This inducement never led to the making of sugar so desired by Governor King, and Norfolk Island became a place of punishment and dread for convicts, not agricultural endeavour by free settlers. The Quaker missionary James Backhouse observed sugar cane still growing near the house ruins of the first settlers when he visited the penal settlement in 1835.[8]

Port Macquarie

The first attempt to promote sugar production on the Australian mainland commenced in December 1819, when the missionary John Gyles, who had spent time in the Society Islands (French Polynesia), advised Governor Lachlan Macquarie that some settlers at Port Macquarie had sought his assistance to establish a sugar plantation. Gyles wondered if Macquarie was interested in employing him to undertake this endeavour on behalf of the government.[9] Macquarie asked Gyles for a report on the suitability of the district for sugar cane cultivation, since he was to be a member of Surveyor-General John Oxley's tour of inspection to the area in January 1820. On his return from this journey, Gyles informed Macquarie that the land around Port Macquarie was 'capable and the climate will allow of the growth of the sugar cane, coffee, cotton and tobacco'.[10] After considering the report, Macquarie asked Gyles to visit London, bearing letters of introduction to Earl Bathurst, with whom he could discuss the project. In his despatch to Bathurst, Macquarie declined to recommend the establishment of a sugar plantation at Port Macquarie as a government measure, but suggested the government could offer some encouragement to wealthy individuals in England to undertake the project, especially as it could lead to a cheaper supply of sugar and spirits than presently obtained from India. Gyles proceeded to England, but was unable to persuade Earl Bathurst to support his venture. He was offered a grant of land in New South Wales, but declined to accept the offer, instead deciding to make his home in London.[11]

After Surveyor-General Oxley's 1820 inspection of the Hastings River district, Governor Macquarie decided to establish a settlement at Port Macquarie for persons convicted of fresh crimes in the colony. During early 1821, the settlement's first Commandant, Captain Francis Allman, supervised the erection of the prison buildings and established gardens where the prisoners could raise their own vegetables.[12] Allman also planted sugar cane cuttings in these gardens and reputedly a small crushing mill was erected in Port Macquarie's main street (see Figure 3.1). The source of Allman's sugar cane cuttings is unclear, but they may have been secured from the Sydney Botanic Gardens where a large collection of plants was being developed.[13] Allman had expanded his original stock to 7 000 sugar cane plants by March 1823, after receiving assistance from a Negro convict, James Williams, a native of Antigua who had been transported to Sydney in 1820. According to the historian Colin Roderick, Williams undoubtedly played a significant part in propagating sufficient plants from Allman's original stock to establish the sugar plantation at Port Macquarie, and it is he who should be recognised as the founder of the Australian sugar industry.[14]

Given such a promising start, why was sugar cane cultivation abandoned at Port Macquarie by the early 1830s? The answer to this question requires the introduction of another character, Thomas Alison Scott, who has acquired perhaps incorrectly the title 'Founder of the Australian sugar industry'. Colin Roderick has argued that Scott was no more than as opportunist, with little regard for the truth, and probably incompetent, either through laziness or lack of knowledge and experience which he claimed to possess. Scott was born in Glasgow in 1777, and he commenced managing his father's Antiguan estates in 1797.[15] He eventually moved to Sydney in early 1823. In August that year, Scott offered his services to Frederick Goulburn, the Colonial Secretary, for the purposes of manufacturing sugar, rum and tobacco at Port Macquarie. Goulburn did not immediately accept Scott's offer, but his repeated solicitations prevailed. In November 1823, Scott was given leave to go to Port Macquarie, without any salary, although he was provided rations. Scott applied to Goulburn in May 1824 to receive a salary, but was advised

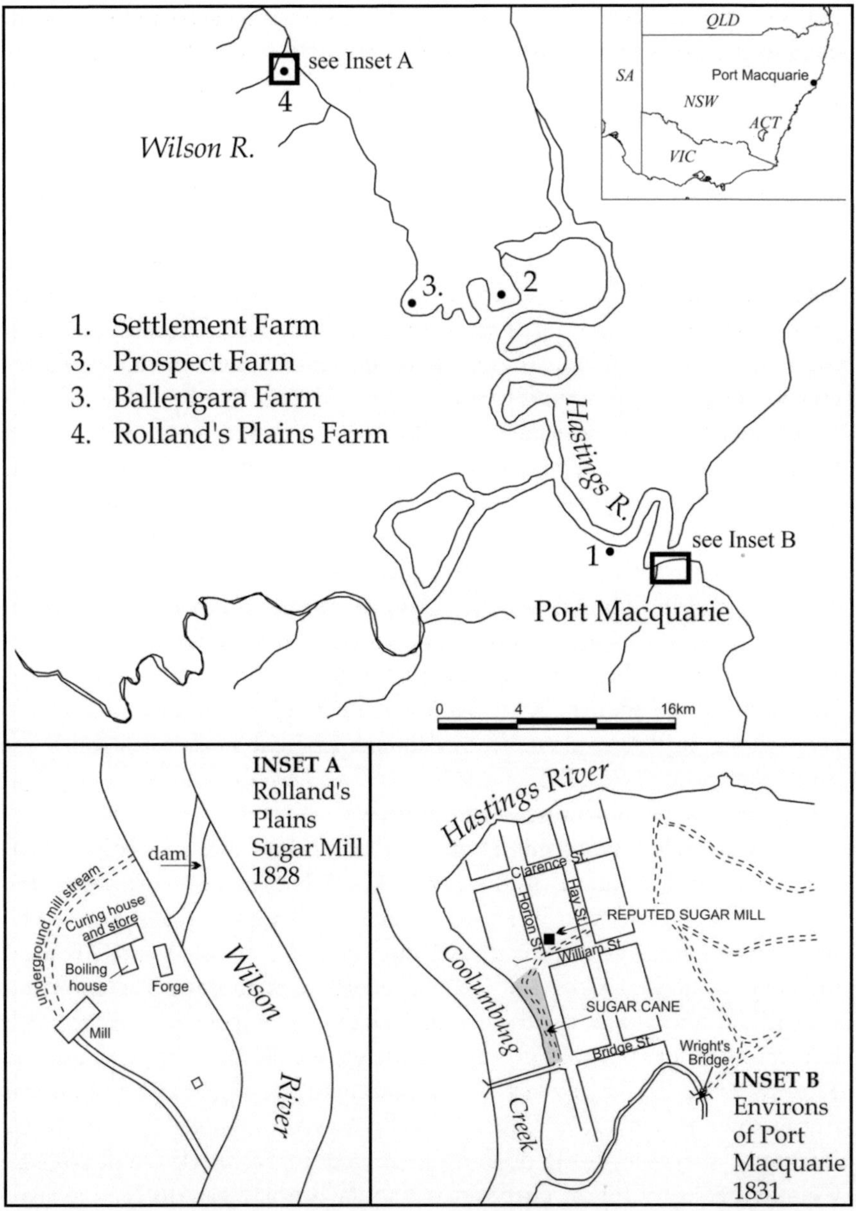

Figure 3.1. Localities around Port Macquarie where sugar cane was cultivated in the 1820s.

Source: Based upon 'Sketch Map of the Port Macquarie District, c. 1828' and 'Map of Port Macquarie, 1831, compiled by Surveyor F. D'Arcy', CSRR, Z/303, Box 36, File C.1.0., Folder 4.3, Doc. 10.

that his income would be derived from the crops of tobacco and sugar cane turned over to the Commissariat.[16] Scott's activities at Port Macquarie during 1824 are unclear. He may have been responsible for the samples of sugar made at Port Macquarie and forwarded to Colonial Secretary Golburn during the latter months of 1824.[17] What is certain, however, is that by February 1825, the Colonial Secretary advised the Commandant at Port Macquarie to interfere as little as possible in Scott's management of the sugar plantation.[18]

Scott initially concentrated on developing the sugar plantation at Prospect Farm, but by June 1825 he had started another sugar estate at Rolland's Plains (Figure 3.1). At the end of the year, 25 ha were under sugar cane. By the winter of 1826, he had erected a water-powered crushing mill at Rolland's Plains and almost 36 ha were under sugar cane.[19] Throughout 1825 and 1826, Scott continued sending samples of sugar to the Colonial Secretary. Eventually in December 1827, Scott supplied three and a half tonnes of sugar to the Commissariat Store in Sydney.[20] Only 220 kg of this sugar was sold at its first offering at auction, with the remainder being auctioned again in March 1828 along with biscuit dust, damaged biscuits and flour and wheat sweepings.[21] Scott's sugar looked suspiciously inferior. Moreover, doubts had been growing about his ability to manufacture sugar successfully.

Scott was dismissed from his position in early 1828 and not replaced. Shortly after Scott's dismissal a Commission inquired into the state of the penal establishment at Port Macquarie, including its sugar plantations. The investigation concluded that Scott had failed to sufficiently cultivate the sugar cane growing on the plantations, but more significantly the Commissioners observed that the sugar cane suffered severely from frosts.[22] Peter Cunningham, the surgeon superintendent of convict transportees, recognised the unsuitability of the district for sugar cane cultivation when he wrote in 1828 that the climate at Port Macquarie was 'too cold' for sugar cane plants.[23] Governor Ralph Darling, however, was optimistic that once the Port Macquarie district was thrown open to occupation, some settlers would cultivate sugar cane, as it had been proven that the crop could be grown in the area. He also advised London in April 1828 that 'two settlers, one of them a sugar planter from the West Indies, are preparing to cultivate the cane and manufacture sugar on the banks of the Manning River'. Nothing was ever heard again of the two Manning River settlers. Three years later, the British government ordered Governor Richard Bourke to 'immediately break up any establishment' at Port Macquarie that was devoted to sugar production, claiming that policy of producing sugar in the colony was 'very questionable'.[24] Consequently, Australia's first sugar plantation was abandoned. Moreover, settlers who tried cultivating sugar cane around Port Macquarie during the late 1860s were also defeated by frost, proving Peter Cunningham to be correct in his assessment that the district was not suited for sustained sugar production.

Official indifference

After the cessation of sugar production at Port Macquarie, the British government hardened its attitude towards the Crown manufacturing sugar in the new Australian colonies. In March 1826, Captain Patrick Logan, the Commandant of the penal establishment at Moreton Bay (site of today's Brisbane), asked for a large supply of sugar cane, tobacco and cotton seeds to form plantations producing tropical products.[25] Colonial Secretary Alexander McLeay replied that the government did not intend to carry out extensive cultivation at the Moreton Bay Penal Settlement. Only enough grain for local needs was to be grown; tobacco might be tried, but not sugar cane.[26] Yet the idea of growing sugar cane in the Moreton Bay district persisted amongst government officials. The settlement's Commandant in 1840, Major B. Walker, proposed establishing a sugar plantation using convict labour, but his scheme was not approved by Lord John Russell, the Secretary of State for War and the Colonies.[27]

Twelve years later, Sir Gordon Bremer, the Commandant at Port Essington (located in today's Northern Territory), sought guidance from Sir George Gipps, Governor of New South Wales, about the policy of allowing permanent settlement around Port Essington, following a request to establish a sugar plantation from a James Jones. Gipps replied that Bremer should not allow the establishment of a sugar plantation or any other kind of cultivation requiring considerable capital outlay by granting free title to any land.[28] Once again official reluctance to encourage sugar cane cultivation stopped the establishment of a sugar plantation. Nevertheless, the suitability of the area for the crop was highlighted in 1841 when Port Essington's Commandant, Captain John McArthur, reported that the sugar cane was 'thriving' and again in 1843 when he remarked, 'the valleys (there) were admirably adapted for the growth of sugar'.[29] Port Essington, however, never became a major trading port and the settlement was abandoned in 1849.

Despite official indifference to the possibilities of sugar cane cultivation around Brisbane, private investors became increasingly aware that the region which was to become Southern Queensland showed potential as a locality where tropical crops might be cultivated successfully. In 1833, Benjamin Sullivan, a police magistrate in New South Wales and a retired major in the British army, proposed the establishment of a joint stock company with a capital of £1 million, to acquire 405 000 ha between the mouth of the Macleay River in New South Wales and the Tropic of Capricorn for the purpose of growing sugar cane. The Colonial Office ignored this scheme and it progressed no further.[30]

The Rev. John Dunmore Lang, a fervent supporter of Protestant immigration to the northern districts of New South Wales, proposed a new colony named 'Cooksland'. It would include part of the existing colony of New South Wales from the Clarence River district (30th parallel south) to the Tropic of Capricorn (27th parallel south). He claimed this area was suitable for growing sugar cane, highlighting that Louisiana was two to four degrees further away from the equator than 'Cooksland', and suffered more severe winters than Australia, yet it was a sugar-producing region.[31] Lang believed sugar production was possible in 'Cooksland', but he needed someone who could advise new settlers on how to cultivate tropical crops. In early 1849, he arranged for Thomas Bowden to visit Brisbane Town. Bowden was definitely a former manager of a Jamaican sugar estate and probably the operator of a short-lived sugar refinery in Sydney between 1847 and 1849 (see Chapter 4). Within a month of arriving, Bowden had called a meeting of those interested in 'establishing a sugar plantation and manufactory'. This meeting resolved to form the Moreton Bay Sugar Co., and sought a capital of £3 000 to carry out the cultivation of sugar cane on the banks of the Brisbane River.[32] Bowden also outlined his scheme at a meeting in Sydney, resulting in number of Sydney-based merchants taking shares in the company.[33] The scheme, however, failed to get off the ground due to insufficient financial support from the Sydney public.

These repeated failures before 1850 to establish permanent sugar production in Australia did not deter settlers in northern New South Wales and Southern Queensland during the late 1850s from again experimenting with sugar cane cultivation. In 1858, a Frenchman named Eleonard Adam purchased nearly 210 ha near Southgate in the Clarence River district, with the aim of forming a sugar plantation. The Rev. John Dunmore Lang observed sugar cane being grown on this estate during a visit to it in 1861, but the venture failed to produce any sugar. Nevertheless, Adam proved that the Clarence River district was suited to sugar cane cultivation, and some other small farmers along the river had planted experimental patches of sugar cane by 1865.[34] Further north in 1859, sugar cane was observed growing in Edgar Thomas Aldridge's gardens at Baddow House, near Maryborough.[35] Four years later, plots containing numerous cane varieties had been developed by John Eaton at his Rosehill Estate (Maryborough district) and Captain Louis Hope on Ormiston Estate, Cleveland (near Brisbane). Walter Hill, the first Colonial Botanist of Queensland, had also established an extensive cane nursery at the Brisbane Botanic Gardens. In addition, settlers along the Caboolture and Richmond Rivers and around Port Macquarie were cultivating small crops of sugar cane by 1863.[36]

Therefore, by the early 1860s, the suitability of the environment for the cultivation of sugar cane in parts of Southern Queensland and northern New South Wales had been well and truly established. Entrepreneurs, however, were reluctant to invest in the erection of the necessary buildings and associated machinery that would make raw sugar production viable, although the domestic demand for sugar was rapidly increasing. Instead, as the discussion in the next chapter will show, businessmen had built Sydney's first sugar refineries, preferring to convert imported raw sugar into the white sugar demanded increasingly by Australian consumers, rather than pursue the cultivation of sugar cane and the production of raw sugar.

Notes

1 *HRNSW*, Vol. 1, Governor Phillip, pp. 1–4, 423–429 & 435–438.

2 D. Collins, *An Account of the English Colony in New South Wales*, Vol. 1. (1789; rpt. Sydney 1975), p. 187.

3 *HRNSW*, Vol. 2, The Norfolk Journal of Lt. King, 24 March 1788, p. 559.

4 *HRNSW*, Vol. 1, p. 420.

5 P. G. King to Lord Glenville, 15 January 1792 and P. G. King to Mr Secretary Dundas, 8 May 1792, in P. King's Norfolk Island Letterbooks, 1788–1789, ML Microfilm No. C 187, pp. 90 & 97.

6 P. King to Sir Joseph Banks, 3 March 1793, Papers of Sir Joseph Banks, Botanical and Horticultural 1789–1796, vol. 18, folio 150, A 81, ML.

7 P. G. King, Norfolk Island Journal, 1791–1794, entry for May 1793, p. 130, A 1687, ML.

8 *Sydney Gazette,* 22 January 1804, p. 1; J. Backhouse, *A Narrative of a visit to the Australian Colonies* (1843; rpt. New York 1967), pp. 268–269.

9 John Gyles to Governor Macquarie, 13 December 1819, *HRA*, Series 1. Vol. 10, p. 258.

10 Report of Mr John Gyles to Governor Macquarie, 18 February 1820, *HRA*, Series 1. Vol. 10. p. 261.

11 Governor Macquarie to Earl Bathurst, 24 February 1820, *HRA,* Series 1. Vol. 10, p. 258; F. Eipper, 'The birth of the sugar industry', *J. of the Scone and Upper Hunter Historical Society*, Vol. 2 (1961), pp. 110–111.

12 Captain Francis Allman to Colonial Sec. Golburn, 24 March 1823, AONSW, NSW Colonial Sec. In-letters, Port Macquarie, 1821–24, 4/1815, Reel 6068, pp. 145–147; B. Levick, 'Scott's sugar mill discovered at Rolland's Plains, Port Macquarie,' CSR, *Newsletter*, No. 99 (May 1967), p. 22 (CSRR, Z/303, Box 36, File C1.0, Folder 4.3, Doc. 41).

13 In the history of the Royal Botanic Gardens, Sydney, no specific mention is made of sugar cane growing in the gardens until 1871, but it is reasonable to assume that sugar cane was cultivated in the gardens much earlier (see L. Gilbert, *The Royal Botanic Gardens, Sydney. A History 1816–1985* (Melbourne, 1986), p. 103).

14 C. Roderick, 'T.A. Scott and his works at Port Macquarie', *J. of the Royal Australian Historical Society*, Vol. 44 (1958), pp. 23–29.

15 V. Parsons, 'Scott, Thomas Alison 1777–1881', *ADB*, Vol. 2 (1967), pp. 430–431; C. Roderick, 'T. A. Scott and his work at Port Macquarie 1823-28', *J. of the Royal Australian Historical Society*, Vol. 44 (1958), pp.1–47; J. Jervis, 'Thomas Alison Scott and the genesis of the sugar industry', *J. of the Royal Australian Historical Society*, Vol. 26 (1941), pp. 387–396.

16 Thomas Scott to Colonial Sec. Golburn, 24 March 1823, 9 August 1823, 14 October 1823 and 29 December 1823, AONSW, Colonial Sec. In-letters, 4/1815, Reel 6068, pp. 235–249 & 325–7; Roderick, 'T.A.Scott', p. 4.

17 Commandant, Port Macquarie to Colonial Sec., 4 September 1824, 10 October 1824 and 3 November 1824, AONSW, Colonial Sec. In-letters 4/1815, Reel 6068, pp. 437–438, 457–458 & 515–516.

18 Colonial Sec. to Commandant, Port Macquarie, 28 February 1825, AONSW, Colonial Sec. Out-Letters, 4/3864, Reel 6019, p. 225.

19 Report of the Commissioners of Inquiry at Port Macquarie, *HRA*, Series 1, Vol. 15, p. 493.

20 Roderick, 'T.A. Scott', pp. 10–13; Advertisements on page one, *Sydney Gazette*, 19 to 26 December 1827.

21 Advertisements on page one, *Sydney Gazette*, 25 February to 5 March 1828.

22 Report of the Commissioners of Inquiry at Port Macquarie, *HRA*, Series 1, Vol. 15, p. 493.

23 P. Cunningham, *Two Years in New South Wales* (1827; rpt. Sydney 1966), p. 219.

24 Governor Darling to Under Sec. Hay, 5 March 1828 and Governor Darling to the Rt Hon. W. Huskisson, 10 April 1828, *HRA*, Series 1, Vol. 14, pp. 15 & 135; Viscount Goderich to Governor Bourke, 29 September 1831, *HRA,* Series 1, Vol. 16, p. 395.

25 Logan to Colonial Sec. McLeay, 19 March 1826, Logan Letterbook, JOL.

26 Colonial Sec. McLeay to Logan, 10 May 1826, AONSW, Colonial Sec.'s Letters to Moreton Bay, 4/ 3794, Reel 749, p. 34.

27 Major B. Walker to Sec. of State for War and the Colonies, 1 October 1840, London Documents Relative to Moreton Bay, Vol. 1, p. 426.

28 Spillet, *Forsaken Settlement*, p. 50

29 Governor Gipps to Lord Stanley, vol. 40, 1842, p. 272, A1229, ML; Governor Gipps to Lord Stanley, vol. 42, 1843, p. 1374, A 1231, ML.

30 Major Benjamin Sullivan to Sec. of State for War and the Colonies, 16 September 1833, London Documents Relative to Moreton Bay, Vol. 1, p. 341.

31 *Moreton Bay Courier*, 5 May 1849, p. 2.

32 *Moreton Bay Courier*, 26 May 1849, p. 2 & 2 June 1849, p. 2.

33 *Sydney Morning Herald*, 24 August 1849, p. 3; see also A. Barnard, *Visions and profits; studies in the business career of Thomas Sutcliffe Mort* (Melbourne, 1961), p. 43. Mort was a Sydney member of the provisional committee of the Moreton Bay Sugar Co.

34 C. Tindal to T. Lloyd, 10 March 1858 in N. Mackey (ed.), *Tindal Letters: source material* (Grafton 1991), p. 53; J.D. Lang, *Queensland, Australia* (London, 1861), pp. 25–26; R. Davies, 'The Clarence River and the City of Grafton: economic and social development'. MA thesis, Uni. of Sydney, 1953, p. 140.

35 J. Kerr, *Sugar at Maryborough Sugar: 120 Years of Challenge* (Maryborough, 1987), p. 3.

36 *MC*, 15 October 1864, p. 2; for sugar cane in the Brisbane Botanic Gardens see *BC*, 14 January 1862, p. 2; Hope's nursery is reported in *BC*, 18 October 1862, p. 5 & *J. of The Society of Arts*, 15 January 1864, p. 140; sugar cane cultivation at Port Macquarie is described in *Sydney Morning Herald*, 9 November 1863, p. 5 & *Sydney Mail*, 30 April 1864, p. 2; for early sugar cane growing along the Richmond River see L. Daley, *Men and a River. Richmond River District 1828–1895* (Sydney 1966), p. 137.

Chapter 4

The Rise of the Colonial Sugar Refining Company

The failure of the new settlement in New South Wales to manufacture its own sugar between 1788 and 1860 meant that the product had to be imported. Sydney was a long distance from Java, Bengal, the Philippines and Mauritius, the nearest established centres of sugar production. Because trading links with the outside world were few and irregular, sugar supplies at times were limited or even unavailable. Once more regular trading networks became established, these constraints were slowly eliminated, but supply and demand was sometimes mismatched, especially as the amount of sugar available for export from the centres of production was not consistent or necessarily abundant.[1] Moreover, sugar was a difficult product to transport by sea during the nineteenth century, because it spoils when wet and the casks of poor quality sugar sometimes leaked molasses. Not all the colony's first merchants were interested in importing such a risky product.

The earliest supplier of sugar to the new colony was the Commissariat. Campbell, Clarke & Co. of Calcutta was awarded several contracts between 1792 and 1820 to supply the Commissariat with sugar from Bengal.[2] Despite these contracts, the Crown's attempts to meet the demand for sugar were not always successful. In 1797 the supply of sugar had been exhausted, and the Commissariat was forced to purchase sugar from a ship that had just arrived from Rio de Janeiro. The sugar supply was again exhausted in late 1798, and those dependent on the Commissariat were issued with wheat instead of sugar. In 1819 the government stockman at Bathurst complained that he had been without sugar for fourteen weeks because supply by the authorities was irregular. As early as 1792 a group of resourceful officers sought to ease this shortage of sugar (and other products) by commissioning the *Britannia* to proceed to Cape Town to buy goods, including sugar.[3]

To counter the Commissariat's monopoly on imports of essential items, independent traders began landing goods in Sydney. Some of the main suppliers of sugar included William Tough representing Chace, Chinnery & Co.

of Madras, Simeon Lord and William Baker. Despite their presence, the market according to Ellis Bent, Governor Lachlan Macquarie's Deputy Judge Advocate, was 'either overstocked or completely empty.'[4] The merchants may have been unable to estimate the next scarcity of sugar, given its popularity. The Calcutta merchants, however, compounded the problem by their irresponsible behaviour, exporting goods to Sydney on speculation, without taking any account of the possible demand. In the early 1810s, for example, there was a glut of sugar. Alexander Riley reported in 1812 that 8 000 bags of sugar were present in the colony or two and half years supply for the residents.[5] The pioneering importers of sugar give way to new commercial figures such as Daniel Cooper of Cooper and Levey (sometimes also know as Waterloo & Co.) and Stuart Donaldson, who both established trading links with Mauritius, and Robert Towns & Co., importers of sugar from Manila, China and Singapore. Other merchants imported sugar from the English refineries.[6]

Calculating the approximate amount of sugar consumed per capita between 1800 and 1830 is not possible because the colonial authorities used tons, bags, mounds, tubs, jars, casks and hogsheads to record the amount of sugar imported into New South Wales and Tasmania, making any calculations of the total weight imported impossible. Governor Sir Ralph Darling, however, tried to calculate the amount of sugar consumption in 1828 by dividing the yearly value of sugar imports by the average cost for half a kilogram of sugar. He concluded that around 20 kg of sugar was being consumed per capita in New South Wales. Sugar import and export statistics were uniformly collected for New South Wales from 1830 onwards, enabling the calculation of the apparent sugar consumption per capita.[7] During the 1830s, the average annual apparent sugar consumption per capita in New South Wales was at least 45 kg. Therefore, the historian John B. Hirst was probably right in his assertion that the annual per capita sugar consumption in early New South Wales was outstandingly high and exceeded that of Great Britain.[8] Sugar permeated every aspect of colonial life. Australians used sugar to sweeten their tea, as an ingredient in many recipes, to make jams, chutney and relishes, to brew their beer and distil their rum and as a medicine. Many rural workers received a standard weekly ration that contained 0.9 kg (two pounds) of sugar. Sugar was also part of the convict rations by 1795, to be given when labour was compliant and withheld as a form of punishment. 'Sugar was power', asserted the historian Robert Dare in his assessment of the relationship between food and authority in colonial Australia.[9]

The Australian Sugar Company

From 1820 onwards, British capital was invested in the New South Wales pastoral industry and other colonial manufacturing activities, particularly flour milling, shipbuilding and boiling down works (i.e. tallow production).[10] However, given the growing demand for sugar in the Australian colonies, it is not surprising that a group of London businessmen decided to invest in a venture that would try and meet this demand. Francis Kemble, London company promoter, and William Knox Child, a magistrate and Inspecting Director of the London and County Joint Stock Bank, headed the investors who formed the Australian Sugar Co. in 1839. Lured by Kemble's grandiose predictions that the venture would make a profit of £40 000 a year, Child sold his estate to raise the capital to invest in a plan to build a sugar refinery in Sydney which would convert imported raw sugar into white sugar.[11]

Steam engines, machinery used to refine sugar and boilers valued at £20 000 arrived safely in Sydney in July 1840. Skilled workers were recruited in Britain to erect the refinery. A block of 25 ha at Canterbury on the Cook's River, about 10 km from inner Sydney, had been purchased from Robert Campbell Jnr., a Sydney merchant. Construction of the refinery was undertaken during the latter half of 1840; stone was quarried from the ground for the buildings. By September 1841, the new sugar refinery was described as a building of 'one hundred feet in length, sixty feet in width and sixty feet high, with a fine chimney shaft, one hundred feet from the ground'. The factory was capable of processing 50–60 tonnes of raw sugar a week.[12]

The promising beginnings to the business venture, however, ended following the partnership split on some matter of dissension between Child and Kemble. The sugar refinery under the original management never began operations. Kemble and Child reconciled their differences and persuaded several Sydney gentlemen and merchants to invest in a new company, the Australasian Sugar Co. (ASC). The new company that took over the assets of the Australian Sugar Co. was eventually formed in March 1842 with a capital of £23 000 provided by sixty shareholders. Another six months elapsed before the venture finally commenced, using imported raw sugar from the Philippines. Towards the end of 1842, the ASC advertised that it could supply all the Australian colonies with loaves of sugar, lump sugar and molasses. Orders were received from Hobart, Launceston, Port Phillip and Adelaide, leading the directors to remark: 'the demand has upon the whole much exceeded any expectations previously entertained'.[13]

Demand for the product may have been brisk, but the ASC was not being managed very competently. The quarrel between Child, the refinery manager, and Kemble flared again in late 1842, with 'charges and counter-charges of deceit, interference and negligence' being exchanged between both men. Ensuing legal action resulted in Child being forced to resign from the company and banned from entering the sugar refinery; his attempt to bring a lawsuit involving charges of libel against Kemble was also unsuccessful. Against this backdrop, the refinery was short of working capital and had been forced to borrow from the Australian colonial banks. The shareholders were in heavy arrears and the refinery was also not generating enough income. Instead of processing the 20 to 50 tonnes of raw sugar a week needed to make the business pay, the refinery had been working only 16 tonnes of raw sugar.[14]

In August 1843, the ASC advertised for a manager to take Child's place. The twenty-four year old Edward Knox was appointed to the position. Knox had arrived in Sydney on 26 January 1840, and by May 1843 he was the manager of the Australian Auction Co.[15] Upon joining the ASC, Knox found the company in poor financial shape; a debt of £10 000 had to be met. However, the company's shareholders were reluctant to advance any further funds for the struggling refinery. The company's directors agreed to shed some workers and raised new loans, using the company's existing raw sugar stocks as collateral. Twenty-three shareholders were also asked to take 30 cwt (1 500 kg) of sugar and sell it through their own businesses. From June 1844 onwards, Knox began to improve the fortunes of the company, adopting the principle of 'always working for the market'. Profits returned, the debt was slowly reduced and the company survived the threat of bankruptcy. Knox was confident enough of the company's future to become one of its directors. In 1847, Knox resigned, explaining that his other duties as managing director of the Commercial Banking Co. of Sydney prevented him from any longer being the ASC's manager. However, he remained an ASC director.

Eliminating rival refiners and sugar importers

Knox's position with the ASC was taken by Archibald Ashdown, who was immediately faced with the challenge presented by the establishment of two rival sugar refineries in Sydney. Thomas Bowden, probably a former Jamaican sugar plantation manager, erected his Sydney Sugar Works in 1847.

In the same year, Robert Cooper, a Sydney publican, brewer and distiller, built his Brisbane House Sugar Refinery, alongside his distillery in Paramatta Street, Ultimo, inner Sydney. The ASC decided to eliminate these rivals and commenced offering its products at a lower price than the opposition, even if it meant occasionally doing so at below the cost of production. The strategy was successful, for by July 1849 both rival refineries had closed. A decade later Sydney merchants alleged Bowden was 'ruined' (and probably went to Queensland at the urging of the Rev. John Dunmore Lang) and Cooper had become insolvent because of the determined opposition from the ASC. To ensure that no other competitors could use the buildings and machinery, the ASC obtained the leases over the properties. In 1852, the directors decided to close the sugar refinery at Canterbury, due to difficulty in securing both sufficient workers to conduct the operation and reliable transport for the refinery's output. Arrangements were made to concentrate operations at the Brisbane House Refinery in Ultimo.[16]

In 1848, Ashdown was also asked to implement the Board's decision to expand the company's business by selling larger consignments of sugar in other Australian colonies. By 1850, this aim had now become complete domination of the Australian market in refined sugars. Ashdown advised Messrs Griffiths, Fanning & Co., the Melbourne agents for the ASC, that the directors wanted to secure the 'entire' Victorian market. In the same year, he told Messrs T. Chapman, Hobart merchants, that the ASC sought the 'entire possession' of the Tasmanian market. Again this strategy involved underselling the merchants who imported refined sugars from Great Britain, Java or Mauritius by offering the ASC's refined sugar at reduced prices. The ASC's directors realised such a strategy posed considerable financial risks, but they argued that the firm was willing to 'submit to a small loss' in order to procure the permanent supply of sugar to customers and to 'drive out the English sugars altogether'. Melbourne merchants Messrs Griffiths, Fanning & Co. were instructed to sell the company's sugar 'at all times at less than the price of English refined, whatever it may be'.[17]

Establishing the Colonial Sugar Refining Company

Dissension amongst the shareholders of the ASC erupted again in 1854, forcing the company into liquidation. On 1 January 1855, the Colonial Sugar Refining Co. (CSR) was established to take over some of the assets of the ASC and to continue the business of sugar refining. Edward Knox, by then Chairman of the Commercial Banking Co., held a third of the capital of £150 000 raised to operate the new company. Other investors included prominent Sydney merchants Ralph Robey, Walter Lamb and William Walker, and the brewer Edwin Tooth. Operations were concentrated in the Brisbane House Refinery, and the factory's machinery was upgraded during 1856 by the installation of new steam-driven sugar-making equipment. CSR also purchased Bowden's old refinery in 1856, with a view to safeguarding their activities against the threat of fire and to ensure that the company had the capacity to meet the growing demand for sugar in New South Wales.[18] Like its predecessor, CSR aimed at a monopoly over all the refined sugar consumed in the Australian colonies and was also considering expanding operations to New Zealand.

CSR flourished for two years. In 1855, a dividend of 20% was declared; this amount was increased to 50% in 1856.[19] Moreover, the growing demand for refined sugar in Victoria prompted CSR's partners to form a subsidiary company in that colony. Edward Knox visited Melbourne and persuaded several Melbourne merchants to invest in the new venture. The Victoria Sugar Co. (VSC) was formed in June 1857 with a capital of £150 000. CSR shareholders owned half its shares. This new organisation spent £90 000 building a large sugar refinery at Sandridge, now Port Melbourne. An early newspaper description of the refinery reported that the main building was as a 'quadrangular structure of bluestone, 94 feet in height, the roof being of metal, and somewhat in the style of the Tudor roof'. The refinery became operational in mid-1859, relying on raw sugar imported from the Philippines to initially produce 250 tonnes of refined sugar a week.[20]

The decision to establish operations in Victoria had been preceded six months earlier by the unpleasant discovery that James Robey, CSR's refinery manager and shareholder, had been planning the establishment of a rival refinery in Sydney. After his dismissal from CSR, James Robey in partnership with his brother Ralph and George Dibbs, the son-in-law of Ralph, formed Messrs Robey & Co. and began operating a sugar refinery on the North Shore at Oyster Cove.[21] A greater challenge, however, emerged towards the end of 1857. The world price of sugar suddenly fell, catching CSR with a large stock of raw sugar on its hands, and imports of sugar at ever lower prices began flooding the Australian market. In December 1857, imported refined sugar from Mauritius sold in Sydney at £7 a ton less than the price offered by CSR. By August 1858, CSR had suffered a loss of £85 000. Moreover, James Robey was threatening to dissolve the company.[22] Knox, who had returned to England in late 1857 after selling his house in Sydney and some shares in CSR, began receiving many letters, advising him that he was ruined and his fortune in jeopardy.

Upon returning from overseas in early 1858, Knox commenced putting the company's affairs in order. He vowed never again to pay such high dividends, but instead insisted on building up capital reserves, especially before embarking upon any expansion. Before his departure from England, Knox had secured a loan from the English shareholders to maintain operations until the price of sugar recovered. After legal advice, Knox assured shareholders that a single shareholder could not dissolve the company against the wishes of all the co-partners.[23] Knox also turned his attention to the new competition provided by Messrs Robey & Co. CSR again lowered its sugar prices to undersell the rival refiner, although Knox in November 1858 instructed Joseph Grafton Ross, CSR's new refinery manager, to try and come to 'any arrangement' with the Robey brothers to reduce opposition to CSR's sugar and treacle. Initial negotiations failed to achieve a satisfactory outcome, and Messrs Robey & Co. tried in February 1859 to form the North Shore Refinery Co. They aimed to raise £90 000, and expand operations from a weekly output of 30 tonnes to 100 tonnes. Financial support for this scheme failed to materialise. Threatened with ruin by CSR's continued and aggressive price reduction policy, Messrs Robey & Co. sought safety from insolvency by agreeing in September 1859 to sell the plant and machinery to CSR for £30 000. Sir Daniel Cooper, CSR's Chairman of Directors, told shareholders that although Messrs Robey & Co.'s opposition was 'currently weak', the business had the potential to grow and seriously interfere with CSR's share of the refined sugar market in New South Wales.[24]

In 1860, CSR's share of the New South Wales market was challenged again when Isaac Peck opened a small sugar refinery at Alexandria, an inner Sydney suburb. Peck manufactured about 4 tonnes of white sugar a week, relying on raw sugar imported from the Philippines. CSR decided to ignore this rival, preferring to tackle the strong competition from unnamed importers of sugar, described as 'local merchants'. Perhaps these merchants were Robert Towns & Co. and Thacker Daniel & Co. who published a pamphlet in 1862 criticising CSR's business strategies. Whoever these rival merchants may have been, CSR decided in 1859 to tackle what the firm perceived as the unfair advantage they had in the business of importing and selling sugar in Australia. CSR argued that its refining operations were not on equal footing with the merchants who imported refined sugars. The refiners were required to hand the government 3–6s per ton for the privilege of importing the sand and clay which made up 5–10% of the raw sugar. They passed on the cost of removing these impurities to consumers, making their sugars dearer than the refined sugars imported by the merchants. Moreover, if CSR or the VSC exported their refined sugars to another Australian colony, both firms paid import duties on refined sugar, further adding to the cost of their sugar. CSR successfully petitioned the governments of Victoria (1860) and New South Wales (1862) to set up bonded sugar houses. As a result, Australian refiners were able to import their raw sugar, extract the contaminants, and only pay the duty on the refined sugar sold to consumers.[25] This arrangement then put them on an equal footing with the merchants and set the scene for a prolonged struggle to control the supply of refined sugar filling the country's sugar basins during the late nineteenth century (see Chapter 9).

Notes

1 Galloway, *The Sugar Cane Industry*, pp. 210–212, 216 & 220.

2 M. Steven, *Merchant Campbell, 1769–1844. A Study in Colonial Trade* (Melbourne, 1965), pp. 27, 62, 67 & 254.

3 J. Cobley, *Sydney Cove 1795–1800* (Sydney, 1986), p. 150; J. Cobley, *Sydney Cove 1793–1795* (Sydney, 1983), p. 49; *HRA*, Series 1, Vol. 2, p. 211; J. Ritchie (ed.), *The Evidence of the Bigge Reports*, Vol. 1 (Melbourne, 1971), p. 178.

4 D. Hainsworth, *The Sydney Traders. Simeon Lord and His Contemporaries, 1788–1821* (Melbourne, 1981), pp. 88 & 109; for William Baker's advertisements see *Sydney Gazette*, 9 March 1806, p. 1.

5 Governor King to Lord Hobart, 16 April 1804, *HRA*, Series 1, Vol. 4, p. 615; Hainsworth, *The Sydney Traders*, p. 99.

6 J. Davidson, 'Daniel Cooper, 1785–1853', *ADB*, Vol. 1, 1966, pp. 244–245; A. Atkinson & M. Aveling (eds.), *Australians 1838* (Sydney, 1987), p. 176; F. Broeze, *Mr Brooks and the Australian Trade. Imperial Business in the Nineteenth Century* (Melbourne, 1993), p. 210.

7 Calculated by adding together the amount of sugar imports and any sugar production, and subtracting the amount of exports, divided by the total population.

8 *HRA*, Series 1, Vol. 4, p. 615; J. Hirst, *Convict Society and its Enemies. A History of Early New South Wales* (Sydney, 1983), p. 419; P. Griggs, 'Sugar demand and consumption in colonial Australia, 1788–1900', in R. Dare (ed.), *Food, power and community. Essays in the history of food and drink* (Adelaide, 1999), p. 77.

9 Griggs, 'Sugar demand and consumption in colonial Australia, 1788–1900' and R. Dare, 'Introduction', in R. Dare (ed.), *Food, power and community. Essays in the history of food and drink* (Adelaide, 1999), pp. 12 & 82–87.

10 G. Linge, *Industrial Awakening. A Geography of Australian Manufacturing 1788 to 1890* (Canberra, 1979), pp. 105–115.

11 A. Birch, 'The origins of the Colonial Sugar Refining Company 1841–1855', *Business Archives and History*, Vol. 5, no. 1 (February 1965), pp. 22–23.

12 Birch, 'The origins of the Colonial Sugar Refining Company 1841–55', p. 23; *The Australian*, 7 September 1841, p. 3 & 28 September 1841, p. 2.

13 Birch, 'The origins of the Colonial Sugar Refining Company 1841–55', pp. 23–24.

14 *The Australian*, 17 September 1842, p. 2; Birch, 'The origins of the Colonial Sugar Refining Company 1841–55', pp. 24–25.

15 A. Lowndes, 'Knox, Sir Edward (1819–1901)', *ADB*, Vol. 5 (1974), p. 39 & A. Birch and J. Blaxland, 'The Historical Background' in A. Lowndes (ed.), *South Pacific Enterprise. The Colonial Sugar Refining Company* (Sydney, 1956), pp. 14–15.

16 Proceedings of the Annual General Meeting of the shareholders of the Australasian Sugar Co., 27 July 1847 & Extracts of Board Minutes, Australasian Sugar Co., 15 September 1852, CSRR, Z/303, Box 13, File A 5.0, Folder 3, Doc. 31 & 32; Towns & Co., Brown & Co. & Daniel Thacker & Co., *Reply to a pamphlet published by the Colonial Sugar Refining Company*, 8pp pamphlet, 1862, pp. 6–7, CSRR, Z/303, Box 9, File A 3.0, Folder 5, Doc. 11.

17 A. Ashdown, Sydney, to Messrs W. Hall & Son, Melb., 16 December 1848; A. Ashdown, Sydney, to Messrs Griffiths, Fanning & Co., Melbourne, 14 December 1849; A. Ashdown, Sydney, to Messrs Griffiths, Fanning & Co., Melb., 8 October 1850, all in CSRR, 142/2764; A. Ashdown, Sydney, to Messrs Griffiths, Fanning & Co., Melb., 28 February 1851, p. 69 & A. Ashdown, Sydney, to Messrs T. Chapman, Hobart, 16 August 1851, p. 192, CSRR, 142/2765.

18 Lowndes, 'Knox, Sir Edward (1819–1901)' p. 39; *The Argus* (Melbourne), 18 November 1856, p. 6; Birch and Blaxland, 'The Historical Background', p. 12.

19 CSR, Half Yearly Reports, 1855 & 1856, CSRR, Z/303, Box 3, File A 1.1, Folder 1, Doc. 1.

20 *Melbourne Argus*, 14 March 1859, p. 5; Birch and Blaxland, 'The Historical Background', p.18.

21 S. Edgar, 'Robey, Ralph Mayer (1809–1864)', *ADB*, Vol. 6 (1976), pp. 47–48; *Sydney Morning Herald*, 27 January 1859.

22 CSR, Half Yearly Reports, 1858, CSRR, Z/303, Box 3, File A 1.1, Folder 1, Doc.1; Birch and Blaxland, 'The Historical Background', p. 19; J. Grafton Ross, Sydney, to E. Knox, 10 December 1857, CSRR, Z/303, Box 2, File A 1.0, Folder 1, Doc. 45.

23 Birch and Blaxland, 'The Historical Background', p. 19; Extracts from CSR Board Minutes, 30 August 1858, CSRR, Z/303, Box 3, File A1.1, Folder F2, Doc. 4.

24 E. Knox, Sydney, to J. Grafton Ross, Manila, 30 November 1858, CSRR, Z/303, Box 70, File M 1.0, Folder 3; Towns & Co., Brown & Co. & Daniel Thacker & Co., *Reply to a pamphlet published by the Colonial Sugar Refining Company*, 8pp pamphlet, 1862, p. 7, CSRR, Z/303, Box 9, File A 3.0, Folder 5, Doc. 11; CSR, Extract from Directors' Report to Shareholders for Half-Year ended December 31, 1859, CSRR, Z/303, Box 14, File A 5.0, Folder 6, Doc. 9; Prospectus of the North Shore Sugar Refinery Co., 7 February 1859, CSRR, Z/303, Box 14, File A 5.0, Folder 6, Doc. 6.

25 E. Knox and I. Peck, 'Minutes of evidence before the Select Committee on the Bonded Distilleries and Sugar-Houses Bill', *NSWVP*, Vol. 2 (1862), pp. 669–671 & 702; G. Patterson, *The Tariff in Australian Colonies* (Melbourne, 1968), pp. 25–26.

Part Two

1864 TO 1914: BECOMING ESTABLISHED

Sewing bags of sugar at Bingera Mill, near Bundaberg, 1902. Bingera, one of the largest sugar mills in the Bundaberg district, was established in the mid-1880s during the rapid expansion in the Australian sugar industry.

Source: Picture 298, Fryer Memorial Library. Image reproduction courtesy of The University of Queensland Libraries, Brisbane.

Chapter 5
The Spread of Sugar Cane Cultivation

In 1863, Queensland parliamentarians asked Robert Herbert, the Premier of Queensland, if his government intended to encourage the widespread cultivation of sugar cane by offering a bonus to those who grew the crop. They were advised that the Queensland government had no such plans.[1] However, the lack of interest shown by Queensland's earliest legislators in the possibility of sugar cane cultivation in their new colony was not shared by all parliamentarians. During the early 1860s, Captain Louis Hope – the seventh son of the Earl of Hopetoun and a member of the Legislative Council – had been extending the amount of sugar cane cultivated on his property at Ormiston, on the eastern outskirts of Brisbane.

To advance his endeavour, Hope had employed the immigrant John Buhôt in mid-1862 as the estate's overseer. Buhôt had arrived in Brisbane in April 1862, after gaining experience on Barbados sugar plantations as both a sugar boiler and estate manager. He had impressed the local agricultural fraternity in June 1862 by converting some juice extracted from the sugar cane found in the Brisbane Botanic Gardens into granulated sugar that was described as being 'very suitable for human consumption'. Hope, however, had dismissed Buhôt by late 1862, probably because of incompetence – one Queensland parliamentarian described Buhôt in 1874 as a 'terror of sugar growers' because his cultivation methods caused crop failure.[2] Undeterred by Buhôt's departure, Hope decided in 1863 to erect the colony's first sugar mill with all the necessary equipment to make raw sugar on a commercial scale. Operations at Ormiston Mill commenced in early June 1864, under the supervision of the engineer James Strahan. Hope's first batch of raw sugar was successfully sold at auction in Brisbane on 17 June 1864, marking the beginning of permanent commercial sugar production in Australia.

Hope's success induced a sudden change in official attitude towards sugar cane cultivation in Queensland. In August 1864, Premier Herbert decided to stimulate tropical agriculture in the colony, by introducing Sugar and Coffee Regulations. This legislation allowed intending sugar or coffee planters to lease one block of land of between 320 and 1280 acres (130–520 ha) in size, limited to within 17 km (10 miles) of the coast. Lessees were required to pay annual rentals of 1s per acre for a three-year lease only, which would not be renewed. To freehold the land at the price of £1 per acre, the lessee had to convince the government that one-twentieth of the block had been cultivated with either sugar and/or coffee and that a sum of 20s or more for each acre leased had been spent on improvements. These initial regulations were amended in 1866 when annual rentals were increased to 2s 6d per acre and annual leases were extended to eight years. The government took this step to make them comparable with the *Leasing Act* of 1866, which gave settlers eight years to purchase their block of land. The Sugar and Coffee regulations were incorporated into the *Crown Lands Alienation Act* of 1868, and functioned until new land legislation in 1876 abolished their existence.[3]

Other circumstances also contributed to the growth of an Australian sugar industry during the late 1860s. The Queensland government's decision to encourage sugar cane cultivation came shortly after the initial exploration of some of the new colony's northern districts. The explorer George Dalrymple (later the Commissioner for Crown Lands, Kennedy Pastoral District) had reported favourably in 1859 and 1860 upon the potential for tropical agriculture in the Port Denison (later Bowen) and Lower Burdekin districts, while the pastoralist John Mackay in 1860 had discovered the Pioneer River valley (west of today's city of Mackay) which looked a very suitable place to commence sugar production.[4] Consequently, new northern districts with environments ideally suited to sugar cane cultivation were being opened to European occupation.

Larger investors and small farmers in coastal New South Wales and Queensland were also seeking alternatives, following the failure of staple crops. The early excitement over the potential of cotton in Queensland had been premature, as yields were disappointingly low. The hopes of many Queensland cotton companies had been ruined by a combination of the Queensland climate, their lack of expertise and the improvement of the American cotton supply towards the end of the Civil War. Consequently, Queensland's cotton growers sought new crops. Sugar cane seemed an ideal alternative, and it had replaced cotton as the most significant crop in Southern Queensland by 1876, except in the Ipswich district. In coastal central New South Wales, small farmers also sought alternative crops, following the failure of the wheat crops due to stem rust and the continuing low maize prices.[5] Again, sugar cane was considered a possibility, especially after 1870, following the erection of milling facilities, particularly those central sugar mills built by CSR (see Chapter 6 for more details). Moreover, CSR assisted farmers expand sugar cane plantings, by advancing them £6 for every acre of cane planted, free of interest.[6]

Labour for Queensland's expanding agricultural sector was in short supply by the mid-1860s, and became even harder to secure after the Gympie gold rush of 1867. Pacific Islanders had been brought to Queensland since 1863 to work on the pastoral properties and sugar and cotton plantations, but without the approval of the Queensland government. In 1868, Robert Mackenzie's government, despite some opposition, assisted the colony's sugar planters by sanctioning and regulating the introduction of Pacific Islanders into Queensland. In the following year, the Queensland government, now under the leadership of Charles Lilley, decided to maintain the duty of 5s per ton on raw sugar imported into the colony. This duty had been in existence since 1855, when Queensland was still part of New South Wales.[7] Thus, Queensland's pioneer canegrowers were protected from sugar imports and cheap labour was made available to them, which may not have been obtainable if the colony had decided to rely solely on the immigration of Europeans for its workers.

Pioneering experimentation and speculative expansion, 1870–1884

During 1865 and 1866, the Queensland Department of Public Lands approved the selection of 15 000 ha under the Sugar and Coffee Regulations, with 60% of this land being located in the Mackay and Beenleigh districts.[8] As a result of the selectors having to meet the conditions of their leases, sugar cane was simultaneously cultivated in several coastal districts throughout Queensland during the late 1860s. By 1874, 59 000 ha were under sugar cane in Queensland, compared to less than 40 ha a decade earlier (Figure 5.1). This expansion of sugar cane cultivation in Queensland was halted in 1875 by the outbreaks of a disease known locally as 'rust' (see Chapter 7). Yields were reduced substantially in the paddocks of cane affected by the disease, and in the most severe cases, the disease killed entire paddocks of cane.

In 1874, most sugar cane cultivated in Queensland was found in districts south of the Tropic of Capricorn (Figure 5.2). Nearly 20% of this total amount was located in the Brisbane district, being grown along the alluvial flats fronting the Brisbane, Mooloolah and Caboolture Rivers, at Cleveland and Tingalpa on the eastern outskirts of Brisbane and at the St Helena Island Penal Establishment in Moreton Bay.[9] West of Brisbane, sugar cane was being planted by farmers around Ipswich, despite the threat of frost.[10]

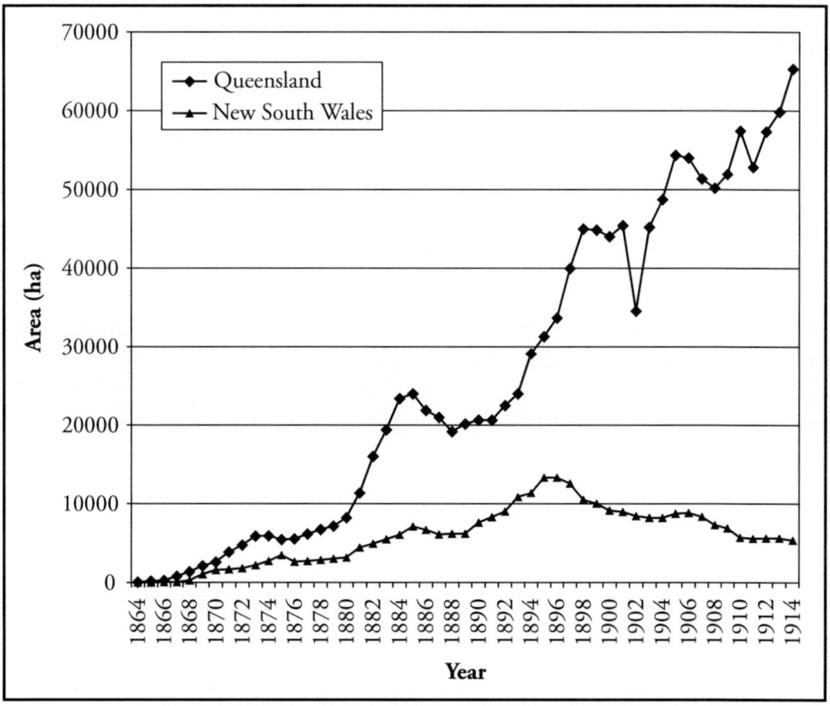

Figure 5.1. Area (ha) under sugar cane in Queensland and New South Wales, 1864–1914.

Source: Based upon data in the *Statistical Registers of New South Wales, Statistics of the Colony of Queensland* and *Statistics of the State of Queensland.*

Just under a quarter of the sugar cane cultivated in Queensland in 1874 was found in the coastal country south of Beenleigh. Here some of Queensland's earliest cotton planters had switched to growing sugar cane on their estates along the Logan, Albert, Coomera and Nerang Rivers. Another quarter of the colony's sugar cane was grown to the north of Brisbane along the Mary River. Nearby, sugar cane growing had just commenced along the Burnett River, so this area was still a minor sugar-producing district in 1874.[11] Beyond the Burnett River, the next sugar cane growing region was Mackay, some 700 km to the north, the intervening country being overlooked for sugar cane cultivation, probably because it was considered too dry for cropping. The Mackay sugar planters had established this locality as the principal sugar-producing region in the colony, with 2 020 ha under sugar cane or 40% of the total area under the crop in Queensland. The most northerly sugar cane cultivated was the 200 ha found in the Herbert River district.[12]

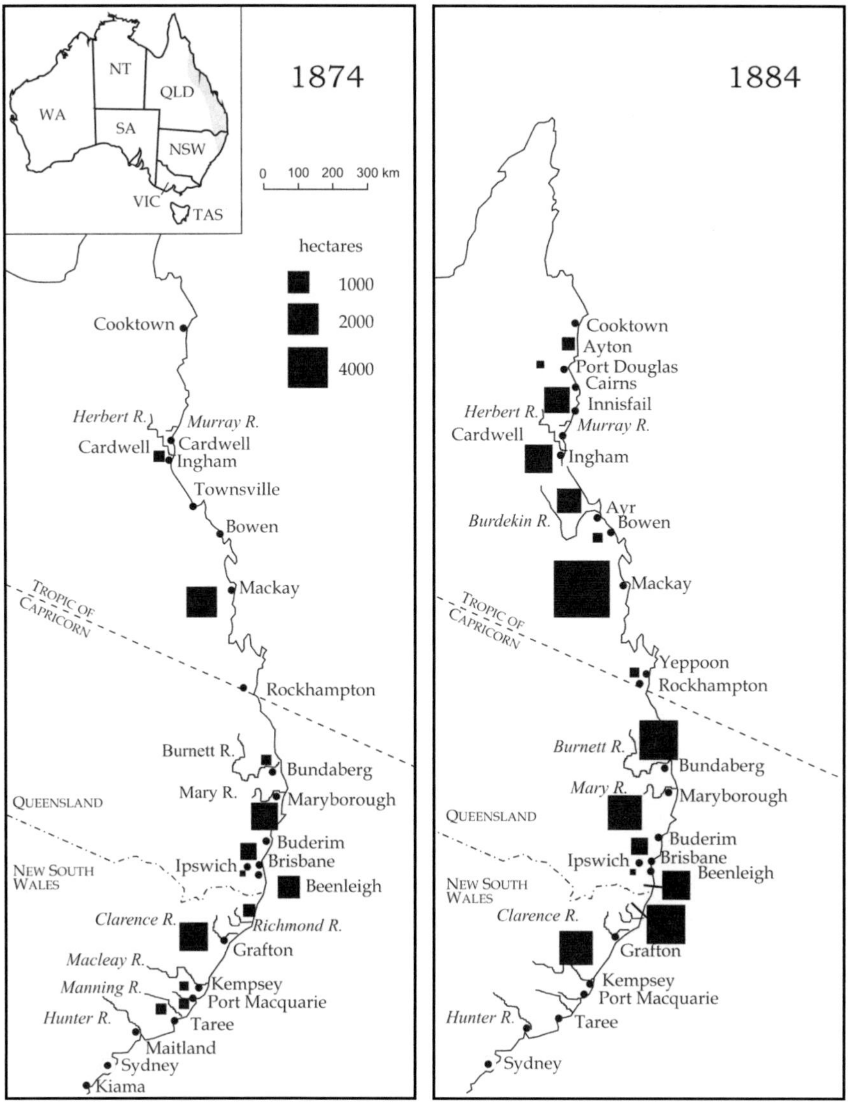

Figure 5.2. Distribution of sugar cane cultivation in Australia, 1874 and 1884.

Source: Based upon figures in the *Statistics of the Colony of Queensland*, 1874 and 1884, and the *Statistical Registers of New South Wales*, 1874 and 1884.

Encouraged by the success in Queensland, settlers also commenced growing sugar cane in New South Wales during the late 1860s, although not on the scale undertaken in Queensland and not with indentured Pacific

Islanders. By 1874, just under 2 700 ha were cultivated with sugar cane in New South Wales (Figure 5.1), stretching from the Manning River district in the south to the Tweed River district in the north of the colony (Figure 5.2). Three quarters of this sugar cane was situated in the Clarence River district between Grafton and Harwood Island, but contemporary commentators considered the more productive land to be downstream closer to Harwood Island.[13] By 1874, small patches of sugar cane were found along the Richmond River as far inland as Lismore, while farmers beside the Tweed River had grown the crop only since 1872, following a promise by a local settler to have a sugar mill operational by 1874.[14] A few hectares of sugar cane were also cultivated on properties at Port Macquarie and along the Manning and Macleay Rivers, although each was a minor sugar-producing district in 1874.[15]

Geographically, three-quarters of the sugar cane cultivated in Australia in 1874 was found south of the Tropic of Capricorn or in districts not ideally suited for the cultivation of the crop (Table 5.1). In New South Wales, the growing of sugar cane at Kiama and along the Hunter River had been attempted in the late 1860s, but had ceased by 1870 due to the crops being repeatedly damaged by frost.[16] During the 1870s, sugar cane cultivation along the Manning River and around Port Macquarie contracted because of the severity of winter frosts and had ceased altogether by 1884 (Figure 5.2). Frosts also forced the majority of the Macleay River district farmers to abandon sugar cane in favour of other crops, although one settler still cultivated sugar cane in 1884. A meeting of Macleay River district farmers in September 1887 discussed reviving sugar cane growing in the district, but even the low maize prices could not induce them to try sugar cane again.[17] By the mid-1870s, settlers in the Clarence, Richmond and Tweed Rivers districts had also

Table 5.1. Regional percentage of the total area cultivated with sugar cane in Australia, 1868–1914.

Region	1868	1874	1884	1894	1904	1914
Queensland – North of Tropic of Capricorn	14.0	23.8	57.2	36.6	51.1	63.8
Queensland – South of Tropic of Capricorn	71.4	39.5	19.5	34.5	33.3	29.5
New South Wales	14.6	36.7	23.3	28.9	15.6	6.7

Source: Calculated from figures in the *Statistics of the Colony of Queensland, Statistics of the State of Queensland* and the *Statistical Registers of New South Wales.*

become aware that the climate in their localities was not ideal for sugar cane, for its growth stopped altogether in winter and frosts damaged the crops. After a series of dry seasons and severe frosts during the late 1870s repeatedly reduced the size of the annual sugar cane harvests, Alexander Meston, a Clarence River canegrower and journalist, predicted in 1881 that sugar cane growing along the Clarence River would eventually fail. His suggestion proved to be incorrect, but at the time of Meston's prediction, the future of sugar cane cultivation in the Clarence River district was very uncertain.[18]

By 1874, environmental problems had also led to sugar cane cultivation being abandoned in four separate Queensland localities. During 1866, the Scottish immigrant John E. Davidson had established the Bellenden Plains plantation on the Murray River, north of Cardwell. Neither he nor the property's two successive owners managed to erect a sugar mill, and after repeated flooding and attacks from Aborigines, the plantation was abandoned in the early 1870s.[19] Further south at Townsville, sugar cane cultivation ceased after Robert Towns' sugar estate was destroyed for a second time by a cyclone in March 1870, although this area has annual rainfall that is marginal for the growing of sugar cane.[20] At Bowen, lack of water plagued the Bowen Sugar Co.'s plantation on the banks of the Don River. The Mauritian sugar planter Joseph Kelly – the main investor in the venture – failed to appreciate that Bowen with its annual rainfall of 1 000 mm was too dry for successful sugar cane growing without irrigation. At Rockhampton, the sugar-producing venture of Peter Murray, commenced in 1870, was abandoned after the sugar cane crop was ruined by frost in 1873.[21]

In late 1873, the Palmer government in Queensland funded an expedition to assess the resources of the colony's north-east coast. Leadership was entrusted to George Dalrymple, a former Commissioner for Crown Lands in the Kennedy District (1861–63) and parliamentarian (1865–7). Dalrymple's investigations of the country between Cardwell and the Endeavour River (site of today's Cooktown) resulted in the discovery of several fertile coastal valleys. He optimistically concluded that 121 500 ha in this region were fit to be cultivated with sugar cane.[22] Dalrymple's claims were possibly exaggerated, but he did encounter environments in which sugar cane would flourish. The Crown, however, delayed opening these coastal lands to occupation because of the 'rust' disease crisis of 1874–1876. Outbreaks of this disease had ended by 1878, due to the planters switching to 'rust' resistant sugar cane varieties, thereby ensuring a return to healthy yearly crops (see Chapter 7). As a result, the Crown commenced opening new, northern coastal lands to occupation. This country was rapidly selected; so great was the demand for land, that

Premier Sir Thomas McIlwraith concluded that it was being disposed of too cheaply. In early 1881, the price for first class agricultural lands was trebled from 5s to 15s per acre and the maximum area that could be selected was reduced from 5 120 acres (2075 ha) to 1 280 acres (518 ha). Even so, some 97 200 ha of agricultural land were sold between October 1881 and March 1882.[23] Undoubtedly speculators purchased much of this land, but some proprietary companies – especially from the colony of Victoria – invested in forming sugar plantations in North Queensland.[24] These factors led to a short period of rapid growth in the area devoted to sugar cane cultivation in Queensland (see Figure 5.1).

This interest in sugar production also led to both Adelaide- and Victorian-financed proprietary companies and individual investors turning their attention to the Northern Territory, which had been annexed to South Australia since 1863. The South Australian government had since 1874 offered a reward of £1 000 for the production of the first 100 tons of sugar from sugar cane grown in the Northern Territory.[25] By April 1881, the South Australian government had approved the apportionment of 40 500 ha of land to 30 applicants under the *Northern Territory Sugar Cultivation Act* of 1880. Most investors received land grants ranging from 1 215 ha to 2 025 ha in size, with the biggest blocks going to the Delissa Pioneer Sugar Co. and the Palmerston Plantation Co., each awarded 4 050 ha. Sugar cane plants from the Botanic Gardens at Darwin were supplied free to the successful applicants to assist them commence operations.[26]

One main feature of this period, therefore, was the extension of sugar cane cultivation to regions north of the Tropic of Capricorn in Queensland and the Northern Territory (see Table 5.1 and Figure 5.2). By 1884, approximately 200 ha were cultivated with sugar cane in the Northern Territory. This amount consisted of the small areas (under 20 ha) grown by Chinese and European settlers around Darwin and the equally small areas on the plantations situated at Delissaville, Cox Peninsula, and Beatrice Hills in the Adelaide River district. The most northerly sugar cane now grown in Queensland was along the Bloomfield River, near Ayton. To the south, small patches of sugar cane were found beside the Daintree and South Mossman Rivers by 1884, but a small sugar mill had only just been erected in the former district.[27] Further down the coast, nearly 1 620 ha were cultivated with sugar cane along the Mulgrave, Johnstone and Tully Rivers (Cairns Region) by 1884.[28] The Lower Burdekin district had also been settled in the late 1870s and early 1880s, after being overlooked during the late 1860s as a potential sugar-producing locality due to its low annual rainfall and difficult river accessibility. Approximately 1 400 ha of sugar cane were grown in the district by 1884, but the pioneering planters resorted to irrigation, using the abundant sub-artesian water supplies of the Burdekin River delta to achieve yearly crops. By 1884, sugar cane was also being planted along the Proserpine River (area included with Bowen figures in Figure 5.2), in anticipation of the Crystal Brook Sugar Co. erecting a sugar mill.[29] In addition, this upsurge of interest in sugar manufacture had led to sugar cane being grown again at Bowen and Rockhampton.[30]

In other sugar-making districts, a considerable expansion occurred in the amount of sugar cane being cultivated. At Mackay, the area under the crop by 1884 reached 8 100 ha, or quadruple the amount in 1874. Mackay sugar planters and small farmers now grew more sugar cane than their counterparts in all other sugar-producing districts in Southern Queensland or New South Wales. This growth had occurred following the formation of an additional ten new sugar plantations, while existing planters and farmers extended the area planted with sugar cane.[31] At Bundaberg, the amount of sugar cane being grown had trebled since 1874. This locality now boasted the second largest area under sugar cane in Australia.[32] In the Beenleigh district, the extent of sugar cane cultivated doubled, partially due to new sugar planters and small farmers, including some inland at Mt Tamborine, commencing the cultivation of sugar cane.[33]

The most spectacular growth, however, had taken place along the Richmond and Tweed Rivers and in the Herbert River district, where the amount of sugar cane grown in each region was nine times greater than 1874. In the latter district, expansion on existing plantations and the formation of new estates (e.g. Victoria; Ripple Creek; Hamleigh) contributed to the significant upsurge in the area devoted to the crop.[34] Along the Richmond and Tweed Rivers, CSR had encouraged the planting of sugar cane following the erection of its Condong Mill near Murwillumbah in 1880, and its Broadwater Mill near Woodburn in 1881 (see Chapter 6 for more details). The Tweed River district had also attracted the attention of William Julius, who had considerable experience in growing sugar cane in the West Indies, but his first sugar-producing venture in New South Wales on the Macleay River had failed due to frosts. He established Cudgen plantation during the late 1870s, using 200 Pacific Islanders who had completed their indenture contracts in Queensland. In addition, settlers, including some who had moved away from the Clarence River district, began cultivating sugar cane in the 'Big Scrub', a heavily forested elevated tract of relatively frost-free country with rich volcanic soils lying north of the Richmond River between Lismore and Brunswick Heads.[35]

Expansion in other sugar-producing regions was less spectacular. In the Ipswich district, the amount of sugar cane cultivated grew by just 40 ha between 1874 and 1884, as the crop was grown only in small patches by the farmers near Marburg, who mostly engaged in mixed farming.[36] Around Maryborough, the modest increase in the area under sugar cane resulted from the crop being cultivated initially in the Pialba district, on the coast north of the Mary River, and new or existing settlers along the Mary River commencing sugar cane growing for the first time or expanding the amount of cane grown on their properties.[37] In the Brisbane district, the area under sugar cane between 1874 and 1884 remained relatively static. Investment in new sugar plantations had occurred at Buderim, but the abandonment of sugar cane growing around Brisbane offset this expansion. Settlers in the Brisbane district switched to other crops after the 'rust crisis' and repeated frosts during the early 1880s.[38] Drought and severe frosts between 1880 and 1882 had also led to a contraction in sugar cane growing along the Clarence River. Yet by 1884, after a return of favourable seasons, the area under sugar cane along the Clarence River was slightly bigger than the amount found in 1874, although the cultivation of the crop was now concentrated along the lower reaches of the river, downstream from Ulmarra.[39]

Depression, recovery and northern expansion, 1885–1914

In August 1885, the Western Australian government offered Mackay-businessman, George Smith, a free grant of 40 500 ha of land in the Kimberley district of northern Western Australia, if he could produce sugar in the region by 1887. This offer by the Western Australian government aimed to attract canegrowers to its tropical districts.[40] The timing was unfortunate, for the world price of sugar had collapsed by a third in mid-1884, primarily due to massive dumping of subsidised European beet sugar on the British and American markets, and the Australian sugar industry had entered a short period of contraction (Figure 5.1). Consequently, Smith never attracted financial support for his scheme, and the tropical regions of Western Australia remained without a sugar industry until the mid-1990s. Moreover, the economic crisis affecting Australia's canegrowers was magnified in Queensland after 1885, following the decision by Samuel Griffith's Liberal government to prohibit further importation of Pacific Islanders to the colony after 31 December 1890, and requiring that the employment of

Pacific Islanders in the sugar industry would cease three years later. Thereafter the industry was to rely on European labour only.

The changed circumstances under which the Australian sugar industry operated led to a contraction in the amount of sugar cane being grown throughout most sugar-making districts. Around Cairns, local plantation closures resulted in sugar cane cultivation being abandoned on 200 ha between 1884 and 1888. In the Lower Burdekin district, 400 ha of sugar cane disappeared between 1885 and 1888, following plantation closures, the district's small farmers switching to growing maize and the surviving sugar planters reducing the size of their crops due to labour shortages.[41] At Bowen and along the Proserpine River, sugar cane growing ceased altogether by 1887, while around Mackay the area under sugar cane in 1888 was 1 460 ha less than the amount grown in 1884. This substantial decline had occurred after the closure of nine sugar-producing estates.[42] In Southern Queensland, the extent of the contraction was even greater. At Maryborough, the area under sugar cane in 1888 had fallen to 1 170 ha, half that cropped in 1884. Heavy frosts in 1884 and 1888, drought in 1887 and exhausted land had also contributed to the decline along the Mary River.[43] In the Beenleigh district, a severe frost in 1885 and drought in 1887 had prompted small farmers to replace sugar cane with maize, potatoes and arrowroot. As a consequence, only 530 ha of sugar cane were left in 1888 or 1 620 ha less than the amount grown in 1885. Moreover, the growing of sugar cane had ceased altogether along the Nerang River.[44]

Sugar cane cultivation had also been abandoned entirely around Brisbane by 1889. The properties that were once cultivated with sugar cane along the Brisbane River and at Cleveland had been sub-divided into housing allotments or converted into fruit orchards.[45] Sugar cane growing was also nearly abandoned in the Northern Territory during the late 1880s. White ant damage to the sugar cane, poor selection of plantation sites and unsatisfactory rains had combined with low sugar prices to defeat the majority of canegrowers in the Northern Territory.[46] In New South Wales, the contraction was not as great (Figure 5.1), and mainly confined to the Clarence River district, where canegrowers had switched to dairying or cultivating maize.

Sugar cane cultivation, however, did not contract in all regions during the mid-1880s. At Rockhampton, the area under sugar cane had doubled by 1888, due to the Yeppoon Sugar Co. bringing additional land under cultivation to stave off bankruptcy. Extended plantings on CSR's properties in the Innisfail and Herbert River districts had contributed to the expansion in the amount of sugar cane grown in those localities.[47] Sugar cane cultivation had also commenced near Coffs Harbour in 1885. At least three settlers had erected small, inexpensive horse-driven mills by 1886, and were crushing the sugar cane crops of their neighbours.[48] The biggest increase, however, occurred around Bundaberg, where the area under sugar cane in 1888 was nearly 1 000 ha more than in 1884. The Bundaberg district was still seen as favourable locality to invest. Sugar planters continued making a profit, partly because they paid less interest, having not secured more land than they could successfully manage.[49] Moreover, they invested less in sugar-making machinery, relying on Millaquin Refinery to convert their juice into sugar (see Chapter 9). Despite the expansion in these districts, the area under sugar cane in Australia stood at 25 3100 ha in 1888 or 5 670 ha less than 1885 (Figure 5.1).

Griffith's Liberals were defeated at general elections in May 1888, and Sir Thomas McIwraith, the new Queensland Premier, reacted to the sugar planters' misfortune by appointing a Royal Commission in late 1888 to inquire into the problems being faced by the colony's sugar industry. Its Chairman was William Henry Groom, a staunch opponent of any arrangement which would favour the sugar industry over his wheat-growing constituents on the Darling Downs. He was assisted by Alfred Cowley, the previous manager of Hamleigh plantation and recently elected parliamentary member for the Herbert electorate, and Henry Edward King, a former parliamentarian who had retired to a Brisbane legal practice. During January and February 1889, the three royal commissioners toured coastal Queensland from Brisbane to Mossman, and according to the historian Geoffrey Bolton they 'sought out evidence diligently', interviewing 154 witnesses, including many sugar plantation owners and managers. The Groom Commission concluded that the depressed conditions in the sugar industry were caused by a combination of the following factors: mismanagement, extravagance and inexperience of some sugar planters; the high levels of indebtedness amongst the sugar planters and small canegrowers who operated using capital borrowed at high interest rates; the low price of sugar owing to the competitive impact of subsidised European beet sugar; losses through the impact of a series of dry seasons in the mid-1880s; and a loss of confidence in the sugar industry due to uncertainty over the industry's future labour requirements. The Royal Commissioners recommended the continued recruitment of Pacific Islanders under government control, government funding of irrigation schemes in the sugar-producing districts, further state support for central sugar mills and reciprocal tariff arrangements with other Australian colonies in order to provide a more reliable market for the colony's sugar. McIlwraith had resigned due to ill health by the time the Royal

Commissioners compiled their report, so leadership of the government had passed to Boyd Morehead. Efforts to implement the Groom Commission's recommendations, particularly the resumption of Pacific Islander recruitment, failed to be supported by the Queensland Parliament.[50]

Recovery from the short depression in the Australian sugar industry commenced after 1888, as the area under sugar cane in Australia rose again (Figure 5.1). In New South Wales, the major part of this growth occurred in the Tweed River district. CSR's payment of higher prices for sugar cane to its Condong Mill suppliers after 1888, led to bigger areas along the Tweed River being cropped with sugar cane. Farmers between Murwillumbah and Byron Bay also commenced growing sugar cane, in anticipation of the completion of the New South Wales government-built railway from Lismore to Murwillumbah. Finally opened in 1894, this railway allowed sugar cane grown at Byron Bay to be carted to Condong Mill for crushing.[51] In Queensland, much of the growth can be attributed to the spread of sugar cane cultivation to the Isis district, near Maryborough. Here Robert Cran & Co. convinced small farmers to grow sugar cane for their newly erected Doolbi Mill, which commenced crushing in December 1890.[52] Some surviving planters also extended their plantings, using whatever labour was available.

Not all Queensland sugar producers shared the optimism of Robert Cran & Co. CSR in particular was very concerned about possible future labour shortages in Queensland and considered closing its three Queensland sugar plantations. Other Queensland sugar planters continued vocally opposing the Queensland government's Pacific Islander labour policy during 1890. Griffith – again Premier but in a coalition ministry with McIlwraith – became increasingly aware of these concerns. To investigate the issues first hand, Griffith and his cabinet visited North Queensland during December 1890 and January 1891. They received a hostile reception from deputations of canegrowers who, according to the historian Geoffrey Bolton, 'bemoaned their hardships and urged the restoration of Pacific Island labour'. On his return from this trip, Griffith, without consulting his cabinet colleagues, announced that Pacific Islander recruitment to the colony could re-commence in February 1892. Griffith retired as Premier in March 1893 to become Chief Justice of Queensland. His place was again taken by McIlwraith, who promised that the Queensland government would provide £500 000 towards the construction of new, co-operative central mills throughout the colony.[53] Together these two actions restored the confidence of Queensland's sugar producers and led to another period of rapid growth in the area cultivated with sugar cane (see Figure 5.1).

By 1894, sugar cane cultivation had commenced around Gin Gin in anticipation of a government-funded central sugar mill being erected in that district, the area under sugar cane had doubled in the Beenleigh district following the decision to build a central sugar mill near Nerang, and sugar cane growing had re-commenced in the Maroochy district where a central sugar mill was being constructed at Nambour (see Chapter 6 for more details). Just over 2 000 ha were now under cane in the Lower Burdekin district, or 600 ha more than 1884. Small farmers had re-commenced growing sugar cane and the two surviving sugar planters were bringing new land under cultivation. At Mackay, the area under sugar cane had returned to the figure reached in 1884, following an upsurge of cane growing amongst small farmers.[54] A resurgence of sugar cane cultivation had also taken place along the upper reaches of the Mary River, following the decision to erect a central sugar mill at Mt Bauple. Nearby in the Isis district, sugar cane cultivation continued to expand, especially after CSR announced it would have a sugar mill near Childers ready for the 1895 crushing season.[55]

This new confidence in the sugar industry's future, however, did not lead to a return to sugar cane cultivation around Brisbane, nor along the Bloomfield River, near Ayton, where Vilele plantation – the colony's northernmost sugar plantation – was abandoned after 1890 (Figure 5.3). Moore's Shoal Bay sugar plantation – the last sugar plantation in the Northern Territory – also closed after the 1892 season. Inquiries into the agricultural potential of the Northern Territory in 1895 and 1896 maintained that the region was suited to sugar cane cultivation, but neither private investors, the South Australian government or the Commonwealth government after it assumed control over the Northern Territory in 1911 sought to re-establish sugar production in the Northern Territory.[56]

During the late 1890s and early 1900s, the area under sugar cane in Australia continued expanding (Figure 5.1), but a significant northward shift in the spatial distribution of sugar cane cultivation had taken place by 1904 (Table 5.1; Figure 5.3). Just over 2 400 ha of sugar cane were now grown around Mossman, while five times more sugar cane was cultivated in the Cairns district in 1904 compared to 1894. Double the amount of sugar cane was now found in the Innisfail and Herbert River districts compared to 1894. Almost 1 600 ha of sugar cane were now cultivated in the Proserpine district, while at Mackay an additional 3 240 ha had been placed under sugar cane between 1894 and 1904. Two main factors contributed to this extraordinary growth of sugar cane cultivation in North Queensland between 1894 and 1904. First, uncultivated sugar plantation lands were leased or sold to small settlers, who turned them into farms where sugar cane was the principal crop.

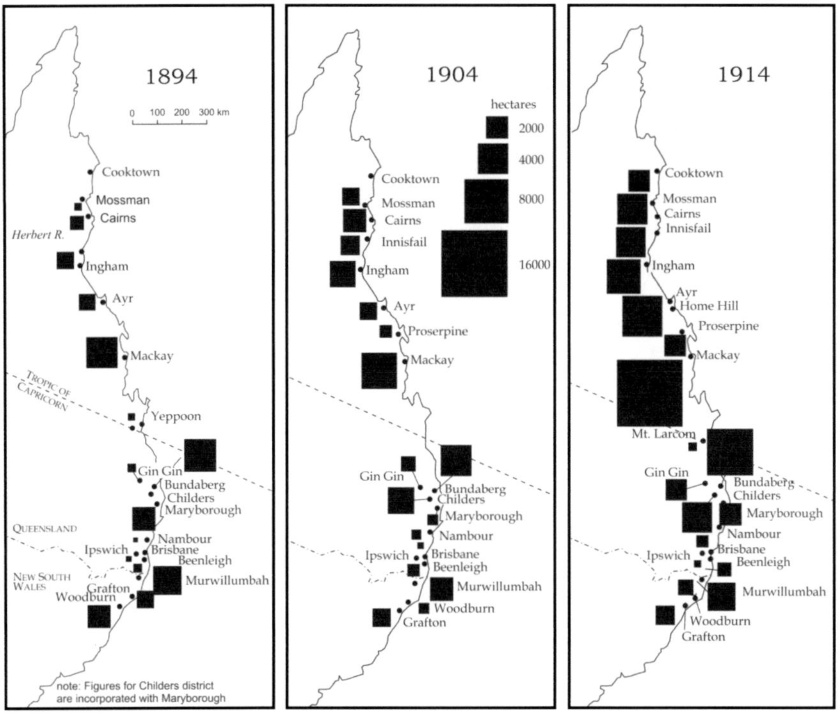

Figure 5.3. Distribution of sugar cane cultivation in Australia, 1894, 1904 and 1914.

Source: Based upon entries in the *Statistics of the Colony of Queensland*, 1894, *Statistics of the State of Queensland*, 1904 and 1914 and the *Statistical Register of New South Wales*, 1894, 1904 and 1914.

Second, sugar cane growing by small settlers was stimulated following the erection of central sugar mills in the Mossman, Cairns and Proserpine districts during the mid-1890s (see Chapter 6 for more details).

South of the Tropic of Capricorn, increases in the area under sugar cane had been confined to the Maroochy and Beenleigh districts, where the central milling companies had stimulated extra plantings of sugar cane by increasing their price for cane by 1s per ton. In contrast, sugar production had ceased at Yeppoon by 1904, following the closure of the local sugar mill, after its owner encountered difficulties in securing labour.[57] The previous expansion in sugar cane growing throughout the Bundaberg, Gin Gin and Isis districts had also halted because of gumming disease outbreaks that severely affected yields (see Chapter 7 for more details). Any recovery in the amount of sugar cane cultivated in these districts was further curtailed by a severe drought and frosts at the turn of the century. Nearby along the Mary River, sugar cane growing again contracted during the late 1890s, as farmers switched to dairying, following gumming disease outbreaks in the district between 1893 and 1895.[58]

Sugar cane cultivation also contracted in all New South Wales districts after 1895. By 1904, the area under sugar cane in New South Wales was 3 240 ha less than 1894 (Figure 5.1). Moreover, only 15% of all the sugar cane grown in Australia was now found in New South Wales (Table 5.1). Such a contraction was a result of the uncertainty about the future viability of sugar production in New South Wales, following outbreaks of gumming disease after 1893 and the announcement by the New South Wales government in 1895 that it would abolish the protective sugar import tariff gradually from July 1897 onwards (a policy which was eventually not implemented). The latter action would have exposed New South Wales canegrowers to imports of sugar from Queensland. New South Wales canegrowers argued that Queensland sugar could be sold more cheaply than the locally produced product, as Queensland canegrowers relied on indentured Pacific Islander workers who were paid considerably less than the European field labourers in New South Wales.[59] This uncertainty led to some New South Wales canegrowers switching to dairying entirely or moving to Queensland sugar-producing districts. Others adopted mixed farming: their best land was utilised for sugar cane cultivation, while the rest was converted to pasture for dairy cattle.[60]

By 1904, the Australian sugar industry operated under very changed political conditions. The new Parliament of the Commonwealth of Australia established in 1901 was dominated by spokespersons for 'white Australia', desiring 'to be one people, and remain one people, without the admixture of other races', as expressed by Alfred Deakin.[61] In response to these sentiments, the Commonwealth government led by Prime Minister Edmund Barton had introduced the *Pacific Islanders Act* of 1901, which prohibited the recruitment and introduction of Pacific Islanders into Australia after 31 March 1904, and allowing for certain exemptions, required those Pacific Islanders in Australia to be repatriated after 1906. To protect the country's sugar industry against a flood of imported sugar which had been produced more cheaply, a protective tariff of £6 and £10 per ton was placed on cane sugar and beet sugar, respectively, via the *Excise Tariff Act* of 1902. In addition, an excise of £3 per ton was imposed upon all sugar manufactured in Australia until 1 January 1907. These funds were used to encourage Australian canegrowers employ European workers only. From 1 July 1902, all Australian canegrowers who harvested sugar cane with European labour only after 28 February 1902 received a rebate of 4s per ton of cane. These arrangements were altered the following year after the introduction of the *Sugar Rebate Abolition Act* of 1903 and *Sugar Bounty Act* of 1903, which together abolished the rebate method and gave a bounty on sugar produced by European workers only. Those Australian canegrowers employing Europeans only to grow and harvest their sugar cane in the previous twelve months were awarded a bounty ranging from 4s to 5s per ton of cane, depending upon geographic location. However, not more than a quarter of Queensland's sugar output was produced by European labour in 1905, so the bonus to canegrowers using European labour only was raised to between 6s and 7s 6d per ton of cane under a new *Sugar Bounty Act* of 1905. The *Excise Tariff Act* of 1905 extended the imposition of the excise on sugar manufactured in Australia until 1 January 1913 and increased the excise on sugar from £3 to £4 a ton, thereby raising additional funds to pay for the increased bounty.[62]

Considerable opposition to the new legislative arrangements was expressed by some Queensland sugar producers and groups of Pacific Islanders. George Drysdale, a partner in the company running Pioneer Mill, for instance, predicted that the Australian sugar industry would 'collapse'.[63] Such a sentiment became so widespread amongst other Australian canegrowers and sugar mill-owners that William Kidston's government in Queensland appointed a Royal Commission in 1906 to inquire into the number of Pacific Islanders to be deported at the end of 1906. Chaired by Robert Ranking, the Brisbane Police Magistrate, the inquiry concluded that sufficient labour would be available to maintain sugar production in Queensland after the deportation of the Pacific Islanders, although the establishment of a Labour Intelligence Bureau would enhance the coordination of its supply, particularly in North Queensland. The Ranking Commission also recommended that the Pacific Islanders unfit for life on their home islands by reason of age or infirmity, those who had been resident for twenty years or more in Australia, and those who had been living for twelve months or more with partners with whom they could not return should be exempt from deportation. Following pressure from the Queensland government, the Commonwealth government relaxed the strict provisions of the legislation requiring repatriation of the Pacific Islanders. As a result of these exemptions, an estimated 1 600 to 2 000 of the approximately 9 300 Pacific Islanders living in Australia in 1901 remained in Queensland when deportations were officially completed on 31 July 1908.[64]

The concerns raised by some Queensland sugar producers about possible labour shortages were mostly ill founded. The incentives provided by the Commonwealth government to use European workers were successful, with the country's canegrowers employing Europeans as field labourers, albeit slowly initially. By 1906, 70% of the sugar cane in Queensland was grown and harvested by European labour. Four years later, less than ten percent of the state's sugar cane was harvested by non-European labour.[65] Canegrowers and mill-owners in Queensland, however, found that waged and increasingly unionised European workers demanded better working conditions and higher wages than indentured Pacific Islanders. The Queensland government had led the way by introducing the *Sugar and Shearers' Accommodation Act* of 1905, which stipulated that workers were to be supplied with 10.2 m^3 per person in respect to sleeping accommodation, proper cooking and cleansing materials and receptacles for rubbish. In addition, the Commonwealth government (Alfred Deakin; Protectionist) in 1907 fixed a field labourer's wage during the non-harvest season at a minimum of 22s 6d per week as a condition to receive the bonus for using European labour.

Despite these measures, there was an upsurge in unionism amongst Australian sugar industry workers, who complained about poor accommodation, poor provisions for hygiene and poor quality food. This trend was opposed strongly by the Australian Sugar Producers' Association (ASPA), formed in 1907 to represent the interests of the country's canegrowers and mill-owners, and CSR, by then the country's largest producer of raw and refined sugar. Both CSR and the ASPA consistently refused to negotiate with the Amalgamated Sugar Workers' Union (after December 1910 the Amalgamated Workers' Association following a merger with the union representing miners in North Queensland). Moreover, CSR during 1907 – at the urging of the Cane Farmers' Association of North Queensland – had recruited 113 Britains, 104 Spaniards, 81 Scandinavians and 41 Austrians. Attempts at securing Italians failed after the Italian government refused to approve the contract labour scheme. Ostensibly this recruitment was undertaken to safeguard against possible labour shortages, but these immigrant workers meant CSR and any sympathetic North Queensland canegrower could avoid employing unionised Australian workers.[66]

Minor strikes occurred in 1909 and 1910, and the united efforts of the mill-owners and the ASPA thwarted the efforts of the Amalgamated Sugar Workers' Union to achieve improved conditions. However, Prime Minister Andrew Fisher's Labor Commonwealth government responded to the workers' concerns by introducing new regulations under the *Sugar Bounty Act* of 1910, which stipulated that the Minister of Trade and Customs might

Plate 5.1. A gang of European canecutters in the sugar cane fields near Ingham, 1904.

Source: Image no. 24680, John Oxley Library. Image reproduction courtesy of the State Library of Queensland, Brisbane.

withhold the bounty from Australian canegrowers if the rates of wages and conditions of employment of workers in the sugar industry were below the standard rates and conditions of employment prescribed by the Commonwealth or State Industrial Authority.[67]

Before these regulations could be implemented, serious industrial unrest occurred between June and August 1911, during a general strike in all Queensland sugar-producing districts by mill workers who were supported by field labourers and waterside workers. Canegrowers, assisted by their families and non-unionists, were not enough to keep the sugar mills working at full capacity. Violence flared in many areas and prosecutions were brought against the strikers for offences such as trespass, unlawful assembly, riot and arson. Some canegrowers claimed they were pressured by the mill-owners to work in the sugar mills with threats that their cane would not be crushed. Strike action, which gained increasing levels of support from interstate unionists, some mill-owners and sympathetic canegrowers, finally ceased after the Queensland government (Denham; Liberal) intervened and negotiated an increase in mill worker wages to 30s with keep for a forty-eight hour working week; field labourers were left to make private arrangements with canegrowers.[68]

The publicity during the strike revealed the poor working conditions faced by both mill workers and field labourers. Prime Minister Andrew Fisher was also concerned about the Commonwealth government's involvement in the affairs of the Australian sugar industry, the prices being paid for cane by the mill-owners (see Chapter 8), and CSR's dominant position in the Australian sugar industry (see Chapter 9). Hence, the Commonwealth government appointed a Royal Commission in October 1911 to inquire into and report upon the conditions of the Australian sugar industry, including whether or not the Commonwealth should persist with its bounty regulations. The political scientist Diana Shogren, however, has suggested that to a large extent the Commonwealth government established the Royal Commission as a means to inquire into the affairs of CSR.

The inquiry was chaired by Sir John Gordon of the South Australian Supreme Court, until ill-health forced his retirement; he was then succeeded by William Jethro Brown, Professor of Law at the University of Adelaide. The other members of the Commission were Albert Hinchcliffe, M.L.C., Robert Anderson, Martin Shannon and Thomas Crawford. The majority of Commissioners had strong ties to the labour movement. Hinchcliffe was the editor of the *The Worker* (Brisbane), Shannon had been an unsuccessful Labor candidate in North Queensland, and Anderson was said to have Labor sympathies and later became one of William Morris (Billy) Hughes' closest associates. Gordon who had Liberal sympathies, however, soon established friendly and confidential relations with Hughes. Crawford was the odd man out: he had actual experience of the sugar industry, having grown sugar cane in the Mossman district and had been associated with the running of the Mossman Central Mill Co. Ltd.[69]

In August 1912, even before the Royal Commission had completed its task, the Commonwealth government introduced new regulations under the *Sugar Bounty Act* of 1912 that required canegrowers to have paid field workers 36s a week for a forty-eight hour week if they wished to claim the bonus on using European labour. Eventually, the Royal Commission reported in December 1912 and concluded that the excise and bounty regulations had successfully engineered a transformation in the fields and sugar mills from non-European to European labour and that the excise had become a source of revenue for the Commonwealth, although there was no justification for this tax on the sugar industry. The Royal Commission recommended that the excise and bounty system be abolished, as long as the States continued to promote the Europeans only labour policy. Digby Denham, the Queensland Premier, had already advised the Commonwealth government that if the excise and bounty regulations were repealed, his government would legislate to exclude non-European labour in the state's sugar industry. Legislation repealing the excise duty on sugar and the bounties were passed by the Commonwealth Parliament in late 1912. In the following year, the Queensland government introduced *The Sugar Cultivation Act* of 1913 which prohibited non-European labour in the production of sugar and *The Sugar Growers' Act* of 1913 which provided for the payment by the sugar mills to the canegrowers of what was the old bounty, plus an additional 2s 2d per ton of cane supplied.[70]

Hence, the transformation from indentured labour to waged European labour in the Australian sugar industry was accomplished without too much disruption to production, despite the industrial unrest between 1909 and 1911. The area under sugar cane cultivation in Australia did drop by 3 600 ha between 1906 and 1909 (Figure 5.1), although this trend cannot be entirely attributed to concerns about the removal of Pacific Islanders or shortages of European workers. Some canegrowers actually extended their areas under sugar cane so as to collect an increased amount of bonus. Moreover, a world-wide improvement in sugar prices following the abolition of the European beet sugar bounty system by the Brussels Convention in 1902 improved confidence amongst canegrowers and mill-owners about the overall future of the Australian sugar industry.[71] By 1914, 70 875 ha were under sugar cane in Australia, the highest to date in the industry's history (Figure 5.1).

The northern expansion in the area under sugar cane cultivation continued during the late 1900s, with just under 64% of Australia's sugar cane being grown north of the Tropic of Capricorn by 1914 (Table 5.1). The biggest contribution to this expansion came from increased plantings of sugar cane in the Lower Burdekin and Mackay districts. In the former region, the area under sugar cane had trebled between 1904 and 1914. Part of the explanation for this trend was a result of the Queensland government opening the Inkerman Estate properties – 3 700 ha in the vicinity of Home Hill and the Haughton River – to closer settlement in 1911. By 1914, sixty Inkerman Estate settlers near Home Hill were growing sugar cane for Inkerman Mill, which was erected in the midst of their farms in 1913. Twenty-five settlers along the Haughton River had also commenced growing sugar cane for Pioneer Mill. In addition, elsewhere in the Lower Burdekin district, several large properties, including the former Airdmillan plantation lands, were subdivided into smaller blocks and occupied by farmers who turned to cane growing.[72] At Mackay, the area under sugar cane in 1914 was recorded as being 16 400 ha or 5 000 ha more than 1904. This massive expansion had come about mostly from the spread of cane growing up the Pioneer River valley to Finch Hatton, where the new Cattle Creek Mill operated from

1906 onwards, and additional estates being subdivided and occupied by farmers who commenced sugar cane cultivation.[73]

In other northern sugar-producing regions, the increased areas under sugar cane were not as great. Around Cairns, the spread of sugar cane growing up the Freshwater Creek valley and southward along the Cairns – Mulgrave Tramway towards Babinda, meant an additional 1 215 ha were under the crop in 1914 compared to 1904.[74] In the Innisfail district, the additional 1 620 ha placed under sugar cane between 1904 and 1914 resulted from the crop being finally cultivated in the Daradgee locality, just north of Innisfail, and the Liverpool and Maria Creek localities, south of Mourilyan. In the former district, settlers commenced growing sugar cane after CSR built a bridge across the North Johnstone River in 1908 to bring the crops to its Goondi Mill, while the latter districts were occupied by small farmers for the first time following the Queensland government opening the country to occupation in 1911. These new settlers began growing sugar cane for their produce could easily be transported to nearby Mourilyan Mill, which had been connected to the Liverpool district by tramline since at least 1900. Another 1 620 ha had also been added to the lands cropped with sugar cane in the Herbert River district, after tramline extensions, particularly along the Stone River, ensured additional settlers could get crops of sugar cane to the district's two mills.[75] The only decrease occurred around Mossman, where 810 ha of sugar cane disappeared between 1904 and 1914. Difficulties in securing European workers for such a remote locality meant canegrowers reduced their annual plantings, particularly on hillside areas and poorer quality land.[76]

A mixture of changes occurred between 1904 and 1914 in the areas cropped with sugar cane in Queensland districts south of the Tropic of Capricorn. A new sugar cane growing district emerged at Mt Larcom, 175 km north of Bundaberg. Here the Queensland government released 250 blocks of land to occupation in 1911, and many of the new settlers planted sugar cane, railing their crops to Millaquin Mill for crushing.[77] Around Bundaberg, sugar cane cultivation continued to grow in popularity, with an additional 2 025 ha of the crop being present in 1914 than in 1904. Part of the explanation for this expansion was the subdivision of several former plantations (e.g. Kalbar, Windemere). The settlers who acquired these properties commenced cultivating sugar cane. In addition, small farmers along Baffle Creek, 50 km north of Bundaberg, also commenced planting sugar cane following the transfer of a closed sugar mill from near Beenleigh to their locality in 1910.[78] In the Maryborough district, the area under sugar cane had doubled between 1904 and 1914. The European labour bonus had proved attractive to the district's small farmers who increased their plantings of sugar cane, while fallow portions of estates were replanted with sugar cane after a lapse of fifteen years.[79] Sugar cane growing throughout the Maroochy district had also increased during the mid-1900s, following the expansion of tramlines to new localities in the Blackall Range, near Nambour. Yet by 1914, the area under sugar cane along the Maroochy River was 405 ha less than 1904, as local farmers converted some or all of their properties into pastures for dairy cattle or commenced cultivating tropical fruits. Dairying also proved attractive to the small farmers in the Beenleigh district, with the amount of sugar cane grown falling by 325 ha between 1904 and 1914.[80] In contrast, the area under sugar cane in the Ipswich district remained static at 200 ha between 1904 and 1914, as the district's small dairy farmers continued growing only two to five hectares of sugar cane as a sideline (see Figure 5.3). This amount of cane was barely enough to keep the small sugar mill at Marburg operational.[81]

Sugar cane cultivation contracted in all New South Wales districts throughout the 1900s and early 1910s. Small farmers switched solely to dairying or mixed farming, where sugar cane was only one of a number of agricultural pursuits on a property. They pursued this strategy, claiming the following reasons: dairying provided quicker financial returns (a crop of sugar cane in New South Wales often took two years to mature); CSR paid too little for their crops; rising labour costs made it too expensive to grow sugar cane on the stony hill country where all the cultivation had to be done by hand; and satisfactory field workers could not be found to load and cart harvested cane in isolated spots.[82] As a consequence, the area under sugar cane in New South Wales fell to 4 455 ha in 1914 or half the amount reached in 1904 (Figure 5.3). Moreover, the distribution of sugar cane growing within the Richmond and Tweed Rivers districts changed during the 1900s and early 1910s. The alluvial flats beside the upper reaches of the Richmond River, and the stony hill country of the 'Big Scrub' were no longer considered suitable for the growing of sugar cane, and the crop was cultivated only along the lower reaches of the Richmond River, downstream from Woodburn. In contrast, throughout the Tweed River district, sugar cane growing had contracted along the river frontages because of frost and poor drainage and had moved to the hill country at Terranora and Duranbah, 5–8 km back from the river.[83] Consequently, northern New South Wales had become a minor sugar-producing region, having just under 7% of the sugar cane grown in Australia in 1914 (Table 5.1).

The spread of sugar cane cultivation throughout the coastal valleys of northern New South Wales and Queensland left an enduring legacy – the disappearance of the native forests. However, on the eve of World War 1, the rapid loss of so much forest in coastal eastern Australia between Grafton and Mossman and its replacement with fields of sugar cane went mostly unnoticed amongst the Australian public or their parliamentarians. Moreover, the remaining coastal rainforests of North Queensland were under serious threat, as most investors in sugar-producing ventures agreed that the coastal country from Mackay northwards was more climatically suited to the cultivation of sugar cane, compared to Southern Queensland and northern New South Wales. Inactivity or limited support by the South Australian and Western Australian governments meant that sugar production had failed to become permanently established in the Northern Territory and northern Western Australia, despite these areas having a climate suited to sugar cane cultivation, albeit with irrigation. Consequently, the geographical centre of sugar production was North Queensland and two distinct production zones had emerged by the early 1910s. In the coastal districts north of Maryborough, sugar cane was the main crop planted by most small farmers. Throughout the Maryborough district and cane growing areas southwards, sugar cane was often grown as a sideline by small farmers, who combined it with dairying, tropical fruits, vegetables or maize. The presence of so many small farmers in the tropical and sub-tropical regions of Queensland, however, was no accident. Government intervention, as will be shown in the next chapter, ensured that the production system whereby sugar cane was cultivated by indentured Asians or Pacific Islanders on large estates had been replaced with central sugar mills supplied with cane grown on small family farms owned by Europeans.

Notes

1 *QVP*, Vol. 2 (1863), pp.115–116.

2 *BC*, 6 June 1862, p. 2; *Sydney Mail*, 30 April 1864, p. 2; C. Wood, 'Buhot, John (1831–1881)', *ADB*, Vol. 3 (1969), pp. 292–293.

3 *QGG*, Vol. 5 (1864), p. 868; *QGG*, Vol. 7 (1866), p.1162; C. Bernays, *Queensland Politics During Sixty (1859–1919) Years* (Brisbane, 1919), pp. 314–315.

4 J. Farnfield, *Frontiersman. A Biography of G.E. Dalrymple* (London, 1968), pp. 17–25; J. Kerr, *Pioneer Pageant. A History of Pioneer Shire* (Mackay, 1980), pp.1–8.

5 J. Farnfield, 'Cotton and the search for an agricultural staple in early Queensland', *Queensland Heritage*, Vol. 2 (1971), pp. 20–25; M. Robinson, 'The decline of wheat-growing in coastal districts of New South Wales', *Australian Geographical Studies*, Vol. 8 (1970), pp. 50–52; *ATCJ*, 14 January 1871, p. 45; C. Lyne, *The Industries of New South Wales* (Sydney, 1882), p. 256.

6 *Sydney Mail*, 13 November 1869, p. 8.

7 C. Moore, *Kanaka. A History of Melanesian Mackay* (Port Moresby, 1985), pp. 200–202; *QVP*, Vol. 2 (1870), p. 445.

8 Deduced from figures in *QVP*, Vol. 1 (1865), p. 1213 & *QVP* Vol. 1 (1866), p.1461.

9 *QVP*, Vol. 1 (1868–69), pp. 650–654; *QVP*, Vol. 2 (1871), p. 994; E. Thorne, *The Queen of the Colonies, or Queensland as I Knew it, by and eight years' resident* (London, 1863) p. 61; Y. Reynolds, 'Sweet surrender: sugar production at St Helena Penal Establishment', in M. Johnson (ed.), *Brisbane: Moreton Bay Matters*. Brisbane History Group Papers No. 19 (Brisbane, 2002), pp. 19–26.

10 *Q*, 8 July 1871, p. 5, 23 March 1872, p.11 & 17 October 1874, p. 9.

11 J. Kerr, *Sugar at Maryborough:120 Years of Challenge* (Maryborough, 1987), pp. 12–31; J. Kerr, *Southern Sugar Saga. A History of the Sugar Industry in the Bundaberg District* (Bundaberg, 1983), pp. 7–11.

12 J. Kerr, *Pioneer Pageant. A History of the Pioneer Shire* (Mackay, 1980), pp. 26–34 & 43–50; *BC*, 25 October 1873, p. 5; *Q*, 24 October 1874, p. 10.

13 *ATCJ*, 18 March 1871, pp. 333–334, 25 March 1871, p. 356, 14 June 1873, p. 746 & 5 December 1874, p. 901; *Clarence, New England, Macleay and Gwydir Almanac*, 1872, pp. 2–3.

14 L. Daley, *Men and a River. Richmond River District* (Melbourne, 1966), pp. 137–142; *Q*, 16 March 1872, p. 10, 10 August 1872, p. 11 & 28 November 1874, p. 9.

15 *Sydney Morning Herald*, 2 December 1870, p. 2; *ATCJ*, 4 February 1871, p. 141; W. Birrell, *The Manning Valley* (Brisbane, 1987), pp. 139–143.

16 *Weekly Herald and Queensland Agricultural, Mining and Pastoral Advertiser*, 6 May 1865, p. 13; *Kiama Independent*, 30 March 1865, p. 2; *ATCJ*, 29 November 1873, p. 685.

17 Lyne, *The Industries of New South Wales*, p. 257; *ATCJ*, 29 March 1884, p. 602; *NS*, 12 March 1884, p. 2 & 14 September 1887, p. 4.

18 A. Meston, 'Report on the Sugar Industry on the Clarence and Richmond Rivers', *QVP*, Vol. 2 (1882), p. 1119.

19 *Cleveland Bay Express*, 15 December 1866, p. 2; *PDT*, 12 January 1867, p. 2; *ATCJ*, 29 April 1871, p. 523; *Q*, 11 January 1873, p. 5; J. Beardmore to the Hon. Minister for Lands, 22 July 1873, In-letter 5378B in Sugar Selection File No. 4B, LAN P/216, QSA.

20 *PDT*, 16 March 1867, p. 3 & 12 March 1870, p. 2.

21 *PDT*, 12 March 1870, p. 2; *Rockhampton Morning Bulletin*, 19 November 1870, p. 2; L. McDonald, *Rockhampton: A History of City and District* (Brisbane, 1981), p. 63.

22 G. Dalrymple, 'Narrative and Report of the Queensland North-East Coast Expedition, 1873', *QVP*, Vol. 2 (1874), pp. 615–667.

23 T. A. Coghlan, *Labour and Industry in Australia*. Vol. 3 (1918; rpt. Melbourne, 1969), pp. 1373–1374; *ATCJ*, 10 December 1881, p. 1124; *Q*, 8 April 1882, p. 432.

24 *QVP*, Vol. 4 (1889), p. 94; A. Serle, *The Rush to be Rich: A History of the Colony of Victoria, 1883–1889* (Melbourne, 1977), p. 50; T. Parsons, 'Melbourne money and Queensland sugar. The experience of Swallow and Ariell in the 1880s', *Victorian Historical Magazine*, Vol. 44 (1973), pp. 32–39.

25 P. Donovan, *A Land Full of Possibilities. A History of South Australia's Northern Territory* (Brisbane, 1981), pp. 120–121.

26 *SAPP*, Vol. 4 (1881), Paper No. 179: Land Selections in the Northern Territory for Sugar Cane Cultivation; *SAPP*, Vol. 4 (1882), Paper No. 147: Quarterly Report by Govt. Resident on the Northern Territory; *Northern Territory Times*, 23 April 1881, p. 3 & 23 April 1883, p. 3.

27 *BC*, 11 July 1883, p. 3 & 22 October 1883, p. 3; J. Kerr, *Northern Outpost* (Mossman, 1979), pp. 21–26.

28 *Q*, 29 January 1881, p. 135; *The Planter and Farmer*, February 1882, p. 41; *Cairns Post*, 7 August 1884, p. 2 & 29 May 1884, p. 2 (Cairns district); *The Planter and Farmer*, January 1882, p. 16; *The Planter and Farmer*, September 1883, p. 605; *Q*, 21 June 1884, p. 991; *Sydney Mail*, 1 November 1884, p. 595 (Johnstone River); *BC*, 5 September 1883, p. 2 (Tully).

29 P. Griggs, 'Plantation to small farm: A historical geography of the Lower Burdekin sugar industry, 1880–1930'. PhD thesis, Uni. of Qld., 1990, pp. 99–115 & 122–125; *PDT*, 12 August 1882, p. 2, 3 March 1883, p. 2 & 23 August 1884, p. 3.

30 *PDT*, 18 November 1882, p. 3 & 28 April 1883, p. 2; *The Planter and Farmer*, November 1882, p. 307; *Rockhampton Morning Bulletin*, 21 April 1883, p. 2.

31 Kerr, *Pioneer Pageant*, pp. 69–79.

32 A. Goodwin, *The Sugar Industry of the Lower Burnett in 1882* (Bundaberg, 1882); Kerr, *Southern Sugar Saga*, pp. 12–30.

33 *ATCJ*, 7 January 1882, p. 26, 14 January 1882, p. 61 & 21 January 1882, p. 115; *Q*, 25 February 1882, p. 231 & 22 April 1882, p. 488.

34 *Queenslander Supplement*, 20 May 1882, p. 3 (inserted in the *Queenslander* of that date); *Sydney Mail*, 19 October 1884, p. 791.

35 Meston, 'Report on the Sugar Industry', p. 1120; *RRE*, 9 October 1880, p. 2; *Sydney Mail*, 6 December 1884, p. 1155; D. Jeans, *An Historical Geography of New South Wales to 1901* (Sydney, 1972), p. 246 (Richmond River); *ATCJ*, 10 September 1881, p. 519; *Q*, 29 July 1882, p. 152 & 5 August 1882, p. 183; *Sydney Mail*, 22 November 1884, p. 1052; J. Boileau, *Caldera to the sea: a history of the Tweed Valley* (Murwillumbah, 2006), pp. 110–113 (Tweed River district).

36 *Q*, 15 April 1882, p. 472; *BC*, 17 August 1883, p. 2.

37 *MC*, 20 March 1880, p. 2; *BC*, 13 November 1883, p. 5; Kerr, *Sugar at Maryborough*, pp. 101–116.

38 M. Gaylard, 'Economic development in the Maroochy district until 1915'. B.A. thesis, Uni. of Queensland, 1967, pp. 46–48; *BC*, 27 September 1882, p. 5 & 4 December 1882, p. 6.

39 Meston, 'Report on the Sugar Industry', p. 1119; *RRE*, 11 September 1880, p. 2, 9 July 1881, p. 2, 12 August 1882, p. 4 & 13 December 1884, p. 6; Jeans, *An Historical Geography of New South Wales*, p. 246.

40 Proposed Scheme for the Establishment of the Sugar Industry in the Tropical Territory of Western Australia, *Votes & Proceedings of the Western Australia Legislative Council*, 1885.

41 *QVP*, Vol. 4 (1889), pp. 123–128 & 157–159; Griggs, 'Plantation to small farm', pp. 115–119.

42 The plantations and their year of closure were as follows: Cedars (1885); Cassada (1886); Coningsby (1886); Mount Pleasant (1886); Dumbleton (1887); Marian (1888); Te Kowai (1889); Victoria (1889); and River (1891); for details see Kerr, *Pioneer Pageant*, pp. 99–101.

43 *Sydney Mail*, 8 November 1884, p. 947; *Q*, 20 March 1886, p. 469 & 13 October 1888, p. 668; Kerr, *Sugar at Maryborough*, pp. 101–106.

44 *QVP*, Vol. 4 (1889), pp. 362, 377 & 378; *Q*, 15 November 1884, p. 783, 21 November 1885, p. 839 & 24 August 1889, p. 342; *BC*, 26 November 1885, p. 2.

45 P. Fletcher, 'The sugar industry of Queensland', in P. Fletcher (ed.), *Queensland: Its Resources and Institutions* (Brisbane, 1886) p. 9; *QVP*, Vol. 2 (1886), p. 1087; *Q*, 4 September 1886, p. 391 & 8 December 1888, p. 1055.

46 *The Planter and Farmer*, December 1885, pp. 258–259; *SAPP*, Vol. 3 (1886), Paper No. 53: Half Yearly Report on the Northern Territory, 5; W. Haslam *et al.*, *Report and Minutes of Evidence of the Northern Territory Commission* (Adelaide 1895), passim.

47 *The Capricornian*, 19 June 1886, p. 11; *Townsville Daily Bulletin*, 6 February 1888, p. 2.

48 *CRE*, 8 December 1885, p. 4, 26 January 1886, p. 4, 1 May 1886, p. 8 & 23 October 1886, p. 3.

49 *QVP*, Vol. 4 (1889), pp. 79–80.

50 W. Groom *et al.*, 'Report of the Sugar Industry Commission Appointed to Inquire into the General Condition of the Sugar Industry' *QVP*, Vol. 4 (1889), pp. 62–75; Bolton, *A Thousand Miles Away*, pp. 198–201.

51 Farm circular, E.W. Knox to Tweed River Contractors, 17 September 1888; Farm circular, E.W. Knox to Tweed River Contractors, 17 April 1889, CSRR, 142/3511; *Q*, 12 August 1893, p. 310; *SJTC*, 15 September 1894, p. 178.

52 B. O'Neill, *Taming the Isis* (Childers, 1987), pp. 215–217; *Q*, 6 April 1889, p. 631; *QVP*, Vol. 4 (1889), p. 317.

53 E. Knox to R. Philp, Townsville, 23 April 1890, CSRR, 142/480, pp. 368–369; Bolton, *A Thousand Miles Away*, pp. 201–203.

54 Griggs, 'Plantation to small farm', pp. 163–166 & 173; Kerr, *Pioneer Pageant*, pp. 116–122.

55 *Bundaberg Mail*, 14 June 1893, p. 2; *MC*, 26 May 1894, p. 2 & 15 September 1894, p. 3; *Q*, 22 September 1894, p. 559 & 15 December 1894, p. 1136 (Maryborough); *Bundaberg Mail*, 19 May 1893, p. 2; *Q*, 27 January 1894, p. 176 (Childers).

56 *SAPP*, Vol. 3 (1893), Paper No. 158: Government Resident's Report on the Northern Territory for 1892, p. 7; L. Brackenbury, *Northern Territory. Report on Agricultural and Other Lands* (Adelaide, 1896), pp. 1–4; Haslam, 'Northern Territory Commission', p.vii.

57 *SJTC*, 15 August 1901, p. 28; L. Carpenter, *Livingstone. A History of the Shire of Livingstone* (Brisbane, 1991), p.199.

58 Kerr, *Southern Sugar Saga*, pp. 63–66; Kerr, *Sugar at Maryborough*, pp. 149–151.

59 *SJTC*, 15 October 1892, p. 190; *Q*, 21 January 1893, p. 119; *Q*, 13 August 1893, p. 310; *SJTC*, 15 October 1894, p. 202 & *Q*, 5 January 1895, p. 33 (gumming disease outbreaks); Bank of New South Wales Insp. to Mang., Perpetual Trustee Co. Ltd, Sydney, 15 November 1897, Bank of New South Wales, Grafton, Manager's Letterbook 1893–1900, pp. 359–360, Clarence River Historical Society.

60 Accounts of this contraction are detailed in *ATCJ*, 16 October 1897, p. 31 & 23 October 1897, p. 22; *Tweed Herald and Brunswick Chronicle*, 21 January 1900, p. 9 & 30 August 1901, p. 5; T. Coghlan, *The Wealth and Progress of New South Wales, 1900–01* (Sydney, 1902) p. 713; & *NS*, 26 September 1900, p. 8; for New South Wales canegrowers moving to Queensland see *Q*, 15 September 1894, p. 512 & *Geraldton Mirror*, 12 August 1896, p. 2.

61 P. Courtenay, *Northern Australia. Patterns and problems of tropical development in an advanced country* (Melbourne, 1982), p. 20.

62 A. Birch, 'The implementation of the White Australia Policy in the Queensland sugar industry', *Australian J. of Politics and History*, Vol. 11, no. 2 (1965), pp. 198–210; H. Easterby, *The Queensland Sugar Industry. An Historical Review* (Brisbane, c. 1933), pp. 28–30.

63 Birch, 'The implementation of the White Australia Policy in the Queensland sugar industry 1901–1912', p. 207; Moore, *Kanaka*, pp. 274–278; George Drysdale to Robert Tullis, 28 August 1902, PMR/CT 1, Box 7, Article 83, JCUA.

64 R. Ranking *et. al.*, 'Report and Evidence of the Royal Commission Appointed to Inquire into and Report Regarding the Number of Pacific Islanders to be Deported from Queensland at the end of 1906', *QVP*, Vol. 2 (1906), pp. 461–464; Bolton, *A Thousand Miles Away*, pp. 250–251; Moore, *Kanaka*, pp. 287–291; P. Corris, 'White Australia in action: the repatriation of Pacific Islanders from Queensland', *Historical Studies*, Vol. 15 (1972), pp. 237–50.

65 D. Hunt, 'Exclusivism and Unionism: Europeans in the Queensland Sugar Industry' in A. Curthoys and A. Markus (eds.), *Who Are Our Enemies?: Racism and the Australian Working Class* (Sydney, 1978), p. 92; R. Norris, *The Emergent Commonwealth* (Melbourne, 1975), p. 86; Griggs, 'Plantation to small farm', p. 213.

66 Easterby, *The Queensland Sugar Industry*, pp. 42–43; W. Douglass, *From Italy to Ingham. Italians in North Queensland* (Brisbane, 1995), pp. 80–85; *Q*, 20 July 1907, p. 22; J. Armstrong, 'The sugar strike, 1911', in D. Murphy (ed.), *The Big Strikes, Queensland, 1889–1965* (Brisbane, 1983), pp. 102–104; Bolton, *A Thousand Miles Away*, p. 308.

67 *ASJ*, 7 October 1909, p. 247–248; Easterby, *The Queensland Sugar Industry*, p. 32.

68 Hunt, 'Exclusivism and Unionism', pp. 91–93; Armstrong, 'The sugar strike, 1911', pp. 105–115; for more detailed descriptions of the strike action see *North Queensland Register* 27 May 1911, p. 45, 5 June 1911, p. 78, 19 June 1911, pp. 29, 50 & 83 & 12 August 1912, p. 26.

69 L. Fitzhardinge, *William Morris Hughes: A Political Biography. Volume 1: That Fiery Particle 1862–1914* (Sydney, 1964), p. 266; D. Shogren, 'The politics and administration of the Queensland sugar industry to 1930', PhD thesis, Uni. of Queensland, 1980, p. 174; for details about Crawford's life see J. Kerr, *Northern Outpost* (Mossman, 1995), pp. 67–68.

70 Easterby, *The Queensland Sugar Industry*, pp. 33–35; for a detailed examination of the 1912 Royal Commission see Shogren, 'The politics and administration of the Queensland sugar industry to 1930', pp. 160–195.

71 P. Courtenay, *Northern Australia: patterns and problems of tropical development in an advanced country* (Melbourne, 1982), p. 29.

72 Griggs, 'Plantation to small farm', pp. 225–242.

73 J. Kerr, *Top Mill in the Valley. Cattle Creek Sugar Mill, Finch Hatton, 1906–1990* (Brisbane, 1991), pp.12–17; Acting Land Commissioner, Mackay, AR 1907, *QPP*, Vol. 2 (1908), p. 1018.

74 Acting Land Commissioner for the Carins, Herberton, Ingham and Geraldton districts, AR 1909, *QPP*, Vol. 3 (1910), p. 94.

75 Land Commissioner, Cairns, Herberton, Ingham and Geraldton districts, AR 1907, *QPP*, Vol. 2 (1908), p.1020 & Land Commissioner, Cairns, Herberton, Ingham and Geraldton districts , AR 1911, *QPP*, Vol. 3 (1912), p. 72; E. Forrest, Brisbane, to Revd Mother Sister S. Patrick, All Hallows Convent, Brisbane, 22 September 1908, CSRR,

142/3274; J. Armstrong and G. Verhoeven, *The Innisfail Tramway* (Brisbane, 1973), pp. 28 & 38–39; *ASJ*, 6 May 1909, p. 15.

76 *Cairns Post*, 30 July 1907, p. 2; *CPP*, Vol. 4 (1913), p. 1174 (Ev. G. Muntz, Chairman of Directors, Mossman Central Mill Co. Ltd., Q., 1174) & p. 1224 (Ev. R. Jones, Mossman Mill supplier, Q. 1484); *ASJ*, 5 August 1909, p. 167 & 8 December 1910, p. 414; Sec., Mossman Central Mill Co. Ltd., to The Hon. W. Kidston, Premier, 26 December 1910, Letterbook No. 10, p. 541, Mossman Mill.

77 *ASJ*, 6 July 1911, p. 223, 4 April 1912, p. 14 & 4 June 1914, pp. 204–205.

78 *ASJ*, 3 March 1910, p. 474, 4 August 1910, p. 195, 9 November 1911, p. 465 & 6 June 1912.

79 Kerr, *Sugar at Maryborough*, pp. 153 & 155; *ASJ*, 7 December 1911, p. 497.

80 *ASJ*, 6 May 1909, p. 21; B. Alcorn and R. Dunn, *Moreton Sugar Mill. Sweet Heart of Nambour* (Nambour, 1997), pp. 49–50; *CPP*, Vol. 4 (1913), p. 648 (Ev. John Paddle, Acting Mang., Nerang Central Mill, Qs. 17637 &17711).

81 *ASJ*, 9 November 1911, p. 453 & 9 January 1913, p. 754; *CPP*, Vol. 4 (1913), p. 963 (Ev. W. Gibson, Managing Director, Marburg Sugar Co., Q. 26187).

82 *CPP*, Vol. 4 (1913), p. 620 (Ev. Albert Biggs, farmer Teven Creek, Qs. 16971–16977); p. 629 (Ev. Robert Browne, ex-Excise Officer, Broadwater, Q. 17225; p. 641 (Ev. N. Joubert, Condong Mill supplier, Qs. 17505–17508); p. 637 (Ev. E. Marks, Condong Mill supplier, Qs. 17400–17405); *ASJ*, 3 June 1909, pp. 98–99; A. Barry, Broadwater Mill Mang., to E. W. Knox, 6 October 1908, CSRR, 142/877, p. 10; A. Barry, Broadwater Mill Mang., to E. W. Knox, 17 May 1911, CSRR, 142/878, pp. 144–145; Russel Dowling, Cane Insp., Condong Mill, to E. W. Knox, 3 April 1913, CSRR, 142/1218, Letter No. 282.

83 G. Kottmann, 'Chemical Report on the 1903 season', pp. 13–14, CSRR, Z/109, Box 376; R. Dowling, Broadwater Mill Cane Insp., to E.W. Knox, 22 January 1914, p. 2, CSRR, Z/109, Box 459 A; *Q*, 3 November 1906, p. 33; *ASJ*, 6 May 1909, p. 14; J. Armstrong, 'The sugar tramways of northern New South Wales', *Australian Railway Historical Society Bulletin*, Vol. 27 (1976), p. 126; *Tweed Times and Brunswick Advocate*, 13 June 1906, p. 3.

CHAPTER 6

The Production Unit: Planters, Central Millers and Small Farmers

During the nineteenth century, three main arrangements existed for the production of sugar. Plantations or vertically integrated agro-industrial units combining both the growing of the sugar cane and its processing into sugar were still found in many parts of the world. Some sugar plantations, mostly in Brazil, still relied on slaves; the majority in the English-speaking colonies employed indentured labour, especially Indians. A second alternative was metayage or sharecropping, whereby labourers shared proceeds with the planters in lieu of wages. Under this system the labourers cut the cane, did all the cultivation and provided the manual work associated with the milling of the crops. The sugar plantation owner provided the land and all the equipment, stock and milling facilities. The sugar made was usually divided in halves, with the planter generally purchasing the metayager's share of sugar. A third alternative was the central sugar milling system in which the planters concentrated on growing the sugar cane and the central sugar millers devoted their resources to the industrial section. This arrangement became known as the *central-colono* system, although the *colonato* grew to incorporate a range of cultivators and types of landholdings. Colonato could be former mill-owners who continued growing cane, ex-slaves or others who leased plots of land from the planters, or nearby smallholders who were persuaded to plant sugar cane on land they owned or leased.[1]

Identifying which late nineteenth century properties were Australian sugar plantations is fraught with difficulties. Reports in contemporary Queensland newspapers referred to any landholding with a sugar mill as a plantation. The Groom Royal Commission into the Queensland Sugar Industry (1889) adopted this classification; other properties were simply labelled farms.[2] This author argues, however, that the existence of a sugar mill on a late nineteenth century Australian sugar-producing property should not automatically confer plantation status. During the late 1860s and early 1870s, small, horse-powered sugar mills could be secured in Queensland for between £200 and £800. Hence, their presence on a property sometimes only indicated the owner had slightly more capital than a small farmer. It said little about the size of the property, number of workers or the division of labour and management.[3]

Given that the size of Australian sugar-producing units during the late nineteenth century ranged between 20 ha and 4 050 ha, a decision was made that any sugar-producing property greater than 130 ha in size would be considered a plantation. This figure was decided upon because it was the size of the smallest blocks issued to selectors in Queensland under the Sugar and Coffee Regulations of 1864, which aimed to encourage plantation agriculture in the colony. Some settlers owning less than 130 ha, with a sugar mill on their property, and employing a few indentured workers could have been included in this analysis as plantations, and would have possibly been considered a plantation in the West Indies during the late nineteenth century. These operations, however, have more in similarity with the 'small millers' category developed by the historian Barry Higman in his study of New South Wales sugar-producing units in the nineteenth century. Most Queensland and New South Wales small millers were men of moderate capital, operating small sugar mills and relying mainly on family labour or a few paid workers. The scale of their operations was not on a size consistent with what could be expected on a sugar plantation.[4]

Plantations

Numbers and location

By 1885, at least 116 sugar plantations operated in Australia (Table 6.1). Nearly half these properties had been developed between 1864 and 1875; the remainder were founded during the next decade. Initially sugar plantations were only established in Queensland; none were founded in New South Wales. Difficulty in securing large blocks of land fronting the Clarence, Richmond and Tweed Rivers deterred investors interested in pursuing large-scale agriculture. Most riverside land had been alienated as small blocks during the 1860s. Moreover, legislative restrictions on the size and shape of holdings (i.e. one side could not be more than twice the length of another) made it difficult to assemble blocks with large river frontages.[5] Some large estates greater than 130 ha in size (e.g. Woodside, Taree and Woodland Estates) along the Macleay, Manning and Hastings Rivers could have become sugar plantations, but their owners had abandoned sugar cane cultivation by 1875.

Table 6.1. The number of sugar plantations in Australia's sugar-producing regions, 1870–1915.

Region	1870	1875	1885	1895	1905
Far Northern*	1	0	8	4 (2)+	4 (4)
Herbert R.	0	3	5	3 (2)	3 (3)
Lower Burdekin	2#	0	5	3 (3)	2 (2)
Mackay	5	17	23	7 (4)	4 (3)
Rockhampton	0	0	2	1 (1)	0
Bundaberg-Childers	0	3	26	19 (9)	10 (7)
Maryborough	6	9	13	3 (0)	0
Brisbane-Ipswich	0	4	4	3 (0)	1 (0)
Beenleigh	12	15	16	3 (1)	0
New South Wales	0	0	13	5 (3)	1 (1)
Northern Territory	0	0	1	0	0
TOTAL	26	51	116	51 (25)	25 (18)

Notes: * Includes the Bloomfield River, Mossman, Cairns, Innisfail & Tully districts.
One plantation each was located at Townsville and Bowen. + The figure in brackets is the number of plantations known to have tenants.

Source: Based upon Appendix 1 in P. Griggs, 'Sugar plantations in Queensland, 1864–1912: Origins, characteristics, distribution and decline', *Agricultural History*, Vol. 74, no. 3 (2000), pp. 644–647.

Severe frosts in these districts probably forced the proprietors to stop the development of their estates into sugar plantations, although Woodside Mill was destroyed by fire and the proprietor decided not to rebuild the factory, thereby eliminating the need to grow any more sugar cane.[6]

The second phase of sugar plantation establishment in Australia occurred between 1878 and 1885. The majority of these new plantations were founded in the Far Northern, Lower Burdekin and Bundaberg-Childers regions of Queensland (see Table 6.1 and Figure 6.1). In the Northern Territory, only Delissaville estate was sufficiently developed by 1885 to be categorised as a plantation. Labour had been employed to plant nearly 80 ha with sugar cane and a sugar mill erected on the property. Holders of other land grants under the *Northern Territory Sugar Cultivation Act* of 1881 had failed to develop their properties beyond establishing small sugar cane plant nurseries of 10–12 ha in size.[7] Investors in New South Wales finally assembled enough blocks of land

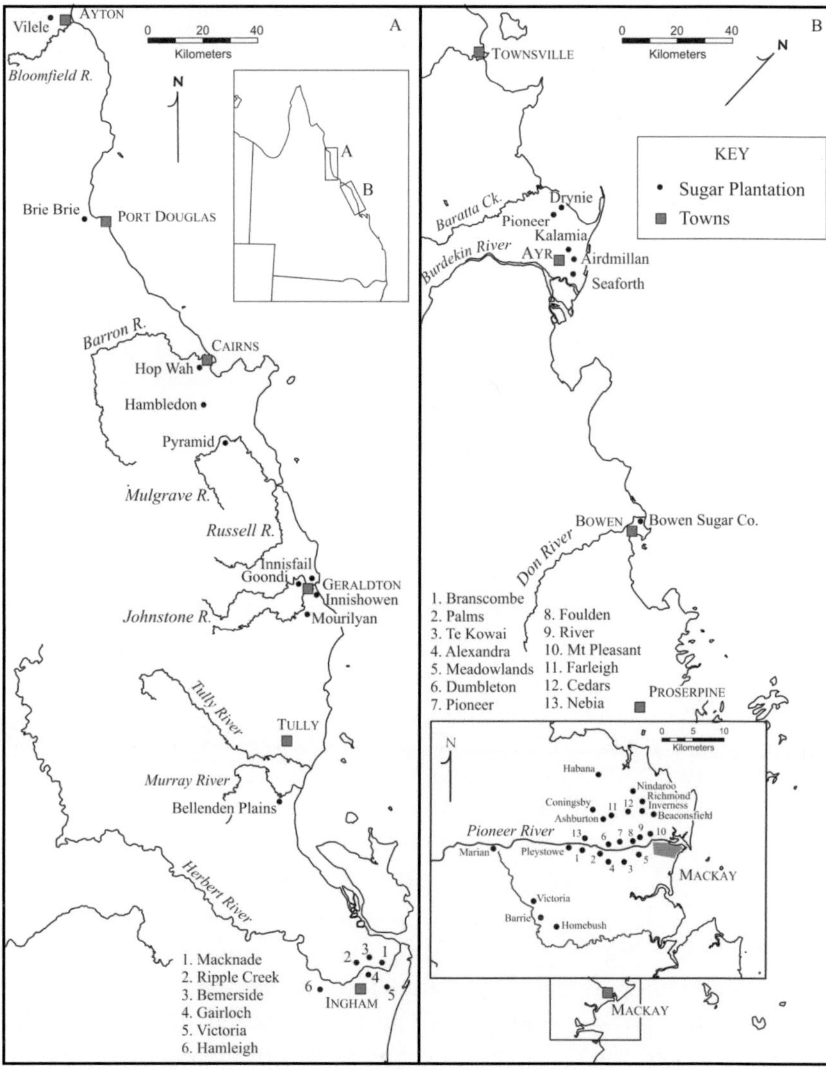

Figure 6.1. The approximate locations of North Queensland sugar plantations.

Source: Based upon JG1, Sheet 5, 1897; JG1, Sheet 8, 1910; and NG1, Sheet 6, 1911, QSA; and C. Moore, *Kanaka. A History of Melanesian Mackay* (Port Moresby, 1985), pp. 105 & 111.

during the late 1870s and early 1880s to create thirteen plantations (Figure 6.2). CSR, even though it had pioneered the establishment of central sugar mills in New South Wales (see discussion later in the chapter), operated two plantations in the colony by 1890. Its Condong estate, however, was never intended to be a plantation, but a tenanted property. CSR had purchased 1 200 ha in the Tweed River district by 1878, but the scarcity of suitable tenants forced the firm to cultivate several hundred hectares of this land in order to keep Condong Mill fully supplied during the 1880s.[8] In contrast, CSR was seeking a large property in the Richmond River district where it could grow sugar cane. As early as 1869,

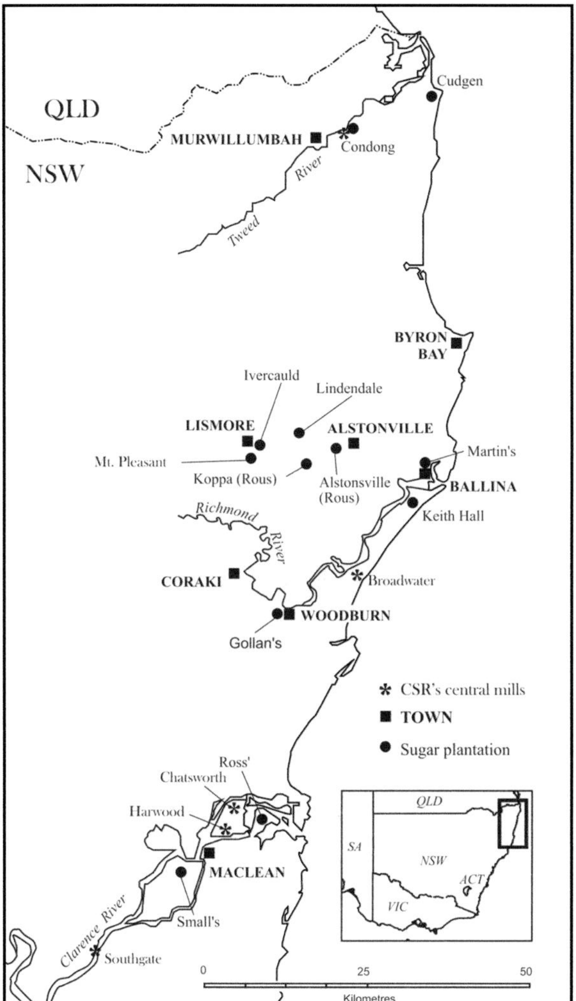

Figure 6.2. The approximate locations of the New South Wales sugar plantations and four of CSR's central sugar mills.

Source: Based upon Figure 2 in B. Higman, 'Sugar plantations and yeomen farming in New South Wales', *Annals Association of American Geographers*, Vol. 58 (1968), p. 703.

Edward Knox advised Thomas Scott that CSR wished to be less dependent on small farmers and had been searching for 400 ha of land in the Richmond River district. In 1882, CSR was offered the Tomki Estate – 3 442 ha of mostly riverfront land upstream from Coraki – but declined to purchase the property because of concerns about frost. Instead, the firm settled for Keith Hall Estate, consisting of 160 ha of riverfront land in the lower reaches of the Richmond River. CSR grew sugar cane on Keith Hall Estate during the late 1880s, but a sugar mill was never sited on this property. By 1895, it had become primarily an experimental farm.[9]

Sugar plantations in Queensland mostly had navigable river or creek frontages (see Plate 6.1). The distribution of sugar plantations in North Queensland showing this locational feature is illustrated in Figure 6.1. The advantages of this arrangement were three-fold. Firstly, the sugar planters secured the rich, alluvial soils fronting the colony's coastal rivers. Secondly, in frontier districts having no railways and limited road networks, river carriage was often the only means of transport. Australian sugar planters established riverside wharfs to receive goods and machinery and to export their raw sugar

Plate 6.1. Eatonvale plantation, Copenhagen Bend, Mary River, c. 1890. Many Queensland sugar plantations established before 1880 had river frontages.

Source: Photograph No. 112, Sidney Richardson Collection, No. 243, Fryer Memorial Library. Image reproduction courtesy of The University of Queensland Libraries, Brisbane.

to urban markets. CSR in particular instructed its officers to ensure that any land being contemplated for purchase was accessible to navigable water, but also free from lengthy flood inundation.[10] Third, a regular supply of fresh water was needed for a sugar mill's boilers and vacuum pans. Riverside locations, however, suffered damage from flooding. Eatonvale plantation, for example, lost a 40 metre strip of riverside land during the 1870 flood of the Mary River, while flooding of the Herbert River in 1873 washed away all the newly planted cane on Macknade plantation.[11]

A third of Australian sugar plantations were not situated on navigable creeks or rivers, but there were mostly sound reasons for their location. In New South Wales, half the plantations were found inland near Lismore, away from the Richmond River. The majority of riverside land had been selected by small farmers during the 1870s, so the intending sugar planters were forced to select their blocks of land further inland, although the soil in this locality was mostly fertile (Figure 6.2). In Queensland, Redland Bay plantation near Brisbane was sited adjacent to Moreton Bay, its owners taking advantage of the fertile red-volcanic loam in this vicinity. Inverness, Cedars and Richmond plantations were found north of the Pioneer River (Mackay) in heavily timbered, but fertile hill country (Figure 6.1). These plantations were developed in the early 1870s, after the land fronting the Pioneer River had been sold or leased to the district's pioneering sugar planters. To the north-east of Bundaberg, a dozen sugar plantations were situated between the Burnett River and the ocean in an area known as the Woongarra district, a particularly fertile plain, consisting of red-volcanic loam. Cudgen plantation was also sited to take advantage of volcanic soils, its original owners foregoing river frontage to the Tweed River and possible river transport (Figure 6.2). In contrast, Woodlands plantation, near Marburg, was situated 65 km inland from the coast and developed in a frost-prone area not entirely suited to sugar cane growing. Thomas L. Smith, the estate's owner, established the sugar plantation in this locality as he needed yearly crops to supply his sugar mill, which had been erected as a central mill. The surrounding small farmers who had promised to supply his factory, however, failed to deliver the necessary cane.[12]

Size of the sugar plantations

The size of Australian sugar plantations created between 1864 and 1884 varied enormously. New South Wales sugar plantations ranged in size from 130 ha (Lindendale plantation) to 870 ha (Condong plantation), with the average size being 310 ha. In Queensland, the average size of the sugar plantations formed between 1864 and 1878 was 208 ha, although Yengarie, Macknade, Magnolia, and Pimpama plantations were much larger, all exceeding 810 ha in size. Between 1878 and 1885, half the sugar plantations founded in the three most northern Queensland sugar-producing regions were over 2 025 ha in size. Smaller sugar-producing estates under 610 ha in size were most prevalent in Southern Queensland.[13] An abundance of virgin land which could be easily secured by intending sugar planters and the practice of dummying (i.e. persons acting on behalf of another, usually a pastoralist or planter) enabled the creation of such large properties.[14]

Less that 80 ha of sugar cane were cultivated on many of the Queensland sugar plantations during the late 1860s and early 1870s. However, some newspaper reports recording the area under cultivation on these sugar plantations occurred soon after their establishment, so the figures probably reflected the early stages in an estate's development. Nevertheless, the owners of Yengarie, Macknade, River and Nerada plantations, to name a few, cultivated only a fraction of their land holdings. Labour shortages (see discussion later in the chapter) limited the area which could be cleared and placed under crop, and the restrictions imposed on the size of the yearly harvest by the capacity of the plantation's mill probably explain why so much land was under-utilised on some Australian sugar plantations.[15]

More comprehensive information on this topic was given to the Royal Commission into the Queensland Sugar Industry of 1889. This data can be used to effectively gauge the size of operations before 1885, because immediately afterwards Queensland's sugar planters were forced to curtail their operations due to labour shortages. In 1888, less than 200 ha of cane were cultivated on the majority of the sugar plantations in Queensland. However, larger areas between 500 and 800 ha were cultivated on some Australian sugar plantations, mostly in the northern sugar-producing regions. Yet these figures hide variations in the utilisation of sugar plantation lands. Between 40 and 50% of Cudgen, Palmyra, Ripple Creek, Sharon, Fairymead and Bingera plantations, to list a few, was under sugar cane. In contrast, CSR grew over 800 ha of sugar cane on each of its Queensland sugar estates in 1888, utilising less than a quarter of its land holdings in the colony.[16] Nevertheless, non-cultivated lands on some Australian sugar plantations were devoted to cattle grazing or placed under small crops of maize and potatoes to feed the estate's horses and Pacific Islanders. Thomas Swallow and Frank Derham utilised part of Hambledon plantation for a fruit orchard, growing enough ginger, pineapples, oranges and lemons to supply their Melbourne biscuit factory.[17]

Many Australian sugar plantations had a succession of owners, suggesting a speculative investment in this pioneering agricultural endeavour. Some original owners were absentee-owners, while others resided on their estates, enduring the difficulties of their establishment. Arthur Neame, for instance, lived in a tent and suffered malaria during the development of Macknade plantation, while John E. Davidson was forced to spend Christmas Day 1866 up a tree due to flooding of his Bellenden Plains plantation.[18] Some of the earliest Australian sugar planters came from wealthy or aristocratic backgrounds. John E. Davidson was the son of a wealthy Scottish merchant, Francis Tyson Amhurst of Foulden plantation (Mackay) was the son of the High Sheriff of Norfolk, and Frederick and Arthur Neame of Macknade plantation (Herbert River) were sons of an English gentleman, Frederick Neame of Macknade Estate, Faversham.

Few of Australia's earliest sugar planters had direct experience of growing sugar cane. Those who did such as the MacDonald Brothers (Inverness plantation), Robert Muir (Benowa plantation) or Hugh McCready (Palmyra plantation) had gained this experience in the British West Indies or Demerara. In contrast, John Spiller (Pioneer plantation, Mackay) had come to Queensland from Surabaya (Java), while Samuel Staines had been an engineer and the general manager of some Hawaiian sugar mills in the 1870s, before forming Koppa plantation (Alstonsville, New South Wales). Others did have some agricultural experience. Alexander Mackenzie (Gairloch plantation) had been a coffee planter in Wynaad, southern India, William Hyne (Balmoral and later Meadowlands plantations) was the son of a Devonshire farmer and a graduate of the Royal Agricultural College, Cirencester, England, while John Davey and Francis Gooding of Beenleigh plantation came from farming backgrounds in Devon. This background, however, was no guarantee they would succeed as agriculturalists.[19]

Close links existed between Australia's sugar plantation owners and other economic activities. Some of Queensland's earliest sugar planters were pastoralists, which is understandable given that pastoral pursuits pre-dated the establishment of sugar production in most districts. In the Maryborough district, Robert Tooth and Robert Cran Snr operated several pastoral properties as well as the Yengarie Boiling Down Works. They diversified into sugar production in the late 1860s, converting Yengarie into the colony's first sugar refinery (see Chapter 9). A similar pattern was found at Mackay. William Hyne and George Bridgman founded the Walkerston Boiling Down Works in 1868, as well as investing in Meadowlands and Balmoral plantations.[20] For John and James Toohey, the Sydney brewers, it was a case of backward integration, as they were able to supply their brewery with sugar from Nashua plantation. George Wilkie Gray, the main investor in the company operating Innishowen plantation, also had brewing links as he was a member of the firm Quinlan, Gray & Co. which operated Castlemaine Brewery (Brisbane).[21] For CSR and its subsidiary the Victoria Sugar Co. (VSC), the investment in Queensland sugar plantations was also backward integration, but in response to competition. The two companies had become concerned about their market share being eroded by Queensland-produced white sugars being sold to merchants who retailed it to consumers in southern markets in direct competition with CSR and the VSC. Hence CSR's move to Queensland expanded its own raw sugar production and enabled the firm to purchase more raw sugar from other Queensland sugar mill-owners for its Melbourne and Sydney refineries.[22]

Australian colonial banks funded both individuals and joint stock companies to form sugar plantations. The Queensland National Bank supported the establishment of at least twenty Queensland sugar plantations, making it the single largest provider of investment capital to the Australian sugar industry. Finance was also secured from other lending organisations. William Sloane & Co., the Melbourne-based woolbrokers and general merchants, for instance, supplied the start up funds in the 1870s for Te Kowai, Nebia and The Cedars plantations (Mackay district), and the initial loans to establish Seaforth, Kalamia and Airdmillan plantations (Lower Burdekin district) during the early 1880s. The money lent to the sugar planters was often secured by mortgage and bill of sale over the estate's land and crop liens. If the sugar planters were unable to meet their repayments, the lending institutions foreclosed, taking control of the plantations.[23]

Before 1878, few Australian sugar plantations were owned by joint stock companies. The most noted were the Pimpama Sugar Co., the Maryborough Sugar Co. and the Bowen Sugar Co., which all attracted start-up investment from local businessmen.[24] After 1878, there was an upsurge in the ownership of plantations by joint stock companies. Melbourne-based companies formed to develop new sugar plantations included the Burdekin Delta Sugar Co. (Airdmillan plantation), Bloomfield River Sugar Co. (Vilele plantation), the Mackay Sugar Co. (Victoria plantation, Mackay), Oakenden Sugar Co. (Oakenden plantation) and the Alstonsville Sugar Co. (Alstonsville plantation).[25] New Melbourne-based companies such as the Pleystowe Sugar Co. acquired the already operational Pleystowe plantation (Mackay district), while the Melbourne-Mackay Sugar Co. was floated with a capital of £500 000 to take over the five sugar plantations held by William Sloane & Co., John E. Davidson and George H. King. Victorian businessmen were also the major investors in the Palmerston Plantation Co. and the Adelaide River Land and Sugar Co., which were founded to develop sugar plantations in the Northern Territory.[26] For Victorian capitalists, investment in Queensland sugar plantations was attractive, given the increasing Victorian demand for sugar driven by rapid population growth.

Sugar plantation ownership in Australia became more concentrated between 1878 and 1885. CSR had the most extensive holdings of approximately 18 000 ha. The firm controlled three sugar plantations in Queensland – Victoria (Herbert River district), Goondi (Innisfail district) and Homebush (Mackay district) – in addition to its Condong plantation and Keith Hall Estate in New South Wales. The Melbourne-Mackay Sugar Co. owned 3 240 ha, consisting of Alexandra, Te Kowai, Branscombe, Palms and Nebia plantations (Mackay district). Edward Fanning and Thomas Nankivel through their investments in various companies had effective control over Gairloch, Bemerside, Macknade and Hamleigh plantations, all in the Herbert River district. One family who also developed a diverse portfolio of sugar plantation investments from this period was Frederic, George and Charles Buss. In addition to their ownership of a major department store in Bundaberg, they invested in/and or acquired together with various partners the following Bundaberg properties: Ashfield (1882); Millbank (1883); Glenmorris (1885); Pemberton (1886); Duncraggan (1886); Kalkie (1890); and Knockroe (1893).[27] This trend in plantation ownership allowed managers to make both the cultivation of cane and the manufacture of sugar more efficient through economies of scale. Their administration was streamlined, sugar cane acreage extended, some methods of scientific cultivation adopted (see Chapter 7) and manufacturing technology improved, often through the acquisition of the latest sugar-making machinery from Scottish engineering firms (see Chapter 8).

Sugar plantation workforce

From the outset, Queensland sugar planters relied heavily on indentured Pacific Islanders, recruited mostly from Vanuatu and the Solomon Islands. Pacific Islanders cost less to employ than Europeans, but it was also widely believed that Europeans were incapable of labouring in the tropics. In addition, labour shortages arose in the late 1860s after the discovery of gold at Gympie. By the early 1880s, over 200 Pacific Islanders worked on large sugar plantations such as

Plate 6.2. A gang of Pacific Islanders hoeing young sugar cane in the Herbert River district, possibly on Macknade plantation, c. 1902.

Source: Image no. 16956, John Oxley Library. Image reproduction courtesy of the State Library of Queensland, Brisbane.

the Palms, Te Kowai, Homebush, Victoria or Gairloch.[28] The indentured Pacific Islanders laboured for long hours, primarily in the fields, but some worked in the sugar mills. They generally worked five and a half days each week, from 6.00 am until about 5.30 pm, with an hour's break at mid-day. During the crushing season, they sometimes toiled longer. They performed exhausting manual toil, including clearing forested land, cutting firewood for the mills, preparing the fields for cultivation, trashing and weeding cane crops, harvesting the cane, loading harvested cane into wagons and carts and unloading it at the mills, and feeding the harvested cane into the sugar mills' crushing rollers.[29]

During the late 1860s and 1870s, some Maryborough and Mackay sugar planters also employed Aborigines to perform a variety of seasonally-determined tasks, such as weeding, gathering firewood and loading harvested cane into wagons. In addition, Europeans worked on the estates, but in more skilled positions such as ploughmen, sugar boilers in the mills, bookkeepers or estate managers. Exceptions to this general pattern occurred on Pimpama plantation, where the field labourers were mostly German immigrants, due to the influence of the Brisbane merchant Johann Heussler, one of the directors of the Pimpama Sugar Co. Management at Malungavel, Dulverton and Loganholme plantations also attempted growing sugar cane using Europeans only. The owners of these three estates, however, had lost control over their properties by 1880, although this insolvency cannot be entirely attributed to the use of European labour only.[30]

After 1878, different ethnic groups made up the workforce on Australian sugar plantations. Delissaville plantation in the Northern Territory relied on Chinese and Aborigines.[31] Strong parliamentary and worker opposition to non-European labour in New South Wales ensured that Europeans were the main plantation labourers. The exception was Cudgen plantation, where time-expired Pacific Islanders (i.e. Pacific Islanders who had elected to stay in Queensland after their first contract of indenture had expired) from Queensland found ready employment.[32] Queensland's sugar planters, however, faced labour shortages during the early 1880s, despite the introduction of measures in 1880 which restricted Pacific Islanders to employment in the colony's sugar industry and record levels of Pacific Islander recruitment to the colony over the years 1881 to 1883. They augmented their labour forces by employing Chinese who had left the colony's mining industry and recruited indentured Cingalese, Chinese, Malays and Maltese. Some Queensland sugar planters, however, found employing workers from alternative nationalities unsatisfactory. Many Cingalese brought to Queensland repudiated their contracts as most had no agricultural skills, while the majority of the Malays employed on the Lower Burdekin sugar plantations absconded shortly after their arrival.[33] Moreover, these workers did not always prove to be cheap labour. Time-expired Pacific Islanders cost North Queensland sugar planters 12s per week to employ, but Chinese cost between 25s and 28s per week.[34]

The class relationship or social hierarchy on a Queensland sugar plantation was expressed in its settlement patterns and this feature is well illustrated in the map of buildings on Pioneer plantation, near Ayr (see Figure 6.3). The sugar boiler's house and other huts, presumably accommodating the European staff, were set apart from the Malay and Chinese camps and buildings housing the Pacific Islander or 'kanaka' workers. Other houses and offices were located further away from the sugar mill and were most likely occupied by the managerial staff. The quality of housing also reflected one's status on a Queensland sugar plantation. Pacific Islanders lived in huts of reeds and thatch work or sometimes wooden barracks. These dwellings were

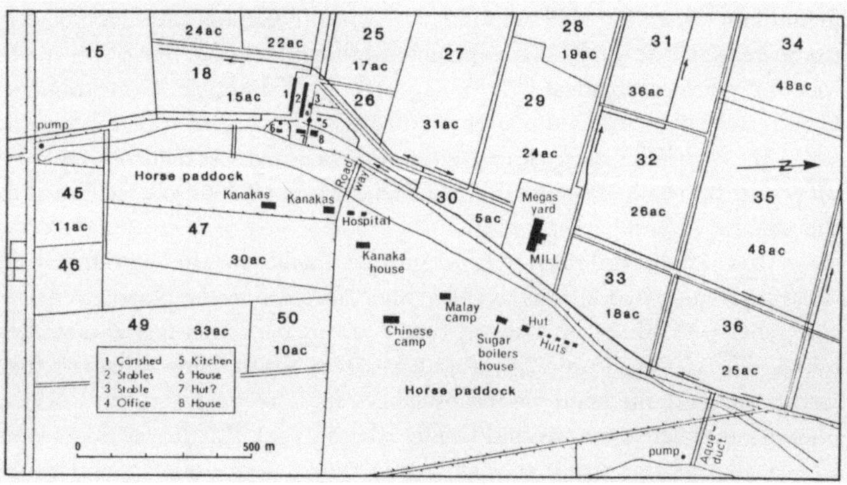

Figure 6.3. Part of a sketch plan of Pioneer Sugar Estate, Lower Burdekin, 1888.

Source: 'Sketch Plan of Cultivation Paddocks, Pioneer Sugar Estate, 1888', PMR, [M/P] Box, Folio 1, Article 1, JCUA. Image reproduction courtesy of James Cook University.

often surrounded by coconut trees and food gardens, containing traditional crops such as paw paw, yams, pineapples, taro and bananas. In contrast, the homesteads of the plantation owners and managers were large, single- or two-storey wooden houses, sometimes with broad verandahs both upstairs and downstairs. These buildings were often set apart from the sugar mill and occasionally situated on the tops of hills, to catch not only the cooling breezes, but enabling better scrutiny of the daily activities on the sugar plantation.[35]

Ending the sugar plantations

In August 1885, the continued existence of sugar plantations in at least Queensland was put in jeopardy. Samuel Griffith's Liberal government introduced legislation banning Pacific Islander recruitment after December 1890. This shock announcement was compounded by the rising cost of Pacific Islander recruitment and a sudden decline in the price of sugar.[36] These three factors began a chain of events that led to 45% of Australia's sugar plantations being closed or abandoned by 1895 (Table 6.1). Sometimes, however, there were a number of stages before the eventual cessation of plantation agriculture. Moreover, a few new sugar plantations were formed in Queensland during the late 1880s. Robert Cran & Co., for example, established Duncraggan plantation (Bundaberg) and Island plantation (Maryborough) in 1886.[37] Consequently, Australian sugar plantations disappeared gradually over the next twenty-five years.

When confronted with the threat of labour shortages, rising costs for labour and falling returns from sugar, Australian sugar planters adopted one or more of the following strategies: do nothing; continue growing sugar cane, but secure alternative labour; swap to another crop or agricultural pursuit; substitute machinery for workers; or shift the growing of sugar cane to small farmers via sharecropping or tenanted plantations. Australian sugar planters could not readily mechanise their cultivation practices. Gasoline-powered tractors only began replacing men and horses in the Australian sugar industry during the late 1910s (see Chapter 11), while practical, workable mechanical cane harvesters were not available until the late 1950s (see Chapter 16). No evidence could be located to indicate that Australian sugar planters engaged in sharecropping. However, in Southern Queensland where alternative crops could be cultivated, the owners of Ageston, Bundall, Benowa and Birribi plantations ceased sugar cane growing completely, closed their small factories and switched to cultivating lucerne (for dairying), oats and maize. Alfred H. Lloyd (Dumbleton plantation, Mackay) also abandoned sugar cane growing in favour of horse breeding, while Colin Munro of Drynie plantation (Lower Burdekin) ceased growing sugar cane, but converted his sugar mill into a condensed milk factory supplied by the herd of dairy cattle grazing on his plantation lands. Other Australian sugar planters immediately subdivided their estates. The owners of Redland Bay plantation on Brisbane's outskirts, for example, took advantage of the estate's favourable location to split it into housing allotments and small farms, while Yatala plantation near Beenleigh had been split into 4 ha small farms by 1888.[38]

Most Australian sugar planters or plantation-owning companies, especially in the northern regions, continued growing sugar cane, for alternative crops which could utilise the milling machinery were limited and many planters believed that the cultivation of coffee, tobacco or tropical fruits also required cheap labour.[39] Indeed, no North Queensland sugar planter switched to growing a different crop during the crisis. To maintain plantation agriculture and prepare for being denied Pacific Islanders, many Australian sugar planters simply sought alternate labour supplies. Several larger companies, including CSR, recruited Javanese, although CSR discovered it was considerably more expensive, for it cost the company £25 to bring Javanese workers to Queensland as opposed to an outlay of £8 for the recruitment of Pacific Islanders. This strategy, however, had to be halted when the Dutch government prohibited the trade at the end of 1886. Robert Smellie of Mourilyan plantation recruited forty

English labourers in 1886, but they disliked working in the sugar mill, creating as many difficulties as they could. Eventually their contracts were cancelled.[40] Other Queensland sugar planters turned to Aborigines, while the recruitment of Italian labourers for Queensland's sugar plantations was sanctioned by Griffith's Liberal government and commenced in 1891.[41]

By 1888, Innisfail, Mt Pleasant, Hamleigh, Mourilyan, Kalamia, Seaforth, Pioneer, Ashburton and Pyramid plantations, to name a few, had been taken over by the mortgagee, when their owners defaulted on repayments.[42] Most mortgagees, particularly the banks, were uncertain what to do with these properties. Some attempted to sell the estates at auction, but sugar plantation assets were virtually unrealisable in the late 1880s. The London Chartered Bank, for example, tried selling Hamleigh plantation at auction in February 1887, but the highest bid received by the vendor was £11 300. Later in the year, the property that cost its investors £110 000 to establish, was sold to one of the original partners for a mere £12 500. Pleystowe plantation attracted no bids when auctioned for the first time in November 1888.[43] The Union Bank of Australia Ltd. tried unsuccessfully in 1889 and 1898 to convince CSR that Mourilyan plantation was a worthy acquisition. Officers from the Union Bank of Australia Ltd. considered Mourilyan plantation a 'troublesome business', with the estate remaining the Union Bank's property until 1907. Similarly, the Bank of Australasia decided to operate Pioneer and Ashburton plantations (Mackay), despite one of the Bank's Inspectors acknowledging that it was a business 'so foreign' to the Bank's usual concerns. Eventually in 1893, the Bank of Australasia disposed of these properties to Sir John Lawes, owner of nearby Farleigh plantation. Similarly, the Australian Estates and Mortgage Co. Ltd, a successor company to William Sloane & Co., decided to operate Seaforth, Kalamia, Te Kowai and the Palms plantations, although the firm made losses on these estates during the mid-1890s.[44] Other mortgagees, however, had resolved by 1890 to close their sugar plantations. Vilele (Bloomfield River), Pyramid (Cairns), Airdmillan (Lower Burdekin), Avoca (Bundaberg), and Cassada, Coningsby and Richmond (Mackay) plantations all suffered this fate. The land was left vacant, the mortgagees giving up even trying to sell the properties.[45]

Creating tenanted sugar plantations

CSR's plantation managers had concluded by 1881 that the firm could not grow sugar cane as cheaply as its tenants on their Condong plantation.[46] Flora Shaw, *The Times* (London) journalist who visited Queensland's sugar-producing districts in 1893, summarised the reasons why small farmers produced sugar cane more cheaply than the planters. She observed that the usual proportion of labour to land on a Queensland sugar plantation with 400 ha under cane was one Pacific Islander to every 2 ha, but a small farmer cultivating 40 ha would employ one Pacific Islander to every 4 ha. Hence, the labour bill would be halved and supervision would be eliminated entirely. Therefore, small farmers could afford to sell their sugar cane to the sugar mills more cheaply than the planters could grow for themselves. An example of this difference was found on the Lower Burdekin sugar plantations of the Australian Estates and Mortgage Co. Ltd. The company spent 16s 6d producing a ton of sugar cane, while their small farmers who received 13s for a ton of sugar cane made an adequate living.[47] Moreover, the sugar planters recognised that Europeans who laboured for wages on estates would work harder, longer and with more care while working for themselves and thereby produce bigger crops, especially as farmers and their families were so dependent on the crop for their entire subsistence.[48]

The realisation that sugar cane cultivation could be undertaken more economically on a small scale prompted a few Australian sugar planters to create tenanted plantations during the late 1880s. Some of the land on Innishowen, Macknade, Mourilyan and Ripple Creek plantations was leased to small farmers who provided part of the cane supplies for the sugar mills.[49] This decision to subdivide the estates into small blocks for rental, however, was considered risky. Previous attempts in Queensland during the late 1870s and early 1880s by the owners of Loganholme, Bemerside, Gairloch and Habana plantations to lease their land to tenants had not attracted much interest from prospective small farmers. CSR had also experienced difficulties in getting tenants for its land along the Tweed River.[50] The proposition that European farmers would grow and harvest sugar cane in the Australian tropics was also considered unsubstantiated. Moreover, once the planters leased their land to small farmers, they lost control over the production of sugar cane for their factories. Guarding against this eventuality entailed subdividing the plantations in stages to gauge the effectiveness of the process and/or binding the small farmers through written agreements to supply the sugar mills with cane for a set period, usually five or seven years.

Initial efforts to get Europeans to lease subdivided sugar plantation lands in the late 1880s and early 1890s were not very successful. Mourilyan plantation had the most success, for its owners sponsored German families from South Australia to come to the Innisfail district. Mostly Chinese and Pacific Islanders were the only persons willing to become tenants, particularly in North Queensland.[51] However, after the Griffith government allowed the resumption of Pacific Islander recruitment to Queensland in 1892, some Queensland sugar planters once more announced their willingness to lease or sell small blocks of land to interested settlers on very liberal terms (see discussion later in the chapter). Consequently, half Australia's surviving sugar plantations were tenanted by 1895 (Table 6.1). Examples of tenanted sugar plantations included Hambledon (Cairns), Macknade (Herbert River), Pioneer and Kalamia (Burdekin), Habana, Nindaroo and Farleigh (Mackay), Yeppoon (Rockhampton), Bingera, Windemere and Invicta (Bundaberg), Knockroe (Isis) and all CSR's estates.[52]

CSR disposed of the largest area, reporting in 1906 that it had sold 16 000 ha of its holdings to small farmers since the late 1880s and currently still had 600 ha under lease. Initially, CSR was reluctant to lease its land in Queensland to small farmers, due to difficulties it had with tenant farmers in the Tweed River district during the early 1880s. Labour shortages on its Queensland sugar plantations, however, led the firm to conclude that the disposal of its land holdings was the only option available.[53] A similar trend occurred in New South Wales, where CSR re-commenced leasing parts of its Condong plantation to small farmers, while sections of Rous plantation had been let to tenant farmers since 1887.[54] In addition, vacant land on closed Queensland sugar plantations such as Barrie, Marian, Dumbleton and Pleystowe (all Mackay), Pyramid (Cairns) and Nerada, Charleville and Magnolia (all Maryborough) had been subdivided into small farms and offered to settlers interested in becoming small sugar cane farmers.[55]

Despite this trend, owners of tenanted sugar plantations still undertook large-scale cultivation of sugar cane in the 1890s. On Cudgen plantation (Tweed River), the proprietor had 450 ha under sugar cane in 1898. In Queensland, CSR cultivated 930 ha of sugar cane on its Goondi estate in 1895, while 530 ha and 425 ha were under sugar cane in 1896 on Habana and Nindaroo plantations (Mackay), respectively. Moreover, some Queensland sugar plantations had been expanded in size, when their owners bought nearby smaller estates that were too weak to survive as separate units, leading to a further concentration in plantation ownership.[56] Labour to sustain such extensive agriculture in Queensland came from time-expired Pacific Islanders, supplemented after 1892 by Japanese workers, who were recruited mostly by CSR, the Union Bank of Australia Ltd. (for Mourilyan plantation), and the Australian Estates and Mortgage Co. Ltd. (for Kalamia and Seaforth plantations). In addition, the owners of Fairymead and Yeppoon plantations secured 200 Gilbert and Ellice Islanders in 1895 for their estates.[57] In New South Wales, John Robb & Co. used a mix of Sikhs and Pacific Islanders on Cudgen plantation, while the Richmond River Sugar Co. employed Sikhs on its Rous plantation. New

Plate 6.3. Australian workers objecting to Indians working on New South Wales sugar plantations.
Source: *The Worker*, 25 June 1898, p. 4.

South Wales workers complained about the increasing use of non-European workers on New South Wales sugar plantations and formed the North Coast Anti-Alien Society in 1895 (see Plate 6.3). During the late 1890s, this organisation campaigned against the use of non-European workers in both the Queensland and New South Wales sugar industries.[58]

The last sugar plantations

By 1900, small farmers supplying sugar cane to central sugar mills were replacing sugar plantations in Australia (see discussion later in the chapter). Yet many Queensland small canegrowers and those sugar plantation owners engaged in extensive cultivation still relied on indentured Pacific Islanders. With the introduction of the Commonwealth government's *Pacific Islander's Act* of 1901, the sugar plantation owners had to decide if they would maintain operations under a system which denied them cheap labour after 1906. By 1904, CSR, the Australian Estates and Mortgage Co. Ltd. (Kalamia and Palms plantations), John and Joseph Woods and Robert Boyd (Ripple Creek plantation) and the Farleigh Estate Sugar Co. Ltd. (Farleigh plantation) had all concluded that growing sugar cane on a large-scale with Europeans would not be practical. They subdivided the remainder of their estates for sale or lease to small farmers who then supplied their factories, which were operated as proprietary central mills.[59] Drysdale Bros. & Co. tried growing sugar cane with Europeans on

Pioneer plantation during 1903 and 1904, but found the experiment unsatisfactory. The firm ceased growing sugar cane during the late 1900s, and leased parcels of land to small farmers.[60] In contrast, the owners of Yeppoon plantation decided in 1904 to cease operations altogether because of the perceived labour difficulties, despite having some small farmers settled on their plantation lands. The Union Bank of Australia Ltd. also decided that it was unprepared to operate Mourilyan plantation with any form of labour, and that the property would be shut after the cessation of crushing in 1906, unless a buyer could be found for the estate. A syndicate of local businessmen did purchase the property in 1907, and subdivided the plantation into small farms.[61]

A few Australian sugar planters, however, resolved to grow sugar cane on a large-scale after 1906. In 1910, John A. Robb still cultivated 1 166 ha of sugar cane at Cudgen plantation, using whatever labour he could muster. By then, however, he too had decided to abandon large-scale agriculture. The following year, Robb sold his property to CSR, and it was subsequently subdivided into small farms.[62] In contrast, the bigger Bundaberg district operators of A., H. & E. Young or Young Bros. (Fairymead plantation), Gibson & Howes Ltd. (Bingera plantation), Buss Bros. (Invicta, Bonna and Pemberton plantations), H. Blisset and W. Hart (Goodwood plantation) and the Queensland National Bank (Waterview, Oakwood, Sharon and Mon Repos plantations, which became known as Millaquin) all continued extensive sugar cane cultivation during the late 1900s and beyond. Europeans did perform satisfactory in the sugar cane fields in this district. The Australian Estates and Mortgage Co. Ltd. also resumed plantation agriculture on the Palms and Kalamia estates during the late 1910s and early 1920s, respectively, following the firm's dissatisfaction with the performance of tenanted plantations. Moreover, the Bundaberg sugar plantations have been particularly enduring, with the Bundaberg Sugar Co. still operating Fairymead, Millaquin and Bingera plantations in 1995.[63]

Small millers

Small millers originated during the late 1860s for a variety of reasons. In Queensland, the East Moreton Farmers' Association sought government funds in 1868 to erect a central sugar mill that would be supplied by the small farmers located around Brisbane. Premier Robert Mackenzie would not support this scheme, so some of the small farmers established their own small sugar mills, equipped mainly with horse-powered machinery. The farmers along the

Manning and Richmond Rivers also belong to this category. Attempts to form central sugar milling companies in these districts during the late 1860s and early 1870s were failures, so some farmers erected their own small sugar mills to handle their crops.[64] Other settlers were located in isolated districts, where not enough farmers existed to form a central sugar milling company. Tweed River district farmers Messrs Pringle, Skankie and Byrne fall into this category. They erected Abbotsford Mill in 1874, after realising they would have to crush their own sugar cane, especially after CSR refused an approach by the Tweed River district small farmers to erect a factory in the district.[65] Isolation was also a problem for the Clarence River district farmer J.E. Chapman, who was located upstream from Grafton, and 30 km from the nearest sugar mills. The proprietors of these factories would not purchase his crop because of the cost of bringing the cane to the factory, so he erected his own small mill.[66]

In the Clarence River district, the farmers who became small millers wished to be independent of CSR's central sugar mills. Between 1871 and 1882, at least twelve small farmers became small millers, using the profits accumulated from growing sugar cane for CSR to fund the erection of their own small mills. Some were unhappy with CSR, claiming they were not receiving a fair price for their cane and that the firm restricted the cane varieties they could plant and would not accept frost-affected cane. Others complained about CSR's harvesting gangs cutting their cane too early, harvesting the best cane only, leaving them with half-harvested crops, or cutting the cane so poorly that they were unable to get ratoon crops. CSR was also accused of engaging in systematic robbery by fraudulently measuring the weight of harvested cane in the punts.[67] Dr J.C. Neild, a Port Macquarie doctor and small miller, wrote that the Clarence River district small farmers had a duty to become independent millers, for those who persisted in supplying CSR's sugar were nothing more than 'plantation labourers'.[68]

Given the reasons for the emergence of small millers, their distribution in New South Wales was necessarily widespread in 1870 and 1875, although the biggest concentration was along the Clarence and Richmond Rivers. In Queensland during the 1870s, the small millers were mainly located in the Brisbane district, although most of these operations were short-lived. This production system also emerged at Mackay briefly in the early 1870s, but the three small millers had all failed by 1875. Later changes in the distribution of the small millers reflected fluctuations in the spread and contraction of sugar cane growing. By 1885, small millers were no longer found along the Manning, Hastings and Macleay Rivers. The establishment of the small miller system at Bundaberg and Coffs Harbour by 1885 followed the expansion of sugar cane growing in these areas. Yet the movement of sugar cane growing to new North Queensland districts and the continued expansion in the area under sugar cane in the Herbert River and Mackay districts was not automatically accompanied by the emergence of small millers. These districts remained localities where the plantation dominated, with the small farmers being unwilling to invest in sugar milling machinery. In contrast, some small selectors in the Beenleigh and Maryborough districts were prepared to invest in crushing machinery, and the small miller system emerged in these districts for the first time by 1885, even though sugar cane had been cultivated along these rivers for almost twenty years.[69]

To become a small miller, the property owner had to acquire milling equipment. In Australia during the late 1860s and 1870s, the purchase and erection of crudely constructed small sugar mills, equipped with horse-powered crushing rollers, cost between £200 and £800. Small sugar mills with steam-driven machinery cost between £1 500 and £2 000. Small millers around Bundaberg in the early 1880s were even spending up to £4 000 on their milling machinery.[70] To utilise this machinery effectively, some small millers purchased cane from neighbouring small farmers. Samuel and George Grimes, the owners of Coongoon Mill, even advertised in the local newspapers their willingness to crush the crops of other settlers.[71] Suppliers were mostly paid a flat rate for their cane or the share or 'halves' system was adopted, whereby the mill-owner shared half the sugar produced with the canegrower. James Knox, CSR's Inspector of Mills in the early 1880s, noted that small millers in New South Wales survived because they bought sugar cane 'independent of age, or quality. Anything is taken'.[72]

Small millers adopted various strategies to fund the purchase of the milling equipment and to operate their properties. Some funds came from pooling of resources. Along the Tweed, Clarence and Richmond Rivers, 25% of the small sugar mills were operated by partnerships, generally based on familial relationships. A good example of such a partnership existed between William Miller, a storekeeper at Ulmarra, and his brother-in-law William Meston, who together operated a small sugar mill at Woodburn from 1876 until 1905.[73] The major external source of capital was the Australian colonial banks. Loans were negotiated through regional branches. Samuel Johnston, for example, received the funds to establish his Binbirrum Mill from the Maryborough Branch of the Commercial Banking Company of Sydney. Nearby in the Bundaberg district, the local branch of the Queensland National Bank supplied the capital to Walter Adams (Summerville), Frederick Gladwell (Mabbro) and Richard Lee (Sherwood). The capital was borrowed on the security of the farms.[74] Merchants also advanced funds to small millers.

Biggingee Poochee, an Indian selector at Walloon near Ipswich, mortgaged his land to the local Ipswich merchant, John Foote, to gain the capital for his small sugar mill, while the Clarence River small miller, Joseph Edmonson, was financed by the merchants Messrs Gibbs, Bright & Co.[75]

Small millers came from a variety of backgrounds. For New South Wales, evidence could be found about the backgrounds of one third of the small millers. Nearly half these small millers were Scottish. Included amongst their ranks was the Rev. Edward Holland, a Presbyterian clergyman who supplemented his stipend by manufacturing sugar in his primitive mill at Port Macquarie during the late 1860s.[76] No clear pattern existed in the nationality of small millers in Queensland, with Scots, English, Danish and German immigrants to the colony trying their hand at sugar milling. Generally, small millers came primarily from the farming class, rather than a wealthy background. Yet some tradesmen, small businessmen and professionals also engaged in this type of sugar production. William Clement, for example, was a building contractor before investing (and losing all his funds) in his small sugar mill near Ballina, while Biggingee Pochee was an Ipswich photographer before turning to sugar production.[77]

The widespread failure of the small miller system dates from 1885, the immediate cause being the collapse in sugar prices. Half the small mills closed between 1885 and 1895. Edward W. Knox had predicted this failure as early as 1871, observing that the Clarence River small millers would come to grief, as they had 'no idea of buildings or machinery or indeed anything but hard work'. A Special Correspondent for the *Sydney Morning Herald* who visited northern New South Wales in 1884 claimed that many small millers were 'almost wholly without experience and labour under the delusion that anyone can run a sugar mill'.[78] Lack of engineering and manufacturing knowledge may have been a contributing factor to the failure of some small millers, but the biggest weakness of this production system was that many small mills operated at less than full capacity because of inadequate sugar cane supplies. By 1886, most small millers crushed for only short periods of time as they processed on average less than 40 ha of cane. J. MacDonald's Mill on the Clarence River, for example, only had seasons of ten and three weeks in 1898 and 1899, respectively. CSR's New South Wales sugar mills, on the other hand, operated between five and six months year.[79] In addition, Charles Lyne, who published a book on the industries of New South Wales in 1882, claimed that small millers manufactured sugar of 'an inferior description, or charged with too large a proportion of molasses'. This sugar was increasingly shunned by consumers, who could obtain high quality white refined sugars at affordable prices. Edward W. Knox had observed this trend in 1879, noting that some of the sugar made at the small mills along the Clarence River was 'unsaleable'.[80]

In 1895, there were still thirty small millers, located mostly in New South Wales. Yet over the next decade, the majority of these last small millers also closed their factories. Some like J. MacDonald (Maclean) just gave up trying to make a profit from small cane supplies when sugar prices remained so low; another small miller, William Newby (Maclean), died in 1902 and his nephews declined to work the sugar mill any longer. By 1905, only four small millers were left in New South Wales, and the last – George Morrison of Goodwood Island – closed his small sugar mill in 1907.[81] In Queensland, three small millers remained in 1905. Two of these mill-owners – William Heck (Rocky Point Mill) and Carl Rehfelt (Alberton Mill) – made the transition to being proprietary central mill-owners and were operational in 1914. Rehfelt's mill eventually closed in the early 1930s, but Rocky Point Mill still operated in 1995, being Queensland's only privately owned sugar mill.

Central sugar mills and small farmers

CSR's first central sugar mills

In New South Wales, the first scheme to establish a central sugar mill was proposed in September 1867, when a public company sponsored by Horace Dean announced its intention to build a new sugar mill on the Manning River if the investors could be guaranteed a supply of cane. The local farmers were unprepared to enter into a seven year contract to supply cane to the sugar mill, and the project proceeded no further. Nine months later, CSR also announced that it would erect central sugar mills in northern New South Wales, if farmers guaranteed to grow sufficient cane to keep the factories supplied.[82] CSR, however, had never intended to become a producer of raw sugar, aiming instead to monopolise the Australian refined sugar market (see Chapter 4). Yet a consumer market existed for unrefined sugars manufactured locally, thereby avoiding the £5 import duty on sugar. Moreover, mill whites could be manufactured in sugar mills equipped with more sophisticated machinery. Hence, CSR's directors decided in 1867 that if raw sugar could be manufactured in New

South Wales, 'it should be made of a quality suitable for refining and not of a description suitable for sale to grocers and others'.

To control the manufacture of raw sugar in New South Wales during the late 1860s, CSR encouraged the manufacture of 'concrete' (i.e. undried sugar in a solid state) by offering to import concretors for prospective millers and agreeing to purchase the product.[83] However, the New South Wales small millers were unwilling to cooperate with CSR. Many equipped their small sugar mills with machinery that manufactured raw sugar, not 'concrete'. To stop the growth of this competition, CSR erected central sugar mills, hoping to slow the construction of additional sugar mills. By 1875, enough small farmers had signed contracts with CSR, so the firm had built four sugar mills in the colony: Harwood, Chatsworth and Southgate Mills along the Clarence River and Darkwater Mill on the Macleay River (Figure 6.2).

Farmers agreeing to sell their cane to CSR, contracted to supply the company's sugar mills for three, five or seven years. CSR employed gangs of canecutters to harvest the sugar cane, and no crop of cane was allowed to remain uncut for more than two years. Farmers were paid 10s per ton for sound, ripe sugar cane, from which all the side shoots and dead leaves had been removed; for inferior lots they were paid a little less. The weight of the cane was determined by measuring the draught of the punts when they were loaded and ready to be towed to the mills. Farmers received another 1s a ton for carting the cane to CSR's punts. To encourage the small farmers to grow more cane, CSR was prepared to advance them £6 per acre of cane planted, free of interest, as long as they sold the entire crop to CSR.[84] A journalist for the *Northern Star* (Lismore) concluded in 1884 that the income after deducting expenses from growing cane for CSR was 'not handsome, but with the aid of farm produce it would be sufficient for a small family to live on'.[85]

CSR initially found operating central sugar mills a difficult task. Its Darkwater Mill on the Macleay River worked for only three seasons (1870–1872), before being transferred to the Clarence River where it became Harwood Mill in 1873. The firm's plans were interrupted by the erection of small sugar mills by contractors who reduced the cane supply to Darkwater Mill. The cost of manufacturing sugar at Darkwater Mill also proved to be higher than at the Clarence River sugar mills, probably because of the inferior cane supplied to the factory.[86] Some Clarence River farmers, as noted above, accused CSR of a range of offences, and established independent small mills, thereby eating into CSR's cane supply. Most New South Wales small cane farmers cultivated 8–12 ha of sugar cane, employed poor cultivation techniques (see Chapter 7), and were reluctant to expand their areas under sugar cane.[87] Declining sugar cane supplies along the Clarence River, eventually prompted CSR to close Southgate and Chatsworth Mills in 1885 and 1887, respectively, thereby centralising manufacturing at Harwood Mill (Figure 6.2). Edward W. Knox had hinted at the eventual closure of Southgate Mill as early 1871, observing that the locality was:

> On the whole very unsuitable for the growth of canes …. the bulk of the land is under water at every fresh; frost is very severe. Drainage would improve the land, but it is almost hopeless to expect the farmers to attempt any improvement of the sort.[88]

Despite these initial problems, CSR erected additional central sugar mills in New South Wales during the late 1870s. CSR officials had inspected the Tweed and Richmond River districts in 1872 and 1873, respectively. They reported unfavourably on the Richmond River district, highlighting that the land along the Richmond River was considered unsuitable for sugar cane growing because it needed extensive drainage. On the other hand, the Tweed River region was described as 'a promising district, with both respect to climate and soil, for the successful cultivation of sugar cane'. Initially, CSR decided not to proceed with erecting a sugar mill in either district.[89] However, CSR's directors rescinded their earlier decision in 1877, and began purchasing land along the Tweed River as a safeguard against the possibility of sugar cane growing contracting so much on the Clarence River as to force the closure of one of the firm's sugar mills. This newly acquired land was to be leased to small farmers who would supply cane to Condong Mill, which would be ready for crushing in July 1880. As mentioned earlier, this plan was initially defeated by a lack of suitable tenants and CSR was forced to cultivate the land itself for much of the 1880s. Farmers along the Richmond River in 1877 also asked CSR to erect a central sugar mill in their district. Following another report into the extent of milling power in the Richmond district, CSR decided in 1878 to erect another central sugar mill at Broadwater, in readiness for the 1881 crushing season.[90]

Queensland's earliest central sugar mills, 1865–1885

One of the earliest, but unsuccessful attempts at establishing a central sugar mill in Queensland, was undertaken by the Mackay Sugar Crushing Co. Ltd. Formed in February 1869 with a capital of £5 000, the company was dissolved by the end of the year without a sugar mill being erected. In early 1869, the small farmers throughout the Brisbane and Beenleigh districts who could not afford their own sugar mill had more success achieving crushing facilities. Their approach was novel for they agreed to support a floating central sugar mill. The *Walrus*, a flat-bottomed paddle-steamer, was equipped with sugar-making machinery that could turn out around a tonne of sugar a day. Over the next three years, the *Walrus* travelled along the Brisbane River and the rivers in the Beenleigh district, stopping alongside small riverside farms to crush the annual crops of cane. The mill's machinery, however, was unreliable and the boat gained a reputation for being no more than a mobile and illicit distillery. The boat's milling functions ceased after the 1871 crushing season. Another floating central sugar mill was proposed for the Mary River in 1869, but not enough canegrowers were willing to invest in such a scheme. Attempts to operate Maryborough Central and Myrtle Grove Mills as central sugar mills in the Maryborough district during the early 1870s also ended in failure. Their promoters lost control over the associated properties, and the small farmers received nothing for their crops, which hardly encouraged small cane farming in the colony.[91]

Central sugar mills were again established in Queensland during the early 1880s. John Howard and Benjamin DeLissa aimed to get all of their cane supplies for their Riverside Mill from the small farmers along the Coomera River. Shortages of cane, however, forced Howard and DeLissa to lease land and grow their own sugar cane for the mill. In 1889, John Howard advised the Royal Commission into the Queensland Sugar Industry that the operation had never returned a profit. During the early 1880s, an attempt by the Queensland Central Sugar Factory Co. to obtain all the cane for its Marian Mill (Mackay) from small farmers only, ended in bankruptcy. The Richmond Sugar Co. suffered a similar fate, its Richmond Mill (Mackay) also being undersupplied by the local small farmers.[92] In contrast, two syndicates of small farmers at Buderim in the Maroochy district and three groups of small farmers in the Beenleigh district did successfully operate central sugar mills between 1880 and 1888, until reduced profits from low sugar prices forced their closure. The particular success of the central sugar mills in the Beenleigh district was most likely due to the willingness of the German partners to cooperate in all activities associated with sugar production, including growing the crop and its harvest and transport to the mill.[93]

In most Queensland coastal districts during the 1870s, the numbers of small farmers increased following the introduction of land legislation in 1868 and 1876 encouraging settlement by small selectors. The lack of central sugar mills, however, did not deter these small farmers from growing sugar cane. They relied on the sugar planters to crush their yearly crops, although such arrangements were not always cordial. The colony's small canegrowers claimed the planters refused to purchase their entire crops in good seasons, because they had too much of their own cane to crush, and they often deferred crushing their cane until late in the season, spoiling the chance of a ratoon crop the following year.[94] Eventually some Mackay canegrowers petitioned the Queensland government (Griffith; Liberal) in July 1885 for a loan to build a central sugar mill. Small farmers in the Herbert River, Lower Burdekin and Maryborough areas supported their Mackay counterparts with their own requests for funds to erect central sugar mills.[95]

As early as 1876, the Queensland government had investigated the economics of central sugar mills supplied by small farmers. Angus Mackay, the Queensland Commissioner to the Philadelphia Exhibition in 1876, visited sugar-producing locations in North America and the Caribbean, and reported that the central sugar mill system successfully operated on the French-controlled islands in the West Indies.[96] Despite this report, the Queensland government had refused to advance the cause of central sugar mills in the colony. In 1885, however, Griffith's Liberal government decided to encourage small-scale sugar cane farming by allocating £50 000 towards the building of co-operatively owned central sugar mills. To obtain this finance, interested farmers had to form a company and mortgage their land to the government as security. The farmers would then be allocated shares in the company in proportion to the value of their land. As added security, the government would hold a mortgage over any property and machinery owned by the company. Such funds, however, would only be advanced if the cultivation of the sugar cane for the sugar mills was undertaken entirely by Europeans.[97] The parliamentary opposition in Queensland opposed Griffith's central sugar mills experiment, pointing out that the colony was already well endowed with sugar mills.[98]

Griffith's central sugar mill experiment

Despite community concerns, the Griffith government in early 1886 ordered an investigation into the most suitable places to locate the central sugar mills. William Hodgkinson – the Gympie Mining Warden – conducted the survey, visiting districts north of Rockhampton, because he noted the legislature favoured the funds being allocated to the colony's tropical districts. Yet such a decision precluded consideration being given to the small canegrowers along the Logan River in Southern Queensland who were anxious to establish a central mill in their district.[99] This inquiry, however, found that the small farmers in the Herbert River, Mossman and Lower Burdekin districts rejected the government's offer because of the restriction concerning labour. Moreover, in the recently settled northern parts of the colony, the majority of the settlers were unable to mortgage their land to the government, as they had not obtained its freehold title. Only the small farmers at Mackay were willing and able to accept the state's terms.[100]

The Queensland government accepted Hodginkson's recommendations that the funds be allocated an association of settlers in the vicinity of the Eton terminus of the Mackay Railway and a group of settlers near the racecourse at Mackay. As a consequence, North Eton and Racecourse Mills were erected, and crushed for the first time in 1888 and 1889, respectively. From the outset, however, both sugar mills were troubled by lack of cane and low sugar prices and had difficulties meeting their repayments to the Queensland Treasury. A government inquiry in late 1889 discovered that some shareholders were not growing sugar cane for their respective factories, thereby forcing the directors to purchase cane from non-shareholders who had used Pacific Islanders to cultivate the crops. Clearly this practice had violated the purpose for which the funds had been advanced, although the Queensland government took no action against the companies.[101] Warnings to shareholders to grow cane or face the possible forfeiture of their shares had the desired impact. By 1895, both companies were in a sounder financial position, and reducing their debt to the Queensland Treasury.[102]

Initially, Premier Griffith's experiment with central sugar mills in Queensland had not been very successful. Yet Griffith's successor as premier, Sir Thomas McIlwraith, announced in 1893 that the Queensland government would fund the construction of additional central sugar mills, and no restriction would be imposed on the type of labour to be employed in cultivating the sugar cane for these mills. Parliamentary opponents of the legislation again raised the issue of the State favouring the sugar industry over other colonial enterprises, and concern was expressed about the State stifling private investment in the sugar industry. CSR also opposed the scheme, arguing that expanded sugar production in Queensland would cause a further fall in the domestic market price for sugar.[103]

The mechanism by which public funds were made available to establish new central sugar mills in Queensland during the 1890s was *The Sugar Works Guarantee Act* of 1893. Under this legislation, the Queensland government guaranteed debentures issued by companies formed to operate central sugar mills. Before any monies were advanced, the Queensland Treasurer had to be satisfied that sufficient land was available to keep a sugar mill fully supplied and that 200 ha of sugar cane had been planted for the proposed factory. To guard against any company loss, the government had first mortgage over the sugar mill and freehold land that had been offered as security. If the company defaulted, the Treasurer could take possession of the sugar mill and manage it until the whole liability had been fully discharged. An amending Act in 1895 provided for the Queensland Treasury to take up the debentures rather than merely guaranteeing them, so the government assumed a real liability, and limited the amount that could be advanced for this purpose to £500 000.[104] A decade later, Dr Walter Maxwell, the Director of the Queensland Bureau of Sugar Experiment Stations, and the author of a 1904 report on the poor financial performance of some of the state's central sugar mills, was highly critical of this legislation. He observed that the legislation contained no provision for enforcing the growing of sufficient quantities of sugar cane to keep the sugar mills fully supplied. In addition, the legislators had failed to ensure that the sugar mills were placed only in localities that were relatively frost free, and where the average annual rainfall was assured by records from previous years, thereby ensuring that the localities were capable of producing the necessary crops.

The new legislation was greeted enthusiastically by the colony's small farmers. At least twenty-seven companies were formed during the 1890s to erect central sugar mills using government funds.[105] To assess their competing claims, the Queensland Government appointed R.W. McCulloch as its Valuator and Inspector of Central Mills. Throughout 1894 he toured the colony, assessing the proposed sites for the sugar mills. The companies paid for this inspection in advance.[106] By early 1895, thirteen companies had met the conditions set by the McIlwraith government and were allowed to commence planning to erect a sugar mill. Official supervision continued during the planning process, with McCulloch being involved in the design of the sugar mills. The directors then paid a government-appointed engineer to supervise the mill's erection and its subsequent initial operation. Edward W. Knox of CSR also vetted the plans of at least Marian and Nerang Mills.[107] The locations of the central sugar mills finally erected are shown in Figure 6.4.

Creating proprietary central sugar mills

The state's funding of central sugar mills clearly benefited localities where existing groups of small farmers had limited or no access to milling facilities. In the major Australian sugar-producing regions, however, much of the best

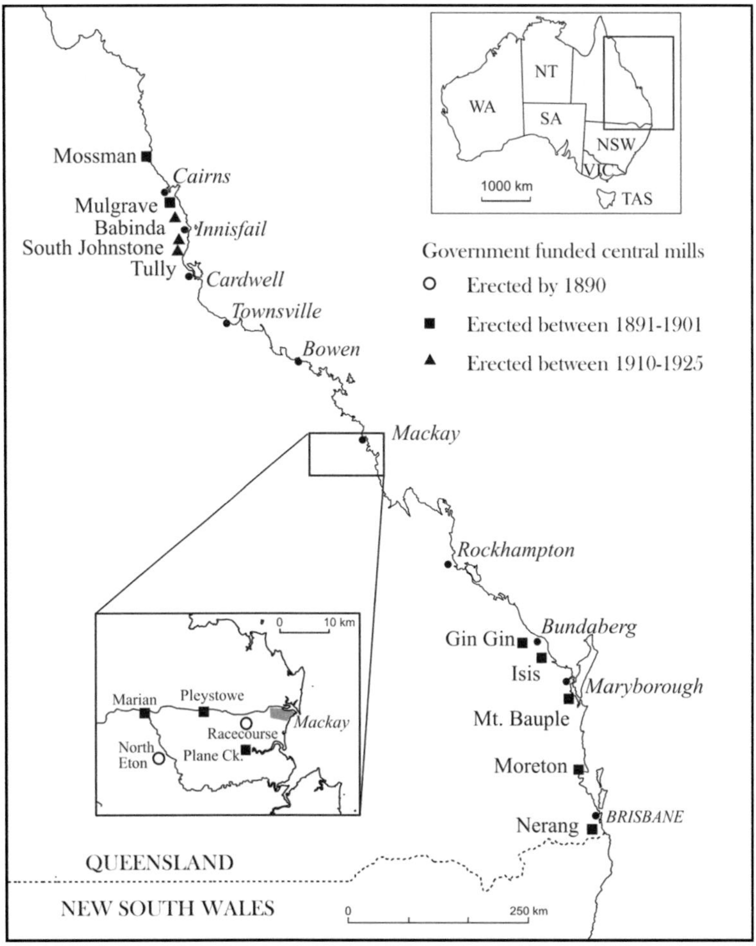

Figure 6.4. The location of the government-funded central sugar mills in Queensland.

Source: P. Griggs, 'The origins and early development of the small cane farming system in Queensland, 1870–1915', *J. of Historical Geography*, Vol. 23, no. 1 (1997), p. 53.

agricultural land was still owned by the sugar planters. Despite their previous lack of success in disposing of this land, the Australian sugar planters again announced their willingness in the early 1890s to lease or sell small blocks of land to interested settlers, thereby creating either tenanted plantations or proprietary central sugar mills supplied by small farmers. The Queensland government assisted intending small cane farmers by publishing details about the terms offered by the colony's sugar planters in a brochure which was distributed throughout the colony and Great Britain, where it was hoped to attract immigrant farmers.[108]

Terms offered by the Australian sugar planters varied, but some generalisations are possible.[109] All Australian sugar planters leased blocks of land from 8–40 ha to tenant farmers for between five and ten years, although CSR's lessees initially were allocated 12 ha only. If this area was thoroughly cultivated it was possible to acquire a further 12 ha, with 24 ha being the maximum amount secured by any lessee.[110] In Queensland there was no government restriction on the size of blocks leased to tenants during the 1890s and 1900s, even non-Europeans, and the sugar planters willingly leased larger areas to approved applicants (see Figure 6.5). This action by the Queensland sugar planters needs some explanation. In the mid-1890s, Maurice Hume Black, Queensland's Special Immigration Agent in London, estimated an unmarried small farmer who did not hire labour or have family assistance could cultivate up to 8 ha only, while married farmers with active wives and

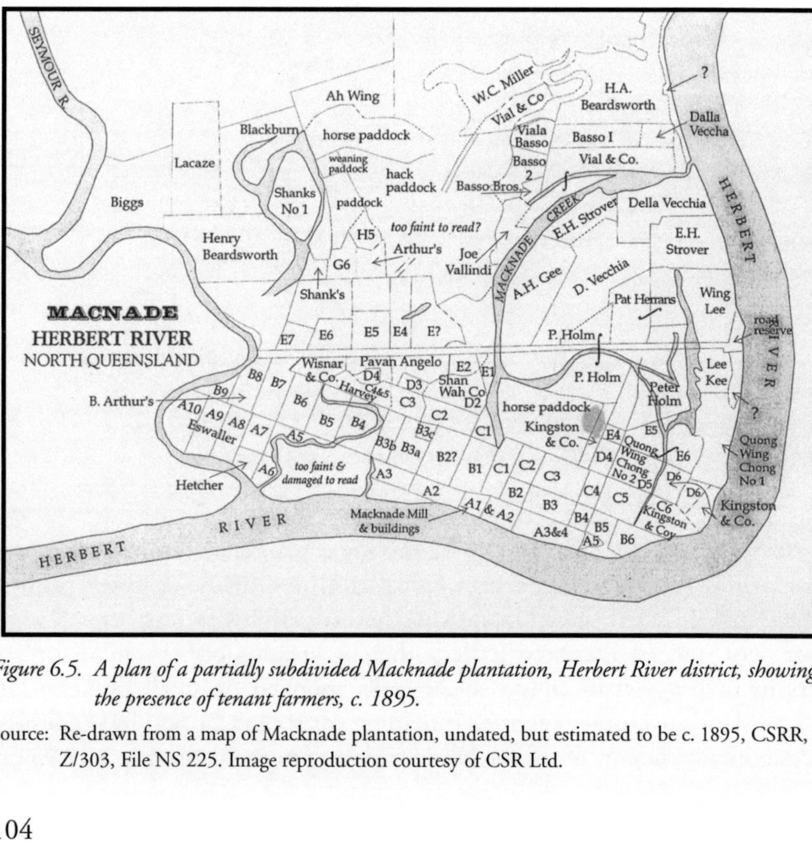

Figure 6.5. A plan of a partially subdivided Macknade plantation, Herbert River district, showing the presence of tenant farmers, c. 1895.

Source: Re-drawn from a map of Macknade plantation, undated, but estimated to be c. 1895, CSRR, Z/303, File NS 225. Image reproduction courtesy of CSR Ltd.

four or five children could cultivate up to 16 ha only.[111] By the early 1900s, some small canegrowers in Queensland relied solely on family labour or co-operated with other farmers to plant and harvest their crops. German immigrants particularly favoured co-operation or the use of family labour only.[112] However, many small canegrowers in Queensland during the 1890s and early 1900s hired non-European workers, mostly Pacific Islanders, Japanese or Chinese, to assist them on their farms. Moreover, some mill-owners recruited and hired out Pacific Islanders or in some cases British immigrant workers to their small farmers. This labour enabled them to cultivate larger areas.[113] Such assistance ceased after 1906, as the Australian mill-owners commenced having difficulties in obtaining workers for their factories.[114]

Some Australian planters such as CSR, John Robb & Co. and Drysdale Bros. & Co. encouraged their employees to become tenant farmers, thereby trying to insure themselves against loss by inexperienced men.[115] This policy led to a diversification in the ethnicity of the colony's small cane farmers during the 1890s and early 1900s. Maltese, Italians, Chinese, Japanese and Pacific Islanders occupied plantation lands. Pacific Islander canegrowers were located particularly throughout the Mackay, Proserpine and Isis districts, although a small group farmed in the Tweed River district. Chinese canegrowers occupied properties mostly around Cairns and Innisfail, a dozen Japanese canegrowers were found each in the Proserpine and Mackay districts, and five to ten Indians cultivated sugar cane at Mackay and in the Richmond River district.[116] CSR and Gibson & Howes Ltd. also willing leased land to applicants with some farming experience, accepting German farmers from Southern Queensland as tenants during the 1890s.[117] These lessees were charged varying rentals depending upon the type of land leased. Uncleared land could be rented for between 1s and 3s per acre; cleared land was available at rentals from between 5s and 10s per acre.[118]

Australian plantation owners had different attitudes to land disposal. Gibson & Howes Ltd. would not sell its land to tenants, although the firm announced that tenants had the pick of their sugar cane growing lands, with the company cultivating the poorer country. However, Rudolph Helms, manager of CSR's Childers Mill, reported to Edward W. Knox in 1911 that this policy had 'not really worked; gradually the farmers leave as poor as they come and gradually the firm's cultivation increases'. Drysdale Bros. & Co. and the Australian Estates and Mortgage Co. Ltd. were also unwilling to sell their land to the tenants. Australian Estates thought the idea of sub-dividing their plantation lands 'altogether too risky' and adopted a leasing policy only to ensure the firm retained control over the production of sugar cane for its factories.[119] In contrast, CSR opposed its tenant farmers remaining lessees permanently and encouraged them to settle, announcing that the company would compensate tenants who left only for improvements such as drainage, fencing and clearing additional land, but not houses or other buildings.[120] CSR tenants could purchase their blocks on five year terms, or for cash after three years, but could not credit the annual rental towards the purchase price, as CSR believed its tenants secured their farms very cheaply anyway. Prices for blocks varied. Uncleared blocks were valued at between £1 and £3 per acre, cleared, but unstumped land cost between £5 and £9 per acre, and land ploughed and under cane was valued at £12 per acre.[121]

Securing a block of land did not ensure immediate success as a small cane farmer. Many of Queensland's small canegrowers needed annual financial assistance to achieve a crop for harvest. Cash advances to European and non-European small growers came from the mill-owners. These funds were often spent on hiring labour or purchasing fertiliser. Generally money was advanced on the basis of the amount of cane planted and after the miller had obtained a crop lien.[122] Some mill-owners also aided their small farmers by purchasing wood, maize and horses from them, while CSR supplied rations to 'trustworthy men' for up to a year.[123] Local businesses also supplied credit or funds. At the 1912 Royal Commission into the Australian Sugar Industry, one Mackay farmer claimed that the liberal terms given to the small canegrowers by the Mackay businessmen ensured their survival during the 1900s, while another farmer claimed that the local Mossman storekeeper had provided farmers with credit totaling £20 000.[124] By the 1900s, however, some Queensland sugar milling companies had reduced their advances to small farmers following the establishment of banks in nearby centres, including the Queensland Agricultural Bank set up by the Queensland government in 1902.[125]

Some Queensland milling companies were particularly generous with their advances. Racecourse Central Sugar Co. Ltd. carried advances to farmers of between £10 000 and £15 000 in the late 1900s. By 1914, Drysdale Bros. & Co. had accumulated farmer debts of £35 000. Suppliers to the two sugar mills operated by Drysdale Bros. & Co. borrowed money without adequate security; merely an instruction from the firm's Managing Director, John Drysdale, to one of the local banks was sufficient to open an account for the farmer, guaranteed by the company.[126] In contrast, Charles Young who managed Kalamia plantation for the Australian Estates and Mortgage Co. Ltd. during the 1890s believed that it was 'a farmer's chief aim and end in life to get into debt as far as possible anywhere and everywhere'. The

Australian Estates and Mortgage Co. Ltd., therefore, lent more cautiously to its tenant farmers. Similarly, CSR was careful in the amount of support it offered to tenant farmers, believing that large advances made farmers thriftless, as they worked for the advance rather than for the return from the crop.[127]

Eliminating non-European small canegrowers

By 1901, the central milling system had become the dominant production arrangement in New South Wales following the demise of the small millers. In Queensland, the central milling system was emerging as the dominant production system, with 2 610 small farmers cultivating sugar cane, compared to 202 in 1889 when the Royal Commission into the Queensland Sugar Industry assessed the extent of sugar production in the colony.[128] This transformation, however, had not been achieved without the reliance on non-European labourers who were employed at low wages. After the introduction of the Commonwealth government's legislation prohibiting the recruitment of Pacific Islanders to Australia after 1904, and the financial incentives to employ only 'whites', many Australian canegrowers and mill-owners ceased employing non-European workers (Table 6.2). In addition, the Australian Workers' Union (AWU) – the organisation representing sugar industry workers following a merger with the Amalgamated Workers' Association in 1912 – conducted a campaign against the presence of non-European workers in the Australian sugar industry during the 1910s. Most canegrowers complied and employed only Europeans; some mill-owners, however, were not as co-operative, refusing to be dictated to by the trade unions. Drysdale Bros. & Co. (Pioneer Sugar Mills Ltd. after 1914), for example, employed Japanese workers at Inkerman and Pioneer Mills until at least 1930, much to the disdain of the local unionists.[129]

Discrimination against non-European small canegrowers also progressively forced them out of the Australian sugar industry during the 1900s and early 1910s. CSR refused to make any new contracts with Indian canegrowers on the Richmond River during the late 1890s, and had ceased leasing land to Chinese tenants on its Goondi and Hambledon plantations by 1900. Dr Walter Maxwell, Comptroller of the Queensland Bureau of Central Mills, instructed all the State's central sugar mills in 1904 to stop providing financial assistance to non-European canegrowers. In 1910, John Drysdale declined to make any contracts with the Lower Burdekin Chinese canegrowers for the purchase of their crops. Moreover, the Commonwealth's

Table 6.2. Sugar production (tonnes) in Queensland and New South Wales by 'white' and 'non-white' labour, 1902–1912, and numbers of New South Wales canegrowers using 'white' and' 'non-white' labour, 1906–1912.

	Queensland		*New South Wales*			
Year	'White' labour	'Non-white' labour	'White' labour	'Non-white' labour	No. of canegrowers ('white' labour)	No. of canegrowers ('non-white' labour)
1902	12 450 (16)	66 630	19 744 (92)	1 550		
1904	40 034 (27)	107 305	18 096 (91)	1 867		
1906	129 579 (70)	55 492	22 153 (93)	1 638	1 405	122
1908	134 191 (88)	18 615	14 580 (92)	979	1 387	164
1910	194 468 (92)	16 028	18 222 (95)	906	1 206	138
1912	107 785 (94)	6 556	16 746 (98)	243	932	79

Notes: (a) The figure in brackets refers to a percentage of the total. (b) The figures on numbers of growers employing 'white' or 'non-white' labour were only available for New South Wales from 1906 onwards.

Source: Based upon figures in *CPP*, Vol. 3 (1912), p. 1022, *Yearbook of New South Wales*, 1913, p. 804 & *Yearbook of Australia*, No. 8, 1915, p. 340.

bonus for those canegrowers who used only European labour penalised non-European canegrowers financially, for growers were themselves classified as labour used in sugar production. In addition, some Pacific Islander canegrowers were deported after 1906, despite having leased land to stave off deportation.

Further regulations targeting non-European canegrowers were introduced under *The Sugar Cultivation Act* of 1913, which required 'aliens' (i.e. non-Australian, non-European and 'non-white' American citizens) to obtain a certificate of authority to cultivate and harvest a crop of sugar cane which they had planted in 1913 for a period not exceeding three years. To secure such a certificate, the applicant had to pass a dictation test in any language determined by the Under-Secretary of the Queensland Department of Agriculture and Stock. At least 188 certificates of exemption were granted to non-European canegrowers: two-thirds of these certificates were given to Pacific Islanders, mostly resident around Mackay. Some Queensland canegrower organisations complained about the ease which 'aliens' could secure certificates of exemptions, especially open-ended certificates that did not state the holder had to cease cane growing after 31 December 1915. However, these concerns were really unfounded, as the legislation forced additional non-European canegrowers out of the industry, so Australia's canegrowers by 1914 were overwhelmingly European.[130]

The departing non-European canegrowers in North Queensland were replaced by ex-miners from the declining mines at Chillagoe, Charters Towers and Ravenswood.[131] Former European canecutters, including new Italian and Maltese immigrants, also commenced cane growing during the 1900s and early 1910s.[132] Two groups of German immigrants who settled in the Bundaberg district, and a party of Finnish immigrants who made their new home in the Maroochy district, also commenced supplying sugar cane to local sugar mills during the mid-1900s.[133] By 1906, 3 300 small cane growers were present in Queensland (Table 6.3), or nearly 700 more than the turn of the century. Their numbers continued expanding during the mid-1900s, reaching 4 238 by 1911, but dropped by 200–300 during the early 1910s, following the exodus of small farmers from sugar cane growing in Southern Queensland sugar-producing districts. On average, these Queensland canegrowers cultivated around 14 ha in the mid-1900s, or three times more than their New South Wales counterparts, who often combined sugar cane growing with dairying (Table 6.3).

Table 6.3. Average area (ha) of sugar cane cultivated per small canegrower in Queensland and New South Wales, 1906–1914.

	Queensland		New South Wales	
Year	No. of growers	Ha	No. of growers	Ha
1906	3 300	16.3	1 527	5.5
1907	-	-	1 579	4.6
1908	3 727	13.4	1 551	4.3
1909	3 947	13.2	1 553	3.7
1910	4 232	13.6	1 344	4.1
1911	4 238	12.4	954	5.9
1912	3 901	14.6	1 011	5.5
1913	3 982	15.0	887	6.3
1914	4 130	15.9	713	7.5

Note: Figures were not available for Queensland in 1907.

Source: Calculated from figures in *QVP*, Vol. 2 (1911–12), p. 932, *Statistics of Queensland, Statistical Registers of New South Wales,* 1907–1916, *Yearbooks of New South Wales,* 1913–1916 & Queensland Department of Agriculture and Stock, ARs, 1912–1916.

The relatively smooth transition to the central milling system in the 1900s, however, had to be accompanied by efforts to improve the financial performance of Queensland's government-funded central sugar mills, in order to ensure that the experiment of 'white' small farmers growing sugar cane remained viable. By 1898, the companies operating Moreton, Nerang, Proserpine and Gin Gin Mills had defaulted on their repayments.[134] The Queensland Treasury assumed control of Moreton and Proserpine Mills in 1899 and 1900, respectively, and advanced the companies additional funds to pay their suppliers.[135] A severe drought and heavy frosts at the turn of the century also meant Gin Gin Mill did not crush between 1901 and 1903, Nerang and Moreton Mills received only enough cane to work the factories at half their capacity in 1901 and 1902, and Mt Bauple Mill did not operate in 1902.[136] In 1904, Dr Walter Maxwell, the Director of the Queensland Bureau of Sugar Experiment Stations, was asked to salvage the situation. Together with officers of the newly formed Bureau of Central Sugar Mills, Maxwell ensured more effective operations at the factories by extending their tramline networks, employing analytical chemists and ceasing to pay non-shareholders less for their cane.[137] By 1914, Pleystowe, Racecourse, Marian and Moreton Mills had finalised their loan repayments, and the other central sugar mills had reduced their arrears in repayments. The exception was Nerang Mill, which was described as having a 'hopeless' financial position in 1912.[138]

The arrangement of financially struggling central sugar mills in Queensland being run by government-appointed officers did not stop other groups of canegrowers lobbying the Kidston government during the late 1900s for public funds to build new central mills. In particular, Premier William Kidston was being pressured by farmers from the Russell River and Daradgee districts, just north of Innisfail. They had been promised funds in 1900 to build a central sugar mill, but had been denied the monies when the bill proposing its allocation had been denied Royal assent.[139] To deal with the increasing number of demands to build new central sugar mills, the Kidston government appointed a Royal Commission in November 1910 to examine where these new central sugar mills should be located in Queensland. The Commission consisted of Robert Ranking, a police magistrate, John Paddle, Comptroller of the Queensland Bureau of Central Mills and Dr Arthur Gibson, a chemist associated with Bingera plantation. During 1911, the Ranking Royal Commission gathered evidence from groups of farmers at twenty locations throughout Queensland, mostly in the north of the state.

After its investigations, the Ranking Royal Commission recommended that the two sugar mills to be erected in readiness for the 1913 crushing season should be one capable of producing 10 000 tonnes of sugar and one of 5 000 tonnes capacity. The larger of the sugar mills should be erected at Babinda Creek, and the smaller at Daradgee, on the North Johnstone River. A third sugar mill of 5 000 tonnes capacity should also be built at Jarvisfield, near Ayr, in readiness for the 1913 crushing season, if no definite assurances had been received by the Queensland government before 31 December 1911 that Drysdale Bros. & Co. would erect a sugar mill on the Inkerman Estate (south side of the Burdekin River, near today's Home Hill). The Commission also suggested that in preparation for the 1914 crushing season, a sugar mill with a capacity of 8 000 tonnes should be erected on the South Johnstone River. Abundant land suitable for growing sugar cane was situated at each locality, but the farmers suffered from an absence of milling facilities or were hindered by the lengthy distance from existing sugar mills.[140]

In response to the Ranking Royal Commission's recommendations, William Kidston's government in Queensland introduced legislation which once again enabled public funds to be allocated to farmer cooperatives seeking to erect central sugar mills. Under the *Sugar Works Guarantee Act* of 1911, land owned by those forming the company had to be offered as security, valued and certified suitable for growing sugar cane. Once enough suitable land had been offered as security for the proposed loan, the Queensland government gazetted a sugar works area. Within this area, prospective suppliers of the proposed sugar mill agreed to participate fully in the scheme, as they were liable for making up any deficiency in the yearly repayments to the government. The new sugar mills were to be controlled by the Bureau of Central Sugar Mills. However, the Denham government which had assumed power in February 1911, moved slowly to implement the Ranking Royal Commission's recommendations and no new central sugar mills were ready to operate in the 1913 crushing season. Eventually, after considerable agitation from the Babinda district farmers, the Babinda Mill was partially erected during 1914, and it crushed for the first time in 1915. The Daradgee proposal dropped out of consideration, and instead the funds were allocated to a central sugar mill at South Johnstone, although its erection during 1915 was hindered due to delays in the delivery of machinery caused by World War I. Crushing finally commenced in September 1916. Overall, the Queensland Treasury by mid-1916 had advanced £560 000 towards the erection of the Babinda and South Johnstone Mills. Additional funds were not required for a Lower Burdekin sugar mill, as Drysdale Bros. & Co. had offered the Queensland government the necessary assurances about its intention to erect a sugar mill to serve the Inkerman Estate lands. Inkerman Mill was built during 1913 and early 1914; its first crushing took place later in 1914.[141]

Institutional adjustment and equilibrium

Between 1884 and 1914, the institutional arrangements for sugar production in Australia went through a period of adjustment, before reaching what the economic historian Ralph Shlomowitz termed 'equilibrium'. This 'equilibrium' became particularly enduring, with the central milling system still being the basis of the sugar industry in 2009. Sugar cane continues to be grown on small, family-owned farms and sugar produced exclusively by European labour only. Moreover, the outcome of this transformation, as mentioned in the book's introduction, had also made the Australian sugar industry in 1914 unique amongst the world's sugar cane industries. Sir Samuel Griffith's bold experiment to create a 'white man's' industry had succeeded.

In 1927, the sugar technologist Francis Maxwell noted, however, that such institutional arrangements were artificial, as they had only come about because the Australian sugar industry received considerable protection from imports of more cheaply produced sugar.[142] This protection, as will be seen in Chapter 18, was not welcomed by all members of Australian society, especially the jam manufacturers, who argued for its abolition during the 1920s and 1930s. Cheap sugar, not white-produced sugar on small family farms, mattered more to these industrialists and some members of the Australian public who were not swayed by the arguments that large numbers of European canegrowers in coastal Queensland were important in the defence of the country.

Notes

1 P. Griggs, 'The origins and early development of the small cane farming system in Queensland, 1870–1915', *J. of Historical Geography*, Vol. 23, no. 1 (1995), pp. 46–47.
2 *QVP*, Vol. 4 (1889), pp. 340–344.
3 *BC*, 25 September 1869, p. 7; *Slater's Queensland Almanac*, 1870, p. 22.

4 B. Higman, 'Sugar plantations and yeoman farming in New South Wales', *Annals Association of American Geographers*, Vol. 58 (1968), pp. 713–717.

5 Higman, 'Sugar plantations and yeoman farming', pp. 703–705.

6 *ATCJ*, 17 December 1870, p. 13 (Woodside); *ATCJ*, 24 December 1870, p. 10 (Taree); Higman, 'Yeomen and plantation landscapes', pp. 172 & 174.

7 *Northern Territory Times*, 1 April 1882, p. 3; *SAPP*, Vol. 4 (1882), Paper No. 147A: Quarterly Report by the Govt. Resident on the Northern Territory, p. 1 & *SAPP*, Vol. 3 (1884), Paper No. 53 A: Quarterly Report by the Government Resident on the Northern Territory, p. 3; the lack of development on other Northern Territory sugar estates is detailed in *Northern Territory Times*, 19 May 1883, p. 2 & 14 July 1883, p. 2.

8 E.W. Knox, AR 1877 season, p. 29; E.W. Knox, AR 1878 season, pp. 29–30; E.W. Knox, AR 1881, p. 77, all in CSRR, Z/109, Box 374; W. Issacs, Condong Mill Mang., 'Notes on Tweed River farmers, c. 1880,' CSRR, Z/303, Box 37, File C 1.0, Folder 6, Doc. 11c.

9 E. Knox to T. Scott, Brisbane Waters, 1 February 1869, CSRR, 142/2333, p. 57; E.W. Knox to W. Issaacs, Condong Mill Mang., 4 February 1882, CSRR, 142/806, p. 376–377; A. Fairgrieve, Insp. of Mills, to W. Isaacs, Condong Mill Mang., 14 February 1882, CSRR, 142/806, p. 389; W. Rothe, Acting CSR Mang., to J. Wyness, Broadwater Mill Mang., 1 April 1882, CSRR, 142/806, p. 485.

10 E. Knox to E. B. Forrest, Brisbane, 24 December 1880, CSRR Z/303, Box 40, File D 1.0. Folder 6, Doc. 4; E.W. Knox to A. Fairgrieve, Brisbane, 26 May 1881, CSRR 142/2339, p. 12.

11 S. May (ed.), 'The Diary of Arthur Neame, 1870–1897', p. 53; Kerr, *Sugar at Maryborough*, p. 49.

12 *BC*, 3 August 1882, p. 4 (Cudgen); *QVP*, Vol. 4 (1889), p. 356 (Woodlands).

13 *NSWVP*, Vol. 6 (1890), p. 67, Evidence of Louis Armbruster, owner of Lindendale; Higman, 'Yeomen and plantation landscapes', p. 28; P. Griggs, 'Sugar plantations in Queensland, 1864–1912: origins, characteristics, distribution and decline', *Agricultural History*, Vol. 74, No. 3 (2000), p. 622.

14 For examples of planters dummying see Kerr, *Pioneer Pageant*, pp. 27 & 32.

15 Griggs, 'Sugar plantations in Queensland', p. 622.

16 *QVP*, Vol. 4 (1889), p. 132.

17 For cattle raising see *Sydney Mail*, 19 October 1884, p. 791; *Q*, 29 July 1893, p. 217; & *Rockhampton Morning Bulletin*, 22 January 1900, p. 6; crops of maize and potatoes on Benowa plantation are described in *The Planter and Farmer*, December 1885, p. 267; for Hambledon's orchards see *Q*, 4 February 1888, p.190.

18 S. May (ed.), 'The Diary of Arthur Neame, 1870–1897', pp. 30–31; C. Wood, 'The Queensland sugar industry as depicted in the Whish and Davidson Diaries', *J. of the Royal Historical Society of Queensland*, Vol. 7, No. 3 (1964–65), pp. 567 & 577–579.

19 Griggs, 'Sugar plantations in Queensland, 1864–1912', pp. 623–624; *NS*, 10 May 1884, p. 2 (Staines).

20 Kerr, *Sugar at Maryborough*, pp. 32–40 (Tooth & Cran); Waterson, *Biographical Register*, pp 15 & 57 (Black and Fitzsimmons); Kerr, *Pioneer Pageant*, p. 18 (Hyne & Bridgeman).

21 T. Parsons, 'Melbourne money and Queensland sugar. The experience of Swallow and Ariell in the 1880s', *Victorian Historical Magazine* Vol. 44, Nos. 1 & 2 (February–May 1973), pp. 32–39; for Swallow sending molasses to his biscuit factory see *Q*, 7 November 1885, p. 759; & *NS*, 26 June 1889, p. 3.

22 CSR purchases of Queensland mill outputs are mentioned in the letterbooks Head Office, CSR, to W.H. Paxton & Co., Mackay, Queensland (OUT) June 1880 to October 1885, CSRR, 142/3171 & Z/109, Box 318.

23 Griggs, 'Sugar plantations in Queensland, 1864–1912', pp. 625–626 & 632–633.

24 Memorandum of Association of the Bowen Sugar Co. Ltd., c. 1866, LAN/214, File Sugar Selection No 32, QSA; *BC*, 4 May 1865, p. 3; Pimpama Sugar Co. File, A/21276, 3/1, QSA; Maryborough Sugar Co. Ltd. File, A/21276, 4/1, QSA.

25 Burdekin Delta Sugar Co. Ltd., Defunct Co. File No. 582, VPRO, 932/PO/34; Alstonsville Sugar Co., Defunct Co. File No. 808, VPRO, 932/PO/49; Oakenden Sugar Co., Defunct Co. File No. 737, VPRO, 932/PO/46; Bloomfield River Sugar Co. Ltd., Defunct Co. File No. 741, VPRO, 932/PO/46; The Mackay Sugar Co. Ltd., Defunct Co. File No. 572, VPRO, 932/PO/33.

26 Melbourne-Mackay Queensland Sugar Co. Ltd., Defunct Co. File No. 581, VPRO, 932/PO/34; Melbourne-Mackay Sugar Co. Ltd., Defunct Co. File No. 753, VPRO, 932/PO/47; Pleystowe Sugar Co. Ltd, Defunct Co. File No. 671, VPRO, 932/PO/41; Kerr, *Pioneer Pageant*, pp. 72–73; Donovan, *A Land Full of Possibilities*, pp. 135 &138.

27 J.A. Hone, 'Fanning, Edward (1848–1917)', *ADB*, Vol. 4 (1972), pp. 151–152 (Fanning & Nankivell); J. Nolan, 'Buss, Frederick William (1845–1926)', *ADB*, Vol. 7 (1979), pp. 499–500; Kerr, *Southern Sugar Saga*, pp. 28 & 31 (Buss family).

28 Q, 26 July 1884, p.159; *Sydney Mail*, 13 September 1884, p. 531, 27 September 1884, p. 635 & 19 October 1884, p. 791.

29 K. Saunders, *Workers in Bondage. The Origins and Bases of Unfree Labour in Queensland 1824–1916* (Brisbane, 1982), p. 62; C. Moore, *Kanaka. A History of Melanesian Mackay* (Port Moresby, 1985), pp. 140–144; R. Evans, K. Saunders and K. Cronin, *Race Relations in Colonial Queensland* (Brisbane, 1988), pp. 181–185.

30 For German employees see D. Nagel, 'Johann Christian Heussler - Father of Queensland', in J. Fletcher, J. Voigt and J. Moses (eds.), *New Beginnings. Germans in Queensland and New South Wales* (Stuttgart, 1983), p. 126; *BC*, 16 April 1866, p. 4; for European workers see *Q*, 2 August 1873, p. 6 & *Queensland Express*, 9 January 1869, p. 3; for Aboriginal workers see Manning, *In Their Own Hands*, pp. 48–49; *MM*, 26 July 1873, p. 2; Kerr, *Sugar at Maryborough*, 13 & 29; & *Q*, 18 November 1871, p. 5.

31 *Northern Territory Times*, 2 October 1880, p. 2 & 1 April 1882, p. 3.

32 *NS*, 20 December 1884, p. 5 & 30 November 1892, p. 2.

33 Griggs, 'Plantation to small farm', pp. 112–113; Graves, *Cane and Labour*, pp. 36–37; B. Swan, 'Sinhalese emigration to Queensland in the nineteenth century', *J. of the Royal Australian Historical Society*, Vol. 67, no. 1 (June 1981), pp. 55–62; B. York, *Empire and race: the Maltese in Australia 1881–1949* (Sydney, 1990), pp. 22–26.

34 *QVP*, Vol. 4 (1889), pp. 125, 129, 164, 184 & 199.

35 *Australasian*, 5 August 1882, p. 183; *MC*, 28 October 1884, p. 3.

36 Shlomowitz, 'Markets for indentured and time-expired Melanesian labour', pp. 78–81; Graves, *Cane and Labour*, pp. 28–33.

37 Kerr, *Sugar at Maryborough*, p. 117; Kerr, *Southern Sugar Saga*, pp. 35–36; *The Planter and Farmer*, August 1887, p. 193.

38 *QVP*, Vol. 4 (1889), pp. 365, 378 & 419; A. F. Munro, *The Sugar Fields of Mackay* (Mackay, 1895), p. 29; *Townsville Daily Bulletin*, 17 November 1887, p. 2; *Q*, 4 September 1886, p. 391.

39 *QVP*, Vol. 4 (1889), pp. 132, 159, 180, 207 & 397.

40 *QVP*, Vol. 4 (1889), p. 176; *MM*, 31 March 1886, p. 2 & 19 August 1890, p. 2; E.W. Knox to J. Knox, Victoria Mill, Ingham, 9 February 1886, CSRR, 142/2857, p. 70; *Q*, 12 June 1886, p. 927; *QVP* Vol. 3 (1889), pp. 215–221 (Javanese); D. Dignan, 'Chiaffredo Venerano Fraire, 1852–1931', in M. Brandle (ed.), *The Queensland Experience. The Life and Work of Fourteen Remarkable Immigrants* (Brisbane, 1991), pp. 49–75 and F. Galassi, *Sotto La Croce del Sud (Under the Southern Cross). The Jumma Immigrants of 1891*. Studies in North Queensland History No. 16 (Townsville 1991) (Italian recruitment); for Smellie's plan see Jones, *Hurricane Lamps and Blue Umbrellas*, p. 154.

41 *QVP*, Vol. 4 (1889), pp. 131, 172, 179 & 186; Griggs, 'Plantation to small farm', pp. 113 & 131.

42 A.P. Ashdown, Manager, Mackay Branch of Commercial Bank of Sydney, to E.W. Knox, 29 March 1887, CSRR, Z/303, File D 2.0, Folder 1, Doc. 9 (Mt Pleasant); E.W. Knox to Goondi Mill Mang., Geraldton, 6 March 1894, CSRR, 142/1460, p. 57 (Mourilyan); *J. of Commerce*, 27 January 1887, p. 3 (Pyramid); Mang., Brisbane Branch of BOA, 'Report on Ashburton and Pyramid plantations, 21 June 1888,' BOAR, A/300/1, p. 7, ANZA; *ADB*, Vol. 4 (19), p. 179 (Innisfail).

43 A.P. Ashdown, Mang., Mackay Branch, CBCS, to E.W. Knox 29 March 1887, CSRR, Z/303, D 2.0, Folder 1, Doc. 9; Superintendent, BOA, Melb. Office, to Sec., London Office of BOA, 18 February 1887, BOAR, A/8/61, Letter No. 2436, ANZA; *J. of Commerce*, 24 February 1887, p. 4 & 30 November 1888, p. 8.

44 General Mang., Melb. Office, UB, to Mang., London Office, UB, 12 April 1889, UBR, U/103/35, Letter No. 324; General Mang., Melb. Office, UB, to Mang., London Office, UB, 16 January 1890, UBR, U/103/37, Letter No 481, p. 14; E.W. Knox to D. Finlayson, General Mang., UB, Melb., 28 January 1898, attached to General Mang., UB, Melb., to Chairman, Standing Committee, London, 2 February 1898, UBR, U/60/7, Letter No. 475; General Mang., UB, Melb., to Chairman, Standing Committee, London, 2 January 1907, UBR, U/60/15, Letter No. 1648, all at ANZA; Insp. C.R. Cowper, BOA, 'Report on Ashburton and Pioneer Sugar Plantations, 18 November 1891,' p. 9, attached to Superintendent, BOA, Melb. Office, to Sec., London Office of BOA, 24 November 1891, BOAR, A/8/70, Letter No. 2976 & Superintendent, BOA, Melb. Office, to Sec., London Office of BOA, 22 August 1893, BOAR, A/8/74, Letter No. 3157, all at ANZA.

45 Kerr, *Northern Outpost*, p. 25; Kerr, *Pioneer Pageant*, pp. 69 &100; Kerr, *Southern Sugar Saga*, pp. 31 & 61; *Townville Herald*, 12 February 1887, p. 19; Mang., Melb. Branch of BOA to Superintendent, BOA, Melb., 15 July 1891, BOAR, A/268/4, ANZA.

46 E.W. Knox to W. Issaacs, Mang. Condong Mill, 8 November 1881, CSRR, 142/1143, p. 448; E.W. Knox to Condong Mill Mang., 7 March 1893, CSRR, 142/1150, p. 390.

47 F. Shaw, 'The sugar industry in Queensland', *The Times* (London), 9 January 1893 (reprinted in M. Black, *The Sugar Industry of Queensland* (London, 1897), p. 7; Sec., AEMCo., Melb., to Sec., AEMCo., London, AER, London Office Book, No. 3, p. 200, MUA.

48 Munro, *The Sugar Fields of Mackay*, p. 3; Royal Commission on the West Indies, *British Parliamentary Papers* Vol. 50 (1898): Appendix C, p. 135; Q, 3 August 1895, pp. 225–226.

49 *QVP*, Vol. 4 (1889), pp. 186, 207 & 212.

50 *Q*, 2 August 1873, p. 6, 30 June 1877, p. 29 & 15 May 1880, p. 616; *Sydney Mail*, 27 September 1884, p. 635; *QVP*, Vol. 4 (1889), p. 388.

51 *QVP*, Vol. 4 (1889), pp. 171, 201, 207 & 212; E.W. Knox to C. Adams, Goondi Mill Mang., Geraldton, 8 February 1888, CSRR, 142/2886, p. 247; *QVP*, 3 (1890), p. 989 (Chinese selectors); E.W. Knox to C. Adams, Goondi Mill Mang., 7 August 1890, CSRR, 142/1457, p. 206 (Germans).

52 Munro, *Sugar Field of Mackay*, passim; *Q*, 27 January 1894, p. 176, 24 February 1894, p. 3 & 30 November 1895, p. 1043; *SJTC*, 15 February 1894, p. 9 & 15 March 1894, p. 36; *Geraldton Advocate*, 6 November 1893, p. 2; *MC*, 15 March 1894, p. 3; Griggs, 'Plantation to small farm', pp. 173–181.

53 *QPP*, Vol. 2 (1906), p. 878; E.W. Knox, to A. Cowley, Mang. Victoria Mill, Ingham, 12 February 1884, CSRR, 142/1547, p. 133.

54 *Q*, 2 March 1889, p. 422 (Advt. for CSR land sale on Tweed River); *ATCJ*, 23 July 1887, p. 178; *Tweed & Brunswick Advocate*, 25 December 1889, p. 3 (Advt. for CSR land sale on Tweed River).

55 *MC*, 11 April 1894, p. 2, 17 May 1894, p. 2 & 15 September 1894, p. 3; *Q*, 25 July 1896, p. 178; Munro, *Sugar Fields of Mackay*, pp. 28, 29, 59, 61 & 64; *MM*, 1 September 1894, p. 4.

56 *Tweed Herald and Brunswick Chronicle*, 16 September 1898, p. 4; *Geraldton Advocate*, 6 November 1895, p. 2; H. Easterby, *The Queensland Sugar Industry. An Historical Review* (Brisbane, c. 1933), p. 83; Kerr, *Southern Sugar Saga*, pp. 34, 36, 78, & 84; Manning, *In their own hands*, p. 75; Griggs, 'Plantation to small farm', pp. 163–169; *The Capricornian*, 11 November 1893, p. 35.

57 J. Armstrong, 'Aspects of Japanese immigration to Queensland before 1900', *Queensland Heritage*, Vol. 2, No. 9 (November 1973), pp. 3–5; F. Galassi, 'From the land of the rising sun to the sugar-growing districts of North Queensland 1892–1901', *J. of the Royal Historical Society of Qld.*, Vol. 17, No. 4 (February 2001), pp. 427–432; D. Munro, 'Gilbert and Ellice Islanders on Queensland Canefields, 1894–1899', *J. of the Royal Historical Society of Qld.*, Vol. 14, No. 1 (May 1992), pp. 449–465.

58 *NSWVP*, Vol. 7 (1897), p. 1; *ATCJ*, 11 September 1897, p. 22; *Tweed Herald and Brunswick Chronicle*, 1 September 1899, p. 5 & 7 June 1901, p. 5; M. de Lepervanche, *Indians in a White Australia* (Sydney, 1984), pp. 85–89 & A. Potts, 'I am a British subject, and I can go wherever the British flag flies: Indians on the Northern Rivers of New South Wales during Federation Years', *J. of the Royal Australian Historical Society*, Vol. 83 Part 2 (1997), pp. 97–105.

59 *SJTC*, 15 December 1904, p. 90; *North Queensland Register*, 2 July 1906, p. vii; Griggs, 'Plantation to small farm', pp. 252–254.

60 *North Queensland Herald*, 18 December 1905, p. 18.

61 *QPP*, Vol. 2 (1911–1912), p. 1180; Report on the Half Yearly Accounts, Queensland Division, Union Bank, 31 August 1906, pp. 56–58, UBR, U/220/1, ANZA; Mourilyan Syndicate Ltd., Minute Book, Vol. 1 (107–1914), entry for 30 March 1907, p. 27, Innisfail Historical Society.

62 R. Dowling, Cane Insp., Condong Mill, to E. Knox, 3 April 1913, CSRR, 142/1218, Letter No. 282.

63 *QPP,* Vol. 2 (1906), pp. 480–486; Qld. Govt. Intelligence and Tourist Bureau, *Queensland Sugar Industry* (Brisbane 1913), p. 110; Kerr, *Southern Sugar Saga,* p. 78; Anon, *Kalamia Sugar Mill. The First 100 Years* (Ayr?, 1984), p. 5; *ASYB* 1997, pp. 182–183; Ev. M. Clapham, Mang, Palms Estate, p. 557, in 'Volume of Evidence before the Piddington Royal Commission, 1920', CSRR, Z/109, Box 45.

64 *Q,* 8 August 1868, p. 10; *Manning River News,* 31 August 1867, p. 2.

65 *Q,* 28 November 1874, p. 9; *BC,* 24 April 1875, p. 7.

66 *CRE,* 25 September 1880, p. 2.

67 *CRE,* 1 August 1871, p. 2, 19 November 1872, p. 2, 26 September 1882, p. 2 & 21 November 1882, p. 2; *ATCJ,* 14 June 1873, p. 746; Meston, 'Report on the Sugar Industry etc'., p. 3; *Q,* 18 March 1882, p. 328; E.W. Knox, AR 1872, p. 23, CSRR, Z/109, Box 374.

68 *CRE,* 28 November 1871, p. 2.

69 Higman, 'Yeoman and plantation landscapes', p. 88; Kerr, *Pioneer Pageant,* pp. 44–45.

70 *The Planter and Farmer,* January 1884, p. 20; *CRE,* 9 September 1868, p. 3; *Slater's Queensland Almanac,* 1872, p. 21; Meston, 'Report on the Sugar Industry etc.,' p. 921; Lyne, *The Industries of New South Wales,* p. 260; E.W. Knox, Clarence River, to J. Grafton Ross, Sydney, 16 July 1872, CSRR, 142/2323, p. 177; *QVP,* Vol. 4 (1889), pp. 402–412.

71 *RRE,* 5 April 1879, p. 2 (Prospect Mill); A. Francis, *Then and now: the story of a Queenslander* (London 1936), p. 17 (Arthur Francis); *BC,* 16 September 1871, p. 5 (Clydesdale Mill); *Port Denison Times Almanac,* 1868, p. 42 (Ormiston) & Diaries of John Mackinnon, entry for 28 August 1885, p. 8, Richmond River Historical Society; for G.& S. Grimes advertising see *Q,* 3 August 1872, p. 1 & 10 May 1873, p. 1.

72 *ATCJ,* 24 December 1870, p. 10; *CRE,* 18 January 1876, p. 2 & 5 September 1882, p. 2; *NS,* 9 August 1879, p. 2; J. Knox, Broadwater Mill, to E.W. Knox, 28 November 1882, CSRR 142/2872, p. 186.

73 Higman, 'Yeoman and plantation landscapes', p. 84; W. Meston, 'Notes about the Meston family, the sugar industry and about Woodburn, after 1876', 8pp typescript, no date, p. 1, Richmond River Historical Society.

74 *Q,* 14 January 1888, p. 71; QNB, Half Yearly Balance Books, 1887–1889, Bundaberg Branch, QNBR; William Walker, Mang., Maryborough Branch, CBCS, to General Mang., CBCS, Sydney, 2 November 1881, CBCS, Outward Letterbook, Maryborough Branch, Vol. 2, p. 357, CBCSR.

75 LAN/AG 386, Selection File No. 535, QSA; *CRE,* 1 February 1885, p. 1 (Advt. for sale of Edmonson's Sugar Mill).

76 Higman, 'Yeomen and plantation landscapes', p. 86; Daley, *Men and a River,* p. 136.

77 Daley, *Men and a River,* p. 138; *Q,* 23 March 1872, p. 11.

78 E. W. Knox, Southgate Mill, to E. Knox, 25 October 1871, CSRR, Z 109, File C10. Folder 4, Doc. 34; *NS,* 10 December 1884, p. 3.

79 *Clarence River Advocate,* 19 September 1899, p. 2.

80 Lyne, *The Industries of New South Wales,* p. 260; E.W. Knox, Sydney, to W. Issaacs, Mang. Condong Mill, 1 September 1879, CSRR, 142/1143, p. 60.

81 *Clarence River Advocate,* 11 July 1902, p. 5 & 18 July 1902, p. 5; Higman, 'Yeomen and plantation landscapes', p. 88 & Appendix 2, p. 183: entry for George Morrison.

82 *Manning River News*, 7 September 1867, p. 1; Birrell, *The Manning Valley*, p. 140; Extracts from CSR's Board Minutes, entry for 30 June 1868, CSRR, Z/303, Box 3, File A 1.1, Folder 2, Doc. 4; *CRE,* 14 July 1868, p. 4 (Advt).

83 Extracts from CSR's Board Minutes, entries for 26 March 1867, 7 May 1867 & 3 March 1868, CSRR, Z/303, Box 3, A 1.1, Folder 2, Doc. 4; *Manning River News*, 6 July 1867, p. 2.

84 *Sydney Mail*, 13 November 1869, p. 8; *ATCJ*, 14 June 1873, p. 746; Lyne, *The Industries of New South Wales*, p. 258; *The Sugar Cane*, 1 November 1870, p. 650; E.W. Knox, AR 1872 season, p. 22, CSRR, Z/109, Box 374.

85 *NS*, 15 November 1884, p. 5.

86 Extracts from CSR's Board Minutes, 14 November 1871 and 9 April 1872, CSRR, Z/303, Box 3, File A 1.1, Folder 2, Doc. 4; Higman, 'Yeomen and plantation landscapes', pp. 46–47; E. W. Knox, Darkwater Mill, to Messrs Forsyth & O'Dangar, Macleay River, 6 May 1872, CSRR, 142/1783, pp. 244–246.

87 *Q*, 2 February 1884, p. 191; Lyne, *The Industries of New South Wales*, p. 921.

88 E.W. Knox, AR 1871 season, p. 7, CSRR, Z/109, Box 374.

89 Report on the Tweed River by Messrs E. Hayley and C. Stephens, June 1872, p. 5, CSRR, Z/303, Box 37, File C 1.0, Folder 6, Doc. 10.

90 E.W. Knox, AR 1876 season, p. 23; E.W. Knox, AR 1877 season, p. 20; E.W. Knox, AR 1878 season, pp. 29 & 31, CSRR, Z/109, Box 374.

91 *Q*, 17 April 1869, p. 10, 2 September 1871, p. 11 & 27 July 1872, p. 1 (Advt.); Kerr, *Sugar at Maryborough*, pp. 26–31, 54–55, 57–83; Kerr, *A Century of Sugar*, pp. 11–12.

92 *QVP*, Vol. 4 (1889), pp. 371–373 (Riverside) & 393 (Marian); *Sydney Mail*, 20 September 1884, p. 583 & Kerr, *Pioneer Pageant*, p. 69 (Marian).

93 Alcorn & Dunn, *Moreton Sugar Mill*, pp. 7–9 (Buderim Mt Sugar Co. & Maroochie River Sugar Co.); *Q*, 3 July 1880, p. 27 & 24 October 1885, p. 678, & *Pugh's Almanac*, 1888, entry for Beenleigh (Excelsior Mill); *QVP*, Vol. 4 (1889), pp. 361–362 & *The Planter and Farmer*, February 1888, p. 352 (Eagleby); *Pugh's Almanac*, 1884–1888, entry for Beenleigh (Waterford Sugar Co.).

94 *Q*, 21 November 1874, p. 6; P. Fletcher, 'The sugar industry of Queensland', in P. Fletcher (ed.), *Queensland: Its resources and institutions* (Brisbane, 1886), p. 12; *QVP*, Vol. 1 (1885), p. 1155; *QVP*, Vol. 4 (1889), pp. 272 & 275; *BC*, 3 October 1882, p. 6.

95 *MM*, 18 July 1885, p. 2; *QVP,* Vol. 1 (1885), pp. 1155–9.

96 *QVP*, Vol. 3 (1878), pp. 1129–1138.

97 *QPD*, Vol. 47 (1885), pp. 1442–1443.

98 *QPD*, Vol. 47 (1885), pp. 1444–1471; *BC*, 27 October 1885, p. 4 & 6 November 1885, pp. 4–5.

99 *Logan Witness*, 29 May 1886, p. 2.

100 *QVP*, Vol. 2 (1886), pp. 1–28.

101 *QVP*, Vol. 3 (1890), pp. 847–66; Racecourse Central Mill Co., Directors Minute Book No. 1, p. 108.

102 QAG, AR 1891–92, *QVP*, Vol. 2 (1892), p. 53; QAG, AR 1892–93, *QVP*, Vol. 2 (1893), p. 24; QAG, AR 1894–95, *QVP*, Vol. 2 (1895), p. 470.

103 *QPD*, Vol. 70 (1893), pp. 947–963; E. Knox, Sydney, to Hon. E. Forrest MLC, Brisbane, 9 July 1894, CSRR, Z/303, Box 40, File D 1.0, Folder 4, Doc. 2.

104 Queensland, 57 Vic. No 18, 1893; Queensland, 59 Vic. No 14, 1895.

105 Deduced from an examination of the Indexes to the Register of Letters to the Under-Sec., Qld. Dept. of Public Works, WOR H/9 and WOR H/10, QSA.

106 Under-Sec., Qld. Dept. of Public Works, to R.W. McCulloch, 10 October 1894, WOR G/51, QSA; *Q*, 24 February 1894, p. 367, 28 July 1894, p. 175 & 8 August 1894, p. 320; Kerr, *Northern Outpost*, p. 33.

107 *Cairns Morning Post*, 30 December 1897, p. 2; Under-Sec., Qld. Dept. of Public Works, to E.W. Knox, Sydney, 5 June 1894, WOR G/50, p. 746, QSA; Under-Sec., Qld. Dept. of Public Works, to E.W. Knox, 10 November 1894, WOR G/51, p. 507, QSA.

108 Qld. Dept. of Agric. & Stock, *Sugar Planting in Queensland* (Brisbane 1893); *QVP*, Vol. 3 (1893), p. 918; this brochure was also reproduced in the *Mackay Standard*, 18 January 1895, p. 3 and in M.H. Black, *The Sugar Industry of Queensland* (London, 1897), pp. 3–6.

109 Qld. Dept. of Agric. & Stock, *Sugar Planting in Queensland* (Brisbane, 1893) & *MM*, 18 May 1895, p.3; Griggs, 'Plantation to small farm', pp. 186 & 225; *Q*, 15 June 1895, p. 1138.

110 *The Sugar Cane*, 1 October 1891, pp. 528–529; E.W. Knox to Mang. Goondi Mill, Geraldton, 20 August 1891, CSRR, 142/1458, p. 99; E. Knox to Mang. Victoria Mill, Ingham, 11 May 1892, CSRR, 142/1551, p. 122; *Q*, 17 February 1894, p. 323.

111 Black, *The Sugar Industry of Queensland*, p. 14.

112 *CPP*, Vol. 4 (1913), pp. 1386 (Ev. G. Forest, Victoria Mill Mang., Q. 6152), 1474 (Ev. J. Moule, farmer, Homebush, Q. 8936), 1484 (Ev. O. Madsen, farmer, Mackay, Q. 9242), 1571 (Ev. T. Hodgett, farmer and chairman of directors, Cattle Creek Sugar Co., Q. 11614) & 2133(Ev. W. Gibson, Mang. Director, Marburg Sugar Co., Q. 2133); *Q*, 24 October 1885, p. 678 & 23 December 1893, p. 1233; *QVP*, Vol. 4 (1889), p. 362; *QPD*, Vol. 47 (1885), p. 1465.

113 E.W. Knox to Victoria Mill Mang., Ingham, 24 August 1893, CSRR, 142/ 1552, p. 75; Mossman Central Mill, Directors Minute Book, , No. 6, p. 12; *SJTC*, 15 April 1901, p. 221; *CPP*, Vol. 4 (1913), p. 1176 (Ev. G. Muntz, Chairman of Directors, Mossman Central Sugar Mill Co., Qs 133–134); Marian Central Mill, Directors Minute Book, No. 2, p. 35; Racecourse Central Mill, Directors Minute Book, No. 4, p. 252; *QPP*, Vol. 2 (1906), p. 879; *Port Denison Times*, 19 February 1901, p. 4; *Geraldton Advocate*, 13 November 1895, p. 2; *The Sugar Cane*, 1 April 1896, pp. 201–202.

114 E.W. Knox to Victoria Mill Mang., Ingham, 20 December 1906, CSRR, 142/1561, p. 351; *QPP*, Vol. 2 (1906), p. 293.

115 *Q*, 26 March 1892, p. 582 & 3 June 1905, p. 34; *CPP*, Vol. 4 (1913), pp. 1366 (Ev. D. Hunter, farmer Goondi, Qs. 5525 & 5567–5575) & 1369 (Ev. J. Doherty, farmer, Goondi, Q. 5660); Griggs, 'Plantation to small farm', p.184; S. May (ed.), 'The Diary of Arthur Neame, 1870–1897', p. 131; E.W. Knox to Under-Sec., Qld. Dept. of Agric. & Stock, 13 June 1891, CSRR, 142/483, p. 108.

116 *SJTC*, 15 July 1894, p. 128; *Q*, 25 January 1896, p. 178; Mossman Central Mill, Financial Ledger, p. 305; Manning, *In their own hands*, p. 262; York, *Empire and race*, pp. 62–63; Galassi, *Sotto La Croce del Sud*, pp. 186–187; *QPP*, Vol. 2 (1906), p. 477; P. Griggs, 'Alien agriculturalists: non-European small cane farmers in Australia, 1880–1920', in P. Ahluwalia, W. Aschroft & R. Knight (eds.), *White and Deadly: Sugar and Colonialism* (New York, 1999), pp. 135–142.

117 *Mackay Standard*, 31 January 1894, p. 2 & 20 February 1895, p. 3; *Bundaberg Mail*, 14 January 1895, p. 3.

118 *SJTC, Supplement*, 15 October 1893; *MM*, 18 May 1895, p. 3; *Q*, 15 June 1895, p. 1138; Qld. Dept. of Agric. & Stock, *Sugar Planting in Queensland*, pp. 18–19.

119 R. Helms, Childers Mill Mang., Childers, to E. Knox, 15 September 1911, CSR, 142/1410, Letter No. 911; W. McLure Smith, AEMCo., Melb., to Gen. Man., AEMCo., London, 11 November 1897, AER, Letterbook No. 2, p. 340, MUA; *QPP*, Vol. 2 (1906), p. 507; *CPP*, Vol. 4 (1913), p. 1975 (Ev. W. Gibson, Q. 22354).

120 E.W. Knox to Victoria Mill Mang., Ingham, 23 May 1895, CSRR, 142/1553, p. 443.

121 *The Sugar Cane*, 1 October 1891, p. 528; E.W. Knox to Victoria Mill Mang., Ingham, 11 May 1892, CSRR, 142/1551, p. 122; *Q*, 17 February 1894, p. 323.

122 *The Sugar Cane*, 1 October 1891, p. 529 (Homebush); May, 'The Diary of Arthur Neame', p. 131 (Macknade Mill); Mourilyan Syndicate, Directors Minute Book, 1907–1914, passim (Mourilyan Mill); Cattle Creek Mill, Register of Crop Liens, 1905–1941, passim; *QPP*, Vol. 1(1905), p. 102 (Marian Central Mill); Racecourse Central Mill, Directors Minute Book, 1893–1908, passim; Mossman Central Mill, Directors Minute Books, 1897–1905, passim; North Eton Central Mill, Directors Minute Books, 1897–1907, passim.

123 *Mackay Standard*, 13 March 1895, p. 3; Racecourse Central Mill, Directors Minute Book No. 2, p. 44; Griggs, 'Plantation to small farm', pp. 147 & 188; *CPP*, Vol. 4 (1913), p. 1298 (Ev. D. Morgan, farmer, Gordonvale, Q. 3499); E.W. Knox to Under-Sec., Qld. Dept. of Agric. & Stock, 13 June 1891, CSRR, 142/483, p. 108.

124 *CPP*, Vol. 4 (1913), pp. 1203 (Ev. R. Rex, farmer, South Mossman, Q. 918) & 1491 (Ev. J. O'Riordan, farmer, Mackay, Q. 9470).

125 *CPP*, Vol. 4 (1913), pp. 1190 (Ev. G. Muntz, Chairman of Directors, Mossman Central Mill Co., Qs. 507–508), 1454 (Ev. F. Meynill, accountant, Prosperpine, Q. 8353) & 1873 (Ev. F. Kinne, Sugar miller, Maryborough, Q. 19451); R. Fitzgerald, *From the Dreaming to 1915. A History of Queensland* (Brisbane, 1982), p. 184.

126 Racecourse Central Sugar Mill, Annual Balance Sheets, 1901–1910; See lists of 'Advances to Farmers' in PMR, IAL/6, pp. 52–53, JCUA; Connolly, *John Drysdale and the Burdekin*, p. 185.

127 W. McLure Smith, AEMCo., Melb., to General Mang., AEMCo., London, 20 October 1896, AER, London Office Books No. 1, p. 146; W. McLure Smith, AEMCo., Melb., to General Mang., London, 21 June 1904, AER, London Office Book, No. 9, p. 395, MUA; E. Knox to A. Farquhar, Insp. of Mills, Macknade Mill, 26 August 1897, CSRR, 142/2859, p. 264.

128 *QVP,* Vol. 4 (1889), pp. 236, 241 & 340–44; *SJTC*, 15 September 1890, p. 43.

129 K. Saunders, 'The workers' paradox: indentured labour in the Queensland sugar industry', in K. Saunders (ed.), *Indentured Labour in the British Empire* (Canberra, 1984), pp. 238–245; for Drysdale Bros.' insistence on employing non-European workers see POL/J 40, Bundle 1336M, QSA; *QPP*, Vol. 3 (1925), p. 48; *North Queensland Register*, 26 December 1927, p. 24 & Sec., Pioneer Sugar Mills Ltd., to Under-Sec., Qld. Dept. of Agric. and Stock, 24 January 1930, PMR/LB/50, p. 248, JCUA.

130 P. Griggs, 'Alien agriculturalists: non-European small cane farmers in Australia, 1880–1920', in P. Ahluwalia, W. Aschroft and R. Knight (eds.), *White and Deadly: Sugar and Colonialism* (New York, 1999), pp. 143–148.

131 Bolton, *A Thousand Miles Away*, p. 303; *CPP*, Vol. 4 (1913), p. 1363 (Ev. C. Jodrell, farmer, Innisfail, Q. 5425).

132 *SJTC*, 15 March 1903, p. 178; *CPP*, Vol. 4 (1913), pp. 1276 (Ev. C. Hives, farmer, Cairns, Q. 3023) & 1283 (Ev. H. Niven, farmer, Aloomba, Q. 3203); *QPP*, Vol. 2 (1914), pp. 857–858; *ASJ*, 5 November 1914, p. 549.

133 Corkhill, *Queensland and Germany*, pp. 92–94 & 99; O. Koivukangus, *Sea, gold and sugar cane. Finns in Australia, 1851–1947* (Turku, 1986), p. 240.

134 QAG, AR 1897–98, *QVP*, Vol. 2 (1898), pp. 34–45; QAG, AR 1898–99, *QVP*, Vol. 2 (1899), pp. 29–30; QAG AR 1899–1900, *QVP*, Vol. 4 (1900), p. 31.

135 QAG, AR 1898–99, *QVP*, Vol. 2 (1899), p. 30; QAG, AR 1899–1900, *QVP*, Vol. 4 (1900), p. 31; Sec., Proserpine Central Mill Ltd., to Under-Sec., Qld. Dept. of Agric. & Stock, 11 March 1899, In-letter 02111 of 1899, A/2093, QSA; *SJTC*, 15 February 1900, p. 180; *PDT*, 19 February 1901, p. 4.

136 *QPP*, Vol. 3 (1902), p. 139; Kerr, *Sugar at Maryborough*, p. 144; Dignan, *The Story of Kolan*, p. 83.

137 BCSM, AR 1904–1905, *QPP*, Vol. 2 (1905), pp. 612–613; BCSM, AR 1905–1906, *QPP*, Vol. 2 (1906), pp. 280–83 & 297; BCSM, AR 1907–1908, *QPP*, Vol. 3 (1908), p. 195; *Nambour Chronicle*, 19 August 1904, p. 4.

138 BCSM, AR 1905–06, *QPP*, Vol. 2 (1906), p. 279; BCSM, AR 1906–07, *QPP*, Vol. 2 (1907), pp. 821–22; QAG, AR 1906–07, *QPP*, Vol. 1 (1907), p. 115; QAG, AR 1909–10, *QPP*, Vol. 2 (1910), p. 1912; QAG, AR 1911–1912, *QPP*, Vol. 1 (1912), p. 486.

139 *BC*, 8 August 1907, p. 10 & 30 April 1908, p. 4; *ASJ*, 4 August 1910, p. 186 & 195, 8 September 1910, pp. 228–229 & 8 November 1910, pp. 365–366.

140 Easterby, *The Queensland Sugar Industry*, pp. 44–46; *QPP*, Vol. 2 (1911–1912), pp. 1027–1043.

141 Easterby, *The Queensland Sugar Industry*, pp. 46–48; QAG, 1916–1917, *QPP*, Vol. 2 (1916–1917), p. 1202.

142 R. Shlomowitz, 'The search for institutional equilibrium in Queensland's sugar industry 1884–1913', *Australian Economic History Review*, Vol. 19, no. 2 (1979), p. 110; F. Maxwell, *Economic Aspects of Cane Sugar Production* (London, 1927), p. 10.

CHAPTER 7
Science and the Canegrowers

The earliest commentators who described sugar cane cultivation methods in Australia were highly critical of the country's canegrowers. In 1880, Henry L. Roth concluded that cane cultivation in Queensland was 'very rude'. George Craig in his 1885 essay about the Australian sugar industry observed that 'cane growers, as a rule, depend upon chance and the productiveness of virgin soil'. Walter Campbell, a politician who toured New South Wales in the mid-1880s and reported upon the state of agriculture in the colony, wrote that cane farming in the Richmond River district was 'wretched and primitive in the extreme' and the chief product was 'weeds and these grow to perfection'. A similar report from the Special Correspondent of the *Sydney Morning Herald* claimed that the Richmond River district cane farmers engaged in a 'careless profligate system of farming'. Jean Despeissis, an advisory field officer with the New South Wales Department of Agriculture, wrote in 1891 that the system of cane cultivation practised in northern New South Wales had 'nothing particularly commendable in itself'. As late as 1900, Dr Walter Maxwell, the Director of the Sugar Experiment Station of the Hawaiian Sugar Planters' Association, in his report on the Queensland sugar industry concluded that the handling of the land by the colony's canegrowers had been 'crude and superficial'.[1]

Field creation was one agricultural practice by the country's first canegrowers which contributed to this poor opinion of their farming capabilities. Numerous reports exist of paddocks containing tree stumps.[2] Many of the country's earliest canegrowers simply cut down the vegetation on their blocks of land, allowed it to dry and then the material was burnt. Bigger tree stumps survived this initial firing and, in most instances, just rotted away, usually in three to five years.[3] Stones and boulders littered other paddocks. At Cedars plantation (Mackay), a visitor in 1882 noticed that field labourers moved the stones in the fields to one side, creating spaces for the cane to be planted. Walter Clarke, a reporter for the *Tweed Herald and Brunswick Chronicle*, described one paddock near Alstonsville in 1900 as 'two great undulating slopes being covered with boulders varying from the size of one's fist to about a foot square'. The cane grew amongst the boulders.[4]

Preparation of the soil to receive the cane plants varied enormously. Initially on many Australian properties where stumps and/or boulders filled the paddocks, canegrowers were unable to use the plough, and relied on hoes to make troughs or oblong holes for the setts.[5] Canegrowers could break up the soil even further when the paddocks had been stumped or cleared of stones and boulders. By the 1870s, some Queensland sugar planters utilised teams of horses or bullocks to pull ploughs. A steam plough facilitated deep ploughing of the soil at Ormiston plantation in 1865, but the use of these devices before 1880 was exceptional due to the expense associated with their purchase.[6] In contrast, Walter Campbell complained in 1888 that there was still not a subsoil plough in the Richmond River district, while along the Clarence River 'very shallow ploughing is the rule; the land merely skimmed over'. Generally, deep ploughing did not occur; the reputed motto of the country's first canegrowers was 'tickle the soil with a hoe and it will laugh a harvest'.[7]

Uncertainty prevailed over the correct method to plant sugar cane. Some Australian canegrowers adopted a system known as 'planting in squares'. They dug holes about 45 cm^2 and about 20 cm deep in a regular pattern, as far as stumps and boulders would permit. Setts were then laid in the holes and covered with soil. They tried almost every conceivable distance between the holes. Archibald Liversidge, Professor of Chemistry at Sydney University and author of a report into agricultural practices in the Maryborough district in 1875, strongly condemned this practice, claiming excess water would pool and stagnate in these holes, thereby damaging the root system of the plants.[8] Moreover, Australian canegrowers using this technique planted as much cane as possible in any given area, irrespective of quality. This practice of over-planting in New South Wales resulted in as many as 1 000 stools of cane per hectare, whereas 600–800 stools per hectare were considered ample.[9] On other Australian properties with teams of horses, deeper furrows could be created with the plough. The setts were then placed in these furrows by teams of workers, usually Pacific Islanders in Queensland, and covered by soil (see Plate 7.1). Australian canegrowers also experimented with various distances between furrows, although no-one really knew which method contributed to the best yields or the optimum distance between furrows. Eventually, CSR conducted trials into the matter and showed conclusively in 1895 that planting in continuous rows gave better stands of cane, higher yields and reduced the expense of weeding.[10]

Some of Australia's earliest canegrowers carefully cultivated the paddocks as the young cane grew, trying to reduce the growth of weeds. At Malungavel plantation in 1869, the fields were kept in a 'remarkably clean state by means of stalwart ploughmen, who use ploughs, horse hoes and grubbers', while at Noyea plantation in 1874, a combination of ploughing with horse-drawn cultivators and hand hoeing kept down the weeds.[11] In

Plate 7.1. Hand planting of cane by Pacific Islanders in the Mackay district, c. 1870.

Source: Image No. 21683, John Oxley Library. Image reproduction courtesy of the State Library of Queensland, Brisbane.

most instances, however, cultivation of the paddocks as the cane grew was neglected. A journalist from *The Queenslander* described the cane fields around Maryborough in 1874 as 'a compact mass of weeds and grass, with a few stunted canes here and there. All attempts at ploughing or anything but scratching are ignored'. A reporter from *The Brisbane Courier* visited the Marburg district in 1883 and found the cane 'in some places overrun with weeds' and no attempt had been made to 'trash or otherwise attend to the cane since it had been planted'. After an inspection of Mourilyan plantation in 1891, the Union Bank of Australia Ltd.'s Inspector wrote that 'on some fields the edges have been kept clean, while in the interior the cane has been well nigh choked by the trash and weeds which have been allowed to accumulate around it'. He had also seen two fields in which he was 'really unable to decide which predominated – cane or jungle'. Reports from the Brunswick River district in 1900 described cane paddocks as being full of the weed 'stinking roger' (*Tagetes minuta*) standing as 'thick as a crop of wheat feet higher than the cane'.[12] The practice of over-planting in New South Wales meant canegrowers could not use horse-drawn cultivators to keep down weeds, and had to rely on the hoe. The colony's canegrowers frequently neglected such expensive and labourious work, leading to paddocks being full of weeds.

Considerable debate existed about the proper care that should be given to the growing cane plants. Some Australian canegrowers advocated trashing or stripping the cane of dead leaves during the hotter growing months. This activity removed any dead leaves, thereby exposing the matured part of the cane stalk to more sunlight which enhanced sucrose production. Other Australian sugar planters such as William Fryar (Loganholme plantation) and Samuel Grimes (Rockholme plantation) opposed the practice on the grounds it made no difference to the sugar yields.[13] Yet the small farmers in New South Wales, especially those contracted to CSR, generally trashed their cane once, because CSR paid a higher price per ton for trashed cane. However, the practice was often done solely to satisfy the mill-owner, without any thought given to its purpose. Writing for the *Queensland Review*, George Craig concluded that those Queensland canegrowers who trashed 'guessed at random' the proper time for completing the task.[14] Moreover, in New South Wales where cane took two years to mature, trashing the cane paddocks twice added considerable expense to a farmer's cultivation budget. Double trashing cost £4 per acre, and Australian canegrowers had difficulty finding workers to undertake what was considered a very unpleasant task. Some small cane farmers in New South Wales attempted to avoid trashing by burning their fields prior to harvesting, thereby saving on labour costs, but CSR refused to accept burnt cane.[15] Other New South Wales canegrowers stopped supplying CSR in preference to the small millers who generally accepted untrashed cane. Trashing remained a contentious issue throughout the 1880s, until CSR conducted experiments into the practice during 1887 and concluded that it was not necessary.[16]

Australian canegrowers often assessed soil fertility by the thickness of the vegetation, with jungle-covered alluvial soils close to rivers believed to be the most fertile.[17] Soil analysis to determine fertility was unheard of in Australian sugar-producing regions before 1880, except on Beenleigh plantation, where the owners had their soils scientifically tested in 1876 to determine its chemical deficiencies. The results guided them in the correct selection and application of fertilisers. A few other sugar planters also fertilised their crops. William Canny, owner of Iveragh plantation (Maryborough), used guano and lime on his fields; meatworks manure – a fertiliser made from meat processing products – was applied at Coongoon plantation; the field overseers at Fisherfield, Koorooroo and Clydesdale plantations added bagasse to their paddocks. Cattle penning was practised at Ferney plantation; the manure was mixed with bagasse before being adding to the fields.[18] Generally, however, Australia's earliest canegrowers cared little for their soil. They cropped the land until it showed signs of declining fertility, before taking any action to restore its natural fertility.[19]

Some Australian canegrowers failed to fertilise, no doubt believing the various reports about the inexhaustibility of the country's soil. Richard

Daintree, Agent-General for Queensland from 1871–1876, wrote that the alluvial soils of coastal Queensland covered with dense jungle were of 'robust description, and capable of repeated cropping with exhausting crops without manure'. Maurice Hume Black, Special Immigration Agent for the Queensland government during the mid-1890s, made similar claims in 1894 when he reported that land at Mackay was still giving 'good yields after being cropped for 20 years, without any artificial aid'.[20] Other Australian canegrowers just appeared to be lazy when it came to this most basic of all agricultural practices. In 1884, the Special Correspondent for the *Sydney Mail* reported that some New South Wales mill-owners had suggested to the farmers that the drays filled with harvested cane, once emptied at the sugar mill, could return to the farms loaded with bagasse. The canegrowers, however, were too indifferent to the matter even to do take away free material which could fertilise their farms.[21]

Design of the paddocks was often not given much thought. The owners of Helensvale and Noyea plantations left square belts of timber around the fields to protect the cane from damaging icy winds.[22] Their approach was exceptional. Shallow surface drains to reduce flooding of paddocks were built on a few plantations, mostly located in the low-lying parts of the Beenleigh district (e.g. Pimpama, Fisherfield, Yahwillpah and Benowa plantations). A more elaborate system operated on Ormiston plantation Diamond-shaped drains built from hardwood to depths of between 0.7 m and 1.8 m and filled with coral rubble criss-crossed the property.[23] Yet these planters were in the minority. Archibald Liversidge observed in 1875 that most sugar planters in the Maryborough district had no system of artificial drainage whatever upon their lands. The *Sydney Morning Herald's* survey of 105 canegrowers in New South Wales in 1870 found that the majority had no drainage, with ten having surface drains, and only two having underground drainage. In the Tweed River district during the late 1870s, CSR's Cane Inspector noted that many farmers planted cane in the hollows or on parts of their farms where the land was wet from want of drainage. They had made no attempt to establish drains to rid their properties of excess water. Dr Gustav Kottman, CSR's Inspecting Chemist during the 1880s and 1890s, claimed to have observed on his travels around the cane growing districts of New South Wales 'fields of cane which for weeks had been literally swamped; little had been done towards draining the low parts of the farms'.[24]

The criticism of the poor agricultural practices used by Australia's earliest canegrowers was harsh, especially in the context of virgin soil cultivation. Initial limited clearing of the fields can be explained by the need of the sugar planters to get a crop into the ground quickly in order to obtain some return on the large amount of capital being expended upon the formation of the estate. Moreover, clearing paddocks of all their stumps cost between £15–£20 per acre if done immediately after clearing the vegetation, compared to £5–£7 per acre after three or four years. In addition, Australian canegrowers faced labour shortages by the early 1880s, so it was not easy to secure workers to undertake such an arduous task. Even CSR with its huge financial backing did not immediately clear all the stumps from the paddocks during the formation phase of its Goondi plantation because of the cost associated with removing so much timber.[25] Another large expense in the 1880s – farm drainage – cost around £5-£10 per acre if canegrowers installed underground pipes. Such expense was beyond the means of most small farmers, and there was the added difficulty of the absence of main outlet drains to natural water-courses to which farmers could connect any surface drains.[26] Additionally, Australian canegrowers planted the first crops in virgin soil, which responded well to little or no cultivation.[27]

The prevalence of poor agricultural practices among Australia's earliest canegrowers may be attributed to the fact that few had farming backgrounds. James Knox, CSR's Inspector of Mills, observed in 1887 that most of CSR's contractors were 'never brought up as farmers', while Henry Roth claimed the majority of Queensland's pioneer canegrowers were not 'agriculturalists'. Dr Walter Maxwell on his visit to Queensland in 1899–1900 concluded that the colony's canegrowers who were farmers in other countries had experience that was no preparation for cultivating sugar cane, while for others 'land work with them is a very recent thing'. He believed there was 'not one farmer in their number who even pretends to know anything of the special principles of the science and practice upon which the continuous and economic production of cane must depend'.[28]

Scientific knowledge that could have improved agricultural practices for growing sugar cane was known by the 1860s. In the West Indies, some sugar planters applied chemical and natural fertilisers to the land to replenish its fertility, although sometimes indiscriminately due to the absence of proper soil analysis. Agricultural experts preferred planting setts horizontally rather than vertically in the soil. Scientists advocated the use of animal-drawn implements (e.g. ploughs, harrows and horse-hoes) for breaking up the soil thoroughly and to keep the paddocks free of weeds.[29] Such advice was available to Australian canegrowers, being printed regularly in contemporary newspapers and specialist journals such as *The Australian Sugar Planter* (Maryborough) and *The Sugar Journal and Tropical Cultivator* (Mackay). Publications on the practical cultivation of sugar cane by Thomas de Keating, John Hincliffe, Angus Mackay, Melmoth Hall and Frederick Bell were also available.[30]

Determining how many Australian canegrowers accessed this information, however, is difficult. In New South Wales, the *Sydney Morning Herald* tried surveying the colony's canegrowers in 1869 and 1870 to determine their cultivation practices, but concluded that 'many of them were illiterate'. Hence, the published material may not have been read by many Australian canegrowers. However, the meetings of the Wide Bay Farmers' and Planters' Association in 1868 and 1870 contained discussions about cane growing, while the Agricultural Society of Southern Queensland organised a conference in 1875 about sugar cane cultivation.[31] Therefore, some Queensland canegrowers had exposure to details about scientific cultivation methods. Nevertheless, promotion of this information did not occur via government agricultural extension programs, as the Departments of Agriculture in Queensland and New South Wales did not exist until 1887 and 1890, respectively. In addition, formal training for prospective farmers did not commence in Queensland until the establishment of the Gatton Agricultural College in 1896.[32]

The impact of a non-scientific approach to cane cultivation became apparent in some Australian sugar-producing districts by 1885. During the mid-1880s, CSR observed that continuous cropping, shallow ploughing and limited fertilising had greatly diminished soil fertility throughout the Clarence River district. Individual farm yields were often half what they were in the mid-1870s. Farm land that could have been purchased for £30–£35 per acre in the past now cost only £15 per acre. In the Maryborough district, a reporter for *The Queenslander* in 1885 concluded that under the present system of cultivation, soil exhaustion was occurring rapidly, and yields of two to three tonnes of sugar per acre a few years earlier had fallen to only 0.25 tonnes.[33] Poor cultivation methods lowered the potential yield and markedly increased the vulnerability of sugar cane to attack by disease and pests. Inappropriate agricultural practices adopted by Australia's pioneer canegrowers also contributed to the 'rust' and gumming disease outbreaks of the 1870s and 1890s, respectively (see discussion later in chapter).

Improving cultivation practices

CSR, the first organisation to actively promote the need for more scientific cultivation methods amongst Australia's canegrowers, decided in 1870 that its Macleay River contractors needed encouragement to improve their land and cultivation methods. Edward Knox Snr, CSR's Chairman of Directors, offered two prizes of £50 each for the best four hectares or more of cane, and two prizes of £25 each for the best two or four hectares of cane supplied to Darkwater Mill in the 1871 season. Three years later CSR announced it would distribute £1 000 worth of prizes, in sums varying from £5 to £50 each among its Clarence River contractors who supplied the firm's sugar mills with cane that had been carefully cultivated. CSR required its suppliers pay particular attention to fertilising, drainage, trashing and keeping the crop free of weeds. Farmers were to keep continuous records of these aspects, so CSR could make an assessment at the end of the crushing season.[34] Such incentives, however, did little to raise the overall level of cultivation among the bulk of CSR's New South Wales contractors. Edward W. Knox observed in 1876 that 'our efforts to obtain better cultivation of the cane are not meeting with much success; farmers are still not ploughing'. Nevertheless, CSR kept trying to improve the cultivation methods of its New South Wales contractors, lending money without interest to farmers who purchased pipes to drain their properties and encouraging the use of superphosphate by supplying it at very reasonable rates.[35]

After CSR commenced cane growing for its sugar mills during the 1880s, Edward W. Knox allowed no discrepancy between the farming methods practised on its plantations and the advice to its tenants and contractors. He counselled the mill manager at Condong Mill that 'if we set a bad example we can hardly expect our tenants who have not the same facilities for hiring men as we have to keep their crops in proper order'. Therefore, steam ploughs at Victoria and Homebush plantations ensured properly prepared paddocks for cultivation. Horse-drawn implements under the control of European ploughmen undertook the annual tillage on each estate. Drains at Victoria and Condong plantations removed excess water from the fields. After the dry seasons of 1885 and 1888, irrigation facilities were established at Victoria and Homebush plantations, ensuring crop survival during future periods of drought.[36] Ash from the mill furnaces, bagasse or molasses were used as fertiliser at Victoria and Homebush plantations. Trashing was practised at Goondi and Condong plantations, but CSR discovered, like its contractors, that labour shortages did not always mean the task could be done once, let alone twice.[37]

To improve their own agricultural practices and those of their contractors, CSR realised the organisation needed to keep up-to-date with the best overseas agricultural practices. In late 1887, Dr Gustav Kottman, CSR's inspecting chemist, travelled to Java and gathered information on Dutch sugar milling and cultivation practices. Kottman concluded that CSR's cultivation methods were 'much behind our Java competitors and though we cannot hope with our dear labour to bestow the same amount of care on cane fields, there is no doubt that closer attention to the fields will improve our cane supply'. Edward W. Knox responded to this report by instigating scientific investigations into growing sugar cane under Australian conditions. On two to four hectares set aside at each estate, CSR tested the effectiveness of green manuring and fallowing on restoring soil fertility and subsequent cane yields. Different chemical fertilisers, including the firm's own fertiliser manufactured at the Pyrmont Refinery since 1886, were despatched to each plantation with instructions to test them on different cane varieties, in order to determine the optimum rate of application and which varieties responded best to chemical fertilisers.[38]

Kottman acted as a clearing house for the new information. As he received results from the plantation trials, he compiled agricultural circulars and sent them to the estates with advice on best practices.[39] Because of these trials, green manuring using Mauritius beans, cowpea and lupin seeds had commenced at Homebush and Victoria plantations by 1891. These crops covered the land during the wet season, thereby reducing soil erosion, and when ploughed back into the fields returned essential nitrogen to the soil. By the turn of the century, a quarter of the land cultivated by CSR at its Victoria plantation annually was under green manure or else fallowed.[40] Subsoiling at all plantations after 1893 improved yields, following the earlier adoption of the procedure at Condong plantation. Further investigations into drainage, planting in squares versus planting in rows, fertilising and ploughing were carried out at Keith Hall estate, CSR's field experiment station in New South Wales. Edward W. Knox hoped CSR's methods at Keith Hall would 'show to farmers that clean and careful cultivation paid'. A booklet issued to CSR's New South Wales contractors in 1895 summarised the conclusions from the trials at Keith Hall. CSR encouraged its contractors to abandon planting in squares in favour of continuous planting in rows, as the latter method gave better yields. Canegrowers were asked to cease ploughing between plant cane in favour of using shallow cultivation with scarifiers, as this approach did not damage the roots of the young cane as much. Better drainage, subsoiling, green manuring and crop rotation were all promoted. CSR issued another booklet outlining the latest thoughts on cultivation methods to its Queensland contractors in 1914.[41]

CSR also undertook investigations into possible ways to reduce frost damage, realising that losses from frost caused New South Wales farmers to abandon growing sugar cane in favour of other activities, particularly dairying. Some CSR suppliers also attempted to reduce their exposure to the risk of frost by harvesting their cane annually. CSR objected to this practice, because in the milder climate of northern New South Wales most cane varieties did not mature enough in twelve months. Thus, during the winter months of 1895 and 1896, CSR and several Clarence River canegrowers under the firm's direction, assessed if surrounding the paddocks with very smoky fires reduced frost damage to the cane crops. Thermometers placed in the paddocks registered the nightly temperatures. This approach was based upon the premise that frost did not occur on cloudy nights, so if the paddocks could be surrounded by clouds of smoke the incidence of frost should decrease. The results, however, showed that temperatures were just as low at spots either a very short distance from the fires or where no fires were lit at all. Moreover, the cane on CSR's experimental farm near Harwood Mill was killed by the frost, despite being surrounded entirely by fires. Such results were not encouraging, and the practice never gained much acceptance in New South Wales. In 1901 and 1902, the Isis district canegrowers were troubled by severe frosts and CSR once more supported frost reduction trials using smoky fires. Again, the results from the trials indicated that this method would not reduce the incidence of frost. Interestingly, a decade later, a similar practice was trialed by a group of canegrowers in the Maroochy district. The results were not encouraging and the practice appears to have been abandoned.[42]

As CSR began settling farmers on its plantation lands, the firm realised that many of its tenants had little experience of tropical agriculture. Moreover, the organisation's future depended upon its sugar mills receiving expanded supplies of cane from these tenants. Hence, CSR tried to improve the farming techniques of its tenant farmers based upon the results from its own investigations. After 1895, CSR decided to forego rent on tenanted land under green manure at its Victoria, Homebush and Goondi plantations. In 1901 and 1902, CSR offered its Homebush plantation tenants a bonus of 10s per acre of cane cultivated if they practised green manuring, installed surface drains, and ploughed the paddocks appropriately. Walter Farquhar, CSR's Inspector of Mills, remarked in 1903 that this strategy at Homebush plantation had led to better drained fields and cowpea crops were becoming 'the rule rather than the exception now'.[43] CSR purchased fertilisers in bulk and allowed its tenant farmers to procure them at cost price, free of interest, with payment to the company coming from the proceeds of the fertilised crop. From 1897 onwards, CSR encouraged its tenants at Victoria plantation to pay more attention to drainage, by either supplying them with pipes which they installed or the firm drained the fields and added the cost to the rental or purchase price of the farms. After 1909, CSR even advanced up to £3 per acre free of interest to farmers in the

Tweed River district to assist in clearing stones so they could plough the land. During the 1900s, CSR began welcoming Italians as contractors for its North Queensland sugar mills. The Italians were favoured because the Macknade Mill Cane Inspector observed:

> There is no doubt that these Italians will grow more cane than the Britishers and succeed where the latter failed, for they live more cheaply, are exceedingly hard workers, keep the young cane thoroughly clean by both horse cultivation and hand weeding (which the British do not), are strong believers in artificial and green manures and procure more labour as they are usually three to four to a company.[44]

Other companies operating Australian sugar plantations during the 1880s also demonstrated more scientific cultivation techniques. With their larger financial resources, these organisations stumped their estates and employed steam ploughs to break up the soil thoroughly. Teams of European ploughmen using horse-drawn implements ploughed the paddocks and kept the growing cane free of weeds. These organisations fertilised their crops, using bagasse or other mill refuse. Soil analysis on Mourilyan plantation guided the owners in their choice of fertilisers to be applied on the property.[45] However, they did not universally practise trashing the cane. The Lower Burdekin district sugar planters claimed not to have the labour for trashing, so they fired their cane to clear the trash before harvesting.[46] Some Australian sugar planters undertook drainage on their properties, with the most elaborate system of channels in North Queensland reputedly existing on Victoria plantation (Mackay). The Mackay Sugar Co. spent an estimated £15 000 to construct and connect 80 km of smaller surface channels to larger surface drains encircling the property.

Irrigation was found to be essential on the Lower Burdekin sugar plantations because of the district's variable annual rainfall. In September 1883, George Drysdale, managing director at Pioneer plantation, advised a London partner in the sugar-producing venture that the estate had endured months of no rainfall, and the cane was dying. In desperation, he ordered the sinking of a well on the property, and the saw bench engine was re-assigned to pump water onto the withering cane. A month later, a visitor to the plantation observed that the paddocks were being irrigated with water obtained from fourteen wells scattered about the estate. However, the failure of the summer rains in early 1885 meant the Lower Burdekin sugar planters were again faced with dying crops by the middle of the year. George Drysdale instructed that the waste water from the sugar mill be diverted onto the cane paddocks, but he realised this measure was only a temporary solution to what was becoming a recurring problem. In November 1885, Pioneer plantation began to be irrigated with water sourced from Sheep Station Creek which ran through the property. By mid-1886, irrigation had been adopted on nearby Airdmillan, Kalamia and Seaforth plantations. The fields of cane were flooded with water pumped from the permanent lagoons on these properties.[47]

Some of the larger, Australian sugar-producing firms that survived into the 1890s adopted the system of scientific agriculture introduced by CSR. Meatworks manure replenished the soil at Nindaroo and Pioneer plantations, while the Farleigh plantation manager imported 610 tonnes of fertiliser (possibly guano) directly from Chile in 1892. By the turn of century, Young Bros. of Fairymead plantation practised crop rotation, with about three-fifths of the property being under cane each year, leaving the remainder covered with cow peas or fallow. They also allowed sheep to graze on the cow peas, assessing if animal manure benefited the soil.[48] In the mid-1890s, Gibson & Howes Ltd. of Bingera plantation carried out extensive underground drainage of low lying lands, making the pipes at the estate's pottery. The mill chemist tested different artificial fertilisers to see which benefited the soil the most, and analysed the estate's soils to determine where they were deficient, thereby ensuring correct fertiliser applications. In 1897, Gibson & Howes Ltd. employed steam ploughs at Bingera plantation to 'purge the ground of roots and stumps'. Soils at Bingera by 1900 were treated with lime and superphosphate made on the property and the firm had adopted green manuring by using cow peas. Soil fertility on Goodwood plantation was restored through the adoption of green manuring and the application of meatworks manure and the refuse from the property's piggery.[49]

The above advances, however, virtually ignored irrigation, except on the Lower Burdekin. By 1890, the Lower Burdekin district contained probably the largest irrigation system in Australia. Over 65 km of canals, some 2.5 m wide, spread throughout the district. Five pumping stations, delivering up to 18.1 million litres of water daily, were built at Kalamia plantation; nearby on Pioneer plantation another three pumping stations were operational. Enormous losses of water by soakage occurred in the canals because of the porous nature of the soil. Hence, the planters built galvanised iron fluming to assist in delivering the water to the main channels, thereby improving a pump's effective output (see Plate 7.2). Tenants on Pioneer and Kalamia sugar plantations took advantage of their landlords' irrigation facilities, which were substantially expanded throughout the 1890s.[50] Moreover, John Drysdale of Pioneer plantation realised during the mid-1880s that plenty of underground

Plate 7.2. Part of a wood engraving showing the irrigation facilities of Pioneer plantation, near Ayr, 1889. Note the main irrigation channel and an example of galvanised iron fluming.

Source: *Australasian Sketcher*, 11 July 1889. Image reproduced from the La Trobe Collection, courtesy of the State Library of Victoria, Melbourne.

water existed throughout the Lower Burdekin district. Iron pipes inserted into the ground, however, quickly clogged with sand, making it difficult to maintain a flow of water. Drysdale, in conjunction with Henry Braby, the plantation's consulting engineer, succeeded in tapping this underground supply by the development of the spear system. In early 1886, they sank iron pipes with one end perforated by holes and covered with wire, while inside the pipe was an inner lining of steel or copper gauze to exclude the finer gravel. Subsequently, multiple spears became the basis of independent pumping units that allowed irrigation to be adopted by the Lower Burdekin canegrowers. By 1914, 145 irrigators, overwhelmingly small canegrowers, operated on the Lower Burdekin, accounting for 65% of the total area irrigated in Queensland.[51]

Elsewhere, a series of dry seasons at the turn of the century forced other Australian sugar-producing companies to devise ways to irrigate their properties. The Australian Estates and Mortgage Co. Ltd. erected a pumping station on the banks of Pioneer River in order to provide water for its Palms plantation. Bores and spears tapped the underground water supplies on Fairymead and Qunaba plantations (Bundaberg district). In contrast, the Millaquin and Yengarie Sugar Co. irrigated its fields with refuse water from the Millaquin Refinery. Gibson & Howes Ltd., who already irrigated the sugar cane on Bingera plantation, spent another £30 000 on expanding their irrigation facilities. Owing to the prevalence of floods in the Burnett River, their plant had to be above flood reach. Consequently, the water passed through a tunnel in the river bank to the main shaft. Pumps then forced the water through 457 m of 76 cm steel piping up the river bank into a reservoir, 61 m above the river level. Coal, delivered by rail to coal hoppers, fuelled boilers for the four pumping station engines that lifted the water to the reservoir. The sugar planters estimated the annual operational cost of irrigation facilities at £3–£5 per acre, so throughout the 1900s and 1910s irrigation remained out of the reach for most small farmers, except in the Lower Burdekin district.[52]

As some Australian sugar planters ceased growing sugar cane, they became proprietary central sugar mill-owners. Nevertheless, these firms still assisted their tenant farmers and other suppliers to improve cultivation methods. Drysdale Bros & Co. provided fertilisers to their suppliers and allowed them to use the fertiliser-distributing machines free of charge. The farmers paid for the fertilisers after the harvest. Gibson & Howes Ltd. of Bingera plantation rented their steam plough on reasonable terms to their tenant farmers, ensuring the land was thoroughly broken up on the rented farms.[53] The central sugar milling companies formed in Queensland during the late 1890s also encouraged their suppliers to adopt more scientific cultivation practices. To promote fertilising, management at Mulgrave, Nerang and Moreton Mills arranged for their contractors to have abundant supplies of lime, artificial fertilisers and mill by-products. Payment was not required until the cane from the fertilised areas had been crushed at the sugar mills. During the mid-1890s, the Mt Bauple Central Mill Co. offered its farmers use of the factory's horses in the non-crushing season to facilitate greater ploughing of the fields. They could even rent up-to-date implements such as disc ploughs from the company. A similar scheme operated at Mossman Mill during 1911. In 1910, the Moreton Central Sugar Mill Co. Ltd. tried to stimulate good cultivation practices by offering five prizes of £100 to the farmers producing the best cultivated crops of cane.[54]

Forming the Queensland Bureau of Sugar Experiment Stations

By 1900, CSR and other Australian sugar milling companies had shown that scientific investigations improved cultivation methods. Governments in New South Wales and Queensland, however, were slower to become involved in promoting such investigations. In 1893, the New South Wales government (Dibbs; Protectionist) established the Wollongbar Experiment Farm, near Lismore. Subsequently, the main trials there during the 1890s concentrated on determining the suitability of different sugar cane varieties to the frost-prone environment of northern New South Wales and their resistance to gumming disease (see discussion later in the chapter), while leaving research on cultivation methods to CSR. State Nurseries were established at Mackay and Kamerunga, near Cairns, by the Queensland government (Morehead; Conservative) in 1889. Trials to determine the disease-resistance and sucrose-richness of cane varieties dominated the research program at these nurseries. Such activities, although appreciated by Queensland canegrowers, did not fully satisfy their needs. They complained that the State Nurseries did not address the sugar industry's requirements for scientific investigations into soil types and the benefits of fertilisers and irrigation. The colony's canegrowers pressured the Queensland government to establish sugar experiment stations and laboratories similar to those established in Java (1888), Louisiana (1885) and Hawaii (1895).[55]

This pressure succeeded. A laboratory addition to the Mackay State Nursery in 1898 allowed experiments to determine the effect of various artificial fertilisers and green manuring on the growth of cane. This development, however, appeased only temporarily the Queensland canegrowers' requests for sugar experiment stations. Because of these continued demands, James Chataway, the Queensland Minister for Agriculture and Stock, announced in 1899 the conversion of the Mackay State Nursery into a Sugar Experiment Station. At the same time, the Bundaberg Planters' and Farmers'

Association persuaded the Queensland government to invite Dr Walter Maxwell, then Director of the Sugar Experiment Station of the Hawaiian Sugar Planter's Association and an agricultural scientist with an international reputation, to prepare a report upon the colony's sugar industry.[56]

Maxwell's report in January 1900 analysed the state of the soils in the colony's sugar-producing districts, causes of declining yields, extent of irrigation and the types of cultivation methods. He determined that continuous cropping and indifferent cultivation methods had greatly diminished soil fertility. Maxwell concluded that the colony's canegrowers needed advice and assistance in maintaining and restoring soil fertility and in the cultivation of crops of sugar cane. His main recommendations included the establishment of three additional sugar experiment stations: one in the neighbourhood of Mulgrave Mill, near Cairns; one at Mackay; and another at Bundaberg. A director aided by a staff of analytical chemists and assistant directors would have overall responsibility for the program of scientific research at these stations, and farmer education in each sugar-producing district.[57]

Based upon Maxwell's report, the Queensland government passed *The Sugar Experiment Stations Act* of 1900 and offered Maxwell the position of Director of the Queensland Bureau of Sugar Experiment Stations (hereafter BSES). Maxwell accepted and returned to Queensland in November 1900. He took over the existing laboratory at Mackay, and arranged for the erection of another laboratory at Bundaberg, which became his headquarters. He established substations on properties throughout the different sugar-producing districts and instigated experimental trials on a wide range of subjects under varying soil and climatic conditions. Financing the BSES came from a levy, not exceeding 1d per ton of cane delivered to the State's sugar mills, paid by the sugar mill-owners and canegrowers, in equal proportions. CSR, however, was outraged when asked to fund the BSES, complaining that improvements at its sugar mills and plantations resulted from their own expensive research program. Edward W. Knox predicted that the proposed investigations by the BSES would be of little benefit to CSR, yet the firm was required to contribute to the funding of these investigations.[58] In what appears as a deliberate retaliatory action, Knox banned Maxwell from gaining access to any figures about the operation of CSR's sugar mills and plantations, although the firm generally did not allow outsiders to view the results of their operations.[59]

Under Maxwell's direction, BSES staff at the substations, aided by local canegrowers, conducted experiments into cultivation methods, ways to restore soil fertility and the impact of irrigation upon yields. By 1907, the results from these experiments confirmed that yields from crops grown on ground subsoiled were a third higher than yields from land ploughed to a depth of 20 cm only, and that irrigation could increase yields by up to 15%. In addition, the BSES had analysed 8 140 soil samples from different cane growing districts by 1908, thus undertaking the first comprehensive assessment of soil fertility in Queensland.[60] Along with experimentation, Maxwell and his staff worked hard at disseminating their findings to the state's canegrowers. They travelled to Queensland's different sugar-producing districts, informing meetings of canegrowers about their results. Some examples will illustrate the type of advice offered by Maxwell. In September 1902, Maxwell advised canegrowers at Mossman to use more lime, as soil analyses indicated the district's soils were deficient in this nutrient. In 1903, he suggested to farmers at Proserpine that their soils needed more nitrogen. Canegrowers could easily obtain this element by burying their trash; the common practice of burning the trash led to the release of nitrogen into the air. During a visit to the Lower Burdekin in 1903, Maxwell assessed irrigation practices throughout the district. He found that damage to some fields on Pioneer plantation had been caused by the application of vast quantities of water. Maxwell advised Drysdale Bros. & Co. not to irrigate with so much water or as often.[61]

Some Queensland canegrowers, however, criticised Maxwell's emphasis on soil analysis. John Campbell, a local parliamentarian for the Morteon electorate, said during a speech in the Queensland Parliament that 'the practical farmers who use the central mills scorn his advice. They tell me they can do better without him'. George Muntz, Chairman of Directors of the Mossman Central Sugar Mill Co. Ltd., claimed to have observed Maxwell's methods and listened to his teaching, but thought he did not 'add to the knowledge of the cane growing community one iota'. Maxwell grew tired of the complaints and the heavy workload of running the BSES and overseeing the operation of the state's central sugar mills, a task he assumed in 1904. He eventually left Queensland in 1909. Ernst Scriven, his successor at the BSES and Under-Secretary for the Queensland Department of Agriculture and Stock, decided that greater interaction with canegrowers would lead to a better appreciation of their problems. In 1912, he appointed field instructors, who commenced moving from farm to farm, giving advice to canegrowers about cultivation methods and the best cane varieties to be grown in their district. Despite this change, the general emphasis in Maxwell's program – soil analyses, improving soil fertility and green manuring – continued into the early 1910s.[62]

Major changes in the farm implements used to cultivate sugar cane did not accompany the improvements in field preparation and care for soil fertility that occurred in Australia after 1880. The plough had replaced the hoe on many cane growing properties, but most Australian canegrowers did not seriously entertain mechanised cane cultivation and cane harvesting before 1914. Two main reasons explain this lack of progress in substituting machinery for labour. Firstly, most Australian canegrowers had access to cheap, indentured workers until the deportation of the Pacific Islanders between 1906 and 1908, thereby reducing incentives to seek out mechanical replacements. Secondly, replacing the hoe with animal-drawn implements was far easier and cheaper than mechanising the cultivation and harvesting processes entirely. Nevertheless, some Australian canegrowers attempted to do so in the first decade of the twentieth century, but not always successfully.

Harvesting was the most laborious and bothersome feature of cane growing during the nineteenth century. Inventors found it difficult to design a machine that mechanised cane cutting. In March 1887, Thomas Tomlinson approached CSR with details about an apparatus that could mechanically harvest sugar cane. CSR was initially willing to advance Tomlinson money so he could complete the machine. After negotiations, however, CSR decided against the project, because Edward W. Knox concluded that 'Mr Tomlinson could not adequately explain how the cane stalks were to be cut off close to the ground or how the cane was to be topped'. CSR never again sought to trial mechanical harvesters until the late 1950s, despite being presented with numerous opportunities during the 1890s and 1900s when at least another twenty inventors took out patents on designs for cane harvesting machines.[63] In 1914, most of these early machines were too bulky and unwieldy. Australia's canegrowers were still forty-five years away from replacing canecutters with mechanical harvesters (see Chapter 16).

Australian inventors achieved more success in devising a mechanical cane planter to replace the usual method of hand planting, where dozens of workers walked behind the plough, filling the furrows with setts. In January 1898, a machine known as Crowe's cane planter was trialled in Bundaberg. Setts were hand fed through a hopper that delivered them into a furrow made by an attached plough. The arrangement of the mouldboards forced a sufficient quantity of soil to fall back into the furrow where it covered the setts. Pyrce Trevor, an inventor of a similar mechanical planter in 1900, claimed that the machines saved up to 10s per acre planted by considerably reducing the number of workers needed during the planting phase. CSR began systematic tests of planting machines in 1898, although it is not clear whose invention they were testing. By 1904, CSR had concluded that two men with a mechanical planter could plant 1.2 to 1.6 ha per day, whereas two men under the traditional system of hand planting could only plant 0.8 ha per day.[64] Improved versions of the cane planting machine were developed during the 1900s and early 1910s. These newer machines were built with a larger capacity (i.e. could hold more setts in their trays), and modified shutes, enabling a better delivery of the setts into the furrows. Even where hand planting was continued, the BSES observed in 1914 that the use of a cultivator with its teeth removed had replaced the old method of covering up the setts by hand.[65] Yet teams of horses continued to pull these new machines; draught power had still not been mechanised.

The adoption of the scientific approaches and new implements by Australian canegrowers during the 1890s and 1900s had more positive benefits than just making the fields of cane look cleaner and tidier. Yields improved, with the average tonnage per acre of sugar cane harvested in Queensland and New South Wales between 1897–1899 and 1910–1914 rising by about two tonnes in Queensland (15.4 to 17.4) and 6.6 tonnes in New South Wales (19.7 to 26.3). In addition, the average weight of sugar cane used to make a tonne of sugar fell by about a tonne during this period, although this reduction could be partially attributed to the adoption of the new, sucrose-rich cane varieties such as Badila, Goru and M. 1900 (see discussion in next section). A lack of surviving detailed financial records for Australian canegrowers from this period also makes it impossible to assess if they benefited by improved incomes. A century ago, it is even doubtful if many Australian small canegrowers could say that adopting the scientific cultivation methods had improved their profits. 'I am not able to state the profit I make on cane growing, but I am satisfied I am making a profit; there is no thorough system of bookkeeping', claimed one small cane grower – representative of many of his colleagues – when he gave evidence to the Royal Commission into the Australian Sugar Industry in 1912.[66] Indeed, the Central Sugar Cane Prices Board set up in 1915 to assess the prices to be paid for sugar cane each season in Queensland encountered difficulties during the late 1910s in determining the profit made by the state's canegrowers – a widespread lack of book keeping amongst the farmers hindered their calculations (see Chapter 17 for more details).

Searching for disease-resistant and sucrose-rich cane varieties

During the late 1860s and early 1870s, Australian canegrowers often grew numerous cane varieties, particularly as they were uncertain about which variety performed best in the different regional environments. In 1873, John Hinchcliffe, for example, had thirty-three cane varieties growing on his plantation fronting the Logan River.[67] Yet by 1870, some favourite cane varieties had emerged in the different sugar-producing districts. The China (or Tahitian) and Ribbon cane varieties appealed to many New South Wales canegrowers, for both varieties were considered frost-tolerant. In Southern Queensland, the Meera cane variety had gained some popularity, for it was relatively sucrose-rich and displayed an ability to survive extremes of weather. The Bourbon cane variety, however, was the most widely grown, especially in districts north of Maryborough. It was quick to mature, sucrose-rich, and tolerated poor soils, but was very susceptible to cane diseases and lodging.[68]

Bourbon's popularity in Australia ended abruptly in the mid-1870s, following what became known as the 'rust' disease outbreaks. Initially, when the cane was attacked by the 'rust' disease, the leaves went pale and became dotted with brown or reddish spots. Later, the stalks of cane rotted near the surface of the ground, and eventually toppled over due to the death of the plant's root system. Within two weeks of the disease appearing the cane could be dead. Cane that survived the disease yielded substantially less sucrose than crops not affected by the disease. 'Rust' disease outbreaks first appeared in Southern Queensland sugar-producing districts in 1871; canegrowers in New South Wales reported its occurrence by 1873. The crisis peaked during 1875 and 1876; outbreaks then petered out, with the last mention of the disease occurring at Grafton in 1881.[69]

The 'rust' disease affecting canegrowers in the 1870s was probably not caused by the rust pathogens *Puccinia melanocephala* H. & P. Syd (common rust) or *Puccinia kuehnii* Butler (orange rust). The first true rust outbreaks did not occur in Australia's sugar cane crops until 1978 (see Chapter 15). Instead, it was most likely that the disease affecting Queensland's sugar crops during the 1870s was caused by the fungus red rot, *Glomerella tucumanensis* (Speg) Arx and E.Mull. In August 1876, Dr Joseph Bancroft, a Brisbane medical practitioner with an interest in the problems affecting livestock and crops in Queensland during the 1870s, had identified the red fungus and white mites or acari on diseased cane. Bancroft concluded that the bite of the acari caused the growth of white blisters, which afterwards became red and shrunken. These were then surrounded by the fungus with black spores. In 1975, L. Jeppson *et al.* suggested that feeding injuries by the mite *Steneotarsonemus bancrofti* Michael had allowed the cane plants to be infected by the spores of the fungus red rot. The mite and fungus had been imported with the first cane cuttings brought to establish the Australian sugar industry. Uncontrolled movement of cane from one district to another had led to the mite's rapid spread and the very favourable conditions after an unusually wet summer season (as in early 1875) resulted in the mites reaching plague proportions.[70]

In the absence of substantial technical advice on how to combat the 'rust' outbreaks, Australian canegrowers tried various methods to eliminate the disease. Some Maryborough planters added lime and salt to their paddocks, but concluded that this strategy offered no relief from the disease. John Davidson of Alexandra plantation (Mackay) and Archibald Meston, manager of Pearlwell plantation (near Brisbane), both claimed to have produced disease-free crops of cane in 1875 after adding the ashes of burnt bagasse to their paddocks. Eliminating the Bourbon cane variety, however, became the commonest approach to stopping the disease outbreaks. By 1875, the canegrowers observed that Bourbon consistently succumbed to 'rust' and in desperation they replaced it with other cane varieties. Yet Bancroft suggested that this strategy was short-sighted and that Bourbon should be preserved in stock, for the disease could be defeated by producing disease-free setts by soaking them in a solution of 4 litres of water to 1 kg of quicklime. His view was supported by Karl Staiger, the Queensland Government Chemist, who argued that under a system of clean and careful cultivation, Bourbon had never been surpassed for weight of yield and that it could still be successfully grown in Queensland. Bourbon was reintroduced into Queensland at the Mackay BSES station in 1901, but it again developed a rust-like disease and was subsequently abandoned.[71]

The economic repercussions of the 'rust' outbreaks for the Australian sugar industry were severe, although concentrated over a few years. In 1875, the colony lost approximately £115 500 in external earnings from reduced sales of raw and refined sugar and rum. At least fourteen sugar mills closed between 1875 and 1877, including three of the four operational sugar mills in the Herbert River district. At Mackay, southern financiers dislodged the resident owners in ten of the eighteen plantations; three of the eleven plantations at Maryborough had also changed owners by 1880 because of 'rust'-related financial difficulties.[72] The McAlister government in Queensland was shocked by the 'rust'-induced setback to the colony's main agricultural industry, and ordered imports of new cane varieties for the colony's canegrowers. Angus Mackay, the Queensland Government's Commissioner at the Philadelphia Exhibition in 1876, dispatched a collection of cane varieties from mainland USA to Queensland, and Hawaiian cane varieties were imported by the Queensland government in 1877.[73]

Trials determining if these new, imported cane varieties were suited to the Queensland environment occurred at the New Farm, St Helena Island and Oxley experimental nurseries (see Figure 7.1), although the choice of locations is peculiar, given that most of the colony's sugar cane grew in districts north of Maryborough. By mid-1881, these trials had concluded that the Creole variety showed promise as a replacement for Bourbon, and the Queensland government began distributing cuttings of this variety to interested canegrowers. Despite government promotion, Creole never found favour with many canegrowers. Robert Boyd of Ripple Creek plantation, for example, tried Creole but concluded it was not worth planting as it yielded 'too light a crop'.[74]

By the mid-1880s, Australia's canegrowers had repeated the same mistake as their pioneering colleagues: relying too much on a single cane variety. The Rappoe cane variety (also known as Rose Bamboo or Light Cheribon) had become favoured by many canegrowers throughout all the sugar-producing districts of eastern Australia. It displayed no signs of attack from 'rust', survived extremes of weather, tolerated all kinds of soil and was sucrose-rich. Striped Singapore, another sucrose-rich cane variety, was also widely grown.[75] Other cane varieties had more localised appeal. Cheribon (or Striped Tana) was held in high repute by the sugar planters along the Mary River. Meera (also known as Black Cheribon or Moore's Purple) was cultivated predominantly throughout the sugar-producing districts north of Mackay, for it had similar characteristics as Rappoe. The Black Java cane variety (sometimes known as Black Tanna), with its high sucrose content and resistance to 'rust', became a staple variety of the Mackay sugar planters who cultivated the hillsides north of the Pioneer River.[76]

The folly of relying on so few cane varieties became apparent in 1892 when entire crops of cane began dying in northern New South Wales. Gumming disease (or gummosis) was identified as the cause of this crop failure and over the next fifteen years outbreaks occurred throughout the following sugar-producing districts: Bundaberg and Maryborough (1894–1896); Isis (1898–1904; 1908–1912); Mackay (1898–1904); and Herbert River (1903–1908). Gumming disease is caused by a bacterium destroying the stalk cells in sugar cane plants. The outstanding characteristic of the disease is the production of a yellowish, slimy gummy fluid that fills the cavities emerging from the death of the stalk cells. Harvesting the sugar cane before it dies from the disease results in reduced sucrose yields, and the gummy fluid interferes in sugar production. Crops of cane grown from diseased setts display poor growth and markedly reduced yields. CSR estimated that gumming disease reduced the yield of the

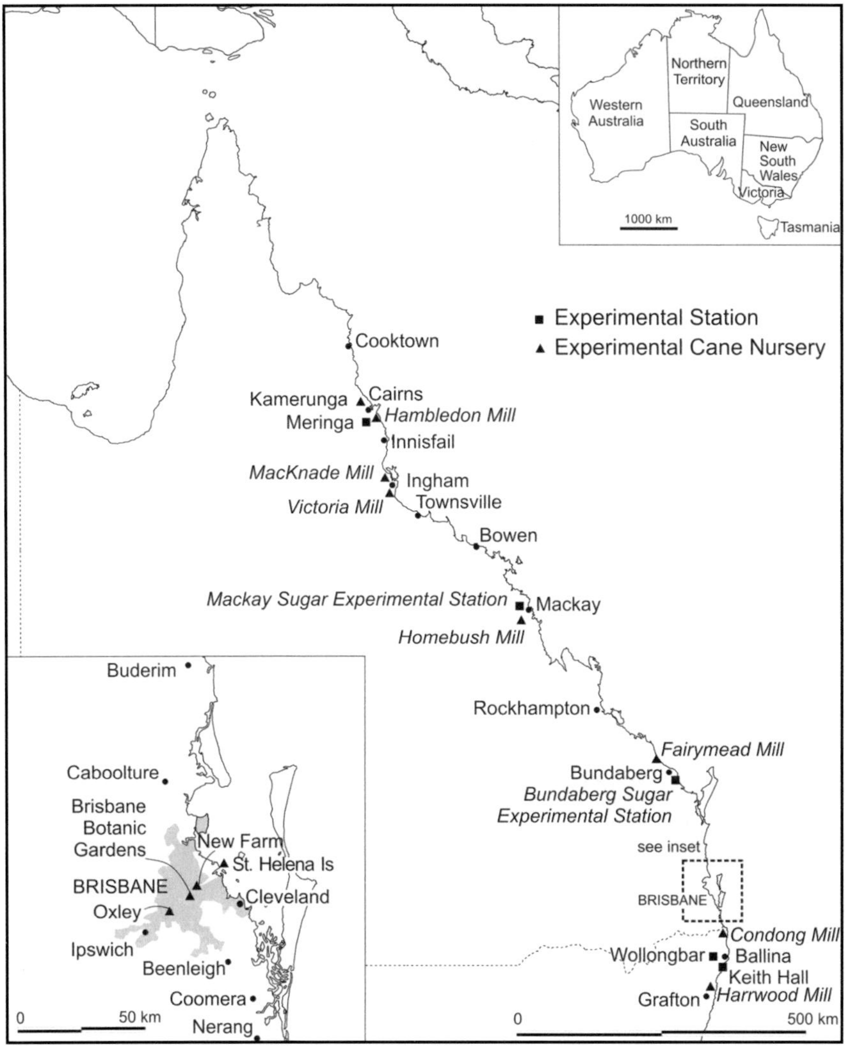

Figure 7.1. The location of experimental sugar cane nurseries and stations in Australia, 1870–1914.

Source: P. Griggs, 'Australian scientists, sugar cane growers and the search for new gummosis-resistant and sucrose-rich varieties of sugar cane 1890-1920', *Historical Records of Australian Science*, Vol. 14 (2000), p. 294.

crops supplied to Harwood Mill from 1893–1895 by about 40% in tonnage and 17% in sucrose content, and that at Broadwater Mill from 1893–1899 by about 30% and 9%, respectively.[77] Moreover, the disease struck the Rappoe and Striped Singapore cane varieties, the industry's mainstay.

Two strategies were developed to combat gumming disease. CSR's technical field staff quickly concluded that the impact of gumming disease could be reduced by better cultivation methods. They urged CSR's contractors to fertilise more regularly, improve the drainage upon their properties, reduce the number of ratoon crops and plant disease-free material.[78] Some cane-growers responded to this advice, contributing to the overall improvement in the cultivation methods throughout Australia's sugar-producing districts and the gradual elimination of the disease. The other strategy involved replacing gumming disease susceptible cane varieties by local, imported or specially bred hybrid cane varieties which were gumming disease resistant. Crucial to this research were the experimental cane nurseries and stations developed in both colonies, by either the colonial governments or sugar-producing organisations such as CSR and Young Bros. of Fairymead plantation (Figure 7.1).

Breeding hybrid cane varieties using cane seedlings had only commenced in Barbados in 1887, so the techniques were still in their infancy and had not been tried in Australia. However, as the gumming outbreaks progressed in New South Wales during the mid-1890s, C.N. Stephens, CSR's Harwood Mill Manager, Harwood Mill supplier Charles Lazar, and David Buchanan, overseer at the Mackay State Nursery, attempted unsuccessfully to raise cane seedlings.[79] Continued investigations into cane variety breeding during the late 1890s and early 1900s by CSR employee James Clark, first at Keith Hall Estate and then at the Hambledon Cane Nursery, eventually led to the successful raising of cane seedlings (see Plate 7.3). By 1902, James Clark had bred the sucrose-rich HQ. 426 (Clark's Seedling), Australia's first hybrid cane variety.[80] In the meantime, those officials, both public and private, who were concerned about the future of the country's sugar industry, relied upon the traditional approach to varietal selection. They arranged for the importation of new cane varieties and their testing to determine if any were sucrose-rich, gumming resistant and suited to local conditions.

At least 235 different cane varieties were imported into Australia between 1890 and 1905 (Table 7.1). Almost half were obtained by the Queensland Department of Agriculture and Stock. The process had commenced in December 1890, when the Queensland Department of Agriculture and Stock received a cane variety known as Kewensis from the Royal Botanic Gardens, Kew, as part of that organisation's hybrid cane variety distribution program to all the English-speaking sugar-producing colonies. The bulk of the new imported cane varieties, however, came from Ebenezer Cowley's and Henry Tryon's cane collecting expeditions to British New Guinea in 1893 and 1896, respectively.[81] The New South

Plate 7.3. Seedling cane trials at the Hambledon Cane Nursery, c. 1902.

Source: A. Birch and J. Blaxland, 'The Historical Background', in A.G. Lowndes (ed.), *South Pacific Enterprise. The Colonial Sugar Refining Company Limited* (Sydney, 1956), opposite p. 30. Image reproduction courtesy of CSR Ltd.

Wales Department of Agriculture shared the expenses associated with these expeditions in return for receiving duplicate sets of the new cane varieties, which were added to their collection of imported cane varieties from Jamaica and Lord Howe Island at the Wollongbar Experimental Farm.[82] Young Bros., stirred into action by the gumming outbreaks at Bundaberg in 1894, were the second largest importer of new cane varieties, obtaining two-thirds of their overseas varieties in 1895 (Table 7.1). In contrast, CSR relied on the Queensland and New South Wales Departments of Agriculture and the Queensland Acclimatisation Society to supply it with imported New Guinea and West Indian cane varieties. CSR, however, finally became an importer of new cane varieties in 1901, securing thirty-nine Mauritian varieties, known collectively as the '1900 Seedlings'. Edward W. Knox advised the Childers Mill Manager that the Mauritius seedling canes were 'being obtained at great expense, so take care with them'.[83]

Each organisation involved in the importation of new cane varieties favoured different source regions. Peter McLean, Under-Secretary for the Queensland Department of Agriculture and Stock, selected British New

Table 7. 1. The number of cane varieties imported into Queensland, 1890–1914, classified according to country of origin and importer.

Importer	Country of Origin	Year of importation and number of varieties
Queensland Department of Agriculture	England New Guinea South Africa Hawaii	1890 (1) 1893 (11); 1896 (72) 1900 (1) 1895 (1)
Queensland Bureau of Sugar Experiment Stations	Trinidad Mauritius Louisiana Hawaii New Guinea	1901 (1) 1902 (7); 1909 (9) 1902 (1); 1910 (4) 1906 (1) 1912 (117)
Young Bros., Fairymead plantation	Demerara Louisiana Barbados Java	1895 (32); 1898 (7); 1905 (1) 1895 (12) 1899 (12) 1912 (1)*
Queensland Acclimatisation Society	Barbados Demerara	1898 (6); 1908 (6) 1898 (4)
CSR	Mauritius Fiji New Guinea	1901 (38) 1909 (4); 1914 (2); 1915 (4) 1914 (130)
NSW Department of Agriculture	Jamaica Lord Howe Island	1893 (16) 1894 (1)

* This variety was imported by Fairymead Sugar Co. Ltd., the company floated by Young Bros.

Sources: H. Easterby, *Varieties of Sugar-Cane in Queensland.* BSES General Series Bulletin No. 2, 2nd edition, 1926, p. 13; *Q*, 7 July 1894, p. 31; *SJTC*, 15 November 1895, pp. 268–269.

Guinea, although the reason guiding this choice is unclear. Perhaps McLean concluded that the absence of a local commercial sugar industry in British New Guinea may have ensured that the cane varieties were not tainted by diseases brought from other cane growing countries. Young Bros. selected hybrid cane varieties from Louisiana, Demerara and Barbados. Precisely why these countries were chosen could not be determined, but Young Bros. did receive West Indian hybrid cane varieties from the Queensland Acclimatisation Society in 1895, so this exchange may have guided the firm's choice.[84] Certainly, the West Indies had become a major centre for breeding new cane varieties by the

1890s. CSR, on the other hand, looked to Mauritius. Again the precise reason for this choice of location could not be determined, but CSR may have been informed about the extensive hybrid cane variety breeding program established on the island during the 1890s.[85]

The approaches to determining whether any of the new, imported cane varieties were resistant to gumming disease and sucrose-rich varied amongst the field technical staff and agricultural scientists assigned the task. CSR initially sought to involve its canegrowers in the research process. Between 1892 and 1895, CSR officials distributed cuttings of the New Guinea cane varieties to their New South Wales suppliers. They were asked to observe the growing habits and disease-susceptibility of the new cane varieties under field conditions. This research, however, yielded little positive and had been abandoned by 1900. On the other hand, careful observations about the growing habits of the different cane varieties and their disease-susceptibility and sucrose-content were assembled at CSR's cane nurseries. A similarly careful approach was adopted by H. Jackson, the officer-in-charge at the Wollongbar Experimental Farm. He tabulated observations about the cropping habits, disease susceptibility and the sucrose content of each cane variety.[86]

In contrast, a disorganised and unscientific approach was initially followed by the Queensland Department of Agriculture and Stock. Under-Secretary McLean issued no instructions for testing the suitability of the imported cane varieties. Tryon, the scientist involved in the collection of the New Guinea cane varieties, was not involved in their assessment. Instead, this task was left to Ebenezer Cowley and David Buchanan, the overseers of the Kamerunga and Mackay State Nurseries, respectively. At both localities, the imported cane varieties were propagated, and within eighteen months setts were distributed to interested canegrowers. This practice continued throughout 1894 and 1895. Vague descriptions on the characteristics and performance of the new cane varieties were reported to McLean. Buchanan wrote in mid-1895 that Chenoma produced a 'heavy crop', while Arabora 'grows a great crop'. Anecdotal feedback from canegrower trials appeared in the Annual Reports of both nurseries, although this information was of little value.[87] This haphazard approach prompted an unidentified journalist from *The Queenslander* to write in October 1896 that the distribution of the imported cane varieties had been 'premature' and that some canegrowers 'had been put to the expense of planting and cultivating utterly worthless canes'. Eventually, during the late 1890s and early 1900s, Alexander Ramsay and then Harry Easterby, both trained chemists, systematically tested the characteristics of the imported cane varieties at the Mackay BSES Station.[88]

Results from these investigations showed that many of the imported cane varieties displayed inadequate growth after several years of cultivation under Australian conditions or died because of attacks from unidentified cane diseases. Yet some imported cane varieties had the potential to replace Rappoe. By November 1898, Jackson at Wollongbar had concluded that Mahona (N.G. 22) (sometimes spelt Mohona) was probably best suited to New South Wales conditions. Sucrose-rich and gumming disease resistant, Mahona was nevertheless a brittle cane variety, making it susceptible to wind damage and difficult to trash and harvest because of extensive lodging.[89] By 1900, Young Bros. had identified D. 1135 as the best of the imported cane varieties trialled on their estate. This cane variety appeared gumming disease resistant, very frost-tolerant and drought-resistant, making it ideally suited to the Bundaberg district, which suffered from both frost and periodic drought. Yet when CSR's scientists finally tested samples of D. 1135 in the mid-1900s, they expressed disappointment at its 'sweetness', but acknowledged that the cane variety resisted frost 'fairly well' and withstood damage from native rats and cane grubs.[90] The most promising discovery, however, had occurred at the Hambledon Cane Nursery. Here James Clark and Dr John Hastings Reed, CSR's Hambledon Mill Manager, had concluded by October 1900 that N.G. 15 (Badila) was the best of the imported cane varieties, being very sucrose-rich. They had also identified that N.G. 24 or Goru, Moore's Purple and Kewensis were sucrose-rich, although the latter variety was considered 'a weakly and tender cane, quite unfit for a standard variety'.[91] By 1905, these findings had been repeated at the Mackay BSES Station, where Easterby concluded that the most promising varieties were N.G. 15 (Badila), N.G. 24 (Goru), N.G. 24 A (Striped Goru) and N.G. 24 B (Green Goru).[92]

Badila was a dark purple to black coloured cane variety with very distinctive white bands across its stalks (see Plate 7.4). It was very erect growing, a good germinator and ratooner, and very resistant to gumming disease. Badila also matured quickly, often in around eight months, yet maintained its peak sucrose content for a protracted period. By the mid-1910s, however, canegrowers realised Badila was susceptible to damage from native rats and cane grubs, and performed best on moist, better class soils or well-fertilised soils. Badila also posed mill-owners a problem because its tough fibres hindered the smooth passage of the cane stalks through the first set of crushing rollers. Such a sucrose-rich cane variety could not be ignored because of an engineering difficulty, so the mill-owners altered their crushing trains by adding preparatory devices (i.e. revolving knives or shredders) to smash up Badila's tough fibres. In contrast, the North Queensland mill-owners favoured

Plate 7.4. A twelve month old crop of Badila cane in the Innisfail district, 1915. Note the distinctive banding on the thick stalks of cane.

Source: *ASJ*, 8 July 1915, photographic section, no pagination.

the Goru cane varieties, which were ground up easily by the crushing rollers. The Goru cane varieties varied in colour from a greenish brown to yellowish green. Each variety was sucrose-rich, gumming disease resistant, produced many stalks of cane per stool and favourable ratoon crops. Unlike Badila, they had a tendency to lodge easily.[93]

During the early 1900s, CSR organised farm trials to test if Badila and Goru were suitable replacements for Rappoe and Striped Singapore. Results from these trials indicated that Badila and Goru were best suited to cane growing districts north of Mackay, not the drier and colder Southern Queensland regions. By 1904, CSR was encouraging the cultivation of these cane varieties amongst its North Queensland suppliers. Four years later, BSES trials confirmed CSR's findings and BSES field staff began promoting the growing of Badila and Goru.[94] Throughout the early 1900s, CSR's staff also evaluated the Mauritian cane varieties imported in 1901.

By 1906, CSR's field officers had concluded that Malagache and M. 1900 were the two cane varieties that performed well under North Queensland conditions. Malagache was a purplish to black cane, a heavy stooler and good ratooner, but only moderately sucrose-rich. CSR's

Homebush suppliers first planted Malagache in 1906, and it was spread to other sugar-producing districts during a BSES cane distribution program throughout 1909.[95] M. 1900 was a stout, blue to black cane, a good germinator and ratooner, very erect growing and sucrose-rich, but suffered wind damage easily due to its soft, brittle stalks. Both CSR and the BSES distributed it to canegrowers during the late 1900s, with neither organisation having identified its high susceptibility to Fiji and gumming diseases.[96] During the 1900s and early 1910s, CSR also continued tests to determine if any other New Guinea cane varieties were suitable to be grown widely in northern New South Wales. These investigations revealed N.G. 14 to be erect growing, a good ratooner and stooler and sucrose-rich. CSR began recommending its cultivation in 1904.[97] By 1910, the cane varieties N.G. 16 (sometimes known as Garven's 16) and N.G. 23 were also identified as being good germinators and stoolers and sucrose-moderate, with N.G. 23 having the added advantage of being particularly frost-tolerant. In 1914, CSR commenced recommending that these two varieties should be cultivated.[98]

As the first gumming disease outbreaks occurred in New South Wales, some canegrowers switched immediately to the Tanna and China cane varieties, much to the chagrin of CSR. All were frost-resistant and showed no signs of gum, but were only poor or moderately sucrose-rich. In 1895, CSR instructed its New South Wales mill managers to allow the planting of no more than 650 ha of the Tanna cane varieties in their respective districts, as the firm sought to identify more sucrose-rich cane varieties that were gumming disease resistant.[99] By 1902, the new, imported cane varieties, particularly the sucrose-rich Mahona, were hardly grown at all in New South Wales. In Queensland, lack of any varietal data from most sugar-producing districts makes it difficult to assess the change in the cane varieties cultivated during the late 1890s. Data for CSR's mills indicate that Rappoe and Striped Singapore remained the favourite cane varieties, except at Homebush Mill where the canegrowers tried the New Guinea cane variety Chenoma. Most CSR mills, however, were in North Queensland where gumming disease did not strike until 1903, so the conclusions drawn from this data should not automatically be applied to Southern Queensland, where gumming disease had occurred in the late 1890s.[100]

Rappoe and Striped Singapore's importance in the North Queensland sugar-producing districts, however, declined rapidly during the mid-1900s. The sucrose-rich and gumming disease resistant cane varieties such as Goru, Badila and B. 208 were cultivated more widely, following their promotion to the canegrowers in at least the localities where CSR operated and the

Table 7.2. The estimated percentage of various cane varieties in a selection of Queensland's sugar-producing districts, 1912.

	Cane Varieties						
District	Badila	Goru varieties	Malagache	Rappoe	D.1135	B.147	Otamite & Cheribon
Mossman	17	50			20	10	
Cairns	60	37					
Lower Burdekin		86*		14			
Proserpine	10	30	10	10			5
Mackay							60
Bundaberg					90		

* The figures also include the small amount of Badila grown in this district.

Source: Based upon statistics in BSES, AR 1912–1913, *QPP*, Vol. 2 (1913), pp. 585–587.

appearance of gumming disease in the Herbert River district in 1903.[101] By 1912, the Goru cane varieties were most popular in the Proserpine, Lower Burdekin and Mossman districts (see Table 7.2). They still attracted a considerable following amongst the Cairns district canegrowers, although Badila had now become the preferred cane variety in this locality. Some Herbert River district canegrowers cultivated the Goru cane varieties, but here farmers were also attracted to Badila and B. 208, although the latter cane variety was rapidly abandoned after 1912, due to its susceptibility to a newly identified disease called yellow stripe (known now as mosaic). The popularity of the Goru cane varieties throughout North Queensland waned during the late 1910s, as canegrowers switched to Badila and HQ. 426. They retained their popularity longest in the Lower Burdekin district, still making up 76% of the crop in 1917. Yet two years later, the amount of Goru grown had halved, its place also taken by Badila.[102]

A more complicated varietal pattern existed throughout the Mackay district during the late 1900s and early 1910s. Gumming disease outbreaks between 1898 and 1904 had resulted in some canegrowers switching to the gumming-resistant Cheribon and Otamite cane varieties, which comprised 60% of the annual harvest in 1912 (Table 7.2). Both cane varieties, however, were sucrose-poor, but most local sugar mill-owners, except CSR, paid for cane on the basis of weight, not sugar content.[103] CSR had refused to allow its Homebush Mill suppliers to cultivate Cheribon and Otamite, so by 1907 some had switched to the Goru cane varieties. Yet five years later Malagache had ousted the Goru cane varieties. In 1910, Walter Farquhar, CSR's Inspector of Mills, advised the Homebush Mill Manager that this trend should be reversed, as Malagache was sucrose-poor compared to the Goru cane varieties. This advice was relayed to farmers, but went unheeded.[104] Malagache was only ousted in the Mackay district during the late 1910s and early 1920s, following the introduction of regulations which paid canegrowers on the basis of the sucrose-content of their cane (see Chapter 17 for more details).

Canegrowers in the sugar-producing districts south of Mackay favoured different cane varieties to their northern counterparts. In Southern Queensland, the canegrowers resisted cultivating Badila and Goru, claiming they were unsuited to the colder climates. By 1912, D. 1135 had become the most important cane variety in the Bundaberg district (see Table 7.2). In addition, the BSES listed D. 1135 as occurring in the Maryborough-Bauple, Maroochy, Beenleigh and Marburg districts in 1912, although provided no indication as to its importance throughout these areas. Yet by the mid-1910s, CSR drew its suppliers' attention to the superiority of M. 1900 over D. 1135 in sugar content.[105] Some Isis district canegrowers had switched to this cane variety by 1914. In New South Wales, canegrowers were slower to take up the newer gumming-resistant and sucrose-rich cane varieties, despite CSR's urging. To rectify this situation, CSR announced in October 1904 that it would not sanction any further plantings of the China cane variety because of its low sucrose-content.[106] As a result, some suppliers to Harwood and Broadwater Mills grew N.G. 14 or Mahona (N.G. 22) by 1907. Others persisted in cultivating Rappoe, Striped Singapore or Malabar. Yet by 1914, Rappoe and Striped Singapore had been abandoned by most New South Wales canegrowers. Suppliers of Harwood Mill had switched to N.G. 14 and N.G. 16, while Mahona had attracted the favour of canegrowers in the Broadwater Mill supply area. Malabar, however, remained popular amongst the Condong Mill suppliers.[107]

By 1910, some Australian canegrowers had switched to the new gumming-resistant and sucrose-rich cane varieties. This trend, together with improved cultivation methods, ensured gumming disease was gradually brought under control in Australia's sugar-producing districts.[108] The new cane varieties saved canegrowers and millers from potentially substantial financial losses if gumming disease outbreaks had persisted during the 1900s and 1910s. Moreover, the new cane varieties provided financial gains to the millers, for they were also sucrose-rich. Hence, millers recovered more sugar from crushing the same amount of cane. In addition, the incomes of Australia's canegrowers benefited from these new cane varieties. Before 1900, Australian canegrowers were paid so many shillings per ton of cane. They aimed to produce heavy crops of cane, having little incentive to cultivate sucrose-rich cane varieties. W.H. Rothe, financial assistant to CSR's General Manager, Edward W. Knox, summed this situation up in 1886 when he observed that CSR's cane suppliers 'do not care as we pay the same price for good or bad canes'.[109] Yet during the 1900s this lack of incentive to cultivate sucrose-rich cane varieties changed as CSR and the government-controlled central sugar mills in Queensland began paying canegrowers by individual analysis (i.e. the sucrose-content of their cane (ccs), determined by chemical analysis). In addition, Mulgrave Mill offered its suppliers a bonus of 1s 6d above the usual price between 1907 and 1918 if they grew Badila.[110] Thus, for the first time real financial benefits accrued to some Australian canegrowers if they produced sucrose-rich cane.

The threat of insect pests

Insect pests troubled Australia's canegrowers shortly after the industry's establishment. Termites attacked the young cane crops along the Macleay River in 1870. CSR advised its Macleay River suppliers to rid the paddocks of any tree stumps which provided food for the insects and to apply small amounts of kerosene to the paddocks. In the same year, caterpillars were reported to be 'boring the heart out of the cane' throughout the Maryborough district; the local canegrowers claimed to have no idea how to stop these depredations.[111] Swarms of grasshoppers appeared in the Herbert River district during 1882, 1883 and 1884, the Cairns district in 1884 and 1885, and throughout the Mackay, Bundaberg and Isis districts over 1903 and 1904. In the mid-1880s, teams of Pacific Islanders on the sugar plantations throughout the Herbert River district drove swarms of grasshoppers into deep, grass-filled trenches, which were set alight, killing the insects. In the Cairns district, Chinese collected the grasshoppers and put them into bags. These were dipped in boiling water to kill the insects. Throughout the Isis district, teams of men on horseback drove the grasshoppers into molasses-filled trenches where they perished. In addition, Alfred Cowley, the Manager of Hamleigh plantation, decided to boost the Herbert River district bird population, hoping they would bring the grasshoppers under control. In 1883, he arranged for a consignment of Indian myna birds (*Acridotheres tristis*) caught in Melbourne to be delivered to the Herbert River district. Most of the birds were released on the Herbert River sugar plantations, some in the Innisfail district and a small batch at Townsville.[112] Today, the descendants of these birds have become a major avian pest in Queensland.

The most troublesome and destructive insect pest, however, was the cane grub or the larvae of the cane beetle. Entomologists have identified nineteen species of native cane grubs which are known to attack sugar cane in eastern Australia. The greyback cane beetle, *Dermolepida albohirtum* Waterhouse, is considered the most damaging of all Australian cane beetles. It occurs in all sugar-producing districts between Mossman and Mackay, although is less common around Mackay. The Frenchi cane beetle, *Lepidiota frenchi* Blackburn, is also particularly widespread, being found in all sugar-producing areas south to Bundaberg. In contrast, the Childers cane beetle, *Antitrogus parvulus* Britton, and the Bundaberg cane beetle, *Lepidiota crinite* Brenske, are more restricted in their distribution and are capable of causing localised damage in the Bundaberg and Isis districts where they are most prevalent.[113] All species of native beetles affecting the cane crops, however, complete a life-cycle that begins when the female lays her eggs in the soil, proceeds through a grub stage before the grubs pupate and transform into beetles (see Plate 7.5).

This life-cycle commences when adult cane beetles are first seen at dusk, generally between October and December. Their emergence is governed chiefly by a period of soaking rain, which moistens the soil enough to allow them to develop from their pupal phase and burrow upwards to escape from the surface. On emergence they fly to the adjacent forest or cane paddocks where they feed upon the foliage. Edmund Jarvis, one of Queensland's earliest entomologists, provided a description of such an event on a summer's evening in 1917:

> Standing among the cane stools, one seems to be encompassed by an immense swarm of beetles – thousands being in view at the one time – which, in their erratic and ill-directed flight, are constantly striking against the cane leaves, the clapping more produced by the sudden impact being plainly audible at a distance of several yards. In addition to this oft-repeated sound, the air, so still before, vibrates loudly with a continuous hum, due to the accumulated buzzing of countless numbers of these insects.[114]

Within a month of the adults emerging and feeding on foliage, the female cane beetle re-enters the soil and deposits her eggs. Under moist soil conditions, incubation proceeds and within a month the small grubs will hatch. The newly hatched grubs live at first by ingesting large quantities of soil and some organic matter. During January-February, these small grubs will enter the second stage of their life-cycle. They will feed on organic matter and some

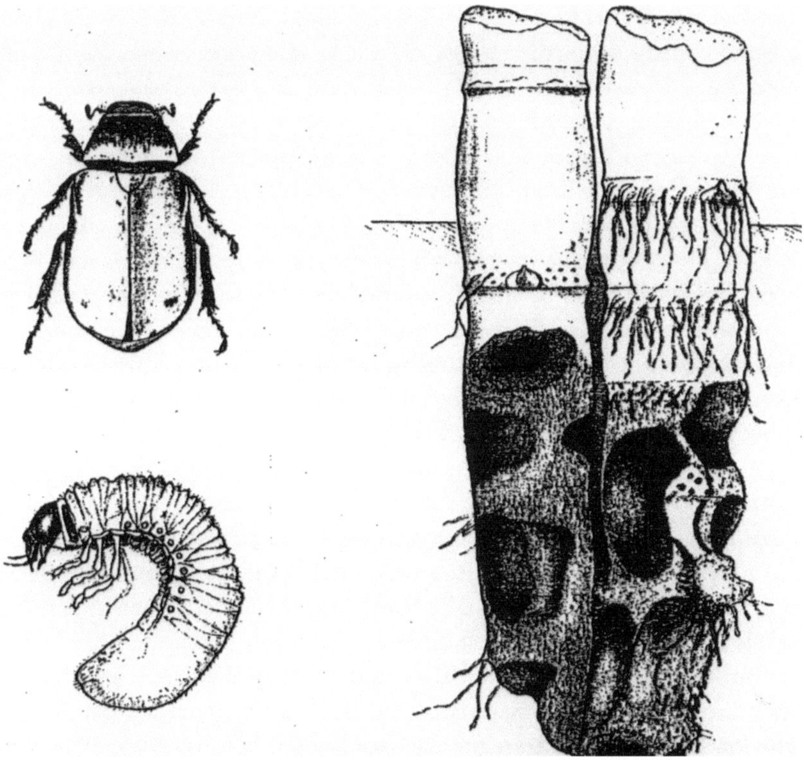

Plate 7.5. Part of Edmund Jarvis' drawing of the metamorphosis of the greyback cane beetle (Dermolepida albohirtum *Waterhouse) and the damage caused to the roots of stalks of sugar cane, 1916.*

Source: E. Jarvis, *On the Value of Poison Bait for Controlling Cane Grubs*. BSES, Division of Entomology, Bulletin No. 4, 1916, p. 3. Image reproduction courtesy of BSES Limited.

cane roots and grow into medium-sized grubs. In those species with a one-year life cycle, the grubs enter their third and final growth phase (large-sized grubs) between February and May. In contrast, those species with a two-year life cycle (e.g. Childers cane grub; Frenchi cane grub) will spend about ten months in the first two stages. All species have a resting phase (i.e. pupal stage) when the large grubs burrow deep into the soil and transform into beetles.

The cane grubs attack and feed upon the roots of the cane (Plate 7.5), causing the leaves to become brown or yellow and withered. The growth of the cane is arrested, and eventually the cane plant dies. As the roots are consumed, the plant is susceptible to lodging during any windy weather. When stools of grub-affected cane are drawn out of the ground, they are easily removed without much tugging, and exhibit few roots. If cane is drawn out of the ground during the earlier stages of attack, many grubs may be discovered amongst the root mass. Grub-affected cane that is harvested will generally have reduced sugar content. Well-grown cane generally suffers damage from beetle species with a one-year life-cycle. Newly planted and smaller cane plants are more affected by beetle species with a two-year cycle because the grubs will feed during the winter months – the growing period for cane planted from May to July. Such cane may be killed outright, or recover from root loss and grow, but it will often be smaller in size than cane unaffected by grubs.[115]

Temporally, there is a clear association between high summer rainfall and severe grub attacks, a relationship initially determined by the nineteenth century entomologist, Henry Tryon.[116] Ameliorative actions taken by cane-growers (e.g. collecting cane beetles; soil fumigation) also reduced the severity and extent of the cane grub attacks. Canegrowers throughout the Cairns, Innisfail and Herbert River districts endured regular and some of the most severe attacks from the cane grubs. Canegrowers in these regions are particularly susceptible to grub attacks because these locations receive high summer rainfall that moistens the soil, making it easily penetrated by the beetles. Friable, red-volcanic soil that is very easily penetrated by the beetles is also found in parts of the Cairns and Innisfail districts. In addition, the cane farms in these regions are often situated near large areas of rainforest that provide abundant feed trees for the cane beetles.

The earliest record of severe damage by cane grubs occurred around Mackay in 1872. Throughout the 1880s, the cane crops in the Herbert River, Innisfail and Cairns districts suffered damage from cane grub attacks, but not on such a scale to cause general alarm amongst sugar producers. During the 1890s, however, extensive damage from cane grubs were reported around Mackay (1893–1895), throughout the Innisfail district (1895–1897) and in the Isis district (1897).[117] Severe outbreaks then ceased, probably a result of the combination of control methods employed during the 1890s and the prolonged drought conditions in most cane growing districts between 1899 and 1902. However, with the return of normal summer rainfall totals in the mid-1900s, numbers of cane beetles rebounded and considerable damage from cane grubs was again prevalent in the North Queensland sugar-producing districts between 1906 and 1911. A major outbreak also occurred in the Isis district in 1911.[118]

Australian canegrowers and mill-owners affected by the grub attacks became alarmed at the scale of damage and losses they suffered in the 1890s and 1900s. Henry Tryon calculated that cane grub attacks around Cairns during the mid-1900s had cost the industry between £40 000 and £50 000 annually. Edward Swayne – the parliamentarian for the Queensland electorate of Mirani – in his evidence to the 1916 parliamentary inquiry into whether to gazette a compulsory canegrower levy to combat the cane grubs, estimated that about 10% of the sugar cane throughout the Mackay district had been lost each year since the grub attacks in the mid-1890s.[119] North Queensland's canegrowers feared that the cane grubs could put them out of business unless the pest was controlled more effectively. For the Queensland government, this matter was of some importance as the State had promoted the idea that European farmers could make a living from cane farming in the tropics, as opposed to the plantation system relying on indentured workers. Any challenge that lessened the likelihood of this bold experiment succeeding led to institutional support for the small canegrowers. Thus, these losses were serious enough for individual sugar plantation owners, groups of canegrowers, and eventually the Queensland government, to devote considerable resources towards the destruction of the grubs via a variety of strategies.

Queensland's canegrowers had limited scientific expertise at their disposal when the cane grub attacks began causing crop losses during the mid-1880s. CSR employed analytical chemists in its sugar mills, but their foremost responsibility was ensuring the effective recovery of sugar during the milling process, not agricultural investigations. The closest person to an official colonial entomologist was the twenty-six year old Henry Tryon, who had been appointed to look after invertebrates at the Queensland Museum in September 1883. Tryon has been described as 'an erudite and versatile scientist, with a brilliant brain, a sarcastic tongue and a cantankerous nature'. His services were much in demand after the formation of the Queensland Department of Agriculture and Stock in 1887, although there is no indication that he conducted investigations into cane grubs anytime during the 1880s or early 1890s.[120]

In the absence of much technical expertise during the 1880s and early 1890s, canegrowers resorted to the hand collection of the beetles and grubs. Initially, individual sugar planters organised the hand collection of beetles, deploying their Pacific Islanders in this task. Walter Paget of Nindaroo plantation (Mackay), for example, claimed to have spent £400 in 1894 on the collection of beetles, paying 4s per pound of beetles. A party of forty men shook the beetles out of the trees on Nindaroo plantation. Each man killed about 1 200 beetles a day, so approximately 50 000 beetles were dispatched daily during the swarming season. In the mid-1890s, small groups of canegrowers in the Mackay and Innisfail districts voluntarily raised funds by levies on the amount of cane they harvested. These funds supported the efforts of local collectors, sometimes children keen to earn extra cash from beetle collecting. Numbers of beetles collected, if the reports were accurate, appeared quite extraordinary. At Habana and Nindaroo plantations, 7.1 tonnes of beetles were collected during 1894; another 9.6 tonnes of beetles were destroyed in 1895. Walter Farquhar, CSR's Inspector of Mills, estimated that 12 million beetles were killed in the Herbert River district alone during 1895.[121]

Grubs were harder to reach and needed to be brought to the soil's surface, where they could be scorched by the sun or eaten by flocks of birds, often ibises, which followed the ploughmen. This outcome could be achieved by more intensive ploughing of the fields. Management at Mourilyan plantation in 1889 decided upon this course of action, but had to first fully stump their land. By 1892, CSR followed the same strategy at its Goondi plantation, bringing extra land under the plough as quickly as possible.[122] Hand collection of the grubs also boosted the numbers destroyed after ploughing. At Pyramid plantation (near Cairns) in the mid-1880s, Chinese workers followed the plough, picking up the grubs; Pacific Islanders were assigned the task at Innisfail plantation. At Goondi plantation in 1891, field labourers filled a hole in the ground 1.8 m in diameter and 3 m deep (6 811 litres) with grubs collected when 12 ha were put under the plough. Boiling water then finished off the grubs. Five years later, 35 000 grubs were gathered by workers from half a hectare of ploughed land at the same plantation.[123]

Some canegrowers explored other avenues to reduce the impact of beetles and grubs. In the mid-1880s, the field overseers at Hambledon, Goondi, Mourilyan and Homebush plantations added lime to the paddocks, but this had negligible impact upon the severity of the grub attacks. At the Hummock plantation, near Bundaberg, William Farquhar applied a light dressing of salt to his fields in 1888, but this strategy did not reduce the impact of the grubs. Edward Long applied 'Paris Green' – an emerald green crystal powder consisting of a mixture of copper and arsenic – to the beetles at Habana plantation in 1894, but to his surprise found that it was not very effective in killing large numbers of the beetles.[124] CSR tried to artificially boost the numbers of insectivorous birds on its Goondi plantation during the mid-1890s by importing more Indian myna birds (source region not given) and crows from Mackay. This strategy achieved little; the birds were observed taking to the nearby hills and showed little interest in the beetles or grubs.[125] In 1897, the Innisfail district canegrowers began planting deep-rooting cane varieties such as Dupont, Black Fiji or Cheribon. The extensive root system of these cane varieties withstood the grub attacks better than Rappoe and Striped Singapore. CSR, however, allowed this measure to be used for three years only, because the Dupont cane variety had to be cut exactly when ripe, as before and after maturity the sucrose content was low.[126]

By the early 1890s, other canegrowers realised that paddocks of cane near vegetated areas were more liable to be damaged by grubs than fields surrounded by cleared land, and commenced ridding their properties of remaining vegetation, particularly the feeding trees such as the Moreton Bay Ash (*Eucalyptus tesselaris*), Bloodwood (*Eucalyptus corymbosa*), Tea Trees (*Melaleuca* spp.) and Scrub Figs (*Ficus* spp.) that were preferred by the beetles. Walter Paget claimed to have reduced the grub attacks at Nindaroo plantation by clearing 30 km^2 of headlands of 'beetles bushes' or the trees favoured by the cane beetles. In 1894, Edward W. Knox ordered that the swamps and watercourses on Goondi plantation be thoroughly cleaned of all undergrowth in order to reduce the threat from the beetles.[127]

The annual losses associated with the grub attacks during the 1880s eventually prompted CSR to commence research into ways of controlling the insect pest. In 1889, Dr Gustav Kottmann reported that the firm's chemists (unnamed) had tried killing grubs using kerosene and chloride of ammonium, but neither was very effective. Three years later, Kottmann reported that CSR had imported *Botrytis tenella* (later *Beauveria tenella* Vuill.), a parasitic fungus found in France. The firm's chemists at Homebush and Goondi Mills were asked to investigate if it killed grubs. Experiments with the fungus produced mixed results. Grubs on Goondi plantation succumbed to the fungus; it had no impact upon the grubs at Homebush plantation. This avenue of investigation was then abandoned.[128] The idea of biological control, however, re-surfaced in 1895 when CSR decided to bring moles to Queensland in the hope these burrowing animals would eat the grubs. The Queensland Department of Agriculture and Stock was not convinced that the introduction of these animals would be in the colony's best interests and refused to sanction their importation. The use of moles was raised again in 1902 by Major A.J. Boyd, a former sugar planter near Beenleigh, but his proposal to bring the creatures to Queensland was once more refused by the Queensland government.[129]

CSR's research into reducing the impact of grub attacks had not ceased during the early 1890s, despite the failure of the investigations into the parasitic fungus yielding useful results. In July 1891, Johannes Brünnich, a Swiss-trained chemist employed by CSR at Pyrmont Refinery, was transferred to Homebush Mill.[130] Here he began experimental work on ways to control cane grubs. Unfortunately, the inward correspondence from Homebush Mill to CSR's Head Office has not survived, so reports on these experiments are absent. From other accounts, we know that Brünnich during 1894 and 1895 had measured the mortality amongst cane grubs in paddocks impregnated with at least carbon bisulphide, salt and iguano – a fine earthy substance strongly impregnated with napthol. Brünnich concluded that iguano had 'no appreciable effect' on the numbers of cane grubs. Salt in an aqueous solution of about 10% applied to paddocks had no marked influence on the grubs, but affected the cane plants 'prejudicially'. However, carbon bisulphide injected into the soil killed the grubs. Brünnich never completed large-scale soil fumigation trials using carbon bisulphide, as he left CSR in March 1897 to accept the position of government agricultural chemist with the Queensland Department of Agriculture and Stock.[131]

Henry Tryon, who had been appointed the Queensland government entomologist in 1894, must have known about Brünnich's experiments with carbon bisulphide. In May 1896, he advised canegrowers that soil fumigation with carbon bisulphide killed the cane grubs. Within a month, Tryon's advice had been taken up by Edward Long, who arranged for carbon bisulphide to be applied to the paddocks at Habana plantation, although the success of this approach is not mentioned in the extant historical records.[132] Gauging the subsequent extent of the adoption of soil fumigation with carbon bisulphide after this initial trial is difficult due to the absence of any statistics on the topic. By the mid-1900s, a handful of Mossman canegrowers added carbon bisulphide to their fields in preference to beetle collecting, and found that it effectively reduced grub attacks. During the late 1900s, Young Bros., also adopted soil fumigation with carbon bisulphide, and concluded that it greatly reduced the impact of grub attacks on their Lynwood Estate, near Childers. Yet according to Edmund Jarvis, the use of carbon bisulphide to combat the cane grubs was 'not widespread' in 1921, so is reasonable to assume that the practice had not become widely adopted in 1914.[133]

Several reasons can be advanced to explain why Australian canegrowers resisted soil fumigation with carbon bisulphide. First, fumigating paddocks with carbon bisulphide during the 1900s was expensive, costing about £2 10s per acre for material and another £1 for labour. Second, carbon bisulphide is a highly flammable liquid, making it not ideally suited to an environment where the annual crop could be lost by an uncontrolled fire. Third, carbon bisulphide had to be injected into the soil manually using hand injectors, a particularly labourious and unpleasant task. Field workers could only treat about 0.13 ha daily, making it a slow process. Moreover, canegrowers had difficulty attracting workers who were prepared to complete this unpleasant task. Finally, Brünnich's failure to complete field trials on the procedure, and lack of follow-up investigations, even by CSR, meant

Queensland's canegrowers had no idea about the correct dosages for effective treatment or the best time to fumigate.[134]

The limited use of soil fumigation during the 1900s and early 1910s meant most Australian canegrowers relied on the hand collection of cane beetles and cane grubs as their main control methods. Voluntary beetle or pest destruction boards were formed throughout the Mossman, Cairns, Innisfail, Herbert River and Isis districts. Funds raised from levies on the amount of cane harvested and subsidies from the sugar mill-owners were matched by the Queensland government subsidy scheme, fixed at 50% of the total funds raised. These funds were used to pay the beetle and grub collectors. Interested parties in the Mackay district had, for example, collected 22 tonnes of beetles over the 1909–1910 swarming season. Further north, collectors in the Cairns district gathered 14 tonnes of beetles and 3 tonnes of grubs during the 1910–1911 swarming season; three years later they gathered another 23 tonnes of beetles during the collecting season.[135] Yet as early as 1897, James Cran, part-owner of Duncraggan plantation (near Bundaberg), suggested that the hand collection of grubs was a 'very expensive remedy and only fairly successful'. Dr Gustav Kottmann also expressed concern in early 1908 about the effectiveness of hand collecting beetles. Kottmann wondered if all the beetles caught and paid for by the pest destruction funds were actually injurious to the crops of cane. In order to make a more informed opinion about this matter, CSR commissioned Walter Froggatt, the New South Wales Government Entomologist, to examine specimens of beetles caught during the swarming season in CSR's North Queensland mill supply areas. Froggatt concluded that all the beetles examined had grubs that were destructive to cane roots. This evidence, together with the unwillingness of CSR's canegrowers to embrace soil fumigation, forced the company to continue supporting the hand collection of beetles and grubs.[136]

Queensland's canegrowers had hoped that the formation of the BSES would have led to investigations into ways to destroy the cane grubs more effectively, but an entomologist was not amongst the first staff hired by the BSES. Therefore, they continued relying on advice from Tryon, who could devote only limited time to this issue. In 1909, Henry Tryon initiated experiments into the possibility of the propagation and utilisation of the green soil fungus *Metarhizium anisopliae* Metschnikoff as a way to control the cane grubs. Tryon discovered that cane grubs would die when confined in vessels filled with soil contaminated by *M. anisopliae*. However, he abruptly abandoned this line of inquiry and never explained why such promising research should have been halted.[137] His efforts were not unappreciated by the state's canegrowers, but by the late 1900s they had grown tired of BSES inaction in this area and the ASPA contemplated hiring its own entomologist. Consequently, the BSES agreed in late 1911 to appoint American-trained Alexandre Girault as its first full-time entomologist. Over the next two and half years he focused upon describing the taxonomy and life-cycle of the cane beetles, much to the annoyance of the state's canegrowers. Girault provided no new clues to the control of the grub attacks.[138]

On the eve of the outbreak of World War 1, the agricultural scientists recommended that soil fumigation with carbon bisulphide reduced the impact of cane grub attacks, although canegrowers were reluctant to adopt this control measure, preferring to rely on the hand collection of the cane beetles and cane grubs. In contrast, Australia's canegrowers willingly adopted the new, sucrose-rich and gumming-resistant cane varieties identified by the agricultural scientists, thereby defeating a disease which had the potential to destroy the Australian sugar industry. Financial benefits flowed to the mill-owners, who secured more sucrose per tonne of cane crushed, and some canegrowers who were paid on the basis of the sucrose content in their cane. Agricultural scientists had also shown the country's canegrowers how to improve their cultivation methods and they had adopted elements of this expanding body of scientific knowledge on cultivation methods. A considerable increase in green manuring had taken place since 1880, while crop rotation and the application of artificial and organic fertilisers had been taken up by some Australian canegrowers. Improved types of implements were now more widely used. Mechanical cane planters were gaining favour, harrows saved labour in some northern cane growing districts and lime and fertiliser were spread mechanically by modified cane planters.[139] Ernst Scriven, the BSES Director, concluded in 1914 that overall sugar cultivation practices in Queensland were generally on the 'upgrade'.[140]

Notes

1 H. L. Roth, *A Report on the Sugar Industry in Queensland* (Mackay, 1880), p. 58; G. Craig, 'The sugar industry in Australia', *Queensland Review*, August 1885, p. 59; J. Despeissis, 'Sugar industry in north coast district', *Agricultural Gazette of NSW*, Vol. 2 (1891), p. 36; *NS*, 29 November 1884, p. 7; W. Campbell, *Extracts Reports on Certain Agricultural Districts of New South Wales* (Sydney, 1888), p. 3; W. Maxwell, *Report upon an Investigation into the Condition of the Sugar Industry in Queensland* (Brisbane, 1900), p. 11.

2 For examples see *ATCJ*, 28 Januray 1871, p.110; *Q*, 15 February 1871, p. 5; *BC*, 24 July 1883, p. 3; P. Clarke, *The 'New Chum' in Australia, or the Scenery, Life and Manners of Australians in Town and Country* (London, 1886), p. 279; T. Dexter, 'Memoirs of Bingera', p. 2, 6pp typescript, date unknown, Gibson & Howes Mss, F1085C, Fryer Library & S. May (ed.), 'Diary of Arthur Neame, 1870–1897', p. 41, JOL.

3 A. Liversidge, 'Report of the sugar cane disease in the Mary River district, Queensland', *J. of the Agricultural Society of NSW*, Vol. 2, no. 8 (1876), p. 128; *QVP*, Vol. 4 (1889), pp. 177, 388 & 389.

4 *The Australasian*, 5 August 1882, p. 183; *Tweed Herald and Brunswick Chronicle*, 5 January 1900, p. 9.

5 *ATCJ*, 28 January 1871, p. 110; *BC*, 4 November 1885, p. 5; May, 'Diary of Arthur Neame', p. 41; Roth, *A Report on the Sugar Industry in Queensland*, p. 58; Campbell, *Extracts from Reports on Certain Agricultural Districts*, p. 5.

6 For some examples see Q, 9 August 1873, p. 6 (Koorooroo plantation); *Q*, 14 December 1876, p. 3 (Iveragh plantation); Q, 25 March 1876, p. 21 (Rice's plantation); and *MM*, 2 June 1877, p. 3 (Pioneer plantation); *Weekly Herald and Queensland Agricultural, Mining and Pastoral Advertiser*, 22 July 1865, p. 13.

7 Campbell, *Extracts from Reports on Certain Agricultural Districts*, pp. 10 & 29; *RRE*, 9 October 1880, p. 2; *Q*, 8 November 1884, p. 767.

8 *MC*, 8 October 1864, p. 2; *Slater's Queensland Almanac*, 1870, p. 23; *Sydney Morning Herald*, 19 September 1884, p. 13; *Q*, 26 May 1883, p. 831; A. Liversidge, 'Report of the sugar cane disease in the Mary River district, Queensland', *J. of the Agricultural Society of NSW*, New Series, Vol. 2, no. 9 (1876), p. 144.

9 *CRE*, 18 June 1881, p. 3; *CRE*, 21 July 1883, p. 6.

10 For descriptions of hand planting cane see J. Strachan, 'Sugar cultivation in Queensland', *Transactions of the Philosophical Society of Queensland*, 1859–1872 (no pagination; reprinted from the *Queensland Daily Guardian*, 30 March 1867) & *Bundaberg Mail*, 22 November 1895, p. 3.

11 *Queensland Express*, 9 January 1869, p. 3; *Q*, 7 March 1874, p. 10.

12 *Q*, 31 January 1874, p. 3; *BC*, 9 July 1883, p. 5; *Tweed Herald and Brunswick Chronicle*, 5 January 1900, p. 9; G. Nankivell, Branch Inspector, 'Report upon Mourilyan Estate, 14 October 1891,'pp. 6–7. Attached to a letter from Gen. Mang., UB, Melb., to Gen. Mang., UB, London, 31 October 1891, Letter No. 845, U 103/40, UBR, ANZA.

13 For views supporting trashing see *Q*, 9 August 1873, p. 5 & 19 September 1873, p. 3 & *Sydney Mail*, 5 August 1871, p. 708; for opposition to trashing see *Manning River News*, 12 October 1867, p. 1; *Q*, 2 August 1873, p. 6 & *QVP*, Vol. 4 (1889), p. 375 (Ev. S. Grimes, Qs. 9437–9439).

14 Despeissis, 'Sugar industry in the north coast district', p. 40; Craig, 'The sugar industry in Australia', p. 60; for CSR paying more for thrashed cane see *QVP*, Vol. 4 (1889), p. 342 (Ev. E.W. Knox, Qs. 8228–8229) & E. W. Knox, AR 1874 season, p. 23, CSRR, Z/109, Box 374.

15 A. Meston, 'Report on the Sugar Industry on the Clarence and Richmond Rivers', *QVP*, Vol. 2(1882), p. 922; for farmers in New South Wales burning their cane see *CRE*, 18 June 1881, p. 3; Lyne, *The Industries of New South Wales*, p. 258; & Despeissis, 'Sugar industry in the North Coast District,' p. 40; for CSR's refusal to accept burnt cane see E. W. Knox, AR 1874 season, p. 25, CSRR, Z/109, Box 374.

16 E. W. Knox, AR 1876 season, p. 22, CSRR, Z/109, Box 374; *The Sydney Mail*, 27 December 1884, p. 1306; G. Kottmann, 'Circular to All Mills: Trashing Cane, 4 February 1887', CSRR, Z/303, File NS 306.

17 For examples of this view see *Slater's Queensland Almanac*, 1869, p. 22 & R. Daintree, *Queensland, Australia. Its Territory, Climate and Products etc. etc.* (London, 1873), p. 20.

18 *Q*, 26 August 1876, p. 8 (Beenleigh); *Q*, 14 December 1876, p. 3 (Iveragh); *Q*, 15 April 1876, p. 21 (Clydesdale); *Q*, 29 January 1876, p. 6 (Coongoon); *Q*, 9 August 1873, p. 6 (Fisherfield); Q, 9 August 1873, p. 6 (Koorooroo); *Sydney Mail*, 8 November 1884, p. 947 (Ferney).

19 For comments about the absence of fertilising and fallowing see: *Q*, 25 June 1870, p. 5; *Sydney Mail*, 28 October 1871, p. 114; Roth, *A Report on the Sugar Industry in Queensland*, p. 59; Despeissis, 'Sugar industry in the north coast district', p. 38, *NSWVP*, Vol. 6 (1889), Report of the Inquiry into the Grafton to Tweed Railway, p. 25 (Ev. C. Brady, farmer, Q. 957); & J. Knox, AR 1882, p. 7, CSRR, Z/109, Box 374.

20 R. Daintree, *Queensland, Australia. Its Territory, Climate and Products Agricultural, Pastoral and Mineral etc. etc.*, p. 20; M.H. Black, *The Sugar Industry of Queensland* (London, 1894), p. 5.

21 *The Sydney Mail*, 27 December 1884, p. 1305.

22 *Q*, 9 September 1876, p. 30 & 2 December 1876, p. 28.

23 *Weekly Herald and Queensland Agricultural, Mining and Pastoral Advertiser*, 22 July 1865, p. 13; A. Mackay, *The Sugar Cane in Australia* (Sydney, 1883), p. 155; *Q*, 5 March 1870, p. 5 & 7 August 1873, p. 3; *Sydney Morning Herald*, 26 May 1873, p. 5; Half Yearly Report of Beenleigh Branch of QNB, 31 December 1874, p. 54 (entry for Yahwillpah plantation), QNBR.

24 A. Liversidge, 'Report of the sugar-cane disease in the Mary River district, Queensland', *J. of the Agricultural Society of NSW*, Vol. 2, No. 8 (1876), p. 128; R. Dowlings, Cane Inspector, Condong Mill, to E. Knox, Sydney, 29 November 1880, CSRR, Z/303, Box 37, File C 1.0, Folder 6, Doc.11d; G. Kottman, 'Circular to CSR's New South Wales farmers: The Sugar Cane Disease, 1892', p. 1, CSRR, 142/3512; *Sydney Morning Herald*, 13 July 1870, pp. 8–9.

25 *QVP*, Vol. 4 (1889), pp. 169 (Qs. 2491 & 2493a), 185 (Q. 3009), 309 (Q. 7362) & 374 (Q. 9396); Graves, *Cane and Labour*, pp. 26–28.

26 *SJTC*, 15 October 1896, p. 226; *Mackay Standard*, 4 November 1895, p. 2; E.W. Knox to Sec., Mossman Central Milling Co., 19 October 1911, CSRR, Z/109, Box 316, Letter book No,. 21, p. 116.

27 For this view see E. Denman, 'The Sugar Industry', *QAJ*, Vol. 9 (1901), p. 100.

28 J. Knox, Harwood Mill, to E.W. Knox, 23 September 1887, CSRR, 142/2890, p. 23. Roth, *A Report on the Sugar Industry in Queensland*, p. 58; Maxwell, *Report Upon An Investigation into the Condition of the Sugar Industry in Queensland*, p. 14 .

29 R.W. Beachey, *The British West Indies Sugar Industry in the Late Nineteenth Century* (Oxford, 1957), pp. 87–88 and R. Ormond, 'The evolution of soil management practices in early Jamaican sugar planting', *J. of Historical Geography* 5, no. 2 (1979), pp. 157–170.

30 T. de Keating, *A Practical Treatise on the Cultivation of the Sugar Cane and Manufacture of Sugar Adapted to New South Wales and Other Australian Colonies* (Sydney, 1865); J. Hinchcliffe, *The Sugar Cane in Queensland* (Brisbane, 1868); F. Bell, *Handbook of*

Practical Directions for Sugar Cane Planting, Sugar Making and the Distillation of Rum (Sydney, 1866); A. Mackay, *The Semi-Tropical Agriculturalist and Colonists' Guide* (Brisbane, 1875); for Melmoth Hall's 'Plain Hints to Plain Men respecting the Cultivation of Sugar Cane' see *Sydney Mail*, 12 March 1870, p. 8.

31 *Sydney Morning Herald*, 13 July 1870, p. 8; *BC*, 9 July 1868, p. 3; *Q*, 8 January 1870, p. 5; Agricultural Society of Southern Queensland, *Cultivation of Sugar Cane, Manufacture of Sugar, Diseases of Cane, Mixed Farming and the labor question* (Brisbane, 1875).

32 P. Skerman, A. Fischer and P. Lloyd, *Guiding Queensland Agriculture 1887–1987* (Brisbane, 1987), pp. 5–7 & 42–43; P. Mylrea, *In Service of Agriculture: a Centennial History of the New South Wales Department of Agriculture 1890–1990* (Sydney, 1990).

33 *RRE*, 13 December 1884, p. 6; *Q*, 7 November 1885, p. 759; J. Knox, Harwood Mill, to E.W. Knox, 23 September 1887, CSRR, 142/2890, p. 21.

34 CSR Board Minutes, Meeting 14 June 1870, CSRR, Z/303, File Box 3, A1.1, Folder 2, Doc. 1; *CRE*, 29 July 1873, p. 2.

35 E.W. Knox, AR 1872 season, p. 12 & E.W. Knox, AR 1876 season, p. 13, CSRR, Z/109, Box 374; Campbell, *Extracts from Reports on Certain Agricultural Districts*, p. 31; E.W. Knox, 'Circular to All Mills - Manures, 6 April 1886' & E.W. Knox, 'Circular to All Mills- Superphosphate, 10 May 1886', CSRR, Z/303, File NS 306.

36 E.W. Knox to W. Issaacs, Condong Mill Mang., 14 April 1881, CSRR, 142/1143, p. 338; for use of steam ploughs see *QVP*, Vol. 4 (1889), p. 221 (Ev. W. McLean, Acting Mang., Victoria Mill, Q. 4357) & *MM*, 13 November 1882, p. 12; drainage is discussed in A. Fairgrieve, Insp. of Mills, Broadwater, to E.W. Knox, 30 September 1881, CSRR, 142/2869, p. 95 & A. Fairgrieve, Insp. of Mills, to E. Cowley, Victoria Mill, 13 May 1884, CSRR, 142/2867, p. 285; for irrigation see E.W. Knox to C. Forster, Victoria Mill Mang., 10 July 1888, CSRR, 142/1548, p. 339.

37 For fertilising see: *QVP*, Vol. 4 (1889), p. 343 (Ev. E.W. Knox; Q. 8259); E.W. Knox to Victoria Mill Mang., 11 April 1892, p. 49, CSRR, 142/1551, p. 49; & Munro, *The Sugar Fields of Mackay* (Mackay, 1895), p. 33; for trashing see *QVP*, Vol. 4 (1889), p. 171 (Ev. G. Adams, Goondi Mill Mang., Qs.2560–61) & A. Fairgrieve to W. Issaacs, Condong Mill Mang., 6 May 1884, CSRR, 142/1445, 139.

38 E.W. Knox, 'Circular to All Mill Managers, 8 March 1889', p. 1; E.W. Knox, 'Circular to All Mill Managers, 8 March 1889', pp. 2–3; G. Kottman, 'Circular to All Mill Managers, 7 May 1889'; G. Kottman, 'Circular to All Mill Chemists, 18 December 1889', all in CSRR, 142/3511.

39 G. Kottmann, *On the application of artificial manures* (Sydney: CSR, 1890), in CSRR, Z/109, Box 369; CSR, 'Agricultural circular – Manuring: phosporic acid, potash and nitrogen, 28 July 1892' & CSR, 'Agricultural circular – Manuring of Cane Fields, Superphosphates, 16 July 1892', in CSRR, 142/3512.

40 E.W. Knox to Victoria Mill Mang., 22 March 1892, CSRR, 142/1551, p. 412; *Q*, 6 June 1901, p. 1110.

41 E.W. Knox, 'Circular to All Mill Managers, 30 October 1893', CSRR, 142/3512; CSR, 'Cane cultivation on Keith Hall Estate, Richmond River, New South Wales', 23 pp. pamphlet, 1895, in CSRR, Z/303, Box 18, File A.6.0, Folder 2, Doc. 38; for observations of cultivation trials at Keith Hall see *SJTC*, 15 July 1895, p. 153 & 15 November 1895, p. 254; CSR, 'Notes on Cane Cultivation in Queensland, 1914' , 12 pp booklet, in CSRR, Z/109, Box 369.

42 C. Lynne, *The Industries of New South Wales*, p. 259; *SJTC*, 15 July 1901, p. 8, *ASJ*, 7 July 1910, p. 163 & 8 June 1911, p. 155; CSR, 'Circular re: frost prevention, c. 1900' in CSRR, 142/1423, pp. 50–51.

43 For the policy of foregoing rent see E.W. Knox to Homebush Mill Mang., 24 October 1895, CSRR, 142/1235, p. 48; E.W. Knox to Victoria Mill Mang., 20 December 1900, CSRR, 142/1556, p. 432; A. Farquhar, Insp. of Mills, Homebush plantation, to E.W. Knox,13 January 1903, CSRR, 142/2894, p. 138; E.W. Knox to Homebush Mill Mang., 12 July 1900, CSRR, 142/1237, p. 287; & *SJTC,* 15 September 1900, p. 58.

44 For purchasing manure in bulk see *Cairns Post,* 8 July 1897, p. 3 (Hambledon estate) & *CPP,* Vol. 4 (1913), p. 219 (Ev. G. Forrest, Victoria Mill Mang., Q. 6189); for the drainage policy see E.W. Knox to Victoria Mill Mang., 8 July 1896, CSRR, 142/1554, p. 472 & *Q,* 5 June 1897, p. 1249; for advancing funds to clear stones from paddocks see *ASJ*, 9 September 1909, p. 203; favouring Italian contractors is discussed in Macknade Mill Cane Insp., AR 1914, p. 6, CSRR, Z/109, Box 467.

45 P. Griggs, 'Improving agricultural practices: science and the Australian sugarcane growers 1864–1915', *Agricultural History*, Vol. 78, no. 1 (2004), p. 17; *Planter and Farmer*, October 1882, p. 272; *QVP*, Vol. 4 (1889), p. 177 (Ev. R. Smellie, Q. 2802).

46 *Townsville Daily Bulletin*, 17 November 1887; *QVP*, Vol. 4 (1889), p. 236 (Ev. Charles Young, Qs. 4878–4881; for cane burning on the Lower Burdekin in the 1890s and 1900s see *North Queensland Register*, 19 October 1898, p. 19; *Q,* 3 December 1898, pp. 1088–89; & J. Critchell, 'The sugar industry in Queensland', *International Sugar Journal*, Vol. 5 (1903), p. 188.

47 Griggs, 'Plantation to small farm', pp. 112 & 121.

48 *SJTC*, 15 March 1892, p. 25 & 15 April 1893, p. 64; Connolly, *John Drysdale and the Burdekin*, p. 110; W. Lees, *Districts of Queensland No. 15: Bundaberg* (Brisbane, 1901), p. 6.

49 *MC*, 2 February 1894, p. 3; Q, 18 April 1896, p. 755; Lees, *Districts of Queensland No. 15: Bundaberg*, pp. 7–8; *Bundaberg Mail*, 11 November 1896, p. 24, 4 October 1897, p. 2 & 7 February 1900, p. 2.

50 For the expansion of the irrigation facilities on Pioneer, Kalamia and Seaforth plantations see PMR/IAL/2, pp. 26–32 JCUA; *North Queensland Herald*, 24 December 1892, p. 20; Charles Young, Kalamia, to Henry Young, Cambridge Downs Station, 18 January 1895, Young Private Letterbook, p. 89, JCUA; & Notes of Interview with Mr John Drysdale, 11 September 1900, p. 3, AER, 165/305, NBAC; for the tenant farmers using the planters' irrigation facilities see PMR/JA/1-JA/3, passim, JCUA; *SJTC*, 15 November 1892, p. 214; & *Q,* 8 June 1901, p. 1108.

51 Griggs, 'Plantation to small farm', pp. 121–125 & 348.

52 *SJTC*, 15 September 1902, p. 49, 15 October 1902, p. 72, 15 January 1903, pp. 134 & 142 & 15 March 1903, p. 187–188; Anon, 'Irrigation at Bundaberg', *QAJ*, Vol. 12 (1903), pp. 373–380; *ASJ*, 8 December 1910, p. 380; BSES, AR 1902, *QPP*, Vol. 2 (1903), p. 814.

53 Sec., AEMCo., Melb., to Sec., AEMCo., London, 19 January 1897, London Office Letterbook, No. 1, p. 321, AER, MUA; *MC,* 7 January 1896, p. 2.

54 *SJTC*, 5 April 1901, p. 221; *CPP*, Vol. 4 (1913), pp. 11 & 1823 (Ev., G. Muntz, Chairman of Directors, Mossman Central Mill Co. Ltd., Q. 239 & Ev. J. Paddle, Nerang Mill

Mang., Q. 17794); *North Queensland Register*, 2 July 1906, p. 17; BCSM, AR 1905, *QPP*, Vol. 2 (1906), p. 300; *ASJ*, 9 June 1910, p. 89.

55 N. King, 'The foundation and development of the Bureau of Sugar Experiment Stations', in BSES, *Fifty Years of Scientific Progress* (Brisbane, c. 1950), p. 5.

56 A. Ramsay, Chemist, Sugar Experiment Station Mackay, AR 1898, *QVP*, Vol. 2 (1899), p. 616; for Maxwell's career before coming to Australia see K. Mason, 'Dr Walter Maxwell and the Queensland sugar industry 1899–1908'. B.A. Hons thesis, Uni. of Qld., 1987, pp. 29–32.

57 Maxwell, *Report upon An Investigation into the Condition of the Sugar Industry in Queensland*, passim.

58 King, 'The foundation and development of the Bureau of Sugar Experiment Stations', p. 7; E. Knox to All Editors, Queensland newspapers, 5 November 1900, CSRR, 142/1334, pp. 366–368; *BC*, 1 November 1900, p. 20.

59 For an example of this instruction see E.W. Knox to Goondi Mill Mang., 20 June 1901, CSRR, 142/1464, p. 250.

60 Summarised from details in H. Easterby and G. Patten, *Summary of some experiments conducted with sugar cane and soil and other chemical analyses from 1902 to 1923*, BSES, General Series Bulletins, No. 4, 1924, pp. 1–30; BSES, AR 1909, *QPP*, Vol. 2 (1909), p. 497.

61 *SJTC*, 15 September 1902, p. 48 & 15 June 1903, p. 255; Anon, 'Tropical industries: Irrigation on the Burdekin Delta', *QAJ*, Vol. 12 (1903), p. 50.

62 *QPD*, Vol. 98 (1906), p. 438; *CPP*, Vol. 4 (1913), p. 8 (Ev. G Muntz, Q. 170); BSES, AR 1912–13, *QPP*, Vol. 2 (1913), p. 585.

63 E.W. Knox to J.E. Knox, Victoria Mill, 1 March 1887, CSRR, 142/2858, p. 12; E.W. Knox, Memo re: Mr Tomlinson's cutting machine, 20 May 1887, in CSRR, 142/2888, p. 142; for descriptions of Tomlinson's machine see *Townsville Herald*, 21 April 1888, pp. 17–18; for other early cane harvesters see B. Kerr and K. Blyth, *They're All Half Crazy. 100 Years of Mechanical Cane Harvesting* (Brisbane, 1993), pp. 11–22.

64 *SJTC*, 15 January 1898, p. 155; *QAJ*, Vol. 7, part 3 (1 September 1900), pp. 270–271; E.W. Knox to Goondi Mill Mang., 18 April 1901, CSRR, 142/1464, p. 215; G. Forrest, Victoria Mill Mang., to E.W. Knox, 23 June 1904, CSRR, 142/1592, Letter 183, p. 3; Victoria Mill, 'Cultivation Overseer's Annual Report, 1904', p. 3, CSRR, 142/1592.

65 *ASJ*, 6 April 1911, p. 41, 8 October 1912, p. 519, 9 January 1913, p. 747, 6 August 1914, p. 383 & 3 December 1914, p. 647.

66 For more analysis see P. Griggs, 'Improving agricultural practices. Science and the Australian sugarcane grower, 1864–1915', *Agricultural History*, Vol. 78, no. 1 (2004), p. 26 & *CPP*, Vol. 4 (1913), p. 656.

67 A survey of cane varieties in Queensland is provided in *QVP*, Vol. 3 (1876), pp. 1026–1029; for Hincliffe's plantation see *Q*, 26 July 1873, p. 10.

68 *CRE*, 28 July 1874, p. 3; *Q*, 27 September 1873, p. 5; *ATCJ*, 10 August 1875, p. 263; A. Mackay, *The Sugar Cane in Australia* (Brisbane, 1870), p. 158; *CRE*, 28 July 1874, p. 3; *Slater's Queensland Almanac*, 1868, p. 30; Q, 4 April 1874, p.5; *Sydney Mail*, 12 March 1870, p. 8.

69 P. Griggs, 'Rust disease outbreaks and their impact on the Queensland sugar industry, 1870–1880', *Agricultural History*, Vol. 69, no. 3 (Summer 1995), pp. 421 & 427–429.

70 Griggs, 'Rust disease outbreaks and their impact on the Queensland sugar industry, 1870–1880', pp. 426–427.

71 Griggs, 'Rust disease outbreaks and their impact on the Queensland sugar industry, 1870–1880', pp. 429 & 435; Paper by Dr Bancroft, reproduced in *QVP*, Vol. 3 (1876), p. 1024; H. Easterby, *Varieties of Sugar Cane in Queensland.* BSES General Series Bulletin No. 2, 2nd edition, 1926, p. 17.

72 Griggs, 'Rust disease outbreaks and their impact on the Queensland sugar industry, 1870–1880', pp. 433–435.

73 A.C. Gregory, 'Second Annual Report of the Board Appointed to Acquire into the Causes of Diseases Affecting Livestock and Plants in Queensland', *QVP*, Vol. 2 (1876), pp. 1037–1038; *Q*, 6 October 1876, p. 4, 13 October 1877, p. 21 & 31 August 1878, p. 699.

74 For a list of cane varieties grown at the Oxley experimental garden see J. Bonwick, *The Resources of Queensland* (London, 1882), p. 70; for the distribution of the Creole variety see *Q*, 26 November 1881, p. 969 & 1 April 1882, p. 408; *ATCJ*, 10 December 1881, p. 1124; *MC*, 9 January 1883, p. 2; 'Applications for Sugar cane Plants, 1881–1883', in LAN/AK 97, QSA; & *QVP;* Vol. 4 (1889), p. 211 (Ev. R. Smellie, Q. 4000).

75 Bonwick, *The Resources of Queensland*, p. 70; *RRE*, 9 January 1891, p. 3; Easterby, *Varieties of Sugar-Cane in Queensland*, p. 17; *Sydney Morning Herald*, 1 November 1884, p. 895; Angus Mackay, *The Sugar Cane in Australia* (Sydney, 1883), p. 36; H. Tryon, *Gumming of Cane* (Brisbane, 1895), p. 24.

76 For Meera's popularity see *MM*, 26 May 1883, p. 2 (quoting the *Townsville Daily Bulletin*); *Sydney Morning Herald*, 19 September 1884, p. 13; & *Townsville Daily Bulletin*, 17 November 1887; for Black Java's popularity see Manning, *In their own hands*, pp. 19–20 & 40 & *Q*, 16 June 1883, p. 952; for Cheribon's appeal see Mackay, *The Sugar Cane in Australia*, p. 36.

77 Tryon, *Gumming of Cane*, pp. 9–12; David North, *The Gumming Disease of the Sugar Cane. Its Dissemination and Control* (Sydney, 1935), p. 13.

78 N. Cobb, *Plant Diseases and their Remedies: Diseases of the Sugar-Cane* (Sydney, 1893), pp. 12–17.

79 D. Buchanan, Mackay, to P. McLean, Under-Sec., Qld. Dept. of Agriculture & Stock, 15 May 1890, AGS/N72, QSA; Overseer Mackay State Nursery, AR 1891–92, *QVP* Vol. 4 (1892), p. 660; C. Stephens, Harwood Mill Mang., to E. W. Knox, 12 October 1893 & C. Stephens, Harwood Mill Mang. to E.W. Knox, 26 February 1895, CSRR, 142/1787, pp. 79 & 690–691.

80 During the mid-1890s, James Clark had been the field overseer at Keith Hall estate, CSR's experimental farm. He was transferred to the Hambledon Cane Nursery in 1898. In 1904, Clark was sent to Fiji to open a cane breeding station at Rarawai. Seedling raising was discontinued at the Hambledon Cane Nursery in 1906 (Staff Register, Officers Service Series 1, 1891–1911, pp. 60 & 335, CSRR, Z/109, Box 306); for his experiments at Keith Hall Estate see J. Clark, Keith Hall Estate, to E.W. Knox, 8 February 1896 & 10 May 1897, in CSRR, Z/303, Box 38, File C 2.0, Folder 1, Doc. 9; for details about his Hambledon Cane Nursery experiments see J. Clark to E.W. Knox, 29 June 1900 & 24 August 1901, CSRR, Z/395, Box 45.

81 D. Norris, Royal Botanic Gardens, Kew, to Thomas Archer, Queensland Agent General, London, 7 October 1890, AGS/N94, QSA; E. Cowley, 'Report on Economic Plant Collecting Trip to British New Guinea, 17 January 1893,' AGS/1, QSA; C.G. Hughes, 'The introduction of Badila into Queensland', *CGQB*, Vol. 11, no. 2 (1947), pp. 97–100.

82 Under-Sec., NSW Dept. of Agriculture, to P. McLean, Qld. Dept. of Agriculture & Stock., 11 November 1892 and 25 January 1895, AGS/1, QSA.

83 For CSR being supplied with imported cane varieties see E. Cowley, Kamerunga, to P. McLean, 14 June 1893, AGS/1, QSA and Ev. of Henry Tryon, pp. 512–514, in 'Piddington Royal Commission Evidence', CSRR, Z/109, Box 45; C. G. Hughes, 'Farewell Nineteen Hundred,' *CGQB*, Vol. 21, no. 4 (1 April 1958), p. 116; E.W. Knox to Childers Mill Mang., 9 November 1901, CSRR, 142/1423, p. 471.

84 *Q*, 19 October 1895, p. 756.

85 J. Galloway, 'Botany in the service of Empire: The Barbados cane-breeding program and the revival of the Caribbean sugar industry, 1880s–1930s', *Annals of the Association of American Geographers*, Vol. 86, no. 4 (1996), pp. 692–693; A. North-Coombes, *A History of Sugar Production in Mauritius*, revised edition (1937; rpt. Mauritius, 1993), pp. 25–26.

86 Harwood Mill Mang. to E.W. Knox, 17 August 1894, CSRR, 142/1787, pp. 449–450; E.W. Knox to Harwood Mill Mang., 6 January 1900, CSRR, 142/1716, p. 397; E.W. Knox to Homebush Mill Mang., 2 November 1901, CSRR, 142/1238, p. 116; *SJTC*, 15 November 1895, pp. 268–269; H. Jackson, 'Wollongbar sugar-cane experiments', *Agricultural Gazette of NSW*, Vol. 9, no. 11 (1898), pp. 1289–1295.

87 Overseer Mackay State Nursery, AR 1894–95, *QVP* Vol. 3 (1895), p. 1041; Overseer Kamerunga State Nursery (hereafter KSN), AR 1894–95, *QVP*, Vol. 3 (1895), pp. 104 & 1043; Overseer KSN, AR 1895–96, *QVP*, Vol. 4 (1896), p. 474; E. Cowley, Kamerunga to P. McLean, 16 May 1894; E. Cowley to P. McClean, 28 September 1894, AGS/70, QSA; *Australian Tropiculturist*, 23 December 1895, p. 347.

88 *Q*, 10 October 1896, p. 707; Mackay Sugar Experiment Station, AR 1899–1900, *QVP*, Vol. 2 (1900), p. 758; BSES, AR 1905, *QPP* Vol. 2 (1906), pp. 332–344.

89 Jackson, 'Wollongbar sugar-cane experiments', pp. 1289–1295; Condong Mill Mang. to E.W. Knox, 31 May 1904, CSRR, 142/1210, Letter No. 179; CSR, 'Planting Circular Broadwater Mill, 1911', CSRR, 142/878; *ASJ*, 2 July 1914, p. 275.

90 Hambledon Cane Nursery Supervisor, 'Report on Variety Trials, 1906,' CSRR, 142/909, p. 432; Condong Mill Mang., 'Notes on Cane Varieties, 30 May 1905', CSRR, 142/1213.

91 BSES, AR 1905, *QPP*, Vol. 2 (1906), p. 332; BSES, AR 1908, *QPP* Vol. 2 (1909), p. 529; James Clark, 'Interim Report on Cane Varieties at Hambledon Cane Nursery, October 1900', CSRR, Z/395, Box 45; W. Farquhar, Mill Insp., to Goondi Mill Mang., 16 September 1904, CSRR, N126/149; G. Kottman, 'Chemical Report on the 1909 season', p. 95, CSRR, Z/109, Box 377.

92 BSES, AR 1905, *QPP*, Vol. 2 (1906), p. 332.

93 P. Griggs, 'Australian scientists, sugar cane growers and the search for new gummosis-resistant and sucrose-rich varieties of sugar cane, 1890–1920', *Historical Records of Australian Science*, Vol. 14 (2003), p. 300.

94 W. Farquhar, Mill Insp., to Goondi Mill Mang., 16 September 1904, CSRR, N126/149; G. Kottman, 'Chemical Report on the 1909 season', p. 95, CSRR, Z/109, Box 377; BSES, AR 1905, *QPP*, Vol. 2 (1906), p. 332; BSES, AR 1909, *QPP*, Vol. 2 (1909), p. 529; Childers Mill Mang., AR Childers Mill, 1907, CSRR, 142/1408, Letter No. 205A.

95 Hambledon Cane Nursery Supervisor to G. Kottman, 7 November 1906, CSRR, 142/909, p. 344; Homebush Mill Mang. to E.W. Knox 17 September 1906, CSRR, 142/1255, Letter No. 458; BSES, AR 1910, *QPP*, Vol. 3 (1910), p. 345.

96 'Planting Circular to Victoria Mill Canegrowers, 7 February 1907', CSRR, Z/395, Box 78; C.G. Hughes, 'Farewell Nineteen Hundred', *CGQB*, Vol. 21, no. 4 (1958), p. 116–117; H. Easterby, *Varieties of Sugar Cane in Queensland.* BSES General Series Bulletin No. 2 (revised 2nd edition, 1926), p. 25.

97 *SJTC*, 15 May 1904, p. 227; E.W. Knox to Harwood Mill Mang., 22 September 1905, CSRR, 142/1721, 482.

98 Broadwater Mill Mang. to E.W. Knox, 1 August 1908, CSRR, 142/876, 474; 'Planting Circular, Broadwater Mill, 1910', CSRR, 142/877, 412–143; 'Planting Circular, Harwood Mill, 1914', CSRR, 142/1727, 150–151; Easterby, *Varieties of Sugar Cane in Queensland*, p. 21.

99 G. Kottmann, Cane Report 1897, p. 19, CSRR, Z/109, Box 376; N. Deerr, *Sugar and the Sugar Cane* (Manchester, 1905), pp. 14 & 22; Easterby, *Varieties of Sugar Cane in Queensland*, p. 19; *SJTC*, 15 October 1895, 202; E.W. Knox to Broadwater Mill Mang., 22 March 1895, CSRR, 142/822, p. 30; E.W. Knox to Broadwater Mill Mang., 14 June 1895, CSRR, 142/822, p. 176.

100 Griggs, 'Australian scientists, sugar cane growers and the search for new gummosis-resistant and sucrose-rich varieties of sugar cane, 1890–1920', pp. 302–303.

101 For an example see 'Circular to Cane Contractors, Victoria Mill, 7 February 1907', CSRR, Z/395, Box 78.

102 CSR, 'Circular to All Mills – Yellow Stripe Disease, 15 January 1913', CSRR, Z/395, Box 78; J. Wright, 'Badila in the Burdekin', *CGQB*, Vol. 31, No. 2 (July 1967), p. 71; Griggs, 'Australian scientists, sugar cane growers and the search for new gummosis-resistant and sucrose-rich varieties of sugar cane, 1890–1920', p. 302.

103 Easterby, *Varieties of Sugar Cane in Queensland*, p. 19; Homebush Mill Mang. to E.W. Knox, 13 February 1904, CSRR, 142/1254, Letter No. 146; J. Kerr, *Top Mill in the Valley. Cattle Creek Sugar Mill, Finch Hatton, 1906–1990* (Mackay, 1991), pp. 24–25.

104 J. Kerr, *A Century of Sugar. Racecourse Sugar Mill, Mackay* (Mackay, 1988), p. 69; *ASJ*, 9 April 1914, p. 30; Griggs, 'Australian scientists, sugar cane growers and the search for new gummosis-resistant and sucrose-rich varieties of sugar cane, 1890–1920', pp. 303–304.

105 CSR, 'Circular to Canegrowers at Childers, October 1914', CSRR, 142/1436, p. 189; *ASJ*, 3 February 1914, p. 626.

106 *SJTC*, 15 July 1901, p. 8, 15 September 1902, p. 49, 15 May 1904, p. 227 & 15 October 1904, p. 55.

107 Griggs, 'Australian scientists, sugar cane growers and the search for new gummosis-resistant and sucrose-rich varieties of sugar cane, 1890–1920', p. 305.

108 North, *The Gumming Disease of the Sugar Cane. Its Dissemination and Control*, pp. 11–13.

109 W.Rothe to Condong Mill Mang., 28 August 1886, CSRR, 142/1146, p. 353.

110 *CPP,* Vol. 4 (1913), p. 902, Q. 24744; Mulgrave Central Mill Co. Ltd., *Fifty Years in Retrospect 1895–1945* (Gordonvale, 1947), p. 11.

111 J. Ross, Sydney to E. Knox, 22 January 1870, CSRR, Z/303, Box 18, File A 6.0, Folder 2, Doc. 10; *Q*, 9 December 1871, p. 5.

112 *Sydney Mail*, 18 October 1884, p. 791; Q, 7 February 1885, p. 207; *Cairns Morning Post*, 9 June 1897; Qld. Government Entomologist, AR 1904–1905, *QPP*, Vol. 2

(1905), pp. 568–569; *SJTC*, 15 December 1904, pp. 87–88; A. Chisholm, 'Introduced birds in Queensland', *The Emu*, Vol. 19 (July 1919): 61; *Q*, 7 February 1885, p. 207.

113 For recent accounts see J. Agnew, *Australian Sugarcane Pests* (Brisbane, 1997), pp. 12–22 & P. Allsopp, P. Samson and K. Chandler, 'Pest Management', in M. Hogarth and P. Allsopp (eds.), *Manual of Cane Growing*. (Brisbane, 2000), pp. 301–310.

114 *ASJ*, 8 March 1917, p. 917.

115 P. Griggs, 'Entomology in the service of the State: Queensland scientists and the campaign against cane beetles', *Historical Records of Australian Science*, Vol. 16 (2005), pp. 2–4.

116 H. Tryon, *Grub Pest of Sugar-Cane (*Lepidiota squamulata*) of the Mackay District* (Brisbane, 1896), pp. 8 & 26.

117 H. Roth, 'On the animal parasites of the sugar cane', *The Sugar Cane*, 2 March 1885, p. 118; *SJTC*, 15 January 1893, pp. 150 & 15 February 1895, p. 8; *MC*, 15 December 1897, p. 3.

118 Qld. Govt. Entomologist, AR 1910–1911, *QPP*, Vol. 2 (1911–12), p. 862; *ASJ*, 6 May 1909, p. 11, 6 April 1911, pp. 9 & 45 & 8 June 1911, p. 131.

119 *Cairns Morning Post*, 28 October 1907, p. 2; *QPP*, Vol. 2 (1916–17), p.1255 (Ev. E. Swayne, Q. 84.).

120 G. Logan, 'Tryon, Henry (1856–1943)', *ADB*, Vol. 12 (1990), pp. 272–273; P. Mather, *A Time for a Museum. The History of the Queensland Museum, 1862–1986* (Brisbane, 1986), pp. 45–47 & 177–178; I. Mackerras and E. Marks, 'In Retrospect: The Insects and Entomologists', in Entomological Society of Queensland Inc., *Changing Patterns in Entomology. Jubilee Publication of the Entomological Society of Queensland.* (Brisbane, 1974), p. 7; for Tryon's earliest agricultural work see H. Tryon, *Report on Insect and Fungus Pests* (Brisbane, 1889).

121 *North Queensland Herald*, 5 December 1894, p. 25; *Geraldton Advocate*, 2 October 1895, p. 2; *The Mirror* (Geraldton, Qld.), 29 July 1896, p. 3; *SJTC*, 15 November 1895, p. 257 & 15 June 1896, p. 117; *Q*, 29 February 1896, p. 419.

122 *QVP*, Vol. 4 (1889), p. 177 (Ev. R. Smellie, Q. 2808); *QAJ*, Vol. 5 (1899), p. 133; E. Knox, Sydney, to P. McLean, Brisbane, 18 August 1892, CSRR, 142/485, p. 180; *SJTC*, 15 June 1896, p. 118; Ag. & Pastoral Conference, Qld., June 1899, session on cane grubs, *QAJ*, Vol. 5, no. 2 (August 1899), p. 133.

123 *QVP*, Vol. 4 (1889), pp. 127 (Ev., W. Langdon, Q. 1046), 164 (Ev., J. O'Halleran Q. 2304) & 412 (Ev. J. Cran); J. Davidson, 'Report on Mourilyan Estate, 1 May 1891', p. 3; attached to a letter from Mang., UB, Melb. to Mang., UB, London, U103/39, UBR, ANZA; *The Mirror* (Geraldton), 29 July 1896, p. 3; Ag. & Pastoral Conference, Qld., June 1899, session on cane grubs, *QAJ*, Vol. 5, no. 2 (August 1899), p. 134.

124 *QVP*, Vol. 4 (1889), pp. 130 (Ev., T. Swallow, Qs. 1173–1174); 169 (Ev., G. Adams, Q. 169); 394 (Ev., R. G. Smith) & 402 (Ev., W. Farquhar); *Geraldton Advocate*, 2 October 1895, p. 2; details on Paris Green courtesy of Queensland Health Scientific Services – Pesticides Division; G.Kottman, 'Circular to All Mill Managers, 7 May 1889', CSRR, 142/3511; *QVP*, Vol. 2 (1896), p. 708 (Ev. W. Farquhar, Q. 4630); Ag. & Pastoral Conference, Qld., June 1899, session on cane grubs *QAJ*, Vol. 5, no. 2 (August 1899), p. 135.

125 *QAJ*, Vol. 5, no. 2 (August 1899), pp. 133 & 136; E.W. Knox to Goondi Mill Mang., 30 March 1892, CSRR, 142/1458, p. 330 & E.W. Knox to Goondi Mill Mang., 25 July 1893, CSRR, 142/1459, p. 266.

126 A. Ramsay, 'The sugar industries of the north', *QAJ*, Vol. 5, no. 1 (1899), pp. 49–50; *Cairns Morning Post*, 8 July 1897, p. 2.

127 *SJTC*, 15 June 1894, p. 106, 15 December 1894, p. 247 & 15 February 1895, p. 8; *Q*, 22 June 1895, p. 1186; E.W. Knox to Goondi Mill Mang., 31 May 1894, CSRR, 142/1460, p. 144.

128 G. Kottman, 'Circular to All Mill Managers, 7 May 1889', CSRR, 142/3511; E.W. Knox to Goondi Mill Mang., 15 June 1892, CSRR, 142/1458, p. 395; *SJTC*, 15 June 1894, p. 104.

129 E.W. Knox to Goondi Mill Mang., 16 May 1895, CSRR, 142/1460, p. 447; P. McLean, Under-Sec., Qld. Dept. of Ag. & Stock, to E.W. Knox, 31 May 1895, CSRR, Z/303, Box 18, File A.6.0, Folder 2, Doc. 50; A. Boyd, 'Extermination of the cane grub', *QAJ*, Vol. 10, no. 6, (June 1902), pp. 468–469.

130 Staff Register, Officers Service Series 1, 1891–1911, CSRR, Z/109, Box 306, p. 19; T. Beckmann, 'Brunnich, Johannes Christian (1861–1933)', *ADB*, Vol. 7 (1979), pp. 463–464.

131 Tryon, *Grub pest of sugar-cane (*Lepidiota squamulata*) of the Mackay district*, pp. 38 & 42; *QVP*, Vol. 2 (1896), p. 708 (Ev. W. Farquhar, Q. 4630); E.W. Knox to Homebush Mill Mang., 4 June 1896, CSRR, 142/235, p. 267; Ag. & Pastoral Conference, Qld., June 1899, session on cane grubs, *QAJ*, 1 August 1899, p. 135.

132 *SJTC*, 15 May 1896, p. 91 & 15 June 1896, p. 117.

133 Qld. Govt. Entomologist, AR 1906–07, *QPP*, Vol. 2 (1907), p. 718; Qld. Govt. Entomologist, AR 1909–1910, *QPP*, Vol. 3 (1910), pp. 205–206; *ASJ*, 6 May 1909, p. 29; *ASJ*, 3 February 1910, p. 412–413; *ASJ*, 3 March 1910, p. 481; E. Jarvis, *Notes on Queensland Cane-Insects and Their Control*. BSES, Division of Entomology, Bulletin No. 17 (1923), p. 53.

134 Griggs, 'Entomology in the service of the state', p. 20.

135 *SJTC*, 15 June 1896, p. 117; Ag. & Pastoral Conference, Qld., June 1899, session on cane grubs *QAJ*, Vol. 5, no. 2 (August 1899), pp. 133–134; *Cairns Morning Post*, 6 December 1901, p. 4; B. Egan, 'Origins and development of the sugarcane pest and disease control board system in Queensland', *Proc. ASSCT*, Vol. 24 (2002), pp. 276–277; R. Helms, Childers Mill Mang., to E.W. Knox, 11 April 1911, CSRR 142/1410, Letter No. 798; Qld. Govt. Entomologist, AR 1910–11, *QPP*, Vol. 2 (1911–12), p. 862; *ASJ*, 9 February 1911, pp. 496–497.

136 *QVP*, Vol. 4 (1889), p. 412 (Ev. J. Cran); G. Kottman, 'Chemical Report on Work of 1908 Season, Part 1', p. 27, CSRR, Z/109, Box 377.

137 *ASJ*, 9 February 1911, p. 534 & 3 December 1914, pp. 631–632; Ev. H. Tryon in A. Piddington *et al.* , *Report of the Royal Commission on the Sugar Industry together with Evidence* (1920), pp. 515–516 (located in CSRR, Z/109, Box 45).

138 *ASJ*, 6 August 1914, p. 374; BSES, AR 1913–14, *QPP*, Vol. 2 (1914), p. 912.

139 *ASJ*, 3 June 1909, p. 95, 10 July 1913, p. 232, 7 August 1913, p. 297, 8 October 1914, p. 468, 5 November 1914, p. 550 & 3 December 1914, p. 645–648; Broadwater Mill Cane Insp., AR 1914, p. 1, CSRR, Z/109, Box 459A.

140 BSES, AR 1913–14, *QPP*, Vol. 2 (1914), p. 859.

Chapter 8
The Milling Sector

As the growing of sugar cane spread in New South Wales and Queensland during the late 1860s and early 1870s, the number of sugar mills in Australia increased rapidly, reaching 147 in 1874: 71 in Queensland and 76 in New South Wales (Table 8.1). An absence of official figures on the amount of sugar manufactured in Australia during the 1860s precludes the calculation of the average annual output per sugar mill until the early 1870s. By then, the average annual output per Queensland sugar mill was 107.7 tonnes or nearly three times more than the average output of a New South Wales sugar mill. The gap had narrowed slightly by the late 1870s, although the average annual output per sugar mill in Queensland was still double that of the New South Wales sugar mills (Table 8.2). Overall, the average yearly output was quite small (under 200 tonnes per annum), reflecting the low level of technology employed in many Australian sugar mills. Yet during this period of rapid change, the type of machinery in these sugar mills was not static. Some sugar mill-owners upgraded their equipment following improvements in personal circumstances, while seven sugar mill-owners in the Maryborough district ceased manufacturing raw sugar entirely, converting their factories to juice mills by concentrating only on the extraction phase of the milling process. Despite such constant change in the machinery used in the country's first sugar mills, some generalisations can be made about their levels of technology.

At one end of the technological spectrum were the juice mills, where juice extraction from the harvested cane was the focus of operations. Some Australian sugar mill-owners found this arrangement advantageous as they did not have to find a market for their raw sugar and they could concentrate on cultivation, not manufacture. During the mid-1870s, the Queensland refiners Robert Tooth and Robert Cran Snr pioneered this arrangement at Maryborough. They built two small mills containing only crushing rollers on Irrawarra and Yerra plantations. The juice was treated with lime to stop its deterioration and then delivered to Yengarie Refinery, either in tankers mounted on rail trucks, which travelled over the estate's tramlines, or through underground pipes. By 1879, Tooth and Cran had convinced seven nearby mill-owners to abandon the manufacture of raw sugar in favour of

Table 8.1. The number of sugar mills in Queensland and New South Wales, 1864–1914.

Year	Queensland sugar mills	Queensland juice mills	NSW sugar mills	NSW cattle-driven mills
1864	1	*	1	*
1870	39	*	27	*
1874	71	*	32	44
1880	83	*	50	24
1884	168	*	86	12
1890	110	*	31	2
1895	64	24	23	1
1900	58	10	8	0
1905	49	6	5	0
1910	47	1	4	0
1914	46	1	3	0

* Data not collected.

Source: Based upon figures in the *Statistics of the Colony of Queensland, Statistics of the State of Queensland,* Queensland Department of Agriculture and Stock, ARs, 1890–1915 (juice mill details) and *Statistical Registers of New South Wales.*

Table 8.2. The average output per Queensland and New South Wales sugar mill (tonnes of sugar), 1870–1874 to 1910–1914.

Period	Queensland	New South Wales
1870–1874	107.6	36.6
1875–1879	182.8	93.9
1880–1884	195.6	119.0
1885–1889	379.0	321.7
1890–1894	949.0	867.9
1895–1899	1 906.4	1 271.7
1900–1904	2 203.3	2 799.5
1905–1909	3 248.7	4 491.8
1910–1914	4 056.6	5 899.4

Source: Assembled from data in *Statistics of the Colony of Queensland, Statistics of the State of Queensland* and the *Statistical Registers of New South Wales.*

juice extraction only (Figure 8.1). Yengarie's owners assured its suppliers that limed juice could be kept for 'an almost indefinite period' without any deterioration. Paddle steamers pulled iron punts or barges equipped with two large tanks filled with the juice along the Mary River to the Yengarie Refinery wharf. The juice was then pumped from the tanks up to Yengarie Refinery for conversion into sugar.[1]

Sugar mills without steam-powered machinery flourished mostly in New South Wales during the mid-1870s. By 1874, forty-four sugar mills in New South Wales were described as 'cattle mills' or sugar mills that were powered by either horses or oxen (Table 8.1). Chambers' sugar mill near Coraki was typical of one of these animal-powered New South Wales sugar mills. In 1871, two horses drove the three, 0.8 m hardwood crushing rollers. The centrifugal was under-driven from a treadmill worked by another pair of horses, although the requisite speed was not reached and the basket wobbled too much. Edward W. Knox was scathing of these horse-driven sugar mills, claiming that their machinery was 'faulty and ill-arranged' and prone to falling to pieces in 'a short time'.[2] Official statistics on animal-powered sugar mills in Queensland were not collected, but newspaper descriptions suggested that horses solely powered at least seven Queensland sugar mills during the 1860s and 1870s. Major A. Boyd of Ormeau plantation, near Beenleigh, operated one such sugar mill. Its crushing rollers were driven by four horses walking around a circular horse path. The work was very heavy and at least eight other horses were needed to maintain crushing operations all day (see Plate 8.1). The technology was so primitive in another two of these horse-powered Queensland sugar mills – Birribi and Benowa – that the wooden cooling troughs consisted of hollow logs.[3]

This reliance on horse-driven crushing rollers reduced the amount of juice recovered from the cane and limited the size of the crop that could be harvested annually due to the slowness of crushing the cane. Major A. Boyd noted that juice could be still squeezed out of the cane that had passed through the single set of rollers at his Ormeau Mill. He also reported that producing sugar using the draining process instead of a centrifugal or relying on a horse-powered centrifugal produced unsatisfactory sugar. Boyd claimed that the main characteristic of the drainers was their 'admirable slowness' and a 'propensity to drown the sugar-house with molasses'. Moreover, Boyd claimed he needed 20 tonnes of cane to produce 1.0 tonne of sugar, instead of the 13–15 tonnes of cane required if steam-driven crushing rollers were used in the extraction phase.[4] Therefore, it is not surprising that such disadvantages prompted half the owners of the Queensland sugar mills erected with horse-driven crushing rollers to upgrade their mill machinery during the 1870s by installing steam-driven crushing rollers.[5]

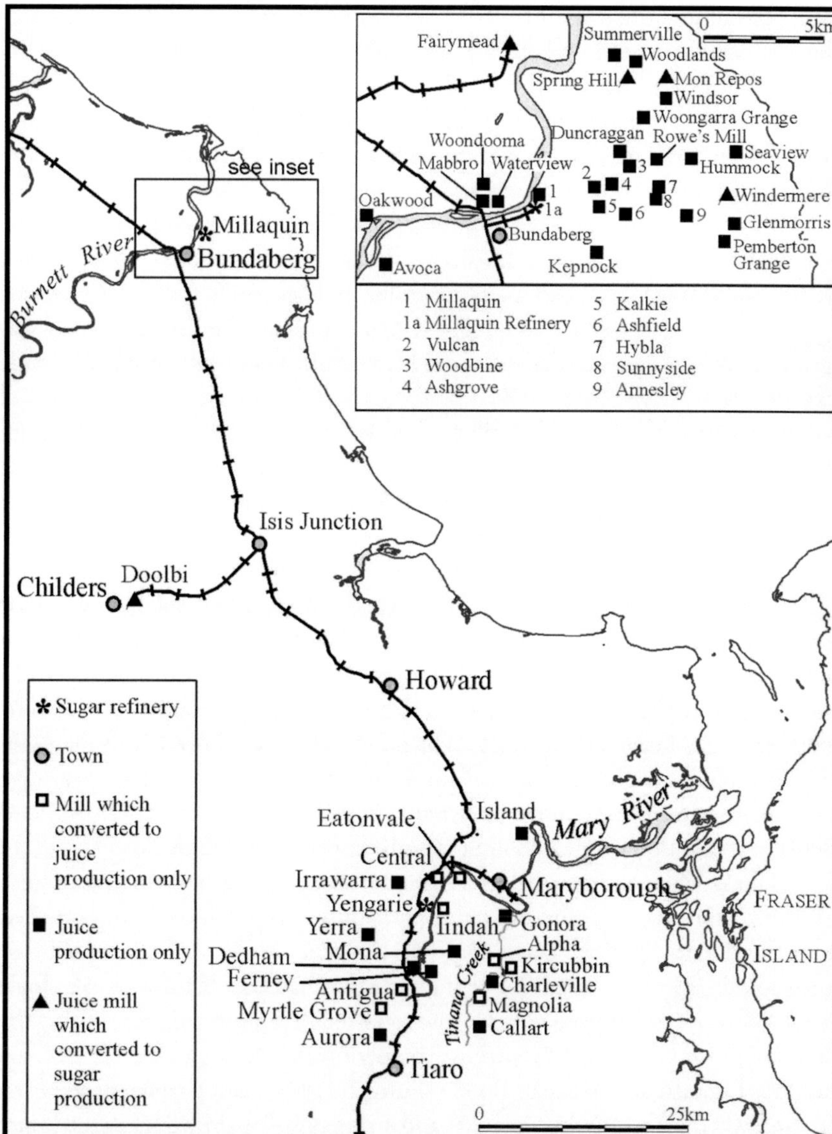

Figure 8.1. The juice mill system that operated in the Maryborough and Bundaberg districts, 1870–1914.

Source: Based upon details in J. Kerr, *Southern Sugar Saga. A History of the Sugar Industry in the Bundaberg District* (Bundaberg, 1983), p. 8 & J. Kerr, *Sugar at Maryborough: 120 Years of Challenge* (Maryborough, 1987), p. 7.

Plate 8.1. Major A.J. Boyd's sugar mill at Ormeau plantation, near Beenleigh. He erected this horse-powered mill c. 1869.

Source: William Boag Photograph Albums, GS-66, John Oxley Library. Image reproduction courtesy of the State Library of Queensland, Brisbane.

Sugar mills relying on steam power to drive the crushing rollers were very common in Australia during the late 1860s and 1870s. Juice extraction may have improved in these mills, but the sticks of harvested cane were still fed manually into the crushing rollers, often by Pacific Islanders (see Plate 8.2). Moreover, a lack of co-ordination between the amount of cane harvested and its feeding through the crushing rollers often led to delays in crushing the cane. A visitor to Ormiston Mill, near Brisbane, in 1867 observed 'a tremendous heap of cane' that had been carted in from the fields on the two previous days, but it was still waiting to be fed into the crushing rollers. In 1881, a journalist describing Cedars Mill, near Mackay, observed that heaps of sugar cane emptied from the many drays that constantly came in from the fields had piled up at one end of the mill building, awaiting feeding into the crushing rollers.[6] Loss of sucrose content occurred during these delays, as harvested cane must be milled within a day to maximise the amount of sucrose recovered during the manufacturing process. This loss could be overcome by using mechanical cane carriers which sped up the process of feeding the cane through the crushing rollers (see Plate 8.3). Yet newspaper descriptions from the 1870s only mention the presence of mechanical cane carriers in CSR's Darkwater Mill and Te Kowai

Plate 8.2. Pacific Islanders feeding sticks of harvested cane into the crushing rollers at Alexandra Mill, near Mackay, c. 1868.

Source: Image reproduction courtesy of Mackay Regional Council Libraries, Image no. 02414.

Plate 8.3. The interior of CSR's Goondi Mill, near Innisfail, showing the mechanical cane carrier and crushing rollers, 1905. Many of the larger and more technologically advanced Australian sugar mills had installed three sets of crushing rollers by 1900.

Source: *The Weekly Times*, 27 May 1905, p. 10 & CSRR, 142/3835, Negative Number 5652. Image reproduction courtesy of CSR Ltd.

Mill. It is highly likely, however, that additional Australian sugar mills, including CSR's other three sugar mills, were equipped with this machinery.[7]

In some Australian sugar mills, steam engines powered both the crushing rollers and centrifugals, but evaporation still occurred in open pans heated by direct fires (i.e. they were semi-mechanised). The machinery in Richard Meares' sugar mill on the Hastings River was typical of these late nineteenth century Australian semi-mechanised sugar mills. A journalist visiting the sugar mill, shortly after it commenced crushing for the first time in June 1868, left the following description of its machinery:

> The juice, or saccharine, is carried from the bed of the mill by duplicate gutters into the clarifiers, which are set within five feet of the mill. There are two clarifiers of four hundred gallons each, with flat bottoms; and the battery consists of four pans, each capable of containing 400, 300, 200, and 100 gallons. The juice, when boiled to the necessary consistency, is discharged through a large tap into galvanised iron coolers. In these it takes about twelve hours to crystallise; it is then passed through the centrifugal machine on the second day, and from the machine it is taken ready to bag for market.[8]

Some Australian semi-mechanised sugar mills with open pans heated by direct fires, however, used a steam-heated pan during the crystallisation process or the final boiling of the juice. Steam-heated pans could be a Fryer's concretor. This apparatus consisted of a series of wrought iron trays, which in their arrangement resembled the tables of a quartz-crushing machine. The clarified juice travelled across the steam-heated trays until it became a thick, viscous syrup. It was then discharged into copper buckets, allowed to cool and solidify, forming a substance known as 'concrete' in the sugar trade. Fryer's concretors were not particularly common in Australia before 1880, with only half a dozen sugar mills, including CSR's Chatsworth and Darkwater Mills, being equipped with this machinery.[9] Instead, many mill-owners favoured either Wetzel or Bour pans, where the juice was boiled by copper pipes or discs heated by steam until crystallisation. Gibson and Sons' Clydesdale Mill, near Brisbane, was typical of one of these semi-mechanised sugar mills equipped with steam-heated evaporating pans. Final evaporation of the juice until crystallisation occurred in a Wetzel pan. A visitor to Clydesdale Mill in 1876 described its operation:

> The Wetzel pan itself is half a cylinder (this one was about 8 feet long and a curve of about 30 inches radius); in this is mounted a plunger in the shape of a large reel, the span connecting the two ends being hollow tubes containing steam supplied from the boiler. This apparatus continues to revolve in the pan, plunging through the liquor therein, raising its density to that required. An experienced hand can tell at once when sufficient density has been arrived at, and will then at once remove the charge to the coolers….. [10]

The most advanced crystallisation equipment available in the 1860s and 1870s was a vacuum pan, which exploited the well-known relationship between the boiling point of liquids and atmospheric pressure. Under artificial conditions of reduced pressure, it lowered the boiling point of cane juice, thereby greatly saving on fuel costs. Moreover, vacuum pans produced less molasses per tonne of sugar, so a mill-owner could manufacture a better quality product with reduced amounts of an unwanted waste product. In Queensland, A. & A.H. Brown pioneered the use of steam vacuum technology in 1872 by including a vacuum pan amongst the original machinery installed in Antigua Mill. Henry Roth in his account of the Queensland sugar industry in 1880, reported that fourteen Queensland mill-owners now employed vacuum pans, instead of bour or wetzel pans. In New South Wales, all CSR's sugar mills were upgraded in the mid-1870s, after the concreters were abandoned in favour of vacuum pans.[11] Yet John E. Davidson, a sugar planter at Mackay, had deliberately delayed installing a vacuum pan in his Alexandra Mill until 1879, believing this apparatus belonged only in sugar refineries. Moreover, installing a vacuum pan in sugar mill did not always guarantee a better quality product. In 1875, the Queensland National Bank manager at Beenleigh advised his superiors that neither Ageston nor Loganholme Mills made 'good sugar', even with a vacuum pan. Both sugar mills had found it difficult to get a confident sugar boiler with knowledge of vacuum pan technology.[12]

Most Australian sugar mill-owners during the 1860s and 1870s relied upon the traditional clarification technique of adding approximately 0.45 kg of lime to 3 784 litres of juice during its gentle heating in the clarifiers. A few Australian sugar mill-owners, however, experimented with different clarification techniques during the 1870s. By 1880, at least sixteen Queensland sugar mills and one New South Wales sugar mill employed the monosulphite of lime process, developed in 1870 by Dr Icery of Mauritius.[13] This process required the creation of a mixture of sulfuric acid and finely sifted lime and then its addition to the juice being heated in the clarifiers. The decolourising effect of the sulfuric acid continually set free under the action of the heat was very marked during the evaporation of the liquor in the pans, where the juice rapidly became a pale lemon colour. Freed of its albumin, the clarified juice produced a marked improvement in the colour of the final sugar.[14] In 1875, Robert Walker, manager at Foulden Mill, near Mackay, went one step further during the clarification phase, pioneering the production of mill whites in Queensland. The juice was passed through iron cylinders filled with charcoal. Most impurities were removed from the juice, resulting in the manufacture of a brilliant white sugar. CSR experimented with adding sulfur dioxide fumes directly into the juice during clarification, although Edward W. Knox was not convinced of the effectiveness of this procedure, noting that it improved the colour of the sugar, but markedly increased lime usage.[15]

Edward W. Knox's uncertainties about clarification procedures stemmed from his and CSR's lack of experience in sugar production. Knox, the second son of Edward Knox Snr, CSR's founder, had been appointed as CSR's superintendent of sugar mills in 1870. He admitted many years later that the only positive aspect associated with him taking up the position was that no one else in the firm 'knew more, or much less' than he did about the manufacture of raw sugar. CSR's main business was sugar refining, not milling. Yet the younger Knox was very intelligent, able and determined. After five years in the position, Knox convinced CSR's directors that the company needed to investigate the latest developments abroad in sugar milling technology, as he believed that their ignorance of the approaches adopted in other countries was contributing to losses in the manufacture of raw sugar. He was sent to the West Indies in 1876 with a view 'to enabling him to compare the results on the Clarence River with those of the West Indian Islands'.[16] Later when Knox was CSR's general manager from 1880 to 1920, he dispatched many CSR officers overseas, charged with the task of reporting upon latest international developments in the refining and milling of sugar and the cultivation of sugar cane so that CSR could be up-to-date as possible in its practices.

Edward W. Knox returned from the West Indies in early 1877. On one hand he was bitterly disappointed, writing that he witnessed nothing that would improve the clarification process in CSR's sugar mills. The West Indian sugar mill-owners still relied overwhelmingly upon the use of skimming and lime during the clarification phase. On the other hand, he observed the benefits of double crushing and triple effets. Knox's visit to the West Indies convinced him that juice evaporation in open, steam-heated evaporating pans consumed too much fuel, the juice was damaged by prolonged exposure to high temperatures while in contact with air (i.e. the inversion process) and that the exposure to air of clarified juice also had a darkening effect due to the result of oxidation. Double crushing also appeared very beneficial, enabling more juice recovery.

Immediately upon Knox's return from overseas, CSR's three sugar mills were equipped with a second set of crushing rollers and maceration during late 1877 in a move designed to improve juice extraction. In 1878, CSR installed an imported a triple effet at Harwood Mill; the following year another was installed at Chatsworth Mill. These two sugar mills became the first Australian sugar mills to become fully mechanised, utilising steam in all stages of production and eliminating open-pan boiling. Edward W. Knox greatly admired the triple effets, noting that the saving in fuel was considerable and fewer workers were now needed in this stage of the manufacturing process. However, CSR continued its clarification trials, experimenting with Icery's monosulphite of lime process and the use of magnesia (or magnesium oxide) during 1877 and 1878. Neither gave Knox much satisfaction, particularly as the magnesia caused the impurities in the juice to rise to the surface, instead of sinking. More labour was then needed to skim the impurities away before the juice could be evaporated.[17]

To secure its triple effets, CSR imported the machinery from France (possibly manufactured by the famous Parisian engineering firm Fives Lille Co.), as this type of specialised equipment was not manufactured in Queensland until 1895. Other pioneering Australian sugar mill-owners also sourced their machinery from overseas, mostly Scotland or England. Maryborough pastoralists A. & A.H. Brown, for example, had ordered machinery worth £10 000 for their Antigua Mill from the Glasgow-based Mirrlees, Tait & Watson (Mirrlees, Watson & Co. after 1882). Another in this category was the immigrant Irish engineer Thomas Fitzgerald, who spent £15 000 on machinery manufactured by W. and A. McOnie for Te Kowai Mill. This machinery included a mechanical cane carrier, vacuum pan and distillery.[18] These pioneering Australian mill-owners who turned to McOnie and Mirrlees, together with other Glasgow engineering firms such as A. & W. Smith and Co. and Duncan Stewart & Co. and the English firms Manlove, Alliot, Fryer & Co. (Nottingham), Fawcett and Preston (Liverpool) and George Fletcher & Co. (Derby and London) because they led the world in the production and export of sugar-making machinery. Basically, Scottish engineering firms produced strong and reliable machinery that suffered less risk of breakage compared to the lighter machinery made by their American or French competitors. Its makers could lavish so much metal upon their machines due to the low cost coal and cheap cast-iron obtained from around Glasgow.[19]

The earliest Australian sugar mill-owners who imported their sugar-making machinery were often seeking more specialised and advanced machinery such as vacuum pans. Other Australian sugar mill-owners who erected semi-mechanised mills with more basic machinery relied on local engineering firms. In New South Wales, the Sydney-based Chapman & Co., the Atlas Engineering Co., and P.N. Russel & Co., who also had foundries in Brisbane and Melbourne, specialised in producing sugar machinery for small sugar mills equipped with horse-driven crushing rollers.[20] The most noted earliest sugar machinery maker was the Brisbane-based Robert Smellie & Co., a partnership between two immigrant Scottish engineers, Robert Smellie and James Sinclair, and the pastoralist Alexander Raff. Other Brisbane engineering firms making sugar machinery included the Queensland Foundry, operated by Alexander Cameron, and Sutton's Northern Copperworks (later J.W. Sutton & Co.). At Mackay, the Scottish immigrant William Robertson opened his Victoria Foundry in 1872, advertising he would repair and manufacture basic machinery. Three years later, R. Hughes & Co. established the Vulcan Foundry at Maryborough, although this firm's advertisements only mention its willingness to manufacture battery pans, clarifiers and copper stills.[21]

The most celebrated Queensland engineering firm to manufacture sugar machinery, however, was Walkers Ltd., iron founders, engineers and shipbuilders. The firm originated as John Walker & Co. on the Ballarat goldfield, but had set up a branch foundry in Queensland because of the demand for both sugar milling and gold mining machinery following the discovery of gold at Gympie. John Walker contemplated expanding to Mackay in the early 1870s, but decided against this move following the opening of the Victoria Foundry at Mackay in 1872. In 1884, the business was floated as a public company, John Walker & Co. Ltd. Four years later it was reorganised as Walkers Ltd. From the outset, the company sought to be competitive with overseas foundries by reducing its labour costs. The firm equipped the foundry with the latest labour-saving devices, and advertised extensively in the colonial newspapers that their sugar machinery was equal in quality and cheaper in price than imports. John Walker, reflecting on the first years of operation in Queensland, claimed his firm manufactured their first orders for sugar machinery at the same price as they could be imported, so as to overcome the prejudice against colonial manufacturers. To remain competitive in the late 1880s, Walkers also paid a Glasgow-based engineer to supply information about recent improvements undertaken by the largest Scottish engineering firms. Walkers manufactured the first Australian-made triple effet in 1895, attempting to stop orders for this specialised machinery being placed with British firms.[22]

New sugar mills and machinery upgrades, 1878 to 1884

Both local and overseas sugar machinery makers were kept busy with orders from Australian sugar mill-owners during the early 1880s, as the number of sugar mills operating in Queensland and New South Wales rose rapidly (Table 8.1). By 1884, 266 sugar mills were found in Australia: 168 in Queensland and 98 in New South Wales. This figure represented the maximum number of sugar mills ever reached in Australia. However, the proliferation of so many new mills in the early 1880s ensured that the average annual output from both Queensland and New South Wales sugar mills rose only slightly compared to the late 1870s (see Table 8.2). Yet this figure is an under-estimation, given that the official statistics were calculated by dividing the total sugar output by all operational sugar mills, including the juice mills producing no raw sugar. Moreover, the average figures hide the great variation in output and level of technology found in the different types of Australian sugar mills by 1884. Sugar mills with horse-driven crushing rollers manufactured six or seven bags of low quality raw sugar (i.e. ration sugar) a day, semi-mechanised sugar mills produced 4–5 tonnes of varying grades of raw sugar daily, while larger, fully mechanised sugar mills could turn out 20–30 tonnes of raw sugar or mill whites daily.[23]

Amongst the new sugar mills erected between 1880 and 1885 were at least two dozen that could be categorised as fully mechanised. Overwhelmingly located in Queensland, these new sugar mills included Hambledon, Mourilyan, Victoria, Airdmillan, Pioneer, Homebush, Palms, Kalamia, Goondi, Fairymead and Bingera, to list a few. Unlike the small, single-storey wooden structures established in the 1870s, the machinery of these fully mechanised sugar mills was often housed in two-storey, cathedral-like buildings (cf. Plate 8. 1 and Plate 8.4). The main building of Palms Mill, near Mackay, for example, was a very substantial structure of 55 m in length by 26 m in width. This building was clad in galvanised iron and the machinery rested on cement floors. The boiler house was separate from the main building, with the steam generated being carried to the machinery via pipes.[24] A large brick chimney was a feature of the Palms Mill and many of the other fully mechanised sugar mills erected in Australia the early 1880s.

These new fully mechanised sugar mills contained most of the more recent technological developments in milling. A majority of them were equipped with mechanical cane carriers, thereby eliminating the cumbersome system of manually feeding the cane into the crushing rollers. Double crushing was employed in some of these sugar mills. Maceration was used in Vilele and Mourilyan Mills and CSR's Victoria, Homebush, Condong and Broadwater Mills. Advanced clarification procedures involving the addition of sulfur fumes, sulfuric acid (Icery's monosulphite of lime process), superphosphate or magnesia to the juice, its filtering through charcoal filters or a combination of both, were employed in at least eleven of these sugar mills.[25] Each of these sugar mills was equipped with triple effets and vacuum pans. Gas or electric lighting enabled a night shift at Mourilyan Mill and all CSR's sugar mills. These sugar mills with their continuous processing technologies had expanded capacities enabling the production

Plate 8.4. Horse-drawn carts loaded with harvested cane beside the Palms Mill, near Mackay, 1895. Palms Mill, erected in 1881, was one of the largest sugar mills built in the Mackay district during the 1880s.

Source: Image reproduction courtesy of Mackay Regional Council Libraries, Image no. 03659.

of between 1 500 and 3 000 tonnes of sugar annually. All or a substantial part of the machinery for these sugar mills was imported, mostly from the Scottish engineering firms A. & W. Smith & Co., Mirrlees, Tait & Watson or W. & A. McOnie.[26]

Many other new sugar mills erected in Australia during the early 1880s were semi-mechanised. All had steam-driven crushing rollers and open pans heated by steam or direct fires, but crystallisation occurred in a variety of arrangements. Pandora Mill, near Rockhampton, was equipped with a concretor, obtained from Manlove, Alliot, Fryer & Co. Pandora Mill was unique, being the only new Australian sugar mill erected in this period to have such machinery.[27] Other new semi-mechanised Australian sugar mills operated with Bour or Wetzel pans only, suggesting limited amounts of capital were available for the enterprise. Vacuum pans were present in at least thirty new Australian sugar mills erected during the early 1880s, highlighting the growing importance of this equipment.[28] Three Australian sugar mills, all located in Southern

Queensland, contained a Sutton's Atmospheric Evaporating Pan, one of the earliest, possibly the earliest Australian sugar milling invention. This apparatus was an open-finishing pan in which heated air was forced by means of a fan blast through a coil of pipes perforated with small holes, facilitating evaporation of the juice until granulation was reached. It allowed the concentration of the boiled juice in little over half the time required by the more costly Wetzel or Bour pans, and produced well-coloured grocery sugar almost equal to that made in a vacuum pan, but at one-tenth the cost.[29] Overall, much of this machinery, except some vacuum pans, was made locally by either J.W. Sutton & Co. or Walkers Ltd. Indeed, William Harrington, the Managing Director of Walkers Ltd., complained to the Royal Commissioners investigating the condition of the Queensland sugar industry in 1889 that many of their recent customers had ordered 'only little sugar mills and small appliances'.[30]

The reliance on horses or bullocks to drive a sugar mill's crushing rollers waned in New South Wales during the early 1880s, with the number of 'cattle mills' being halved between 1880 and 1884 (see Table 8.1). In Queensland, individuals or groups of canegrowers with limited capital continued to erect horse-driven sugar mills. The Maroochie River Sugar Co. is a good example of one such venture. By 1884, this loose alliance of local canegrowers had built a small sugar mill with hardwood rollers powered by horses and were manufacturing 'concrete' for the Brisbane market.[31] Amongst this category of mill-owner were also at least five Southern Queensland canegrowers who erected what was known locally as a 'Victor mill'. This machinery consisted of horse-driven vertical crushing rollers, combined with a Cook's sugar evaporator (i.e. an open boiling arrangement resembling a down-sized Fryer's concreter) that had been manufactured by the Cincinnati-based Blymyer Manufactory Co. Its operation had been demonstrated throughout Southern Queensland during the late 1870s by J. Vincent, Blymyer's Queensland agent, and Blymyer's travelling representative.[32]

Some existing Australian sugar mills were also upgraded during the early 1880s. An analysis of newspaper reports from this period indicated that vacuum pans were added to the machinery in at least twenty-one Queensland sugar mills and eleven New South Wales sugar mills.[33] By 1880, many Australian mill-owners had concluded that vacuum pans were 'an established and a necessary portion of first-class sugar making appliances'. They reduced fuel consumption, were made locally and cost considerably less than triple effets, especially as Mirrlees, Tait & Watson supplied cast-iron vacuum pans costing £300–£500, or at half the cost of copper vacuum pans of similar size.[34] However, a vacuum pan's biggest advantage was that it allowed the manufacture of high quality raw sugar. Mill-owners trying to sell sugar in southern Australia's main cities had to produce a better quality product if they were to break into the markets increasingly dominated by white sugars imported from Mauritius or supplied by CSR. However, not all of these upgrades were successful. Edward W. Knox reported in 1880 that a small vacuum pan installed in one of its rival New South Wales sugar mills (un-named) was sitting idle, as its owner could not find a boiler who could operate the equipment.[35]

Despite the growing emphasis on steam vacuum technology, at least thirty-five juice mills were erected in Queensland during the early 1880s. Most crushing rollers for these juice mills were built locally, either by John Walker & Co. or at Tooth & Co.'s, Vulcan Foundry. A third of the juice mills were erected around Maryborough, their owners sending limed juice to Yengarie Refinery via large iron tanks situated on barges. The other juice mills were built near Bundaberg and supplied juice to Robert Cran & Co.'s new Millaquin Refinery, which first operated in October 1882. From Millaquin, Robert Cran & Co. constructed a 15 cm cast-iron main pipe for a distance of 40 km through the Woongarra district. Various branches led from the main pipe to the local juice mills, where large wrought-iron tanks were erected to store the juice. Steam pumps placed at the bottom of each tank forced the liquid to Millaquin. Juice mills upstream of Millaquin (e.g. Avoca and Mabbro) punted their juice to Millaquin. At the height of the juice mill system in the late 1890s, approximately 416 000 litres of limed juice was delivered daily to Millaquin Refinery.[36] However, during the 1880s at least four juice mills in the Bundaberg district were converted into factories which could manufacture raw sugar (Figure 8.1).

Andrew Fairgrieve, CSR's senior chemist during the mid-1880s, suggested that a juice mill arrangement could be adopted in the Clarence River district, with CSR purchasing juice from the local small mill-owners. Edward W. Knox firmly rejected this suggestion, arguing that such a policy could encourage more canegrowers to erect their own sugar mills, further reducing CSR's cane supply, but more significantly the valuable juice would be lost through improper crushing. Almost a decade later Flora Shaw, the correspondent for *The Times* (London), expressed a similar sentiment when she observed that the juice mill system in Queensland left the 'very important item of crushing still in the hands of the unscientific farmer'.[37]

As increasing numbers of Australian sugar mills relied on steam power during the 1870s and early 1880s, the issue of fuel economy attracted the attention of some mill-owners. Most Australian sugar mills used bagasse combined with firewood or coal, although Balmoral Mill (Mackay) operated the entire 1875 season by using only bagasse. Firewood was plentiful before 1880, but the mill-owners still paid for it to be gathered, even if they relied on their own Pacific Islanders to undertake the task. Bagasse was free and immediately available, being produced as part of the manufacturing process, but had one main disadvantage: the product needed to be dried before it could be used in the boiler furnaces. In 1875, a journalist described the process of bagasse or megass drying at Inverness Mill, near Mackay:

> The works, as is usual in this district, stand in the centre of a clear space used as the megass drying grounds. Here the megass is spread thinly immediately on leaving the rollers; a few hours suffice to dry it, when it is gathered into rows with a large horse rake, and then carted to the battery stoke-hole for use as firing. All surplus megass is carefully stored in a large building specially erected for the purpose, from which in damp or wet weather the supply is obtained.[38]

Clearly wet weather hindered the above process, which was time consuming and labour-intensive, using according to one report an 'enormous amount of kanaka labour'. In the mid-1880s, Drysdale Bros. & Co., for example, allocated twenty-four workers to this process at Pioneer Mill. Moreover, sheds filled with combustible material ran the risk of fire. One such shed burnt down at Te Kowai Mill in December 1880, resulting in the mill not having any stored bagasse for the commencement of the following year's crushing season.[39] Therefore, when maceration commenced in Australian sugar mills during the 1880s, the bagasse emerged from the crushing rollers wetter than previously, thereby taking even longer to dry.

The sun-drying of bagasse persisted in Queensland because mill-owners employed cheaper, indentured workers who could be used for the task. In New South Wales, CSR did not have access to indentured workers, so bagasse drying was costly using European workers, although burning larger amounts of coal instead of bagasse proved an expensive alternative. During the 1870s, CSR tried eliminating the sun-drying of bagasse at its New South Wales sugar mills by using a bagasse-drying oven, but found that this equipment actually consumed 20% of the bagasse during the drying process. A bagasse-drying machine using heated air was tested at Condong and Broadwater Mills in 1881 and 1882, but CSR concluded that the machine could not cope quickly enough with all the bagasse being produced during the crushing phase, so it piled up around the sugar mills.[40] Henry Braby, a Mackay-based mill engineer, eventually developed a practical solution to the mill-owners' dilemma in 1888. As the bagasse left the last set of crushing rollers it travelled on an elevator made of perforated iron slats and was exposed to waste heat from the boiler flues. By the time the bagasse reached the mill's boiler furnaces it had been dried sufficiently for use as a fuel, without the need for any sun-drying.[41]

Coping during the sugar price downturn, 1885 to 1899

The rapid growth in the number of sugar mills in Australia came to an abrupt halt in 1884. At least thirty-one new sugar mills were erected in Australia during the late 1880s and 1890s, but their presence could not offset the widespread number of sugar mill closures throughout both colonies. By 1890, only 110 sugar mills were left in Queensland; this number had fallen to fifty-eight by 1900. A similar trend was observed in New South Wales, where the total number of sugar mills fell rapidly from ninety-eight in 1884 to eight in 1900 (Table 8.1). The immediate cause of this decline was the world price of sugar in mid-1884 fell by about a third. This decline resulted in an identical fall in Australian sugar prices. By the early 1890s, Australian sugar mill-owners received between £20 and £25 per tonne for white sugar sold in Australian capital cities, compared with £30 to £35 per tonne in the early 1880s.

Some Australian mill-owners tried surviving this price downturn by producing more sugar, often from the same amount of cane. This strategy could be implemented by installing the new technologies, although raising the capital during such uncertain times was difficult.[42] A few Australian sugar mill-owners upgraded their factories by the addition of double crushing, in order to extract more juice from the harvested cane. Included amongst this group were Springhill and Woodlands, Bundaberg juice mills, and Oakwood, Kalkie, Rubyana, Windsor and Seaview Mills, five semi-mechanised sugar mills in the Bundaberg district, and Beenleigh and Yatala Mills, located in the Beenleigh district.[43] Most of these sugar mills were owned by individuals with limited capital, so it is reasonable to assume they sought the cheapest way of expanding sugar output.

Other Australian sugar mill-owners, including both individuals and joint stock companies, undertook more extensive upgrades, converting their sugar mills from semi-mechanised operations to full mechanisation by the addition of triple effets, additional sets of crushing rollers and/or preparatory devices (i.e. shredders or revolving knives). These Australian sugar mill-owners not only recovered more sugar from the harvested cane, but reduced fuel consumption. The Richmond River Sugar Co.'s Rous Mill, near Lismore, is good example of an Australian semi-mechanised sugar mill fully mechanised after 1884. A second set of crushing rollers, a triple effet, two new vacuum pans and three new boilers were added to Rous Mill during 1889 and 1890. Maceration was introduced and a night shift became possible following the addition of electrical lighting to the sugar mill.[44]

The Australian owners of the mechanised sugar mills established prior to 1885 did not rest on their laurels during this price downturn, hoping to survive because of their superior technology. CSR adopted new milling arrangements during the late 1880s and 1890s. A shredder which prepared the cane before it entered the crushing rollers was installed at Harwood Mill in 1889. By 1892, all CSR's Australian sugar mills were equipped with shredders, except Victoria, which was upgraded with a set of revolving knives. Knox, however, was uncertain about the effectiveness of preparatory devices, but CSR persisted with their use during the 1890s. CSR also adopted a new innovation known as treble crushing or passing the crushed cane fibres through a third set of crushing rollers.[45] From 1891 onwards, conical subsiders, a CSR invention, were installed in all CSR's sugar mills. This new equipment, which converted clarification from a batch to continuous process, consisted of a shallow inverted cone, through which the dirty juice flowed, being guided in a spiral path by baffles. The settled impurities from the juice were removed at the bottom of the subsider; the clear juice was taken off at an overflow weir at the periphery. Triple effets were upgraded to quadruple effets in CSR's sugar mills. In 1899, CSR even began experimenting with a quintuple effet at Macknade Mill.[46] Some of this new machinery was ordered from the Scottish sugar machinery makers; in other instances it was secondhand and obtained cheaply from the Australian sugar mills that had closed as a consequence of the depression in sugar prices in the late 1880s.[47]

Other Australian sugar mill-owners either independently kept abreast of advances in milling technology or copied CSR's developments. At Bundaberg, Young Bros. upgraded Fairymead Mill during the 1890s by adding a shredder, a third set of crushing rollers, hydraulic pressure attachments which ensured better feeding of the cane through the crushing rollers and a Yaryan quadruple evaporator – an 1886 invention in which the liquor to be boiled flowed through a series of tubes that formed one continuous coil. This machinery was imported from Mirrlees, Watson & Co. In 1892, their neighbours, Gibson & Howes Ltd., installed a new electricity plant at Bingera Mill, enabling crushing at night. A shredder and a third set of crushing rollers built by Walkers Ltd. were added to the sugar mill in 1894 and 1896, respectively. In 1897, Gibson & Howes Ltd. imported a Molіereux pan from Duncan Stewart & Co. of Glasgow, thereby introducing crystallisation by motion into Australia. The principal of this process was to crystallise by slow rotary movements, the slower the revolutions the bigger the crystals.[48] In the Lower Burdekin district, Drysdale Bros. & Co. began upgrading Pioneer Mill in 1892 by installing the following machinery: a contrivance for drying, screening and sifting sugar to reduce lumps, developed by the engineer Henry Braby; a second set of crushing rollers obtained from the idle plant at the nearby closed Airdmillan Mill; and hydraulic pressure attachments for the crushing rollers, imported from Mirrlees, Watson & Co. Further additions in the mid-1890s included revolving knives, the conversion of the triple effet into a quadruple effet, electric lighting and a third set of crushing rollers.[49] Richard Kingsford, who acquired Hambledon Mill in the early 1890s, authorised its upgrade through the addition of a third set of crushing rollers, a second triple effet, a continuous subsider and maceration. His investment was rewarded in 1897, when CSR purchased Hambledon Mill as part of its expansion in North Queensland.[50]

Thirty-one new sugar mills were also erected in Australia between 1885 and 1899. Almost a quarter were semi-mechanised, but contained vacuum pans, and most were located in the Kolan River district. Machinery sourced from closed New South Wales sugar mills was shipped northwards and re-used in two of these sugar mills – Waterloo and Greenwood.[51] These new sugar mills, however, were no match technologically for the fully mechanised, state-funded Queensland central sugar mills which made up half the new sugar mills built in Australia during the late 1880s and 1890s. Between £30 000 and £50 000 had been spent on acquiring the latest machinery for these sugar mills. The most advanced technologically were Mulgrave, Proserpine, Isis and Moreton Mills, with each having preparatory devices, double crushing and continuous subsiders. Marian Mill was erected with three sets of crushing rollers, instead of any preparatory devices, while Mossman Mill had continuous subsiders, but no preparatory devices at the time of its construction.[52] Walkers Ltd. secured the orders to build the machinery in Marian, Mt Bauple, Gin Gin, Plane Creek, Isis and Proserpine Mills. Its main rival, A. & W. Smith & Co. of Glasgow, constructed all the machinery in five other Queensland central sugar mills, despite the colony's parliamentarians campaigning to ensure that all machinery contracts for the central sugar mills were awarded to Queensland engineering firms only.[53]

Six sugar mills erected during the late 1880s and early 1890s were juice mills, mostly situated in the Bundaberg or Isis districts. The most noted of these juice mills was Doolbi, built to improve the dwindling supply of juice to Yengarie Refinery, following the closure of several juice mills along the Mary River and Tinana Creek during the late 1880s. Robert Cran & Co. had purchased a large property in the Isis district in the late 1880s and commenced leasing sections of it to farmers on the condition they supplied cane to the new juice mill. The juice was moved by tankers mounted on rail trucks from Doolbi to Isis Junction and then to Yengarie Refinery (see Figure 8.1). Doolbi's double crushing plant commenced in December 1890 and produced 75 000 litres of juice daily or enough to fill ten rail tankers. By 1893, seven-eighths of all the juice dealt with at Yengarie came from Doolbi. Moreover, Doolbi was upgraded by the installation of electric lights in 1894, enabling crushing at night. Juice then began being sent to Millaquin Refinery, and the supply subsequently directed according to whichever refinery had the capacity to deal with the juice.[54] Yet this move to the Isis district could not stave off the inevitable decline of the juice mill system, which had a serious defect – the fibre in the cane was lost to the refiners as fuel, forcing them to purchase more expensive coal. Yengarie Refinery eventually closed in 1900 and some of its machinery was used in 1901 to convert Doolbi Mill into a complete raw sugar mill with treble crushing. A dwindling number of juice mills supplied Millaquin Refinery during the 1900s, with only the Millaquin juice mill being left in 1914.[55]

Fuel use became a concern of some Australian sugar mill-owners during the 1890s as they sought to reduce manufacturing costs. In 1900, two Queensland sugar mill managers advised Dr Walter Maxwell that bagasse provided only one-half of the fuel needed to produce a ton of sugar; an additional 16 cwt (0.8 tonne) to 20 cwt (1.01 tonnes) of firewood was required to manufacture that ton of sugar. Consequently, Australian sugar mills were often surrounded by huge piles of firewood which had been created during the slack season in preparation for when the wood was needed during the crushing season. The extent of one such stockpile was provided by Walter Farquhar, CSR's Inspector of Mills, who reported in April 1895 that 8 200 tonnes of firewood had been stored at Victoria Mill and this amount would enable the mill to operate without purchasing coal during the forthcoming crushing season.[56]

To meet this demand for firewood, Australian sugar mill-owners initially used timber cleared from their own plantations or purchased timber from other landowners who were clearing their properties and establishing cane farms. Shortages of firewood, however, emerged by the early 1890s. In 1893, Arthur Neame, the co-owner of Macknade Mill, noted in his diary that 'latterly, we had a very considerable distance to go for the wood'. Charles Young, Manager of Kalamia Mill and plantation, complained to the estate's London-based owners in 1895 that the shortage of firewood had caused him to clear nearby mangroves, but this practice was becoming unsustainable as most of the local mangrove forests had been cleared. John Drysdale, Managing Director of Pioneer Mill and a neighbour of Charles Young, also faced firewood shortages in the early 1900s. Eventually, as firewood shortages became more acute during the late 1900s, both Kalamia and Pioneer Mills relied increasingly on coal sourced from Bowen, approximately 100 km south of the Lower Burdekin district.[57] In the Mackay district where ten sugar mills competed for firewood in the early 1900s, the Local Land Commissioner noted in 1908 that 'the lands in the vicinity of the mills have been almost cleared, and fuel is becoming scarce; the mills are running portable lines out to tap new country'. In 1911, CSR's Chief Engineer advised the firm's engineers in a circular that 'in North Queensland each year it is more difficult and costly to obtain good wood fuel'. CSR sanctioned the expenditure on economisers or heat exchange devices that heated the water up to, but not normally beyond its boiling point. They were fitted to the boilers to improve their efficiency. However, coal sourced from Newcastle, nearly 2 500 km south of CSR's northern-most sugar mill, had to be used more frequently as supplies of suitable firewood dwindled. By the early 1910s, Mourilyan Mill relied on coal sourced from Ipswich, after its owner, the Australian Sugar Co. Pty. Ltd., concluded that firewood in the Innisfail district had become 'inferior'.[58]

The introduction of chemical analysis

Prior to 1890, manufacturing sugar in Australia was often more art than science. Cane was generally not weighed as it entered the factories and analytical chemists were not employed before 1880. Consequently, most Australian sugar mill managers did not know how much cane it took to manufacture a tonne of sugar and they often had little idea if their factory operated efficiently.[59] Questions were not even asked about how much sugar was lost in waste products such as bagasse, molasses and filter muds. For many Australian sugar mill-owners this question was not crucial, as sugar prices were high at over £30 per tonne, leaving a good margin of profit.

All Australian sugar mills before 1880 relied on sugar boilers to oversee the manufacture of sugar. A sketch of an unidentified Australian sugar boiler standing beside a vacuum pan is shown in Plate 8.5. Percy Clarke in his book, *The New Chum in Australia,* observed that the sugar boiler in the mill he

Plate 8.5. A sugar boiler testing the syrup in a vacuum pan at an unidentified Australian sugar mill, c. 1886.

Source: A. Garran, *Picturesque Atlas of Australasia,* Volume 1. (Sydney, 1886), p. 116.

visited in 1886 used the 'thumb test' to determine the stage of crystallisation. The sugar boiler felt the 'grain' of the sugar by dipping his fingers into the hot syrup. He separated his fingers while holding them up to the light. If specks of sugar crystals were present in the thin film of syrup, it was ready for the coolers or centrifugals.[60] Late nineteenth century sugar boilers, however, did have a testing instrument to aid them in their work, so it was not all based on observation. The instrument, known as the saccharometer, was simply a hydrometer so graduated as to indicate the amount of sugar contained in a saccharine solution. Sugar boilers could read off the specific gravity of the solution on the graduated scale of the saccharometer, each degree indicating the presence of 0.019% of sugar in the liquor. If the saccharometer floated in the liquor so as to indicate 10° Beaume′, for example, there would be 19% of sugar in the liquid.

By the early 1870s, Australian sugar mill-owners also had the polariscope to assist them measure the quantity of crystallisable sugar and other impurities present in the cane juice. However, a Queensland journalist observed in 1879 that it had 'not come into favour' amongst the colony's millers. The use of the instrument demanded too much chemical and analytical skill, whereas the results of tests with a saccharometer were understood by all. In 1885, James McHenley, the analytical chemist at Mourilyan Mill during the early 1880s, supported this observation, claiming that he had never seen a polariscope in a Queensland sugar mill during his travels throughout the colony.[61]

Sugar boilers were highly regarded by Australian sugar mill-owners because of their special skills. Two Australian sugar boilers who gained almost celebrity status at Mackay during the 1870s were the Mauritius-trained Frenchmen, Louis Duval and Daniel Dupont. In 1877, Duval was responsible for producing sugar at River plantation using the Icery monosulphite process, before moving northwards to manage Innishowen Mill, which was equipped with machinery imported entirely from the French engineering firm Fives Lille & Co. Dupont operated as the sugar boiler at Te Kowai. They may have been responsible for training Alfred and Frederick Knobel, who initially gained experience at Foulden Mill in the mid-1870s from two 'unidentified Mauritius-trained Frenchmen'. Alfred Knobel later worked as a sugar boiler at several Mackay sugar mills before going to Pioneer Mill, where he remained for twenty-three years.[62] Moreover, the high demand for experienced sugar boilers by the early 1880s, especially with an understanding of the operation of vacuum pans, led Edward W. Knox to complain that they were being lured away from CSR by other Australian sugar mill-owners who offered them higher rates of pay. In 1883, CSR decided that it would no longer train any men in the art of vacuum pan or triple effet boiling unless they signed an agreement to stay with the company for a set period of time, as a considerable expense was devoted to this training.[63]

By the late 1870s, Edward W. Knox recognised that the strength of the European beet-sugar industries and their increasing superiority in the international sugar market derived from 'the organised application of scientific principles and scientific manpower to sugar production'. His views led to CSR's first investment in specialist scientific expertise in 1879, when sugar refinery chemist Andrew Fairgrieve from Greenock, Scotland, joined the firm. Fairgrieve was sent to Harwood Mill to gain some experience at a sugar mill. Edward W. Knox's reflection on this initial attempt at using a chemist suggested the experiment was not a success. In 1890, he recalled: 'We did not know clearly what we wanted nor did the man we engaged, so the start was not a success; but it showed us we were on the right track[…]'. Undeterred by these initial difficulties, CSR during the early 1880s hired several other analytical chemists including Thomas Walton (1850–1917), another Scottish sugar refinery chemist, and Dr Gustav Kottmann (1852–1936), Assistant Director of the Sugar Institute in Berlin.[64]

The chemists quickly established a major presence as a recognised group within CSR. In 1890, CSR employed approximately thirty chemists and trainee chemists. By 1910, this number had expanded to ninety, reflecting the steady growth in the number of CSR-operated sugar mills and refineries and the expanded experimental program at the head office laboratory in Sydney. The chemists, however, were not a homogenous group in regard to their experience or their functions. A definite hierarchy emerged amongst the chemists. Walton emerged as CSR's head chemist, establishing the head office laboratory in Sydney and assuming responsible for training the young chemists within CSR. With his background in refinery chemistry, he also oversaw innovation and chemical control in CSR's refinery operations, but also played an important part in developing and implementing a system of chemical control in the sugar mills. Initially, Fairgrieve assumed the role of inspecting chemist, but he was replaced by Kottmann in 1885 because of his failure to create a clear program of research which was relevant to the company, and his inability to draft reports that were understood by management. Kottmann then became 'CSR's main technical trouble-shooter'. As CSR's inspecting chemist he lived a peripatetic existence, visiting the sugar mills, assessing their performances and carrying out elaborate and lengthy experiments on various sugar milling problems. The next layer consisted of the chemists with some training and experience, who could take charge of the chemical work at a sugar mill or refinery, be responsible for chemical reports from that unit and be able to conduct experiments, either independently or under instruction from the leading chemists. Johannes Brünnich, who worked at Homebush Mill in the mid-1890s, was typical of this category of CSR's chemical staff. The largest group of chemists functioned entirely as routine analysts.[65]

Walton and Kottmann initially set about to determining if bagasse contained recoverable sugar and if so how much? Their experiments revealed that anywhere up to 10% of the recoverable sucrose was left in CSR's bagasse, even after double crushing. CSR's mill managers accused the chemists of talking nonsense. However, Edward W. Knox, CSR's general manager since 1880, realised that the chemists held the key to improving sugar production at the firm's expanding number of sugar mills. In June 1884, he instructed CSR's sugar mill managers that the company was introducing a system of chemical book-keeping at all its factories and that they should co-operate with the chemists who were joining their mill staff. Knox believed this course of action was essential, given the decline in sugar prices troubling the industry.

CSR's system of chemical control involved measuring, by chemical analysis, the amount of sucrose in the cane as it entered the mill, the sucrose in the juice and syrups at the various stages during the manufacturing process, and the amounts of sucrose leaving the mill as either raw sugar or in the bagasse, filter muds and molasses. Each factory's chemical returns were forwarded to the central laboratory in Sydney weekly, for further analysis. This system was refined further in June 1886, when Kottmann sent each sugar mill a table, identifying thirty-one stages in the manufacturing process. It indicated the tests to be done at each stage and their frequency. To provide a standard for measuring the performance of a sugar mill processing cane of varying quality, Kottmann devised the pocs (pure obtainable cane sugar) formula after trials throughout 1885 and 1886. Kottmann's original pocs formula was defined as 'the amount of sugar in the cane, minus half the impurities in the cane'. It enabled the calculation of the amount of cane sugar capable of theoretically being recovered by milling, if very high standards of milling performance could be reached. The pocs formula later evolved into the ccs (commercial cane sugar) formula, and was used to calculate the price paid for a canegrower's cane.[66]

Tensions soon developed between CSR's sugar mill managers and the chemists over the question of the authority to be exercised by the latter. In May 1886, Edward W. Knox resolved this matter in an unambiguous and decisive way by instructing the sugar mill managers to concentrate on supervising expenditure, leaving the complete responsibility for the manufacture of sugar to the chemists and wherever possible they were to be guided by the chemists' advice about the treatment of the juice. During the late 1880s, additional Head Office chemical circulars were issued to each sugar mill on topics such as minimising losses of sugar during the manufacturing process, chemical book-keeping and chemical control in sugar mills. Throughout the 1890s a well developed system of sugar mill inspections emerged, with inspecting chemists visiting each of CSR's seven Australian sugar mills regularly, and reporting to Head Office about the manufacturing work being carried out at the sugar mill and areas which needed improvement. In 1912, Edward W. Knox claimed that CSR now spent £20 000 annually on sugar mill chemical control. He stated that if the work of all the sugar mills in 1910 had been carried on as in the 1900 season, the company would have made about 8 600 tonnes less sugar. If a comparison was made with 1885 when chemical control was first introduced by CSR, then the company would have made 29 300 tonnes less sugar.[67]

CSR's analytical chemists, particularly Kottmann, investigated other aspects of sugar production. In the late 1880s, Kottmann was responsible for testing CSR's diffuser manufactured by the Fives Lille Co. Erected at Broadwater Mill in 1887, the sugar exhaustion from the pulped slices of cane stalk was good, but the machinery's fuel consumption proved to be very high. Moreover, the diffusion process left a sodden mass of cane stalk slices which had to be crushed again before they could be used as fuel. Consequently, CSR never switched to diffusion, unlike John Robb & Co. at Cudgen, which was the only Australian sugar mill equipped with a diffusion unit during the late nineteenth century.[68] During the late nineteenth century, Kottmann also explored the efficiency of maceration. The descriptions of these experiments are confusing as they refer to Kottmann testing his 'diffusion drum' or 'diffusion apparatus', which is not the same as the diffuser. Kottmann tested the effectiveness of dragging the bagasse through a shallow tank filled with water, located between the sets of crushing rollers. He measured how much sucrose was dissolved from the bagasse soaked in water for varying lengths of time and at different temperatures.[69] Experiments determined the influence of the crushing speed and feed regularity of the cane into the crushing rollers on the amount of sugar recovered, the most effective clarifying agents, loss of sugar from delays in milling harvested cane, the economical consumption of the bagasse and the benefits gained by using shredders and continuous subsiders.[70]

Some Australian mill-owners observing CSR's progress also recognised the need to employ chemists. As early as 1884, the Mourilyan Sugar Co. employed an analytical chemist to oversee the manufacture of mill whites at Mourilyan Mill. Another mill-owner who acknowledged the importance of scientific observations in sugar production was Edward Long. In 1887, he established a chemical laboratory at his Habana Mill. Long even convinced Edward W. Knox to allow Habana's new chemist to spend two weeks at CSR's Homebush Mill during 1887, a practice rarely allowed as CSR was generally not particularly anxious to impart chemical knowledge to others, having acquired it at what Knox termed 'considerable expense'. From 1889 onwards, the operation of John Robb & Co.'s diffuser at Cudgen was supervised by an analytical chemist.[71] During the 1890s, some of the more astute proprietary and central milling companies hired analytical chemists. By 1900, Rous, Pioneer, Fairymead, Bingera, Marian and Mulgrave Mills, to list a few, used chemists, mostly during the crushing season, although at some sugar mills (e.g. Fairymead) the chemist was a full-time member of staff.[72]

Mostly, however, analytical chemists were not employed at all or were expendable staff in the 1890s and early 1900s. Racecourse Mill, for example, operated without a chemist in the early 1900s, after the farmers objected to such an appointment on the grounds that a chemist would 'oppress' them. Management at Marian Mill in 1901 believed that an analytical chemist in the off-season was a luxury and his services were dispensed with at the conclusion of the crushing season. In the following year, a motion at a meeting of the Directors of the Marian Central Milling Co. to employ a chemist for the 1902 crushing season lapsed for want of a seconder.[73] However, chemical control assumed higher priority in Australian sugar mills as the 1900s progressed. In 1904, Dr Walter Maxwell assumed control of the debt-ridden Queensland central sugar mills, and ordered that all the state's central sugar mills would be staffed with analytical chemists. Maxwell understood that reducing the debt of these factories could be achieved by introducing chemical control which would lead to less waste and more sugar being recovered from the harvested cane. CSR also pressured Australian sugar mill-owners into employing chemists. Many Australian sugar mills by the early 1900s had ceased manufacturing mill whites in favour of producing refining crystals for CSR's refineries (see Chapter 9 for a full discussion). As part of their agreement with CSR, Australian sugar mill-owners had to produce sugar with initially 88 net titre (n.t.) strength (later 94 n.t.). If they failed, CSR could refuse to accept the sugar. Chemical analysis was the only way mill-owners could determine whether they met CSR's requirements. Yet in the early 1910s, a few Queensland sugar mills still had no chemist. James Bartholomew, a member of the firm operating Goodwood Mill, for example, advised the 1912 Royal Commission into the Australian Sugar Industry that he did not know the percentage of juice recovered from the crushed cane at his sugar mill as they did 'not keep a chemist there'.[74]

Milling developments, 1900–1914

As a result of expanded capacities, chemical control and mill closures, the average annual output of individual sugar mills in New South Wales and Queensland rose considerably between 1885 and 1899. By the late 1890s, this figure stood at nearly 1 930 tonnes per mill in Queensland, and just under 1 320 tonnes in New South Wales (Table 8.2). The majority of Australian sugar mills, except the juice mills and some semi-mechanised mills, had reached levels of technological advancement comparable to the factories in at least Cuba, Demerara and Louisiana, where mill-owners during the 1880s had converted rapidly to multiple crushing, multiple effet evaporation and the use of vacuum pans.[75] Edward W. Knox had concluded in 1900 that the system of manufacture adopted by CSR, in so far as they could ascertain, was 'better than that in vogue in any other country where sugar cane is produced'. Dr Walter Maxwell suggested in his 1900 report on the Queensland sugar industry that the best of the state's central sugar mills were equal in juice extraction to the best Hawaiian sugar mills. Yet he was critical of the heavy use of maceration in Queensland sugar mills, believing it caused increased fuel consumption due to the need to evaporate the excess water added during the process with little return in extra sugar. Maxwell also argued there was an urgent need for the introduction into Queensland sugar mills of more thoroughly trained managers who had an improved knowledge of the chemistry of sugar juices and clarification procedures.[76]

Despite Maxwell's assessment of the performance of Queensland's sugar mills, the earliest BSES officers were not concerned about the state's milling arrangements, concentrating instead on farming practices. Yet the formation of the BSES did have a direct impact upon milling practices in Queensland. Following the introduction of *The Sugar Experiment Stations Act* of 1900, all Queensland sugar mill-owners had to install a reliable weighbridge or other weighing instrument in their factory. All cane entering the sugar mills was to be weighed and records kept of the tonnage of cane delivered by each canegrower so that the levy to fund the BSES could be accurately assessed.[77] In addition, a few Queensland sugar mill-owners during the 1900s and early 1910s modified their factories by catching up with technological developments from the 1890s. Some examples of these changes included the following: sets of triple effets being converted to quadruple effets at Marian Mill in 1910 and quintuple effets at Bingera in 1908; electricity finally being installed in Racecourse Mill in 1911; and Goodwood and Mulgrave Mills being converted to treble crushing in 1912 and 1913, respectively.[78] Over this fifteen year period, CSR focused upon the following: ensuring all its sugar mills had preparatory devices and treble crushing; experimenting with different types of mechanical pushers to assist the flow of cane through the crushing rollers; and trying to reduce fuel usage.[79]

Throughout the 1900s and early 1910s, crushing power became concentrated in fewer Australian sugar mills. Mill closures continued in Queensland, particularly around Mackay and Bundaberg. CSR became the only sugar mill-owner in New South Wales, after the last of the smaller, independently-owned sugar mills shut in the mid-1900s and CSR purchased Cudgen Mill in 1911.[80] By 1914, fifty sugar mills operated in Australia, twenty less than at the turn of the century. As a consequence of these closures and growing chemical control over milling operations, the average annual output per sugar mill nearly doubled in Queensland and did double in New South Wales between 1900–1904 and 1910–1914 (Table 8.2). Yet this improvement was not lauded by industry officials. In 1914, CSR's management expressed disappointment with the milling performance of its sugar mills, observing that the conversion to treble crushing had not enabled the firm to 'extract any more sugar by crushing alone'. John Newman, sugar boiler and superintendent of sugar production at Marian Mill, advised the Royal Commission into the Australian Sugar Industry in 1912 that the harvested cane was rushed through the crushing rollers in many Queensland sugar mills, without the slightest technical control. Engineers at Queensland's central sugar mills were satisfied to extract as low as 88% of the juice from the cane, and seldom achieved more than 91%. This level was very poor in comparison to CSR's mills which extracted about 96% of the juice. Thus, many Queensland sugar mill-owners were burning valuable sugar in the bagasse which was used as a fuel in their boiler furnaces. Harry Easterby, the general superintendent of the BSES concurred, remarking in 1912 that the majority of Queensland's sugar mills were 'not doing the work that they should do'. He suggested the introduction of 'better methods of milling'.[81] However, it was a decade before Easterby turned his attention to the state's sugar mills (see Chapter 17).

From field to market: moving harvested cane and sugar

Late nineteenth century photographs of the exterior of some Australian sugar mills often show wagons or drays filled with harvested cane, waiting to be unloaded (see Plate 8.4). Drays were a two-wheeled vehicle or a cart with fixed sides. They were generally pulled by one horse. A wagon was a four-wheeled vehicle, pulled by teams of horses or bullocks. Mostly, the harvested cane was loaded vertically or upright into the drays or wagons. In Southern Queensland, especially throughout the Marburg and Beenleigh districts, the preferred method of transporting harvested cane was the German system of light wagons. Pulled by two horses, the harvested cane was loaded horizontally into the wagon which had a long body and high side frame. These wagons could handle 25–30 cwt (1.2–1.6 tonnes) of cane. Moreover, the design of some German wagons sped up the unloading process as the wagon box sat upon rollers in front and behind, and were fastened into place by an iron catch. After the catch was lifted, the box, with its load of cane was easily pushed back, the tailboard being removed, and when the centre of gravity was passed the whole box tipped up, emptying its contents onto the ground.[82]

Transporting harvested cane by wagons and drays, however, had several disadvantages. Delays were common, especially during rainy weather. Habana plantation's tenant farmers, for example, found that the constant wet weather during 1886 made the carting of harvested cane by wagons and drays to the sugar mill impossible. Much of their cane remained unharvested at the end of the crushing season. Moreover, wagons and drays worked most effectively over short distances on plantations, which were compact, self-contained units. However, as the central milling system emerged, suppliers became increasingly located further away from a sugar mill, so it was not possible to transport their cane over long distances by wagons or carts. The Richmond River Sugar Co. reported in 1887 that the arrangement of hauling cane to its Rous Mill by wagons and drays was far too expensive, especially when it had to be brought in from the out-lying farms, as the roads were 'for the most part in a most primitive state, and in a dreadfully shocking condition'.[83] Moreover, such arrangements were not desirable, as the more technological advanced, fully mechanised sugar mills needed a regular supply of large amounts of cane to maintain continuous operations, and this supply could not always be assured by the use of wagons or carts. The solution in some sugar-producing districts was moving harvested cane by wagons and drays over much shorter distances to riverside loading points or narrow gauge (60.9 cm or 106 cm) tramway sidings where the cane was placed either into punts and then moved to the sugar mills by water or loaded onto rail trucks which were pulled to the mills by teams of horses or steam-powered locomotives.

The use of water transport to move harvested cane emerged simultaneously in Southern Queensland and New South Wales during the late 1860s and 1870s. CSR's New South Wales mills were supplied with cane from many different canegrowers scattered along the Clarence, Richmond, Tweed or Macleay Rivers and their tributaries. Coordinating the movement of so many wagons or drays over poor roads to ensure a reliable cane supply would have been impossible, so CSR relied on water transport. A small fleet of steam-powered tugs pulled red-painted punts or droghers, sometimes chained together, to its riverside sugar mills.[84] In Southern Queensland, some sugar mill-owners crushed the cane of neighbouring canegrowers without processing facilities, taking advantage of the navigable rivers throughout the region. Punts loaded with harvested cane were moved along the Brisbane, Mary, Albert and Logan

Rivers, discharging their loads at riverside wharves. The cane was then loaded onto railway trucks, and hauled, sometimes by steam engines, to the sugar mills over short tramlines running from the wharves to the mill's cane carrier.[85]

Punting of harvested cane ceased along the Brisbane River during the 1880s following the closure of Pearlwell and Coleridge Mills, the main riverside sugar mills which relied on punts to deliver their cane. However, punts loaded with cane continued moving along the Albert River in the early 1890s, being discharged at the Albert River Milling & Distilling Co.'s Beenleigh Mill. At this sugar mill, the mechanical cane carrier had been lengthened from the factory to the riverside wharf, enabling the cane to be unloaded from the punts directly onto the cane carrier, thereby avoiding double handling.[86] In the mid-1890s, punts were used along the Kolan River, supplying cane to Invicta Mill. The Maryborough Sugar Factory also relied on punts. Harvested cane was derricked off the punts and dropped onto the cane carrier, which extended from the river to the mill's crushing rollers. In 1909, the directors of the Moreton Central Sugar Mill Co. Ltd. expanded Moreton Mill's cane supply by accepting cane that had been punted along the Maroochy River from downstream suppliers. Two years later, seventy farmers in the Rosedale district, north of Bundaberg, agreed to support a sugar mill erected at Baffle Creek. Most canegrowers used the Baffle Creek Sugar & Trading Co.'s motor boat and punts to get their cane to the mill.[87]

In New South Wales, the punting system became more sophisticated during the 1890s and 1900s. On busier loading spots along the Richmond and Tweed Rivers, and to a lesser extent the Clarence River, permanent tramlines extended by portable tramlines led away from the rivers. By 1897, 16.3 km of permanent track throughout the Tweed River district connected farms to Condong Mill, while the length of permanent tramline associated with Broadwater Mill had increased from 4 km in 1882 to 21 km in 1897. Railway trucks loaded with harvested cane were lugged over these tramlines by horses to the riverside loading points. Canecutters moved and re-laid the portable tramline as harvesting shifted during the crushing season. Cane arriving at Condong Mill by punt had sometimes reached the river by unorthodox methods. During the early 1900s, harvested cane from the high ridges along the north side of the Tweed River had descended to the riverside loading spots by either aerial ropeway or dry flumes or chutes. The most elaborate system, however, was the funicular established at Terranora. Two tramlines ran from Joubert's Wharf to the high ridges on the compensation plan, the loaded railway trucks coming down serving to pull the empty railway trucks on the other rails.[88]

Certain disadvantages were associated with punting cane. Firstly, the harvested cane was double-handled. Wagons, drays or railway trucks had to be unloaded at the riverside loading points. The canecutters shouldered the cane across two parallel planks known as cane slips. Accidents were common, as workers slipped and fell at the bank, off the slip or in the punt. At the sugar mill, the punt then had to be unloaded, again initially manually. Eventually, CSR established mechanical grabs at its mill-side wharves, enabling the faster unloading of the harvested cane (see Plate 8.6). Secondly, CSR claimed there was some difficulty in maintaining a regular supply of cane to its New South Wales sugar mills from punts, especially at night, compared to its Queensland sugar mills where the entire supply of cane came via railway trucks. Thirdly, poor stowage of the harvested sticks of cane on the punts often led to delays at unloading or losses when the sticks of cane dropped overboard. Recovering sticks of cane lost overboard into a river was much harder than retrieving sticks of cane that had fallen from rail trucks. Fourthly, tidal conditions and changing depths of rivers and creeks influenced the amount of cane loaded in punts. During the early 1890s, shoaling had made it increasingly difficult for CSR's

Plate 8.6. Mechanical cane grab unloading punts filled with harvested cane at the Broadwater Mill wharf, 1907.

Source: CSRR, 142/3749, Negative No. 5650. Image reproduction courtesy of CSR Ltd.

cane punts to move along North Creek to the Richmond River, from where they could reach Broadwater Mill. To overcome this difficulty, a canal was cut in 1896 from a point about 1.6 km west of Ballina northwards towards North Creek, creating a new channel between North Creek and the Richmond River. Punting of cane along Baffle Creek also quickly became uneconomical during the mid-1910s because shifting sandbanks reduced the amount which could be loaded in a punt from 60 tonnes to only 10 to 15 tonnes.[89]

Narrow gauge tramlines or railways avoided many of the disadvantages associated with punting harvested cane. The Mackay sugar planter, John Spiller, is credited with pioneering the use of tramlines in Australia's sugar-producing districts. In 1879, he commenced building a tramline on Pioneer plantation for the purpose of conveying firewood and harvested cane to his sugar mill. A year later he built another tramline on his River plantation. A locomotive equipped with a vertical boiler pulled the loaded wagons. It had been built locally at the Victoria Foundry, Mackay, but was eventually replaced in 1881 by a locomotive from the famous engineering firm, John Fowler & Co. of Leeds.[90] Spiller's initiative was soon matched by other Australian sugar mill-owners, as several of the new mechanised sugar mills established in the 1880s were connected to tramlines. Mourilyan Mill is a good example of an Australian sugar mill erected in the early 1880s with a tramline from the outset. By 1884, it was connected to approximately [illegible] km of permanent tramline of 60.9 cm gauge (two foot gauge) and 4 km of portable tramline. Two locomotives pulled the sugar mill's sixty rail wagons. In contrast, horses hauled the cane trucks over the 60.9 cm gauge Decauville rail network at Innishowen Mill. Young Bros. also relied on horses during the 1880s to move loaded rail trucks along Fairymead Mill's 60.9 cm gauge tramline network. They claimed that performing a similar task without a tramline would have necessitated the employment of ten drays, each pulled by two horses, costing £5 daily, as opposed to the 10s spent daily on the tramway operation.[91]

During the 1880s, successive Queensland governments also commenced building railways that benefited canegrowers. In October 1882, the Queensland Parliament approved plans for the Mackay Railway, to run from Mackay westwards to Mirani, with a southern branch to Eton. The railway was designed to assist small canegrowers get their harvested cane to the sugar mill of their choice, except initially the railway avoided any contact with Meadowlands, Te Kowai, Palms, Pleystowe and Victoria Mills. This arrangement was amended in 1887 when the Queensland Department of Railways agreed to build short spur lines to the local sugar mills, their owners agreeing to repay the costs over five years. By the early

1890s, the Mackay Railway was busy conveying cane for small farmers who sold their crops to the sugar mill offering the best terms. The Queensland government's Bundaberg to Mt Perry Railway also benefitted canegrowers, although its construction was not originally intended for that purpose. Built in the early 1880s for the copper mine at Mt Perry, the railway when completed was of little use to the Mt Perry copper mine which had all but ceased operations by 1884. Harvested sugar cane, however, became a major freight item over the railway following the opening of Bingera Mill in 1885. Gibson & Howes Ltd. built their own spur lines, one to the sugar mill, and the other to the Watawa locality for cane.[92]

After the collapse of world sugar prices in 1884, Australian sugar mill-owners realised that greater mill throughput would improve their income, although this improvement could only be achieved by obtaining increased cane supplies, often grown on lands further away from the sugar mills. Some Australian sugar mill-owners who had erected tramlines over short distances in the early 1880s expanded the size of their tramline networks. Farleigh Mill, for example, had the length of its tramline enlarged from 6 km in 1891 to 24 km in 1896.[93] Other Australian sugar mill-owners built their first tramlines in the late 1880s and 1890s. Edward Long of Habana plantation commenced building a narrow gauge tramline in 1887, after being unable to process all of his tenants' cane in late 1886 because persistent wet weather had made its carting to the sugar mill impossible. By 1896, this property was criss-crossed by 27 km of narrow gauge tramline. The Richmond River Sugar Co. also fits into this category. In 1891, a 60.9 cm gauge tramline was built from Rous Mill to Alstonsville, a distance of 8 km. The cumbersome system of moving the harvested cane by bullock wagons to the sugar mill was eliminated. Drysdale Bros. & Co. was another mill-owner to engage in tramline building during this period, constructing their first 5 km of tramline for Pioneer Mill in 1893; this tramline network had been extended to a distance of 25 km by 1899. Drysdale Bros. & Co. adopted the standard Queensland Railways gauge of 106 cm (three feet six inches), anticipating the eventual railway link between the Lower Burdekin district and Townsville.[94]

During the period 1895–1914, several local authorities in Queensland's sugar-producing districts built tramlines to take advantage of the growing trend towards moving harvested cane and bagged sugar by rail. The Cairns Divisional Board was one of the first local authorities to engage in this activity. In 1896, the Cairns-Mulgrave Tramway was opened from Cairns to Nelson (now Gordonvale), and over the next decade it was extended southwards, reaching Babinda by 1910. Farmers located beside the Cairns-Mulgrave

Plate 8.7. A rake of loaded cane trucks being hauled over a narrow gauge tramline to Isis Mill, 1896.

Source: Image no. 410, John Oxley Library. Image reproduction courtesy of the State Library of Queensland, Brisbane.

Tramway sent their harvested cane by rail wagon to either Mulgrave Mill at Gordonvale or CSR's Hambledon Mill at Edmonton. In the same year, the Pioneer Shire Council commenced extending the Mackay Railway westwards from its terminus at Mirani. By July 1911, the line had reached the foot of the Eungella Range at Netherdale. Sugar cane and firewood for the central sugar mills near Mackay and the Cattle Creek Mill at Finch Hatton provided the bulk of the line's traffic. In 1897, the Hinchinbrook Divisional Board also secured a Queensland Treasury loan to extend the East Ingham to Lucinda tramline by another 27 km westwards along the Stone River. The Stone River canegrowers benefited from being able to rail their cane to CSR's Victoria Mill. Three local authorities – the Townsville Municipal Council and the Ayr and Thuringowa Divisional Boards – combined resources in 1900 to construct the Ayr Tramway. This tramline was primarily built to move bagged sugar from the Lower Burdekin sugar mills to the Port of Townsville. However, farmers in the Haughton River district were using the tramline by 1906 to send their cane to Pioneer Mill for crushing, an arrangement that lasted until Invicta Mill was moved from the Kolan River district to Giru in the Haughton River district in 1918.[95] The Woongarra Shire Council also constructed a tramline in 1912 from Woongarra Junction (near Bundaberg), eastwards to Bargara on the coast and then southwards towards Elliott Heads. The line served the transport needs of Qunaba, Windemere and Pemberton Mills.[96]

The building of tramlines resulted in different arrangements being developed for the loading of harvested cane onto rail trucks. Suppliers of

Pioneer Mill in the 1890s brought their harvested cane in drays or wagons to elevated loading sites. The loads of cane were tipped straight down into the rail trucks, in an attempt to avoid manual loading of the stalks of cane. At other localities, cane derricks were constructed to facilitate easier lifting and loading of bundles of cane onto the rail trucks.[97] One of these overhead, circle boom type derricks on the Ayr Tramway is shown in Plate 8.8. In addition, arrangements at some Australian sugar mills were modified to ensure a quicker unloading of the rail trucks, obviating the dropping of the stalks of cane onto the ground, to be picked up afterwards by hand and placed onto the cane carrier. At Mourilyan Mill, for example, the cane carrier ran through an underground pit over which were three lines of rail. Three rail trucks at a time were discharged onto the cane carrier. A similar arrangement existed at Bingera Mill. Rail trucks were tipped sideways, discharging their loads onto the cane carrier.[98] Despite these innovations, many Australian sugar mill-owners during the late 1890s still relied on teams of Pacific Islanders, Japanese or Chinese to unload the cane manually from railway wagons onto the cane carrier, a time-consuming process which sometimes contributed to a bottleneck in the milling process.

The impetus to eliminate as many workers as possible at the cane carrier came in 1900, when the Commonwealth government announced the deportation of the Pacific Islanders from Australia after 1906. Australian sugar mill-owners were faced with either the prospect of out-laying more in their wages bill to employ Europeans who would take the place of the departing non-Europeans unloading the harvested cane from rail trucks at their sugar mills or

Plate 8.8. An overhead, circle boom type cane derrick at Hodel Siding, Ayr Tramway, 1911.
Source: *ASJ*, 6 April 1911, p. 27.

reducing the number of mill workers at this station by the installation of new labour-saving equipment which sped up the process of unloading. Peter Macgregor, the engineer in charge at Hambledon Mill, had been experimenting with such a device in the late 1890s.[99] In 1898, he advised Edward W. Knox of his intention to have two sprocket wheels running across the cane carrier. Chains with short projecting forks would revolve around the sprocket wheels, and the feeding down of the sticks of cane from the wagons could be stopped and started by a small hand winch. Unusually, Knox showed limited interest in the potential of this device, but the evidence suggests that Smith William Davids, the General Manager at nearby Mulgrave Mill, was very interested in these devices.

Davids had served as an apprentice with A. & W. Smith & Co., the Glasgow sugar machinery makers, before coming to Australia and gaining engineering experience in several sugar mills. He was appointed General Manager of Mulgrave Mill in 1898.[100] Davids and his co-inventor, Matthew Muir, registered provisional specifications for a mechanical cane unloader of railway trucks in October 1901. He received a patent for this invention in February 1902. The design of the mechanical cane unloader looked very much like the device operating at Hambledon Mill, and CSR unsuccessfully challenged the validity of the patent. CSR eventually came to an agreement with Davids, paying him £500 in 1903 for the right to use his unloaders at their sugar mills. Other Queensland sugar mill-owners also readily adopted the new innovation during the 1900s, ordering the machinery from Walkers Ltd. which held the exclusive rights to manufacture Davids mechanical cane unloaders in Queensland. An analysis of newspaper reports showed that at least half Australia's sugar mills had installed this equipment by 1914.[101] Mechanical cane unloaders not only sped up the manual unloading of sticks of cane at the carriers, but slashed the number of workers at the carrier, so that unloading could be undertaken with three or four men instead of eight to twelve workers, depending upon the size of the sugar mill.[102] They performed best, however, where cane was delivered to the sugar mill by rail trucks, not wagons or carts (see Plate 8.9).

Moving harvested sugar cane to the sugar mills was not the only transport challenge faced by Australian sugar mill-owners during the late nineteenth century. Once the raw sugar or mill whites had been manufactured, it had to be sent to a commission agent, wholesaler, retailer or a refiner who further processed the sugar into white sugar. Whatever type of sugar was produced at Australian sugar mills during the late nineteenth and early twentieth centuries, it mostly ended up filling jute bags, each weighing 32–45 kg.

Plate 8.9. Davids mechanical unloaders in operation at Bingera Mill, 1909. Note the hand winch that controlled the chains and the short, projecting forks which dislodged the stalks of cane into the carrier.

Source: *ASJ*, 7 October 1909, p. 256.

A limited quantity of bagged sugar could be stored in the sugar mill's storeroom, and it was not uncommon for Australian sugar mill-owners to stockpile sugar for short periods due to shipping delays. Some mill whites produced at the Maryborough Sugar Factory during the mid-1900s were even stockpiled on occasions for as long as twelve months in order to fetch a better price on the market.[103] Generally, however, the sugar mill-owner had to sell most of the factory's output as the six month crushing season progressed in order to generate income to sustain milling operations.

Before 1880, some Australian-produced sugar was sold locally to wholesale merchants or grocers in nearby towns or cities. In the early 1870s, William Clement of Alpha Mill (Richmond River district), for example, delivered his sugar to John Grime, a storekeeper located 25 km away at Casino.[104] Therefore, this sugar was being transported by wagons only over relatively short distances. By the early 1870s, however, shipments of sugar from some North Queensland and New South Wales sugar mill-owners were directed increasingly to Brisbane or Sydney. As the decade progressed, other Queensland sugar mill-owners also supplied the Melbourne market. Colin Munro, owner of Fisherfield Mill, even tested the overseas market by sending samples of sugar to England and Scotland for sale.[105] Sea transport and ports, therefore, became vital in the movement of this sugar to markets.

Many Australian sugar mills established between 1864 and 1890 were sited on navigable rivers or creeks. In the pre-railway era, the sugar mill-owners had built riverside wharves to receive goods and machinery and to export their sugar to market. Bagged sugar was moved by bullock wagons to these riverside wooden jetties. To overcome the difficulties associated with moving wagons laden with bagged sugar over unsealed roads, several short tramlines were constructed between some sugar mills and the estuarine wharves. The Bloomfield River Sugar Co. in the early 1880s built a 3 km tramline from their Vilele Mill to the Ayton Wharf on the Bloomfield River. In the Cairns district, Thomas Swallow and Frank Derham erected a 5 km tramline from Hambledon Mill to a wharf at Trinity Inlet. A similar arrangement existed in the Lower Burdekin district, where the Burdekin Delta Sugar Co. built 5 km of tramline from its Airdmillan Mill to a wharf on Plantation Creek, which was navigable only to small steamers. CSR, however, was required to build a much longer tramline of almost 13 km to ensure Victoria Mill was linked to its riverside wharf on the Herbert River at Halifax.[106]

These river sugar ports had one main disadvantage: lack of deep water. In many instances, the lighterage process enabled the bagged sugar to be transported from the upstream shipment points to freighters anchored down-river in deeper water or larger ports nearby. Two examples will illustrate this process. In the Lower Burdekin district, the bagged sugar was loaded onto small steamers in Barratta or Plantation Creeks. These steamers then took the bagged sugar approximately 100 km to Townsville, where it was transferred to larger ships for despatching to southern markets. This process cost on average 14s: 4s to load the sugar onto the small steamers and then another 10s in freight charges to Townsville. A similar process occurred at Mackay, where the Pioneer River often runs dry at low tide. Launches and cargo lighters ferried waterside workers, general cargo and bagged sugar from the Mackay wharves, about 6.5 km from the bar at the entrance to the Pioneer River, to bigger ships anchored at Flat Top Island where 7 m of water was available (Plate 8.10). Trying to synchronise the movements of bullock wagons or rail wagons (after 1888) conveying bagged sugar from the district's sugar mills with ever changing tidal times and the loading and leaving of sugar lighters at the wharves was a difficult task. After 1897, the ships anchored at Flat Top Island also received sugar from Proserpine Mill. This sugar had initially travelled 7 km by tramline from the sugar mill to a wharf on the Proserpine River, and then had been lightered a distance of approximately 100 km to Flat Top Island.[107]

In northern New South Wales, the Clarence and Richmond River Steam Navigation Co. catered for the growing traffic in coal demanded by the sugar mills and the export of their sugar to Sydney. However, CSR decided in 1872 to purchase its own vessel after the Clarence and Richmond River Steam Navigation Co. announced that it was unwilling to devote one of its ships

Plate 8.10. Hand-loading bagged sugar into a lighter at Mackay for transfer to a southern ocean-going steamer anchored at Flat Top Island, 1911. This cumbersome system of moving bagged sugar persisted until the opening of a deep water harbour at Mackay in 1939.

Source: *The Queenslander*, 27 May 1911, p. 22.

to maintain a regular service of coal deliveries to CSR's Clarence River sugar mills. CSR then became the only Australian sugar mill-owner to develop its own fleet of ships. By 1880, CSR had acquired the vessels SS *Keera*, SS *Fiona,* the paddlewheel *SS Terranora* and the smaller, iron steam paddlewheel tugs *Chatsworth, Wollumbin, Darkwater, May Queen* and *Iluka* which pulled the cane punts along the rivers in northern New South Wales. The larger vessels moved coal from Newcastle to CSR's New South Wales sugar mills and returned with bags of sugar, barrels of molasses and hogsheads of rum from the distillery at Harwood Mill. In 1882, the original SS *Fiona* was wrecked when she struck Seal Rock Point on the New South Wales coast. She was replaced with a second SS *Fiona* in 1883, and this vessel remained in the service of CSR until 1909. The third SS *Fiona* was delivered in 1909 and had been built with special modifications: tanks to carry molasses.[108]

Along the Queensland coast, the growing demand for shipping services to move sugar southwards was initially met by the major established coastal shipping companies such as the Australasian Steam Navigation Co. (ASN), Howard Smith and McIlwraith McEacharn Ltd. In 1887, after a period of intense competition and amalgamations between some shipping companies, a series of agreements were reached between the main competitors about 'spheres of influence' along the Australian coastline. The Australasian United Steam Navigation Co. (AUSN), the successor to the ASN, agreed not to trade past Melbourne, nor to trade in coal past Sydney, in return for a promise from Howard Smith not to sail north of Mackay and to share its sugar trade with the AUSN. Other northern ports were also served by the Queensland Steam Shipping Co. (QSS) and Burns, Philp & Co. This arrangement, however, was dramatically altered in April 1893 when the Adelaide Steamship Co. (ASC) announced the conclusion of a contract with CSR for the transport of all its Queensland sugar.[109] Moreover, Queensland Railways from 1893 onwards began carrying bagged sugar from the Isis district mills to Brisbane; the freight rates quoted were most attractive and gave the ships little hope of competing. Consequently, a period of competition between the major shipping companies commenced, with eventually negotiations concerning the movement of general cargo, including sugar, achieving another arrangement in July 1902: the ASC carried CSR's sugar only, while the AUSN and Howard Smith shared the remainder of the sugar that needed moving by sea, including the output from Millaquin Refinery.[110]

By 1914, the spate of railway and tramline building in Queensland during the 1890s and 1900s mentioned earlier had contributed to the abandonment of several river sugar ports or their movement seawards in the direction of deeper water. In the Cairns region, the bagged sugar from Hambledon and Mulgrave Mills now travelled via the Cairns-Mulgrave Tramway to the Port of Cairns. The Trinity Inlet wharf had been abandoned after 1897. Sugar from Mossman Mill, however, was still lightered down the coast to Cairns for reloading onto larger ships. In the Herbert River district, small paddle steamers pulling oaken barges loaded with bagged sugar no longer moved along the Herbert River to Dungeness, where the sugar was re-loaded onto larger ships anchored in the deeper water of Hinchinbrook Channel. Instead, the sugar was sent over the tramline built to connect Ingham with the new wharf and jetty erected at Lucinda Point in 1896.[111] Further south, sugar from the Lower Burdekin sugar mills could now be railed directly to the Port of Townsville after the completion of the Ayr Tramway in 1901, thereby eliminating the use of coastal lighters. Proserpine Mill's sugar was no longer lightered southwards to Mackay, but travelled northwards to the deep water harbour at Bowen, after a tramline from Proserpine was extended to Bowen in 1910. Yet the lighterage system still persisted at Mackay, and would not be eliminated until the erection of the Mackay Outer Harbour in 1939. River sugar ports were also still in existence at Bundaberg and Maryborough. Shoaling in the Mary River, however, had led to the decision to move Maryborough's port to deeper water. Maryborough was linked by a rail extension to a deep water jetty constructed at Urangan, although sugar was not removed from this wharf until 1917.[112]

These newer arrangements may have sped up the movement of bagged sugar from the sugar mills to local ports, but the task was still cumbersome and involved considerable man-handling. The raw sugar was bagged by teams of men at the mills, then manually loaded onto the rail wagons, before being taken to the wharf, put into cargo slings, swung on board and stacked in a ship's hold. On arrival at the sugar refinery or the wharves of the shipping or commission agents in the capital cities of the southern Australian colonies (later states), the process was repeated in reverse order. The work was heavy and unpleasant. Yet in 1914, the arrangement for moving sugar to the ports and refineries was to persist for another forty years, until the introduction of mechanical bulk handling of sugar from mill to port (see Chapter 18 for more details).

Competing for cane

Most Australian sugar mills established before 1885 were supplied by sugar cane grown on the property associated with the mill. Hence, no competition existed between the sugar mill-owners. Sometimes, however, these pioneering sugar mill-owners supplemented their cane supply by purchasing from outside canegrowers. Occasionally, millers competed for this cane. In 1870, sugar mill-owners rowed along the Mary River from one small farm to another, bidding for cane. This practice had disappeared by 1874, after the sugar mill-owners had divided the river into zones and agreed to purchase only in their own area.[113] In contrast, competition between New South Wales sugar mill-owners before 1885 was more common. CSR bore the brunt of this competition. By the late 1870s, CSR's cane supply throughout the Clarence River district had been reduced, as former suppliers erected their own small sugar mills. CSR required its contractors to trash their cane and offered between 13s and 15s per ton of cane, although it arranged for harvesting and delivery of the cane to the sugar mills. The small millers offered up to 18s per ton of cane delivered to the factory, permitted their suppliers to leave their cane untrashed and purchased cane refused by CSR and at full price. To remain competitive, CSR offered a bonus of £ 2 per acre in 1877 to all canegrowers who renewed their contracts with the company for another decade. Following repeated defections by small farmers, the bonus was raised to £3 for new contractors in 1879.[114]

CSR believed that this competition would eventually force the firm to move its milling operations elsewhere. However, CSR emerged as the only miller in New South Wales, after all the small sugar mills shut following the sugar price decline in 1884.[115]

After 1885, many Queensland sugar mills handled larger volumes of cane due to improved process technology (e.g. double or treble crushing; multiple evaporators). Moreover, it was an imperative that these sugar mills processed more cane, in order to reap the benefits of the capital outlay on this new technology. Hence, some sugar mill-owners increasingly sought ways to improve their cane supplies. In the Proserpine, Mossman, Maroochy and Yeppoon districts where only one sugar mill existed, the local sugar mill-owners extended tramlines to distant localities in order to encourage settlers to grow cane and to assist these settlers with the transport of this cane to the sugar mill. Competition with other sugar mill-owners was not an issue. Other sugar mill-owners in the Mackay, Bundaberg, Lower Burdekin and Isis districts faced greater difficulties in expanding their cane supplies because of the presence of rival sugar mills. Young Bros., for example, avoided competition with existing sugar mill-owners around Bundaberg by purchasing estates 50 km away in the Kolan River and Isis districts. They grew their own cane on these properties, railing it to Fairymead Mill. At Mackay, the Racecourse Central Sugar Co. Ltd. realised in 1894 that as time progressed, its sugar mill would face competition for suppliers with nearby Marian and Pleystowe Mills. To avoid competition, the Racecourse Central Sugar Co. Ltd. purchased large blocks of nearby land in 1894 and 1903 and subdivided them into smaller farms. Settlers could acquire these farms via purchasing leases on the condition they grow cane only for Racecourse Mill.[116]

Other Queensland sugar mill-owners improved their cane supplies by competing with nearby mill-owners for the canegrowers' annual crops, persuading them to switch sugar mills by offering higher prices than their competitors. The North Eton Central Sugar Co. Ltd. directors secured the cane from some Marian Mill suppliers in 1888 by offering them 13s per ton of cane. Marian Mill offered only 10s per ton. North Eton suppliers, however, were also willing to switch mills. In 1894, one of North Eton's biggest cane suppliers was induced to transfer his crop to Meadowlands Mill, after receiving an offer of a higher price for his cane. A decade later, more North Eton suppliers switched allegiances by sending their cane to Pleystowe Mill, which offered a higher price for cane than all its competitors.[117] Doolbi Mill, near Childers, actively competed with the Maryborough Sugar Factory and Mt Bauple Mill for cane from the Pialba canegrowers, who refused to sign long-term supply agreements with any mill-owner or to sell all of their cane to one sugar mill. The bulk of the Pialba cane available in 1903, for instance, went to Doolbi Mill, which offered the best price. Five years later, Mt Bauple Mill outbid the other two sugar mills for the Pialba-grown cane.[118]

Competition for cane was reduced or eliminated in some sugar-producing districts when sugar mill-owners realised it would be counter-productive and reached arrangements amongst themselves to reduce price wars and/or establish 'zones of influence'. From the mid-1900s onwards, the Australian Estates and Mortgage Co. Ltd. and Drysdale Bros. & Co., the two Lower Burdekin mill-owners, agreed to pay the same price for cane, thereby removing an incentive for canegrowers to switch sugar mills.[119] Various restrictive arrangements also existed in the Bundaberg district. The Millaquin and Yengarie Sugar Co. and Angus Gibson of Bingera Mill agreed not to compete for cane during the mid-1900s, but the arrangement expired in 1909. George Buss of Pemberton Mill agreed not to take cane from former Ashfield Mill suppliers, following the purchase of that factory by the Millaquin and Yengarie Sugar Co. CSR also sought to avoid competition in the districts where its factories operated. By 1912, 'understandings' existed between CSR and other milling companies in the Cairns, Innisfail and Isis districts. In each area, zones of influence had been established, with CSR's managers agreeing not to purchase cane outside their zones.[120]

The most elaborate arrangement to reduce competition emerged at Mackay. By the turn of the century, eight sugar mills existed south of the Pioneer River. Each was competing for cane, especially from canegrowers who were located close to the railway that ran through the district. In 1900, representatives of Marian, Racecourse and North Eton Mills met and agreed to pay the same flat rate per ton of cane. In 1902 they were joined by the representatives of Pleystowe and Palms Mills, and resolved to form the Mackay Sugar Manufacturers' Association (MSMA). Members again agreed to pay a common flat rate per ton of cane and not to skirt the pact by making any undisguised payment for cane, such as transport allowances. In 1906, the Cattle Creek Sugar Co. Ltd. which had erected Cattle Creek Mill at Finch Hatton, also joined the organisation. The arrangement between the six Mackay mills lasted until 1908, when Pleystowe Mill withdrew from the MSMA and offered suppliers higher prices than its competitors. The arrangement brokered by the MSMA finally collapsed in 1912, when Marian Mill offered transport allowances to suppliers loading cane onto wagons themselves. The Mackay sugar mill-owners, however, reformed the MSMA in late 1913, agreeing to purchase cane on the basis of sucrose content, thereby eliminating the contentious issue of paying a flat rate per ton of cane. Canegrowers before 1 June each year had to tell the sugar mill-owners which factory they intended to supply throughout the year and then no other sugar mill had the right to receive cane from them.[121]

In New South Wales, CSR's virtual monopoly over milling after 1895 and growing farmer dissatisfaction with the prices they received from the firm for their cane, led to hostile editorials in local newspapers. The Editor of the *Tweed Herald and Brunswick Chronicle,* for example, wrote in 1898 that CSR was a 'blood sucking octopus' and the 'farmer must suffer' so its shareholders could be paid enormous dividends.[122] A group of canegrowers in the Lismore district petitioned the New South Wales government in 1894, asking for funds to build a farmer-owned central mill in either the Richmond River or Tweed River district, so they would be independent of CSR. This scheme remained unfunded, as did an 1898 proposal for a central sugar mill from the farmers along the lower reaches of the Clarence River.[123] The most dissatisfied canegrowers in New South Wales simply left the industry, and as mentioned in Chapter 5, the area under sugar cane contracted substantially.

In Queensland, the issue of cane prices, zones of influences and lack of competition amongst the sugar mill-owners was first aired in 1908, when a group of Lower Burdekin canegrowers refused to sign new contracts to supply cane to Pioneer Mill. Instead, they formed the Macdesme Co-operative Mill Association, with a view to erecting their own central sugar mill, aiming to provide some opposition to the established Lower Burdekin sugar mill-owners who worked in combination by paying the same price for cane, which the farmers considered far too low. Drysdale Bros. & Co., the owner of Pioneer Mill, caved into the farmers' demand and raised the price paid for cane by 1s per ton.[124] Other Queensland canegrowers, however, were not receiving sympathetic hearings from sugar mill-owners about the price of cane, so they decided to lobby their national parliamentarians. On 21 October 1909, a petition was received by the Commonwealth Parliament. In this petition, the Queensland canegrowers claimed that the low prices received for cane was insufficient for them to carry out the required labour conditions. The Fusion ministry led by Alfred Deakin was sympathetic, and in March 1910, shortly before general elections, he appointed a Royal Commission to inquire into a range of issues associated with conditions in the Australian sugar industry, including the continued operation of the sugar bounty regulations. However, the Commission was in limbo because its Chairman, Judge Cohen of the New South Wales Arbitration Court, had resigned shortly after his appointment due to illness. Moreover, Deakin's government was defeated at the general elections on 13 April 1910 and the new Fisher (Labor) Commonwealth government shut down the Royal Commission.[125]

During 1910, canegrowers throughout the Bundaberg, Maroochy, Innisfail and Herbert River districts complained to the mill-owners about the low prices they received for their cane and lack of competition between the millers. All the sugar mill-owners, especially CSR, steadfastly refused to advance the price they paid for cane. Eventually, as noted in Chapter 5, the Fisher Commonwealth government appointed a Royal Commission into the Australian Sugar Industry in late 1911. During the hearings associated with this inquiry, several canegrowers and representatives from cane growing organisations complained that CSR paid its farmers too little in comparison to the grower-operated central sugar mills, and mostly these canegrowers were unable to switch their supply of cane to any other sugar mill. Other Queensland central milling companies were reputedly discriminating against non-shareholder suppliers by paying them less for their sugar cane than shareholders, but local 'zoning' agreements meant it was difficult to supply an alternate sugar mill that would offer a better price for the cane.[126]

Agitation for legislation to regulate the price paid for sugar cane by Queensland mill-owners intensified after the 1912 Royal Commission. On 8 July 1913, a deputation of Labor parliamentarians led by Thomas Ryan, then Leader of the Opposition, urged Premier Digby Denham to introduce legislation establishing cane price-fixing boards. Lt-Col. Colin Rankin (Burrum; Country Party) who was championing such legislation finally introduced his private members' bill in October 1913. During the parliamentary debate associated with 'The Sugar Cane Price Boards Bill', Rankin noted that Queensland canegrowers suffered from 'an absence of competition' and that tribunals would ensure a fair distribution of profits between miller and farmer. He stressed, however, that this measure was not an attempt to establish state control over cane processing. Premier Denham opposed the legislation, fearing it would lead to the 'nationalisation of production distribution and exchange'. Other parliamentarians announced that Queensland's sugar mill-owners, especially CSR, opposed the bill. Labor party support, however, ensured the bill was passed by the Legislative Assembly, but Rankin then withdrew his bill; it never reached the Legislative Council. Shortly afterwards, Rankin became Minister for Railways.[127]

Bitterly disappointed with their failure to achieve cane price-fixing boards, some canegrowers in the Mackay, Isis and Innisfail districts refused to harvest their crops for part of the 1914 crushing season as a protest against the prices received for their cane and the lack of competition between the sugar mill-owners. Yet towards the end of the crushing season they relented and harvested their crops when it became clear that the sugar mill-owners would not accede to their demands.[128] These protesting canegrowers, however, were probably unaware that the 1914 crushing season was about to become quite unique. Their protests certainly made the crushing season distinct, but it was also the last time that the Australian sugar milling sector would be fully unregulated for another seventy-five years. Within twelve months of the last Queensland sugar mill stopping its crushing rollers in 1914, the Commonwealth and Queensland governments had introduced legislation formalising and regulating the informal milling arrangements that existed in 1914, thereby heralding the beginning of a new and regulatory epoch in the history of the Australian sugar industry.

Notes

1 Kerr, *Sugar at Maryborough*, pp. 89–91 & R. Parsons, *The Port of Maryborough. A History* (Magill, SA, 1976), p. 30; for this process see *Q*, 6 September 1873, p. 5, 6 February 1875, p. 5 & 15 June 1878, p. 6.

2 *ATCJ*, 29 April 1871, p. 523; E.W. Knox, AR 1872 season, p. 27, CSRR, Z/109, Box 374.

3 For descriptions of mills with horse-driven crushing rollers see *BC*, 19 September 1873, p. 3 (Biribi); *Q*, 5 August 1871, p. 1 (Yawillpah); *BC* 30 October 1869, p. 5 (St Helena); *The Planter and Farmer*, March 1882, p. 59 (Clydesdale); *Q*, 23 September 1871, p. 4 (Benowa); *Q*, 16 September 1869, p.10 (Fisherfield); & *BC*, 3 October 1873, p. 7 (Otmoor).

4 *BC*, 27 August 1879, p. 6; A. Boyd, 'Early sugar mills of Queensland', *QAJ*, Vol. 14 (1920), p. 285; for CSR's Chatsworth Mill making a ton of sugar from 13 tons of cane see E.W. Knox, Harwood, to E. Knox, 3 November 1871, CSRR, 142/2323, p. 45.

5 For descriptions of mills being converted from horse-driven to steam-powered crushing rollers see *Q*, 26 November 1870, p. 5 (St Helena); *Q*, 20 September 1873, p. 5 (Ageston); *Q*, 14 October 1876, p. 26 (Coleridge); *Q*, 16 September 1871, p. 5 (Clydesdale); *Q*, 18 November 1871, p. 5 (Cassada); & *Q*, 31 May 1873, p. 1 (Benowa).

6 *BC*, 2 August 1867, p. 3; *Q*, 8 October 1881, p. 471; *The Planter and Farmer*, January 1883, p. 370.

7 *ATCJ*, 28 January 1871, p. 110; *MM*, 16 June 1877, p. 3.

8 *The Manning River News*, 13 June 1868, p. 3.

9 For a description of a concretor in operation at Belvidere Mill see *ATCJ*, 7 January 1871, p. 12; for other concretors see *BC*, 15 September 1875, p. 3 (Loganholme), Kerr, *Sugar at Maryborough*, p. 58 (Myrtle Grove), E.W. Knox, AR 1872 season, p. 34, CSRR, Z/109, Box 374 (Chatsworth) & *ATCJ*, 28 January 1871, p. 110 (Darkwater).

10 *Q*, 15 April 1876, p. 21; for descriptions of steam-heated evaporating pans in other sugar mills see *Q*, 19 November 1870, p. 5 (Meadowlands); Kerr, *Sugar at Maryborough*, pp. 18 & 45 (Maryborough Sugar Co. & Iindah); *Q*, 26 July 1873, p. 10 (Tygum); & *Q*, 18 October 1873, p. 10 (Logan Sugar Mill).

11 Kerr, *Sugar at Maryborough*, p. 83; for a selection of Queensland mills being upgraded or built with vacuum pans during the 1870s see *BC*, 16 February 1876, p. 6 (Macknade); *Q*, 1 September 1877, p. 6 (Meadowlands, River & Te Kowai); & *Q*, 20 September 1874, p. 5 (Ageston); H. Roth, *A Report on the Sugar Industry in Queensland* (Brisbane, 1880), p. 62.

12 J.A. Mills, 'Davidson, John Ewen (1841–1923)', *ADB*, Vol. 8 (1981), p. 228; *MM*, 19 April 1879, p. 2; Half Yearly Report, 30 December 1875, Beenleigh Branch of QNB, QNBR.

13 Roth, *A Report on the Sugar Industry in Queensland*, p. 61; those mills identified using the Icery process were as follows: St Helena (*Q*, 14 June 1873, p. 5); Pearlwell (*Q*, 6 December 1873, p. 5); Helensvale (*Q*, 9 September 1876, p. 30); Redland Bay (*Q*, 15 July 1876, p. 4), Alexandra and Meadowlands (Kerr, *Pioneer Pageant*, p. 57); Ormiston and Eatonvale (*Q*, 4 January 1879, p. 27); River, Pioneer & Branscombe (*Q*, 5 January 1878, p. 26); Pleystowe (*Q*, 26 July 1884, p. 159; Otmoor (*LW*, 26 July 1879); Bundall (J. Elliot (ed.), *Letters to Bundall, 1872–1879 and Lina Cooper's Manuscript* (Brisbane, 1993), p. 112); & Small's (*MC*, 6 April 1880, p. 4 (quoting from the *Lower Clarence Advocate*).

14 *Q*, 9 October 1875, p. 21; A. Mackay, *The Sugar Cane in Australia* (Sydney, 1883), pp. 144–146.

15 *MM*, 25 September 1875, p. 2; Kerr, *Pioneer Pageant*, pp. 58–59; E. W. Knox, AR 1873 season, p. 27, CSRR, Z/109, Box 374.

16 A. Birch and J. Blaxland, 'The historical background', in A. Lowndes (ed.), *South Pacific Enterprise. The Colonial Sugar Refining Company Limited* (Sydney, 1956), pp. 24–25.

17 Birch and Blaxland, 'The historical background', pp. 25 & 35; E.W. Knox, AR 1877 season, pp. 1, 4 & 13 & E.W. Knox, AR 1878 season, pp. 4 & 16, in CSRR, Z/109, Box 374.

18 *QVP*, Vol. 4 (1889), pp. 396 & 414; Mirrlees Watson & Co., Order Books, UGD, 118/2, 7/1, p. 54, BRC; *MM*, 31 October 1874, p. 2; *Q*, 21 September 1872, p. 5 & 26 October 1872, p. 5.

19 For nineteenth century exports of Scottish-made sugar machinery see R. Beachey, *The British West Indies Sugar Industry in the Late Nineteenth Century* (Oxford, 1957), pp. 175–176 & A. Mattei, 'The role of the Scottish sugar machinery manufacturers in the Puerto Rican plantation system', *Scottish Industrial History*, Vol. 8 (1985), p. 21; for the durability of Scottish-made sugar machinery see M. Moss and J. Hume, *Workshop of the British Empire: engineering and shipbuilding in the west of Scotland* (London, 1977), p. 35.

20 For details see B. Higman, 'Yeoman and plantation landscapes in the sugar industry of New South Wales'. BA Hons thesis, Sydney Uni., 1967, Appendix 2; for a historical account of the Atlas Engineering Co. see *CRE*, 5 July 1879, p. 3; advertisements of P.N. Russel & Co. are found in *Pugh's Almanac*, 1864 and 1868 and *Slater's Queensland Almanac*, 1870.

21 For Smellie & Co. see C. Wood, *Sugar Country. A History of the Raw Sugar Industry of Australia, 1864–1964* (Brisbane, 1965), p. 74; for the Mackay foundries see Kerr, *Pioneer Pageant*, p. 49; for the Vulcan Foundry see G. Loyau, *The History of Maryborough and Wide Bay Burnett Districts from the Year 1850 to 1895* (Brisbane, 1897), pp. 339–340.

22 Loyau, *The History of Maryborough and Wide Bay Burnett Districts*, pp. 335–340; Kerr, *Sugar at Maryborough*, pp. 42–43; Ev. John Walker, *QVP*, Vol. 2 (1879), p. 629.

23 For mill outputs see P. Fletcher, 'The sugar industry of Queensland', in P. Fletcher (ed.), *Queensland. Its Resources and Institutions* (Brisbane, 1886), p. 11.

24 *ATCJ*, 7 January 1882, p. 21; *Sydney Mail*, 13 September 1884, p. 531.

25 *QVP*, Vol. 4 (1889), pp. 130, 182 & 212 (Hambledon, Mourilyan and Ripple Creek); *Sydney Mail*, 13 September 1884, p. 53 (Palms); *Sydney Mail*, 4 October 1884, p. 687 (Farleigh); *Q*, 6 September 1884, p. 407 (Vilile); *Q*, 14 June 1884, p. 951. (Innishowen); *Queenslander Supplement*, 29 May 1882, p. 3 (Hamleigh); *MM*, 1 June 1882, p. 2 (Richmond); *Sydney Morning Herald*, 11 September 1884, p. 4 (Airdmillan); *The Planter and Farmer*, August 1882, p. 209 (Cudgen).

26 For examples of such orders see: McOnie Order Book, UGD 118/4, 5/3, p. 281 (Kalamia); McOnie Order Book, UGD 118/4, 5/3, pp. 277–278 (Seaforth); Mirrlees Order Book, UGD 118/2, 4/3, pp. 54–56 (Farleigh); Mirrlees Order Book, UGD 118/1, 7/2, pp. 21–22 (Palms); & Mirrlees Order Book, UGD 118/1, 7/2, pp. 31–32 (Fairymead).

27 *BC*, 30 August 1883, p. 2; *The Planter and Farmer*, October 1883, p. 633.

28 For examples of new sugar mills erected with vacuum pans see J. Kerr, *Southern Sugar Saga. A History of the Sugar Industry in the Bundaberg District* (Bundaberg, 1983), pp. 10–11 (Branyan and Waterview); UGD, 118/1, 2/8, pp. 232, 246, 266, 271 & 311, BRC (Palmyra, Nindaroo, Conningsby, Mt Pleasant and Beaconsfield); *The Planter and Farmer*, February 1882, p. 41 (Hop Wah); & *BC*, 2 December 1885, p. 2 (Brie Brie); for a selection of mills with just Wetzel or Bour pans see *LW*, 1 November 1879, p. 3 (Roseville); *BC*, 31 December 1885 (Rockholme); *The Planter and Farmer*, July 1882, p. 166 (Loebieger's); & *The Planter and Farmer*, March 1882, p. 1 (Anderson's).

29 The sugar mills identified using Sutton's Evaporating Pan were Rehfeldt's Junction (*LW*, 4 September 1880, p. 3), Buderim Sugar Co. (*BC*, 10 March 1880, p. 5), and Bahr & Co.'s factory on the Albert River (*BC*, 10 March 1880, p. 5). This evaporating pan was also used in Riverside and Benowa Mills the early 1880s, but had been replaced with vacuum pans by 1885 (*LW*, 28 January 1882, p. 4, 8 November 1884, p. 3 & 21 January 1882, p. 4; *Q*, 24 October 1885, p. 651).

30 *QVP*, Vol. 4 (1889), p. 325.

31 B. Alcorn & R. Dunn, *Moreton Sugar Mill. Sweet Heart of Nambour* (Nambour, 1997), p. 9.

32 Victor mill trials are described in *Q*, 30 December 1876, p. 27, 26 July 1879, p 122 & 6 September 1879, p. 294; & *BC*, 4 February 1880, p. 5; see also *ATCJ*, 14 January 1882, p. 61 & 21 January 1882, p. 115 (Palk's); Kerr, *Sugar at Maryborough*, p. 11 (Eaves'); & *The Planter and Farmer*, January 1884, p. 20 (Arnold's & Preston's Mills, near Brisbane).

33 For Queensland examples see Kerr, *Pioneer Pageant*, pp. 57–58 (Alexandra and Nebia); Kerr, *Sugar at Maryborough*, p. 85 (Nerada); *LW*, 28 July 1883, p. 3 (Yatala); *Q*, 10 November 1883, p. 765 (Otmoor); & *LW*, 18 June 1881, p. 3 (Beenleigh); for New South Wales examples see *ATCJ*, 1 January 1887, p. 10 (Martin's, Richmond River); *CRE*, 6 August 1881, p. 2 (Belmore Mill, Clarence River); *NS*, 18 September 1884, p. 3 (Bryant's, Richmond River); & *RRE*, 29 July 1880, p. 2 (Gollan's, Richmond River).

34 *The Australasian*, 15 July 1882, p. 87; *BC*, 4 February 1880, p. 6.

35 *BC*, 4 February 1880, p. 6; E.W. Knox, AR 1880 season, p. 28, CSRR, Z/109, Box 374.

36 For the new juice mills around Maryborough see Kerr, *Sugar at Maryborough*, pp. 106–119, *Sydney Mail*, 8 November 1884, p. 947 & *Q*, 10 June 1882, p. 727; for details about the Bundaberg juice mill system see Kerr, *Southern Sugar Saga*, pp. 14–16, 19 & 23, *ATCJ*, 29 March 1884, p. 1602 & H. Turner, *Rural Life in Sunny Queensland* (Bundaberg, 1975), pp. 20 & 25.

37 E.W. Knox to A. Fairgrieve, Harwood Mill, 13 August 1884, CSRR, 142/12856, p. 30; F. Shaw, *Letters from Queensland* (London, 1893), p. 23.

38 *MM*, 13 November 1875, p. 2; *Q*, 11 August 1877, p. 4.

39 *MM*, 24 December 1880, p. 2; *Q*, 28 October 1882, p. 593; Connolly, *John Drysdale and the Burdekin*, p. 72.

40 E. W. Knox, AR 1876 season, p. 5; A. Fairgrieve, AR 1881 season, p. 6; J. Knox, AR 1882 season, p. 13, CSRR, Z/ 109, Box 374; E. W. Knox, London, to Sir E. Knox, Sydney, 13 January 1886, CSRR, Z/303, Box 70, File M 1.0, Folder 5, Doc. 42.

41 Connolly, *John Drysdale and the Burdekin*, p. 72; *Q*, 14 January 1888, p. 70.

42 *QVP*, Vol. 4 (1889), p. 306; G.R. Drysdale to Drysdale Bros. & Co., London, 20 May 1887, PMR/[C.T./1] Box 7, Article 4a, JCUA.

43 For examples of these upgrades see *Q*, 21 November 1885, p. 839 (Helensvale), *LW*, 22 October 1892, p. 3 (Beenleigh and Yatala), *SJTC*, 15 March 1894, p. 37 (Oakwood and Kalkie) & Kerr, *Southern Sugar Saga*, pp. 31–33 (Rubyana, Windsor and Seaview); for juice mills upgrades see *SJTC*, 15 March 1894, p. 37.

44 For Rous Mill see J. Browning, 'Rous Mill tramway', *Light Railways*, No. 88, Vol. 22 (April 1985), p. 4 & *NS*, 28 August 1889, p. 4. & 7 December 1889, p. 2; for other mill upgrades see *Q*, 16 April 1887, p. 632, *SJTC*, 15 February 1893, p. 8 & *SJTC*, 15 July 1894, p. 28 (Nindaroo and Meadowlands); *SJTC*, 15 May 1893, p. 94 & *Q* 18 January 1896, p. 120 (Yeppoon); *SJTC*, 15 March 1894, p. 37 (Goodwood); UGD 118/2, 4/7, Orders 627, 656 & 719, BRC (Sharon, Windemere and Pemberton); *QVP*, Vol. 4 (1889), p. 409 (Mon Repos); & *Bundaberg Mail*, 27 May 1898, p. 3 (Kalbar).

45 For upgrades at CSR's sugar mills see E.W. Knox, 'Notes on the work of CSR's mills, 1891', p. 25, CSRR, Z/109, Box 375 (Goondi), *SJTC*, 15 July 1893, p. 148 (Victoria), *SJTC*, 15 December 1892, p. 238 (Homebush) & CSR, *Notes on the work of the Mills during 1892* (Sydney, 1893), p. 4 (shredders); for doubts about shredders see E. W. Knox to Goondi Mill Mang., 18 November 1891, CSRR, 142/1458, pp. 197–198 & Gen. Mang., AEMCo., London, to Melb. Office, 27 April 1898, AER, London Office Book No. 3, p. 250, MUA.

46 For continuous subsiders see G. Kottman, 'Chemical Report on Work of 1897', p. 54, CSRR, Z/109, Box 376; for quadruple effets see CSRR, 142/1233, p. 431 (Homebush), 142/1462, p. 29 (Goondi) & 142/1010, p. 120 (Hambledon), CSR, *Notes on the work of the Mills during 1892* (Sydney, 1893), p. 17 (Condong) & Mirlees, Watson & Co. Order Book, UGD, 118/2, 4/12, Order 652 of 1893 (Victoria) & 4/13, Order 515 of 1894 (Broadwater), BRC; E. W. Knox to W. Farquhar, Inspector of Mills, Macknade Mill, 6 September 1899, CSRR, 142/2860, p. 323.

47 Kerr, *A Century of Sugar*, p. 71.

48 *SJTC*, 15 August 1893, p. 182; *Q*, 31 March 1894, p. 608; *North Queensland Herald*, 9 January 1895, p. 16; Mirrlees, Watson & Co. Order Book, UGD, 118/2, 4/11, Order 531 of 1892 (Fairymead), BRC; *The Australian Tropiculturist and Stockbreeder*, 25 February 1896, p. 53; Duncan Stewart Order Book, UGD, D52/1/1, p. 124

(Bingera), BRC; for Bingera's milling machinery see Anon, 'Sugar at Bundaberg', *QAJ*, Vol. 1, no. 4 (1 October 1897), pp. 295–296.

49 Griggs, 'Plantation to small farm', p. 163; Mirrlees, Watson & Co. Order Book, UGD, 118/2, 4/11, Order 435 of 1892 (Pioneer), BRC; A. & W. Smith Order Book, UGD, 118/1, 2/33, pp. 171 & 265, BRC; Connolly, *John Drysdale and the Burdekin,* p. 90; Charles Young, Kalamia, to Sec., Union Mortgage & Agency Co., Melb., 17 September 1898, CY/LB/1, p. 135, JCUA; *North Queensland Herald*, 26 November 1900, p. 47.

50 Kerr, *Pioneer Pageant*, pp. 129–130; Manning, *In Their Own Hands*, pp. 75 & 131; Mirlees, Watson & Co. Order Book, UGD, 118/2, 4/13, Order 120 of 1894 (Ashburton Mill); A. & W. Smith Order Book, UGD, 118/1, 2/33, pp. 215–218 (Hambledon Mill), BRC.

51 For descriptions see *Q*, 2 February 1895, p. 226 (Albionville), Kerr, *Southern Sugar Saga*, pp. 33 & 41 (Annesley & Greenwood), *SJTC*, 15 March 1894, p. 37 (Bellview) & *Bundaberg Mail*, 28 August 1895, p. 3 (Waterloo).

52 For descriptions of central sugar mills see *The Australian Tropiculturist and Stockbreeder*, 22 May 1897, p. 136, *Bundaberg Mail*, 7 July 1897, p. 3 & A. & W. Smith Order Book, UGD, 118/1, 2/33, pp. 296–322 (Mulgrave), 335–342 (Moreton) & 382–418 (Mossman), BRC.

53 For these views see *Mackay Standard*, 4 April 1894, p. 2 & 15 April 1895, p. 3, *MC*, 6 April 1895, p. 2 & *QPD*, Vol. 75 (1895), pp. 205–229.

54 Kerr, *Southern Sugar Saga*, p. 35; *Bundaberg Mail*, 13 July 1894, p. 3 & 30 December 1892, p. 2; *Q*, 17 February 1894, p. 322.

55 For Doolbi's conversion see Kerr, *Sugar at Maryborough*, p. 89.

56 J. Kerr, *Only Room for One. A History of Sugar in the Isis District* (Childers, 1996), pp. 52–53; W. Maxwell, *Report upon an Investigation into the Condition of the Sugar Industry in Queensland* (Brisbane, 1900), p. 15; W. Farquhar, Inspector of Mills, Victoria Mill, to E.W. Knox, 19 April 1895, CSRR, 142/2892, p. 2.

57 S. May (ed.), 'The Diary of Arthur Neame, 1870–1897', pp. 41 & 42; Charles Young, Kalamia, to R. J. Jeffray, London, 30 March 1895, Private Letterbook, p. 3, CY/LB/2, JCUA; *Port Denison Times*, 11 June 1907, 5; Griggs, 'Plantation to small farm', pp. 167 & 257.

58 Extract from the Report of the Acting Land Commissioner for the Mackay District, *QPP,* Vol. 2 (1908), p. 1018; C. Petrie, CSR, Sydney, to All Mill Engineers (Circular), 23 February 1911, CSRR 142/3510; E.W. Knox, to Victoria Mill Mang., 30 August 1906, CSRR, 142/1561, p. 194; Australian Sugar Co. Pty. Ltd., Minute Book of Directors Meetings, pp. 32 & 58, Mss 3565/32, ML.

59 For the lack of chemical control in Queensland sugar mills see *Q*, 31 October 1885, p. 720 and G. Craig, 'The Sugar Industry', *The Queensland Review*, August 1885, p. 62; for the absence of weighbridges see *MM*, 16 December 1890, p. 2 & *QVP*, Vol. 4 (1889), p. 299 (Ev. J. Gavin, Secretary, Yeppoon Sugar Co., Q. 6916).

60 P. Clarke, *The New Chum in Australia, or the Scenery, Life and Manners of Australians in Town and Country* (London, 1886), p. 286.

61 *Q*, 29 September 1879, p. 410 & 31 October 1885, p. 720.

62 Kerr, *Pioneer Pageant*, pp. 57–58; Manning, *In their own hands*, p. 115; Connolly, *John Drysdale and the Burdekin*, p. 53.

63 E.W. Knox to Victoria Mill Mang., 29 October 1883, p. 50, CSRR, 142/1547.

64 E.W. Knox, 'An application of chemical control to a manufacturing business', *Proc. of the Second Meeting of the Australasian Association for the Advancement of Science*, Melbourne 1890, pp. 372–379; G. Bindon and D. Miller, 'Sweetness and light: industrial research in the Colonial Sugar Refining Company, 1855–1900', in R. Home (ed.), *Australian Science in the Making* (Sydney, 1988), pp. 177–179.

65 Bindon and Miller, 'Sweetness and Light', pp. 179–181.

66 Bindon and Miller, 'Sweetness and Light', pp. 185–187; E.W. Knox, Circular to All Mill Managers, 19 June 1884, CSRR, 142/3505; G. Kottmann, Sydney, to All Mill Managers, 8 June 1886, CSRR, 142/2004, p. 56; for the pocs formula see Appendix 23 in A. Lowndes (ed.), *South Pacific Enterprise. The Colonial Sugar Refining Company* (Sydney, 1956), pp. 452–454.

67 E.W. Knox to All Mill Managers, 6 May 1885, CSRR, 142/3505; for these chemical circulars see CSRR, Z/303, Box 17, File A 6.0, Folder 1, Docs. 26–36; for the inspecting chemist reports see CSRR, 142/ 2696 & 2697; *CPP*, Vol. 4 (1913), p. 997 (Ev. E. W. Knox, Q. 26844).

68 B. Levick, 'Diffusion at Broadwater Mill, 2pp typescript, 16 April 1958, CSRR, Z/303, Box 20, File A 6.1, Folder 1, Doc. 2; *QVP*, Vol. 4 (1889), p. 341 (Ev. E.W. Knox, Qs. 8200–8210); for the diffusion unit at Cudgen see *The Sugar Cane*, Vol. 21, no. 240 (1 July 1889), pp. 356–357 & Vol. 21, no. 244 (1 November 1889), p. 571, *LW*, 13 February 1892, p. 3 & *Q*, 30 May 1896, p. 1030.

69 For experiments with the diffusion apparatus see E.W. Knox to J. Knox, Victoria Mill, 10 August 1885, CSRR, 142/2856, p. 281; for later trials of the diffusion drum at Condong and Harwood Mills in 1896 and 1900, respectively, see G. Kottmann, 'Chemical Report on 1901 season', p. 65, CSRR, Z/109, Box 376.

70 E. W. Knox, 'Notes on the Work of the Mills during the season of 1889', pp. 9 & 42–43, CSRR, Z/303, Box 15, File A 5.1, Folder 2, Doc.18; E.W. Knox, 'Chemical Knowledge Applied to Sugar Producing in Queensland', *The Sugar Cane*, Vol. 22 (April, 1890), pp. 212–215; Bindon and Miller, 'Sweetness and light', pp. 183–188.

71 *The Planter and Farmer*, November 1884, p. 209 (Mourilyan); E. W. Knox to E. Long, Mackay, 22 August 1887, CSRR, 142/477, p. 457; *Q*, 14 January 1888, p. 71 (Habana); *QVP*, Vol. 4 (1889), p. 130 (Hambledon).

72 *NS*, 7 December 1889, p. 2 (Rous); *QVP* Vol. 2 (1898), p. 71 (Nerang); *Cairns Post*, 10 May 1899, p. 7 (Mulgrave); Kerr, *Pioneer Pageant*, p. 122 (Marian); Kerr, *Sugar at Maryborough*, p. 141 (Mt Bauple); *SJTC*, 15 August 1896, p. 178 (North Eton and Pleystowe); *Q*, 28 January 1893, p. 150 (Fairymead); Kerr, *Southern Sugar Saga*, p. 59 (Bingera and Fairymead); *Bundaberg Mail*, 9 October 1895, p. 3 (Goodwood); *Bundaberg Mail*, 27 August 1894, p. 3 (Mon Repos); *Q*, 13 July 1895, p. 82 (Knockroe); Victoria Mill Mang. to E. W. Knox, 9 May 1895, CSR, 142/1553, p. 423.

73 Racecourse Central Milling Co., Minute Book No. 4, 1 August 1900, p. 177, Racecourse Mill Office; Marian Central Milling Co., Minute Book No. 2, 22 March 1902, p. 42, Marian Mill Office.

74 *CPP*, Vol. 4 (1913), p. 716, Q. 19861.

75 F. Garcia, 'The development of capitalism in Cuban sugar production, 1860–1900', in M. Fraginals *et al.* (eds.), *Between Slavery and Free Labour. The Spanish Speaking Caribbean in the Nineteenth Century* (Baltomore, 1985) p. 70; Beachey, *The British West*

Indies Sugar Industry, pp. 62–63; J. Sitterson, *Sugar Country. The Cane Industry of the South, 1753–1950* (Lexington, 1953), pp. 281–284.

76 E.W. Knox to E. Young, Katoomba, 23 March 1900, CSRR, Z/303, Box 40, File D 1.0, Folder 9, Doc. 3; W. Maxwell, *Report upon an Investigation into the Condition of the Sugar Industry in Queensland*, (Brisbane, 1900), pp. 15–17.

77 Under-Sec., Qld. Dept. of Agriculture & Stock, to Dr H. Reed, Hambledon Mill Mang., 22 June 1901, AGS, N/92, QSA.

78 *ASJ*, 8 June 1911, p. 149 (Marian), 7 December 1911, p. 345 (Racecourse) & 6 June 1912, p. 155 (Mulgrave); Kerr, *Southern Sugar Saga*, p. 75 (Bingera); *CPP*, Vol. 4 (1913), p. 716, Q. 19859.

79 CSR, 'Chemical Report on 1911 season', pp. 9 & 11; CSR, 'Chemical Report on 1912 season', p. 8 & CSR, 'Chemical Report on 1914 season', p. 16, CSRR, Z/109, Box 377.

80 Kerr, *Southern Sugar Saga*, pp. 74–75 & 81; Kerr, *Pioneer Pageant*, pp. 130 & 158.

81 CSR, 'Chemical Report on 1914 season', p. 16, CSRR, Z/109, Box 377; *CPP*, Vol. 4 (1913), p. 346, Q. 10121; BSES, AR 1912–13, *QPP*, Vol. 2 (1913), p. 631.

82 *The Planter and Farmer*, January 1883, p. 370; *Q*, 8 October 1881, p. 471 & 14 January 1888, p. 71.

83 *Q*, 14 January 1888, p. 71; *ATCJ*, 8 January 1887, p. 80.

84 G. Burrows and C. Morton, *The Canecutters* (Melbourne, 1986), pp. 188–190; *ATCJ*, 18 March 1871, p. 333, *RRE*, 2 November 1884, p. 7; W. Clarke, 'The industries of the North Coast', *Agricultural Gazette of NSW*, Vol. 10 (1900), p. 1292.

85 Kerr, *Sugar at Maryborough*, pp. 28 & 49; *Q*, 22 October 1870, p. 5 & 14 October 1876, p. 6.

86 Kerr, *Southern Sugar Saga*, p.33; *LW*, 22 October 1892, p. 3.

87 Kerr, *Southern Sugar Saga*, p. 42; Kerr, *Sugar at Maryborough*, p. 129; *ASJ*, 9 September 1909, p. 212 & 9 November 1911, p. 465–466.

88 J. Knox, 'Report on mills, 1882', p. 39, CSRR, Z/109, Box 374; J. Armstrong, 'The sugar tramways of northern New South Wales', *Bulletin of Australian Railway Historical Society*, Vol. 27, no. 464 (1976), pp. 122 & 126; Burrows and Morton, *The Canecutters*, pp. 190–191; *Sydney Morning Herald*, 7 September 1903, p. 5.

89 E. Knox, 'Notes on the work of the mills during 1889', pp. 7 & 9 , CSRR, Z/303, Box 15, File A 5.1, Folder 2, Doc. 18; W. Clarke, 'The Industries of the North Coast', *Agricultural Gazette of NSW*, Vol. 10 (1900), p. 1292; C. Murray, *Across Three Bridges: Island Base for Council 1883–1983* (Ballina, 1983), pp. 139–141.

90 *BC*, 31 May 1879, p. 6; *Q*, 5 June 1880, p. 711; Manning, *In their own hands*, p. 26.

91 *Sydney Mail*, 1 November 1884, p. 595; A. Goodwin, *The Sugar Industry of the Lower Burnett in 1882* (Bundaberg, 1882), p. 2; *The Planter and Farmer*, October 1882, p. 272; J. Armstrong & G. Verhoeven, *The Innisfail tramway: the story of the history and development of the Geraldton Shire tramway and the Mourilyan tramway of North Queensland* (Brisbane, 1973), pp. 37–38.

92 J. Kerr, *Triumph of the Narrow Gauge. A History of Queensland Railways* (Brisbane, 1990), pp. 49–50; Kerr, *Southern Sugar Saga*, p. 38.

93 Manning, *In their own hands*, p. 130.

94 J. Browning, 'Rous Mill Tramway', *Light Railways*, No. 88, Vol. 22 (April 1985), pp. 4–5; Griggs, 'Plantation to small farm', p. 173; J. Kerr, *Black Snow and Liquid Gold. A History of the Burdekin Shire* (Ayr, 1994), p. 113.

95 D. Rapkins, *A Remarkable Achievement. The Story of the Cairns-Mulgrave Tramway* (Cairns, 1997); Griggs, 'Plantation to small farm', pp. 428–432; Kerr, *Triumph of the Narrow Gauge*, pp. 102–103; J. Wegner, 'Hinchinbrook: The Hinchinbrook Shire Council, 1879–1979'. M.A. thesis, Dept. of History, James Cook Uni., 1984, pp. 472–492.

96 Kerr, *Triumph of the Narrow Gauge*, pp. 104 & 108; J. Armstrong, 'Millaquin Sugar Mill tramways', *Bulletin Australian Railway Historical Society*, Vol. 36, No. 573 (July 1985), p. 149 & 151.

97 For loading derricks in the Mackay district see J. Kerr, *A Century of Sugar. Racecourse Sugar Mill, Mackay* (Mackay, 1988), pp. 56, 80 & 89–90.

98 *The Planter and Farmer*, November 1884, p. 209; *ACTJ*, 24 March 1888, p. 589.

99 This discussion is based upon the correspondence, legal opinions and patent documents in CSRR, Z/109, Box 35, File: A Sugar Cane Unloader.

100 Kerr, *Southern Sugar Saga*, p. 24.

101 G. Kottman, 'Chemical report for the 1903 season', p. 4 & G. Kottman, 'Chemical report for the 1904 season', p. 7, CSRR, Z/109, Box 376; for the installation of mechanical unloaders see *SJTC*, 15 January 1904, p. 139 (Mossman) & 15 February 1904, p. 163 (Plane Creek); & *ASJ*, 3 March 1910, p. 453 (Marian), 9 March 1911, p. 587 (Racecourse), 8 June 1911, p. 148 (Cattle Creek) & 4 April 1912, p. 16 (Meadowlands).

102 *SJTC*, 15 December 1901, p. 107; *ASJ*, 7 September 1911, p. 345; for their disadvantage see *International Sugar Journal*, Vol. 7, no. 76 (April 1905), p. 15.

103 *CPP*, Vol. 4 (1913), p. 700 (Ev. F. Kinne, part-owner Maryborough Sugar Factory, Qs. 19315–19316).

104 John Grime, Casino, to William Clement, Alpha Mill, 20 November 1872, in Correspondence Folder labelled Sugar cane and Alpha Mill, Clement Business Papers, Box 691, Richmond River Historical Society, Lismore.

105 For sugar being sent from northern Queensland to Brisbane see *BC*, 17 September 1870, p. 4; for sugar going to Melb. see *MM*, 21 September 1872, p. 2; for Munro's trials see *Q*, 22 July 1871, p. 4.

106 *The Planter and Farmer*, July 1883, p. 569; Jones, *Trinity Phoenix*, p. 167; Griggs, 'Plantation to small farm', p. 98.

107 Connolly, *John Drysdale and the Burdekin*, p. 102; Kerr, *Triumph of the Narrow Gauge*, p. 103; Qld. Dept. of Harbours and Marine, *Port and Harbour Development in Queensland From 1824 to 1985* (Brisbane, 1986), pp. 412–413.

108 D. Byrne, 'A Paper Covering the Colonial Sugar Refining Company (CSR) and a history of its use of water borne transport, including ocean going vessels, in executing the company's business, with notes on bulk sugar cargoes'. M. Litt. thesis, Australian National Maritime Museum Library, pp. 10–18 & 62–64; R. Parsons, *Australian Shipowners and their Fleets*. Vol. 1 (Lobethal, South Australia, 1972), p. 59.

109 G. Lewis, *A History of the Ports of Queensland. A Study in Economic Nationalism* (Brisbane, 1973), pp. 40 & 95; J. Bach, *A Maritime History of Australia* (Melbourne, 1976), pp. 196 & 204; N. McKellar, *From Derby Round to Burketown. The A.U.S.N. Story* (Brisbane, 1977), pp. 35 & 118.

110 McKellar, *From Derby Round to Burketown. The A.U.S.N. Story*, pp. 117–122 & 220–222; M. *Page, Fitted for the Voyage. The Adelaide Steamship Company Limited 1875–1975* (Adelaide, 1975), p. 166; Lewis, *A History of the Ports of Queensland*, p. 148.

111 H. Shaw, *Lucinda North Queensland. Yesterday and Today* (Ingham, 1978), pp. 9–10.

112 Kerr, *Triumph of the Narrow Gauge*, p. 103; Qld. Dept. of Harbours and Marine, *Port and Harbour Development in Queensland From 1824 to 1985*, pp. 385–386.

113 *Slater's Queensland Almanac*, 1870, p. 22; *Q*, 3 October 1874, p. 10.

114 *NS*, 15 September 1877, p. 3; E.W. Knox, AR 1877 Season, pp. 9–10; E.W. Knox, AR 1879 Season, p. 15, CSRR, Z/109, Box 374; *QVP*, 4 (1889), p. 340 (Ev. E.W. Knox, Qs. 8172–8176).

115 E.W. Knox to A. Fairgrieve, Insp. of Mills, Broadwater Mill, 3 December 1881, CSRR, 142/2854, p. 98; E. W. Knox to A. Fairgrieve, Harwood Mill, 13 August 1884, CSRR, 142/2856, p. 30.

116 Kerr, *Southern Sugar Saga*, p. 34; *MC*, 26 October 1896, p. 2; *SJTC*, 15 August 1894, p. 149; Kerr, *A Century of Sugar*, pp. 51–52 & 82–84.

117 *QVP*, Vol. 4 (1889), p. 294 (Ev. A.MacKenzie, Marian Mill, Q. 6780); Sec., North Eton Central Sugar Co. Ltd., to Hon. A. Hawthorn, Treasury, Brisbane, 11 January 1909, A/20292, In-letter 00481 of 1909, QSA; Kerr, *Pioneer Pageant*, p. 158.

118 Kerr, *Sugar at Maryborough*, p. 152; Chief Clerk, BCSM, to Qld. Treasurer, 14 July 1910, A/20288, In-letter 6941 of 1910, QSA; *CPP*, Vol. 4 (1913), p. 703 (Ev. F. Kinne, part-owner, Maryborough Sugar Factory, Q. 19448).

119 *CPP*, Vol. 4 (1913), p. 1435 (Ev. A. Rutherford, farmer, Q. 7702).

120 *CPP*, Vol. 4 (1913), pp. 183 (Ev. W. Desplace, Mang., Mourilyan Mill, Qs. 5160–5163) & 824 (Ev. J. Johnston, Millaquin Sugar Co. Ltd, Qs. 22668–22670); *QPP*, Vol. 2 (1911–12), p. 1195 (Ev. E.W. Knox, Qs. 1657–1660); E. W. Knox, Sydney, to W. Ralston, QN Bank, 16 March 1903, CSRR, 142/2934, p. 116.

121 Directors Minute Book, Vol. 1, p. 184, North Eton Central Sugar Co. Ltd; Directors Minute Book No. 4, p. 70, Racecourse Central Sugar Co. Ltd.; Directors Minute Book, Vol. 2, pp. 225 &239, Cattle Creek Sugar Co. Ltd., J. Kerr, *Top Mill in the Valley. Cattle Creek Sugar Mill, Finch Hatton, 1906–1990* (Brisbane, 1991), p. 24; *CPP*, Vol. 4 (1913), p. 304 (Ev. W. Kenzey, Manager, Meadowlands Mill, Qs. 8924–25).

122 *Tweed Herald and Brunswick Chronicle*, 15 September 1898, p. 6; *RRE*, 24 August 1894, p. 4 & 1 July 1898, p. 5; *Clarence River Advocate*, 19 July 1898, p. 3.

123 *NS*, 4 July 1894, p. 2, *NSWVP*, Vol. 4 (1895), pp. 1131–1139; *Clarence River Advocate*, 4 October 1898, p. 4; *The Tweed Herald and Brunswick Chronicle*, 10 June 1898, p. 6.

124 Griggs, 'Plantation to small farm', pp. 272 & 275–276.

125 *ASJ*, 7 April 1910, p. 6 & 9 June 1910, p. 104; L. Fitzhardinge, *A Political Biography of William Morris Hughes. Vol. 1: That Fiery Particle, 1862–1914* (Sydney, 1964), p. 265.

126 *ASJ*, 6 May 1910, pp. 70–71, 9 June 1910, p. 85, 7 July 1910, p. 162, 8 September 1910, p. 282, 7 October 1909, p. 262 & 8 December 1910, p. 411; *CPP*, Vol. 4 (1913), pp. 1263, 1274, 1317, 1480 (CSR's mill), 1466, 1468–69, 1494, 1530, 1928–29 (other districts), 1469, 1494, 1507 & 1529 (shareholders versus non-shareholders); *QPP*, Vol. 2 (1911–12), pp. 1090 & 1097.

127 *QPD*, Vol. 106 (1913), pp. 1934–1935; 1974–1976; & 2488–2509; Shogren, 'The politics and administration of the Queensland sugar industry to 1930', pp. 230–234.

128 *CPP*, Vol. 4 (1913), p. 833 (Ev. W. Horswill, canegrower, Qs. 22833–22838); *ASJ*, 7 May 1914, pp. 140–141, 2 July 1914, pp. 262–264, 6 August 1914, p. 880 & 3 September 1914, p. 412.

Chapter 9

Competing for the Sugar Basin: Refiners, Merchants and the Australian Sugar Market

During the late nineteenth century the demand for sugar in Australia continued expanding, due to population growth and because Australians eagerly consumed more sugar in their diets. By the late 1860s, apparent annual sugar consumption per capita in Australia had reached 36.2 kg. It slowly increased during the late nineteenth and early twentieth centuries, peaking at 51.1 kg per capita in the late 1900s. On the eve of World War 1, this figure had fallen slightly from its record high and stood at just over 50 kg.[1] Moreover, the type of sugar demanded by many Australians had also changed by the early 1870s. Muscovado or moist brown sugar had fallen out of favour with the increasingly fastidious Australian consumer, who shunned it in favour of better quality and drier raw sugar or white sugar. Indeed, one Queensland parliamentarian observed in 1885 that 'even the working man would take nothing but purely white sugar'.[2] Furthermore, the British authors Charles Lock, George Wigner and Robert Harland in their survey of world sugar production and refining concluded that colonial Australians during the mid-1870s consumed more sugar per capita than anywhere else in the world. By the late 1900s, Australians still had the world's highest sugar consumption per capita (Table 9.1). Therefore, potential Australian suppliers of white sugar were assured of a growing domestic demand for their product, assuming they could survive CSR's attempts to monopolise this market.

In the early 1860s, three sugar refineries operated in Australia: two in Sydney and the other in Melbourne. Two of these refineries were controlled by either CSR or its subsidiary the Victoria Sugar Co. (VSC). These sugar refineries were joined by another in the early 1870s, when Robert Cran Snr and his partner Robert Tooth – one of the partners in the Kent Brewery, Sydney – combined meat preservation and sugar manufacture in a complex at Yengarie, near Maryborough. Initially, Tooth and Cran Snr converted the juice from the cane grown on their own properties into good quality raw sugar. However, five charcoal retorts were added to the factory in 1870, enabling the preparation of char from the bone residue of the meat preserving plant. In the following year, Yengarie operated as a sugar refinery for the first time, the

Table 9.1. Sugar consumption per capita (kg) for selected countries, 1874, 1887 and 1909.

Country	1874	1887	1909
Great Britain	28.5*	32.0	46.3
France	7.0	12.3	20.0
Germany	7.5	8.6	19.5
United States of America	17.1	27.7	38.1
Netherlands	11.3	10.5	17.2
Austria-Hungary	6.8	5.5	10.4
Russia	2.5	4.0	6.4
Australia	38.9	40.4 #	51.7

Notes: * Figure is for 1875 # Figure is for 1889

Source: Based upon figures in T. Coghlan, *The Wealth and Progress of New South Wales 1888–89* (Sydney, 1890), p. 331, *The Sugar Cane,* 1 October 1888, p. 503 and C. Lock, G. Wigner and R. Harland, *Sugar Growing and Refining* (London, 1882), p. 694.

filtered juice being passed through Taylor bag filters and bone char to remove any impurities, leading to the production of fine, white sugar. Tooth and Cran Snr made more additions to Yengarie Refinery during 1872, expanding production to meet the growing demand for white sugar in Queensland. To increase their output, Tooth and Cran purchased raw sugar from some unidentified Mackay mill-owners and juice from the nearby Alford Mill.[3]

Tooth and Cran Snr had also petitioned the Queensland government in April 1872, asking to import foreign raw sugar, refine this sugar in bond, and then pay the duty only on the amount of refined sugar exported. Tooth and Cran Snr argued such an arrangement would place them on the same footing as the sugar refiners in Melbourne and Sydney. Legislation granting Tooth and Cran's request was passed by the Queensland Parliament in August 1872. Known as the *Bonded Distilleries and Bonded Sugar Houses Act* of 1872, the measure did not gain favour with Tooth and Cran Snr, who complained that clause 12 made the legislation unworkable, for it required all products made from imported foreign sugar, molasses or treacle to be removed from the bonded sugar house before any Queensland-produced sugar could enter the premises. Such a cumbersome condition would slow production considerably.

Queensland's parliamentarians considered Tooth and Cran Snr's objections and the *Bonded Distilleries and Bonded Sugar Houses Act* of 1872 was repealed in July 1874 and replaced by *The Imported Sugar Refiners*

Act of 1874. The new legislation allowed the refiner to import sugar; a bond was given to the government for the amount of duty required and that bond could not be released until a corresponding quantity of refined sugar for export had been produced or the duty was paid. Its usefulness, however, was never really tested, for Tooth and Cran Snr, after using imported sugar from Taiwan in 1873 and 1874, decided not to rely solely on imported foreign sugar. Instead, they convinced half a dozen nearby mill-owners during 1875 and 1876 to supply juice to Yengarie, thereby guaranteeing the supply of raw materials for the refinery. By 1882, 200 tonnes of refined sugar was produced weekly, but the refinery's reliance on juice, not raw sugar, meant Tooth and Cran Snr could not manufacture refined sugar all year, which was a serious handicap in trying to build up a market for their product.[4]

During the mid-1870s, Tooth and Cran Snr began to differ on the future direction of their business activities. They had agreed to abandon meat extraction after the 1873 season, in favour of concentrating on sugar production. Cran Snr wanted to expand by duplicating Yengarie at Bundaberg. Tooth was not convinced and instead moved off-shore, establishing refineries in Manila and on the island of Penang. Cran Snr was not part of these ventures and the growing tensions between the two partners led to the dissolution of their partnership on 11 August 1879. Tooth sold his entire interest in Yengarie for £255, a strange arrangement, but according to the sugar historian John Kerr, 'Tooth was no fool'. The debt of £200 000 owing to the Queensland National (hereafter QN) Bank probably exceeded the value of the refinery and its associated plantations. Moreover, Tooth Snr had been a CSR director during the late 1850s and had sanctioned the elimination of rival refiners. He may have grown concerned that CSR would eventually move against its Queensland competitor. In the aftermath of the dissolution of the partnership between Tooth and Cran Snr, Robert Cran Snr and his sons John and Robert Junior formed the new firm of Robert Cran and Co., although later it was referred to as the Millaquin and Yengarie Sugar Co. Continued financial support from the QN Bank enabled them to maintain operations at Yengarie Refinery.[5]

Cran Snr had not abandoned the idea of building a new refinery with double the capacity of Yengarie at Bundaberg. After convincing the QN Bank to advance another £160 000 for the project, Cran Snr left for Europe in September 1880 to visit sugar beet factories and sugar refineries – the design of his new refinery was to be modern and it was to be equipped with the most up-to-date machinery. Before his departure, Cran Snr selected a location for the new refinery that was named Millaquin. The Burnett River site had deep-water frontage, a plentiful supply of fresh water at scarcely any depth and solid rock beneath on which to rest the foundations. Abundant coal supplies could be obtained from the nearby Burrum River coalfields at Howard. The refinery's main, three-storey building was split into four sections which corresponded with the four gables of the roof: one for the sugar house; two for the machinery; and the fourth for the gas and bone char works. Other buildings included a large engine fitting shop, premises for the blacksmith, the lime kiln and a boiler shed. Machinery from Mirrlees, Tait & Watson began filling the building in late 1881. Operations commenced in 1882, with the refinery relying on juice purchased from local mill-owners.[6]

Robert Cran & Co.'s burgeoning debt with the QN Bank did not raise much concern amongst its directors – the Queensland sugar industry was booming and the industry's future looked prosperous. However, when the price of sugar suddenly dropped in mid-1884, E. Drury, the QN Bank's General Manager, became alarmed at the company's large overdraft of £300 000. Drury commenced investigations into ways that the bank could recover its debts. Attempts to form a new limited liability company failed, with potential investors unwilling to support such an enterprise at a time of low sugar prices. Drury continued fretting over Cran & Co.'s liabilities during the late 1880s, but the QN Bank did not foreclose on the Cran family. When Cran Snr died in September 1894, his sons discovered that the debt owing on the sugar refining business exceeded its market value, given the depressed state of the Australian sugar industry. The QN Bank as mortgagee agreed to take over the refineries at a valuation equal to the debt. The new owners eventually concentrated their production at Millaquin Refinery, after closing Yengarie Refinery in 1900. The machinery at Millaquin Refinery was entirely renovated in 1909, enabling an output of around 20 000 tonnes of refined sugar annually, considerably less than the amount that could be produced by CSR's refineries in Sydney and Melbourne. Eventually, the QN Bank incorporated their sugar assets in 1911 into a public company known as the Millaquin Sugar Co. Ltd.[7]

Initially, CSR ignored Tooth and Cran Snr, concentrating on establishing its milling operations in northern New South Wales. Refining matters, however, attracted the attention of CSR's directors in January 1875 when they became aware of a rumour that an unidentified individual had acquired land at Pyrmont, in inner Sydney, and was contemplating erecting a rival sugar refinery. Just what happened next is unclear, but CSR had acquired about 2 ha at Pyrmont by the end of 1875. Such an acquisition provided a new site which offered the benefits of a deep water frontage, thereby eliminating wharfage costs, plenty of space to expand the business and an inner city location close to potential customers. Moreover, it was time to vacate the ageing Brisbane House Refinery on Paramatta Road, its water supply having become polluted through the growth of housing around the refinery. Despite having outlaid a large amount of capital on their New South Wales sugar mills, CSR decided to spend more funds to erect a new refinery at its Pyrmont site. Built during 1876–1877, the Pyrmont Refinery complex consisted of a large wharf-side store which could hold 8 000 tonnes of raw sugar and a nine-storey main building that housed the machinery used during the refining process. The new refinery was capable of turning out 700 tonnes of refined sugar a week. Edward Knox Snr planned a refinery 'as perfect as the best in either England or Scotland' and the general consensus was that he got it, but CSR spent £120 000 to secure such an impressive factory. Construction was placed under the direction of Frederick Poolman, the former manager of the VSC's Sandridge Refinery during the 1860s and early 1870s. Poolman, on behalf of the VSC, had travelled to England during 1875 to investigate refining techniques, so his knowledge was keenly sought by CSR. He subsequently became the refinery's first manager and consulting engineer to CSR, until he was dismissed in February 1885 for 'lax management'.[8]

The VSC, CSR's subsidiary, also faced a serious rival in late 1873, when Joshua Brothers, London- and Melbourne-based distillers, erected Melbourne's second sugar refinery at Yarraville, towards the mouth of the Yarra River (see Plate 9.1). Costing £160 000, the nine-storey building and its 30 m high chimney was described as 'a great structure of brick, resting upon a massive bluestone basement'. Joshua Brothers commenced operations using raw sugar imported from Java, China and the Philippines, but their involvement in this industrial endeavour was short-lived. They sold their factory to the VSC in July 1875 for £200 000, following the destruction of the VSC's Sandridge Refinery by fire a month earlier. Yarraville's capacity was then expanded by the addition of extra machinery, enabling the production of 400 tonnes of refined sugar per week or larger amounts if the factory operated continuously.[9] However, within six months of the VSC taking control of the Yarraville Refinery, the firm was confronted with another, but less organised rival refiner. In 1876, William Murray Ross erected a sugar refinery at Rosstown (today the Caufield area of suburban Melbourne), with the aim of refining raw sugar manufactured from processing locally grown sugar beet. Built of red brick, the Rosstown Refinery was 21 m long, 15 m wide and seven storeys high. The estimated total cost to complete the building and equip it with machinery was £60 000. The refinery, however, never commenced production. Ross had built the refinery without managing to convince the local market gardeners to engage in sugar beet cultivation, so he had no raw material to process.[10]

THE YARRAVILLE SUGAR REFINERY.

Plate 9.1. The Victoria Sugar Co.'s Yarraville Refinery, 1876. Originally erected in 1874 by the Melbourne-based distillers, Messrs Joshua Brothers, Yarraville was purchased by the Victoria Sugar Co. in 1875, after its Sandridge Refinery was destroyed by fire.

Source: *The Australasian Sketcher*, 2 September 1876, p. 216. Reproduced from the La Trobe Collection, courtesy of the State Library of Victoria, Melbourne.

The refiners' stranglehold over white sugar production in Australia was challenged during the late 1870s and early 1880s as new sugar mills were built with the capacity to produce mill whites. Thomas Buckland, a CSR director between 1861 and 1896, urged Edward Knox Snr in February 1874 to prepare for this eventuality, advising him that in addition to Tooth and Cran Snr, CSR 'must expect that some-one will produce a high class sugar under protection of the duty'.[11] Within a year of Buckland raising this possibility, mill whites were manufactured for the first time at Foulden Mill, near Mackay. Foulden's sugar was judged to be equal to that manufactured at the Yengarie Refinery. A decade later, at least a dozen Queensland sugar mills had some refining capacity (i.e. iron cylinders filled with charcoal). Some of the more noted mill white makers included the Mourilyan Sugar Co. (Mourilyan Mill), Thomas Swallow and Frank Derham (Hambledon Mill), Robert Boyd and John and Joseph Woods (Ripple Creek Mill), Sir John Lawes (Farleigh Mill), Gibson & Howes Ltd.

(Bingera Mill), Young Bros. (Fairymead Mill) and Thomas Smith (Woodlands Mill). In New South Wales, Cudgen and Rous Mills commenced producing mill whites during the late 1880s.[12] According to one journalist from *The Queenslander*, most ordinary consumers considered mill whites to be 'refined' sugar. However, the majority of these sugar mills had ceased manufacturing white sugar by 1900 (see discussion later in chapter), although the production of mill whites had commenced at the Maryborough Sugar Factory and Nerang Mill during the 1900s.[13]

Interest in establishing new sugar refineries in Australia continued during the 1880s and 1890s. In Brisbane, Robert Tooth & Co. was registered in April 1882, its main investors being Robert Tooth, Edward Corser, a Maryborough merchant, and the four partners of the Maryborough engineering firm of John Walker & Co. The company raised £20 000 to erect a very small brewery and sugar refinery on the Brisbane River at Bulimba. In the following year, however, the partnership was dissolved and the factory sold to a new syndicate of investors who decided to concentrate on brewing beer.[14] In contrast, Frederick Poolman's Port Melbourne Sugar Refinery, was more enduring. Poolman, as mentioned above, had supervised the erection of Pyrmont Refinery and had managed its operation during the early 1880s, so he was experienced in the art of sugar refining. Built in 1887, the Port Melbourne Refinery was initially a small affair, capable of producing a maximum of 4 000 tonnes per year. Following Poolman's death, his three sons maintained production until 1907, using imported sugars from Java, declining to purchase Australian-produced raw sugars which they considered too expensive compared to the cheaper, overseas manufactured sugars. They built a second refinery in Sydney in 1906.[15] Moreover, the Poolman family was more successful than the Australasia Sugar Refining Co. This firm's main promoters were a group of Melbourne businessmen engaged in the wholesale grocery trade. In 1891, they invested £150 000 in another Port Melbourne sugar refinery capable of producing 400–500 tonnes of refined sugar a week. The Melbourne merchants' dream of being free of CSR's influence was, however, short-lived, for the refinery had closed by December 1894.[16]

Throughout the 1880s and 1890s, CSR's own refining (and business) activities were in a state of flux. CSR's subsidiary the VSC had embarked on an ambitious, but somewhat risky program of expansion during the early 1880s. Just over £40 000 was spent upgrading its Yarraville Refinery. These extensive improvements involved trebling the storage capacity at the refinery and installing new machinery that enabled a weekly output of 800 tonnes of sugar, although demand stood at only 450 tonnes of sugar a week. Another £600 000 was invested in three Queensland sugar plantations and mills in an effort to augment its imports of raw sugar with Australian-produced raw sugar for the Yarraville Refinery.[17] However, after the sudden drop in the price of sugar during 1884, Victorian merchants flooded the Victorian market with cheap imports of sugar. With no capital reserves, the VSC was unable to compete effectively and faced insolvency. Its liabilities stood at £860 000, and assets, if then sold, would probably have realised no more than £350 000. Alarmed at the prospect of losing a presence in the Melbourne and Tasmanian market – Tasmania was supplied from Melbourne – CSR's directors agreed to an amalgamation between the two companies. In 1888, the old Colonial Sugar Refining Co. became the Colonial Sugar Refining Co. Ltd. The headquarters of the former VSC became the Melbourne office of CSR and Edward W. Knox began salvaging the disaster in Victoria.[18]

CSR had just finished finalising the merger with the VSC, when the directors were confronted with the announcement in August 1890 that the Melbourne-based merchants Charles Jacobs and Sons – described as the city's 'leading firm of Mauritius sugar importers' – intended to erect a sugar refinery in Adelaide. CSR immediately decided to establish a presence in South Australia and acquired 1.5 ha of riverside land at Glanville, towards the mouth of the Port Adelaide River. Following CSR's announcement of their intention to establish operations in Adelaide, Charles Jacobs and Sons abandoned plans to build a sugar refinery in South Australia. CSR, however, decided to press ahead with their plans and over the next year, £100 000 was spent on erecting a small refinery. Initially, the Glanville Refinery, which commenced production on 3 November 1891, relied on imports of raw sugar from Java and managed a weekly output of around 300 tonnes of refined sugar. This amount satisfied much of the South Australian demand.[19]

Twelve months later, CSR decided to erect another sugar refinery, but this time in response to an unfavourable political decision made by the Queensland government. In early 1892, the McIlwraith government began taxing imports of golden syrup into Queensland at the rate of 10s per one-gallon packages. CSR had strived to build up a considerable trade in this item throughout the colony, and was not prepared to see its market share diminish. Moreover, the firm realised that its Queensland share of the market in white sugars would only grow if the product was made locally so as not to attract the import duties of £6 6s on white sugars brought into the colony. Thus, CSR in mid-1892 acquired 1 ha of land at New Farm, an inner Brisbane suburb. The site had two particular advantages: closeness to the bulk of the firm's main customers; and riverside frontage where wharves could be constructed. Construction of the

New Farm Refinery began in mid-1892 and after delays caused by the Brisbane River flood of 1893, the building was completed by August 1893 at the cost of £100 000. At the time of construction, the refinery could produce 250 tonnes of refined sugar weekly. CSR also contemplated erecting its fifth Australian sugar refinery in Western Australia during 1897, but did not proceed any further in the matter. The venture was 'unwarranted', claimed Edward W. Knox, as the small demand for sugar in Western Australia did not justify the expense and the operating costs would be considerably more expensive due to the dearer cost of coal in that colony. Later, E.W. Knox advised the Royal Commission into the Australian Sugar Industry in 1912 that the advantages of combining all the refining in one city were obvious, but CSR with its refineries in four Australian state capital cities was able to quickly deliver fresh and undamaged refined sugar to a consumer which was a distinct advantage.[20]

When Glanville Refinery was built in 1890, Edward Knox Snr advised CSR's shareholders that the additional capacity would for a time render unnecessary any further enlargement of the other refineries which were considered equal to meeting the Australian demand for sugar.[21] However, this view was very short-lived, as the Australian demand for refined sugar products continued to grow throughout the early 1890s and 1900s. Consequently, CSR embarked upon upgrades at its refineries during the 1890s and 1900s, enabling the production of larger amounts of white sugar.

Pyrmont Refinery was the first to have its capacity enlarged by the addition of new and bigger boilers in 1891. New bone char driers and granulators were installed in the mid-1890s and a new bone char kiln house was built in 1900. Additional land near to the refinery was also purchased in 1900, 1904, 1910 and 1911, allowing further expansion, including the erection of a distillery (1901), affination station (1901), tablet house (1908), enlarged bone char making plant (1910) and carbonation station (1912).[22] By the early 1910s, Pyrmont was CSR's premier refinery, its annual output of refined sugar having trebled since 1890 (Figure 9.1). Additional land near Yarraville Refinery was also purchased in 1896, as part of that refinery's expansion program which included the erection of new Buchanan pipe kilns to produce bone char and new sugar driers. By the early 1910s, Yarraville was capable of producing close to 70 000 tons of refined sugar each year, almost treble the amount produced in the early 1890s (see Figure 9.1). In addition, within weeks of the New Farm Refinery commencing operations, CSR was forced to admit that 'the capacity of the refinery was insufficient', following the receipt of so many orders. Nearby blocks of land were purchased, allowing the construction of a new hessian bag store and an enlarged raw sugar house. In 1897, the refinery was also

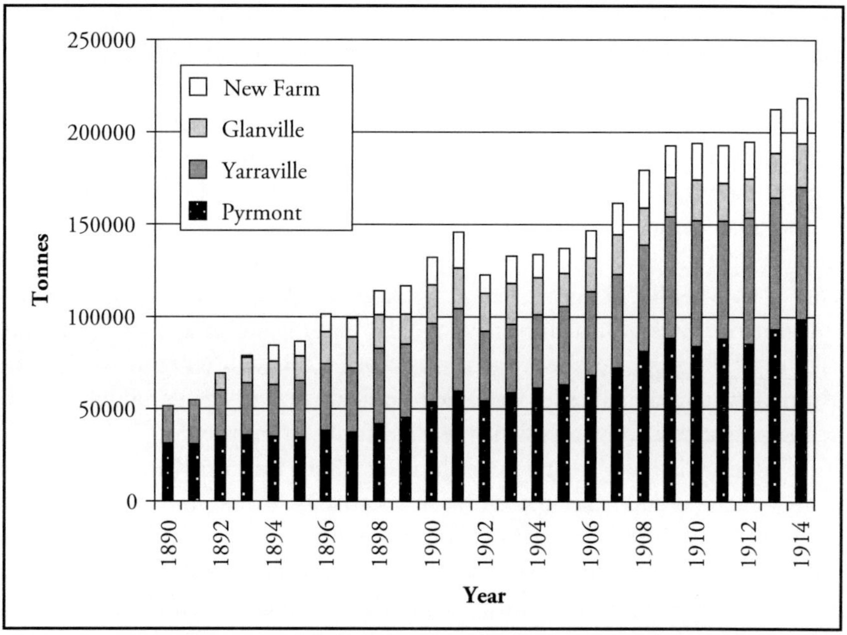

Figure 9.1. Refined sugar production at CSR's Australian sugar refineries, 1890–1914.

Source: Based upon refinery output figures, CSRR, Z/303, Box 9, File A 3.0, Folder 7, Doc. 6.

connected to Queensland's rail network, thereby enabling additional loads of raw sugar to be railed to the refinery from Moreton and Isis Mills, located close to Brisbane. By 1910, the New Farm Refinery occupied 3.6 ha, and the annual output had increased threefold from approximately 7 000 to 24 000 tonnes of refined sugar.[23]

CSR also adopted new refinery processes during this period in order to produce better quality sugar and reduce its burgeoning refining costs. From the beginning of operations, each CSR refinery used a combination of the phosphatation-Taylor bag filter process and bone char paste in the decolourisation of the liquor, before its conversion into refined sugar. This process involved adding phosphoric acid and lime to the raw sugar as it was being melted by hot water in the melting tanks, known as 'blowups' in the nineteenth century (see Plate 9.2). The melter liquor was heated and given time for the defecants to flocculate and to occlude the suspended impurities. The heavy flocculants were then separated out by filtration through Taylor bag filters. The liquor was then sent to the bone char decolorisation plant where

Plate 9.2. The interior of the Victoria Sugar Co.'s Yarraville Refinery, 1881.
(a) Top left: Workers known as 'shot out' men breaking up raw sugar to be dissolved in hot water. (b) Top right: Large circular melting tanks known as 'blow-ups'. The raw sugar was dissolved in hot water which filled these tanks. (c) Centre: Washing the filter bags. (d) Bottom: Emptying the centrifugals.

Source: *The Australasian Sketcher*, 2 July 1881, p. 216. Reproduced from the La Trobe Collection, courtesy of the State Library of Victoria, Melbourne.

it passed through metal cisterns filled with bone char paste to remove additional impurities and to decolorise the liquor before it was boiled in vacuum pans to produce refined sugar. Departure from this arrangement began in 1900 when Pyrmont Refinery adopted affination or the process of washing the raw sugar before it was melted to remove the layer of molasses containing the bulk of the raw sugar's impurities. With a good affination performance, around 85% of the total impurities can be removed, and its use together with the phosphatation-Taylor bag filter process improved product quality. In addition, affination reduced the amount of bone char paste required in the refining process, as the quantity of bone char required to treat 100 tonnes of raw sugar increases as the purity of the melter liquid decreases.[24]

Details on exactly when CSR's other refineries adopted this process could not be reconstructed, but presumably they were all converted during the 1900s in keeping with CSR's practice of generally standardising industrial processes at its factories once the benefits of a particular process were established.[25] A further change occurred over the period 1911–1913, when Pyrmont Refinery was converted to carbonation with diatomite pressure filtration. The carbonation process commences when milk of lime is mixed with the melter liquid. This solution is then run through a series of saturation or 'gassing out' tanks filled with carbon dioxide, obtained from washing boiler flue gases. The process results in the precipitation of large quantities of calcium carbonate in the melter liquid, which is then removed by pressure filtration using diatomaceous earth. Carbonation with the associated filtration plant requires a larger capital investment and higher maintenance costs than phosphatation, yet reduces chemical costs, as the use of phosphoric acid ceases, and the cumbersome system of bag filters is eliminated. Moreover, refineries equipped with phosphatation plants normally burn about 20% bone char on refined sugar output, whereas the corresponding figure for a carbonation refinery is about 10%. Thus, substantial savings can be made in the bone char department using carbonation, which CSR undoubtedly found an attractive proposition.[26]

CSR's capacity to deliver increasing amounts of white sugar to its consumers, however, relied upon refinery workers who during the nineteenth and early part of the twentieth centuries endured jobs that were unpleasant, sometimes dangerous and over time, unhealthy. Machinery in different parts of a refinery generated much heat and high levels of humidity, making working conditions uncomfortable. A picture of the interior of Yarraville Refinery in 1881 showed semi-naked men either breaking up lumps of raw sugar that were to be melted or washing the filter bags (see Plate 9.2). A year later, a visitor to Yarraville Refinery observed that:

> The sugar when released from the baskets or bags is broken up with small axes, or shot out men who are obliged, as in many parts of the premises, to work in a comparatively nude state, on account of the high temperatures caused by the number of steam pipes in the building.

Benjamin Lewis, a union representative, advised the Royal Commission into the Sugar Industry in 1911–1912 that refinery workers suffered:

> the ill-effects of the terrific heat, and the change from that heat into draughts, which may occur at any time, and working in cold and hot water swishing over them all the

time. There are divers weaknesses peculiar to the men who work in different parts of the refinery; they are almost all affected by what is called toe-rot; their toes get in such a condition that they are never dry, and are pulpy and nasty.

In addition to rotting feet, Lewis claimed that the sugar fumes rotted the teeth of employees and that working around hot and cold water, naked, made the workers susceptible to rheumatism and influenza. Boys who worked in the washhouse cleaning the bag filters, he said, 'never reach a proper state of maturity'.[27]

CSR's strategies to dominate the Australian sugar market

As the production of different types of sugar grew in Australia during the 1870s and 1880s (see Figure 9.2), the sugar mill-owners had to market their annual outputs. This task depended very much on the type of sugar-produced and the location and ownership of the sugar mill. Some sugar mill-owners sold their own sugars, as they also operated retail outlets. Samuel Grimes, for instance, sold the sugar from his Coongoon Mill free of commission via his retail store in central Brisbane.[28] Other mill-owners had an association with a particular retail outlet. The Brisbane-based merchant Howes Bros. was the outlet for some of Bingera Mill's mill whites, as they were partners with Gibson Bros. in this venture. Another producer of mill whites in this category was Robert Boyd, part-owner of Ripple Creek Mill. His Newcastle partners, the merchants John and Joseph Woods, sold a third of the mill whites from Ripple Creek Mill.[29] Other sugar mill-owners who produced raw sugar often sold it directly to local storekeepers. In the early 1880s, the sugar from Hilderbrand's Mill, near Bowen, was delivered directly to J. Palmer & Co.'s store in Bowen, while George Martin of Rocky Mouth Mill (Richmond River district) sold his sugar to the Clarence River storekeepers.[30]

Much of the sugar before 1890, however, was dispatched to mercantile and commission agents, who either had customers willing to purchase the sugar or it was sold at auction. For those Australian sugar mill-owners who supplied the Victorian market, many dealt with the sugar brokers Greig & Murray or Godwin Crespin, the manager of Fanning, Nankivell & Co. The latter firm traded extensively in sugar and tea, including the sugar produced by Millaquin and Yengarie Refineries and Macknade and Gairloch Mills, the factories owned by the company during the 1880s. Prominent sugar brokers in

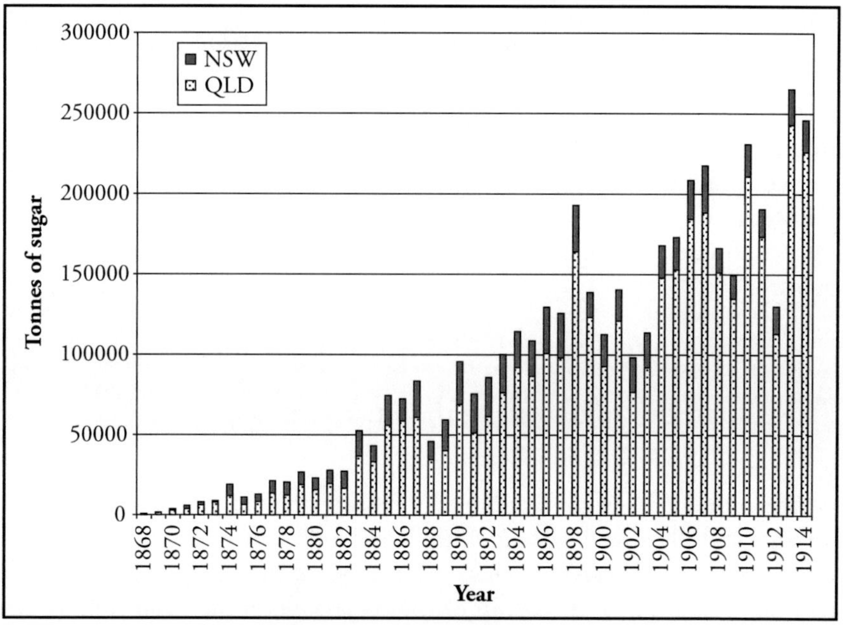

Figure 9.2. Tonnage of sugar produced in Australia, 1868–1914.

Source: Based upon figures in the *Statistics of the Colony of Queensland*, the *Statistics of the State of Queensland* and the *Statistical Registers of New South Wales.*

Sydney during the early 1880s were Perry Boyce & Co., who acted as the agent for a dozen of the small Richmond River millers, while Thomas Barry & Co. and William Miller & Co. handled the sugar from the Clarence River sugar mill-owners.[31] Gibbs, Bright & Co. acquired sugar from several New South Wales and Queensland sugar mills, including the mill whites from Foulden and Pleystowe Mills at Mackay. Moreover, during the early 1880s, this firm attempted to secure more of the output from the smaller New South Wales sugar mills by agreeing to advance their owners funds to buy machinery, in return for the right to acquire their sugar.[32] Southern Queensland sugar mill-owners could deal with several Brisbane-based commission merchants, the most noted being Parbury, Lamb & Co., Brabant & Co. and James Stoddart. The Bundaberg merchant Michael Duffy and the Maryborough merchants Corser & Co. and Graham & Co. sold the sugar from those sugar mills in the Wide Bay-Burnett region that did not supply juice to Cran's refineries.[33] Further north, Walter Reid & Co. of Rockhampton was the central Queensland outlet for River, Nindaroo, Marian and Farleigh Mills, all located at Mackay, while the Mackay-based merchants James Croker and W. H. Paxton & Co.

arranged the sale of the sugar produced by many of the Mackay sugar mills.[34] In North Queensland, Aplin, Brown & Co. and Burns, Philp and Co. handled the sugar from at least nine northern sugar mills, including the mill whites produced by Mourilyan and Macknade Mills.[35]

Some Australian commission agents also imported white sugar, although it attracted import duties of between £3 and £6 per ton in each Australian colony. The most noted sugar importer was the Melbourne-based firm Charles Jacobs and Son, who owned a ship which was devoted entirely to the trade in refined sugar sourced from Mauritius. During the 1880s, John Jacobs claimed that the firm imported 8 000 or 9 000 tonnes of white sugar annually. Other importers of white sugar included the Melbourne-based Fanning, Nankivell & Co. and Fraser and Co. and the Sydney-based Gibbs, Bright & Co.[36] The imported white sugar, together with the mill whites, competed directly with the refined sugar being produced by CSR and the VSC. CSR, however, decided that this upsurge in the volume of sugar being traded in Australia during the 1870s and 1880s made it too difficult to try and entirely dominate the Australian sugar market. Two reasons can be advanced to explain this departure from CSR's previous aim of trying to control the white sugar market by eliminating rivals. Firstly, CSR was in no position financially to engage in widespread underselling of its numerous rivals, having spent so much money on building its own sugar mills and Pyrmont Refinery and then rescuing the VSC. In fact, the practice of underselling appears to have been abandoned during the 1870s and 1880s. Secondly, there were too many rivals, following the upsurge in the number of merchants and commission agents dealing in white sugar across the six colonies. CSR could not possibly influence all these independent traders. Nevertheless, during this period, CSR corresponded extensively with grocers, merchants and mercantile agents throughout Australia, and had established an effective distribution network for its products in the eastern Australian colonies. In particular, the travelling agents of Robert Harper & Co. – a Melbourne-based tea, coffee and spice importer – carried with them from 1887 onwards samples of VSC (and later CSR) sugar and arranged for the firm to fill orders for sugar from inland pastoral stations. Robert Harper & Co. also acted as CSR's agent in Western Australia.[37]

CSR could have eliminated some of the better quality raw sugar or mill whites entering the Australian market during the mid-1870s by purchasing it from the Australian sugar mill-owners for use in its Sydney refinery. Yet CSR preferred to rely on imports of sugar from Java and the Philippines, the firm believing that 'Java sugar is better than the Queensland product for refining'.[38] However, CSR decided in 1877 to purchase some locally produced sugar. The firm arranged with the Mackay-based W. H. Paxton & Co. to acquire all the raw sugar made by open evaporating pans in the Mackay district, because CSR wished to reduce the competition against its own sugar by using Mackay sugar in place of imported sugar from Java.[39] Sugar purchases from the Mackay sugar mill-owners continued throughout the 1880s. CSR paid the Mackay sugar millers a little below the price offered by southern capital city commission agents, but the advantage was payment in cash upon delivery to the wharves at Mackay and no commission fees.

Just why CSR turned only to the Mackay sugar millers, instead of buying the raw sugar produced in New South Wales could not be discovered amongst CSR's records. Maybe CSR was impressed by the refining qualities of the Mackay sugar and preferred to source its refining sugars from one, discrete geographic location, instead of having to try and coordinate the purchase of small batches of raw sugar from many sugar mills in northern New South Wales. On the other hand, actual offers to buy the raw sugar from the New South Wales sugar millers were not found in CSR's archival records, so perhaps Edward W. Knox perceived that there was little point in approaching these sugar producers who had taken up milling in order to be free of CSR's dominance. Anyway, they may not have been willing to sell their sugar to a firm that in their view dictated too many conditions. Certainly this attitude appeared prevalent during the attempt by the New South Wales Sugar Mills Association in early 1884 to convince CSR to purchase the raw sugar manufactured by its Clarence River members. These sugar millers were not prepared to accept CSR's proposal that a minimum quantity of 1 000 tonnes was supplied, all the sugar from the mills had to be sold to CSR and payment was cash on the basis of chemical analyses of the strength of the sugar (i.e. net titre).[40]

CSR's directors, however, never forgot that the company once enjoyed a near monopoly over the Australian refined sugar market, and took action to eliminate rivals when presented with a chance. The first opportunity CSR received to acquire a rival's business occurred in early 1887, when Isaac Peck offered to sell his refinery to CSR. The deal was struck in March 1887, and CSR promptly shut down the refinery.[41] Five years later, Edward W. Knox was now confident enough in CSR's financial situation to resume using the underselling strategy that had been employed so effectively by the company in the 1850s. The Australasia Sugar Refining Co. was targeted, the firm having commenced making inroads into CSR's trade in refined sugars in Melbourne, Newcastle, Tasmania and New Zealand. At Melbourne in September 1893, for example, a ton of CSR's white sugar sold for 5s less than the Australasia

Sugar Refining Co.'s product, while at Newcastle in August 1894, CSR's best white sugar could be purchased for £1 per ton less than the Australasia Sugar Refining Co.'s refined sugars.[42] The Australasia Sugar Refining Co. tried to reach an agreement with CSR in 1893 over prices and market share, but refused to accept CSR's terms that it should limit its New Zealand business to orders already accepted. Edward W. Knox was aware that the demand for refined sugar in Victoria and Tasmania could not sustain two large refineries, so he initiated negotiations in October 1894 with the Australasia Sugar Refining Co. about a possible merger between both firms. Two months later the companies had merged, with the shareholders in the Australasia Sugar Refining Co. acquiring an interest in the amalgamated concern proportionate to their holdings in the old company.[43]

Oddly, Edward W. Knox was initially more tolerant of the challenge from Frederick Poolman (and later his sons after Frederick died), advising the Melbourne office of CSR in 1895 to 'let him go his own ways'.[44] Knox never explained why he chose not to eliminate this rival in the 1890s. As a result of offering lower prices than CSR, the Poolman family slowly increased their share of the Australian market in refined sugars from 3 050 tonnes (1898) to 12 200 tonnes (1906). Large customers like the Melbourne-biscuit makers Swallow & Ariell Ltd. had even been induced to deal with the Poolman family instead of CSR.[45] After 1903, CSR decided that competition from the Poolman family was beginning 'to assume serious proportions'. In response, CSR began underselling Poolman Bros. in Victoria to try and regain their market share. Shortly after Poolman Bros. purchased land in Sydney for their new refinery, CSR dropped its prices by 10s per ton. When the refinery commenced production CSR reduced its prices by another 10s.[46] Poolman Bros. were probably aware that CSR would not tolerate for long their attempt at gaining a foothold in the New South Wales market, and that they would face strong competition. Edward Poolman approached Edward W. Knox in July 1906 with a request that the two Poolman-owned refineries be allowed to sell up to 30 500 tonnes of refined sugar in Australia. Knox declined to enter any such arrangement. A year later Edward Poolman asked Knox if CSR was interested in purchasing the family's refineries. CSR concluded that the acquisition of the two refineries was very advantageous. The price they received for white sugars could rise immediately by 10s per ton and they had an excellent chance of securing most of Poolman's customers, thereby giving them even greater control over the Australian market. CSR accepted Poolman's offer, but buying out the opposition cost the firm £223 000.[47]

As mentioned earlier, CSR did not try to compete with Robert Cran Snr and the makers of mill whites, although Edward W. Knox was aware that they had undersold CSR for years in the Australian market and had occasionally interfered considerably with the firm's business.[48] In 1891, CSR expanded its purchases of sugar beyond the Mackay district by acquiring the annual sugar outputs from some New South Wales and Bundaberg sugar millers.[49] The success of this policy in reducing the amount of sugar being traded in opposition to CSR on the Australian market clearly influenced Edward W. Knox, who decided that it was time in 1893 to take more concerted action against CSR's Queensland competitors. CSR's new sugar refinery in Brisbane needed raw materials and Knox tried convincing all Queensland and New South Wales sugar mill-owners that they would receive better financial returns from manufacturing refining sugars for the Brisbane refinery, instead of trying to make mill whites. If the sugar mill-owners agreed, CSR could eliminate more producers of mill whites, get the refining sugars it needed and bolster its position as the main supplier of refined sugar products in Australia. The company's terms were that it would only purchase the entire output from a factory, the sugar mill-owners had to cease using sulfur fumes in the manufacturing process and that the sugar produced had to be 88 n.t. or stronger, although this latter condition was modified in 1896 when CSR required the sugar mill-owners to manufacture a product with an average strength of not less than 93 n.t. so the company could meet the gradually increasing Australian demand for 'the finest sugar'.[50] As an inducement, CSR commenced agreeing to requests from the Australian sugar mill-owners for loans to meet operating expenses, on the basis of liens over their sugar.[51]

CSR's proposition provided the Australian sugar mill-owners with certain advantages. The offer theoretically could improve their returns. The merchant William Brown of the firm Aplin, Brown & Co. advised the Royal Commission into the Queensland Sugar Industry in 1889 that Australian producers of sugar received reduced returns because of 'injudicious competition', a result of limited coordination in the supply of sugar reaching the Sydney and Melbourne markets.[52] The Australian sugar mill-owners would be able to opt out of the current chaotic marketing arrangements, avoid the yearly worry of engaging a commission agent to sell their product and the associated annual fees charged by the agents to cover the receiving, storage and delivery of the sugar to customers would be eliminated. In return, however, the Australian sugar mill-owners had to manufacture good quality refining sugars of a particular net titre. This restriction appeared not to be too onerous, and the owners of Farleigh, Mourilyan, Palms and Hambledon Mills (all Queensland) and Rous Mill (New South Wales) – all makers of mill whites – agreed to CSR's terms. By 1899, CSR was sold the output from almost half Australia's sugar mills (Table 9.2). However, the Australian sugar mill-owners who signed agreements with CSR found that Edward W. Knox was very serious about the supply of high quality refining sugars. CSR chastised several Australian sugar mill-owners during the mid-1890s about their dirty sugar and urged them to improve its strength, but also offered advice on ways to achieve this goal.[53] Consequently, as mentioned in the previous chapter, Australian sugar millers commenced employing analytical chemists to supervise the manufacturing process. The beginnings of a regulated sugar market and the manufacture of a uniform product had commenced in Australia.

The QN Bank, owner of the Millaquin and Yengarie Refineries, did not welcome CSR's manoeuvres in Queensland, and maintained its policy of underselling CSR. Joseph Johnston, General Manager of the Millaquin Sugar Co. Ltd., outlined the strategy. In the late 1890s (he did not supply the exact date), he left written instructions with newspaper editors that if CSR lowered its price, Millaquin followed suit, so that Millaquin's prices were always less than CSR's prices. They were to put Millaquin's advertisements underneath CSR's, so the customers would know the prices for Millaquin sugars.[54] In addition, the QN Bank resolved in 1894 to increase Millaquin's annual output, but

Table 9. 2. The number of Australian sugar mill-owners selling their annual output to CSR, 1893–1908.

Year	QLD	NSW	No. of CSR-owned sugar mills	% of mill output sold to CSR	Total no. of sugar mills in Aust.*
1893	28	0	6	39	86
1896	27	3	7	43	85
1899	23	3	9	49	71
1902	18	1	9	57	49
1905	20	0	9	54	54
1908	24	1	9	62	54
1911	22	1	9	66	48

* Excludes juice mills, which only crushed cane.

Source: Assembled from acceptance letters in CSRR, 142/482, pp. 375 & 425, 142/484, pp. 19, 20, 22, 25, 77 & 232 & Z/109, Box 312, Letterbooks No. 3 to No. 12 & Box 313, Letterbooks No. 13 to 16.

to do so the refinery needed more raw materials to allow refining to occur all year. Local juice was only available for six months of the year. Therefore, the QN Bank converted Millaquin Refinery into a factory capable of treating raw sugar, and tried convincing some of Queensland's sugar mill-owners to manufacture refining sugars for Millaquin. As incentives, the QN Bank offered the Queensland sugar millers loans in return for their yearly outputs and did not require them to sell their entire output to the bank, unlike CSR. Despite its efforts, the QN Bank only ever succeeded in securing the output from between three and eight Queensland sugar mill-owners each year during the late 1890s and early 1900s. Regular suppliers included Ashfield, Pemberton, North Eton, Marian and Plane Creek Mills. Cran Bros. & Co. also sold the sugar from their Mon Repos Mill, near Bundaberg, to the Millaquin Refinery in 1898 and 1899, before the firm sold the sugar mill to the QN Bank.[55]

Some Australian sugar millers who did not sell to CSR or the QN Bank during the 1890s manufactured only small amounts of sugar in mills equipped with increasingly out-dated machinery. Henry Heinemann of Mt Cotton sugar plantation on the outskirts of Brisbane belonged in this category. He still used open pan boiling in his small sugar mill, turning out raw sugar that was sold to Smith Bros., the Brisbane-based commission agents.[56] About a dozen other Australian sugar mill-owners persisted in manufacturing mill whites at various times during the late 1890s, in defiance of CSR and the QN Bank. The main mill white producers in Southern Queensland included the Maryborough Sugar Factory (from 1897 onwards) and Goodwood, Bingera and Fairymead Mills, while Ripple Creek and Macknade Mills, both in the Herbert River district, were the most noted North Queensland suppliers of mill whites. Each of these sugar millers continued relying on commission agents to sell their sugar, although different sale strategies were put in place. Macknade's mill whites, for example, were handled by Gibbs, Bright & Co. of Sydney and the Townsville-based Messrs Aplin Brown & Co., while Burns Philp & Co. distributed Ripple Creek's mill whites throughout Queensland. The Maryborough Sugar Factory's output was sold locally to the Maryborough merchants Webster & Co., Corser, Graham & Gataker, and Hanley & Sons.[57]

CSR, however, tried eliminating these mill white producers. In 1895, Edward W. Knox approached Young Bros. of Fairymead – one of the biggest Queensland makers of mill whites – to determine if they could together broker a deal to limit the output of mill whites from the Queensland sugar millers who refused to sell to the refiners. His efforts were unsuccessful. Knox had more success in convincing CSR's directors to purchase Macknade Mill in December 1896, after it was offered for sale by its owners, Frederick and Arthur Neame. CSR not only expanded its milling operations in North Queensland, but removed a mill white producer. A chance to eliminate another major mill white producer occurred four months later, when Young Bros. offered to sell Fairymead Mill and plantation to CSR. Rudolph Helms, CSR's Childers Mill Manager, inspected the property, but declared it would 'be a white elephant to the Company' on account of the estate being subject to frost and doubts over its suitability for subdivision into small farms as CSR would not maintain plantation agriculture on the property. The CSR Board, on Helms' advice, declined to purchase Fairymead. Instead, Knox tried again in May 1898 to convince the mill white producers in the Bundaberg district to restrict their sale of mill whites on the domestic market to 35 600 tonnes, in order to reduce what he termed the 'senseless competition' between each mill white producer as they sought higher prices. Knox suggested the remainder of their annual output would have to be exported. The mill white sugar producers around Bundaberg met throughout May 1898, but eventually refused CSR's proposal.[58]

The growing competition between CSR and the QN Bank made some Queensland sugar millers prone to selling their raw sugar to a different refiner each year.[59] To stop this swapping between refiners, CSR and the QN Bank reached an agreement in 1903 concerning the purchase of the sugar mill-owners' annual outputs. The QN Bank agreed not to import any raw sugar and to limit its purchases of raw sugar in Queensland to enable the manufacture of 17 300 tonnes of refined sugar. This raw sugar would be purchased from sugar mill-owners located only in the Bundaberg and Isis districts. The 1903 agreement between CSR and the QN Bank was re-negotiated each year during the mid 1900s and generally allowed some increase in Millaquin Refinery's output. In 1906, the QN Bank convinced the owners of Goodwood Mill to manufacture refining sugars instead of mill whites, so CSR agreed to its output being added to Millaquin's quota, raising it to 18 800 tonnes of sugar annually. Two years later, the two firms agreed to Millaquin Refinery turning out 19 800 tonnes of sugar annually.[60]

These agreements meant that every Queensland sugar miller north of Rockhampton had to sell their sugar to CSR, revert to using commission agents or make mill whites for the domestic market. The sugar mill managers at Pioneer and Kalamia Mills complained privately about 'CSR being the only market available', but took no action. A similar policy was adopted by their counterparts at other Queensland sugar mills, except Mossman Mill.[61] In 1905, the directors of the Mossman Central Mill Co. Ltd. authorised the production of 1 000 tonnes of mill whites for the domestic market at Mossman Mill, believing that it was possible to get £3 a ton more than CSR offered for their raw sugar. However, the financial results from this trial did not justify the continuation of the scheme. Throughout all of these business manoeuvers, Robert Boyd at Ripple Creek Mill manufactured mill whites for his Newcastle partners, but in 1906 he switched to making refining sugars for CSR. In the following year, CSR purchased the sugar mill which had been offered for sale by Boyd, and immediately closed it.[62]

With the northern sugar mill-owners firmly tied to CSR, the firm turned its attention to the Beenleigh district. Here most of the sugar mills were small affairs, owned either individually or co-operatively by German immigrants or their descendants. During the non-crushing seasons in 1907 and 1908, A. Wright, the manager of CSR's Brisbane office, attempted to convince these sugar mill-owners to manufacture refining sugars, but he managed to secure agreements from only half the sugar millers. CSR then abandoned its efforts to secure this sugar, but their output had little impact upon CSR's dominance over the Queensland market, as together they produced only approximately 1 000 tonnes of low-grade sugar annually.[63]

The agreements between CSR and the QN Bank had not resolved the issue of competition in the Bundaberg and Isis districts where eleven sugar mill-owners sold their sugar to different refiners each year or made mill whites. In October 1909, CSR and the QN Bank agreed that the former would no longer offer to buy sugar in the Bundaberg and Isis districts, except from Bingera and Fairymead Mills. In return, the QN Bank would not import sugar, offer to buy sugar from only the millers in the Bundaberg and Isis districts and could manufacture up to 25 500 tonnes of refined sugar.[64] At the signing of this one-year agreement, CSR was able to bid for the annual outputs of just under 70% of Australia's sugar mills; the QN Bank was allowed to make offers to a quarter of the country's sugar mill-owners. The QN Bank, however, was not entirely happy with this agreement, claiming that Millaquin Refinery had the capacity to produce 35 000 to 40 000 tonnes of refined sugar a year.

From 1911 onwards, CSR and the QN Bank appeared not to have any formal written agreement about bidding for the output from particular Queensland sugar mills, and CSR secured the output from several sugar mills in the Bundaberg district (e.g. Pemberton, Bonna, Windemere). Joseph Johnston, the General Manager of the Millaquin Sugar Co. Ltd., advised the Royal Commission into the Australian Sugar Industry in 1911–1912 that his organisation had no arrangement with CSR in 'any shape or form' in connection with the purchase of raw sugar from particular sugar mill owners. Such evidence, while technically correct, could not disguise that the QN Bank and CSR had previously colluded to purchase sugar from specific sugar mills only, and their owners had suspected or known of this arrangement.[65]

Throughout the 1890s and 1900s, CSR also sought to deny its competitors outlets for their mill whites or refined sugars. Brewers, jam makers, aerated water manufacturers and confectioners, who accounted for 25% of CSR's sales of refined sugars by 1912, had to make twelve-month contracts with CSR. These contracts contained a restrictive practices clause that prevented the manufacturer from importing sugar during the life of the contract. As a result, CSR locked manufacturers into its supply network for an entire twelve months. William McGown, a merchant who represented Young Bros. in Sydney, acknowledged that these arrangements CSR had with its manufacturing clients made it difficult for him to sell Fairymead's mill whites.[66]

Australian manufacturers were also encouraged to purchase their sugar exclusively from CSR, which offered discounts on the price their customers paid based upon the tonnage demanded, with the maximum discount in 1912 being a 6% reduction in the price per ton of sugar. Regular buyers were protected against any loss on their stock in the event of prices being reduced, for CSR guaranteed that it would reimburse the wholesalers for the difference, provided the stock was purchased within a reasonable period before the fall. Such generous terms ensured that many merchants and manufacturers preferred to deal with CSR, although the QN Bank from 1899 onwards offered wholesalers a reduction of 5s per ton on orders of five tons or more.[67] In addition, some merchants and commission agents were awarded a CSR agency for a particular geographical area, but on the condition they sold CSR sugars only. CSR agencies in Queensland during the early 1900s, for example, were awarded to Burns Philp & Co. (Townsville and Cairns), W. H. Paxton & Co. (Mackay) and Walter Reid & Co. (Rockhampton). The Queensland mill white producers were then denied wholesale outlets for their product in the main North Queensland centres.[68]

Despite the pressure applied by the QN Bank and CSR to the wholesalers and manufacturers, Nerang, Fairymead and Bingera Mills manufactured mill whites during the early 1910s, offering their sugar to the public at 2s to 7s per ton less than the QN Bank or CSR.[69] The Gibson family at Bingera tried assessing the market each year, manufacturing a combination of mill whites for their wholesale merchant partners in Brisbane and refining sugars that were sold to both refiners. Young Bros. at Fairymead made only mill whites. Yet by 1912, both firms struggled to find outlets for their sugar, despite its lower price. The Tasmanian-based jam manufacturer Henry Jones claimed that Bingera and Fairymead mill whites were not suitable for his business, as they did not make the jam 'sufficiently bright'. The Melbourne sugar-broker Ernest Connell observed in 1912 that there was a limited market for the Queensland mill whites because their colour was 'not quite so good' as CSR's products. Frederick Sharp, the general manager of Thomas Heaslop & Co. Ltd., Brisbane merchants, advised the Royal Commission into the Australian Sugar Industry in 1911–1912 that his firm had tried selling Bingera mill whites, but their customers preferred CSR's sugars - they were regular in grain and in whiteness, and there was no treacly smell or flavour of molasses about them. This sentiment was also repeated in the evidence from Angus Gibson, co-founder of Bingera Mill, when he lamented that 'according to the taste of the people, our sugars are not equal to CSR'. In 1913, Fairymead and Bingera Mills ceased making mill whites and made raw sugar for the refiners, leaving only Nerang Mill manufacturing mill whites.[70] Nerang Mill's annual output in the late 1900s and early 1910s was only a few hundred tonnes, which were handled by John Cardno, a Brisbane-based commission agent. Cardno, however, did not recommend the making of mill whites, as he struggled to sell Nerang Mill's output.[71]

The only other challenge CSR faced during its attempts to dominate the Australian sugar market came from merchants or manufacturers who imported refined sugar. At the turn of the century, at least Gibbs, Bright & Co. and Charles Jacobs & Sons brought sugar to Australia from Mauritius and Java. Macpherson Robertson, a Melbourne-based confectioner, was also importing sugar directly from Hong Kong in the early 1900s, the Chinese merchants initially supplying him white sugar at lower prices than CSR.[72] Imports of sugar into Australia after Federation attracted a protective duty of £6, so the prices offered by the Chinese merchants were quite competitive. W. Rothe, CSR's secretary, suggested to Edward W. Knox that the firm could come to some arrangement with the Hong Kong refiners, whereby they ceased selling to Australian manufacturers and merchants in return for CSR purchasing 10 000 tonnes of their sugar annually. Knox, however, did not welcome this suggestion or interference from the Hong Kong merchants and refiners and ordered an aggressive underselling strategy. From 1902 onwards, CSR reduced its rates for refined sugar by 10s a ton, with a further concession to manufacturers under contract, every time Chinese-sourced sugar arrived in Sydney or Melbourne.[73] Macpherson Robertson advised the Commonwealth government's Royal Commission on Customs and Tariffs in 1906 that this strategy had forced the Chinese refiners to cease sending sugar to Australia during 1905.

In mid-1907, Henry Jones also attempted to break CSR's control over sugar supplies by investigating the possible purchase of Millaquin Refinery, but he concluded that the refinery was too small to supply all his jam factories on a permanent basis. In the following year, Jones took a different approach, taking delivery of a consignment of imported Jamaican refined sugar. Jones told the Royal Commission into the Australian Sugar Industry in 1911–1912 that he did not continue with the practice in subsequent years as the price-margin between CSR and Jamaican sugar was too fine when the import duties were taken into consideration. The unofficial version which quickly circulated at the time of the incident, however, painted a different picture. CSR was furious with Jones for importing sugar and made it clear that if he ever did it again, the supply of Australian-made sugar to his factories would quickly dry up. Only Burns, Philp & Co. defiantly ignored CSR's underselling strategy, maintaining annual imports of 500 tonnes of white sugar from Hong Kong during the 1900s and early 1910s, partly for use by Tooth & Co. Ltd.'s brewery in Sydney.[74]

An Australian monopoly?

From the late 1880s onwards, an oligopoly rapidly emerged in the supply of white sugar in Australia. By 1887, Edward W. Knox advised Sir Daniel Cooper, one of the London partners in CSR, that the company had a 'strong hold on the sugar trade in New South Wales' and that all the Victorian trade circulars admitted 'CSR determined the price of refined sugar in that colony', although the firm still faced competition from the mill white producers who sold their sugar in Melbourne and the importers of white sugar. However, as CSR eliminated Queensland's mill white producers, this competition dwindled. In 1896, the Victorian government's Board of Inquiry into the effect of the Fiscal System of Victoria upon Industry and Production concluded that CSR in Victoria 'held a virtual monopoly on the sugar trade'.[75] During the 1880s, CSR in South Australia faced competition from Charles Jacobs & Son who imported sugar from Mauritius and the small amount of sugar from the Northern Territory being sold in Adelaide. Yet by 1892, CSR controlled the entire South Australian market. Moreover, CSR offered its South Australian customers a bonus of 10s per ton of sugar purchased from the Glanville Refinery between 1896 and 1904, thereby ensuring that no rival suppliers of white sugar could challenge its supremacy in the colony.[76] By 1905, CSR produced 80% of the refined sugar in Australia. This figure rose to nearly 90% in 1907 following the purchase of the Poolman family's refineries. In addition, as shown above, CSR had also forced most merchants or manufacturers who once imported sugar into Australia to cease this trade. By the mid-1900s, CSR effectively controlled the Australian sugar market.[77]

CSR's growing control over the supply and price of the sugar filling the country's sugar basins had not gone unnoticed by journalists, Australian sugar millers and parliamentarians. Editorials in the *Mackay Standard* and *The Bundaberg Mail* during 1895 and 1896, respectively, urged the Queensland government to erect a state-owned sugar refinery at either Proserpine or Gladstone, although they did not specifically state that such action aimed to break CSR's near monopoly over sugar refining. Sugar millers and cane-growers in the Mackay district met on several occasions throughout 1898 to discuss erecting a co-operative sugar refinery, but supporters of the proposal failed to raise the necessary capital for the project.[78] In February 1899, James Chataway, Queensland's Minister for Agriculture and Stock, told a meeting of constituents at Mackay that 'Federated Australia had room for a central refinery and the CSR Company as well', so the Queensland government was considering funding the erection of refinery that could produce 1000 tonnes of refined sugar weekly. Edward Knox Snr wrote in a private letter to Edward Forrest, the principal partner in the Brisbane-based merchants Parbury, Lamb & Co., that he thought Chataway was speaking nonsense, and his assessment proved to be correct: the Queensland government never proceeded to erect a sugar refinery in opposition to CSR.

Perceptions of a CSR monopoly would, however, not vanish. In 1901, Thomas Givens (Labor; Cairns) tried unsuccessfully to convince the Queensland Parliament to fund the erection of a state-owned refinery in order to break CSR's control over sugar refining. In the same year, Givens was elected to the Commonwealth Parliament as a senator for Queensland and his crusade against CSR then abated, albeit temporarily. Givens resumed his attacks on CSR in the Senate on 28 September 1905, when he introduced a motion that stated CSR's control over the refining and wholesale distribution of sugar throughout Australia constituted a monopoly and that this arrangement needed to be broken by nationalising the refineries. The debate on the motion was lengthy, with some speakers supporting the motion and taking the opportunity to criticise CSR's business practices. Other senators, however, were not convinced of CSR's monopoly, highlighting that the Poolman family and the QN Bank still operated refineries in opposition to CSR. Eventually, the question was resolved in the affirmative in

December 1905, but no action was taken by the Deakin (Protectionist) government to implement the Senate's resolution.[79]

CSR's control over the refining and wholesale distribution of sugar in Australia was raised again in 1906 as part of the debate in the Commonwealth Parliament over the Deakin (Protectionist) government's bill that eventually became the *Australian Industries Preservation Act* of 1906. This legislation was associated with the policy of 'new protection' by which tariff encouragement to Australian industry was linked with measures to prevent the growth of injurious monopolies, fair prices for consumers and fair wages for workers. It sought to protect Australia's industries from overseas firms that were dumping cheaply-produced products in Australia with the intention of destroying local industries. The bill contained anti-trust sections, but the Deakin government lacked a serious commitment to an anti-monopoly policy and the anti-trust sections of the bill were added because the unfair competition from overseas about which the government was concerned happened at the time to be mounted by a trust. Indeed, when challenged in parliament about the impact upon local monopolies, Sir William Lyne, the Minister for Trade and Customs who had responsibility for the introduction of the bill, thought the sugar monopoly would be untouched.

Lyne's view turned out to be correct: the mere passage of the anti-trust legislation did not usher in an era of 'trust-busting' in Australia for the anti-trust provisions of the *Australian Industries Preservation Act* 1906 were rendered largely ineffective by a High Court decision in the case of *Huddart Parker v. Moorehead* (1908). The High Court found that Commonwealth power as to certain corporations – sec. 51 (xx) – did not enable it to engage in trust-busting in relation to the intra-state activities of such corporations. Accordingly, the High Court held (Isaacs J. dissenting) that sections five and eight of the *Australian Industries Preservation Act* 1906 were invalid. This decision restricted the scope of the legislation to concerns operating within the field of inter-state and foreign trade, and left all intra-state trade free from federal anti-monopoly control.[80]

Antagonism towards CSR's perceived monopoly over the sugar market continued to be voiced during the late 1900s. In January 1910, Dr Walter Maxwell, who had been asked to report to the Commonwealth Minister for Trade and Customs on the effectiveness of the bounty in promoting the production of sugar by European labour only, raised the issue of collusion between CSR and the QN Bank over the amount of white sugar production and price fixing. Maxwell argued that the free production and free sale of white sugar should be established in Australia, noting it was unusual that so few Australian sugar mills manufactured mill whites, given that greater profits accrued from the sale of white sugar. A few months later in mid-July 1910, a meeting of sugar mill-owners at Mackay and Bundaberg again considered a proposal to erect a co-operative sugar refinery in Queensland. Their aim was to have 'the refining and distribution of their sugar in their own hands'. The Bundaberg sugar millers were lukewarm about the idea; the Mackay sugar mill-owners voted to ask for financial assistance from the State and Commonwealth governments, although no record could be found indicating that the scheme was ever placed before a representative of either level of government.[81]

More vehement attacks on CSR's business activities occurred during the campaign associated with the referendum held on 26 April 1911, when Australians were asked to give the Commonwealth Parliament – at the expense of the States – powers over all matters of trade, commerce and industry within the Commonwealth and to give the Commonwealth government power to acquire and nationalise any business deemed by both Houses of Parliament to be a monopoly. These proposals were put forward largely at the instigation of William Morris Hughes, Attorney-General in the Fisher (Labor) government, but only after a long battle within the Labor Party. Hughes publicly claimed on 13 April 1911 that 'one of the great monopolies injuring the people was the Sugar Company. It was the only buyer of the raw product and the only seller of the finished article'. A few days later on 25 April 1911, Hughes again criticised CSR, challenging 'anyone to land 5 000 tons of sugar in Sydney, and sell it. He could not sell it, because if a storekeeper bought as much as a ton of it, he would be unable to buy any more sugar from any other source'.[82] Both proposals were defeated at the April 1911 referendum.

Anti-CSR sentiment did not abate. In August 1911, Miles Ferricks, a Queensland parliamentarian (Labor; Bowen), proposed that the Queensland government should establish a state-owned sugar refinery which could refine the sugar from the four central sugar mills controlled by the Queensland Bureau of Central Sugar Mills and any new central sugar mills erected in the future. His motion in the Queensland Parliament was defeated, but a month later the idea of a state-owned sugar refinery was again being canvassed, but this time by the press in Victoria. William M. Hughes and *The Age* (Melbourne) suggested that CSR charged too much for sugar, but having attacked the refiner the two stalwarts parted company. The newspaper went further, suggesting that the Commonwealth should establish a sugar refinery in the Northern Territory, Papua or near Jervis Bay, so that non-CSR sugar could be supplied to Commonwealth government public servants.[83] Hans Irvine, the Liberal member for Grampians (Victoria) also launched an attack on CSR in the Commonwealth Parliament on 3 September 1911. He claimed that the company was 'the only trust in all Australia of which I know; and the Government have not attempted to deal with it, although octopus-like, it is crushing the people of this country'. In the same month, Queensland's parliamentarians again debated the issue of establishing a state-owned sugar refinery during the introduction of the legislation to establish new central sugar mills in the State. Despite Miles Ferricks' second attempt to convince his colleagues such a move was necessary, Queensland's legislators again voted not to build a state-owned sugar refinery. Alan Birch and J. Blaxland, two contributors to CSR's official history, claimed that Edward W. Knox 'felt bewildered when it was suggested that Australia needed rescuing from the company' and that the attacks on CSR during the late 1900s had left him 'sincerely distressed'.[84]

As mentioned in Chapter 5, the Fisher (Labor) Commonwealth government eventually appointed a Royal Commission in October 1911 to investigate all facets of the Australian sugar industry's operation. The Royal Commission gathered evidence in Queensland during late 1911 and the first six months of 1912. Its attempts to uncover CSR's profits, however, were frustrated by the actions of the firm, which had objected on the grounds that its internal accounts were unable to distinguish the Australian operations from those in Fiji and New Zealand, and that matters concerning the company's operation outside Australia were not subject to Commonwealth jurisdiction. Moreover, CSR wished to be represented at the Royal Commission's hearing by legal counsel, an arrangement not permitted by Sir John Gordon, the Chairman of the Royal Commission. Various CSR officers had given general evidence about the Company's policies during the early months of 1912, but when Edward W. Knox attended the hearing of the Commission in Sydney on 1 May, he declined to answer the questions of the Commissioners, requesting that he be allowed to read a statement giving details of its trading operations and relationships with its customers. Sir John Gordon refused this request, and after a heated exchange with Knox, he adjourned the hearing.

On the following day, Knox, Henry Kater, CSR's Chairman of Directors, and the other directors all failed to appear at the sittings of the Commission. A test case of prosecution for contempt of court was then brought against one of CSR's directors, Sir Norman MacLaurin, a prominent Sydney doctor. After a thirty-eight day court hearing, he was found guilty and fined, but this judgment was subsequently overruled by appeal to the High Court of Australia and the Privy Council. Essentially, the higher courts ruled that CSR could not be compelled to give evidence or produce books relating to its internal organisation, the wages or salaries it paid, or anything but the interstate or overseas trade in sugar, which the company maintained it could not as this information was not separately distinguished in its books.[85]

As the court cases were heard, the Royal Commission suspended its hearings, but resumed them on the 14 October 1912, although the High Court of Australia had issued an injunction against the Commission limiting its investigations into the financial affairs of CSR. Professor W. Jethro Brown had assumed the chairmanship of proceedings and Edward W. Knox now agreed to appear before the Commission. Brown took a more conciliatory attitude towards CSR, commencing by stating that Knox's 'knowledge of the industry made it desirable that the commission should have the assistance of his counsel in Sydney'. CSR's prepared statement previously offered to the Royal Commission was then accepted, and Knox subsequently cross examined on its contents two days later. Knox admitted to the commissioners that CSR determined the price of refined sugar in this country, subject only to global movements in the price of sugar plus the import duty. In addition, CSR determined the prices paid to the Australian sugar mill-owners for their raw sugar, subject only to the constraint that the prices had to be sufficiently high to avoid squeezing them out of existence or forcing them to make mill whites. This control had contributed to CSR achieving handsome profits; it had paid annual dividends (plus bonuses) in the order of 15% on subscribed capital during the 1900s.[86]

After hearing Knox's evidence, the commissioners retired to prepare their report. The Melbourne representative of the *Australian Sugar Journal* suggested in November 1912 that the commissioners 'did not know exactly what to recommend', as they had failed to discover any serious case of offence or grievance in the general conditions of the Australian sugar industry. Their report was delivered to the Commonwealth government on 3 December 1912. Summing up, they concluded there was a virtual absence of competition in the country's sugar markets and refining sector, but surprisingly argued that Australia's sugar refineries were highly efficient, that nothing would be gained by public ownership of the factories and that CSR's large profits were 'largely due to items other than refining'. Indeed, the commissioners expressed their admiration of the economic efficiency which characterised 'every branch of CSR's business'. They declined to recommend that Australia's sugar refineries needed nationalising, that state-owned sugar refineries should be erected in opposition to CSR or that any anti-trust measures should apply to CSR. Instead, the Royal Commission recommended the following measures: regulating government control over sugar imports, so that the customs duty on imports of raw and refined sugar fluctuated in accordance with foreign prices, falling as those prices rose; the domestic price of sugar should be fixed by an Interstate Commission; and the price of sugar cane should be set for each sugar mill supply area by independent boards consisting of representatives from the canegrowers and mill-owner.[87]

The Commonwealth government accepted the Commission's recommendations about the refining sector and the determination of the domestic price of sugar was referred to the Interstate Commission set up in late 1912. The Commissioners, however, were not appointed until August 1913 and they showed no interest in the domestic price of sugar until October 1914, and only after extensive lobbying from the ASPA. Yet the Interstate Commission still had not determined the domestic price of sugar as 1914 drew to a close.[88] In the meantime, CSR effectively controlled the Australian sugar refining sector and sugar market until late 1914, when its ability to set the domestic price of sugar was assumed by the state-based tribunals appointed to administer the various pieces of price-fixing legislation which had been introduced by the states in the immediate aftermath of the commencement of World War 1. The Food Prices Boards formed in all Australian states were supposed to prevent the artificial inflation of domestic prices by persons taking advantage of wartime conditions. In November 1914, the New South Wales and Victorian Food Prices Boards had fixed the domestic price of sugar at £22, or 5s less than the price during the interval between October 1912 and July 1913, much to the chagrin of CSR and complaints from the ASPA.[89]

Antagonism towards CSR, however, did not abate following the Royal Commission's report in 1912. Within a week of the Commonwealth government receiving the report, William M. Hughes during the debate associated with the Nationalisation of Monopolies Bill again attacked CSR, claiming that if the government were running the industry, the 'profits wrung from the people' would be returned either in 'giving better wages to the workers, and better prices to the canegrowers, or cheaper sugar to the consumers'. The attempt to nationalise monopolies and regulate trusts was defeated in a referendum to amend the constitution on 31 May 1913. On the same day, Fisher's Labor government lost the Federal election and Hughes, CSR's nemesis, was consigned to the opposition benches, where his tirades against CSR ceased. Just over a year later, Andrew Fisher led the Labor Party to another victory at double dissolution Federal elections. Within a month of this victory William Archibald, the new Minister for Home Affairs, had again singled out CSR by saying it had the power 'to tax the community in defiance of Parliament or any authority'. As the Labor Party had campaigned on a platform which included a Commonwealth sugar refinery, this speech suggested that the new government was about to commence implementing its promises.[90] However, events in Europe increasingly attracted the attention of the new Commonwealth government more than the refining of sugar and any move to establish a Commonwealth-operated sugar refinery petered out during late 1914.

Despite the criticisms of CSR, Edward W. Knox had stopped any attempt by the Commonwealth government to interfere directly in the marketing and refining of sugar in Australia, although state-based price fixing tribunals had assumed some of this responsibility by late 1914. Yet 1914 was to be the last year that CSR fully controlled the marketing and refining of sugar in Australia. Despite CSR's vast financial resources and Knox's business acumen, he could not control the geo-political events in Europe and their repercussions in Australia. As will be shown in Chapter 18, CSR's own behaviour in possibly trying to create a sugar shortage in Australia so as to increase the price of sugar finally prompted the Ryan (Labor) government in Queensland and the Hughes (Labor) Commonwealth government to introduce legislation in mid-1915 that would eventually regulate the refining and marketing of sugar in Australia for the next seventy-four years. Prime Minister Hughes, however, despite his antagonism towards CSR, did not remove the firm's almost monopololistic control over the refining sector. Competition in the refining of sugar and its wholesale distribution in Australia remained an elusive feature, even when the state intervened.

Notes

1 H. Roth, *A Report on the Sugar Industry in Queensland* (Brisbane, 1880), p. 93; P. Griggs, 'A Natural Part of Life: The Australian sugar industry's campaign to reverse declining sugar consumption 1980–1995', *J. of Australian Studies*, Vol. 87 (2006), pp. 142–144.

2 *BC*, 24 October 1874, p. 7; *Bundaberg Star*, 22 February 1879, p. 22; Roth, *A Report on the Sugar Industry in Queensland*, p. 62; G. Craig, 'The Sugar Industry', *Queensland Review*, August 1885, p. 63; J. Holland, Bundall Plantation, to W. Miskin, Brisbane, 24 May 1879, reproduced in J. Elliott (ed.), *Letters to Bundall 1872–1879* (Brisbane, 1993), p. 182; *QPD*, Vol. 47 (1885), p. 1456.

3 Kerr, *Sugar at Maryborough*, pp. 32–40; *MM*, 21 December 1872, p. 2; *QVP*, Vol. 1 (1874), pp. 671–676; *ASJ*, 10 October 1936, p. 409.

4 *QPD*, Vol. 27 (1874), pp. 960–961 & 982–984; *Australasian*, 24 June 1882, p. 793; Kerr, *Sugar at Maryborough*, pp. 89– 90; *Q*, 10 June 1882, p. 727.

5 Kerr, *Sugar at Maryborough*, pp. 91–92.

6 Kerr, *Southern Sugar Saga*, pp. 12–14; *MC*, 20 October 1882, p. 2; J. Walker, *The History of Bundaberg: A Typical Queensland Agricultural Settlement* (Brisbane, 1890), p. 212.

7 Kerr, *Southern Sugar Saga*, pp. 43, 72–73 & 77; E. Drury to R. Stodart, Brisbane, 13 June 1884, A/QNB/652, p. 125, E. Drury to J. Sutton, Brisbane, 8 August 1884, A/QNB/652, p. 148 & E. Drury to R. Cran Snr, Bundaberg, 15 February 1886, A/QNB/652, p. 159, QNBR; *ASJ*, 9 September 1909, p. 225.

8 *ATCJ*, 26 January 1878, p. 169; *The Sugar Cane*, 1 May 1878, p. 270; S. Fitzgerald & H. Golder, *Pyrmont and Ultimo. Under Siege* (Sydney, 1994), pp. 48–51; CSRR, Z/303, Box 5, File A 2.0, Folder 8, Doc. 10: Documents about Frederick Poolman.

9 *Melbourne Argus*, 11 February 1874, p. 6; *Australasian Sketcher*, 2 July 1876, p. 84.

10 D. Jowett & I. Weickhardt, *Return to Rosstown. Railways, Land Sales and Sugar Beet Ventures in Caufield* (Mordiallic, Victoria, 1978), pp. 141–150.

11 T. Buckland to E. Knox, 17 February 1874; attached to the 38th Half Yearly Report of CSR's Directors, 21 February 1874, CSRR, Z/303, Box 3, File A1.1, Folder 1, Doc. 1.

12 *J. of Commerce*, 26 July 1886, p. 6; Manning, *In their own hands*, pp. 65 & 71 (Farleigh); Kerr, *Southern Sugar Saga*, p. 42 (Bingera); *Cairns Post*, 7 August 1884, p. 2 & Supplement to the *British Trade Journal*, 1 April 1890, p. 56 (Hambledon); Kerr, *Pioneer Pageant*, p. 69 (Richmond); *CPP*, Vol. 4 (1913), p. 2131 (Ev. T. Smith, Qs. 26121–26125); *Q*, 21 June 1884, p. 991 (Mourilyan); *Sydney Mail*, 20 September 1884, p. 583 (Marian); *LW*, 6 December 1890, p. 3 & CSR, *Notes on the work of the mills during the season of 1892* (Sydney, 1893), p. 9 (Rous); *RRE*, 8 November 1884, p. 8 (Cudgen).

13 *Q*, 12 September 1874, p. 5; W. Maxwell, 'Report upon the conditions of the sugar industry of Australia', *CPP*, Vol. 3 (1910), pp. 1027–1029; Kerr, *Sugar at Maryborough*, p. 147.

14 Kerr, *Sugar at Maryborough*, p. 119.

15 N. U'Ren and N. Turnbull, *A History of Port Melbourne* (Melbourne, 1983), pp. 114 & 144; E.W. Knox to A. Draper, Chairman of Directors, Mulgrave Central Mill, 16 July 1908, CSRR, Z/109, Box 315, Purchase of Raw Sugar Letterbook No. 1, pp. 18 & 162.

16 *The Sugar Cane*, 2 November 1891, p. 562; *J. of Commerce*, 10 November 1891, p. 7.

17 Anon, *Local Industries in Williamstown, Footscray and Yarraville. A Compilation of Articles Published in the* Williamstown and Footscray Chronicle (Williamstown, 1882), pp. 17–22 & 'Improvements at Yarraville Refinery', CSRR, File E 2.0, Folder 4, Doc. 14.

18 Minutes of a conference between Messrs Bligh, Richardson & Smith representing the Victoria Sugar Co. & Messrs Knox & Rothe representing CSR, 12 January 1887, CSRR, Z/303, Box 47, File E 1.0, Folder 2, Doc. 1; Agreement of amalgamation between CSR and the Victoria Sugar Refining Co., 23 July 1888, CSRR, Z/303, Box 47, File E 1.0, Folder 2, Doc. 24.

19 *J. of Commerce*, 19 August 1890, p. 9; CSR, Half Yearly Report, 30 September 1890, CSRR, Z/109, Box 179; *South Australian Register*, 27 October 1891, p. 3; CSR Ltd., *Glanville Refinery from beginning to end: 100 years of sugar refining in South Australia* (Adelaide, 1991), p. 3.

20 CSR, Half Yearly Report, 30 September 1892, CSRR, Z/109, Box 179; CSR Ltd., *1893–1993. Spirit of New Farm. Celebrating a Hundred Years at CSR's New Farm Sugar Refinery* (Brisbane, 1993), pp.4–5; *BC*, 7 October 1893, p. 5; E.W. Knox to

H. Symonds, Melb., 18 August 1897, CSRR, 142/2933, p. 446; W. Rothe to J.S. Richardson, CSR Melb., 6 September 1893, CSRR, 142/55; *CPP*, Vol. 4 (1913), p. 2190 (Ev. E. Knox).

21 CSR, Half Yearly Report, 30 October 1890, CSRR, Z/109, Box 179.

22 S. Fitzgerald and H. Golder, *Pyrmont and Ultimo under siege* (Sydney, 1994), pp. 90 & 118; E. W. Knox, Memo to Board of Directors re: refinery alterations, 27 September 1899, CSRR, 142/2753; Charles Davis, 'The Pyrmont Sugar Refinery with particular reference to Char'. PhD thesis, Uni. of Sydney, 1984, pp. 157, 165, 171, 173, 184 & 190–191.

23 E.W. Knox, Memo to Board of Directors re: visit to Yarraville Refinery, 16 September 1896, CSRR, 142/2753; CSR, *1893–1993. Spirit of New Farm*, pp. 6–8; W. Rothe to J.S. Richardson, CSR Melb., 18 September 1893, CSRR, 142/55.

24 T. Heriot, *The Manufacture of Sugar from Cane and Beet* (London, 1920), p. 391; A. Saranin, 'A flocculation-flotation process for clarification of sugar refinery melt using phosphoric acid and lime'. M. Eng. thesis, Uni. of Qld., 1962, pp. 2–6; Davis, 'The Pyrmont Sugar Refinery with particular reference to Char', pp. 74 & 171.

25 CSR Ltd., *1893–1993. Spirit of New Farm*, p. 8.

26 J. Abram and J. Ramage, 'Sugar refining: present technology and future development', in G. Birch and K. Parker (eds.), *Sugar: Science and Technology* (London, 1979), pp. 65–68; Davis, 'The Pyrmont Sugar Refinery with particular reference to Char', pp. 45–46.

27 Anon, *Local Industries in Williamstown, Footscray and Yaraville*, p. 18; *CPP*, Vol. 4 (1913), p. 1745 (Ev. B. Lewis).

28 *BC*, 24 July 1881, p. 3; *The Planter and Farmer*, March 1883, p. 473 (Advt. for S. & G. Grimes).

29 Kerr, *Southern Sugar Saga*, pp. 26 & 42; *CPP*, Vol. 4 (1913), p. 579 (Ev. Robert Boyd, Q. 15785).

30 *BC*, 30 October 1883, p. 4; *LW*, 14 January 1882, p. 4.

31 *Sands and McDougall Melbourne Directory*, 1884, entry for tea, coffee and sugar brokers, p. 927; J. Smith (ed.), *The Cyclopedia of Victoria*. Vol. 1 (Melbourne, 1903), pp. 312–313; *J. of Commerce*, 23 August 1887, p. 3; for sugar auctions in Sydney see *CRE,* 4 October 1880, p. 2, 23 October 1880, p. 5, 3 December 1881, p. 5 & 20 February 1883, p. 2, *RRE*, 11 December 1880, p. 2 & 19 February 1881, p. 2 & *NS*, 14 January 1880, p. 2, 14 April 1880, p. 2 & 13 August 1881, p. 3.

32 V. Gibbs, Sydney, to A. Gibbs, London, 24 October 1883, MS 1110 A: Private Letters of Vicary Gibbs, MUA; F. Keating, Melb., to P. Hannay, Gibbs, Bright & Co., Brisbane, 3 September 1885, MS 11044, Out Letterbook of F. Keating, p. 289, MUA; Q. Deloitte, CSR Head Office, Sydney, to W.H. Paxton & Co., Mackay, 23 July 1880, CSRR, 142/3171, p. 3; *Q*, 11 March 1871, p. 4; *BC*, 22 January 1870, p. 7 & 26 July 1882, p. 1.

33 For Brisbane commission agents advertising the sale of sugar see Q, 9 January 1869, p. 4, *Pugh's Almanac*, 1874 to 1893, Advertisement section, & *BC*, 13 January 1877, p. 1; for the Maryborough agents see *Pugh's Almanac*, 1875, p. 76, *Bundaberg Star*, 10 March 1875, p. 1 & *MC*, 16 July 1883, p. 1; for Duffy's thoughts on marketing sugar see *QVP*, Vol. 4 (1889), p. 307 (Qs. 7175 to 7182).

34 *Mackay Standard*, 22 August 1892, p. 2 & *The Capricornian* , 26 August 1886, p. 13 & 4 August 1888, p. 16.

35 *Townsville Herald*, 1 August 1876, p. 1 & 10 September 1887, p. 17 & *Wilmett's North Queensland Almanac*, 1883, Advt. (Clifton, Aplin & Brown); *Townville Herald*, 17 July 1886, p. 1 & *NQH*, 13 January 1892, p. 1 (Burns, Philp & Co.).

36 For mention of Victorian merchants importing sugar see *J. of Commerce*, 6 September 1887, p. 3, 16 September 1890, p. 9, 1 March 1892 & 24 May 1892, p. 11; see also *Victorian Parliamentary Papers*, Vol. 2, Part 2 (Session 1895–1896), p. 975 (Ev. John Jacobs, Q. 3433).

37 For examples of this correspondence see CSRR, 142/458–472 & *J. of Commerce*, 26 June 1887, p. 3.

38 E.W. Knox to E. Long, Habana Plantation, via Mackay, 23 March 1887, CSRR, 142/477, p. 227.

39 *Q*, 29 September 1877, p. 1, 6 October 1877, p. 6, 20 December 1879, p. 775 & 5 November 1884, p. 40; the voluminous correspondence associated with CSR's sugar purchases in the early 1880s are to be found in CSRR, 142/3171; for examples of purchases in the early 1890s see CSRR, 142/484, pp. 19 & 110–111 & Q. Deloitte, Acting CSR Mang. to W. H. Paxton & Co., Mackay, 6 June 1882, p. 118, CSRR, 142/3171, p. 119.

40 E.W. Knox to Chairman of the New South Wales Mills Assoc., 1 April 1884 and 10 June 1884, CSRR 142/474, pp. 286 and 386.

41 Extracts from CSR's Board Minutes, 29 March 1887, CSRR, Z/303, Box 3, File A 1.1, Folder 2, Doc. 4.

42 W. Rothe to J. Richardson, CSR Melb., 22 September 1893, CSRR, 142/55; W. Rothe to J. Richardson, CSR Melb., 28 August 1894, CSRR, 142/66.

43 E.W. Knox to J. Richardson, CSR Melb., 25 January 1893, CSRR, 142/45; W. Rothe to J. Richardson, CSR Melb., 24 January 1893, CSRR, 142/48; *SJTC*, 15 November 1894, p. 236; A. Birch and J. Blaxland, 'The Historical Background', in A. Lowndes (ed.), *South Pacific Enterprise. The Colonial Sugar Refining Company* (Sydney, 1956), p. 45.

44 E.W. Knox to J. Richardson, CSR Melb., 7 May 1895, CSRR, 142/42933, p. 391.

45 W. Rothe to E.W. Knox, 2 July 1906, CSRR, 142/173, Letter No. 532; 'Details of Messrs Poolman's annual outputs', CSRR, Z/303, Box 15, File A 5.0, Folder 9, Doc. 17.

46 E. W. Knox to the Acting Mang., CSR Melb., 30 March 1903 & E. W. Knox to the Acting Mang., CSR Melb., 28 April 1903, CSRR, 142/256, pp. 228 & 283; *Australian Cordial Maker* 30 May 1907, p. 32.

47 E. W. Knox, 'Notes of interview with Edward Poolman, 30 July 1906', CSRR, Z/303, Box 15, File A 5.0, Folder 9, Doc. 9; E. W. Knox, 'Notes of interview with Edward Poolman, 23 March 1907', CSRR, Z/303, Box 15, File A 5.0, Folder 9, Doc. 8; 'Details of the purchase of Messrs Poolman's Refineries, 1907', CSRR, Z/303, Box 15, File A 5.0, Folder 9, Doc. 25.

48 Parbury, Lamb & Co., Brisbane, to J. Grafton Ross, CSR Head Office, Sydney, 17 June 1879, CSRR, 142/3357; E. Knox to E. Forrest, Brisbane, 25 July 1881, p. 49, CSRR, 142/2339, p. 49; E. W. Knox to J. E. Davidson, Palms Mill Mang., Mackay, 11 September 1893, CSRR, Z/109, Box 311, Raw Sugar Purchases Letterbook No. 2, p. 292.

49 For CSR's sugar purchases in 1891–1892 see CSRR, Z/109, Box 311, Raw Sugar Purchases Letterbook, 1891–1893.

50 CSR, 'Circular to all Australian sugar millers re: purchase of annual sugar outputs, 6 April 1893', CSRR, Z/109, Box 311, Raw Sugar Purchases Letterbook No. 1, p. 489; CSR, 'Circular to All Millers, 4 March 1896', CSRR, 142/3506.

51 See for example W. Rothe to Messrs Larsen and Broadhurst, Good Hope Mill, via Childers, 13 March 1894, CSRR, Z/109, Box 312, Raw Sugar Purchases Letterbook No. 3, p. 138.

52 During the 1880s, Australian mill-owners were mostly charged a commission of 6.25% on the income from the sale of their sugar. See the evidence of E. Forrest of Parbury, Lamb & Co. and W. Brown of Aplin, Brown & Co. in *QVP*, Vol. 4 (1889), pp. 246 & 344.

53 For examples see CSRR, Z/109, Box 312, Raw Sugar Purchases Letterbook No. 4, pp. 458–460 & 474–475.

54 *CPP*, Vol. 4 (1913), p. 825; W. Rothe to A. Wright, CSR Brisbane, 21 August 1900, CSRR, 142/1334, p. 237; W. Rothe to A. Wright, CSR Brisbane, 28 January 1902, CSRR, 142/13335, p. 171.

55 QNB, Minute Book No. 3, entry for 9 May 1894, p. 309, A/QNB/303; QNB, Minute Book No. 5, entry for 28 April 1897, p. 80 & entry for 7 June 1897, pp. 125–126, A/QNB/305; QNB, Minute Book No. 6, entry for 11 November 1898, p. 133, A/QNB/ 306, QNBR; *SJTC*, 15 April 1898, p. 59, 15 June 1898, p. 264 & 15 January 1903, p. 142; *J. of Commerce*, 26 March 1895, p. 11.

56 *Pugh's Almanac*, 1893, Shipping & Signals Section, Advt. Smith Brothers, opposite p. 17 A; *BC*, 1 August 1895, p. 4; *The Sugar Cane*, 1 March 1893, p. 153.

57 For the Maryborough Sugar Factory making mill whites see *SJTC*, 15 September 1897, p. 52 & Kerr, *Sugar at Maryborough*, p. 147; for Macknade Mill making mill whites see *SJTC*, 15 November 1893, p. 262 & 15 April 1897, p. 56; for the sale of sugar see *NQH*, 15 May 1895, p. 20 (Ripple Creek) & *NQH*, 17 January 1894, p. 1 & *SJTC*, 5 September 1897, p. 52 (Macknade).

58 E.W. Knox to A. Young, Fairymead, 1 March 1895, CSRR, Z/303, NS 529; *Bundaberg Mail*, 23 May 1898, p. 2; Kerr, *Southern Sugar Saga*, pp. 44–45.

59 For mill-owners swapping refiners see *CPP*, Vol. 4 (1913), p. 386 (Ev. T. Whitcomb, Chairman of Directors, Racecourse Central Milling Co., Q. 11221).

60 E.W. Knox to W. Ralston, Gen. Mang., QN Bank, Brisbane, 16 March 1903, CSRR, 142/2934, pp. 116–119; E. W. Knox to W. Ralston, QN Bank, Brisbane, 21 June 1906, E. Knox to J. Johnston, Gen. Mang., Millaquin Sugar Co., Bundaberg, 3 February 1908, CSRR, Z/303, Box 43, File D 2.0, Folder 6, Doc. 9; *CPP*, Vol. 4 (1913), p. 713 (Ev. James Bartholomew, part-owner of Goodwood Mill, Q. 1950–1953).

61 H. Gordon Bell, Mang. Kalamia Mill, to Mang. AEMCo., Melb., 22 June 1907, Letterbook, Kalamia Mill, Ayr, p. 493.

62 W. Rothe, CSR Head Office, Sydney, to Sec. Mossman Central Milling Co. Ltd., 27 May 1905, CSR, Z109, Box 315, Letterbook No. 15, p. 389; Kerr, *Northern Outpost*, p. 139; *Cairns Post*, 30 April 1906, p. 4; for Ripple Creek see CSRR, Z/109, Box 315, Letterbook No. 16, p. 232.

63 A. Wright, CSR Brisbane, to CSR Head Office, Sydney, 27 May 1907, 18 June 1907 & 8 July 1907, CSR, 142/1271, Letter No. 129, 152 & 173; A. Wright, CSR Brisbane, to CSR Head Office, Sydney, 30 May 1908, 5 June 1908 & 15 June 1908, CSR, 142/1271, Letter No. 523, 529 & 538.

64 W. Ralston, QN Bank, Brisbane to W. Rothe, CSR Head Office, Sydney, 20 October 1909, CSR, Z303, Box 43, File D 2.0, Folder 6, Doc. 9.

65 W. Ralston, QN Bank, Brisbane to W. Rothe, CSR Head Office, Sydney, 1 June 1909, CSR, Z303, Box 43, File D 2.0 Folder 6, Doc. 9; *CPP*, Vol. 4 (1913), p. 824 (Ev. J. Johnston, Qs 22673–22675); for CSR's purchases from the Bundaberg district sugar mill-owners see CSRR, Z/109, Box 316, Purchases of Raw Sugar Letterbook, 1911–1912, various pages.

66 *CPP,* Vol. 4 (1913), p. 475 (Ev. Robert Philp); p. 998 (Ev. E. W. Knox) & p. 1750 (Ev. W. McGown, Qs. 15795–15798).

67 *CPP*, Vol. 4 (1913), p. 998 (Ev. E.W. Knox); QN Bank, *Price Lists and Trade Circulars for the Millaquin and Yengarie Sugar Co., 1899–1904*, CSRR, 142/3593; CSR, *Price Lists and Trade Circulars, 1904–1914*, CSRR, 142/3594.

68 A.C. Wright, CSR Brisbane, to W. Rothe, CSR Head Office, Sydney, 8 June 1904, CSRR, 142/1266, Letter 903 of 1904.

69 For examples of price lists see *ASJ*, 6 May 1909, p. 10, 6 October 1910, p. 293 & 8 February 1912, p. 647.

70 *ASJ*, 7 October 1909, p. 260; Kerr, *Southern Sugar Saga*, p. 78; *CPP*, Vol. 4 (1913), pp. 455 (Ev. H. Jones), 541 (Ev. E. Connell), 970 (Ev. F. Sharp), 1948 (Ev. W. Gibson) & 2013 (Ev. A. Gibson).

71 *ASJ*, 3 November 1910, p. 313; *CPP*, Vol. 4 (1913), p. 1819 (Ev. J. Paddle, Acting Mang., Nerang Mill, Q. 17653) & p. 966 (Ev. J. Cardno, mercantile broker, Brisbane); W. Maxwell, 'Report upon the conditions of the sugar industry of Australia', *CPP*, Vol. 3 (1912), pp. 1029–1030.

72 *The Manufacturers and Storekeepers Guide*, 31 December 1900, p. 36; *The Australasian Grocer* 24 February 1903, p. 52.

73 W. Rothe, Memo to E.W. Knox re: competition from imported sugars, 15 July 1902, CSRR, Z/303, Box 2, File A 1.0, Folder 1, Doc. 30; *J. of Commerce*, 21 October 1902, p. 18.

74 M. Robertson, 'Minutes of evidence given to the Royal Commission on Customs & Excise Tariffs', *CPP*, Vol. 4 (1906), p. 819; K. Buckley and K. Klugman, *The History of Burns Philp. The Australian Company in the South Pacific* (Burns, Philp & Co. Ltd., 1981), pp. 202–203; *CPP,* Vol. 4 (1913), p. 1731 (Ev. G. Seymour, Qs. 15348–15351); B. Brown, *I Excel! The Life and Times of Sir Henry Jones* (Hobart, 1991), pp. 68–72.

75 E.W. Knox to Sir D. Cooper, London, 9 July 1887 & E.W. Knox to Sir D. Cooper, London, 18 December 1887, CSRR, 142/2933, pp. 62 & 108; *Victorian Parliamentary Papers*, Vol. 2, Part 2 (Session 1895–1896), p. 732.

76 E. W. Knox to J. Richardson, CSR Melb., 5 October 1892, CSRR, 142/44; CSR, Glanville Refinery Price Circulars, 1896–1904, CSRR, 142/3593; for Northern Territory sugar being sold at auction in Adelaide see *The South Australian Register*, 5 November 1891, p. 4.

77 T. Givens, 'Minutes of the debate on the nationalisation of the Australian sugar industry', *Commonwealth Parliamentary Debates*, Vol. 27 (1905), pp. 2893–2895.

78 *Mackay Standard*, 1 February 1895, p. 2 & 18 November 1895, p. 2; *The Bundaberg Mail*, 4 November 1896, p. 3; *Q*, 9 July 1898, p. 54; Manning, *In their own hands*, p. 157.

79 *MM*, 23 February 1899, p. 4; E. W. Knox to E. Forrest, Brisbane, 13 March 1899, CSRR, Z/303, Box 40, File D 1.0, Folder 8, Doc. 1; *QPD*, Vol. 87 (1901), p. 563,

QPD, Vol. 87 (1901), pp. 1068–1082 & *QPD*, Vol. 88 (1901), p. 2678; *Commonwealth Parliamentary Debates*, Vol. 27 (1905), pp. 2893–2903, 5222–5250 & 7455–7456.

80 G. Sawer, *Australian Federal Politics and Law* (Melbourne, 1956), pp. 40 & 84; A. Hopkins, 'Anti-trust and the bourgeois 1906 and 1965', in E. Wheelwright and K. Buckley (eds.), *Essays in the Political Economy of Australian Capitalism*. Vol. 2 (Sydney, 1978), pp. 89–90, *Commonwealth Parliamentary Debates*, Vol. 30 (1905), p. 6820.

81 W. Maxwell, 'Report upon the conditions of the sugar industry of Australia', *CPP*, Vol. 3 (1912), p. 1030; *ASJ*, 4 August 1910, pp. 196–198; C. Young to E. W. Knox, 22 July 1910, CSRR, Z/303, Box 40, File D 1.0, Folder 8, Doc. 3.

82 For the referendum campaign see F. Browne, *They Called Him Billy. A Biography of The Rt. Hon. W. M. Hughes* (Sydney, 1946), pp. 83–85; for Hughes' attacks on CSR see *The Daily Telegraph* (Sydney), 29 April 1911, p. 13; Deakin's opposition to the referendum proposal are discussed in W. Murdoch, *Alfred Deakin* (Melbourne, 1999), pp. 269–270.

83 *QPD*, Vol. 108, p. 420; *ASJ*, 7 September 1911, pp. 320–321 & 5 October 1911, pp. 362–363.

84 *Commonwealth Parliamentary Debates*, Vol. 60 (1911), pp. 402 & 478; *ASJ*, 5 October 1911, pp. 381–382; Birch and Blaxland, 'The Historical Background', pp. 46–47.

85 Shogren, 'The politics and administration of the Queensland sugar industry to 1930', pp. 198–199; Fitzhardinge, *William Morris Hughes: A Political Biography. Volume 1: That Fiery Particle 1862–1914*, p. 266; *ASJ*, 9 November 1911, p. 440 & 9 May 1912, pp. 85–87 & 103–112.

86 *CPP*, Vol. 4 (1913), pp. 1009–1017 (Ev. E.W. Knox); *ASJ*, 7 November 1912, pp. 559–560.

87 *ASJ*, 5 December 1912, p. 626; W. Brown *et al.* 'Report of the Royal Commission into the Australian Sugar Industry', *CPP*, Vol. 3 (1913), pp. 1077 & 1081–1082.

88 J. La Nauze, 'The Inter-State Commission', in C. Hughes (ed.), *Readings in Australian Government* (Brisbane, 1968), pp. 262–263; *ASJ*, 8 October 1914, p. 464.

89 *ASJ*, 5 November 1914, pp. 529–530 & 3 December 1914, pp. 601–602; T. Bavin and H. Evatt, 'Price-fixing in Australia during the War', *J. of Comparative Legislation and Internal Law*, 3rd series, Vol. 3, no. 4 (1921), pp. 202–204.

90 *Commonwealth Parliamentary Debates*, Vol. 69 (1912), p. 6718; *ASJ*, 5 November 1914, p. 536.

Part Three

1915 TO 1995: EXPANSION, EXPORT-ORIENTATED AND REGULATED

Semi-trailers with twin boxes waiting to unload at the Mackay Bulk Sugar Terminal receiving station, 1963. By 1980, the Mackay Bulk Sugar Terminal was the world's largest storage complex for raw sugar.

Source: Image reproduction courtesy of the National Library of Australia. Image No. nla.pic-vn3294946.

Chapter 10
A Regulated Expansion

In 1916, the Ryan (Labor) government in Queensland decided to hold another inquiry into the suitability of establishing additional government-funded central sugar mills in the state, following repeated requests for such action from groups of small farmers in the sugar-producing districts throughout Queensland. William J. Short, an accountant and public servant, headed this new inquiry into where to locate any new central sugar mills in Queensland. The Short Inquiry received applications for a new sugar mill from thirteen different settler groups, mostly in the state's north. Sites investigated such as the Freshwater Valley, near Cairns, and the Atherton, Tully River and Yeppoon districts had been considered but rejected during the Ranking Royal Commission into the establishment of new central sugar mills in Queensland during 1911. Short uncovered concerns about the possibility of production outstripping Australia's demand for sugar, with little likelihood of sugar exports to Great Britain being established. Consequently, the Short Inquiry recommended that the Queensland government should not build any new central sugar mills in the immediate future, although if the government proceeded to increase the state's sugar milling capacity it should consider the following localities in order of preference: Tully River district; Bailey Creek, 50 km north of Port Douglas; Daradgee; Freshwater, near Cairns; and Long Pocket, near Ingham. Premier Ryan accepted the inquiry's advice and deferred building any new central sugar mills.[1]

New sugar milling facilities were not really needed in Queensland as the total area under sugar cane fluctuated between 65 000 ha and 75 000 ha during the late 1910s (Figure 10.1), a result of two opposite trends. On one hand, the amount of sugar cane grown increased in North Queensland, following additional land being brought under cane in the Lower Burdekin, Gumlu, Babinda and South Johnstone districts.[2] This trend, however, was not repeated in Southern Queensland. The small farmers throughout the Marburg district found dairying more profitable and the extent of sugar cane cultivation dwindled, eventually forcing the closure of Woodlands Mill after the 1917 crushing season. In the Beenleigh district, the amount of sugar cane cultivated contracted following the closure of

Nerang Mill after 1918. The small farmers along the Nerang River no longer had a nearby factory to process their crops of cane, so they switched to dairying. Those small farmers in the Beenleigh district who continued cultivating a few hectares of sugar cane were concentrated near the small Rocky Point, Eagelby and Alberton Mills, just to the south-east of Beenleigh, although they often combined it with dairying or growing arrowroot. The most extensive contraction, however, occurred in the Wide Bay-Burnett region. Many of the Maryborough district's small farmers combined tropical fruit production with sugar cane growing, which became a sideline rather than the focus of their operations. At Bundaberg, the amount of sugar cane cultivated dropped from 11 340 ha in 1914 to 8 100 ha in 1920. This contraction was the largest loss of area devoted to sugar cane in any Australian sugar-producing region during the late 1910s. Sugar mill closures in the Kolan River and Baffle Creek districts, both to the north of Bundaberg, meant the small farmers were forced to return to dairying, pig raising and maize and arrowroot cultivation.[3]

After 1915, control over the area cultivated with sugar cane in Queensland was the responsibility of the Central Sugar Cane Prices Board (hereafter Central Board) located in Brisbane. Established under *The Regulation of Sugar Cane Prices Act* of 1915, it consisted of a District Court Judge as chairman, a growers' representative, a millers' representative, a person experienced in accountancy and audit and an industrial chemist with experience in the manufacture of raw sugar. The Central Board heard disputes about the price of cane fixed by the local sugar cane prices boards (see Chapter 18 for more details) and allocated or 'assigned' canegrowers to a particular sugar mill. The assignment applied to the entire property, not just to the part where the sugar cane was being cultivated. The Central Board could also refuse applications from farmers to be assigned to a particular sugar mill. In essence, it could regulate sugar production by determining which canegrowers received access to crushing facilities.[4]

During the late 1910s, the Central Board did not have to worry about the overproduction of sugar in Queensland. However, this situation changed rapidly during the 1920s. A resurgence of interest in growing sugar cane occurred in all Queensland's sugar-producing districts, leading to another period of rapid growth in the area under the crop during the early 1920s (Figure 10.1). The main factor contributing to this turn around was an improved domestic sugar price. Although sugar production was below Australian requirements during World War 1 and immediately afterwards, the domestic sugar price had been regulated and kept low by the Commonwealth government in order to protect the Australian consumer against wartime profiteering. Internationally, the price for sugar soared to as high as £80 per ton throughout the war years, but it stood at

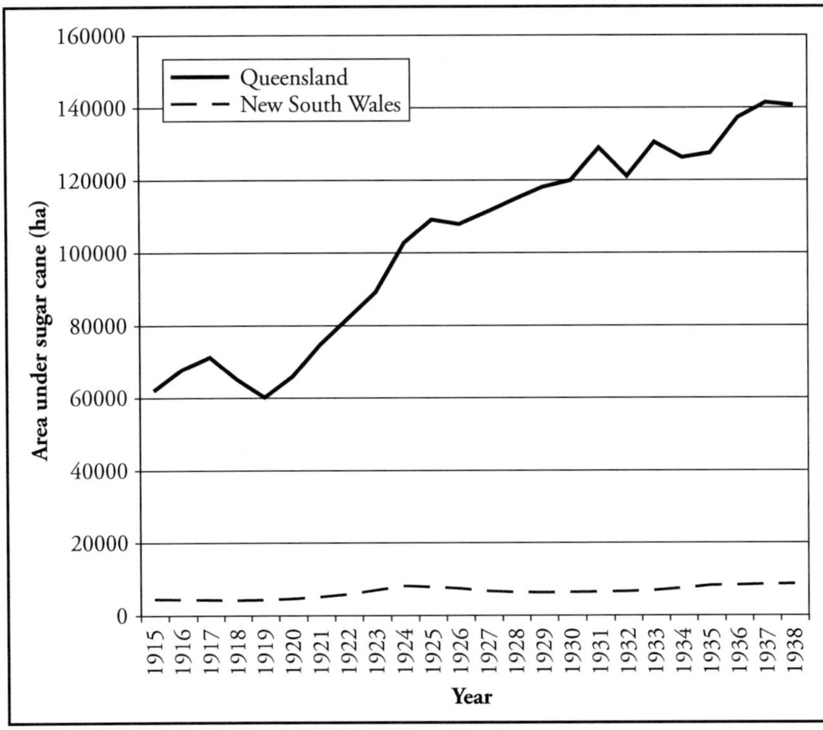

Figure 10.1. The area (ha) under sugar cane in Queensland and New South Wales, 1915–1938.
Source: Based upon figures in the *Year Book of Australia*, various issues.

£21 per ton in Australia during 1919. Australian canegrowers and sugar mill-owners, however, had suffered increased operating costs during the war period, a consequence of labour shortages and increased wages (see Chapter 16 for more details). After lobbying from Australian sugar industry representatives, Prime Minister Hughes raised the price of sugar to £30 6s 8d per ton for the years 1920 to 1922 (see Chapter 18). Australian sugar mill-owners could now afford to pay their suppliers more for their cane.[5]

The rapid growth in the area devoted to sugar cane in Queensland during the early 1920s occurred because of a combination of two processes. Firstly, Ernest Scriven, the BSES Director, claimed some existing canegrowers were encouraged by the higher cane prices and increased the areas planted on their properties, by 'not only putting under cane land that has been lying unproductive, but in many instances bringing new land under cultivation'. Secondly, the number of farms in Queensland where any sugar cane was cultivated steadily increased during the early 1920s, rising from 4 148 in 1918 to 7 062 in 1924.[6] Some farmers, particularly those in Southern Queensland, replanted areas where sugar cane had ceased to be cropped. Fields which had been out of cane for years and heavily stocked with cattle around Mt Bauple, for example, were being replanted with cane by 1922. In the Beenleigh district, a slump in the price of arrowroot coincided with the increased price of sugar, tempting some small farmers to replant sugar cane in preference to arrowroot.[7]

In other sugar-producing districts, settlers in areas already occupied turned to growing sugar cane, often for the first time. Farmers 40 km south of Ingham, between Toobanna and Rollingstone – known as the Ingham Line locality – had commenced growing sugar cane in the early 1920s, following the extension of the railway from Townsville to Ingham which enabled their harvested crops to be railed 150 km south to Invicta Mill, at Giru. At Sarina, the Plane Creek Central Mill Co. Ltd. decided in 1921 to build a tramline from Sarina southwards for 55 km towards Carmila. Transport to a sugar mill was now assured, so the district attracted new settlers who turned to growing sugar cane while the old established settlers were observed in 1921 to be 'rigorously clearing and planting land that was hitherto used only for grazing'.[8] Additional land openings for closer settlement around Bloomsbury, Calen and Mt Ossa, following the building of a missing section of the North Coast Railway from Mackay to Proserpine between 1919 and 1923, attracted settlers who turned to cane growing. Further north, sugar cane also became the preferred crop of the new soldier settlers at El Arish, the district being occupied under the *Discharged Soldiers' Settlement Act* of 1917, while the large blocks of land making up the former Pyramid plantation in the Pyramid and Little Mulgrave localities near Gordonvale were subdivided into small farms and sold to farmers who commenced growing sugar cane. Swamp drainage by two syndicates in the Maroochy district in 1922 also created 400 ha of new land which was devoted to sugar cane for the nearby Moreton Mill.[9]

During the late 1910s, the area cultivated with sugar cane in New South Wales had fallen, although this trend was reversed in the early 1920s, following a resurgence of interest in cane growing (Figure 10.1). However, the small farmers who recommenced cultivating sugar cane were motivated by more than just the improved price they received for cane from CSR. Outbreaks of the disease 'bunchy-top' on the banana farms along the Tweed and Richmond Rivers in the early 1920s caused a reversion to sugar cane planting in the affected areas as the small farmers sought any alternative income. Some dairy farmers also returned to growing sugar cane as a sideline. New areas such as the Bagotville district, near Broadwater, and the Turkey Island Soldier Settlement Estate, near Maclean, were occupied by settlers who cultivated sugar cane. In addition, CSR's funding of extensive drainage activities throughout the Tweed and Richmond River districts ensured a reduction in the length of time that arable land was inundated with floodwaters. Crop damage was reduced and yields improved, thereby convincing the farmers to extend their sugar cane plantings. In addition, better drainage led to peat swamplands being reclaimed on the former Cudgen plantation lands, on parts of the Tweed River floodplain, and in the Mooball and Crabbes Creek district, just north of Brunswick Heads. This land was cultivated with sugar cane.[10]

The rapid growth in the area devoted to sugar cane during the early 1920s was also accompanied by a change in the ethnic composition of the North Queensland canegrowers. By 1914, approximately 3 000 Italian-born immigrants resided in North Queensland. After 1920, new Italian immigrants, especially Sicilians, flocked to North Queensland. Most were drawn to the established Italian communities at Innisfail and Ingham; smaller numbers were found in the Cairns, Proserpine, Mackay and Lower Burdekin districts. Many Italian canecutters, however, aspired to be land owners. They worked cooperatively in small groups, pooling their savings to allow one member to purchase a cane farm. Sometimes they obtained farms as going concerns; on other occasions the Southern Europeans willingly acquired marginal, poorly managed or derelict properties, including abandoned banana farms around Innisfail, vacated by Chinese settlers affected by the White Australia Policy. An Italian landowner then sent for his family from Italy and a new recruit took his place.[11] Greek, Spanish, mostly Basques and Catalans, and Maltese immigrants were also attracted to cutting cane or working in the sugar mills, and like their Italian counterparts some settled permanently in North Queensland after purchasing a share in a farm or a farm outright. The Greeks favoured the Gordonvale, Innisfail, Babinda, and Lower Burdekin districts; the Spanish settled mostly in the Lower Burdekin and Herbert River districts; the Maltese showed a predisposition to the Proserpine and Mackay districts.[12] These non-British canegrowers, however, were not welcomed by some sections of the Australian sugar industry.

Avoiding overproduction

As the area under sugar cane in Queensland grew rapidly during the early 1920s (Figure 10.1), the Queensland government (Theodore; Labor) decided in 1922 to hold another inquiry into the most suitable places to erect additional government-funded central mills in the state. William Harris, a Police Magistrate, assumed the chairmanship, and was assisted by Harry Easterby, the BSES Director, and Matthew Salisbury, the Rockhampton Land Commissioner. Unlike the 1911 and 1916 inquiries into this matter, the Harris Royal Commission did not have to consider direct applications from groups of settlers, but was permitted to investigate any localities which could benefit from the erection of a central sugar mill. The commissioners examined sixteen localities throughout Queensland. Several sites considered had been examined previously by the two earlier investigations into where to locate new central sugar mills (e.g. Atherton; Tully River; Bailey Creek; Long Pocket, near Ingham). The Harris Royal Commission, although not charged with the task of deciding whether or not new sugar mills were warranted, noted that the uncertainty over the sugar price was causing much anxiety to the Australian sugar industry and so recommended that only one, new government-funded central sugar mill should be erected, and that it should be in the Tully River district, an unsettled locality south of Innisfail. The Theodore government accepted the Commission's recommendations and £750 000 were provided under *The Sugar Works Act* of 1922 for the building of Tully Mill, which became the last government-funded central mill erected in Queensland, and the last sugar mill to be built in Australia for seventy years, until CSR erected its Ord River Mill in 1995.[13]

As planning for the new Tully Mill commenced, the marketing arrangements for Australian sugar changed (see Chapter 18). In return for the Commonwealth government's continued embargo on sugar imports, the Queensland government established the Queensland Sugar Board. This organisation assumed responsibility for acquiring and marketing all raw sugar in Australia, CSR agreeing to the output from its New South Wales sugar mills coming under the Sugar Board's control. In addition, the Queensland government agreed that it would bear any losses associated with exporting sugar surplus to domestic requirements. By 1925, all the Australian sugar industry's organisational structure that operated for the next sixty years had been created, except for the Sugar Experiment Stations Board which was formed in 1937. Two organisations – the Queensland Sugar Board (hereafter Sugar Board) and the Central Board – assumed crucial roles in regulating future sugar production in Australia.

Tully Mill's inaugural crushing occurred in late 1925, its suppliers harvesting their first crops during a time of growing industry concern about overproduction. The 1924 crop had been a record and the Australian sugar industry exported, for the first time, a substantial amount of raw sugar – just over 76 000 tonnes or nearly 20% of the country's total sugar output

(Figure 10.2). World market prices, however, had dropped very sharply during the latter half of 1924, so the return to Australian canegrowers fell dramatically from £21 10s to £11 3s per ton of sugar during early 1925, as the surplus sugar had to be exported at prevailing low world prices. In February 1925, the Central Board recognised that the growth in the area cropped with sugar cane in Queensland needed to be halted and announced that it would limit any new assignments to land already cropped with sugar cane or prepared for cane growing. The assignment system, which began as a means of preventing sugar mills from poaching sugar cane from other sugar mills, suddenly became a method of controlling production. Cane harvested from unassigned land produced sugar that could not supply the domestic market. It had to be exported and received lower returns due to the prevailing low export price.[14]

Possible future overproduction was discussed at a conference of Australian sugar industry delegates in April 1925. Chaired by William Forgan Smith, Queensland's Minister for Agriculture and Stock, the conference considered how to deal with the record 1925 crop. Three possibilities were explored: all the cane should be crushed and marketed; immature cane should be allowed to stand over; or all the cane should be crushed and the surplus raw sugar converted into refined sugar and stored. CSR was opposed to standing over an estimated surplus of 203 000 tonnes of cane and the ASPA argued against converting the raw sugar into refined sugar, as Great Britain wanted to import raw sugar for its own refineries. Funds to export the surplus sugar caused the Sugar Board concern – the 1924 crop had required £2 million to finance and export – but CSR offered to find the necessary finance to cover the export of sugar without any charge of interest. This offer was accepted and subsequently became an arrangement that existed until after World War II. With finance secured, the conference agreed to harvest all the Australian cane. Subsequently, 44% of the Australian sugar produced in 1925 was exported (Figure 10.2), and the lower world prices reduced the overall returns from the season's harvest.

Dry seasonal conditions in early 1926 reduced the amount of sugar produced during 1926, so the amount exported fell, and the average price of sugar rose again slightly from £19 5s to £24 5s per ton. The Central Board, however, warned in August 1926 that the area planted since March 1925 had increased substantially, despite its efforts to restrict the assignment of new lands to the state's sugar mills, and the area cropped with cane in 1927 could be a record. Overproduction was looming as a serious possibility.[15]

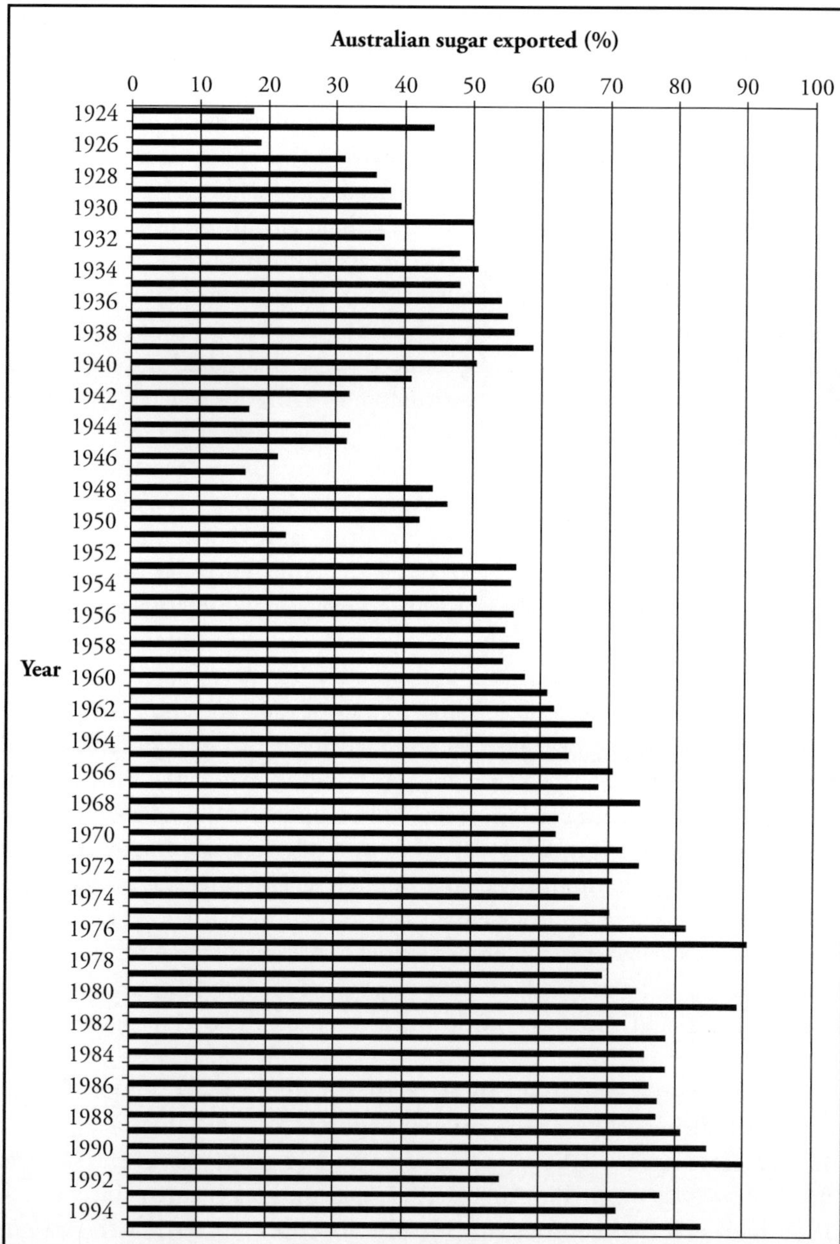

Figure 10.2. The percentage of Australia's raw sugar production exported, 1924–1995.

Source: Calculated from figures in the *ASYB*, various issues.

During the early 1920s, the Theodore (Labor) government in Queensland had introduced legislation providing for the formation of a Council of Agriculture to deal with primary products. Much of this legislation was consolidated under the *Primary Producers' Organisation and Marketing Act* of 1926, but the act also linked the sugar industry to the overall organisation of agriculture in Queensland. Consequently, the United Cane Growers' Association, formed in 1913, was renamed the Queensland Cane Growers' Association, and became a compulsory association for canegrowers. Canegrowers in each mill supply area then elected Mill Suppliers' Committees to handle purely local matters, and these in turn generally provided members for each district's Cane Growers' Executive, which looked after issues affecting canegrowers in a number of mill supply areas. Each Cane Growers' Executive then elected a delegate to the governing body, the Queensland Cane Growers' Council (QCGC), which set policies on behalf of all canegrowers.[16]

At the QCGC's inaugural conference in January 1926, delegates discussed the impact of low sugar prices upon the state's canegrowers and agreed that each Queensland sugar mill should be allocated a quota of sugar to meet Australian requirements, thereby reducing the production of surplus sugar. This idea gained little support amongst other sections of the Australian sugar industry. However, the need to restrict sugar output was discussed again at the annual conference of the QCGC in March 1927. The conference delegates called for action to prevent any further land from being assigned to a sugar mill, other than what was being presently cultivated with sugar cane or being prepared for cultivation with sugar cane. The Central Board agreed to this request and began a review of all the land assigned to the sugar mills throughout the state; in the meantime no new lands would be assigned to a sugar mill.[17] Following the completion of this review, which took two years, the Central Board in April 1930 gazetted the areas on which farmers could cultivate sugar cane and the sugar mill to which they were assigned. All Queensland canegrowers were now limited to a total area they could cultivate with sugar cane. Despite the thoroughness of the process, William Forgan Smith, the Queensland Premier, noted during a parliamentary debate in December 1933 to amend the *Regulation of Sugar Cane Prices Act* of 1915 to 1931 that too many Queensland canegrowers in the late 1920s had been assigned far greater areas than they could ever use. Too much speculation had occurred during the lead up to the limitation of cane growing areas. Such anomalies had the potential to contribute to future overproduction problems.[18]

During the review of assignments, Australia's raw sugar output continued expanding beyond domestic requirements, requiring the sugar to be exported, mostly to Great Britain. Sir Philip Goldfinch, CSR's General Manager, thought this production upsurge had occurred because of a combination of two ideas: firstly, the President of the British Empire Producers' Organisation during a visit to Australia in 1928 had hinted that British preference on Dominions-produced sugars would be increased, thereby ensuring Australia a guaranteed export market for its surplus sugar; and secondly, the wide dissemination of the ill-informed idea that surplus sugar could be profitably converted into power alcohol and that bagasse was needed for the production of celotex, paper and artificial silk.[19] Despite some Australian canegrowers subscribing to such views, there was a growing unease amongst the Australian sugar industry about overproduction and falling prices for exported sugar.

In March 1929, the QCGC's annual conference resolved that the limit of any future year's production for a particular sugar mill would be the highest output of sugar at each Queensland sugar mill in any one year since 1915, or where that figure was insufficient to keep the sugar mill crushing for twenty weeks, a figure based on twenty weeks output. Sugar acquired within mill peaks (later No. 1 Pool) was categorised for home consumption or as being surplus to that requirement. All sugar produced by any sugar mill beyond its mill peak was categorised as 'excess sugar', would be placed in a separate export pool and that sugar would receive an average price realised for all sugar exported (later No. 2 Pool). This scheme became known as the Peak Year Scheme and was discussed in June 1929 at a special conference of delegates from the different organisations representing the Australian sugar industry. Later, during negotiations with the Moore (Country Party) government in Queensland about the introduction of the scheme, the proposal was slightly modified so that any sugar manufactured from sugar cane grown on unassigned lands was also to be placed in the export pool. In addition, the scheme's introduction was to be delayed until 1930.[20]

The Central Board finally announced the peaks for the thirty-six Queensland sugar mills in June 1930. Together, the state's sugar mills could produce a total of 621 210 tonnes of 94 n.t. sugar. CSR agreed that its New South Wales sugar mills would also limit their production in accordance with the principles of the Peak Year Scheme, despite the area under cane in New South Wales actually falling in the late 1920s (Figure 10.1). The New South Wales equivalent of the Queensland peak was considered to be 304 800 tonnes of cane. New South Wales-produced sugar was also taken into the pool on the basis of the '80/20 Agreement' for 20 320 tonnes of sugar, or an allocation of production in the proportion of 80% to home consumption and 20% to export. The percentage for export fell or rose by 1.0% for each 1 000 tonnes of production above or below 20 320 tonnes of output. If the New South Wales output was 40 000 tonnes, for example, 24 000 tonnes or 60% of New South Wales' production was deemed to be home consumption; the remainder had to be exported. CSR also agreed to maintain its land assignment system, introduced in 1925, by confining its suppliers to areas that had been cultivated previously.[21] Further production controls were introduced by the Moore government in August 1931 when the *Regulation of Sugar Cane Prices Act* of 1915 to 1922 was amended to give the Central Board the authority to determine and allot the percentage of the assigned area that could be harvested by any Queensland canegrower. Initially the amount was set at 75% of the gross assigned area.[22]

Challenging the Peak Year Scheme

Immediately following the announcements of the mill peaks, some Queensland sugar mill-owners and canegrowers began requesting increased mill peaks. In August 1930, the Farleigh Cooperative Sugar Milling Association unsuccessfully applied to the Central Board for an increase in the mill peak of Farleigh Mill. Pioneer Sugar Mills Ltd. claimed that the mill peak for its Inkerman Mill was incorrect, for it had diverted 17 800 tonnes of cane from Inkerman Mill to Invicta Mill during 1929, thereby being robbed of the sugar which would have contributed to its mill peak. Canegrowers in the Tully district had a much stronger case for complaint, arguing that they had been disadvantaged during the process as their lands had not been fully developed, and the sugar mill had only five years of output. Delegates representing Tully and Inkerman Mills at the ASPA's annual conference in 1932 failed in their attempt to persuade other delegates to support a resolution calling upon the Sugar Board to review the output for their respective sugar mills. Hugh Henry, one of the Tully Mill delegates to the ASPA's annual conference in 1933, tried again unsuccessfully to convince other delegates to establish an independent tribunal to 'deal with the anomalies as disclosed by the operations of the peak year scheme'. Other delegates were unsympathetic to the Tully case. Charles Young, the miller's representative for Fairymead Mill, for example, blamed the Tully situation on the Queensland government's 'socialistic experiment' and that the rest of the industry should not be asked to bear the burden of overproduction just because the Tully canegrowers had been misled about their prospects by the Queensland government.[23]

As the 1932 crushing season progressed, the Australian sugar industry faced the immediate prospect of a record sugar output in 1933, and the possibility of a still larger crop in 1934. Part of the problem was the attitude of some sugar mill-owners, who quickly assumed that the mill peak was a target to be reached every year. In October 1930, Harry Burstall, Chairman of the Haughton Sugar Co. Ltd., for example, urged Invicta Mill's suppliers to increase their area under cane so as to keep their mill fully supplied in order that its mill peak would be reached regularly. Similar sentiments were expressed by the Moreton Mill Manager, who urged each of the mill's canegrowers to increase their annual crop by 'at least two acres', thereby ensuring the mill peak was met annually.[24] Clearly, the Peak Year Scheme was not working. As a result of half the state's sugar mills exceeding their mill peaks in 1933 (Table 10.1), Premier Forgan Smith called an industry conference on 24 January 1934 to discuss adjusting the mill peaks. Inkerman, Tully, Plane Creek and Proserpine Mills and the Maryborough Sugar Factory all unsuccessfully applied for increased mill peaks. Furthermore, the conference resolved that the Peak Year Scheme should be maintained and all existing mill peaks be confirmed.[25]

The Australian sugar industry's attempts to control overproduction during the early 1930s occurred against a backdrop of growing concern amongst the world's leading sugar producers (e.g. Cuba; Germany; Java) about the unremunerative levels of the world sugar price.[26] During early 1936, the British government unsuccessfully attempted to convene an International Sugar Conference in London to discuss controlling sugar production globally. As a preliminary to this conference, the British government had also arranged a meeting between Great Britain and her sugar-exporting Dominions to review the policy of granting preference to Dominions-produced sugar. This British Empire Conference also failed to materialise. Yet the Australian representatives, William Forgan Smith, the Queensland Premier, and Sir Philip Goldfinch, CSR's general manager, had travelled to England for this conference and used their time in London to meet members of the British government. They received assurances that no quotas would be imposed for the time being on any Dominion sugar producer, and that any alteration in existing British policy would only be made after eighteen months notice had been given about the proposed change. Moreover, the international conference's postponement suited the Queensland Premier, as he discovered that some delegates who had made the trip to London were proposing a 7% cut in sugar output for all sugar-producing countries.[27]

Table 10.1. The number of Queensland sugar mills in excess of the their allocation under the Peak Year Scheme of 1929

Region	No. of sugar mills, 1930–1938	1930	1931	1932	1933	1934	1935	1936	1937	1938
Mossman & Cairns	4	0	2	1	3	0	2	4	4	2
Innisfail & Tully	4	3	4	3	3	1	0	3	4	4
Herbert River	2	0	1	1	1	0	1	2	2	2
Lower Burdekin	4	1	1	3	4	4	2	4	4	4
Mackay & Proserpine	8	1	2	1	6	7	6	8	8	8
Isis-Marybor-ough	3	0	0	0	0	0	1	1	0	3
Bundaberg	5	0	0	0	0	5	4	4	2	5
Moreton & Beenleigh	3	0	0	0	0	0	0	0	0	1
TOTAL	33	5	10	9	17	17	16	25	24	29
% of total		15	30.3	27.2	51.5	51.5	48.5	75.6	72.7	88

Source: W. Webb *et al.*, *Report of the Royal Commission on Sugar Peaks and Cognate Matters* (Brisbane, 1939), p. 14.

Forgan Smith returned to Queensland in August 1936, armed with information about the state of the overseas sugar markets. He was convinced that global restrictions on sugar production were inevitable and decided to prepare the Australian sugar industry for the eventuality. Forgan Smith invited representatives from the main sugar industry organisations to a Premier's conference in Brisbane on 21–22 October 1936 to discuss the industry's production and marketing problems. However, he faced major challenges trying to convince Queensland's canegrowers that they should limit their production even further. The 1935 and 1936 conferences of the ASPA and QCGC had reaffirmed support for the Peak Year Scheme, but the 1935 resolutions contained a rider that sugar production should be reduced to the Peak Year quotas as soon as possible, but no later than the end of the 1937 season.

Yet no progress had been made towards this goal. Moreover, the combined Mill Suppliers' Committees at Mackay had passed a resolution on 19 September 1936 advocating the addition of an extra 40 640 tonnes of sugar in the No. 1 Pool and an upwards revision of all mill peaks. In addition, the Central Board had exacerbated the trend towards overproduction by permitting the area of assigned land in Queensland to rise from 140 928 ha (10 April 1930) to 142 155 ha (28 August 1936). Consequently, more land in Queensland could be put under cane in 1936 than at the beginning of the Peak Year Scheme and the number of properties where cane was being cultivated had risen from 7 228 in 1930 to 7 538 in 1935.[28]

At the Premier's conference, the QCGC tried convincing the other delegates to support a motion raising the amount of sugar production in Queensland to 711 200 tonnes of sugar, and that a competent tribunal should be appointed to review the aggregate quota of 621 210 tonnes in No. 1 Pool together with individual mill peaks. This review aimed to establish new mill peaks for No. 1 Pool, so that together these peaks did not exceed 660 400 tonnes of sugar. This motion was defeated, with delegates eventually deciding to recommend that the *Regulation of Sugar Cane Prices Act* of 1933 be amended to enable 70% of the canegrowers supplying 75% of a sugar mill's cane to enter into an agreement with the mill-owner to cease manufacturing at any time after the sugar mill had reached its peak. The 1929 mill peaks remained unaltered. Premier Forgan Smith, however, decided to ignore the sugar industry's wishes, declining to introduce legislation giving the conference resolutions any legal effect, claiming that he required a more definite expression of the will of the entire industry.[29] Yet the industry appeared to be heading towards disaster: 75% of the state's sugar mills had exceeded their mill peak in 1936 (Table 10.1), and the sugar available for export was almost 420 000 tonnes or approximately 102 000 tonnes above the previous year. Half the country's sugar now had to be exported, mostly to Great Britain, but the average export price for Australian sugar had fallen from £9 4s in 1931 to £7 9s in 1935, even with Empire preferences.[30]

As the leaders of the Australian sugar industry dithered about how to control production, they were overtaken by events in Great Britain. After prompting by the British government, the League of Nations issued invitations to interested governments to attend an international conference in London in April 1937 to consider the better organisation of the production and marketing of sugar globally. Organisations representing the interests of the Australian sugar industry objected when Prime Minister Joseph Lyons announced that Richard Casey, the Federal Treasurer, and Earle Page, the

Federal Minister for Commerce, would represent Australia at this conference, for they clearly wanted Forgan Smith to argue their case in London. Lyons diffused the situation by appointing Forgan Smith as a substitute delegate. Unlike the abandoned meeting in 1936, enough delegates assembled in London to discuss the regulation of the international sugar market and on the 6 May 1937 they agreed to control global sugar production. Twenty-one governments of sugar-producing and sugar-consuming countries, then controlling three-quarters of the world's sugar trade, ratified the first International Sugar Agreement (ISA) which fixed the amount of sugar to be supplied to the free market. As a signatory to the ISA, Australia was awarded an export quota of 406 423 tonnes for the first quota year beginning 1 September 1937. This amount exceeded any previous level of exports, enabling the Australian sugar industry to consolidate the gains made during the early 1930s. In addition, the governments of Great Britain, Australia and South Africa could proportionately increase the export quotas for Australia, South Africa and other sugar-exporting British colonies if warranted by an increase in consumption throughout the British Empire. Failure to fill a quota in any year did not carry a right to exceed a quota in any subsequent year and stocks of sugar at a fixed date in any country could not be allowed to exceed 25% of its production. However, the agreement freed any contracting government that became involved in hostilities, and many of them qualified in this way all too soon. Quota provisions of the agreement were suspended in 1939, under the pressures and dislocation arising from World War II, although the ISA was extended in 1943 for another two years.[31]

Australia's export quota of just less than 407 000 tonnes of sugar under the first ISA presented the Australian sugar industry with major challenges in late 1937. The previous year, 423 075 tons of sugar had been exported from Australia. A slightly larger amount looked like being exported during 1937 and indications in late 1937 pointed to a potential export of a somewhat similar amount from the 1938 crop. If predictions were correct, then Australia could exceed her export quota during the first year of the ISA's operation. In September 1937, a meeting of the executives from the ASPA and QCGC and a representative from CSR met to consider the need to introduce stronger measures to reduce production. The conference recommended the adoption of what became known as the 'Equalisation Scheme'. Under this scheme, the 1929 mill peaks were retained, but the amount paid for any excess export sugar (No. 2 Pool) depended upon No. 1 Pool price. Essentially, if the original value of No. 1 pool did not exceed £17 per ton, there would be no distribution to No. 2 Pool. If No. 1 Pool price exceeded £17 per ton, then the amount would be distributed proportionately to No. 2 Pool (i.e. excess export sugar), although no amount was to be distributed from the No. 1 Pool which would increase the value of excess export sugar beyond £13 10s per ton. For example, if the price of sugar in the No. 1 Pool reached £17 2s per ton and the No. 1 Pool consisted of 600 000 tons, then the extra 2s, multiplied by 600 000 would provide £60 000 for distribution on the tonnage of excess export sugar manufactured. If 100 000 tons of excess sugar was manufactured, then there would be 12s available from the proceeds of No. 1 Pool to be added to the real export price of the excess sugar (i.e. £ 60 000 divided by 100 000). In addition, the meeting concluded that enough sugar cane was being grown in Australia, enabling the country to meet its export quota, and recommended that the Central Board should grant no further assignments and cancel all unused assignments. These proposals were to be considered by a conference of all interested parties by no later than the end of January 1938.[32]

The conference of industry delegates convened by Premier Forgan Smith to consider the means of controlling sugar production met in Brisbane on 7 March 1938. Four schemes to readjust the 1929 mill peaks were considered during the three day conference. The 'Equalisation Scheme' was preferred by the ASPA and QCGC delegates. The Tully Mill delegates proposed the 'Tully Scheme' which sought a reallocation of mill peaks based on the average of the highest outputs of a sugar mill during the past nine years, adjusted to allow an increase of No. 1 Pool sugar to 680 720 tonnes. In no instance would a new mill peak be less than the 1929 mill peak. The proposal from the Mossman Mill representatives called for each Queensland sugar mill to be allocated 20 320 tonnes of No. 1 Pool sugar, or a total state output of 629 920 tonnes of No. 1 Pool sugar. The 'Mackay Scheme', which had been heavily promoted by the Mackay District Canegrowers' Executive and the Mackay Sugar Manufacturers' Association in the lead up to the conference, proposed that the total for No. 1 Pool sugar should be increased to 741 680 tonnes, with a readjustment of the mill peaks in accordance with the most recent highest three outputs, with the proviso that no mill peak would be reduced. In addition, the conference considered applications from Tully and Inkerman Mills for an increase in their mill peaks.

The conference agreed to ask the Central Board to add a further 2 540 tonnes to the mill peak for Inkerman Mill, but again declined to support the request from Tully Mill for a revised mill peak. However, delegates were divided over which sugar production scheme to support and finally with a margin of thirty-six to thirty majority recommended the Equalisation

Scheme. Initially, Premier Forgan Smith declined to implement the conference decision, arguing that support for the scheme did not represent a sufficient majority of delegates. He then referred the scheme to the Queensland Bureau of Industry for investigation. In particular, the Queensland government wanted advice on how Queensland's sugar producers might have fared if the Equalisation Scheme operated during 1937. After receiving the Bureau's report, Premier Forgan Smith announced that the Equalisation Scheme did not warrant adoption. Another joint meeting of the ASPA and QCGC executives in August 1938 voted 12 to 11 to reaffirm the Equalisation Scheme, with the modification that the No. 1 Pool price should be reduced from £17 to £16 16s An attempt to increase the mill peaks by 101 600 tonnes was just lost, eleven to twelve votes. Facing a divided sugar industry which was showing no signs of cooperating in the national interest, Premier Forgan Smith on 5 December 1938 referred the matter of the Peak Year Scheme and the regulation of sugar output to a Royal Commission.[33]

Justice William Webb, Chairman of Central Board, headed the 1939 Royal Commission into Sugar Peaks and Cognate Matters, which was asked to investigate the general operation and effect of the Peak Year Scheme, did the scheme need to be modified, and whether or not should new mill peaks be established? He was assisted by William Short, Chairman of the Sugar Board, Ernest Smith, the mill-owners' representative on the Central Board, William Doherty, the retiring secretary of the QCGC and Frederick Curlewis, secretary of the ASPA. The Royal Commission found that the area under sugar cane in Queensland had slowly increased during the early 1930s, and then grew markedly by almost 14 500 haafter 1935 (Figure 10.1), despite the serious threat of overproduction. Some Queensland sugar mill-owners had tried limiting the amount of cane cultivated in their respective mill supply areas or had limited the acreage harvested in order to meet the mill peak (e.g. South Johnstone). On the other hand, Millaquin Mill, for example, had accepted all the cane grown in its mill supply area between 1934 and 1938, even though the mill's sugar output was in excess of its peak. Tully Mill had deliberately flouted the intent of the Scheme, allowing an increase in the area under sugar cane, arguing that the canegrowers had considered the mill's peak to be inadequate, as the district had come under the Peak Year Scheme as an underdeveloped area, handicapped severely by the number of years used to assess its peak production.[34]

Future arrangements were canvassed from those witnesses appearing before the Royal Commission. CSR was the only Australian sugar mill-owner to ask for the Peak Year Scheme to be saved, un-modified. At the other end of the spectrum, Walter Merrill, Secretary of the Cattle Creek Mill Suppliers' Committee, told the commissioners that his organisation wanted the Peak Year Scheme 'scrapped entirely'. Mostly, the commissioners listened as mill representative after mill representative claimed that they supported the intent of the Peak Year Scheme, but nevertheless they sought an increase in their mill peak, claiming special circumstances.[35] After listening to the evidence, the commissioners were clearly faced with a conundrum. On one hand, CSR and the Queensland government through its funding of the BSES advocated the adoption of scientific farming methods. Improved yields were the consequence of the wider use of fertiliser and more sucrose-rich cane varieties (see Chapters 11 and 13) and a reduction in the deleterious impacts of cane diseases and pests (see Chapters 14 and 15). In addition, more sugar was recovered per tonne of cane because of better milling procedures (see Chapter 17). On the other hand, the State sought to avoid overproduction, but the excess sugar produced in the mid-1930s had ensured that Australia was granted a substantial export quota under the first ISA. The only unanimity amongst the representatives of the Queensland sugar mill-owners was a desire for increased mill peaks.

Not surprisingly, the commissioners were divided in their recommendations which were delivered to the Queensland government on 27 April 1939. Each commissioner agreed that the aggregate of the home consumption quota plus Queensland's share of the annual quota for export as determined by the International Sugar Council was 748 792 tonnes of sugar and that it would be futile and disastrous to produce additional sugar for which no market existed. Commissioners Short, Doherty and Curlewis in the majority report recommended the creation of a single pool for all sugar, with one price for this sugar and revised mill peaks. Justice Webb made similar recommendations, but he devised slightly different mill peaks. Commissioner Smith argued for the retention of a two pool scheme with a price differential between the two pools. All commissioners, except Commissioner Smith, recommended that effective control over production should be achieved by offering a nominal price of 10s only per ton of sugar delivered in excess of the scheduled allocation to any sugar mill. Commissioner Smith thought the penalty price should be 1s per ton. All Commissioners were unanimous in their recommendation that some form of farm production controls were necessary to limit production within mill supply areas.[36]

Premier Forgan Smith quickly considered the recommendations of the Royal Commission and in May 1939 announced the new mill peaks suggested in the Royal Commission's majority report. These provided for an increase in all mill peaks to a total of 748 792 tonnes of sugar. Each

Queensland sugar mill had its peak increased, except Isis Mill, which received a 4% reduction in its mill peak, although this arrangement came about due to the re-assignment of the cane grown at Pialba from the Isis Mill to the Maryborough Sugar Factory. In addition, the Queensland government increased the margin from 5% to 7% on each sugar mill's quota to be paid for at the net export price before the nominal penalty price of 10s per ton, after no doubt being influenced by the addition of 12 200 tonnes to the Australian ISA export quota in May 1939. This arrangement was further amended to 12% in July 1939, following additional purchases of Australian sugar by the British Ministry of Food.[37]

The outbreak of war in September 1939 suddenly changed the international sugar market, disrupting sugar supplies and causing a sharp increase in the world 'free' price for sugar. The British Ministry of Food immediately responded by contracting to buy on behalf of Great Britain, Canada and New Zealand all the exportable sugar produced in the Commonwealth during the war at a price to be negotiated with suppliers. On 20 October 1939, the Sugar Board announced it could acquire all sugar produced in 1939, thereby enabling every Australian sugar mill to process its entire crop in order to produce as much sugar for export as possible. Three days earlier, the Queensland government had also introduced amendments to The *Regulation of Sugar Cane Prices Act* of 1915 to 1938. The new regulations, which were passed by the Queensland Parliament on 31 October 1939, contained provisions for farm peaks or the maximum quantity of sugar cane that could be grown on assigned land and which could be sold to a particular sugar mill. Farm peaks were to be determined each year as part of the mill awards and could be implemented when 70% of the canegrowers supplying a particular Queensland sugar mill agreed to a set of conditions. Mill peaks were to be considered annually by the Central Board during what became known as Section 12A hearings in the early part of the year before the commencement of each crushing season.[38] Following the introduction of these new regulations, the sugar industry became the most regulated of all Queensland's primary industries. The Queensland government, through the operation of the Central Board and Sugar Board, had control over the marketing and production of sugar, through its influence over the land which was allocated (or assigned) for the purpose of growing sugar cane, the tonnage of sugar cane produced on a cane farm (farm peak), the allocation of production quotas to each sugar mill (mill peaks), and the purchase, sale or lease of assigned land.

World War II and its immediate aftermath

In February 1940, just four months after the outbreak of World War II, several Queensland sugar mill-owners applied to the Central Board to overturn the new mill peaks established in 1939. These applications were all rejected on the grounds that the outbreak of war had not brought about a change of circumstances to warrant an alteration in mill peaks, especially as there was no certainty about the amount of shipping that would be available for the export of sugar. Moreover, the Central Board refused any upward adjustment of mill peaks in Queensland for the remainder of the 1940s.

Production controls in New South Wales during late 1939 and 1940 were unresolved due to the unwillingness of the NSW Cane Growers' Association to accept the Webb Royal Commission's recommendation that excess sugar produced in New South Wales be deemed to be that in excess of 40 640 tonnes of sugar, instead of the product of cane in excess of 304 800 tonnes of cane. When Queensland's production was restricted to mill peaks in 1941, the New South Wales industry voluntarily agreed to restrict its crushing to 365 760 tonnes of cane on the basis that it was entitled to an increase of 20% in its '1929 mill peak equivalent', the same as Queensland (i.e. 304 800 tonnes of cane increased by 20% to 365 760 tonnes of cane). Consequently, about 60 960 tonnes of cane was left unharvested in New South Wales at the conclusion of the 1941 crushing season. Some New South Wales canegrowers who were very disgruntled by this arrangement formed a breakaway organisation known as the New South Wales Sugar Promotion League. They made representations to both the Commonwealth and New South Wales governments, alleging unfair treatment by Queensland interests, but failed to achieve their goal of no production restrictions on the New South Wales industry. Eventually, the arrangement that had been followed in 1941 was incorporated into the 1944 agreement between the Queensland government and CSR. The 365 760 tonnes of cane harvested in New South Wales corresponded to a peak of 48 768 tonnes of sugar, of which 40 640 tonnes was covered by the '80/20' arrangement (i.e. 24 384 tonnes at home consumption price and 16 256 tonnes at export price) and 8 128 tonnes attracted the surplus price, less lighterage. By 1944, however, the New South Wales canegrowers were in no position to worry about overproduction, the wartime shortages of labour having a severe impact upon their activities.[39]

During 1940, disruptions caused by the war were not that severe and the amount of sugar produced in Queensland slightly exceeded the state's aggregate mill peaks and almost equaled the mill peak equivalent in New South Wales (Table 10.2). Yet as the war progressed the country's canegrowers suffered shortages of farm workers and fertiliser. Consequently, Queensland's sugar mills only managed to produce two-thirds of their mill peak entitlements in 1943. During 1944 and 1945, the three New South Wales sugar mills produced less than 60% of their mill peak equivalent (Table 10.2). Moreover, at the conclusion of World War II, the area under sugar cane in Australia stood at 136 500 ha or approximately a 10% contraction since 1939. Proportionally, the biggest decline had occurred in New South Wales, where the amount of sugar cane being cultivated in 1945 was 28% less than 1939 (Table 10.2). A severe lack of field labour, especially for harvesting, enforced a reduction in the area planted on farms to an amount that the canegrower could handle efficiently with available assistance – usually the canegrower's own family.[40] In Queensland, the impact on production had not been as great as in New South Wales, with the state's sugar mills producing nearly 90% of their mill peak entitlements in 1945 (Table 10.2).

Table 10.2. Sugar production statistics for New South Wales and Queensland, 1939–1948.

	QLD	QLD	QLD	QLD	NSW	NSW	NSW	NSW
	Area under cane (ha)	No. of farmers cultivating sugar cane	Sugar production (tons)	Percentage of mill peak	Area under cane (ha)	No. of farmers cultivating sugar cane	Sugar production (tons)	Percentage of mill peak equivalent
1939	143 368	7 820	891 738	120.9	8 493	*	36 883	Not operating
1940	142 098	7 554	758 841	102.9	8 408	*	46 977	97.9
1941	135 051	7 538	702 788	95.4	7 580	*	48 361	100.7
1942	125 599	6 945	605 482	82.1	7 666	*	44 259	92.2
1943	129 926	6 871	486 555	66.0	6 323	*	37 407	77.9
1944	126 431	6 601	624 970	84.8	*	687	26 377	54.9
1945	130 367	6 622	641 660	87.0	6 101	597	21 194	44.1
1946	124 823	6 419	511 945	69.5	6 555	558	39 768	82.8
1947	133 036	6 424	571 508	77.5	6 653	584	33 507	69.8
1948	148 247	6 552	922 896	125.2	7 070	580	33 001	68.7

* Data not collected

Source: Based upon figures in the *Queensland Government Statistical Bulletins*, various issues, 1940 to 1953 and *ASYB*, 1950, p. 463.

The difficulties confronting the Australian sugar industry, despite the declaration of peace in 1945, continued over the latter part of the 1940s. Only half the tractors impressed for urgent defence works were returned during 1946, so machinery shortages still impacted upon normal cultivation practices (see Chapter 11). Field workers, especially those willing to harvest the annual crops of cane, remained difficult to procure (see Chapter 16). Australia's canegrowers continued suffering fertiliser shortages. Bad seasonal conditions, particularly in Southern Queensland during 1946, meant Queensland's sugar mills failed to reach their mill peaks in 1946 and 1947.[41] Yet the area under sugar cane in Queensland had recovered by 1948, surpassing the level reached in 1939, and the amount of sugar produced in 1948 exceeded mill peaks for the first time since 1940. Signs of recovery occurred in New South Wales, where the area under cane had risen by almost 1 000 ha since 1945 and the number of farmers cultivating sugar cane had shown a slight increase (Table 10.2).

In the immediate aftermath of World War II, the Queensland government (Cooper; Labor) in partnership with the Commonwealth government (Chifley; Labor) established another soldier settlement scheme to repatriate ex-servicemen. Both governments were keen to avoid the errors associated with the settlement of Australian ex-servicemen following World War 1. In February 1946, the Queensland government appointed a Royal Commission on Soldier Settlement on Sugar Lands. Its aims were to inquire into the practicability and expediency of settling ex-members of the armed forces on sugar cane farms in Queensland, how land for this purpose should be acquired and whether any new sugar mills should be established in Queensland. The three commissioners were John Hutcheon (Chairman), one of Queensland's leading barristers, Hunter Freeman, the technical member of the Central Board, and Frederick Wheeler, the millers' representative on the Central Board. Future purchases of sugar by Great Britain were uncertain beyond 1947, so the commissioners recommended soldier settlement designed to produce an additional 32 500 tonnes of sugar, to be provided by settlement to the extent of 10 200 tonnes within the peaks of certain sugar mills, and by a small extension of 3% to all mill peaks, comprising the output from any new farming areas occupied by 344 ex-servicemen. The commission investigated the possibility of building new central sugar mills at Bingil Bay and Jarra Creek, near Tully, and Abergowrie, near Ingham, but declined to recommend their erection because of the long delay of up to five years in settling ex-servicemen in areas developed around a new sugar mill.[42]

In response to the Royal Commission's recommendations, the Queensland government introduced *The War Service (Sugar Industry) Land Settlement Act* of 1946. This new measure granted an increase of no more than 3% to mill peaks, and permitted the rehabilitation of approximately 400 ex-servicemen. The legislation gave ex-servicemen priority in the purchase of existing cane farms, and left the task of finding new land for any additional assignments in the hands of the Queensland Lands Department and the Central Board. During 1947, Land Commissioners in each Queensland sugar-producing district were instructed to identify unassigned land that could be occupied by ex-servicemen. By June 1948, 237 ex-servicemen buyers had received preference in sugar cane farm sales and 20 ex-servicemen had received an increase in the size of their assignment. However, the Queensland Lands Department reported in December 1948 that its field officers were having great difficulty in finding suitable land for ex-servicemen, as there was practically no suitable Crown land available for cane growing within reasonable distance of the sugar mills, and the Department was having to acquire freehold or leasehold land for the ex-servicemen. By June 1951, only 123 ex-servicemen had been allotted farms and new cane assignments by ballot in six sugar-producing districts, mostly in North Queensland.[43]

Post-War Expansion, 1950 to 1963

During the late 1940s, the size of Australia's population increased, a consequence of the post-war immigration program. The Central Board realised that the modest increase in mill peaks permitted under *The War Service (Sugar Industry) Land Settlement Act* of 1946 would not produce enough sugar to meet this rising domestic demand and Great Britain's expected increased requirements for sugar as the British Ministry of Food considered relaxing wartime rationing. Moreover, the average export price for sugar had risen from £16 7s per ton in 1945 to £28 2s per ton in 1948, so Australian canegrowers could expect enhanced returns if they could meet the demand for sugar. Therefore, the Central Board in June 1949 recommended that the 1939 mill peaks be increased from 748 792 tonnes to 861 568 tonnes, exclusive of the 3% rise for soldier settlement. With provision for New South Wales (48 768 tonnes) and including the 3% increase for soldier settlement, the new Australian mill peaks totaled 936 752 tonnes. To produce the extra sugar, half Queensland's sugar mills were to be granted new assignments; the other sugar mills were to meet their new peaks by expanded tonnages per farm.[44] In mid-1949, the Central Board called for applications for new and increased assignments, but the finalisation of this matter was delayed because of the events in London during the latter half of 1949.

Immediately after World War II, the bulk of Australia's surplus sugar continued to be exported to Great Britain. In the autumn of 1948, the British government extended the bulk purchase of Commonwealth-produced sugar until the end of 1952. This offer safeguarded Britain's immediate needs and was probably as far as the British government wished to go. Australia's sugar producers, however, wanted more certainty. 'Ned' Hanlon, the Queensland Premier, called for a ten year agreement. CSR opposed any future extension of bulk purchases, considering them to be a strictly wartime measure. In particular, CSR feared an unlimited increase in Commonwealth production if bulk purchases continued after 1952, leading to overproduction, and a fall in the 'free' price of sugar. Rather than an extension of bulk purchases, CSR advocated a return to pre-war marketing arrangements, which it believed enabled efficient producers to compete on reasonable terms. Stable prices could be maintained by a renegotiation of quotas under the ISA, and a degree of protection afforded by Empire preferences in the British and Canadian markets. In order to advance the matter of a long-term agreement, the British West Indies Sugar Association asked the Colonial Office to convene a conference in London. This meeting was held in the summer of 1949, and culminated with the Colonial Office announcing on 9 August 1949 that the conditions on which a long-term agreement might be made would be considered in the autumn. Consequently, all Commonwealth sugar exporters participated in the conference, which opened in London in November 1949. Australia's delegation included Premier Hanlon, James Dixon of CSR, and David Muir, Agent-General for Queensland.[45]

Negotiations to reach the British Commonwealth Sugar Agreement (CSA) occurred during three series of talks: November 1949 to January 1950; November 1950 to January 1951; and November to December 1951. In the first round, the British Ministry of Food and the Australian delegates quickly reached a five year agreement for a total outlet to preferential markets of 609 600 tonnes of sugar, of which half would come to Great Britain at annually negotiated prices. This provisional agreement was signed with Australia on 8 December 1949. Discussions over the method of determining the price for the sugar, quotas, any preferential treatment of colonies over the Dominions, the length of the agreement, the future of the Canadian market and the extent of Britain's sugar purchases from non-Commonwealth sources then dragged on throughout 1950 and 1951. Eventually, the final agreement was signed on 21 December 1951, although the introduction of the CSA was backdated to 1 January 1950, and it was to last until 31 December 1959. However, provision was made for annual extensions of one year at a time, from 1952 onwards; this was a compromise with the Commonwealth sugar exporters, including Australia, who had pressed for a ten year agreement. Extensions were made every year from 1952 onwards and the CSA was extended in 1963 until the end of 1971.

Under the CSA, Great Britain was committed to buy for herself and New Zealand, specified quantities of raw sugar at an annually negotiated price, and to import for her own consumption, but not New Zealand's, additional quantities of Commonwealth sugar under a preferential tariff. Great Britain agreed to purchase 609 600 tonnes of sugar from Australia annually for a period of eight years (basic overall quota) and would pay a negotiated price to be decided each year upon 319 024 tonnes. The remainder would be marketed in the preferential markets of Great Britain and Canada at a price equivalent to the world market price plus Empire preferences. Australia could exceed her quota under the CSA, but only in the event of shortfalls in other Commonwealth countries. If Great Britain's demand for sugar was not being met, the quotas could be correspondingly corrected and Australia would benefit proportionately. Great Britain stipulated one condition for its agreement to the CSA: Commonwealth sugar exporters would participate in the ISA when that was renegotiated.[46]

Premier Hanlon was aware during the negotiations to acquire a provisional agreement with the British Ministry of Food that the proposed new export quota exceeded current Australian sugar production. Therefore, Hanlon's government in March 1950 appointed a Royal Commission to ensure that the Australian sugar industry could meet its new international and domestic obligations. The inquiry had three main goals: inquiring into the practicality of establishing new sugar mills in Queensland; should Tully Mill be duplicated to enable the production of an additional 30 480 tonnes of sugar; and preparing an orderly expansion plan for the Australian sugar industry during the next quarter of a century in order to avoid overproduction. A judge of the Supreme Court in Queensland, The Hon. Mr. Justice Alan Mansfield (later Sir Alan and governor of Queensland), headed the Commission.

Over the next seven months, the commissioners investigated eight localities as possible sites for new sugar mills in Queensland, including four sites that had previously been rejected as unsuitable in the 1916 and 1922 investigations into the expansion of sugar milling in Queensland. The commissioners heard that unused land in existing sugar-producing districts was suitable for cane growing and could be developed as part of an industry expansion. Kenneth West, Secretary at Mourilyan Mill, for example, said that 1 012 ha of forested land at Maria Creek could be converted to sugar cane farms. Each sugar mill-owner in the Bundaberg district stated additional land was available for extra production. Other witnesses, however, urged caution. John W. Inverarity, the Manager of Kalamia Mill, advised the commissioners that his mill's owner, the Australian Estates Co. Ltd., thought erecting any new sugar mills anywhere in the state would be 'unwise and unsound economically'. A similar view was held by Walter Davidson, Manager of Pioneer Mill (Lower Burdekin district).[47]

In September 1951, the Mansfield Royal Commission recommended that sugar production in Queensland should not be increased suddenly. Domestic sugar consumption per capita was not predicted to increase substantially, so any production increase only needed to meet the growth in the size of the Australian population. The Commission accepted CSR's view that Australia's exportable surplus would not be much more than 600 000 tonnes of bagged raw sugar each year for the next twenty-five years, and that Australia's main customers would remain Great Britain and Canada (as it turned out, an incorrect assessment of the market). Thus, the Commission estimated that available markets for Australian sugar would be 1.2 million tonnes by 1962 and 1.34 million tonnes by 1975. The State's existing sugar mills were capable of manufacturing such an output – no new sugar mills needed to be built between 1951 and 1962. Moreover, even though the national interest dictated the settlement of the areas near to or north of Cairns, the proposals to establish new sugar mills in the Cooktown, Daintree-Bailey Creek or Atherton Tableland districts could not be recommended by the Commission because of one or other of low rainfall at vital growing periods, high transport costs, including the difficulty of finding a shipment port for the sugar and the very substantial establishment costs of a new sugar mill. Yet the Commission recommended the duplication of Tully Mill, although such a recommendation did not have the support of local canegrowers.

Looking further ahead, the Mansfield Royal Commission also recommended that no new sugar mill should be erected in Queensland between 1962 and 1975, unless economic conditions improved. If economic conditions did change during the 1960s, then a new sugar mill should be erected in the Tully district and supplied from both sides of the Murray River and the Kennedy Valley lands (i.e. undeveloped lands located south of Tully). However, if building this new sugar mill proved uneconomic, then Macknade

Mill (Herbert River district) could be duplicated. The Commission also considered how any future industry expansion could occur, and concluded that new farms were less attractive than expanded existing farms, provided suitable land could be obtained for expanded farms. Yet the Commission warned against using inferior marginal lands to obtain this goal. Mindful of the high capital costs associated with farming, the Commission recommended that no expansion to the area of larger assignments should be made until after the assignments of below 6 ha had been increased and an upward revision made of the Central Board's assessment of the minimum area (or tonnage) sufficient to provide a reasonable living.[48]

By the time the Queensland government had received the Mansfield Royal Commission's report urging cautious growth in sugar production, an expansion program was well underway. Upon the announcement of the provisional export agreement with the British government in December 1949, the Central Board reviewed the size of the mill peaks issued in June 1949. Beginning on 27 February 1950, the Central Board heard evidence from many of Queensland's sugar mill-owners about their plans for increasing mill peaks. The Central Board was particularly interested in schemes enabling the rapid production of extra sugar, and was very attracted to the case put forward on behalf of Victoria Mill for an almost doubling of its mill peak from 45 600 to 83 000 tonnes. CSR assured the Central Board that it could duplicate Victoria Mill in readiness for the 1954 crushing season by importing the key plant and equipment, thereby overcoming local machinery shortages. The sugar mill would be supplied by cane produced in the Abergowrie district, 30 km north-west of Ingham. After hearing evidence for almost a month, the Central Board on 23 March 1950 recommended mill peaks in Queensland totaling 931 570 tonnes or an increase of 69 392 tonnes and prospective 1954/1955 mill peaks totaling 1 061 720 tonnes or an increase of 130 149 tonnes (Table 10.3). Just over half this tonnage was to come from new settlement permitted under the granting of approximately 680 new assignments; the rest from increased existing assignments. Three-quarters of the increase was allocated to canegrowers with small tonnages in Central or Southern Queensland, thereby creating the average farm peak per canegrower of 114 and 88 tonnes of sugar in the Mackay and Bundaberg districts, respectively. The Central Board also supported CSR's scheme to duplicate Victoria Mill and the development of the Abergowrie district. These recommendations were subsequently approved on 28 March 1950 by the Queensland government. By November 1951, the Central Board had granted 1 891 increased assignments and 698 new assignments.[49]

Table 10.3. Sugar production statistics for Queensland, 1949–1963.

Year	Area under cane (ha)	Assigned area (ha)	No. of farms producing cane cut for crushing	Sugar production (tonnes)	Mill peaks (94 n.t. tonnes)	Percentage of mill peaks
1949	148 247	183 624	6 597	910 755	888 593 * (1 061 720)**	102.5
1950	154 438	+	6 578	893 921	956 462* (1 061 720)**	93.5
1951	154 525	+	6 505	712 562	956 462* (1 061 720)**	74.5
1952	157 280	249 417	6 545	949 567	978 489 (1 181 201)**	97.1
1953	170 032	232 858	6 936	1 239 909	1 170 940 (1 188 313)**	105.9
1954	188 923	240 535	7 032	1 322 064	1 189 634	111.1
1955	197 130	239 416	7 018	1 153 855	1 189 634	97.0
1956	195 268	239 808	7 001	1 190 629	1 189 634	100.1
1957	194 446	240 518	6 927	1 276 371	1 223 162	104.4
1958	196 568	240 556	6 971	1 375 199	1 226 718	112.1
1959	197 154	241 331	6 877	1 237 287	1 232 408	98.2
1960	187 063	240 855	6 786	1 340 747	1 233 932	108.6
1961	182 057	+	6 695	1 336 439	1 239 621	107.8
1962	191 319	239 815	6 599	1 798 405	1 239 621	145.0
1963	194 087	239 813	6 612	1 674 645	1 255 369	133.4

Notes: * Includes 3% soldier settlement increase of 25 806 tons of sugar ** Projected ultimate mill peak target by 1954/55 + Figures not recorded in *ASYB*.

Source: Based upon figures in *Year Book of Australia*, various issues, *ASYB*, various issues, and *Queensland Government Statistical Bulletins*, various issues.

In preparation for the hearings about prospective mill peaks for 1952, the Central Board received reports upon the progress being made towards the expansion in newly cropped areas and the upgrade of each sugar mill's crushing capabilities. At the March 1952 meeting of the Central Board, its members agreed that most expansion programs were proceeding according to schedule and that the enhanced milling capacities should be achieved or in some cases exceeded by 1953 or 1954 at the latest. The Central Board noted that there had been a sharp rise in the amount of sugar consumed domestically, so that it now stood at approximately 558 800 tonnes per annum. In order to meet this growing domestic demand and to cover the increased sugar production arising from the new and increased assignments, the

Central Board recommended the mill peaks be increased by 46 700 tonnes, so that the 1952 mill peaks reached 978 489 tonnes. The 1954 mill peaks were projected to be 1.18 million tonnes or a slight increase over the 1950 projections (Table 10.3).

Future mill capacity, however, troubled the Central Board, for they considered that many Queensland sugar mills had reached the limit of their single train crushing capacity (i.e. a sugar mill with only one mill tandem or group of crushing rollers) and suggested that the issue of duplicating Tully Mill should again be considered.[50] Subsequently, the Queensland government approved the mill peak increases and referred the matter of the duplication of Tully Mill to a special committee of inquiry consisting of The Hon. Mr. Justice Alan Mansfield (Chairman), Mr P. Donnollan, Deputy Secretary of the Sugar Board and Harold Lund, Chairman of Directors of the Tully Cooperative Sugar Milling Association Ltd. The Committee met for the first time on 28 July 1952 and presented its report to the Queensland Parliament a little over a year later. On present costs, terms of finance available, and probable returns to canegrowers, the Committee declined to recommend the duplication of Tully Mill, arguing that it was not in the public interest, as the state's other sugar mills had the increased capacity to handle the extra tonnage of cane needed to be crushed to meet the projected demand for sugar in Australia and any overseas markets.[51]

As a consequence of the Central Board's expansionary policy, the total assigned area in Queensland rose from 183 624 ha in 1949 to 249 416 ha in October 1952. Almost 90% of this increase of 6 6015 ha occurred between 1950 and 1952; the remaining 10% was permitted under the *War Service (Sugar Industry) Land Settlement Act* of 1946. Geographically, a third of this increase occurred in the Mackay district; another third was found in the sugar-producing districts north of Townsville. The total number of assignments rose from 7 104 to 8 186, with just over half the assignments being increased in size. New assignments had been granted to 1 082 canegrowers by October 1952, with just on 50% of these new assignments being issued in Far North Queensland.[52]

Two very large land settlement projects were part of this expansion. The Abergowrie Scheme consisted of 123 farms (equivalent to 183 new assignments), which were offered to ex-servicemen during 1951 and 1952. The Queensland government supervised the development of the blocks of land, which represented new assignments totaling 3 804 ha of land. In contrast, the Fairydale and Moore Park Settlement Scheme was the initiative of a private company. During the period 1952 to 1955, the Fairymead Sugar Co. Ltd. spent approximately £500 000 to drain 1 750 ha of coastal freshwater swamps located 20 km, north-west of Bundaberg. The reclaimed land was then subdivided into fifty farms, with eighteen of the farms being occupied by ex-servicemen. Other new localities settled during this expansion included Utchee and Camp Creek, near South Johnstone, Woopen Creek, 16 km south of Babinda and the Syndicate, Riversdale and Jarra Creek districts, near Tully.[53]

In light of the threat to northern Australia during World War II, it is clear that the Hanlon (Labor) government in Queensland used the expansion as a way of increasing the number of settlers in the northern parts of the state. However, the expansion did not proceed without its difficulties. The Queensland Canegrowers' Association reported in 1953 that the implementation of the expansion program was being slowed by credit restrictions, high capital costs associated with developing new farms and labour and material shortages. In particular, difficulties were being encountered in the Abergowrie district, where the clearing of the land covered in dense rainforest proved difficult and time consuming. Of the 120 settlers established by 1954, 40% had left their farms by 1964. Objections were also raised by sugar industry leaders about the proportion of the expansion being granted to new settlers. They argued that it was more economical to increase production on existing farms, since many existing farms had unassigned land already cleared (often former horse paddocks no longer needed after the switch to tractors) and the existing infrastructure and expertise necessary to rapidly and economically increase production.[54] This expansion in production also had an environmental downside. Sloping land, poorly drained sections of farms and freshwater swamps – land considered previously marginal for cane growing – was placed under sugar cane, sometimes after costly modifications. Increased soil erosion and a substantial upsurge in the loss of freshwater wetlands began occurring in the sugar-producing lands of eastern Australia (see Chapter 12).

As the Australian sugar industry embarked on its first major post-war expansion, the main exporters and importers of sugar decided to renew the ISA that had been made largely ineffectual by World War II. The 1953 ISA, which became operative on 1 January 1954, covered a period of five years, with a provision for review in the third year. This took place in 1956. The ISA aimed to assure supplies of sugar to importing countries at equitable and staple prices and guaranteed markets for sugar-exporting countries. Achieving these aims required the participating governments to limit their sugar exports to the free market in each quota year to the prescribed quotas, and that any excess, subject to some margins, was charged to the following year's quota. The ISA was renewed in 1958 and covered the period 1959 to 1963, with meetings to renew the agreement to take place in 1961. By 1960, signatories to the international sugar agreement accounted for 95% of global sugar exports and 65% of the world's sugar imports.[55] The British Commonwealth sugar-exporting countries, including Australia, joined the ISA as a group in 1953. The exporting parties to the CSA agreed to limit their basic overall quota of net exports initially to 2.375 million tonnes. Australia's basic overall agreement quota was 600 000 tonnes of sugar. However, the total amount of net exports allowed to the exporting territories contracted to the CSA was subsequently adjusted upwards in 1956, 1958 and 1960. The extra sugar allocated to the CSA territories, known as the 'International Quota' of the Commonwealth territories, was then distributed among the Commonwealth territories pro rata to their overall agreement quotas. Hence, Australia's export quota continued to rise during the 1950s, reaching 661 416 tonnes by 1961.[56]

The market situation for Australian sugar did not change much during the mid-1950s. The Central Board, perhaps unwisely, allowed slight increases to the state's mill peaks between 1954 and 1958, yet sugar production continued to mostly exceed mill peaks (Table 10.3). Faced with the emerging prospect of production substantially in excess of available markets, the Sugar Board announced in May 1958 that it would accept no more than 7.5% in excess of mill peaks, although following representations from sugar industry organisations, the Queensland government approved the acquisition of sugar up to 12.5% above mill peak on 8 September 1958. Consequently, approximately 1 million tonnes of cane was left unharvested in Queensland at the end of the 1958 season.[57] The state's canegrowers, not unexpectedly, criticised the arrangements for the handling of the 1958 harvest. The sugar mill-owners decided the best way to reduce the volume of stand-over cane was to produce more sugar and applications for increased peaks totaling 147 320 tonnes were received at the Central Board's mill peak hearings in March 1959. The Central Board, after considering probable market requirements for 1959–1960, announced only minor increases to individual mill peaks totaling 5 690 tonnes for the 1959 crushing season. These adjustments were subsequently approved by Otto Madsen, the Queensland Minister for Agriculture and Stock.[58]

In October 1959, the emerging likelihood of Queensland's sugar mills exceeding their mill peaks again, prompted Otto Madsen, Queensland's Minister for Agriculture and Stock, to complain to the ASPA and the QCGC that there was 'no evidence of any concerted or serious effort' to restrict production. Representatives from the main sugar industry organisations were invited to yet another conference on 18 November 1959 to discuss the matter of overproduction. At this meeting, L. Coates, CSR's delegate, claimed that present control measures 'have proved to be quite inadequate, have perhaps not been implemented or there may have been in some cases insufficient effort to implement the controls which exist'. An opposite view was put forward by Ben Foley, President of the QCGC, who claimed that no evidence indicated 'normal production controls have been disregarded'. All submissions from the delegates were against any changes being made to the industry's existing production control measures. At the close of the conference, the delegates resolved that the Australian sugar industry was aware of the dangers of overproduction and of its obligations to honour the current regulated marketing agreements, that the matter be considered by the annual conferences of the industry organisations in March 1960 and that the 1959 crop would not be harvested beyond its established peak controls.[59] At the subsequent annual conference of the ASPA in March 1960, the delegates supported a resolution stating that the current arrangements contained all the necessary provisions to enable each mill supply area to control its sugar production.[60]

During the 1950s, Queensland's canegrowers had capitalised on the shortfalls in production by other Commonwealth sugar-exporting countries and a growing demand for sugar globally. Between 1952 and 1957, the Sugar Board acquired all Australian sugar, so that 540 891 tonnes of sugar had been exported in excess of Australia's quota. However, the decision to accept all the industry's sugar was sometimes an eleventh hour development, and the Australian sugar industry suffered from uncertainty as to the quantity of sugar it could safely produce.[61] Moreover, the Sugar Board was unable to stockpile too much excess sugar under the terms of the ISA. After acquiring so much extra sugar in 1958, the Sugar Board limited its sugar acquisition during 1959 to mill peak only. Consequently, approximately 2.4 million tonnes of cane in Queensland remained unharvested at the conclusion of the 1959 season. Canegrowers in some districts were particularly badly affected. Nearly half the crop grown in the Maroochy district remained unharvested, the highest in Queensland; a third of the cane in the Bingera Mill supply area was left standing. Portions of this cane were subsequently destroyed.[62] In some instances, the crop was harvested and the sticks of cane gathered into heaps and burnt or used for cattle fattening. Elsewhere, the crops of cane were pushed over by tractors equipped with blades and rollers, and the cane allowed to dry out before being burnt or the stalks were shredded by rotary hoes or rotary slashers. The land could then be disced and ploughed, leaving little or no trace of the stand-over crop.

In light of the large amount of stand-over cane at the end of 1959, each of Queensland's thirty-one sugar mills applied for increased mill peaks in 1960. Some mill advocates argued they made the representations on the basis of future adjustments when available markets permitted an increase; several advocates implored the Central Board to display 'cautious optimism' in its recommendations. The Central Board was not swayed by such arguments, noting that a restricted market for sugar sales still existed in 1960, and recommended no significant increase in the aggregate of mill peaks in Queensland. However, the Central Board granted small increases of 510 tonnes of sugar to the mill peaks of Gin Gin and Rocky Point Mills in order to try and improve the financial affairs of these two sugar mills which were considered to be in a 'parlous state'.[63] Given such large amounts of stand-over cane, some of which was processed during 1960, sugar production exceeded mill peaks by 8% in 1960. Yet the Sugar Board acquired all excess sugar to make up the production shortfall caused by disastrous cyclone damage to the cane crops in Mauritius. Nevertheless, surplus cane still existed in many Queensland mill supply areas and an estimated 1.27 million tonnes of sugar cane remained unharvested at the end of 1960.[64]

Overproduction, however, was not being repeated in New South Wales. In October 1948, the NSW Cane Growers' Association decided not to take up the 3% soldier settlement increase in production that was being implemented in Queensland. The number of farmers growing sugar cane in New South Wales dropped below 500 in 1953 and the amount of cane cultivated fell under 6 000 ha, less than the area cultivated during World War II (cf. Tables 10.2 and 10.4). The Tweed River district was the most affected and difficulties were being encountered in keeping Condong Mill fully supplied. A survey of the 170 northern New South Wales farmers who decided not to renew the 1951–1955 cane agreements found that 60% claimed growing sugar cane was unprofitable following the steep rise in the cost of farming since World War II; another 20% had switched to dairying or small crops which paid better.[65] New South Wales canegrowers were also troubled by labour shortages after 1946, with only about 40% of harvesting labour being available locally. This situation made growing sugar cane even more difficult.[66]

A new sugar quota agreement between CSR and the Queensland government in December 1947 had allocated New South Wales a mill peak quota equal to 6.5% of the aggregate Queensland mill peaks, and entitled New South Wales theoretically to pro rata increases in any expansion of the aggregate of Queensland mill peaks.[67] Following the introduction of the new agreement, CSR was faced with a dilemma. Sugar output in New South Wales during the

Table 10.4. Sugar production statistics for New South Wales, 1949–1963.

Year	Area under sugar cane (ha)	No. of farms with sugar cane cut for crushing	NSW Mill Peak Allocation (Tonnes of sugar)	Tonnes of sugar produced	Percentage of NSW sugar allocation
1949	7 070	587	57 758*	41 357	71.6
1950	6 842	569	62 170*	41 921	67.4
1951	6 308	560	62 170*	41 716	67.1
1952	5 880	529	63 601*	14 500	22.8
1953	5 694	424	76 110*	34 546	45.4
1954	6 125	443	77 325*	26 723	34.6
1955	6 122	445	77 325*	36 604	47.4
1956	6 852	491	77 325*	36 491	47.2
1957	7 533	506	79 505*	37 443	47.0
1958	8 381	537	79 736*	59 683	74.9
1959	9 602	594	76 200#	71 655	94.0
1960	10 185	582	76 507#	64 721	84.6
1961	10 372	586	76 856#	75 802	98.6
1962	10 706	583	76 856#	77 536	100.9
1963	11 040	585	77 832#	84 074	108.0

Notes: *6.5% of aggregate Queensland mill peaks as agreed to by CSR and Queensland government in 1947 # 6.2% of aggregate as agreed to by CSR and Queensland government in 1959.

Source: Based upon figures in the *Year Book of Australia*, various issues, and *ASYB*, various issues.

early 1950s was well below its production quota (Table 10.4). CSR decided the situation had to be rectified. Declining cane supplies in the Tweed River district prompted CSR to advise its Condong Mill suppliers in September 1952 that the firm had operated the sugar mill at a loss for the last decade, and that CSR would close Condong Mill if its cane supply did not improve by the end of 1953. Enough additional plantings throughout the Tweed River district had occurred by October 1953 so that CSR was assured of an additional 101 600 tonnes of cane for the 1955 season.[68] Nevertheless, CSR advised its Tweed River canegrowers at the renewal of the cane purchasing agreement in 1955 for the period 1956–1960 that the firm would review the cane supply to Condong Mill in 1958. If the farmers showed no greater interest in growing cane than in the past few years, Condong Mill might still be closed. In addition, CSR advised all its New South Wales suppliers that it could not risk the large capital expenditure required urgently at the three New South Wales sugar mills, without some assurances of substantial additional cane supplies which would give reasonable expectations of operating the sugar mills profitably. CSR was willing to grant assignments to farmers who had never grown sugar cane before, as long as they were reasonably accessible to one of its mill's transport networks. In the Tweed River district, however, CSR pursued a deliberate policy of granting increased assignments to farmers who produced small tonnages, ensuring what the firm termed 'living tonnages' or an annual harvest of at least 1 000 tonnes of sugar cane. This approach encouraged farmers to avoid the tendency of mixing dairying or small crop production with cane growing.[69]

Existing New South Wales canegrowers responded to CSR's frankness by putting more land under sugar cane and many farmers returned to cane growing. By 1963, the Broadwater Mill Cane Inspector reported that canegrowers in his mill supply area were cultivating all available land and that 'fences are being removed to gain extra ground'. Eight farmers in the Dungarubba locality, east of Coraki, had also become canegrowers for the first time by 1957, boosting the cane supply for Broadwater Mill.[70] This turnabout occurred as a result of a combination of five factors. Firstly, dairying returns slumped in the early 1950s making this alternative no longer attractive as it had been in the late 1940s, leading to 85% of dairy farmers in northern New South Wales exiting the industry between 1945 and 1975.[71] Secondly, CSR paid higher prices for its cane and bonuses as an incentive to canegrowers to increase their cane tonnages. Thirdly, the planting of higher yielding and more sucrose-rich hybrid cane varieties bred by CSR's agricultural scientists (e.g. Pindar; Triton) improved crop returns (see Chapter 13). Fourthly, canegrowers paid greater attention to pests and diseases, improved on-farm drainage and invested more heavily in fertiliser use and modern implements, thereby increasing their yields. Fifthly, the introduction of mechanical loaders reduced the difficulties associated with crop harvests due to continued labour shortages since the conclusion of World War II (see Chapter 16 for more details). In addition, CSR specifically subsidised existing and potential Condong Mill suppliers by paying 4s 6d per ton towards freight on cane carried on the New South Wales government railway from the Crabbes Creek district to Condong Mill.[72]

As a result of this upsurge in cane growing, the number of New South Wales canegrowers in 1963 had almost reached the level that existed at the conclusion of World War II, and the area under sugar cane doubled between 1953 and 1963. Moreover, the gap between the actual tonnage of sugar produced in New South Wales and its sugar production quota had been narrowed substantially (see Table 10.4), although the New South Wales sugar industry representatives and the Sugar Board had agreed in 1959 that 76 200 tonnes would be the figure to correspond with Queensland's combined mill peaks totalling 1 226 718 tonnes (i.e. 6.2% of Queensland's aggregate mill peaks, reflecting the absence of the 3% soldier settlement increase in New South Wales).[73]

The recovery of the New South Wales section of the Australian sugar industry occurred at a fortuitous time for all Australian canegrowers. In January 1959, Fidel Castro had established a revolutionary government in Cuba. By July 1960, trade between the USA and Cuba had been suspended, and a great structural change began occurring in the world sugar markets, since the USA had previously relied heavily on Cuba for sugar supplies to the extent of about 3 million tonnes per annum. Under changes to the US sugar legislation in July 1960, Australia had been placed in a very low priority as a source of sugar to replace Cuban supplies, but managed to supply 80 400 tonnes to the USA in 1961. However, under the new US sugar legislation of July 1962, Australia received a quota of 35 714 tonnes per annum until the end of 1964. This quota was subsequently increased to about 165 000 tonnes in 1965 and extended on a firm basis to 1971. In addition, a United Nations sugar conference met in 1961 for the purpose of renewing the ISA. Cuba had decided to export about 1.1 million tonnes of sugar in excess of her export rights under the ISA to other communist countries during 1961. This breach of faith, together with the 'insatiable and ridiculous' demands made by the Cuban delegation during the negotiations, meant that the conference failed to agree upon quotas for 1962 and 1963. Consequently, the ISA collapsed and so from the opening of 1962 there was no limit on Australia's sugar exports to free market importers. Immediately it became apparent that there was a very substantial demand for Australian sugar in Japan.[74]

Even though the international demand for Australia's sugar looked promising and average returns fluctuated around £45 per ton of sugar, Otto Madsen, Queensland's Minister for Agriculture and Stock, accepted the Central Board's recommendations that overall mill peaks in 1961 and 1962 remain unchanged. Queensland's sugar production in 1962 reached a record high, given the effects of favourable seasonal conditions in all sugar-producing districts (Table 10.3). The Sugar Board, by progressive announcements during 1962, acquired all the sugar produced, and no cane had to be left unharvested.[75]

Given such an outcome, the Central Board in March 1963 heard exaggerated claims for increased mill peaks. The Mackay mill-owners, for instance, asked that their mill peaks be increased by 100 000 tonnes. Other applicants, however, urged caution. CSR pointed out that although Australian sugar was currently enjoying buoyant sales, the elements of continuity as to quantity, and stability as to price, appeared lacking. The Central Board's recommendation, accepted by the Queensland government, permitted Queensland's aggregate mill peaks to increase by 15 750 tonnes, bringing the state's overall mill peaks to 1.255 million tonnes (Table 10.4). However, this increase was confined to ten mills, including Rocky Point and Gin Gin Mills, which were again favoured with increases in their mill peaks of 1 000 and 1 500 tonnes, respectively. Moreover, the demand for Australia's sugar was so strong that the Central Board in April 1963 granted canegrowers permission to harvest areas in excess of their net assignments. The Sugar Board acquired all sugar produced by Queensland's sugar mills during 1963.[76]

Reckless Optimism, 1964 to 1974

Average returns per ton of sugar were almost £50 in the early 1960s, so Australian canegrowers and mill-owners were keen to capitalise on the high prices. However, the Australian sugar industry by the early 1960s exported over 60% of its sugar (Figure 10.2), so it was increasingly exposed to fluctuations in the international price of sugar. Moreover, the new markets for Australian sugar were not fully assured as permanent long term markets. This uncertainty presented the Australian sugar industry with two problems: firstly, just how long would these new markets persist; and secondly, a new expansion in output was essential to meet the demand of the new markets, but if these markets should contract or cease to exist, then the livelihoods of many Australian canegrowers could be put in jeopardy. Before committing to another substantial industry expansion, the Queensland government (Nicklin; Coalition of Country and Liberal Parties) appointed a committee of inquiry in June 1963 into the Australian sugar industry, under the chairmanship of Justice Harry Gibbs, a Supreme Court judge, with Norman King, Director of BSES, and Otto Wolfensberger, Chairman of the Sugar Board assisting. The committee of inquiry was asked to investigate the availability and assuredness of markets in the foreseeable future, the sugar requirements of the major markets between 1965 and 1971, measures to increase production, requirements for the equal treatment of present and any new canegrowers and measures to protect canegrowers from any changes in the market.

Over four months during mid-1963, the commissioners heard conflicting views from the sugar mill-owners and canegrowers about the future development of the Australian sugar industry. Firm opinions were presented in favour of and against permitting the annual harvesting of more than 75% of the gross assigned area. Some Australian sugar mill-owners and canegrowers, mostly in the northern sugar-producing districts, favoured mill peaks being fixed at 1.72 million tonnes and that the difference between this level and the production tonnages desired should be subject to secondary peaks. Representatives from the northern mill supply areas argued that they should receive preference in the allocation of expanded mill peaks in any expansion; such claims were strongly challenged by the submissions from the other mill supply areas. The committee considered the issue of canegrowers producing only small crops, with most submissions suggesting that 1 000 tonnes of cane was the most desirable minimum amount that should be produced on a cane farm.[77]

The Gibbs Committee of Inquiry reported in October 1963. It recommended that the Australian sugar industry should progressively increase production from 2.23 million tonnes of sugar in 1965–1966 to 2.54 million tonnes of sugar by 1970–1971. This increase should be permitted through the granting of new assignments sufficient to guarantee a total crop of at least 2.16 million tonnes of sugar in a normal season, but no new sugar mills should be built to cater for this expansion. About 60 900 ha of new sugar cane growing land should be allocated in such a way that the relative importance of the various sugar-producing regions remained unaltered. The recommended split was as follows: Northern region (27 580 ha); Burdekin (7 903 ha); Mackay (1 2519 ha); and Southern Region (12 907 ha). Thus, cane growing areas were to remain unchanged, despite proposals from groups of settlers to expand cane cultivation to new districts (e.g. Atherton Tableland; Gumlu; Bowen). Approximately 45% of the total recommended area for expansion (i.e. 25 920 ha) should be allocated to about 1 000 new canegrowers (i.e. an average of 30 ha per grower). This suggestion attempted to counter the decline in the number of canegrowers in Queensland, which had fallen from 7 900 in 1953 to 6 971 in 1963. The Committee of Inquiry recommended that the Central Board should give preference to ex-servicemen and small producers when granting increased and/or new assignments within a mill supply area and those annual harvest areas for delivery to a sugar mill in any season should be increased to 85% of the gross assigned area.[78]

The recommendations of the Gibbs Committee of Inquiry were rejected by some of the state's canegrowers. The QCGC considered the proposed expansion too rapid, and favoured a more orderly development, especially in light of the dependence upon overseas markets that were not secured by long term contracts. In addition, the QCGC argued too much emphasis was placed upon granting assignments to new farmers. Existing canegrowers were being dealt with unfairly, as they had shown restraint in their production during the 1950s, and now when a large expansion was being considered, they were being denied the chance to undertake the bulk of the new expansion. Moreover, the most economical and efficient way to increase production was on already established farms. Norman King, the BSES Director, worried that the expansion would force cane growing onto country that was 'doubtful agriculturally' and this scenario would not be in the best interests of any new canegrowers. The Bundaberg District Canegrowers' Executive expressed concerns about the unevenness of the expansion, arguing that the northern sections of the state were being favoured.[79] The Nicklin government, however, ignored these concerns and approved the recommendations of the Gibbs Committee of Inquiry.

Implementation began immediately and occurred over three phases. To provide increased sugar output quickly for the 1964 and 1965 seasons, the Central Board removed in January 1964 a long-standing barrier which limited the harvested area to 75% of the gross assignment and permitted the harvest of 85% of the gross assigned area in any one harvest season. The longer term solution to increased sugar production required an expansion in the number of canegrowers, so the Central Board called applications for new and increased assignments on 6 and 7 December 1963. Applicants for new assignments, however, had to be capable of producing 1 000 tonnes of cane in a normal season. Preference would be granted to land that was suitable for mechanical harvesting.[80] In August 1964, the Central Board announced it had issued 1 114 new assignments, totaling 26 295 ha. Additional assignments were granted during 1965, bringing the total area assigned in Queensland to almost 303 500 ha in 1966. Mill peaks were also raised each year between 1964 and 1966, leading to a 75% increase in the aggregate of mill peaks. This increase of almost 944 880 tonnes, the first state-wide expansion since 1953, lifted the state's mill peaks to 2.199 million tonnes of sugar (Table 10.5).

The mid-1960s expansion, however, was not spread evenly across the state's sugar-producing regions. The Northern Region received almost 40% of the total increase or the largest slice of the increases. Yet the canegrowers in other regions had produced evidence before the Gibbs Committee of Inquiry that they had the proven capacity to produce well above their mill peaks, while the sugar

Table 10.5. Sugar production statistics for Queensland, 1964–1974.

Year	Area under cane (ha)	No. of assignments	Area assigned (ha)	Mill peaks (tonnes)	Sugar produced (tonnes)	Percentage of mill peaks
1964	239 257	8 471	266 421	1 716 024	1 884 561	109.8
1965	245 826	8 292	303 206	2 144 776	1 913 497	89.2
1966	253 883	8 047	303 485	2 199 132	2 238 053	101.8
1967	256 574	8 115	303 096	2 200 554	2 249 230	102.2
1968	260 762	8 191	302 616	2 201 367	2 645 987	120.2
1969	259 411	8 025	302 586	2 201 570	2 114 332	96.0
1970	259 055	7 967	302 068	2 201 570	2 375 426	107.9
1971	260 901	7 791	301 393	2 201 570	2 669 621	121.3
1972	267 402	7 578	300 706	2 226 868	2 714 062	121.9
1973	264 018	7 424	299 370	2 226 999	2 405 005	107.9
1974	243 230	7 280	299 058	2 439 999	2 727 532	111.8

Source: Based upon data in *ASYB*, various issues, and BSES, ARs, 1965 to 1975.

mills in the Northern Region had produced far less excess sugar. Clearly, the Queensland government was again promoting northern settlement because it was seen as a desirable goal from a national viewpoint.[81]

In response to these changes, the area under sugar cane in Queensland rose by 17 317 ha between 1964 and 1967 (Table 10.5). Mostly the new land cropped with cane was an extension of areas being already farmed. Sometimes the land had been considered marginal and was brought under cultivation after modifications. A large drainage scheme at Garradunga, near Innisfail, for example, allowed waterlogged land on eleven farms to be reclaimed and cropped for the first time. To a lesser extent, new farms were established, often close to other existing farms. One area which received a major boost in its farm numbers during the mid-1960s expansion was the Herbert River district, where 122 new farms were established between 1964 and 1966. Fifty-four new farms were also created in the Riversdale district, southwest of Tully, thereby continuing the occupation of this locality, which commenced in the 1950s. Often the land that became these new farms was already partially or fully cleared prior to the commencement of cane growing, having been used for cattle grazing. A different scenario occurred in the Lower Burdekin district, where cane assignments were awarded to 200 former tobacco growers at Clare, Millaroo and Dalbeg. These three small settlements had been originally developed by the Commonwealth and Queensland governments in 1952 for the establishment of ex-servicemen on irrigated tobacco farms, except tobacco growing had proved unprofitable when difficulties were encountered selling the tobacco leaf.[82] The New South Wales section of the Australian sugar industry also agreed to participate in this expansion, and an additional 4 000 ha were brought under sugar cane between 1964 and 1966 (Table 10.6). Part of this increase came from landholders, mainly ex-dairy farmers, who commenced supplying cane, sometimes for the first time, to Broadwater and Harwood Mills.[83]

According to the historian Linda Hungerford, the atmosphere of 'expectancy and optimism' that accompanied the 1963–1964 expansion was 'rapidly extinguished' in the mid-1960s. The Nicklin government had acted unwisely by failing to secure contracts before approving any expansion. Australia was not the only sugar-producing country to expand its sugar output during the mid-1960s. Inevitably, surpluses accrued on the world sugar market, unregulated since the lapse of the ISA of 1958. Average export prices fell in 1964 and 1965; they then crashed in 1966, averaging at just under 2 US cents/lb, their lowest post-war level to date (Figure 10.3). Consequently, the average return per ton of sugar received by Australian sugar mill-owners

Table 10.6. Sugar production statistics for New South Wales, 1964–1974.

Year	Area under cane (ha)	Holdings with sugar cane cut for crushing	NSW Mill Peak Allocation (tonnes of sugar)	Tonnes of sugar produced	Percentage of allocation
1964	12 273	594	106 393	96 718	91.0
1965	15 066	595	132 976	71 108	53.5
1966	16 136	603	136 346	142 206	106.0
1967	16 862	685	136 433	122 512	89.8
1968	16 779	672	136 484	122 307	89.6
1969	16 751	655	136 496	103 584	72.7
1970	16 125	605	136 496	149 227	109.3
1971	17 246	619	136 496	123 812	90.7
1972	17 016	570	138 065	102 940	74.6
1973	16 050	554	138 073	121 139	87.8
1974	16 973	549	151 279	121 007	80.0

Source: Based upon data in *New South Wales Official Year Book*, various issues, and *Year Book of Australia*, various issues.

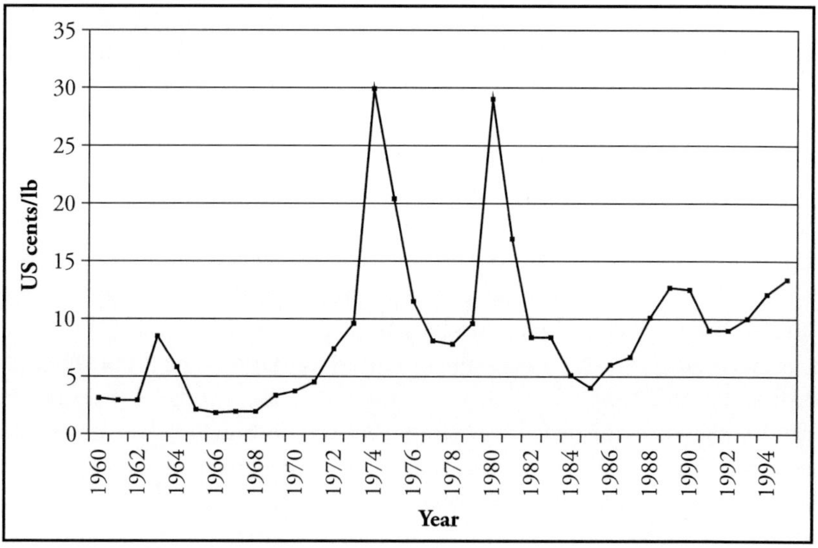

Figure 10.3. World average sugar prices (US cents/lb) by calendar year, 1960–1995.

Source: Based upon figures supplied by the Economic Research Service, United States Department of Agriculture, http://ers.usda.gov/Briefing/Sugar/Data.htm

and canegrowers fell sharply from just under $130 in 1963 to $75 in 1966. This price collapse, together with a serious state-wide drought in Queensland during 1965, left many of the state's canegrowers facing financial ruin, while nine of Queensland's sugar mills approached the Queensland government for a financial rescue operation. The Queensland State Opposition Leader, John Duggan (Labor), accused the Nicklin government of 'reckless optimism' in approving such a large increase in sugar production. Sir Francis Nicklin appealed to the Commonwealth government (Menzies; Coalition of Liberal and Country Parties) for financial assistance to supplement the price paid for raw sugar. Funds borrowed by the Sugar Board in the first instance from the Reserve Bank of Australia, to augment the sales proceeds from the 1966 and 1967 crops, were repaid to the Reserve Bank of Australia from a Commonwealth government loan of $19 million to the Sugar Board, on the condition that it was paid back over the next thirteen years.[84] In view of the financial difficulties being faced by the Queensland sugar mill-owners and canegrowers, the Central Board halted the expansion in 1966 and declined to increase mill peaks, except for some minor adjustments.

Between 1968 and 1974, the international sugar market again experienced significant changes, which all impacted upon the Australian sugar industry. After several years of protracted negotiations, most of the world's major importers and exporters of sugar reached a new ISA in October 1968. It became operational on 1 January 1969 for a period of five years and aimed to stabilise prices between 3.5 and 5.25 US cents/lb. Australia was allotted a basic export tonnage of 1.1 million tonnes.[85] The new arrangements brought some stability to the price of sugar sold on the world free market, and by the early 1970s the average world price had crept above 5 US cents/lb (Figure 10.3), and the average return per tonne of sugar to Australia's canegrowers and mill-owners had climbed above $100.

Because of ISA restrictions, a substantial quantity of sugar cane was left unharvested in Queensland at the closure of the 1968, 1969 and 1970 seasons, as the state's sugar mill-owners tried to produce only their mill peaks. Some of this cane had to be destroyed.[86] The uncertainty over the world sugar price and the future of Australia's markets for sugar prompted the Central Board to leave the mill peaks unchanged between 1968 and 1973, except for minor adjustments (Table 10.5). By 1971, the sugar price had recovered and the world surplus had been eroded to such an extent that a global shortage of sugar loomed. During 1972, the world sugar price rose rapidly, and remained high during much of 1973. During this period, both New South Wales and Queensland canegrowers consistently produced large crops (except for the drought-ravaged crops of 1969 and 1973) and record amounts of sugar (Tables 10.5 and 10.6). Fortunately for Australia, the ISA quotas were suspended following the failure of the ISA to be renewed in 1973, and the Sugar Board exported all this excess sugar. Moreover, the improved sugar prices had assisted some Australian canegrowers reduce their debts accumulated during the mid-1960s. However, the high level of inflation in the early 1970s and rising labour and fertiliser costs meant some canegrowers were forced to take off-farm employment to remain viable; others with substantial debts were forced out of the industry.[87]

Gloom to boom and gloom again – 1974 to 1984

In order to capitalise on the rising sugar prices, Victor Sullivan, Queensland's Minister for Primary Industries, accepted the Central Board's decision in February 1974 to raise mill peaks by 213 000 tonnes to an aggregate of 2.44 million tonnes (Table 10.7). Further increases in the world sugar price during 1974 prompted Minister Sullivan to announce on 8 November 1974

Table 10.7. Sugar production statistics for Queensland, 1974–1984.

Year	Area under cane (ha)	Farms producing cane harvested for crushing	Sugar production (tonnes of 94 n.t. sugar)	Mill peaks (tonnes of sugar)	Percentage of mill peaks
1974	278 413	6404	2 727 533	2 440 000	111.8
1975	321 143	6357	2 751 414	2 490 000	110.5
1976	338 930	6294	3 163 239	2 790 000	113.3
1977	341 985	6191	3 209 255	2 793 500	114.9
1978	330 160	6105	2 748 858	2 793 970	98.4
1979	336 712	6040	2 807 477	2 794 380	100.5
1980	354 874	5929	3 148 955	2 888 200	109.3
1981	372 993	5844	3 250 331	3 080 000	105.5
1982	369 680	5759	3 325 115	3 171 200	104.9
1983	366 937	5701	3 011 590	3 171 760	94.6
1984	368 778	5650	3 349 205	3 181 660*	105.3

* For the 1984 season only, Rocky Point's mill peak was increased by 9 000 tonnes.

Source: Based upon figures in Australian Bureau of Statistics, *Crops and Pastures, Queensland,* Cat. No. 7321.3 & Australian Bureau of Statistics, *Queensland Agricultural Sector: Section 2 – Crops,* Cat. No. 7302.3.

that the time had come to seriously consider another industry expansion. After seeking the industry's views which supported increasing its productive capacity, Sullivan asked the Central Board on 25 November 1974 to increase the assigned area sufficiently to expand sugar production by up to 300 000 tonnes of sugar. Regional assignment committees considered applications for increased assignments from existing canegrowers only, as the aim was to increase production quickly. By April 1975, the Central Board had allocated new assignments, which would increase the assigned area from 297 257 ha to 325 000 ha by 30 June 1976, thereby permitting the production of the entire 300 000 tonnes. Mill peaks were increased by 50 000 tonnes in 1975 on the grounds that the 'sales of sugar on the world free market at reasonable prices appeared bright'. They would then increase by another 300 000 tonnes in 1976. In addition, the Central Board permitted the utilisation of 100% of an assignment, so in essence the net area became the gross area.[88]

As the expansion program commenced, optimism gave way to foreboding. Manufacturers of sugar-based products sought less expensive sweeteners and demand for sugar fell in developed countries.[89] In early 1975, the average world price of sugar plummeted, coming to rest in June at about a third of its 1974 high, and then remained at under 10 US cents/lb for the next two years

(Figure 10. 3). Despite this trend in sugar prices, the Queensland sugar industry proceeded with its planned expansion, producing a record 3.209 million tonnes of sugar in 1977. Average yearly returns from the overseas sale of Australian sugar, however, fell sharply from $267 per tonne in 1975 to $198 per tonne in 1978, given that 80% of the country's sugar was now exported (Figure 10.2). The price fluctuations of the mid-1970s prompted the world's major sugar exporters and importers to try and resurrect the ISA in 1977. Prolonged and difficult negotiations culminated in October 1977 and the new five-year ISA came into force on 1 January 1978. The agreement aimed to stabilise raw sugar prices initially in the target range of 11 to 21 US cents/lb, with a final range being 13 to 23 US cents/lb. Australia did not fare well during the negotiations, having its basic export tonnage (BET) set at 2.35 million tonnes. In addition, during the first three years of the ISA, Australia was required to establish uncommitted buffer stocks of 400 000 tonnes of sugar, which could only be progressively released onto the market when the price reached 19 US cents/lb. The Australian sugar industry felt it could not afford to produce sugar which cost to be stored, but was not necessarily assured a market.[90]

As a result of the new ISA quotas, the Sugar Board restricted its sugar purchases during 1978 and 1979, forcing canegrowers to leave about 4.0 million tonnes of cane unharvested. Yet the ISA had a stabilising influence on world sugar prices; they began rising during 1979, slowly at first, and then sharply in early 1980. In an attempt to curb the rising price, caused by a sugar shortage, the ISA was temporarily suspended. The lifting of quota restrictions, however, failed to slow demand, and during 1980 it became necessary to release the buffer stocks that had been established during 1978 and 1979. Australia sold its entire stockpile of 326 000 tonnes of sugar, providing canegrowers with a financial windfall, since the stocks were sold at considerably higher prices than when they were placed into storage.[91] Average yearly returns from the export of Australian sugar reached $302 per tonne in 1979, prompting the Queensland government to accept the Central Board's recommendation to lift mill peaks by 93 820 tonnes of sugar in March 1980 (Table 10.7).

Following further increases in the world sugar price during early 1980, Victor Sullivan, Queensland's Minster for Primary Industries, directed the Central Board to hold an inquiry into the practicability of further increasing production in 1981 to reap the benefits from the high world sugar prices. Submissions to the Central Board were divided in their views, with some Queensland sugar mill-owners urging only a moderate expansion; others were opposed to any expansion (e.g. Mulgrave and South Johnstone Mills). The Proprietary Sugar Millers' Association suggested the Australian sugar industry should not have 'a knee-jerk reaction to high world prices which follow sudden shortfalls in supplies'. The Central Board recommended an interim 5% increase in the size of the total assigned area, allocated to existing canegrowers only, and adjourned the inquiry until December when it was hoped a clearer picture would have emerged about future markets for sugar. However, this decision was not unanimous, with concern being expressed about the impact upon canegrowers when the ISA quotas were re-imposed. Nevertheless, Minister Sullivan granted the increase in August 1980 and the successful canegrowers were given immediate permission to proceed with planting.

The Central Board inquiry resumed on 15 December 1980 and was held concurrently with the mill peak hearings for 1981. Many submissions suggested an increase of up to 100 000 tonnes of sugar in addition to the 150 000 tonnes rise covering the July expansion. Marketing advice suggested that secure outlets could be obtained for the additional sugar. Opinions were divided over the introduction of new farmers into the industry. In a departure from usual procedures, the Australian Conservation Foundation was granted permission to present evidence to the Central Board for the first time. This organisation argued for a restriction on any expansion of cane growing onto freshwater wetlands in Far North Queensland. The Central Board's recommendations, subsequently approved on 23 December 1980 by the new Queensland Minister for Primary Industries, Michael Ahern, permitted a further 3% expansion in the assigned area.[92] Mill peaks in 1981 and 1982 were increased by 283 000 tonnes to accommodate the increased production that was projected to occur on the newly assigned land (Table 10.7).

Under the 1980 expansion, the assigned area in Queensland had been lifted from 354 809 ha to 364 600 ha by the end of 1981. Existing canegrowers from all mill supply areas were invited to apply for increased assignments and most of the expansion came from increased assignments. Yet to facilitate the expansion, the Central Board agreed to consider applications from potential new canegrowers in the Mossman, Hambledon, Mulgrave, Babinda, Goondi, Mourilyan, Cattle Creek and Qunaba mill supply areas. Sixteen new assignments, mostly located in Far North Queensland and totaling 859 ha, were granted by the Central Board, leading to the spread of cane growing to new localities. Mossman Mill, for example, acquired new suppliers at Thornton, to the north of the Daintree River, and at Julatten on the Atherton Tableland. Mulgrave Mill also acquired a new canegrower on the Atherton Tableland.[93] However, these new canegrowers, the first admitted to the industry since 1964, were to be the last to enter the industry for another seven years. Prices for sugar had declined sharply during the early 1980s, falling below 10 US cents/lb (Figure 10.3). Submissions to the Central Board from some Queensland sugar mill-owners arguing for a continuation of the expansion in order to prepare the industry to capitalise on any price rise were rejected at the mill peak hearings in 1982, 1983 and 1984.[94] Aggregate mill peaks remained unchanged for 1982, 1983 and 1984 (Table 10.7).

In New South Wales, CSR had urged its suppliers during the early 1970s to increase their output by placing more land under cane, replacing assignments not being used or lost to other uses (e.g. residential; road developments) and by improving production per hectare. Areas abandoned for cane growing were partially offset by a revival of interest in cane growing in the elevated country covering the Rous and Alstonsville areas between Ballina and Lismore, while reclaimed land was opened up for cane growing in the Newrybar Swamp region, to the north of Ballina, following the drainage of the freshwater wetlands in this district. Land reclaimed from wetlands drained by flood mitigation works in the Tweed River district during the early 1970s was also placed under sugar cane.[95] CSR's efforts at encouraging new cane suppliers, combined with the higher sugar prices in the early 1970s, contributed to the steady increase in the area under sugar cane in New South Wales during the late 1970s (Table 10.8). Moreover, the pattern of New South Wales sugar production being under its mill peak equivalent was reversed after 1978. Following negotiations between New South Wales sugar industry representatives and the Sugar Board, New

Table 10.8. Sugar production statistics for New South Wales, 1974–1984.

Year	Area under cane (ha)	Number of canegrowers	Sugar production (tonnes)	NSW peak equivalent (tonnes)	% of mill peaks
1974	18 163	549	121 008	151 280	80.0
1975	19 266	584	103 962	151 300	68.7
1976	20 528	584	132 082	151 300	87.3
1977	23 623	581	134 190	151 300	88.6
1978	23 734	584	152 623	151 300	100.9
1979	21 330	567	155 637	151 300	102.9
1980	23 474	622	181 049	161 600	112.0
1981	25 220	636	184 522	177 800	103.8
1982	26 571	639	211 677	183 200	115.5
1983	27 135	635	159 002	183 200	86.8
1984	26 996	579	198 886	183 200	108.6

Source: Based upon figures in *ASYB*, 1986, p. 201 and *Year Book of Australia*, various issues.

South Wales received increased mill peak equivalents in 1980, 1981 and 1982. Further growth in sugar output and the area under sugar cane continued in New South Wales during the early 1980s (Table 10.8).

By the mid-1980s, Australian sugar production had reached record levels, but the average world price for sugar during 1982 was 8.4 US cents/lb; it fell even further during the next twelve months, averaging 7.78 US cents/lb in 1983 (Figure 10.3). Australia's sugar industry leaders hoped that the negotiations to renew the ISA, due to expire in late 1984, would resolve the crisis. Major conferences in 1983 and 1984, however, failed to reach a consensus, and the ISA was pronounced dead, mostly because of the unwillingness of Cuba and the European Economic Community (EEC) to meet international demands for production restraint. Upon the announcement of the ISA's demise, sugar prices fell even further, dipping to 4.2 US cents/lb in December 1984. The Australian sugar industry was deeply disturbed by these developments, fearing unrestricted production would lead to the dumping of surpluses on the market, causing further price falls, and a possible erosion of Australia's markets.[96]

The attitude of the Commonwealth government towards the Australian sugar industry had also changed during the 1970s. Some federal parliamentarians questioned the persistence of the Commonwealth government's embargo on sugar imports and its role in regulating the domestic sugar price (see Chapter 18 for full details). Consequently, the Fraser government (Coalition of Liberal and National Parties) asked the Industries Assistance Commission (IAC) on 12 November 1982 to conduct an inquiry into the level of assistance that should be provided to the Australian sugar industry. The IAC held two rounds of public hearings. The first round in December 1982 considered short term assistance to the industry and the price to be received for sugar sold domestically. Much to the Australian sugar industry's disappointment, the IAC did not recommend any short term assistance to the industry, nor a substantial increase in the domestic sugar price for the remainder of 1983 and 1984. However, the Commonwealth government announced on 7 June 1983 that it would make $11 million available for special concessional carry-on loans to canegrowers – $10 million to the Queensland government and $1 million to the New South Wales government. Additional funding was also provided for the Rural Adjustment Scheme, with the expectation that some of this increase would flow to canegrowers.[97]

The second round of public hearings held by the IAC in March and April 1983 were concerned with longer term assistance and protection for the Australian sugar industry. The IAC concluded that regulations insulated some Australian canegrowers and sugar mill-owners from changing market conditions and restrained others from responding to profitable production opportunities. There were barriers to entry as well as disincentives to leave for those with limited prospects. In particular, the land assignment system encouraged the use of poor land and overuse of non-land inputs such as fertilisers, pesticides and irrigation water. Consequently, the main thrust of the IAC's report was industry deregulation and the main recommendations included the following: termination of the embargo on sugar imports, although the import duty on sugar could be maintained to minimise disruption; cessation of the administrative control over the domestic sugar price; removal of the land assignment system so that no restrictions were placed on the land used for growing sugar cane; the introduction of full negotiability and transferability of farm peaks within mill supply areas; and the abolition of reduced payments for sugar that was produced beyond mill-peaks. Such recommendations were strongly opposed by the QCGC and Australia's sugar mill-owners. Prime Minister Bob Hawke (Labor) announced in September 1984 that the Commonwealth government had no intention of implementing the report's recommendations.[98]

By mid-1984, deregulation ceased to be a priority as industry leaders and both levels of government were faced with an Australian sugar industry suffering considerable financial stress. For most Australian canegrowers experiencing marked declines in their incomes, the slump meant a substantial curtailment in expenditure; others who had borrowed heavily to participate in the expansion struggled to repay their debts. To survive, some canegrowers sold part of their land to fund continued farming operations, and/or to reduce mounting debt, some of which had been acquired expanding their output.[99] Many canegrowers reviewed their production methods, and adopted minimum tillage techniques (see Chapter 11). Queensland canegrowers also showed a greater interest in diversification, particularly in horticultural crops which could be grown on fallow land or on small amounts of unused land.[100] Other canegrowers not equipped to deal with such problems left the industry. The number of farms where cane was cut for crushing in Queensland began declining (Table 10.7).

Deregulation and rapid expansion, 1985–1995

Australian canegrowers, shocked by the IAC proposals and government inaction over financial assistance, met in Brisbane on 22 November 1984. Here they held a rowdy demonstration and a series of meetings with Prime Minister Hawke and sugar industry leaders. At one meeting, Ron Belcher, the QCGC's secretary, told the protesting canegrowers that market prospects in 1985 were so poor that 'even a tonne of gravel or sand had more value than a tonne of cane'. During the meeting with Prime Minister Hawke, he was asked to provide financial assistance to canegrowers by guaranteeing a minimum average price for up-to-peak cane based on a No. 1 Pool sugar price of $240 tonne of 94 n.t. sugar (or equivalent to 7 US cents/lb). Hawke proposed a Sugar Summit, which eventually occurred in Canberra on 1 April 1985. At this meeting the Queensland and Commonwealth governments and the Australian sugar industry agreed to prepare within the next 100 days a Sugar Industry Plan for assistance and restructuring.[101]

Russell Savage, a Babinda district canegrower, chaired a group of sugar industry representatives and senior Commonwealth public servants who made up the Sugar Industry Working Party. Commissioned on 3 May 1985, the Sugar Industry Working Party reported on 8 August 1985. Discussions were held with canegrowers and sugar mill-owners in Queensland and New South Wales, yet the report's main focus and recommendations related to Queensland, although they were still applicable to New South Wales. The main conclusion from the Sugar Industry Working Party's investigations was that by giving priority to equity considerations in determining industry expansion, rather than economic efficiency, the Central Board had encouraged the continued use and development of high production cost land. Hence, the Working Party recommended the abolition of the existing system of assignments and mill peaks and their replacement with a system of 'Peak Entitlements to No. 1 Pool', which would be divisible and freely transferable in whole or part between and within mill supply areas, that production decisions become the responsibility of individual producers and that sugar should not be acquired at penalty rates. However, the embargo against sugar imports should be maintained. In addition, the Sugar Industry Working Party acknowledged that Australian sugar producers had long requested price support equivalent to a No. 1 Pool return of $240 per tonne of sugar (or equivalent to approximately 7 US cents/lb for 1985). Lower sugar prices meant canegrowers slipped further into debt; for example if the price of sugar was $220 per tonne, the returns, based on 1984 costs of production, would cover approximately 43% of a Queensland canegrower's cash operating costs. The Sugar Industry Working Party, therefore, recommended price support to canegrowers in 1985, 1986 and 1987 if the No. 1 Pool price was below $220, $215 and $210 per tonne of sugar, respectively, and additional funding for household support, farm build-up and carry-on purposes.[102]

Following its analysis of the Savage Report, the QCGC advised the Queensland and Commonwealth governments that while Australian sugar producers acknowledged the need for some change, it considered that entire industry deregulation would not improve the financial problems threatening the viability of the country's canegrowers and sugar mill-owners. Moreover, the QCGC was opposed to the main thrust of the report which sought to remove large numbers of canegrowers from the industry. The Commonwealth Minister for Primary Industries, John Kerin, told a meeting of Australian sugar industry representatives in September 1985 that the Commonwealth would not allow such an important industry to collapse, but it could not expect much in the way of financial assistance if it rejected deregulation entirely. A month later, Neil Turner, the Queensland Minister for Primary Industries, promised Queensland canegrowers that the Queensland government would not force them to accept a deregulation package against their wishes. Over the next six months, the Queensland and Commonwealth governments then engaged in bitter and protracted negotiations to achieve a package of government assistance for Queensland's sugar producers. The three-year assistance package of $100 million was eventually agreed upon in May 1986, its centre-piece being price support of up to $54 million from the Commonwealth government to achieve price levels of $230 per tonne of sugar for 1985, and $225 and $220 per tonne of sugar in 1986 and 1987, respectively. The balance of the $100 million was earmarked to assist with farm and mill restructuring. In return, the Queensland government agreed to some modest reforms of the industry's regulatory structure and contributed $27 million to the Sugar Board for interest repayments and loans to cover first delivery advances.[103]

To meet its commitments under the Sugar Industry Rescue Package, the Bjelke-Petersen government (Coalition of National and Liberal Parties) in Queensland amended the *Regulation of Sugar Cane Prices Act* of 1962 to 1981 in September 1986. The legislation provided for some Central Board functions to be abolished or devolved to the local boards, thereby freeing up the previously complex approval mechanisms. Other changes recognised that canegrowers had property rights to their assignments and farm peaks and provided for the transfer of assignments within mill supply areas. All restrictions on cane farm sales were removed and transactions to sell, leave or otherwise dispose of farm peaks within a mill supply area became permissible. Canegrowers were also allowed to expand their assignments through 'roaming'. Under previous arrangements, canegrowers usually followed a crop rotation system that fallowed about 15% of an assignment each year. Effectively, 'roaming' allowed canegrowers to harvest an additional area equal to what they fallowed each year, as long as they sought annual approval from their Local Board to engage in the practice and provided they identified and harvested an area not exceeding the original 100% of their assignment in any one season. Further changes introduced in 1989 included reducing the size of the Central Board from five to three members, permitting the Central Board to reassign cancelled assignments (prior to the 1986 amendments) anywhere in the State, although mill peaks would not follow the assignments, and the introduction of seasonal permits to use unassigned land in other mill supply areas.[104]

The introduction of 'roaming' permits enabled some farmers to take advantage of unused land within or even outside their mill supply areas, although the Australian Bureau of Agricultural and Resource Economics concluded in 1991 that few canegrowers used the 'roaming' provision during the late 1980s. Between June 1985 and July 1988, the Central Board had also re-allocated 2 381 ha of cancelled assignments within the mill supply areas where it originally applied, although in mid-July 1988 it still had 3 534 ha to allocate.[105] These actions contributed to an additional 7 000 ha being brought under cane between 1986 and 1988, although the mill peaks remained unaltered, but the number of canegrowers in Queensland continued falling (Table 10.9).

Some of this reduction in the number of canegrowers occurred as a result of farm amalgamations, assisted by way of interest subsidies and grants under the Sugar Cane Growers' Adjustment Scheme.[106] Other canegrowers, however, sold their properties and the land was converted to different activities. Between 1980 and 1990, 7 550 ha of assigned sugar cane growing land in Queensland was lost to competing activities such as golf courses, hobby farms, residential estates or better paying crops. Far North Queensland was the worst hit. Land used to grow sugar cane for Hambledon and Goondi Mills started being lost to residential developments during the early 1970s, and this loss gathered pace during the 1980s, particularly as Cairns City expanded. Some Innisfail district canegrowers sought better incomes by switching to growing bananas and/or paw paws.[107] In the Mackay district, Farleigh, Racecourse and North Eton Mills were the most affected by the expansion of Mackay City. In Southern Queensland, the largest conversion occurred in the Fairymead and Bingera mill supply areas as canegrowers moved land out of sugar cane into horticultural crops, mostly tomatoes. By 1990, Rocky Point Mill had also lost so much cane growing land to urban expansion that it was considered to be 135 ha short of the minimum requirement to be economically viable.[108]

Table 10.9. Sugar production statistics for Queensland, 1985–1995.

Year	Area under cane ('000 ha)	Assigned area ('000 ha)	Annual increases in assigned area (%)	No. of cane-growers	Sugar production ('000 tonnes of 94 n.t.)	Mill peaks ('000 tonnes of 94 n.t.)
1985	356.6	359.6	+	5 561	3 207.7	3 171.7
1986	354.8	359.6	+	5 999	3 208. 5	3 171.7
1987	359.8	359.5	+	5 913	3 244.6	3 171.7
1988	3668	359.4	+	5 828	3 482.2	3 171.7
1989	370.8	362.6	5.0	5 736	3 617.2	3 171.7
1990	382.6	372.1	8.0	5 672	3 353.3	3 171.8
1991	382.9	395.2	2.5	5 686	2 929.7	3 171.8*
1992	384.3	420.0	5.5	5 956	4 015.7	3 171.8*
1993	404 .1	442.9	5.0	6 063	4 149.6	3 171.8*
1994	414.2#	466.8	3.6	6 237	4 821.0	3 171.8*
1995	428.3#	477. 3	3.4	6 364	4 744.7	3 171.8*

Notes: + Automatic increases under 1991 sugar industry reforms not yet operational. # Estimates due to the agricultural census not being held. * Mill peak system inoperable under sugar industry reforms of 1991.

Source: Based upon figures in the Australian Sugar Milling Council, ARs 1989–1995 and *ASYB*, various issues.

World sugar prices slowly recovered during 1986. The 1985 season No 1. Pool return (i.e. up-to-peak) for Queensland sugar mill-owners was increased by government assistance from $223 to $230 per tonne of sugar. No assistance was needed in 1986 or 1987 as the up-to-peak returns had increased to $282 and $289 per tonne, respectively. However, the average price for sugar was still under 10 US cents/lb during 1986 and 1987 (Figure 10.3). Some Australian canegrowers, despite receiving better returns, continued struggling financially and the QCGC complained that the Commonwealth government (Hawke; Labor) had been very slow in delivering its assistance funds, providing only $5.02 million for grower assistance and $3.779 million for miller assistance during 1986/87.[109] However, concern over the Commonwealth government's tardiness in supplying financial assistance was swept aside in May 1988, when the Federal Treasurer, Paul Keating, announced, without industry consultation, that the Commonwealth government was changing the marketing arrangements for a number of agricultural industries, including the sugar industry, by reducing industry protection. The seventy-four year embargo on sugar imports would be removed on 1 July 1989. In its place, the Commonwealth government proposed ad valorem tariffs which would be reduced to an effective 15% by 1 July 1992.[110]

In response to this shock announcement, the Queensland and New South Wales premiers and Australian sugar industry leaders lobbied the Commonwealth government to delay removing the embargo, claiming it would allow cheap, subsidised sugar to be imported into Australia, causing a profound negative impact upon the regional economies of coastal Queensland and northern New South Wales. After several months of negotiations which dragged on into early 1989, John Kerin, the Commonwealth Primary Industries Minister, reached an agreement with the Australian sugar industry that for the period 1 July 1989 to 30 June 1991, the embargo on sugar imports into Australia would be replaced by a tariff of $115/tonne, applied to both white and raw sugar imports. This tariff was to be reduced to $95/tonne during 1991–1992 and to $70/tonne in the following year. Further levels of protection were to be determined within three years, following an Industry Commission review of the industry. However, these arrangements were altered in March 1991, when the Commonwealth government, again without industry consultation, announced that the tariffs would be reduced to $76/tonne from 1 July 1991 and to $55/tonne from 1 July 1992.[111]

As the Australian sugar industry responded during 1988 to the proposed major changes to its marketing arrangements, the world sugar price fluctuated between 8 and 11 US cents/lb, except for a brief mid-year period when it hit 15 US cents/lb. The better price for sugar and a growing demand for sugar in developing countries prompted the Central Board to conduct a hearing in December 1988 into whether another expansion was warranted in Queensland. In their report to Neville Harper, the Queensland Minister for Primary Industries, the majority of the Central Board favoured a 5% expansion in the assigned area, but without an increase in mill peaks. Mill capacity was sufficient to accommodate the increase by means of little capital expenditure and the use of continuous or overtime crushing arrangements. The Central Board acknowledged that some canegrowers may be unwilling or unable to take part in such an expansion, and this recommendation was prejudicial to them. In February 1989, Neville Harper accepted the Central Board's recommendation and the Queensland section of the Australian sugar industry commenced implementing another expansion.

As this expansion commenced, the world sugar price increased even further, hovering between 11 and 15 US cents/lb during much of 1989. In October 1989, the Australian Sugar Milling Council wrote to Neville Harper requesting another review of the productive capacity of Queensland's sugar producers prior to the commencement of the 1990 season, as there were firm market indications that additional sugar could be exported, if available. Subsequently, the Central Board in April 1990 heard submissions about increasing sugar production in Queensland. The majority of the industry submissions favoured a further significant expansion in the industry's productive capacity up to an annual output of 4.0 million tonnes of sugar. The Central Board accepted this proposition, and calculated it would be made effective by an 8% expansion in the assigned area (or approximately 30 000 ha). Its recommendation of an expansion was approved by Neville Harper on 8 June 1990 on the condition that the expansion was implemented quickly. In both the 1989 and 1990 expansions, mill peaks remained unchanged, so all the additional sugar was destined for No. 2 Pool.[112]

The expansion permitted in 1989 and 1990 contributed to a modest growth in the area under sugar cane in Queensland (Table 10.9). Nearly 250 new farmers commenced growing sugar cane in Queensland (i.e. 15 new growers in 1989 and 232 new growers in 1990), the first significant upsurge in new Queensland canegrowers since 1965. However, the expansion was unequal throughout the state's sugar-producing regions. The biggest growth in the assigned area occurred in the Herbert River, Lower Burdekin and Mackay regions. These districts still had a considerable amount of unused land for development, and new farms suitable for growing sugar cane were being established in the Burdekin River and Eton Irrigation Areas (see Chapter 12). Small crop farmers, particularly in the Maroochy district, diversifying into sugar cane secured the bulk of the new assignments in Southern Queensland. The smallest increase in assigned areas occurred in Far North Queensland. Only one application for a new assignment was received from the existing canegrowers supplying Mulgrave, Hambledon and Babinda Mills, as most land in these mill supply areas was already committed. The small expansion came mostly from ex-rice farmers near Mareeba diversifying into sugar cane when the returns from rice fell below production costs. Their cane, however, had to be carted almost 100 km to Mossman Mill for processing.[113] Yet by late 1991 the allocation of these assignments was too oversubscribed, with new canegrowers taking up only 40% and 26% of their 1989 and 1990 allocations, respectively, so that 34 800 ha of assignments had not been made effective.[114] The drought conditions in some districts and the drop in the average world sugar price to under 10 US cents/lb during 1991 and subsequent reduced average returns had clearly dampened the enthusiasm for outlaying large capital expenditure on developing new land for cane cultivation.

As the expansion commenced during 1989, the Queensland sugar industry was subjected to more upheaval following the election of the Goss (Labor) government in December 1989. Another inquiry was established by Ed Casey, the new Minister for Primary Industries, who was concerned that parts of the Savage Report, especially those dealing with the organisational structure of the industry, had not been implemented. The Sugar Industry Working Party, consisted of E. Fitzpatrick, former Deputy Secretary of the Commonwealth Department of Primary Industries and Energy, Don Watson, Chairman of Directors, Mossman Mill, and Fred Soper, former Chairman of QCGC 1982–1989. They were given two tasks: considering ways to further restructure the Queensland sugar industry; and reconsidering the recommendations of the Savage Report. The Working Party recommended the formation of a new sugar industry organisation, the Queensland Sugar Industry Corporation, operable from July 1991. On the basis of the marketing advice it received, the Working Party concluded that the Australian sugar industry needed to expand if it wanted to exploit future market opportunities. The global sugar market, however, had been characterised by cycles of shortages and surpluses with corresponding peaks and slumps in the world sugar price. In view of this pattern and the probability of a surplus on the world market following the current shortage, the Working Party considered that future expansions should be in the form of small yearly increases, to keep pace with global demand, instead of larger increases every few years which had characterised the Australian sugar industry since 1945. The Working Party recommended the area of assignments in Queensland should be increased by 3% each year, unless the proposed replacement for the Sugar Board, the Queensland Sugar Industry Corporation, after consultation with the industry, decided on a larger or smaller increase in any one year on the basis of available markets. Up to 50% of this expansion should be allocated to new farms within existing mill supply areas, although the Working Party's preferred option was for the Queensland government to release land serviced by irrigation for this purpose. Assignments were to be freely transferable where agreement could be reached between the mills and mill suppliers' committees.[115]

Following the release of the Sugar Industry Working Party's report and further consultation with the industry, the Goss government enacted the *Sugar Industry Act* of 1991, which repealed or replaced nearly all the thirty-seven existing laws pertaining to sugar production in Queensland. The

Central Board was merged with the Sugar Board to form the Queensland Sugar Corporation, which assumed the responsibilities for regulating the production and marketing of sugar in Queensland. The legislation specified that the assigned area in Queensland should increase by 2.5% yearly up to and including 1995, or by such greater proportion as the Minister for Primary Industries determined after consultation with Queensland's sugar producers. This guaranteed expansion provision replaced the time consuming and costly Section 12A mill peak hearings involving the Central Board, sugar mill-owners and canegrowers when an expansion was being sought by the industry. In addition, the 'roaming' provisions were modified so that canegrowers could roam up to 100% of their assignments within the boundaries of their properties, thereby allowing the adoption of less exploitive rotational practices. Sugar produced from cane grown on unassigned land no longer received a penalty price of $1/tonne, but attracted a price determined by the Queensland Minister for Primary Industries. Effectively, the emphasis on production controls shifted from the mill and farm peak system to the land assignment system. Mill peaks were preserved and carried forward from the repealed legislation, but they became effectively frozen. Transfers of assignments and farm and mill peaks were permitted under the new legislation. These new arrangements provided canegrowers and sugar mill-owners increased flexibility to vary production in response to market signals. These changes were also accompanied by the introduction of *The Sugar Milling Rationalisation Act* of 1991 which set out detailed procedures to be followed by the Queensland Sugar Corporation when a Queensland sugar mill was to be closed.[116]

As the above legislative changes were being implemented, the Industry Commission (successor to the IAC) commenced another inquiry in March 1991 into the Australian sugar industry, especially the level of tariff protection that should exist from 1 July 1992. The Industry Commission's Report, delivered in April 1992, acknowledged that the Australian sugar industry competed very successfully on world markets, but its growth and performance was being impeded by one of the most restrictive regulatory regimes of any Australian industry. The assignment system was blamed for encouraging the use of marginal land at the expense of land in other areas better suited to cane growing, and encouraging practices such as greater use of pesticides and fertiliser per hectare and reduced fallowing, which may have had adverse environmental impacts. In contrast, regulations which constrained the activity of many other rural industries (e.g. grains, cotton, peanuts) had been relaxed or removed entirely. Therefore, the Industry Commission strongly recommended further deregulation for the Australian sugar industry, suggesting the following: tariff protection for Australian sugar producers should be gradually reduced from 1992 onwards until its abolition in 1998; the sole marketing powers of the Queensland Sugar Corporation should end in 1997; and the total area of assigned land should be allowed to increase each year by 5% until the cessation of the assignment system following the 1995 season. The Industry Commission considered the principal social consequences of implementing its recommendations would be an expansion of cane growing in the Lower Burdekin, Herbert River, Proserpine and Tully districts.[117]

Australian sugar industry leaders and the Queensland government rejected such radical deregulation, arguing it would have a negative economic and social impact on hundreds of coastal communities in eastern Australia. Simon Crean, the Federal Minister for Primary Industries and Energy, announced he was ambivalent about the thrust of the Industry Commission's report, and that his aim was to develop a growth strategy for the Australian sugar industry. Consequently, Simon Crean announced in July 1992 the formation of a Sugar Industry Task Force to advise him on impediments to sustainable growth, investment and competitiveness at all levels of the Australian sugar industry, ways to overcome these impediments and the appropriateness of future government support for the sugar industry, including tariff arrangements. The Chairman of the Task Force was Brian Courtice, Chairman of the Prime Minister's Country Task Force, member for the electorate of Hinkler and a canegrower. The Sugar Industry Task Force considered submissions from interested parties, but held no public hearings. According to the economist Geoff Edwards, the Task Force's report, delivered to the Commonwealth government in late December 1992, contained a 'shortage of recommendations on key issues' and conflicting views, reflecting the different interests of the task force members. The Task Force could not agree on whether it was in the industry's best interests to maintain the tariff and the continued compulsory acquisition of all raw sugar manufactured in Queensland, but its members did agree that a 'single desk' seller of Queensland raw sugar should be retained and that a significant shift from industry level to local level should occur in the allocation of responsibility for the expansion in the area under cane.[118]

Simon Crean acted very quickly after receiving the Sugar Industry Task Force's report, probably because of the possibility of a Federal election, which did eventuate on 13 March 1993. Even before the Sugar Industry Task Force's report had been made public, Crean, after consultation with the Queensland government and sugar industry representatives, announced on 2 February 1993 a package of initiatives 'designed to ensure the future growth of the Australian sugar industry'. The 'Agreed Sugar Industry Package of 1993' contained provisions for the retention of the tariff on imported sugar at $55 a tonne until mid-1996. To improve the economics of cane growing and transportation of cane, thereby facilitating further industry expansion, the Commonwealth and Queensland governments both agreed to allocate $20 million over four years towards supporting new infrastructure projects. In addition, both levels of government agreed that any future expansion of the Australian sugar industry would be largely determined at the local level. Assignments were to continue, but they were not be used to constrain growth. Harry Bonanno, Chairman of QCGC, noted that the current industry expansion was for the first time 'a bottoms up process fully involving growers and millers at the local level' and that it allowed Queensland's canegrowers and sugar millers to make sounder investment decisions. Previously, there had been a tendency for the state's canegrowers to 'grab every expansion opportunity which became available even if it did not suit them'.[119]

The inquiries into the Queensland sugar industry during the early 1990s were conducted against the backdrop of an industry embarking upon another major expansion. The granting of new assignments associated with this expansion was guided by the *Sugar Industry (Assignment Grant) Guidelines*, which were established under Section 29(b) of the *Sugar Industry Act* of 1991. For the first time since the introduction of cane price-fixing legislation in 1915 (see Chapter 17 for more details), a set of principles guided the local boards in their decision making, after they were delegated much of the responsibility for receiving and determining applications for assignments by the Queensland Sugar Corporation. For assignments to be granted the land had to be capable of producing commercial crops of sugar cane when subjected to correct agricultural procedures, situated in a location from which sugar cane could be efficiently and cost effectively delivered to a sugar mill and be used for the growing of sugar cane without undue damage to the environment.[120]

These more relaxed regulatory conditions also coincided with an increase in the average world sugar price to above 10 US cents/lb and improved average returns to Queensland canegrowers (Figure 10.3), so existing and potential canegrowers showed considerable interest in obtaining new assignments. Demand for new assignments was so strong that every year between 1992 and 1995, Ed Casey, the Queensland Minister for Primary Industries, approved increases in the assigned area in excess of the 2.5% increase permitted under the *Sugar Industry Act* of 1991. The assigned area rose from 395 000 ha in 1991 to 477 300 ha in 1995 – a 21% increase – and was accompanied by a 20% increase in the area under sugar cane. Consequently, raw sugar production also grew by 40%, rising from 3.3 million tonnes in 1990 to 4.7 million tonnes in 1995 (Table 10.9). Yet this rapid growth in the area under cane in the early 1990s attracted the ire of the Australian Conservation Foundation, which criticised the dream of 'unfettered growth in output', noting that it failed to respect the environmental limitations in which the Australian sugar industry operated. The Australian Conservation Foundation suggested a more prudent and responsible approach would have been to focus the industry's efforts on production that was both ecologically sustainable and profitable.[121]

The expansion in the Queensland sugar industry during the early 1990s, however, was concentrated in particular localities, because some mill supply areas had little or no land left for any expansion. The biggest increase occurred in the Lower Burdekin district, and was associated with the release of 100 irrigated farms in the Burdekin River Irrigation Area and existing Burdekin rice growers switching to sugar cane. A 40% increase in the assigned area occurred in the Herbert River district, and was concentrated mostly on country formerly used for grazing or small crops at Mutarnee, 35 km south of Ingham. In the Mackay district, new cane expansion occurred between Koumala and Carmila, south of Sarina, west of Proserpine and to the north of Farleigh.[122] In Far North Queensland, the modest increase in the assigned area was located in quite disparate localities. Some expansion occurred on the Atherton Tableland, and was associated with ex-tobacco farmers situated at Tolga and Walkamin and Arriga, to the west of Mareeba, cultivating sugar cane as their new livelihood. Cane growing also spread to the Murray Upper locality, south of Tully, following the subdivision of the Tully River (formerly King Ranch) cattle station into farms during the early 1990s.[123] The least expansion occurred in Southern Queensland and was concentrated south and south-east of Gin Gin, around Baffle Creek and in the Maryborough area, where some pineapple farmers at Goomoorian near Gympie diversified by commencing the growing of sugar cane.[124] Approximately 500 new farmers started growing sugar cane during this expansion, thereby reversing the slump in canegrower numbers that had occurred during the late 1980s (Table 10.9).

Like their Queensland counterparts, the New South Wales canegrowers during the early 1980s suffered major financial difficulties, with some farmers on small properties seeking off-farm incomes to remain financially viable. In early 1985, the New South Wales Sugar Milling Cooperative Ltd. (hereafter NSW Sugar Milling Coop.), the purchaser of CSR's New South Wales sugar mills, requested financial assistance from the New South Wales and Commonwealth governments. Before committing to a financial rescue package, the New South Wales government appointed a New South Wales Sugar Industry Study Group to assess the long term viability of the New South Wales sugar industry.

This committee of inquiry under the chairmanship of Mr B. Scarsbrick, the Regional Director of Agriculture, North Coast Region, presented its report to the New South Wales Minister for Agriculture and Fisheries on 29 August 1985. Recommendations included short term price support for both the miller and canegrowers equivalent to what was suggested in the Savage Report, Rural Assistance Scheme funds to support adjustment to projected lower sugar prices and a government guarantee for a $2.5 million loan to the NSW Sugar Milling Coop. to assist its operation during 1986. A little over a month later the Commonwealth and New South Wales governments agreed to assist the New South Wales Sugar industry financially. The Commonwealth government (Hawke; Labor) contributed $8 million while the New South Wales government (Wran; Labor) gave $4 million towards rural and infrastructure adjustment and price support equal to future arrangements with the Queensland government. The inquiry also seeded the idea of independence from Queensland by suggesting that the NSW Sugar Milling Coop. should investigate the manufacture of mill whites and/or the packaging and distribution of direct consumption raw sugar.[125]

Relations between the New South Wales and Queensland sections of the Australian sugar industry had also become strained during the mid-1980s. The origins of this dispute arose because of the Sugar Board's refusal to allow the New South Wales marketing quota, or access to No. 1 Pool, to increase at a rate faster than any increase in Queensland's quota and differences over transport charges for sugar moved between New South Wales and Brisbane. In addition, the New South Wales canegrowers opposed the establishment of a new bulk sugar terminal at the Port of Brisbane, about 300 km from the centre of sugar production in New South Wales, arguing instead for its erection near Harwood Mill. Consequently, the New South Wales sugar industry severed its marketing arrangements with the Queensland Sugar Board on 1 July 1989, and established a joint venture sugar refining company with the Manildra group of companies (see Chapter 18 for more details). The NSW Sugar Milling Coop. then began a ten year expansion program, encouraging existing and new canegrowers to increase their crop sizes in order to expand sugar output from around 180 000 tonnes to 240 000 tonnes.

To achieve its goal, the NSW Sugar Milling Coop. had to reverse the loss of a significant area of previously assigned caneland to non-farming activities. This trend had commenced in the mid-1970s, with the loss of land around Ballina to residential developments, and had escalated between 1980 and 1990, when 2 700 ha of caneland was transferred to other uses, particularly prawn farms, hobby farms, roads and urban development.[126] The encouragement of cane growing by the NSW Sugar Milling Coop. led to production area entitlements steadily rising in northern New South Wales during the early 1990s, and the area under cane increased, reaching almost 30 000 ha in 1995 (Table 10.10). Part of this upsurge came from landholders around Codrington and in the Brunswick River valley (near Mullumbimby) switching from raising beef and dairy cattle to growing sugar cane.[127]

On the other side of the continent, the Western Australian government (Dowding; Labor) had clearly been monitoring developments in Queensland and New South Wales, especially the decision by the New South Wales sugar industry to pursue development independent of Queensland. Sugar production in the Ord River district of northern Western Australia had long been considered a possibility by successive Western Australian governments. Trials on growing sugar cane at the Kimberley Research Station, Kununurra, during the 1950s and 1960s had shown that the sugar cane was not affected by major pests or diseases. Yields were similar to those in the Lower Burdekin district, where irrigation was widely used in the cultivation of crops of sugar cane. However, cotton became the preferred crop in the Ord River Irrigation Area, although its failure prompted the Western Australian government (Charles Court;

Table 10.10. Sugar production statistics for New South Wales, 1985–1995.

Year	Area under cane (ha)	Production Area Entitlement (PAE) (ha)	Number of assignments	Sugar production (tonnes)	NSW peak equivalent (tonnes)	% of mill peaks
1985	25 899	**	575	170 010	183 200	92.8
1986	23 702	**	576	162 540	183 200	88.7
1987	26 411	**	574	194 808	183 200	106.3
1988	25 448	29 710	572	196 066	183 200	107.0
1989	25 448	30 398	604	179 630	183 200	98.1
1990	23 878	31 205	599	160 489	*	*
1991	24 958	31 517	620	180 415	*	*
1992	25 546	31 972	626	240 978	*	*
1993	26 324	32 940	634	218 453	*	*
1994	27 348#	34 258	637	258 740	*	*
1995	29 711#	34 214	698	244 350	*	*

Estimates due to the agricultural census not being held. * No longer applied as New South Wales withdrew from the pooling arrangement. ** Concept not introduced until 1988.

Source: Based upon figures in the Tweed River Branch of the NSWCGA, AR 2002–2003, data supplied by the Australian Bureau of Statistics, and figures in the *Year Book of Australia*, various issues.

Coalition of Liberal and National Country Parties) to ask CSR in the mid-1970s to conduct additional trials on the feasibility of growing sugar cane in the Ord River district. Results were promising, but the Western Australian government was unable to attract potential financial backers for the development of a Western Australian sugar industry.[128]

A growing demand for sugar globally eventually prompted the Western Australian government (Lawrence; Labor) in May 1990 to seek expressions of interests from organisations that might be willing to invest in the establishment of a sugar industry in the Ord River Irrigation Area. The Western Australian government received twenty-five expressions of interest, and decided to evaluate five proposals. The Ord River District Co-operative was permitted to proceed to the planning stage and by November 1993 it had convinced CSR to build a sugar mill at Kununarra. The new sugar mill, which cost $30 million, crushed its first cane from twelve farmers in November 1995, and the sugar was shipped through the Port of Wyndham, in bulk, to CSR's Cottesloe Refinery at Fremantle. Moreover, with the commencement of the Ord River Mill in 1995, sugar cane was being grown for the commercial production of sugar in three Australian states, the first time since the 1880s.[129]

In retrospect, the Australian sugar industry expanded its output for most of the time between 1915 and 1995, save for short periods when its production was restricted by the Central Board and Sugar Board or the impact of World War II. Despite the constraints imposed by the regulatory system developed to avoid overproduction, it allowed Australia to move from being a net importer of sugar to one of the world's top four exporters of raw sugar during the 1990s. This shift was achieved by bringing some new land under cultivation, and as will be seen in the next five chapters, by mechanising the cultivation of a crop of sugar cane, more intensive use of fertiliser, pesticides and herbicides, improved drainage and irrigation, better control over cane pests and diseases and the breeding of new sucrose-rich cane varieties.

Notes

1 *ASJ*, 9 April 1914, p. 68 & 4 February 1915, p. 779; *QPP*, Vol. 2 (1916–1917), pp. 835– 836 & 859–860.

2 Griggs, 'Plantation to small farm', pp. 315–316 & 325–330.

3 F. Snars, *German Settlement in the Rosewood Scrub: a pictorial history* (Rosewood, 1997), p. 32; Kerr, *Sugar at Maryborough*, p. 172; Nolan, *Bundaberg. History and People*, p. 199; R. Longhurst, *Nerang Shire. A History to 1949* (Nerang, 1994), p. 153.

4 Bolton, *A Thousand Miles Away*, p. 306; D. Murphy, *T. J. Ryan: A Political Biography* (Brisbane, 1975), p. 120; D. Shogren, 'Agriculture 1915–1929', in D. Murphy (ed.), *Labor in Power: The Labor Party and Governments in Queensland 1915–1957* (Brisbane, 1980) p. 183; for the assignment of a canegrower's land to a particular mill in 1915 see *QGG*, Vol. 105, p. 1971.

5 Easterby, *The Queensland Sugar Industry*, pp. 142–145; *CPP*, Vol. 4 (1920–21), p. 903.

6 Based upon figures in the ARs of the Queensland Government Statistician on Agricultural Production, in *QPP*, various issues.

7 BSES, AR 1919–1920, *QPP*, Vol. 2 (1920), p. 165; *ASJ*, 4 November 1921, p. 473 & 7 April 1922, p. 43.

8 Griggs, 'Plantation to small farm', pp. 376–377; BSES, AR 1922–1923, *QPP*, Vol. 2 (1924), p. 293; Wegner, 'The Hinchinbrook Shire Council, 1879–1979', pp. 162–163; *ASJ*, 8 July 1921, p. 237.

9 BSES, AR 1922–1923, *QPP*, Vol. 2 (1924), p. 295; Ev. C. McKinley (Pleystowe), p. 539, in 'Royal Commission on Sugar Peaks and Cognate Matters Evidence, 1939', in CSRR, Z/109, Box 48; A. Hudson, *By the Banyan* (Brisbane, 2000), pp. 83–85; *ASJ*, 7 July 1922, p. 227; D. Jones, *Cardwell Shire Story* (Brisbane, 1961), pp. 327–328; Mulgrave Central Mill, *50 years in retrospect 1895–1945: a souvenir of the golden jubilee of the Mulgrave Central Mill* (Gordonvale, 1947), p. 51.

10 *ASJ*, 6 January 1922, p. 578; *Year Book of New South Wales,* 1925–1926, p. 457; Harwood Mill Cane Insp., AR 1918, p. 5, CSRR, Z/109, Box 4; *Sydney Mail*, 13 January 1926, p. 36.

11 D. Menghetti, *The Red North. The Popular Front in North Queensland.* Studies in North Queensland History No. 3 (Townsville, 1981), pp. 53–54; Bolton, *A Thousand Miles Away*, p. 332; K. Manning, *In their own hands. A North Queensland Sugar Story* (Mackay, 1983), pp. 265–267.

12 D. Conomos, *The Greeks in Queensland. A History from 1859–1945* (Brisbane, 2002), pp. 325–326, 332 & 341–342; A. Grasby, *The Spanish in Australia* (Melbourne, 1983), p. 51; B. York, *Empire and race: the Maltese in Australia 1881–1949* (Sydney, 1990), pp. 62–64 & 145; Manning, *In their own hands*, pp. 262–265.

13 *QPP*, Vol. 2 (1923), pp. 192, 198, 203 & 211; for more details about Tully Mill see A. Hudson, *By the Banyan* (Tully, 2000), pp. 37–38.

14 Kerr, *A Century of Sugar*, p. 130; *ASJ*, 3 November 1932, p. 406.

15 *QGG*, Vol. 127 (1926), p. 787; W. Webb *et al.*, *Report of the Royal Commission on Sugar Peaks and Cognate Matters* (Brisbane, 1939), p. 9; Shogren, 'The politics and administration of the Queensland sugar industry to 1930', pp. 336–339.

16 D. Shogren, 'Agriculture: 1915–1929', pp. 190–191; Nolan, *Bundaberg. History and People,* p. 119.

17 *ASJ*, 11 February 1926, p. 699 & 12 April 1927, p. 69.

18 *QGG*, Vol. 129 (1927), p. 274; *QGG*, Vol. 134 (1930), pp. 1415–78; W. Webb *et al.*, *Report of the Royal Commission on Sugar Peaks and Cognate Matters* (Brisbane, 1939), pp. 9–10; *ASJ*, 5 December 1933, pp. 443–444.

19 *ASJ*, 9 April 1931, p. 20.

20 *ASJ*, 4 April 1929, p. 15 & 6 June 1929, p. 137; *North Queensland Register*, 28 June 1930, p. 10.

21 *QGG*, Vol. 134 (1930), p. 1961; Anon, 'New South Wales Industry Pools and Prices', p. 1, 4 pp. typescript dated 20 May 1977, in CSRR, Z/109, Box 545; Ev. D. Robins, Pres. NSWCGA, p. 821, in 'Royal Commission on Sugar Peaks and Cognate Matters Evidence, 1939', CSRR, Z/109, Box 48.

22 *ASJ*, 3 September 1931, pp. 307–308 & 5 November 1931, p. 455.

23 Manning, *In their own hands*, p. 277; Hudson, *By the Banyan*, p. 95; *ASJ*, 7 April 1932, pp. 15–16, 6 April 1933, pp. 32–35 & 3 August 1933, p. 233.

24 *North Queensland Register*, 23 August 1930, p. 7; *ASJ*, 5 February 1931, p. 665 & 7 November 1933, p. 1.

25 *ASJ*, 9 January 1934, p. 501 & 6 February 1934, pp. 558–559.

26 A. Dye and R. Sicotte, 'How brinkmanship saved Chadbourne: credibility and the International Sugar Agreement of 1931', *Explorations in Economic History*, Vol. 43 (2006), pp. 223–231; H. Prinsen Geerligs and R. Prinsen Geerligs, *Cane Sugar Production 1912–1937* (London, 1938), p. 29.

27 Geerligs and Geerligs, *Cane Sugar Production*, p. 40–43; B. Carroll, 'William Forgan Smith. Dictator or Democrat', in D. Murphy and R. Boyce (eds.), *Queensland Political Portraits 1859–1952* (Brisbane, 1978), p. 430; BSES, AR 1934–1935, *QPP*, Vol. 2 (1936), p. 785; BSES, AR 1935–1936, *QPP*, Vol. 2 (1936), p. 850.

28 *ASJ*, 14 March 1936, p. 639 & 10 October 1936, pp. 367–368, 406–407 & 427.

29 *ASJ*, 11 December 1936, p. 495 & 11 March 1937, pp. 663–664 & 709.

30 *ASYB*, 1950, p. 15.

31 *ASJ*, 10 November 1936, p. 432; Carroll, 'William Forgan Smith', p. 430; P. Courtenay, *Plantation Agriculture* (London, 1980), p. 166; Geerligs and Geerligs, *Cane Sugar Production*, pp. 44–48.

32 *ASJ*, 10 September 1937, pp. 353–355; Kerr, *A Century of Sugar*, p. 147.

33 Kerr, *A Century of Sugar*, p. 147; *ASJ*, 12 April 1938, p. 94 & 10 September 1938, pp. 335–336; *The Courier Mail*, 8 March 1938, p. 5, 9 March 1938, p. 4 & 10 March 1938, p. 1.

34 Ev. of G. Hudson (South Johnstone), p. 812; Ev. of W. Batchelor (Millaquin) p. 152 & Ev. W. Moran (Tully), p. 930, in 'Royal Commission on Sugar Peaks and Cognate Matters Evidence, 1939', CSRR, Z/109, Box 48.

35 Ev. of P. Goldfinch (CSR), p. 1073; Ev. of W. Merrill (Cattle Creek), p. 461; Ev. G. Moore, (Millaquin), p. 169; & Ev. E. Evans (Farleigh), pp. 427–430, in 'Royal Commission on Sugar Peaks and Cognate Matters Evidence, 1939', CSRR, Z/109, Box 48.

36 *ASJ*, 11 May 1939, pp. 79–80;

37 *ASJ*, 13 June 1939, pp. 143 & 166 & 10 August 1939, p. 311.

38 M. Moynagh, 'The negotiation of the Commonwealth Sugar Agreement, 1949–1951', *J. of Commonwealth and Comparative Politics*, Vol. 15, no. 2 (July 1977), p. 172; *ASJ*, 10 November 1939, p. 432 & 11 January 1940, pp. 540–541.

39 *ASJ*, 13 April 1940, p. 19, 10 June 1941, p. 91 & 11 April 1942, p. 5; *ASYB*, 1960, p. 127; Anon, 'New South Wales Industry Pools and Prices', p. 2, 4 pp. typescript dated 20 May 1977, CSRR, Z/109, Box 545.

40 C. McPhail and C. Palmer, 'Precarious state of sugar industry in N.S.W. A detailed survey of a critical position', *ASJ*, 15 July 1952, p. 279.

41 BSES, AR 1945–46, p. 8; BSES, AR 1946–1947, pp. 6 & 8.

42 J. Hutcheon *et al.*, *Report of the Royal Commission on Soldier Settlement on Sugar Lands* (Brisbane, 1946), pp. 7, 12, 13 & 21–22.

43 *ASJ,* 15 April 1947, pp. 15–16 & 64, 15 June 1948, p. 149 & 15 December 1948, p. 530; for this scheme's operation at the local level see J. Kerr, *Only Room For One. Isis Central Mill* (Childers, 1996), p. 109 & Hudson, *By the Banyan,* pp. 119–120; Queensland Lands Department, AR 1950–1952, *QPP,* Vol. 2 (1951–52), p. 6.

44 *ASYB*, 1950, p. 60.

45 J. Southgate, *The Commonwealth Sugar Agreement 1951–1974* (London, 1984), pp. 13–16; Moynagh, 'The negotiation of the Commonwealth Sugar Agreement, 1949–1951', pp. 172–173.

46 *ASYB*, 1950, p. 59; *ASYB*, 1952, pp. 43–44; Moynagh, 'The negotiation of the Commonwealth Sugar Agreement, 1949–1951', pp. 177–183; International Sugar Council, *The World Sugar Economy: Structure and Politics.* Vol. 2: *The World Picture* (London, 1963), pp. 198–199.

47 A. Mansfield *et al.*, 'Report of the Royal Commission appointed to prepare an orderly plan for the development of the sugar industry (1950)', *QPP*, Vol. 2 (1951–52), p. 873; Ev. of J. Invevarity, Kalamia Mill, Vol. 3, p. 643; Ev. W. Davidson, Pioneer Mill, Vol. 3, p. 584; Ev. W. West, Mourilyan Mill, Vol. 4, p. 967, CSRR, Z/109, Box 56, Mansfield Royal Commission on the Sugar Industry Evidence, 1950.

48 A. Mansfield *et al.*, 'Report of the Royal Commission appointed to prepare an orderly plan for the development of the sugar industry (1950)', *QPP*, Vol. 2 (1951–52), pp. 845–866; *ASYB*, 1952, pp. 165–169; Hudson, *By the Banyan*, pp. 124–125.

49 Ev. E. Pearce, Sec. ASPA, before the Sugar Inquiry Committee, 1952, Vol. 1, pp. 79–81, CSRR, Z/109, Box 56; *ASYB*, 1951, p. 95; *ASJ*, 15 March 1950, pp. 753–755; *ASJ*, 18 May 1951, pp. 154–156 & 15 November 1951, p. 556.

50 *ASJ*, 15 May 1952, pp. 119–120.

51 *ASJ*, 15 October 1953, pp. 447–451; Hudson, *By the Banyan*, pp. 124–125; *ASYB*, 1954, p. 24.

52 *ASYB,* 1953, p. 183; *ASJ*, 15 October 1952, p. 483.

53 *ASJ*, 15 November 1952, p. 575, 15 December 1952, pp. 630 & 679 & 15 January 1953, p. 699–700; L. Hungerford, 'Sugarcane farming in the Bundaberg district 1945 to 1985'. M.A. thesis, Uni. of Qld, 1991, pp. 20–21; Victoria Mill Cane Insp., AR 1952, pp. 24–25, CSRR, Z/109, Box 475; J. Tanzer, 'An investigation of new settlement in the sugar industry as a result of post-war expansions. A case study in the Herbert River district, North Queensland'. B. Econ. thesis, James Cook Uni., 1979, pp. 78–83; Hudson, *By the Banyan*, pp. 118–120.

54 Tanzer, 'An investigation of new settlement in the sugar industry', pp. 38–39 & 82–86; *ASJ*, 15 April 1950, pp. 3–4.

55 Courtenay, *Plantation Agriculture*, pp. 166–167; International Sugar Council, *The World Sugar Economy: Structure and Politics.* Vol. 2: *The World Picture* (London, 1963), pp. 212–214; *ASYB*, 1959, p. 41.

56 International Sugar Council, *The World Sugar Economy: Structure and Politics.* Vol. 2: *The World Picture* (London, 1963), p. 199; *ASYB*, 1963, p. 171.

57 *ASYB*, 1959, p. 29; J. Clark, 'Developments in the Queensland sugar industry', *Quarterly Review of Agricultural Economics*, Vol. 12, No. 1 (1959), p. 25.

58 *ASJ*, 16 March 1959, p. 967; *ASYB*, 1960, p. 81.

59 Dr A. Summerville, Director-General, Qld. Dept. of Agriculture and Stock, to B. Foley, Chairman, QCGC, 16 October 1959 & Minutes of the Conference on Overproduction in the Sugar Industry, 18 November 1959, pp. 17 & 20, in PRV 14721/1/115, Box 47, File: Overproduction, 80/1/1, QSA; *ASJ*, December 1959, p. 716.

60 *ASJ*, 19 April 1960, p. 21.

61 *ASYB*, 1959, p. 119; *ASYB*, 1963, p. 95; Clark, 'Developments in the Queensland sugar industry', p. 25.

62 *Courier Mail*, 17 November 1959, p. 15; *Bundaberg News*, 9 May 1960, p. 3; *ASYB*, 1961, p. 39; L. Chapman and J. Hucknall, 'Destruction of standover cane in the Burdekin district', *CGQB*, Vol. 24, no. 2 (October 1960), pp. 64–65; C. Rehbein, 'Crop disposal', *CGQB*, Vol. 23, no. 2 (October 1959), p. 58.

63 *ASJ*, 15 December 1959, p. 939 & 19 April 1960, p. 79.

64 *ASYB*, 1962, p. 81; AR BSES 1960–61, p. 6; *ASJ*, 15 March 1961, p. 929.

65 Minutes of Meeting 21 October 1948, NSWCGA Minute Book, 1927–1950, at the Office of NSWCGA, Wardell, NSW; *ASJ*, 15 July 1952, pp. 279–281; CSR Mill Insp., Memo re: NSW mill inspection, to R. Harman, CSR Gen. Mang., 15 February 1954, p. 1, in CSRR, Z/109, Box 538.

66 Ev. C. Palmer, Sec., NSWCGA, before the Sugar Inquiry Committee, 1952, Vol. 3, p. 446, CSRR, Z/109, Box 57.

67 Anon, 'New South Wales industry pools and prices', pp. 2–3, 4 pp. typescript dated 20 May 1977, in CSRR, Z/109, Box 545; *ASYB*, 1960, p. 129.

68 Minutes of Special Meeting, Tweed River Branch of NSWCGA, 10 September 1952 & Minutes of the Quarterly Meeting, Tweed River Branch of NSWCGA, 1 October 1953, in Minutes Folder, 31 January 1946–4 November 1976, Office of the Tweed River Branch of NSWCGA, Murwillumbah, NSW.

69 CSR, 'Rehabilitation of the sugar industry in northern NSW', CSR, *Newsletter* No. 64, April 1961, p. 3, CSRR, Z/109, Box 371; Clarence River Branch of NSWCGA, AR 1952–1953, at Office of the Clarence River Branch of NSWCGA, Maclean, NSW; CSR Mill Insp., Memo to CSR Gen. Mang., re: NSW mill inspection August 1961, p. 3, CSRR, Z/109, Box 538.

70 Broadwater Mill Cane Insp., AR 1957, p. 22 & Broadwater Mill Cane Insp., AR 1963, p. 25, CSRR, Z/109, Box 461; G. Smith, *Sweet Beginnings*, p. 329.

71 M. Ryan, *Norco 100. A Centenary History of Norco 1895–1995* (Lismore, 1995), pp. 325–326; J. Bird, *Alternatives in Dairy Industry Adjustment in the Far North Coast Area of New South Wales* (Sydney, 1968), pp. 2–4.

72 CSR, 'Rehabilitation of the sugar industry in northern NSW', CSR, *Newsletter* No. 64, April 1961, pp. 3–4, in CSRR, Z/109, Box 371; CSR Mill Insp., Memo re: NSW mill inspection, to CSR Gen. Mang., 23 May 1955, p. 3, in CSRR, Z/109, Box 538; *ASYB*, 1957, p. 111; Harwood Mill Cane Insp., AR 1954, p. 13, CSRR, Z/109, Box 470.

73 Anon, 'New South Wales industry pools and prices', p. 3, 4 pp. typescript dated 20 May 1977, in CSRR, Z/109, Box 545.

74 *ASJ*, 27 September 1963, pp. 290–305 & 31 March 1964, pp. 863–864; *ASYB*, 1962, p. 33.

75 BSES, AR 1962–63, p. 9; *ASJ*, July 1962, p. 279.

76 *ASJ*, March 1963, pp. 937–938, April 1963, p. 5 & July 1963, p. 269.

77 H. Gibbs *et al.*, *Report of the Committee of Inquiry into Matters Concerning the Expansion of the Australian Sugar Industry* (Brisbane, 1963), passim.

78 *ASJ*, 31 March 1964, pp. 866–874.

79 *ASYB*, 1964, pp. 32 & 43; Hungerford, 'Sugarcane farming in the Bundaberg district 1945 to 1985', pp. 33–34; BSES, AR 1962–63, p. 8.

80 *ASJ*, 27 December 1963, p. 638 & 30 January 1964, p. 711.

81 H. Gibbs *et al.*, *Report of the Committee of Inquiry into Matters Concerning the Expansion of the Australian Sugar Industry*, p. 14; *ASYB*, 1964, p. 181 & *ASYB*, 1967, pp. 169 & 185.

82 Goondi Mill Cane Insp., AR 1966, p. 2, CSRR, Z/109, Box 466; Tanzer, 'An investigation of new settlement in the sugar industry', pp. 100–105; Hudson, *By the Banyan*, p. 122; Kerr, *Black Snow and Liquid Gold*, pp. 246–247.

83 Broadwater Mill Cane Insp., AR 1964, p. 28, CSRR, Z/109, Box 461.

84 L. Hungerford, 'A short history of sugar cane farming in the Bundaberg region', *Central Queensland J. of Regional Development*, Vol. 3, no. 1 (June 1994), p. 49; R. Fitzgerald, *From 1915 to the Early 1980s A History of Queensland* (Brisbane, 1984), p. 396; *ASJ*, September 1966, pp. 375–376; IAC, *The Sugar Industry. Report No. 209* (Canberra, 1979), p. 7.

85 J. Sault and W. Hellier, 'The new International Sugar Agreement and the Australian sugar industry', *Quarterly Review of Agricultural Economics*, Vol. 22, no. 4 (1969), pp. 210–229.

86 *The Producers' Review*, April 1971, p. 43.

87 J. Reid, *The Long View. History of the Sugar Research Institute 1949–1999* (Mackay, 1999), p. 95; *ASYB*, 1973, pp. 163–164.

88 *ASYB*, 1976, p. 23; *ASYB*, 1977, p. 29; *ASJ*, April 1975, pp. 3 & 6.

89 ASYB, 1976, p. 119, Hungerford, 'Sugarcane farming in the Bundaberg District, 1945 to 1985', pp. 73–74.

90 IAC, *The Sugar Industry*. Report No. 209 (Canberra, 1979), p. 22; Hungerford, 'Sugarcane farming in the Bundaberg District 1945 to 1985', pp. 103–104.

91 Hungerford, 'Sugarcane farming in the Bundaberg district 1945 to 1985', pp. 104–105; A. Hannah and D. Spence, *The International Sugar Trade* (Cambridge, 1996), p. 232; Reid, *The Long View*, p. 129.

92 *ASJ*, August 1980, pp. 219–220 & 228–237; *ASJ*, November 1981, p. 431; *ASJ*, January 1981, p. 465–466.

93 *ASYB*, 1981, p. 156; *ASYB*, 1986, p. 38; A. Hudson, *Sweet Success. A Story of South Johnstone Mill* (Brisbane, 1995), p. 267; *AC*, January 1982, p. 72; IAC, *The Sugar Industry 1983: 11 November 1983.* Report No. 332 (Canberra, 1983), p. 36; Kerr, *Northern Outpost*, p. 172.

94 *AC*, January 1983, pp. 8–9, April 1984, pp. 6–9 & January 1985, pp. 6–9; *ASYB*, 1982, p. 148.

95 *ASJ*, November 1974, p. 385; *The Producers' Review*, April 1964, p. 77; Smith, *Sweet Beginnings*, p. 332; State Pollution Control Commission, NSW, *Water Quality in the Tweed-Terranora Estuary* (Sydney, 1987), p. 9.

96 Hannah and Spence, *The International Sugar Trade*, pp. 136–137; Reid, *The Long View*, p. 153.

97 IAC, *The Sugar Industry: interim report, 18 February 1983*. Report No. 314 (Canberra, 1983), pp. 16, 21 & 33; IAC, *The Sugar Industry 1983: 11 November 1983.* Report No. 332 (Canberra, 1983), p. 4; *AC* March 1983, pp. 6–7.

98 IAC, *The Sugar Industry 1983: 11 November 1983.* Report No. 332 (Canberra, 1983), pp. 82–92; *ASYB*, 1984, pp. 123–128; *AC*, October 1983, pp. 16–17.

99 Hungerford, 'A short history of sugar cane farming in the Bundaberg region', p. 50; P. Barnewall and J. Passmore, *Analysis of sugar-cane farm size and the potential for horticultural diversification* (Brisbane, 1988), pp. 1–2; for the impact in the Innisfail district see *AC*, February 1984, p. 10 & February 1985, p. 12.

100 *AC*, November 1984, p. 31 & December 1984, p. 36; N. Rackemann, *The Growing Harvest. Bundaberg Fruit and Vegetable Industry from 1867* (Bundaberg, 1987), pp. 123–127; K. Sedgwick, *Analysis of sugar-cane farm size and potential for horticultural diversification – Burdekin region* (Brisbane, 1988), p. 42.

101 *AC*, January 1985, pp. 22–35 & April 1985, p. 5.

102 J. Cairns, 'Queensland sugar cane industry', *Quarterly Review of the Rural Economy*, Vol. 7 (June 1985), pp. 75–81; R. Savage *et al.*, *Sugar Industry Working Party Report* (Brisbane, 1985), pp. 18–23.

103 *AC*, September 1985, pp. 6–7, October 1985, pp. 4–5, November 1985, p. 6–7; April 1986, pp. 7–8, May 1986, p. 9; *ASYB*, 1987, pp. 103 & 111.

104 *AC*, July 1988, pp. 21–22; *ASYB*, 1989, p. 99; Qld. Dept. of Primary Industries, *A Green Paper on the Regulation of Sugar Cane Prices Act Amendment Bill 1988* (Brisbane, 1987).

105 *ASYB*, 1988, pp. 58 & 99; *AC*, May 1988, p. 16; ABARE, *The Australian Sugar Industry in the 1990s* (Canberra, 1991), p. 23.

106 *AC*, 27 March 1989, p. 9; *ASYB*, 1988, p. 101.

107 *ASJ*, July 1972, p. 209 & November 1972, p. 383; *ASYB*, 1985, p. 154; P. Pak-Poy & Associates Pty Ltd., *Coastal Management Investigation Innisfail to Mossman* (Brisbane, 1976), pp. 105 & 184–185.

108 M. Capelin, 'Planning strategies to resolve land-use conflict between the sugar industry and urban-related development in Australia'. M. Urban & Regional Planning thesis, Uni. of Qld., 1989, pp. 55 & 57; *AC*, 29 January 1990, p. 4.

109 *ASYB*, 1988, p. 100.

110 *ASYB*, 1988, pp. 11–17; *ASYB*, 1990, p. 22.

111 ABARE, *The Australian Sugar Industry in the 1990s*, p. 33.

112 ABARE, *The Australian Sugar Industry in the 1990s*, pp. 8–9 & 23; *AC*, 13 February 1989, p. 6; ASMC, AR 1990, p. 6.

113 *AC*, 27 August 1990, p. 20, 8 October 1990, p. 23 & 31 May 1993, p. 18; P. Connell and P. Whish-Wilson, 'Sugar', *Agriculture and Resources Quarterly*, Vol. 2, no. 4 (1990), pp. 411–412; E. Fitzpatrick, D. Watson and F. Soper, *Report of the Sugar Industry Working Party* (Brisbane, 1990), p. 96.

114 QSC, *The Production and Marketing of Queensland Raw Sugar: Response to the Industry Commission's Draft Report on the Australian Sugar Industry* (Brisbane, 1991), p. 12.

115 E. Fitzpatrick, D. Watson and F. Soper, *Report of the Sugar Industry Working Party* (Brisbane, 1990), pp. 43–45.

116 ABARE, *The Australian Sugar Industry in the 1990s*, pp. 22 & 24; Boston Consulting Group, *Main Report to the Sugar Industry Review Working Party. Analysis of Issues and Identification of Possible Options* (Sydney, 1996), p. 38.

117 Industry Commission, *The Australian Sugar Industry. Report No. 19* (Canberra, 1992), pp. 14–18.

118 G. Edwards, 'Two government failures? A tale of sugar and wool', *Review of Marketing and Agricultural Economics*, Vol. 61, No. 2 (August 1993), pp. 102–103; Sugar Industry Taskforce, *Report of the Sugar Industry Taskforce* (Canberra, 1993), p. 10.

119 ASMC, AR 1992, p. 12; ASMC, AR 1995, p. 3; *ASYB*, 1994, p. 58; Edwards, 'Two government failures? A tale of sugar and wool', pp. 103–104; *AC*, 8 February 1993, pp. 5–7.

120 QSC, *The Production and Marketing of Queensland Raw Sugar*, p. 8; QSC, AR 1991–92, p. 7.

121 L. Geno, *Keeping it sweet: sustainability and sugar cane* (Melbourne, 1996), pp. 16–17.

122 For the Burdekin River Irrigation Area farms see *AC* 16 November 1992, pp. 8 &14, 17 May 1993, pp. 15–16 & 5 September 1994; for Burdekin rice farmers switching to cane see *ASYB*, 1993, p. 47; for the Herbert River and Mackay regions see Chenoweth & Associates, *Nature Conservation in Sugar Cane Areas* (Brisbane, 1995), p. 43 & 56, *AC*, 23 August 1993, p. 17, 27 June 1994, p. 5 & 12 December 1994, p. 4.

123 *AC*, 8 October 1990, p. 23 & 20 September 1993, p. 2 and P. Griggs, 'Changing rural spaces: deregulation and the decline of tobacco farming in the Mareeba-Dimbulah Irrigation Area, Far North Queensland', *Australian Geographer*, Vol. 33, No. 1, pp. 54–56; *AC*, 25 July 1994, p. 6 & 18 September 1995, p. 4.

124 Chenoweth & Associates, *Nature Conservation in Sugar Cane Areas* (Brisbane, 1995), p. 60; *AC*, 26 June 1995, p. 3.

125 NSW Sugar Industry Study Group, *NSW Sugar Industry Study Group Report to the Minister for Agriculture and Fisheries* (Sydney, 1985), pp. 1–7; P. Briody, *Sunshine Sugar. The History of the New South Wales Sugar Milling Co-operative Limited* (Brisbane, 2003), p. 64; *AC*, October 1985, p. 6; *ASYB*, 1989, p. 35.

126 Broadwater Mill Cane Insp., AR 1972, pp. 50–51, in CSRR, Z/109, Box 462; Le Plan Community Research and Planning and Morton Consulting Services Pty Ltd., *The NSW Sugar Industry, Planning for the Future* (NSW, 1991), p. 24; *ASYB*, 1990, pp. 11–12.

127 ABARE, *The Australian Sugar Industry in the 1990s*, pp. 19–20; *AC*, 27 February 1989, p. 9; *ASYB*, 1991, p. 36; *ASYB*, 1993, pp. 45–46.

128 *ASJ*, 15 January 1954, p. 684 & 26 February 1964, p. 775; J. Gallagher, *Summary of Sugar Cane Research, 1950–1985* (Perth, 1986), p. 9; CSR Ltd., *Ord Sugar Industry Feasibility Study. A Report to the Government of Western Australia* (Sydney, 1976); *ASYB*, 1981, p. 150.

129 ASMC, AR 1995, p. 4; *ASYB*, 1991, p. 22; *ASYB*, 1995, pp. 18–19.

CHAPTER 11

Growing The Crop: Cultivation Methods, Implements and Plant Nutrition

The abolition of the trade in Pacific Islanders in 1904 and the repatriation of the majority of the Pacific Islanders working in Australia to their home islands in 1906 forced Australian canegrowers to employ European field labourers. They were encouraged to do so by the award of a £2 per ton bounty to those who used 'whites' only to grow and harvest their sugar cane. However, when faced with the loss of indentured Pacific Islander field workers and the need to pay higher wages for European field workers (see Chapter 16 for more details), Australian canegrowers adopted a combination of two strategies. Some North Queensland canegrowers during the late 1900s and 1910s tried employing Aborigines. This practice was most widespread in the Mackay, Herbert River, Innisfail and Mossman districts. To the canegrowers' annoyance, the Commonwealth government refused to recognise Aborigines as 'white' labour for the purposes of the bounty. Consequently, the Australian sugar industry's use of Aborigines then petered out during the late 1910s.[1] Other Australian canegrowers realised that they could not immediately eliminate European workers in the annual sugar cane harvest. No suitable machine had been developed to replace canecutters and there was little likelihood that a mechanical cane harvester could be developed quickly. Yet Australian canegrowers could reduce the number of field workers in the other phases of sugar cane growing by replacing them with machinery. This strategy radically changed the nature of sugar cane farming in Australia.

The first phase in the annual cycle of growing sugar cane (excluding the original clearing of the land) is soil preparation for a plant crop. By the mid-1900s, most Australian canegrowers relied upon teams of horses to pull the ploughs that prepared the soil for planting. However, ploughing with horses was slow, for the animals had to be spelled at intervals. Canegrowers were lucky to plough 0.2 ha daily. In addition, maintaining horses always involved much labour in feeding and attention, and the farmer had to build and maintain stables to house the animals. By the late 1900s, some Australian canegrowers were keen to reduce their reliance on horses, and wondered if the wheeled internal combustion engine tractor that had first appeared early in the twentieth century could replace the horses. Powered by benzene or power kerosene, the newly invented tractors displayed little elegance, having iron wheels with steel grips. However, they had a haulage capacity equivalent to a team of six to eight horses and enabled the use of multiple disc ploughs, subsoilers and grubbers. Tractors enabled ploughing to be undertaken continuously, even during the hottest spells of weather, whereas horses had to be spelled when the weather was too hot for continuous ploughing, leading to costly and irritating delays. Tractors were also versatile, able to be utilised for purposes other than field work. Equipped with pulleys, they could drive an irrigation pump or drive a chaff-cutter. The earliest tractors, however, suffered two main disadvantages: firstly, they were expensive, costing around £600; and secondly, they broke down and sometimes required mechanics to fix the problem if the farmer was unfamiliar with machinery.[2]

By the early 1910s, a few Australian canegrowers had commenced using tractors; their numbers, particularly in North Queensland, had swelled by the late 1910s.[3] Included amongst the earliest types of tractors used on Australian sugar cane farms were Titan tractors, manufactured by the International Harvester Co., and Mogul tractors, described as 'big clumsy machines'. Nevertheless, testimonials from Australian canegrowers who were early adopters of tractors mentioned the main advantage of the machines was the amount of work that could be achieved daily. Mr. G. Woods of Victoria Estate (Herbert River district), for example, ploughed 2.5 ha in seven and a half hours; R.T. Thomas from the Palms Estate, near Mackay, ploughed 2 ha in an eight-hour day. Moreover, a July 1922 editorial in the *Australian Sugar Journal*, the monthly magazine of the ASPA, endorsed the benefits of tractors.[4]

During the late 1910s and early 1920s, the agricultural machinery makers designed new tractor models which were smaller in size and more manoeuverable than the original versions. Jelbarts had manufactured a lightweight tractor by 1917, and the International Harvester Co. released its McCormick 15–30 in 1921, but it was Henry Ford's Model F Fordson that turned the tractor industry upside down. The mass-produced steel-wheel Fordson tractors, first manufactured in 1917, were an unprecedented success worldwide, although not because they were brilliant as tractors. Indeed, they were difficult to start, rough to ride and had inadequate brakes, but as a tractor they were cheap and readily available, and this feature was frequently often the difference between farmers owning a tractor or not.[5] In Queensland's sugar-producing districts the new models of tractors proved very popular; in the Lower Burdekin district, for example, one hundred Fordson tractors were in use by 1924. For some Queensland canegrowers who had delayed purchasing tractors, the newer versions finally persuaded them to add the machinery to their range of farm equipment. Other Queensland canegrowers replaced their older, cumbersome tractors with the newer models.[6]

Planting setts in Australia varied during the 1910s. Teams of workers still walked behind the plough, filling the furrows with setts that they carried in small hessian bags slung over their shoulders. Drop planters pulled by two or three horses were employed by a few canegrowers during the 1910s, but the Piddington Royal Commission in 1919 concluded that the 'ingenious machines for dropping cane segments in the drill and applying fertiliser at the same time' had been adopted to 'a limited extent only'. Some Australian canegrowers persisted with hand planting, claiming they had a better chance of being able to reject diseased or deformed setts than if a machine was used. Yet a decade later, drop planters, pulled either by two or three horses or tractors had become popular in New South Wales. Harry Easterby, the BSES Director, also claimed in 1930 that machine planting had virtually eliminated hand planting in Queensland.[7] Moreover, by the late 1920s, two types of drop planters were employed: the conventional drop planter that just allowed the mechanical planting of the setts, while the combine mechanical planters were drop planters fitted with attachments for delivering fertiliser into the drills with the setts. Alternatively, mechanical fertiliser distributors were employed to apply a suitable planting mixture in the drill prior to planting.

Mechanisation of other cultivation activities also occurred in Australia's sugar-producing districts during the late 1910s and 1920s. Weeds that were once removed by teams of 'chippers' – field labourers with hoes – were increasingly controlled by machines such as a drill weeder and cleaner that consisted of a revolving wheel fitted with teeth (Plate 11.1) or a Willmann's sugar cane weeder, an implement with centre tines that weeded the bottom of the furrow.[8] Chopping out old cane stools in fields selected for replanting or burying trash in the interspaces was sped up after 1925, following the introduction of Howard's rotary hoe or cultivator, which was readily attached to tractors (Plate 11.2). The revolving blades of this equipment pulverised the plant material, enabling the particles to be readily assimilated into the soil. Other canegrowers used the ratoon plough, the spring tooth cultivator or disc or tine harrows pulled by tractors to deal with the ratoon roots.[9]

The widespread interest shown in new cultivation implements during the late 1910s and 1920s was also accompanied by attempts at soil improvement. BSES agricultural scientists in the early 1900s had identified that the soils of the sugar-producing districts from Mackay northwards were very acidic or deficient in lime. Adding lime to a soil makes its more neutral,

Plate 11.1. Weeding cane mechanically by the use of Hodge Rotary Weeders, Green Hill, near Cairns, 1949.

Source: CSRR, 171/615. Image reproduction courtesy of CSR Ltd.

Plate 11.2. A side view of a Howard Rotary Cultivator, showing the blades and drive casing, attached to an early model tractor.

Source: *ASJ*, 8 May 1925, p. 90.

but calcium, the basic ingredient of lime, is an important plant food and its absence slows normal plant development. BSES scientists had advocated the application of lime, and some North Queensland canegrowers did access lime supplies during the 1900s and early 1910s. At Mossman, dead coral that had washed up on the local beaches was collected, burnt or ground up, and then distributed to farmers; burnt lime from the Chillagoe quarries was also obtained for the cane farms around Cairns and Mossman. Demand for lime created through crushing shells gathered on Hinchinbrook Island was said to be so great by 1914 that 'a sufficient quantity cannot be supplied as yet to fulfill orders'.[10] Queensland's canegrowers could also secure pulversised lime from Brisbane-based suppliers such as the Australian Cooperative Fertilisers Ltd. or the Queensland Cement and Lime Co. Ltd. However, lime was very expensive, costing about £2 per ton in Southern Queensland and up to £4 to £5 per ton in North Queensland. The BSES thought that the high steamship freights associated with the movement of lime from southern Queensland to the northern sugar-producing districts during the mid-1910s made its price too prohibitive, thereby hindering its widespread use.[11]

The price of lime fell during the late 1910s and 1920s, following three major developments. Firstly, the completion of the North Coast Railway in 1924 meant lime could be moved more cheaply to the northern sugar-producing districts from the southern limestone mines at Gore, Tamaree and Degilbo (Figure 11.1). The BSES also negotiated concessional rates with Queensland Railways for the carrying of lime when the distance exceeded 40 km.[12] Secondly, the growing demand for lime in the early 1920s prompted the establishment of new terrestrial limestone mines at Ambrose and Marmoor, just south of Rockhampton, thereby improving the supply of the product (Figure 11.1). Thirdly, the mining of coral and its conversion into lime, which had commenced in the 1900s, became more widespread and organised during the 1920s.[13] A dozen coral reefs off the North Queensland coast were being mined for coral by 1930 (Figure 11.1). Together, these factors ensured a greater supply of lime at cheaper prices for canegrowers. Consequently, the BSES noted a steady increase in the use of lime by North Queensland canegrowers during the 1920s.[14]

The application of lime improved the calcium available to the crops of sugar cane being grown in North Queensland, but the three principal plant foods are nitrogen, phosphorus and potassium. This plant nutrition can be supplied from natural products such as crushed bones and guano or artificial fertiliser mixes. However, potash and nitrate of soda became exceedingly hard to procure during World War 1, inflating the

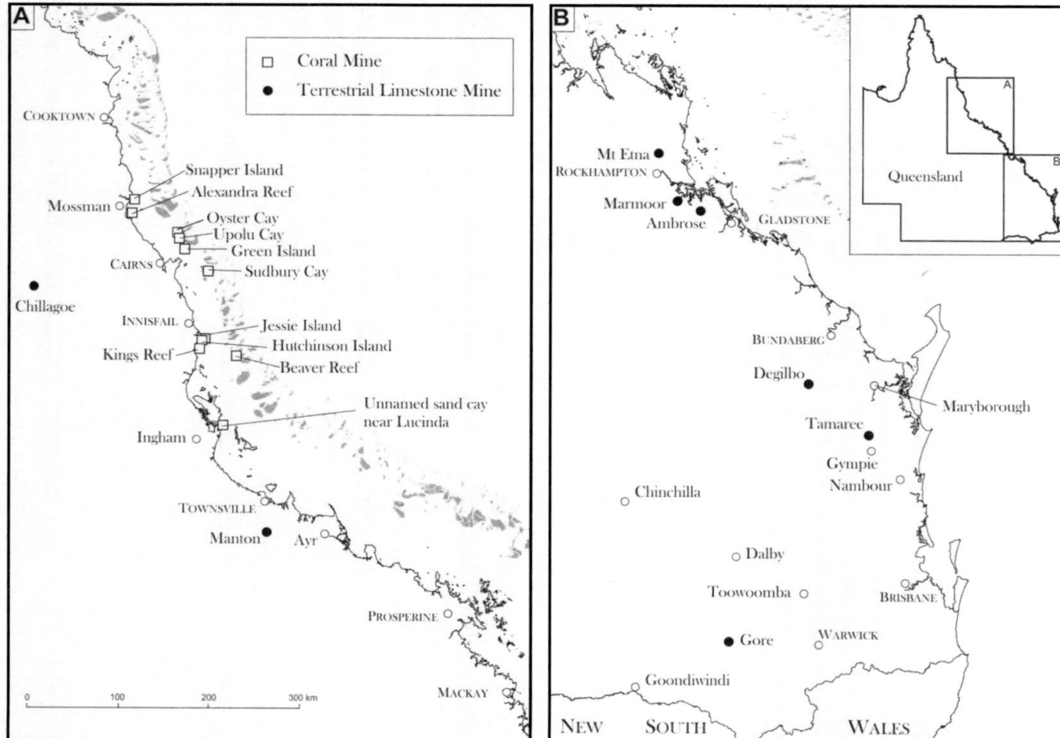

Figure 11.1. The location of terrestrial limestone mines and coral reefs mined to produce agricultural lime in Queensland, 1900 – c. 1945.

Source: Based upon details in the Queensland Department of Mines, various annual reports, and Figure 5 in B. Daley and P. Griggs, 'Mining the reefs and cays: coral, guano and rock phosphate extraction in the Great Barrier Reef, Australia', *Environment and History*, Vol. 12, no. 4 (November 2006), p. 412.

price of inorganic fertilisers. Those Australian canegrowers who applied fertilisers during the wartime period relied on organic fertilisers such as meatworks manures (i.e. blood and bone) and bone dust available from local suppliers or export abattoirs in Brisbane, Gladstone, Rockhampton and Townsville. Bat guano recovered from the Mt Etna caves near Rockhampton was also supplied from 1917 onwards by the Guano Fertilser Co. The Piddington Royal Commission, however, was scathing of the overall limited use of fertilisers in Australia's sugar-producing districts, concluding that associations of canegrowers rendered little, if any, assistance by cooperative effort in ensuring the bulk purchase of fertilisers which would have reduced their cost. Even BSES interest and extension work in plant nutrition suffered during World War 1, as a shortage of field assistants forced a suspension of its farm-based fertiliser and cultivation experiments.[15]

After World War 1, some canegrowers resumed purchasing inorganic fertilisers. Sulphate of ammonia could be secured from the Australian Gas Light Co., Sydney, or the Newcastle works of BHP; superphosphate was sourced from the Port Kembla works of Australian Fertilisers Pty. Ltd. after 1921. A range of artificial fertiliser mixtures were supplied by the Australian Cooperative Fertilisers Ltd. In 1915, the firm built a mixing plant and bone crushing plant at Runcorn, near Brisbane, and three years later it built another mixing plant at Garbutt (a suburb of Townville), blending Houlbourne Island (off Bowen) seabird guano with bone dust and imported sulphate of potash.[16] Gibson & Howes Ltd. even sought a new source of fertiliser when they experimented with molasses in 1922, finding that light applications of molasses in irrigation waters or direct applications of molasses to fields of ratoons on Bingera plantation improved yields and could be made without injuring the soil or plant.[17] Yet the BSES claimed in 1927 that the overall percentage of Queensland canegrowers using fertilisers still remained very low. Moreover, Arthur Gibson, the northern BSES field officer, reported in 1928 that much money was wasted annually in the haphazard use of artificial fertilisers, as most Queensland canegrowers had not used soil analysis to ascertain what plant nutrients were actually required.[18]

Restoring the fertility of cane growing soils was also encouraged by the BSES and CSR's agricultural scientists during the 1920s, although their approach was not always similar. Canegrowers are automatically supplied with a supply of humus which can be added to the soil because after each harvest they are left with trash. Harry Easterby, the BSES Director, reported in 1924 that most trash in Queensland's sugar-producing districts was burnt, and that he favoured this practice as the trash harboured insect pests (e.g. weevil borer) and diseases caused by fungi (e.g. downy mildew). Nevertheless, humus and nitrogen could be restored to the soil at intervals by the growth of a green manure crop such as Mauritius beans and cowpeas. The extent of green manuring in Queensland's sugar-producing districts contracted throughout the war years due to difficulties in obtaining seed, but the practice had resumed during the 1920s. However, the BSES reported in 1927 that the number of Queensland canegrowers using green manures was still far too low and more attention should be paid to this activity.[19] CSR, on the other hand, had a different view of trash. In 1923, an experimental trial in the Broadwater Mill supply area showed that trash from a well grown crop of cane provided double the amount of humus supplied by an ordinary crop of green manure. During 1924, Edward W. Knox advised each CSR mill manager that the burning of trash must have had a seriously deleterious effect on the fertility of the soils throughout Australia's sugar-producing lands, and that the organisation had decided to remedy the situation in the districts where its sugar mills were located.[20]

Many Australian canegrowers had claimed that their implements were not suited to trash conservation (i.e. the ploughing of the trash into the soil) or incorporating green manure crops into the soil.[21] During 1924, James Clark, CSR's field supervisor at Fiji, visited each of CSR's Australian mill supply areas and demonstrated trash conservation by using a modified Secretary Single Disc plough manufactured by the Clyde Engineering Co. In addition, new trials identifying the best kinds of leguminous crops for the New South Wales mill supply areas were conducted in the Clarence River district during 1924, and the results disseminated in a circular to CSR's suppliers in 1926. CSR's officers were instructed to promote the use of the modified Secretary Single Disc plough, trash conservation and green manuring. CSR even hired out its own tractors and the modified disc ploughs to New South Wales canegrowers who still relied on horses.[22] This campaign was assisted by the Massey Harris Co. having released an inexpensive trash plough by 1926, so that the required implement could now be afforded by canegrowers. By the early 1930s, CSR concluded that its campaign had contributed to an upsurge in trash conservation and green manuring throughout its mill supply areas. A quarter of the suppliers to its Macknade Mill, for example, conserved trash around ratoons in 1926; this figure had risen to 57% by 1933. A similar increase was recorded when replanting occurred, with 51% of the farms adopting trash conservation in 1926, increasing to 71% of the farms by 1933.[23]

New implements and improving soil fertility, 1930 to 1939

In mid-1930, Harry Easterby, the BSES Director, outlined the cultivation practices of Queensland canegrowers at a conference of sugar cane technologists in the Dutch East Indies. Fowler steam ploughs were still used to prepare land for planting on the few surviving plantations (e.g. Bingera; Fairymead), but the bulk of the preparation on the small cane farms was done by tractor or horse tillage. Three to four ploughings prepared the land for planting, with drills varying from 1.3 to 1.8 m apart. Cane planting machines were used extensively. Tillage was generally done with shallow cultivating instruments such as the Planet Junior or Cotton Disc harrows, mostly still horse-drawn. In the sub-tropical districts, up to three ratoon crops were harvested; at Mackay, up to second ratoons; in North Queensland generally only one ratoon crop was grown. Trash was generally burnt, and then three to four furrows were ploughed between the cane rows and the land levelled down by the use of tine harrows or cultivators, or the land was treated with a spring tooth cultivator or a grubber instead of being ploughed. Sometimes the trash was left and rolled into each alternate interspace, with every other interspace being cultivated with the plough. Shortly after the above address, Easterby appeared as a witness before the Gunn Royal Commission into the Australian Sugar Industry. He claimed that Australia led the world in the development of implements for the tillage of sugar cane. Representatives from other countries frequently visited Queensland's sugar-producing districts to gather knowledge on cane farming implements.[24]

Easterby's address to the international gathering of sugar cane technologists, however, failed to mention the marked regional differences in cultivation practices that had emerged in the Australian sugar-producing regions. Far Northern Queensland canegrowers during the late 1910s and early 1920s cultivated an average of 20.2 ha, almost double the area grown by their Wide Bay-Burnett counterparts and four times the area of cane found on Southern Queensland sugar-producing properties (Table 11.1). Mechanised cultivation practices were probably the most advanced in the Far Northern and Edgecumbe regions, although the proprietors of Fairymead, Millaquin and Bingera plantations in the Wide Bay-Burnett region had also invested heavily in mechanised agriculture by 1930. Farmers in the Southern Queensland region, growing under 5 ha of cane, were less inclined to mechanise cultivation practices.

Table 11.1. Average area (ha) cultivated on properties growing sugar cane in Queensland, 1917–1919 to 1934–1939.

Region	1917–1919	1920–1924	1925–1929	1930–1934	1934–1939
Far Northern*	21.9	21.5	20.4	21.4	22.3
Edgecumbe**	14.5	15.9	17.6	18.8	20.4
Wide Bay-Burnett	13.9	13.3	11.5	12.4	13.3
Southern	6.0	5.6	3.6	4.4	4.7
Queensland	16.2	15.7	15.3	16.7	17.7

Notes: * Includes the Herbert River district northwards. ** Includes the Burdekin, Proserpine and Mackay regions.

Source: Based upon figures in the ARs of the Queensland Government Statistician on Agricultural Production, in QPP.

New South Wales canegrowers had also been slower to adopt machinery. Arthur Watson, the President of the Clarence River Branch of the NSW Canegrowers' Association, told the Gunn Royal Commission in 1930 that the small areas of sugar cane cultivated by farmers in northern New South Wales had retarded the introduction of labour-saving devices, as the cost and up-keep of a tractor was beyond the reach of the average farmer. In 1932, CSR's Broadwater Mill Cane Inspector painted a similar picture as Watson. Very few farms in the Broadwater Mill supply area had more than three horses, and quite a number had only two horses, and these received little attention. Planting machines were used, but most ploughs were very old and poorly maintained and there was often only one, which was shared amongst many farms. The Broadwater Mill Cane Inspector blamed the farmers for trying to combine dairying with sugar cane cultivation, as the correct working of each industry interfered with the other, and neither was carried out satisfactorily.[25]

From 1933 onwards, CSR encouraged its firm's canegrowers to acquire more modern implements and tractors. Three years later their efforts had yielded some successes, for the Broadwater Mill Cane Inspector reported that 'many new modern implements are in use' and some farmers who cultivated sufficiently large areas had purchased tractors; twenty-five were now in use. By 1939, the total number of tractors in the North Coast Coastal Division in New South Wales stood at 421, although some of these tractors may have been found on non-sugar-producing properties.[26] Partial mechanisation of field activities on the sugar cane farms in the Southern Queensland region had also probably occurred during the late 1930s, although a survey of tractors in Queensland during the Mansfield Royal Commission into Certain Aspects of the Sugar Industry Queensland in 1942–1943 only recorded fifteen tractors in the supply areas for Rocky Point and Eagleby Mills in Southern Queensland, compared, for example, to the 462 tractors found in the Herbert River district, North Queensland.

As the New South Wales canegrowers struggled to acquire modern tillage implements and tractors, some Queensland canegrowers during the 1930s readily adopted the new farming implements and machinery that was being developed both locally and by the international agricultural machinery makers. Caterpillar or crawler tractors, with their greater power and costing around £1 000, appeared on some Queensland cane farms (Plate 11.3). By 1932, Gibson & Howes Ltd., for example, employed fifteen caterpillar tractors on Bingera plantation. In 1934, CSR's Goondi Mill cane inspector reported that 80% of the Goondi Mill suppliers now used tractors, the majority of which was the crawler type. International Harvester's Farmall range of wheeled tractor was adopted by a number of Australian canegrowers. Their high clearance made them ideal for cultivating young plant and ratoon crops of cane. Other Australian canegrowers favoured McCormick Deering's F-30 tractor, which was designed for clean, speedy cultivation between rows as it could be raised for higher clearance.[27]

Plate 11. 3. A cutter planter being pulled by a Caterpillar tractor, 1935.
Source: *ASJ*, 10 October 1935, p. 368.

The adoption of more powerful tractors, including newer models with pneumatic rubber tyres, forced Australian canegrowers to purchase more modern implements as their older ones were not strong enough to stand up to the work required of them. These newer implements contributed to better and deeper cultivation, especially of ratoon crops, as grubbers could be used more extensively.[28] Better ratoon crops were also encouraged by the use of a stubble-shaver, developed by the Onomea Sugar Co. in Hawaii. First imported into Queensland in 1933, this equipment levelled the ground and destroyed the uppermost eyes of the cane stubble, ensuring a ratoon crop that originated from the lower dormant buds and not from those at or above the land surface.[29]

Planting was also sped up on some Queensland cane farms following the adoption of three new types of planting machines. Around 1929, Tom Ross, a North Eton cane farmer, developed a crude cutter planter which used a sharp knife driven by the wheels of the planter to cut the stalks of cane into setts before depositing them into the furrow. Engineer Andy Milne of Mackay modified Ross' design slightly in 1932 and commenced marketing what became known as the Milne Cutter Planter. Sticks of cane were laid horizontally in the wooden box of a cutter planter; the capacity of the machine was approximately half a tonne. The stalks of cane were fed into the centrally placed chute as the machine was drawn across the field (Plate 11.3). A revolving knife driven from the land wheel cut the stalks of cane into setts which then fell directly into the furrow. A single-furrow cutter planter allowed approximately 2 ha to be planted daily. Double-furrow machine planters had also made an appearance by the mid-1930s, enabling farmers to plant two furrows simultaneously. However, although the larger planting units may have sped up the planting operation, they did not save appreciably on manpower, since the double-furrow planting machines required two operators and a larger gang of workers to cut the cane into setts and load them onto the planting unit. Charlie Hodge, a former farmer and implement manufacturer at Mackay, had also been experimenting with machine planters during the early 1930s, before deciding in 1936 upon the design of what became known as a Hodge cane planter (Plate 11.4). In this planter, the stalks of cane were placed vertically in the planter box and supported by steel racks in front of the operator.[30]

Plant nutrition and soil enhancement were not overlooked during the 1930s. In New South Wales, agricultural scientists at the Wollongbar Experiment Farm, near Lismore, commenced a series of trials in 1931 to show canegrowers the benefits of the wider use of inorganic fertilisers. The New

Plate 11.4. A Hodge cane planter in operation, 1953.

Source: Image reproduction courtesy of the National Archives of Australia: Series A1200, Item L16031.

South Wales Department of Agriculture in its 1933 edition of *The Farmers' Handbook* also endorsed the benefits of green manuring and trash conservation during the cultivation of sugar cane.[31] In Queensland, Dr H. (Bill) Kerr who had been the BSES's graduate scholarship holder in soils returned from overseas in 1928, and headed the newly formed Division of Soils and Agriculture, created as a result of the complete reorganisation of the BSES's experimental program. During 1929, Kerr commenced a series of field experiments to determine the influence of fertiliser on crop yields on a variety of soils throughout each Queensland sugar-producing district. Statistical analysis of the figures enabled the significance of the results to be measured. Experiments were also conducted in 1932 on molasses as a fertiliser.

By 1934, the studies of the relationship between the results from field trials and laboratory soil tests had been completed, and the BSES issued details of the chemical composition of six standardised fertiliser mixtures, arranged in pairs – one for planting and one for ratoons (e.g. Sugar Bureau No. 1: Planting; Sugar Bureau No. 3: Ratooning). The essential differences were that the No. 1 mixtures were designed for soils deficient in phosphorous, the No. 3 mixtures targeted soils lacking potash and the No. 2 was a balanced mixture. The ratooning mixtures were all richer in nitrogen. From 1937 onwards, soil samples were also analysed for canegrowers to assist in making fertiliser recommendations, based upon the knowledge gained from the farm fertility trials.[32] Soil mapping by the BSES commenced in the northern sugar-producing districts in 1930; Norman King updated these preliminary maps in 1937. He also completed soil maps for the Bundaberg region in 1935.[33] In addition, Kerr from 1933 onwards began disseminating knowledge about fertilisers and their use and appropriate tillage and cultivation methods in *The Cane Growers' Quarterly Bulletin*, a new BSES-sponsored publication. Just before the outbreak of World War II, all the cultivation and fertiliser advice assembled by the BSES was published in a single volume – *The Queensland Canegrowers' Handbook*.

CSR also assembled information on the effectiveness of various rates of fertiliser application under different environmental conditions from what the company termed 'manure trials' in all its mill supply areas during the 1930s. Details were then incorporated into CSR's annual agricultural circulars to its suppliers. Herbert River district canegrowers, for example, were consistently advised during the 1930s to apply sulphate of ammonia or another nitrogenous fertiliser at the rate of 2–3 cwt (100–150 kg) per acre for ratoon crops, but to avoid applying fertilisers high in nitrogen to plant cane when a green manure crop had been ploughed into the soil. If a green manure crop had not been used, fertilisers applied at planting should be high in phosphates and placed in the drills – meatworks manure or a mixture of sulphate of ammonia and superphosphate were recommended. CSR also continued advocating the benefits of trash conservation, liming and green manuring, and secured seed for legume crops on behalf of its canegrowers. Aborigines from the Palm Island Aboriginal Settlement, near Townsville, assisted by growing green manure crops and gathering the seed which was then distributed to CSR's northern suppliers. A company tractor and trash ploughs were even hired to the canegrowers supplying Broadwater Mill so that trash conservation could be conducted on the farms.[34]

Extension efforts by the agricultural scientists during the 1930s clearly influenced some canegrowers who adopted cultivation practices that improved soil organic matter. Green manuring had become 'firmly established amongst Queensland canegrowers', wrote the BSES's Bill Kerr and Arthur Bell in 1939. [35] Maize, Poona peas and Mauritius beans were favourite green manure crops, although the practice of growing maize on

Table 11.2. Average percentage of the area planted that had been covered by a green manure crop in CSR's mill supply areas, 1927–1945.

Mill supply area	1927–1929	1930–1934	1935–1939
Harwood	36.0	51.0	62.6
Broadwater	*	22.7	44.0
Condong	*	36.7	53.4
Victoria	24.0	38.0	51.6
Macknade	32.5	61.3	51.7
Goondi	68.6	84.0	90.6
Hambledon	*	*	69.5
Childers	46.7	57.0	Mill closed

* Figures not available

Source: Based upon figures in CSR, ARs on Cane Seasons, 1927 to 1939, CSRR, Z/109, Box 387.

cane growing lands was increasingly frowned upon because of its susceptibility to mosaic disease, which could also infect the cane crops. CSR's figures on green manuring in its mills areas confirm Kerr and Bell's claim, as they show a steady increase in the average percentage of the area planted that had been covered by a green manure crop during the 1930s (Table 11.2). In addition, some canegrowers, especially in New South Wales and the Herbert River district, adopted trash conservation (Table 11.3). CSR reported a steady increase in the number of trash ploughs used throughout its New South Wales mill supply areas during the 1930s and rotary hoe contractors had been turning the trash into the soil for canegrowers without the necessary equipment.[36]

Other North Queensland canegrowers, however, had less trash to conserve following the introduction of localised pre-harvest burning of crops during the mid-1930s. This practice was adopted as a health precaution to reduce the incidence of Weil's disease amongst canecutters, thereby avoiding strike action by canecutters (see Chapter 16). Post-harvest burning of crop residues, instead of their conservation, also increased in North Queensland mill supply areas because of concerns that this material harboured rats and weevil borers which damaged the stalks of cane.[37]

Table 11.3. Average percentage of trash conservation on the area ploughed out in preparation for planting in CSR's mill supply areas, 1930–1939.

Mill supply area	1930–1934	1935–1939
Harwood	19.4	24.6
Broadwater	22.3	44.0
Condong	24.5	38.6
Victoria	80.2	84.6
Macknade	84	33.7
Goondi	44	36.0
Hambledon	*	66.0

* Figures not available

Source: Based upon figures in CSR, ARs on Cane Seasons, 1933 to 1940, in CSRR, Z/109, Box 387.

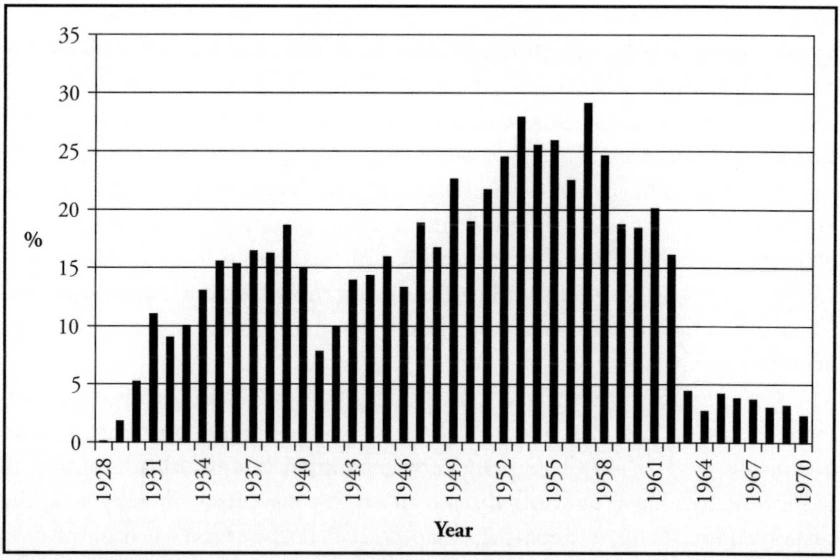

Figure 11.2. The percentage of the yearly molasses production used as fertiliser in Queensland, 1928–1970.

Source: Based upon figures in the ARs of the BSES.

Table 11.4. Fertiliser use by the Queensland sugar industry, 1936–1943.

Year	Tonnes of fertiliser	Tonnes per ha	Year	Tonnes of fertiliser	Tonnes per ha
1936	45 720	0.33	1940	59 944	0.42
1937	50 800	0.36	1941	54 864	0.41
1938	56 895	0.40	1942	35 560	0.28
1939	61 976	0.43	1943	15 240	0.12

Source: Bases upon figures in BSES, AR 1942–43, p. 4.

During the 1930s, the BSES and CSR encouraged Australian canegrowers to pay more attention to plant nutrition. Molasses, which was not recorded in the official statistics as a fertiliser before 1928, grew in popularity as a fertiliser in Queensland during the decade (Figure 11.2). Some canegrowers also purchased filter mud to spread on their farms as fertiliser. Dr H. (Bill) Kerr, who had assumed the role of BSES Director in 1933, advised the Royal Commission into Sugar Peaks and Cognate Matters in 1939, that Queensland canegrowers now appreciated the value of mill by-products as soil fertility builders, leading to the keenest competition between canegrowers wishing to have access to them. He remarked that 'whereas large quantities of molasses were formerly run to waste at many of the northern factories, farmers are now prepared to pay as much as 12s 6d a ton for this by-product, because of its undoubted effects on soil quality and fertility'.[38]

BSES figures on the Queensland sugar industry's fertiliser use first become available in 1936 (Table 11.4). They show a rapid growth in the use of inorganic fertilisers by Queensland canegrowers during the late 1930s. This upsurge may have been related to Queensland's largest fertiliser company, A.C.F. and Shirleys Fertilisers Ltd., reducing their prices in 1931, 1934 and 1935 and introducing a new policy in 1935 to extend credit to farmers, enabling them to purchase fertiliser, subject to adequate security being arranged. The application of fertilisers had also been encouraged from 1933 onwards by a Commonwealth government subsidy of 15s per ton, payable to the users of artificial fertilisers.[39] Not surprisingly, Dr H. (Bill) Kerr advised the Royal Commission into Sugar Peaks and Cognate Matters in 1939 that the use of fertilisers had grown in popularity in Queensland and 'many farmers have embarked on a policy of permanent fertility building, as opposed to the tendency towards soil depletion and crop degeneration of former years'. However, Kerr in his annual report for 1939 sounded a note of warning for at least Queensland canegrowers when he said that 'too many farmers are still content to carry on their fertiliser practices without due regard for the special requirements of their land. As a result, much money is still expanded unwisely in the purchase of artificial manures'.[40]

Wartime interruptions

In late 1939, Australia was heavily dependent on the outside world for its fertiliser supplies. All potash salts for the manufacture of superphosphate, half the nitrogenous fertiliser and a substantial proportion of the sulphate ammonia used were imported. Consequently, just before the outbreak of World War II, considerable amounts of the raw materials required for fertiliser manufacture had been stockpiled by the various Australian fertiliser distributing companies. Therefore, the serious influence which the war exerted later on the continuity of supplies was not at first evident to the sugar industry (Table 11.4). However, interruptions to fertiliser supplies occurred throughout 1940 due to a shortage of ships. In response, the Queensland government (Forgan Smith; Labor) tightened the regulations controlling the sale of fertilisers during 1940, so that the maximum amount of potash sold in general mixtures was fixed at 7.5% and the purchase of potash unmixed was prohibited. Nevertheless, special permission was granted in 1941 to Queensland canegrowers known to farm potash deficient soils – often red-volcanic soils – to purchase mixtures containing up to 25% potash. The use of potash was prohibited on the potash-rich soils of the Burdekin delta.[41]

As the war progressed, fertiliser supplies from abroad during 1941 became more difficult to obtain due to the armed forces' demand for shipping space and the closure of the phosphate works at Nauru (the source of raw rock phosphate) due to enemy action in December 1940. Sulphate of ammonia and other nitrogenous fertilisers, used mostly on ratoons, were restricted to less than one-half of the normal supplies. By early 1942, a general fertiliser rationing scheme had been introduced amongst Queensland canegrowers, in order to ensure an equitable distribution of the available, yet inadequate supplies of fertiliser, and to prevent hoarding by farmers. Calculation of the size of the rations became the responsibility of the BSES, using the knowledge their agricultural scientists had accumulated on soil types and plant deficiencies. Queensland canegrowers were then notified of the amount of blood and bone, nitrate of soda and potash they could receive under the rationing system; canegrowers who had not been consistent users of fertilisers were not able to enter the purchasing field under emergency conditions.

Further disruption to the rationed fertiliser supplies occurred during 1942, as priority of space on the state's transport system was given to military supplies and equipment, following Queensland becoming a large Allied supply base. At the conclusion of 1942, Queensland's sugar industry had been allotted a third of the amount of fertiliser used normally, and the bulk of the fertiliser supplied was meatworks manure, which was considered less effective than the quick-acting nitrogenous fertilisers such as sulphate of ammonia and nitrate of soda.[42] The use of molasses as a fertiliser also fell as farmers struggled to maintain the lorries needed to cart the product between the mill and fields (Figure 11.2). Further fertiliser shortages meant only 0.12 tonnes of fertiliser were applied per ha of sugar cane cultivated in Queensland in 1943, less than a third of the amount used at the outbreak of the war (Table 11.4).

Fertiliser restrictions were also accompanied by a shortage of field labourers. Reduced manpower at the beginning of the war contributed to an upsurge in the purchase of cutter planters amongst North Queensland canegrowers.[43] A lack of labour also reversed the practice of trash conservation, so championed by CSR during the 1930s. Canegrowers immediately reverted to burning the trash. CSR also reported a marked reduction in green manuring throughout its mill supply areas, caused by a lack of suitable seed.[44] Normal pest and disease control practices (e.g. soil fumigation for grub control) were reduced or suspended completely because of a lack of field workers (see Chapter 14). As the war progressed, women and children replaced the absent field workers in some sugar-producing districts. The Macknade Mill Cane Inspector, for instance, noted that much of the planting in the Macknade Mill supply area during 1942 was done by women and girls who drove the tractors and worked the planting machines. Later, weed control was undertaken by women; a few gangs of women 'chippers' were also formed and they accepted contracts to undertake the work on other properties. In New South Wales, women and children were used extensively during 1943 to complete the cultivation and planting work as not enough labour was available for these tasks.[45]

Normal cultivation practices were also interrupted from early 1942 when the Commonwealth government's Allied Works Council and the Civil Constriction Corps impounded tractors to help build defence roads and aerodromes. Heavier type tractors, particularly Caterpillar crawlers, were favoured by the authorities. Overall, 324 tractors were taken, mostly from North Queensland sugar-producing districts, where approximately

20% of the tractors were impressed during 1942.[46] The effectiveness of non-impounded tractors and other farm equipment was impeded by a shortage of spare parts and rubber tyres. Rotary hoe contractors in New South Wales curtailed their operations due to spare parts shortages, thereby delaying or reducing the cultivation activities of those canegrowers who still relied upon their services. Other motive power was limited; many northern canegrowers had reduced their number of horses. Indeed, the Mansfield Royal Commission to Investigate Certain Aspects of the Sugar Industry in 1942–1943 heard from several North Queensland canegrowers who claimed that the use of horses was restricted, even when they were still available, because of the non-availability of experienced ploughmen and owing to the farm implements being converted to or designed for mechanical power.[47]

The impact of the war on cultivation practices eased during late 1943 and 1944. Fertiliser deliveries in 1944 were almost double that supplied in 1943, although fertiliser use was still rationed. The amount of molasses applied as fertiliser returned to almost pre-war levels (Figure 11.2). Additional men released from the Armed Forces and Civil Construction Corps improved the number of available field labourers. Nearly half the impressed tractors were returned throughout 1944 and 155 new tractors were released for sale in the sugar-producing districts. Further improvements to the labour, equipment and fertiliser supply occurred during 1945, and the area fertilised recovered. However, fertiliser rationing in Queensland still existed at the conclusion of World War II.[48] Yet the yields on Queensland's sugar cane farms had made a partial recovery to 19.8 tonnes of cane per ha in 1945–1946, after slumping from 23.4 tonnes of cane per ha in 1939 to 15.2 tonnes of cane per ha in 1943–1944.

Arthur Bell, the Acting BSES Director, claimed in 1945 that the immediate impact of the war was a fairly serious undermining of the 'virility of the industry'. The cumulative effects of acute shortages of labour, machinery and materials, especially fertilisers, had depreciated the fertility levels of soils to a point from which it would take some years to recover. However, the fertiliser shortages caused a renewed interest in the correct use of fertiliser, a development advocated by the BSES since the mid-1930s. Spectacular advances had also occurred in the creation of pesticides, herbicides and fungicides, and their future impacts upon cultivation practices were yet to be determined. Labour shortages stimulated a revived interest in labour-saving devices.[49] Australian canegrowers were also exposed to a piece of new machinery – the bulldozer. This machinery was soon used by Australia's canegrowers as they embarked upon the first post-war expansion during the 1950s.

Bulldozers, trace element deficiencies and herbicides: 1946 to 1974

Traditionally, forested land acquired by Australian settlers intending to cultivate sugar cane was cleared by hand. Brush-hooks were used to remove any vines or undergrowth and the trees and larger shrubs were felled by axe or cross-cut saw. This process was usually conducted during the wetter summer and autumn months, so that during the drier winter and spring months the felled and stacked material could dry out. Walter Figg, an English bricklayer who commenced cane growing in the late 1910s on the Inkerman Estate, near Home Hill (Lower Burdekin district), recalled the countryside each night being 'ringed by fires as each selector burned the timber he had fallen'. Other intending canegrowers used fire to clear the forests. Edwin Brady who toured throughout Queensland in the early 1920s claimed that canegrowers gained great satisfaction using fire for 'a good burn – which leaves only blackened stumps of forests standing – saves money and labour'. Subsequently, the first crop of sugar cane was planted amongst the ashes. Bigger tree stumps that survived any fires just rotted away, usually in three to five years, but sometimes longer.[50]

As the post-war expansion of the Australian sugar industry commenced, contractors using modern earth moving equipment replaced the brush-hook, axe and crosscut saw. Bulldozers or crawler type tractors with a bulldozer blade pushed down the trees, ripping out their root systems. Sometimes stumps were blown out with explosives.[51] After the heavy timber had been shifted to the edge of the cleared fields, grubbers went through bringing any remaining roots to the surface. The land was then cleared of the roots, sticks and any stones brought to the surface by the grubbing action. Tractor-drawn pin wheel scrub rakes, including the Pederick scrub rake developed in Western Australia, and root rakes, larger and more robust versions of a hay rake, swept the roots and sticks into rows or smaller bundles before all the assembled material was burnt. Mechanical rock or stone pickers were also employed to clear the newly emerging paddocks of smaller rocks and stones.[52] Mechanical clearing became favoured because of post-World War II labour shortages and it allowed Australian canegrowers to quickly get the land under crop. Twenty to thirty hectare blocks of virgin forest could be cleared and planted with sugar cane in approximately three to four months. Newly cleared land was also observed being 'ploughed' by bulldozers during the 1950s, although such a practice was risky because of the possibility of bringing subsoil clay to the surface.[53]

As the mechanical clearing of land commenced for the post-war expansion, some canegrowers, mostly in Southern Queensland and New South Wales, strove to catch up with the machinery that had already been adopted by other Australian canegrowers before the war. Instead of relying on contractors to supply rotary hoes, New South Wales canegrowers during the late 1940s purchased this implement for use in trash conservation and plough-outs. By 1958, rotary hoes were the most popular implement for ploughing out cane in the New South Wales sugar-producing districts. Tractors were acquired for the first time by more New South Wales canegrowers during the 1940s and early 1950s. In 1958–1959, nearly 1 600 tractors were found in the northern New South Wales shires where sugar cane cultivation was predominant, compared to less than 500 tractors in 1939 (Table 11.5). A 1958 report on farming activities in the Rocky Point Mill supply area also claimed that it was only in the last few years that mechanical planters had been used and tractors now replaced the horse on most farms.[54]

Australian canegrowers without tractors were not the only purchasers of tractors during the late 1940s and 1950s. After World War II, the Queensland Cane Growers' Association reported a rapid upsurge in tractor purchases as canegrowers replaced their older models with newer tractors and/or acquired additional tractors for the farm. The extent of this upsurge can be seen by comparing the figures on the number of tractors found during the Mansfield Royal Commission into Certain Aspects of the Sugar Industry in 1942–1943 with later tractor figures for 1958–1959. Mansfield recorded 3 431 tractors in Queensland's sugar-producing districts in 1942–1943, although this figure was an underestimate as not all mill supply areas provided figures on the number of tractors. In contrast, Queensland's canegrowers operated 17 857 in 1958–1959 (Table 11.5), so there were at least two tractors per farm. One tractor was used for cultivation and weed control, while the second, heavier and more powerful tractor was devoted to heavy ploughing and grubbing.[55]

Imported Ferguson wheeled and diesel-powered tractors, developed by British Tractor and Implements, emerged as a very popular tractor model in Australian sugar-producing districts by the early 1950s. The International Harvester Co.'s Farmall range of wheeled tractors, both heavy and light models, continued to be favoured by some Queensland canegrowers because of their manoeuvrability and high clearance.[56] By 1958–1959, crawler tractors, mostly supplied by Hastings Deering, made up nearly 16% of the tractors in Australian sugar-producing districts (Table 11.5), although this figure could have been deflated because of a shortage of this type of tractor during the 1950s,

Table 11.5. The number of wheeled and crawler tractors in Australia's sugar-producing regions, 1958–1959, 1962–1963 and 1968–1969.

Region	1958/59		1962/63		1968/69	
	Wheeled	Crawler	Wheeled	Crawler	Wheeled	Crawler
Southern Qld**	3 673	387 (9.5)*	4 667	303 (5.9)	5 963	363 (5.9)
Mackay	4 738	449 (8.6)	5 359	358 (6.2)	6 450	427 (6.2)
Northern Qld#	6 605	2 005 (15.3)	7 928	1 168 (11.9)	10 533	1 767 (14.3)
New South Wales	1 510	83 (5.2)	2055	Data not collected	2 775	Data not collected
TOTAL	16 526	2 924 (15.9)	20 009	2 462 (12.0)	25 257	2 557 (10.0)

Notes: * Percentage of the total ** Incorporates the Queensland cane growing districts south of the Tropic of Capricorn # Incorporates the Queensland cane growing districts north of Bowen.

Source: Based upon figures in Bureau of Census and Statistics, Brisbane, *Tractors on Rural Holdings, Queensland*, 1962/63 and 1968/69 and Australian Bureau of Statistics, *Machinery on Rural Holdings in New South Wales*, Cat. No. 7103.1, various issues.

whereas wheeled tractors had been freely available since 1950. However, a different viewpoint was advanced in 1955 by the Macknade Mill Cane Inspector, who claimed that the Herbert River district canegrowers had been deliberately acquiring diesel-powered wheeled tractors because they were cheaper to maintain than crawler tractors. In addition, crawler tractors were less versatile than wheeled tractors, as their low height damaged young cane, so they could not be used in the cultivation of partially grown crops of cane.[57]

During the 1960s, wheeled tractors remained dominant in all Australian sugar-producing districts. Crawler tractors continued to be most popular in North Queensland where they were suited to conducting operations in wet and boggy soil. Across the entire country, however, their numbers fell, with only 10% of the tractors in Australia's sugar-producing districts being the crawler type in 1968–1969 (Table 11.5). Accompanying the shift away from crawler type tractors was an increase in the size and horsepower of the tractors used by Queensland's canegrowers. Three quarters of the tractors in the Queensland sugar-producing regions in 1962–1963 were under 40 hp

(30 kW), with very few being above 55 hp (42 kW). Six years later, however, the data suggests that some Queensland canegrowers had acquired more powerful tractors during the mid-1960s, with tractors having greater than 60 hp (45 kW) becoming more common.[58]

The acquisition of more powerful tractors provided canegrowers with certain advantages. Tractors with less than 30 hp (22 kW) could only pull two-disc ploughs, enabling fewer drills to be created each day during ploughing operations. Bigger and heavier implements such as three-disc ploughs were easily moved by the more powerful tractors, ensuring faster ploughing of the paddocks and enabling more land to be ploughed overall. These advances were welcomed during the 1950s and early 1960s as the Australian sugar industry embarked upon its major expansions. Bigger tractors and the introduction of mechanical cane harvesters during the 1960s, however, led to the start of soil compaction in the cane paddocks throughout the sugar-producing districts. Australian canegrowers, recognising the need for deep working to restore the soil to good physical condition following its compaction by such heavy machinery, engaged in more deep ripping and grubbing of the soil.

Other Australian canegrowers continued developing new agricultural implements. Fertiliser distributors made their first appearance in the Australian sugar-producing districts. They consisted of either one or two hoppers, which could easily be affixed to or detached from any grubber frame pulled by a light tractor.[59] Some of these fertiliser distributors also enabled the distribution of the insecticide 'Gammexane' which was released in the late 1940s to control cane grubs (see Chapter 14). The identification of pineapple disease, discussed in Chapter 15, and its control by dipping the cane setts in a mercurial solution led to the development of two new types of mechanical planters. In 1954, Ellis Blundell, a canegrower at Daradgee, near Innisfail, developed a cutter-dip-planter that cut the cane stalks into setts, dipped them in a bath of mercurial solution and planted them into the drills. The machine also made the drill, supplied the fertiliser to the sett and covered it with soil. Other cutter planters were equipped with a spraying attachment, so that a pressure operated spray (or bank of sprays) soaked each sett thoroughly as it fell down the chute.[60] Wyper Bros., agricultural machinery makers at Bundaberg, also released their new 'Don' single row cutter planter with a fertiliser attachment in 1958. This cutter planter's capacity had been increased by enlarged cane hoppers and its all steel construction overcame canegrower objections about the previous wooden cane platforms which were so prone to deterioration in the hot and wet climates of the sugar-producing districts. The two-row Nutall cutter planter, developed by

Nuttall's Farm Supplies at Ayr during the 1960s, was an enhanced version of the Wyper Bros.' planter. Its use sped up the planting process, enabling larger areas to be planted in less time.[61]

Advances in mechanical planter design during the 1950s, however, had not entirely mechanised the planting process. Before planting work could commence, teams of workers stripped the trash from the stalks of cane to be planted. Sometimes the canecutters had been called upon to assist with this work during labour shortages. By the late 1960s, planting operations on many Australian cane farms were being prolonged because of insufficient field labour to undertake the tedious and time consuming task of trashing the planting material, especially as many of the canecutters had been replaced by mechanical cane harvesters. By 1969, a mechanical cane stripper operated at Fairymead plantation, but the volume of cane to be trashed on this estate made its operation economical.[62] Such a machine was not the solution to the lack of labour confronting many Australian canegrowers at planting time, as they planted much smaller areas. During 1971 and 1972, some enterprising Lower Burdekin and Mackay canegrowers used chopper cane harvesters to produce billets for planting, thereby eliminating the need for manual stripping of the cane. Cecil Etwell of Walkerston, near Mackay, had developed his Etwell billet planter by October 1971. This planter was equipped with two wire-mesh bins, each holding one tonne of plant material. The arrangement enabled two rows of setts to be planted simultaneously, thereby not only eliminating the need for trashing the plant material, but facilitating the planting of large areas in a shorter time.[63]

Further developments in planter design occurred in early 1972, when Peter Gough, a Herbert River canegrower, developed his trash planter (see Plate 11.5). Planting operations were simplified by a reduction in the handling of the planting material and the elimination of trash removal. Untrashed whole stalks of cane to be planted were loaded mechanically or by hand onto detachable small trailers which were towed by tractors. Several trailers loaded and ready ensured no time was wasted reloading the planter. The small compact planter unit, mounted on the tractor's 3-point linkage, was fed by the operators sitting or standing on the trailer. A chopper box conversion kit also enabled canegrowers to convert their existing cutter planters. By August 1973, the Don Gough trash planter was described as being 'well established' on the cane farms from northern New South Wales to Mossman.[64] However, a resurgence of pineapple disease occurred after the introduction of the Don Gough trash planters, as the Aretan fungicide solution ended up being applied mostly to the trash, instead of wetting the ends of the setts. This problem was overcome by repositioning of the spray nozzles.[65]

Plate 11.5. A Don Gough trash planter in operation, c. 1976.
Source: BSES, AR 1975–76, p. 44. Image reproduction courtesy of BSES Limited.

New implements and cultivation practices were also developed to control weeds during the 1950s and 1960s. Traditionally, weeds were controlled by hand chipping and/or inter-row cultivation. However, during World War II, the synthetic plant hormone 2, 4-D (2, 4-dichlorophenoxy acetic acid) was developed, and proved to be effective at killing broadleaf weeds. BSES trials on the effectiveness of 2, 4-D under Queensland conditions commenced in the spring of 1949. During the late 1940s, 1950s and early 1960s, other synthetic herbicides were developed for either pre-emergence or post-emergence treatment for weeds. These included 2, 4, 5-T, didquat, paraquat, diuron, maleic hydrazide and dalapon, to name only a few. Tests conducted by the BSES showed that herbicides sprayed at correct dosages did not adversely affect crops of cane and were effective at controlling a variety of weeds found in Queensland's sugar-producing districts.[66] After the BSES began recommending their use, Queensland canegrowers developed new pieces of machinery or adapted old pieces of equipment to deliver these herbicides. Knapsack sprayers could be used to apply herbicides over small, isolated patches of specific weeds interfering in the growth of the cane. Moveable-arm boom sprays attached to tractors enabled the herbicides to be sprayed in irrigation ditches or over wider areas, particularly headlands. Tractor-mounted powder misters allowed herbicides to be applied to vines that choked cane crops (see Plate 11.6). Aerial spraying of 2, 4-D had also commenced by 1957.[67]

Interest in weed control, especially in mature cane crops, increased even further during the early 1960s because of the switch to mechanical cane

Plate 11.6. Tractor-mounted powder mister applying herbicides for vine control, 1971.

Source: O. Myatt, 'Vine weeds in the Herbert River District', *CGQB*, Vol. 34. no. 4 (April 1971), p. 122. Image reproduction courtesy of BSES Limited.

harvesting. Canegrowers received price deductions for supplying cane tainted by extraneous matter, mostly weed residue. The main weeds of concern were blue top (*Ageratum* spp.) and rattle pod (*Crotalaria* spp.), which grew mostly late in the season, the introduced and invasive giant sensitive plant (*Mimosa invisia*), and the pink convolvulus vine (*Ipomoea triloba*). Post-emergence herbicide application on some weed species could be done using knapsack sprayers or inter-row tractors and misters. The use of these methods proved more economical than aerial spraying, which was reserved increasingly for controlling the vines that often smothered the crops of cane. Mowing of headlands and applying arsenical preparations for killing weeds around roads and headlands also expanded considerably during the early 1960s, in order to make the way clearer for mechanical cane harvesters. In addition, controlling Para grass (*Brachiaria mutica*) and the common reed (*Phragmites australis*), found growing beside or in the numerous drainage channels constructed during the 1950s and 1960s (see Chapter 12), became a priority. Herbicides were widely applied to these two weeds that quickly clogged the newly constructed drains, thereby ensuring the more efficient movement of water through these channels.[68]

Regional variations, however, had emerged in weed control practices during the 1960s. By 1968, the BSES reported that chemicals were favoured for routine weed control in the wetter areas, especially North Queensland. Aerial spraying of vines to control the giant sensitive plant (*Mimosa invisa*) had also become routine practice in the Herbert River and Tully districts. In contrast, mechanical cultivation means were often considered sufficient elsewhere. In the Mackay district, for example, only about 1.0% of the harvested area in 1968 was treated by herbicides. Mackay district canegrowers had concluded that after treatment failures the cheapest and most effective control method was continued inter-row cultivation. BSES trials had also shown that Queensland canegrowers using pre-emergence herbicides or mechanical cultivation means received an equal financial return.[69]

Reducing the deleterious impact of weeds ensured more vigorous crops of cane, as did the application of inorganic fertilisers. Reduced supplies of inorganic fertilisers troubled canegrowers after World War II until 1949, and a shortage of meatworks manures emerged in Queensland due to fewer cattle being killed at Queensland's meatworks. Blood and bone shortages were still being reported in 1953.[70] Molasses, however, was readily available and a fifth of the molasses produced in Queensland during the late 1940s was sold to canegrowers as fertiliser (Figure 11.2). Despite the inorganic fertiliser shortages, the total amount of fertiliser applied to sugar cane in Queensland slowly increased during the late 1940s, exceeding 50 000 tonnes for the first time in 1949–1950 (Figure 11.3). Queensland canegrowers were applying an average of 0.48 tonnes of fertiliser per hectare during the late 1940s (Figure 11.4).

To guide their fertiliser use, some Queensland canegrowers sent the BSES soil samples from their properties. These samples were analysed at the Brisbane BSES laboratories and fertiliser recommendations developed for the canegrower. BSES agricultural scientists also resumed soil fertility survey work throughout the different Queensland sugar-producing districts in the late 1940s, after its interruption during World War II, and assisted the Tully Cooperative Sugar Milling Association Ltd. carry out soil testing in its mill supply area during the late 1940s and early 1950s. This type of survey involved soil sampling on each farm in the district selected for survey and determining its phosphate, potash and lime requirements. However, Norman King, the BSES Director, claimed in 1953 that fertiliser procedures in Queensland could be improved, as only 10% of the state's canegrowers had taken advantage of the fertiliser advisory service.[71]

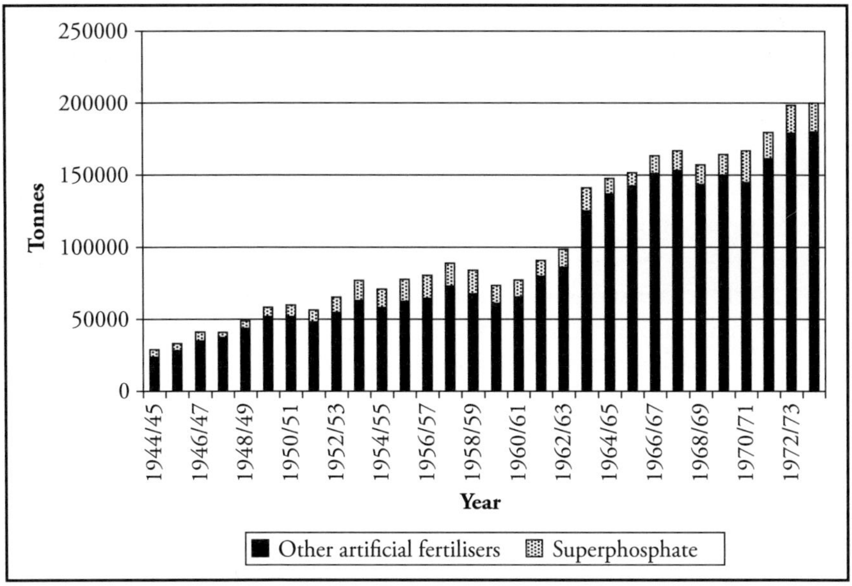

Figure 11.3. Tonnes of inorganic fertilisers applied to sugar cane in Queensland, 1944/1945–1973/1974.

Source: Based upon figures in the *Queensland Year Book,* various issues.

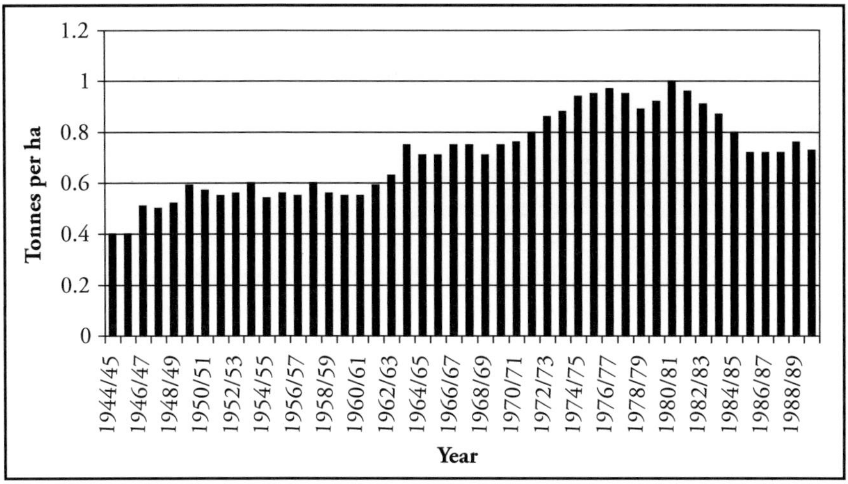

Figure 11. 4. Tonnes of inorganic fertilisers applied per hectare of sugar cane fertilised in Queensland, 1944/1945–1989/1990.

Source: Based upon figures in the *Queensland Year Book*, various issues, and Australian Bureau of Statistics, *Agricultural Land Use and Selected Inputs*, Cat. No. 7411.3, various issues.

The high costs of inorganic fertilisers in the early 1950s prompted some Queensland canegrowers to pay more attention to the correct application of these fertilisers. To achieve this goal, the canegrowers needed to know more about their soils and its deficiencies. Consequently, the number of soil samples tested for fertiliser recommendations by the BSES steadily rose during the late 1950s. Demand for this service was so strong that soil sample analysis also commenced in mid-May 1957 at the BSES's Mackay laboratories. Many samples showed low potash levels, a consequence of fertiliser shortages during World War II or the sample coming from newly assigned lands which were often deficient in this plant nutrient.[72] The number of soil samples from Queensland canegrowers analysed by the BSES fertiliser advisory service fell during the early 1960s, but then rose again as new canegrowers throughout the mid-1960s availed themselves of the service. Farmer interest in soil analysis then dwindled. By the early 1970s, many Queensland canegrowers had established a sound approach to using fertilisers and no longer needed such a service. The soil sampling service ended in 1974 and the function was taken up by commercial laboratories. The BSES continued offering advice about fertilisers, but the new service shifted the emphasis from plant food reserves in the soil to requirements for crop production that took into account the cost of fertilisers and the price obtained for sugar.[73]

Running parallel to the BSES fertiliser advisory service and its experiments on correct rates of fertiliser usage were the activities of CSR. Annual agricultural circulars issued to CSR's canegrowers contained recommendations on the rates of fertiliser use, based upon the results from field trials which were resumed in 1949. Canegrowers supplying CSR's Harwood Mill, for example, were advised in 1963 to apply during planting a mixture low in nitrogen, medium in phosphate and medium in potash at a rate of 2–3 cwt (0.10–0.15 tonnes) per acre.[74] In 1957, CSR further developed its fertiliser advisory service by introducing foliar analysis in the Herbert River district. Foliar analysis involved examining samples of leaves collected from a field of cane at an appropriate time. This examination provided information on the vigour of the cane's growth and whether the right amount of fertiliser had been applied. These trials showed that on average the Herbert River district crops were under-supplied with nitrogen to the extent of the equivalent of 1.00 cwt (0.05 tonnes) of sulphate of ammonia per acre and over-supplied with phosphorous by 0.85 cwt (0.04 tonnes) of sulphate of ammonia per acre. By 1960, CSR used foliar analysis as a basis for fertiliser advice to canegrowers in respect to Pindar ratoon crops grown for Macknade and Victoria Mills. The service was then expanded to CSR's other mill supply areas, but was withdrawn during the mid-1970s following a decline in support for the service by canegrowers.[75]

During the early 1950s, Australian canegrowers obtained their plant nutrition requirements from various traditional sources that had existed since the 1910s. Molasses remained a favourite soil enrichment agent, with a quarter of the molasses produced in Queensland during the 1950s being sold as fertiliser, mostly to canegrowers (Figure 11.2). Rich in potash, molasses was clearly being used by Queensland canegrowers to alleviate the potash deficiency that had been identified immediately after World War II in some of the soils used for growing sugar cane in Queensland. However, molasses begins to be exported following the establishment of the Australian Molasses Pool in 1963, and the product was increasingly used as stock feed by Australian pastoralists.[76] Consequently, the amount of molasses available for sale as fertiliser dwindled, leading to a sharp drop in its use as a fertiliser (Figure 11.2). Phosphorous was delivered by the application of superphosphate, with approximately 12 000 tonnes of the product being applied each year to the crops of sugar cane in Queensland between 1950–1951 and 1955–1956 (Figure 11. 3). Nitrogen continued to be sourced from applications of sulphate of ammonia or leguminous crops (see discussion later in chapter). In addition, figures for CSR's mill supply areas showed that some canegrowers were still treating small areas, usually under 50 ha, with organic fertilisers such as blood and bone.[77] However, major changes occurred in the range of fertilisers available during the late 1950s and 1960s, reducing the proportions of the market held by superphosphate and sulphate of ammonia.

The first new fertiliser product appearing on the market was urea, a white crystalline and very soluble material containing almost double the amount of nitrogen that was found in sulphate of ammonia. Its more favourable price than sulphate of ammonia then ensured urea's use as a top dressing rapidly accelerated during the late 1950s and early 1960s.[78] Urea's dominance, however, was soon challenged by aqua ammonia or water in which ammonia gas is dissolved. It progressively became available in Queensland during 1961 and 1962, following A.C.F. and Shirleys Ltd. completing its bulk terminals at the Ports of Brisbane, Cairns, Townsville and Mackay, enabling the distribution of this cheaper form of nitrogen to the sugar-producing districts. Unlike other fertilisers which were applied as a solid on top of the soil, aqua ammonia was carried on the tractors in small tanks and the liquid injected into the root zone from nozzles fitted behind the cultivator tines. Its popularity, however, had declined by 1972, with ammonium phosphate and ammonium nitrate – two new fertiliser products developed during the 1960s – being preferred by some Australian canegrowers.[79]

Fertiliser was delivered to Australian cane farms during the 1950s in bags weighing 73 kg. Following representations from the canegrowers, A.C.F. and Shirleys Ltd. introduced new, lighter 50 kg polythene lined jute fertiliser bags in October 1960. Further modifications to the delivery of fertilisers occurred after the development of aqua ammonia. Some canegrowers using the product erected bulk tanks with a capacity of either 500 gallons (1 892 litres) or 1 000 gallons (3 784 litres) on their farms. These tanks were situated on stands at a suitable height, enabling the aqua ammonia to gravitate into the smaller tanks mounted on the tractors. Bulk handling of fertiliser was introduced into the Bundaberg and Mackay districts during 1969, and then spread to other sugar-producing districts throughout Australia. In some instances, fertiliser was brought by tip trucks to the cane farms, and subsequently transferred to bins or fertiliser distributors. On other Australian cane farms, the fertiliser arrived in one ton or five ton plastic bags. These were then attached to a tripod or A-frame, so they were held above the fertiliser distributor. Cost savings derived from using bulk handling were substantial. Australian canegrowers in 1971 paid $40.50 to receive five tons of fertiliser in bags. The same amount arriving in bulk cost $24.50.[80]

Handling fertiliser, however, could be eliminated entirely by its aerial distribution. This practice commenced in Australia during 1956, when the BSES successfully applied sulphate of ammonia by aeroplane to a crop of cane on the Bureau's Pathology farm, at Eight Mile Plains, near Brisbane. Successful aerial distribution trials with urea were then conducted by the BSES during mid-1957 and the first aerial fertilising of a canegrower's crop occurred at Mackay in November 1957. New South Wales canegrowers had also adopted the practice by 1964. Aerial fertilising offered the advantage of quick application. Three tonnes of urea could be spread by aerial application during a morning, whereas the equivalent amount would take a week to be spread by a single row, tractor-drawn fertiliser distributor. Crops could also be fertilised from the air when they had become too high for application from a tractor or when the ground was too wet for tractors to enter the fields. Uneven distribution of the fertiliser across the paddocks and drift onto non-cultivated areas, however, was a problem of this method. Despite its advantages, aerial fertilising never became the most widely used method of applying fertiliser to crops of sugar cane – many Australian canegrowers persisted with ground-based distribution methods.[81]

The amount of fertiliser applied on Queensland's cane crops doubled from 50 000 to 100 000 tonnes between 1949–1950 and 1963–1964 (Figure 11.3). Much of this increased use in fertiliser occurred because new lands were being brought under cultivation, as the amount of fertiliser applied per hectare remained at 0.56 tonnes for most of the 1950s (Figure 11.4). This situation, however, changed during the latter part of the 1960s. The total amount of fertiliser applied on Queensland's sugar cane crops doubled from 100 000 tonnes to 200 000 tonnes between 1963–1964 and 1973–1974 (Figure 11.3). Part of this substantial increase can be again attributed to the application of fertiliser to the new lands brought under cultivation during the 1964–1966 expansion, although the amount of fertiliser applied per hectare jumped from 0.59 tonnes in 1961–1962 to 0.88 tonnes in 1973–1974 (Figure 11.4). Fertiliser use figures for New South Wales revealed a similar trend. The total amount of fertiliser applied to sugar cane in New South Wales trebled during the 1960s, rising from just over 2 000 tonnes in 1960–1961 to nearly 7 000 tonnes in 1974–1975. Part of this upsurge can be attributed to the application of fertiliser to the new lands being brought under cultivation, although New South Wales canegrowers were also applying more fertiliser to each hectare cultivated during the early 1970s, the amount rising from 0.5 tonnes per ha in 1964–1965 to 0.6 tonnes per ha by 1974–1975.[82]

Several reasons can explain this upsurge in the amount of fertiliser applied per hectare of sugar cane cultivated during the 1960s. Firstly, the price for inorganic fertilisers supplied to North Queensland canegrowers fell after May 1955, following the commencement of fertiliser manufacture in Cairns by North Queensland Fertilisers and Chemicals Ltd., a subsidiary of the Brisbane-based A.C.F. and Shirleys Fertilisers Ltd. Additional price falls throughout 1956, 1958 and 1959, and late 1961 made fertilisers more affordable at a time when incomes for Australian canegrowers were increasing. Indeed, CSR noted that with more stable incomes, its New South Wales canegrowers throughout the early 1960s were prepared to spend extra on purchasing fertilisers, especially for ratoons.[83] Secondly, cheaper fertiliser would not automatically mean greater use of the product, except the new products were easier to use, and some Australian canegrowers were increasingly over fertilising. Norman King, the BSES Director, reported in 1965 that he knew of many cases where it was not uncommon for canegrowers to apply a ton of fertiliser per acre (2.5 tonnes per ha). In 1970, the BSES reported that it suspected many Lower Burdekin canegrowers had been applying above the recommended rate because they believed some fertiliser might be 'lost in application', and in the Maryborough district there was 'a general tendency amongst growers to apply excess amounts of nitrogen on plant and ratoon crops'.[84] Yet as early as 1957, the BSES had shown that heavy dressings of fertiliser at levels higher than 0.5 tonnes per hectare (or four cwt per acre) of mixed fertiliser plus a top dressing of 0.5 tonnes of sulphate of ammonia per hectare (or four cwt per acre) were uneconomic and did not induce any additional benefits in yields.[85] Thirdly, as the industry expanded, some canegrowers commenced cultivating poorer, nitrogen-deficient soils, and applied increased amounts of fertiliser to obtain appropriate yields on this marginal country.[86]

Restoring soil organic matter was also advocated by the agricultural scientists during the 1950s and 1960s, although the practice was not always undertaken by the country's canegrowers. Trash conservation did not recommence during the late 1940s, even though it had been used by some canegrowers in CSR's mill supply areas before the war.[87] In other Queensland mill supply areas, the expansion in the pre-harvest burning of the crops of cane that occurred during World War II persisted after 1945, thereby reducing the amount of trash available for conservation, and the raking and burning of any crop residues created during harvest was common before ratooning. Results from the BSES's long-range trash conservation trial established at Bundaberg in 1933, and a similar trash trial at Mackay between 1949 and 1959, had also revealed that trash conservation did not improve cane yields.[88] Consequently, the BSES had an ambivalent attitude towards trash conservation, although its staff acknowledged that the practice could prevent soil erosion (see Chapter 12).

In the absence of widespread trash conservation, Australian canegrowers during the 1950s and 1960s were urged to replenish the soil's organic matter through green manuring. Favourite pre-war green manure crops such as Mauritius beans, Gambia pea and Giant Cowpea were gradually replaced during the 1950s with Poona pea and new varieties of velvet beans and cowpeas. To assist Queensland canegrowers, the BSES had continued its experiments, commenced in 1930, on identifying the types of leguminous crops most suitable to the wide range of conditions throughout the sugar-producing districts. From 1955 onwards, the BSES had also initiated a legume breeding program to produce pest- and disease-resistant varieties of beans and peas.[89]

Lime continued being added to cane growing soils, especially in North Queensland, to improve their condition during the 1950s and 1960s. The product was applied by any of the conventional lime-spreaders or increasingly it was applied direct from motor-trucks which had conveyed the lime to the farm. The contents of a few bags of lime were emptied onto the floor of the truck and men with shovels threw the lime over the side, covering a strip 8 m wide with a fairly even application. Canegrowers could obtain lime from two new limestone mines established in the 1940s – the Ryan Lime Co.'s limestone mine at Calcium, near Woodstock, or the Northern

Lime Co.'s limestone quarry just south of Mossman – or the limestone mines shown in Figure 11.1, most of which were still operational during the 1940s and 1950s. Lime sourced from mining coral reefs had probably ceased by the early 1940s, although the idea had not been forgotten. An application by the Cairns District Cane Growers' Executive to restart coral mining for agricultural lime in 1967 was rejected by the local mining warden.[90]

Bagasse was also being used as a soil enhancement agent during the 1950s and 1960s, although the relationship between applying bagasse to paddocks and better yields was less clear than the link between lime and improved productivity. CSR had been treating fields with bagasse in its Victoria Mill supply area since 1948 and found that bagasse-treated sections of fields during drought conditions retained moisture better than untreated sections. Yet CSR did not advance bagasse as a major soil enhancement product. In 1953, BSES agricultural scientists at Mackay concluded much the same as CSR after experiments on adding bagasse to paddocks. By the mid-1960s, the BSES considered that bagasse would be best utilised in the improvement of the physical condition of poorly drained soils, particularly if these had become compacted. The main disadvantages with using the product as a soil enhancer were the economics of carting it from the sugar mill to the field were doubtful over long distances and there were difficulties incorporating the bagasse into the soil. Despite the BSES's ambivalent attitude towards the use of bagasse on fields, some canegrowers adopted the practice, although the BSES noted in 1970 that its application during the past years had been 'haphazard'.[91]

The first trace element disorder in Australian sugar cane crops was identified by BSES scientists during 1949 and 1950 in the Mackay and Maroochy districts. Copper deficiency caused a symptom known as droopy top – young leaves of the cane plants drooped and the stalks became rubbery and flexible. Large areas of sandy coastal land brought under cultivation during the late 1950s and 1960s required applications of copper to enable the production of satisfactory crops. Applications of copper sulphate, usually mixed with fertiliser, overcame the symptoms of droopy top.[92] A magnesium deficiency was also identified in some soils used to grow sugar cane in Queensland during the 1950s. Sugar cane grown on these soils suffered a condition known as orange freckle. Typical freckle symptoms are patches of young cane with yellowing of the lower leaves and a reddish brown freckle on leaf surfaces, leading to losses in cane yield, poor sett root development and poor root growth in mature cane. Applications of magnesium reduced freckling, but there were no consistent yield responses to treatments.[93]

Switching to Minimum Tillage, 1975–1995

By the mid-1970s, Queensland canegrowers cultivated an average of 50 ha per farm; New South Wales canegrowers grew on average about 20 ha less (Figure 11.5). The cropping systems on these farms had several distinctive features. Heavy machinery was used to prepare the soil for the setts, planting of cane, cultivation of the growing plant crop and subsequent harvesting and transport of the harvested cane to the sugar mills. When the final ratoon crop had been harvested, Australian canegrowers generally removed the rootstock by tillage. Assignment restrictions forced canegrowers to fallow some of their land as they were not allowed to harvest the entire assigned area. Fallow planting strategies during the summer period involved growing a legume crop or sometimes the paddocks were left as a bare or weedy fallow. Weeds were controlled mechanically or increasingly through the application of herbicides. Soil productivity was maintained by the increasing use of inorganic fertilisers, which were one of the most expensive costs associated with cane growing, being second only to harvesting.[94]

Costs associated with growing a crop of cane suddenly increased during the mid-1970s. Fertiliser prices rose rapidly between 1974 and 1978: the cost of urea and aqua ammonia doubled and the price of superphosphate trebled. Further price increases for fertiliser occurred during the early 1980s. Wages for field workers in the Australian sugar industry increased by over two-thirds in the two years after March 1974. The price of diesel increased due to the OPEC-induced oil crisis, although fuel prices did decline during the late 1970s after this spike in the mid-1970s.[95] Moreover, the average area cultivated per sugar cane farm between 1974 and 1984 rose by almost 22 ha in Queensland and 13.6 ha in New South Wales (Figure 11.5), thereby adding to overall production costs. Yet farm returns per tonne of sugar fell throughout the mid-1970s, rallied slightly, and then had dropped again by 1985. By the mid-1980s, many Australian canegrowers suffered major financial stress, as they dealt with a substantial cost-price squeeze. The combination of these circumstances forced some canegrowers to question traditional cultivation practices and their impact upon the natural environment.

Increased cultivation costs were not the only factor encouraging a change in agricultural practices amongst some Australian canegrowers during the late 1970s. Mechanical harvesting throughout the 1960s and early 1970s was still accompanied by the pre-harvest burning of the cane crops. This practice assisted the earliest mechanical cane harvesters which struggled to

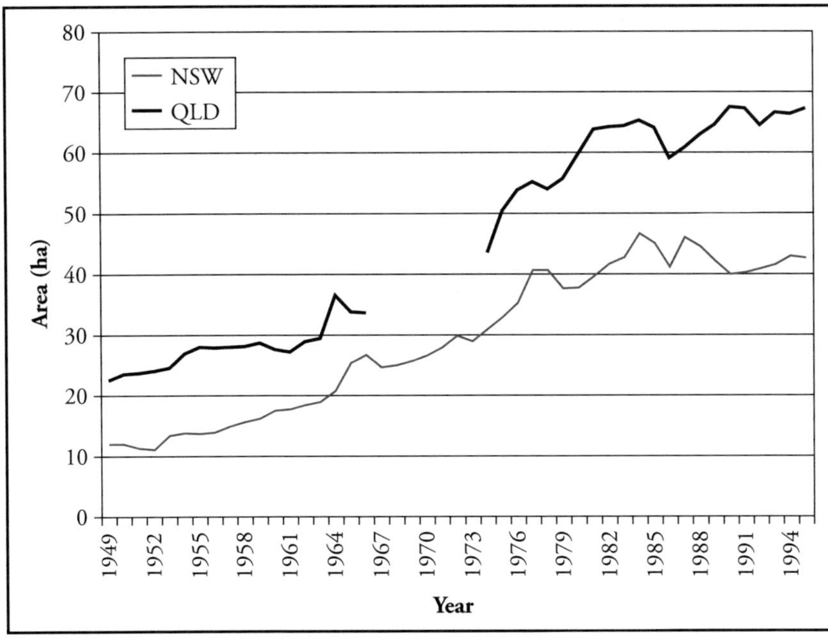

Note: Figures could not be calculated for Queensland between 1966 and 1973 due to incomplete statistics.

Figure 11. 5. Area (ha) cultivated per farm producing cane for harvest in Australia, 1949–1995.

Source: Calculated from figures in the *ASYB,* various issues.

handle large amounts of trash. In addition, any residues generated during the harvest were still raked into heaps and burnt by the canegrowers. The idea of green cane harvesting, however, had been explored during the early 1960s by Mackay brothers, Ron and Dick Creber, who developed a modified version of the Falkiner cane harvester that would mechanically harvest unburnt cane. Remo and Guido Ghidella of Babinda also used a modified Massey-Ferguson chopper harvester to experiment with green cane harvesting in the mid-1960s, following poor burns that are so frequent in the Babinda district. Yet the concept never gained acceptance amongst canegrowers, and the Creber cane harvester cut its last cane in 1973.[96]

The idea of green cane harvesting would have been forgotten about except for severe disruptions to the 1973 and 1975 harvest seasons by prolonged periods of wet weather, which delayed the burning of crops or retrieving burnt cane from wet fields. During the 1976 crushing season, the Rossi family at Aloomba, near Gordonvale, purchased a Claas green cane harvester and demonstrated that it was possible to harvest green cane successfully by machine. In 1985, the practice was still not widely used in Queensland due to the limited green cane capacity of most mechanical cane harvesters. Eventually, the wider adoption of green cane harvesting was made possible during the late 1980s by the release of the Austoft 7000 series mechanical cane harvester and by successful modifications to older model mechanical cane harvesters, enabling them to cut unburnt cane.[97]

Green cane harvesting produced larger amounts of post-harvest residues than burnt cane harvesting and not all canegrowers were inclined to burn this trash, allowing it to cover their fields to slow soil erosion (see Chapter 12). Trials initiated by the BSES in North Queensland in 1977 sought to identify any unexpected problems associated with trash retention. Results from these trials and subsequent investigations during the 1980s and early 1990s, including those conducted by CSR, confirmed that trash retention or blanketing caused no agronomic problems such as reduced ratoon growth, reduced yields or an increased incidence of diseases, although canegrowers were at risk from a heavy trash blanket accidentally burning and the possibility of this fire spreading to unharvested cane.[98] Ironically, the drier years of the early 1980s provided a boost for trash blanketing, as North Queensland canegrowers observed the moisture conserving effects of a heavy trash layer. Later investigations showed that the extra moisture stored under a trash blanket was assumed to be equivalent to two irrigations in the Bundaberg district and a trash blanket at Mackay increased available water for a crop by an estimated 1.8 megalitres per hectare, an extra $35 saving for canegrowers per megalitre of water conserved.[99]

As trash blanketing was adopted by more Queensland canegrowers during the 1980s, the practice became associated with minimum tillage techniques. One of the earliest observed changes was the substantial reduction in the cultivation of ratoon crops. A BSES study in the Mackay district found that canegrowers between 1976 and 1981 averaged just over six cultivations per year in ratoon crops. Cultivation was used to break up crusted and compacted layers following the crop's harvest by heavy machinery, control weeds and to allow fertiliser placement into the soil. Minimum tillage in ratoon crops sought to cut through the trash blanket with minimal disturbance, break up any compacted soil layer and place the fertiliser in the centre of the inter-row space in a single pass. By 1985, a Queensland Department of Primary Industries study at Mackay found that the number of cultivations in ratoon cane had been reduced on average to less than four. Many mechanical operations such as stubble shaving, inter-row cultivating and sub-soiling were used less in the management of ratoons, as their benefits began to be questioned.[100] Moreover, investigations during the mid-1980s highlighted that substantial savings could be made by switching to minimum tillage methods. A survey amongst a group of Bundaberg canegrowers in 1985, for example, revealed that cultivation costs for a 40 ha cane assignment ranged from $15 949 for conventional burnt cane, burnt trash and intensive cultivation, to $13 056 for green trash blanket and no tillage of ratoon crops.[101]

Trash blanketing also reduced weed growth, but those weeds that did grow still needed to be destroyed. Herbicides largely replaced mechanical control methods during the late 1970s and early 1980s. Moreover, the problem of spray drift and its negative effects on the cane leaves were reduced by the invention of new herbicide application equipment. By 1985, innovative El Arish canegrowers, Frank and Tony Costanzo, had developed an adjustable four-row sprayer with hinged droppers that trailed behind the spray boom. The spray nozzles were attached to this trailing leg. The legs were built using a parallelogram system that kept the spray nozzles at a constant height above the ground, thereby applying the spray to the weeds more accurately and the spray pattern from the nozzles could be directed away from the leaves of the growing cane. Herbert River district canegrower, John Irvin, further developed this arrangement, by using two trailing legs per interspace. The legs were curved so that they kept the spray nozzles positioned in the centre of the interspace. The 'Irvin leg' was further modified by the addition of an adjustable T-bar, allowing the adjustment of the nozzle height to cater for varying heights of the growing cane in the fields.[102]

Canegrowers may have used their tractors less as they switched to minimum tillage, but the trend towards increasing tractor size with more powerful engines continued during the late 1970s and 1980s. Large tractors with kilowatt ratings ranging from 75–150 became more common and four-wheel drive articulated steering type tractors made their appearance in the cane fields. These new tractors were fitted with sophisticated inbuilt equipment such as hydrostatic steering, auto-transmission, depth control and weight transference power. Many of these new tractors also improved the working conditions for the operator by being fitted with air-conditioned cabins to reduce dust and noise hazards and to provide protection from the sun. The increased power of these tractors enabled canegrowers to move into multi-row cultivation and to obtain bigger implements such as five disc ploughs, four-furrow mouldboard ploughs and large multi-tined rigid rippers. Fertiliser distributors featuring large twin boxes holding up to 0.5 tonnes of fertiliser became standard equipment as they could be moved by the larger tractors.[103] Australian canegrowers also turned to high clearance tractors equipped with spray equipment, inter-row spray equipment or aerial spraying to apply herbicides to the weeds, especially the vines which continued troubling North Queensland canegrowers, even with trash blankets.[104]

Some machinery designed for traditional cultivation methods on Australian cane farms was not entirely suited to minimum tillage practices and it had to be modified. By 1983, Paul and Laurence Mizzi, two Herbert River district canegrowers, had developed a minimum tillage fertiliser applicator that created minimum soil disturbance when applying fertiliser to the fields covered with a trash blanket. The design of this implement enabled two slits to be made by coulters in the trash blanket of each interspace. Fertiliser was placed in the soil beneath an undisturbed layer of trash on each side of the row behind the coulters and a narrow ripper tine. In the Mackay area during 1984, a centre ripper minimum fertiliser applicator was developed to achieve the goal of applying fertiliser with minimum soil disturbance. Fertiliser was fed into the soil behind a coulter/ripper which moved down the centre of the interspace breaking up the soil in the compacted inter-row zone. By 1987, over 800 implements based on this design were being used on cane farms in the Mackay district.[105]

Rising labour costs also meant that planting operations came under scrutiny during the late 1970s and early 1980s. Don Gough planters required three workers: a tractor driver and two operators to feed the stalks of cane through the cutting unit. By the late 1970s, Peter Populin of Populin Enterprises, Home Hill, and Paul Mizzi had developed automatically-fed billet planters which required only two operators. Bonel Bros. Pty. Ltd. of Bundaberg went one step further by releasing a single person operated automatic billet planter in 1982.[106] These machines, pulled behind tractors, still enabled the setts to be dipped in fungicide and the application of fertiliser at the time of planting (Plate 11.7). However, to ensure successful planting operations, Australian canegrowers were required to supply good quality planting material – the billets needed to be the correct size, with limited eye damage and not fractured or split. Mechanical chopper cane harvesters were modified to reduce damage during the production of the billets. By the late 1980s, the successful introduction of automatic billet planters, rising production costs and shortages of field workers had even prompted some canegrowers in the Richmond River district to abandon individual planting operations and form two planting cooperatives. Three workers using the Populin billet planter handled 8 ha in a day, compared to conventional methods which required six workers to plant 2 ha daily.[107]

The switch to minimum tillage was accompanied by two other major changes in agricultural practices between 1975 and 1995. Firstly, the number of ratoon crops being grown industry-wide almost doubled from 2.2 in 1966 to 4.3 in 1986.[108] Part of this trend was clearly related to cost savings.

Plate 11.7. A billet planter in action, c. 1995.

Source: CANEGROWERS, *Cane Growers' Information Handbook 1995/96* (Brisbane, 1996), p. 57. Image reproduction courtesy of CANEGROWERS, Brisbane.

The variable costs of growing ratoon crops of sugar cane are less than the variable costs of growing a plant crop, until a point is reached where the loss of production returns from a ratoon crop is greater than the cost of replanting. However, the adoption of newer cane varieties (e.g. Q. 124 and Q. 138) with better ratooning capabilities also made this switch possible. Secondly, fallow planting declined. Traditionally, the roots of the final ratoon crop were removed through cultivation towards the end of the crushing season (October/November). Over the summer period, either a legume fallow crop was grown on the paddock, or it was left bare and became covered with weeds. Fallowing had been forced on Queensland canegrowers due to a legislative requirement. Prior to 1965, Queensland canegrowers could only harvest 75% of their gross assigned area. In 1965, this amount was increased to 85% and in 1974 this restriction was removed completely. With the lifting of the restrictions, some Queensland canegrowers moved towards a plough-out/replant cycle as a means of increasing overall farm production. During the 1980s, between 80 and 90% of the assigned area in Queensland was harvested annually. Other canegrowers, however, persisted with a legume fallow crop, although there was a trend during the 1980s towards growing complementary crops such as soybeans and maize which provided an additional source of revenue. However, legume husbandry had been severely downgraded by the early 1990s, resulting in poor legume growth which negated many of their potential benefits.[109]

The switch to minimum tillage did not diminish the need for Australian canegrowers to consider plant nutrition, especially as some of the country's cane growing lands had been in continuous use for a century by the 1970s. BSES advice about the topic was guided by the information its field staff gained from the soil fertility monitoring program commenced in 1975 as a replacement method for studying cane nutrition. Eight fertility monitoring sites were established within assigned cane blocks in each of the thirty Queensland mill supply areas. Chemical and physical analyses of soil samples from each site were made at the start of the program and then repeated every five years. Other BSES trials during the late 1970s and early 1980s gathered more data on soil fertility trends and sugar yield responses from nitrogen, potassium and phosphorous fertilisers.[110]

A review of all the soil fertility trial data from 1964 onwards occurred in 1984. In the following year, the BSES issued new optimum fertiliser recommendations. Nitrogen recommendations were based upon yield response functions developed in different cane growing districts. However, separate nitrogen recommendations for different districts, apart from the Lower Burdekin, were not issued on the grounds that the differences in the mean yield responses calculated for each district were not significant enough at varying rates of fertiliser application. Nitrogen recommendations were lower for plant cane than ratoons and were reduced in special situations such as when dry conditions limited the yield. Advice on the rates and methods of application for other essential elements were based upon soil analysis, where the approach was to maintain soil nutrient levels between a critical level and an upper level where no response to fertiliser can be expected. Consideration was also given to economics in the framing of these recommendations. The amounts of nitrogenous fertiliser to be applied were recommended when the sugar price was above or below $300 a tonne. These guidelines were reviewed and re-issued in 1995.[111]

Critics of the BSES guidelines, however, suggested that they had not changed much over the last fifty years, were too broad and with the exception of nitrogen, did not differentiate for regions, climatic conditions or soil types. Moreover, they had been developed when maximisation of production and short-term economic gains were the key factors driving the industry, not sustainability, especially the creation and maintenance of a high level of soil organic material. A survey of Herbert River district canegrowers in 1995 also found that they considered single fertiliser recommendations for the application of nitrogen, phosphorous and potassium covering the entire district to be unsatisfactory.[112]

The average amount of fertiliser applied per hectare of sugar cane cultivated in Queensland during the 1970s hovered around 0.9 tonnes, but jumped to 1.0 tonne per hectare in 1980–1981 (Figure 11.4). Clearly, the increased fertiliser prices in the mid-1970s did not immediately force Queensland canegrowers to curtail their use of the product. Moreover, a BSES study into fertiliser application to crops of sugar cane harvested from 240 sites across Queensland between 1976 and 1982 revealed that many Queensland canegrowers used above the BSES recommended rates of application for nitrogen, potassium and phosphorus for better economic conditions. CSR's Herbert River technical field staff and the BSES suggested that some Queensland canegrowers believed that cane yields could be increased by simply applying more fertiliser or that the perceived falling productivity of their land, in some instances brought about by soil erosion, could be solved by using more fertiliser.[113] However, after the 1980–1981 season, the average amount of fertiliser applied per hectare cultivated begins falling, reaching 0.73 tonnes per hectare in 1989–1990, the last year that the Australian Bureau of Statistics collected figures on fertiliser application to specific crops in Queensland. This decline presumably occurred in response to low sugar prices forcing Australian canegrowers to limit their fertiliser use.[114]

Types of fertiliser applied to sugar cane crops in Queensland changed during the late 1970s and 1980s. Superphosphate use fell over this period, probably because older cane growing areas had built up high phosphorous levels due to a long history of phosphorous application.[115] Urea became the most popular nitrogen fertiliser used by canegrowers, in preference to sulphate of ammonia (Table 11.6). Urea with its high nitrogen concentration was competitively priced compared with other nitrogen sources. The switch in the late 1970s and early 1980s to bulk handling systems and one tonne bags for solid fertilisers, like urea, made it less labour-intensive to store and apply.[116] The trend to trash blanketing also enhanced urea's popularity as the product could be easily surface-applied, without soil disturbance. Aqua ammonia lost market share on account of the erosion in the cost advantage it once enjoyed over urea. Moreover, the falling demand for aqua ammonia increased its distribution and maintenance costs per tonne of the product. By 1990, aqua ammonia was no longer marketed by Incitec Fertilisers Ltd. in North Queensland and its availability in cane growing districts was restricted to immediately north and south of Brisbane. Blended fertilisers (e.g. Grocane 150 and Grocane 44) increased in popularity during the early 1990s, because they conveniently combined two or more principal plant foods in a single application.[117]

Table 11.6. Tonnes ('000) of artificial fertilisers used on the crops of sugar cane in Queensland, for selected years between 1974–1975 and 1989–1990.

Year	Superphos-phate	Urea	Sulphate of ammonia	Other straight nitrogenous fertilisers	Other or mixed fertilisers
1974/75	20.5	31.1	21.7	65.7	83.6
1976/77	25.5	45.7	15.5	76.5	100.8
1978/79	20.9	40.8	11.9	64.2	87.6
1980/81	31.9	*	*	138.7**	122.1
1982/83	26.8	*	*	129.0**	116.3
1984/85	17.6	61.8	10.9	41.3	105.7
1986/87	16.4	61.8	7.8	28.1	95.5
1987/88	19.2	61.8	6.8	23.8	102.3
1989/90	15.9	*	*	*	119.3

* No details ** Figures include amounts of urea and sulphate of ammonia.

Source: Based upon figures in *Queensland Year Book*, 1974 to 1982 and Australian Bureau of Statistics, *Agricultural Land Use and Selected Inputs*, Cat. No. 7411.3, various issues.

Efficient use of nitrogenous fertilisers, given their escalating costs, became a major concern of some Australian canegrowers during the 1980s and early 1990s. The presence of trash on the soil surface posed a problem for the management and efficient use of urea, the most popular fertiliser. Many Australian canegrowers changed their method of fertiliser placement from banded application beneath the soil surface to banded or broadcast applications on top of the trash. However, Australia's canegrowers were aware of the potential high nitrogen loss that could occur from the surface applications of urea and wondered if their crops were receiving enough nitrogen. Hence, some Australian canegrowers commenced pulling chains behind the fertiliser distributor to disturb the trash and to shake the fertiliser granules through the trash; others mounted coulters beneath their fertiliser boxes, applying the fertiliser beneath the soil surface.[118] BSES studies during the late 1980s, in response to an upsurge in canegrower questions about the placement of urea fertiliser on cane fields with trash blankets and its rates of application, confirmed that anywhere up to 40% of the nitrogen applied to the surface of cane trash was lost via ammonia volatilisation.[119] Additional Australian canegrowers during the early 1990s switched to applying urea in narrow bands beneath the trash using coulters to try and reduce this loss. Yet a joint CSIRO/CSR/Griffith University study on ammonia loss following urea addition to sugar cane trash blankets in 1994 concluded that the majority of canegrowers in districts where green cane harvesting was most advanced (i.e. Mackay northwards) still applied urea or urea blends to the trash surface at the rate of 160–200 kg nitrogen per ha so that a loss of 64–80 kg nitrogen per ha by volatilisation would have little impact upon final yields. The study's authors claimed such irresponsible action constituted a 'great waste of resources on a national scale'.[120]

Soil amelioration continued in Australia's sugar-producing districts during the late 1970s and 1980s. The analysis of soil samples from North Queensland during the 1977 revealed that calcium and magnesium levels were low in substantial areas of all districts between Ingham and Mossman. These deficiencies had emerged after many years of using high analysis fertilisers (i.e. fertilisers with high levels of nitrogen or phosphorous). Applications of calcium and magnesium were initially limited in North Queensland, but such ameliorative practices were more widespread by the early 1990s. Zinc deficiencies also emerged in North Queensland sugar-producing districts during the early 1980s following the adoption of high analysis fertilisers and the decline of superphosphate as a phosphorous source (see above), because zinc was often a contaminant of superphosphate. Applications of zinc sulfate heptahydrate, a soluble granulated product incorporated into the soil prior to planting, overcame this trace element deficiency.[121] By the mid-1980s, some cane crops in the Mossman and Mackay districts had also commenced displaying sulfur deficiencies. Symptoms included leaves that developed a faint purplish tinge, leaves failing to develop their full length and plants lacking overall vigour. Further BSES investigations revealed that approximately 15% of the 18 000 ha of caneland from Mossman to Ingham was potentially sulfur deficient. Applications of low sulfur fertilisers such as urea, aqua ammonia and diammonium phosphate were blamed for the emergence of this problem. Increased application of sulphate of ammonia was recommended to meet the sulfur requirements.[122] Industry expansion during the 1970s and 1980s also resulted in cane being increasingly grown on sodic soils, especially in the Bundaberg, Mackay, Proserpine and Lower Burdekin districts. Sodic soils contain excessive amounts of water-soluble salt. Symptoms of soil salinity often became apparent within five to ten years after clearing the natural vegetation, and range in intensity from slight restrictions on cane growth to complete scalding of the soil surface. Reclamation of sodic soils was achieved by the application of gypsum or lime.[123]

By the early 1990s, the average area cultivated on a Queensland and New South Wales cane farm was around 65 ha and 42 ha, respectively (Figure 11. 5). Moreover, this cane was now grown by a mixture of cultivation practices. At one end of the spectrum were those Australian canegrowers who used minimum tillage. They adopted reduced tillage planting that involved killing the old stools by spraying them with herbicides such as glyphosphate (Roundup®) prior to the wet season and then cultivating a narrow inter-row strip for planting (known as ploughout/replant). Cultivation was done with a ripper-rotary hoe combination, although this method was not suitable for all soils. Green cane trash blanketing with reduced tillage was used on successive ratoon crops. Generally, the main operations involved a single pass with a fertiliser applicator and the spraying of weeds with a contact herbicide if necessary. In some instances, controlled traffic zones were established on properties to reduce the amount of soil compaction and damage to soil structure.[124] At the other end of the spectrum were canegrowers who continued using more conventional agricultural practices. Planting involved one or two offset discings or ploughing to remove old stools, one or two rippings of the paddocks, additional offset discings and rotary hoeing to break up the soil and to create a fine seed bed to receive the setts. Cultivation of ratoon crops commenced with the raking and burning of the residues after harvest, cultivation of the interspace using some combination of a rotary hoe, ratoon discs or tines, the application of the fertiliser and additional cultivations with tines or the use of pre-emergence or contact herbicides to control weed growth. Other canegrowers adopted a combination of methods, including burnt cane trash blanketing with reduced tillage on their ratoon crops. Under this latter method, canegrowers were left with a reduced trash blanket because of pre-harvest burning, but they loosened the soil minimally and controlled weeds by spraying with herbicides.[125]

At the beginning of this chapter, the analysis showed that Australian canegrowers in 1915 were just beginning to mechanise the cultivation of sugar cane to minimise their labour costs. This task had been completed by the early 1950s, although Australia's canegrowers continued designing innovative farm machinery and implements to reduce their labour costs and to permit the switch to minimum tillage agriculture during the 1980s and early 1990s. Mechanisation may have reduced production costs, but growing a crop of cane also required an understanding of soils, their properties and deficiencies and plant nutrition. Increasingly refined information on these topics became available after 1945, although many Australian canegrowers probably used more fertiliser than the recommended application rates, even in the late 1980s. Moreover, growing a crop of sugar cane throughout Australia's sugar-producing regions required more than just an understanding of the soil conditions on each property. Some Australian canegrowers, as will be seen in the next chapter, had also developed a better appreciation of how to manage water, which was also vital to the success of producing a crop of sugar cane.

Notes

1 Bolton, *A Thousand Miles Away*, p. 252; *ASJ*, 3 October 1912, p. 525; Chief Protector of Aboriginals, Qld., AR 1912, *QPP*, Vol. 3 (1913), pp. 1073 & 1079; Kerr, *Pioneer Pageant*, p. 161.

2 A. Barnes, *Agriculture of the Sugar-Cane* (London, 1953), p. 254; Rolleston, *The Defiance*, p. 113; *ASJ*, 3 August 1916, p. 316.

3 *Statistical Register of New South Wales*, 1911, p. 582; *ASJ*, 7 December 1911, p. 499 & 11 January 1912, p. 557; Kerr, *A Century of Sugar*, p. 153; BSES, AR 1917–18, *QPP*, Vol. 1 (1918), p. 1151.

4 *ASJ*, 10 May 1917, p. 132, 10 July 1919, p. 267 & 7 July 1922, pp. 177–178 & 195–196.

5 I. Johnston, *Agricultural Tractors: their evolution in Australia* (Kenthurst, NSW, 1985), pp. 59, 75 & 77; *ASJ*, 10 May 1917, p. 132; Rolleston, *The Defiance*, p. 113.

6 Kerr, *Southern Sugar Saga*, p. 101; Kerr, *Black Snow Liquid Gold*, p. 193.

7 Report of the Piddington Royal Commission, *CPP*, Vol. 4 (1920–21), p. 927; *ASJ*, 7 July 1922, p. 196; Easterby, *The Queensland Sugar Industry*, p. 173; Clarence River Branch of the NSWCGA, AR 1928, in Minute Book of the Clarence River Branch, NSWCGA, 1925–1930, Maclean, NSW.

8 Easterby, *The Queensland Sugar Industry*, pp. 173–174; *ASJ*, 7 February 1918, p. 714; Manning, *In their Own Hands*, pp. 209–210.

9 BSES, AR 1924–25, *QPP*, Vol. 2 (1926), p. 689; BSES, AR 1925–26, *QPP*, Vol. 2 (1927), p. 653; F. Wheelhouse, *Digging Stick to Rotary Hoe: Men and Machines in Rural Australia* (Sydney, 1966), pp. 202–203.

10 BSES, AR 1901–1902, *QPP*, Vol. 3 (1902), p. 235; *ASJ*, 9 April 1914, p. 29; Kerr, *Northern Outpost*, p. 86; *NQR*, 2 July 1906, p. 17.

11 Victoria Mill Cane Insp., AR 1918, p. 3, CSRR, Z/109, Box 474 & *ASJ*, 5 April 1917, p. 47; for views on the high prices of lime see *ASJ*, 12 June 1919, p. 183, BSES, AR 1914–15, *QPP*, Vol. 2 (1915–1916), p. 1175, BSES, AR 1916–17, *QPP*, Vol. 3 (1917), p. 858; & Ev. H. Easterby, p. 151, in 'Piddington Royal Commission Evidence', CSRR, Z/109, Box 45.

12 H. Easterby, *The Cultivation of Sugar Cane in Queensland*. BSES General Series, Bulletin No. 3 (Brisbane, 1924), p. 12.

13 B. Daley and P. Griggs, 'Mining the reefs and cays: coral, guano and rock phosphate extraction in the Great Barrier Reef, 1844–1940', *Environment and History*, Vol. 12, No. 4 (November 2006), pp. 409–424.

14 BSES, AR 1920–21, *QPP*, Vol. 2 (1922), p. 1034; *ASJ*, 13 March 1925, p. 837 & 5 June 1925, p. 159.

15 *ASJ*, 5 April 1917, p. 47 & 6 December 1917, p. 641; Macknade Mill Cane Insp., AR 1918, pp. 14–15, CSRR, Z/109, Box 472; Report of the Piddington Royal Commission, *CPP*, Vol. 4 (1920–21), p. 923; BSES, AR 1917–18, *QPP*, Vol. 1 (1918), p. 1150; BSES, AR 1918–19, *QPP*, Vol. 2 (1919–20), p. 1286; L. Vallance, 'The history of sugar soils investigation and agricultural research', in BSES (ed.), *Fifty Years of Scientific Progress* (Brisbane, 1950), p. 26.

16 *ASJ*, 13 March 1925, p. 794; Macknade Mill Cane Insp., AR 1922, pp. 12, CSRR, Z/109, Box 472.

17 H. Kerr, 'Molasses as fertiliser', *Proc. ISSCT*, 1932, Bulletin No. 56, p. 1; *ASJ*, 7 January 1926, p. 645.

18 BSES, AR 1926–27, *QPP*, Vol. 2 (1927), p. 725; BSES, AR 1927–28, *QPP*, Vol. 2 (1929), p. 869; *ASJ*, 5 June 1930, p. 176; *QAJ*, 1 January 1928, p. 10.

19 BSES, AR 1917–18, *QPP*, Vol. 1 (1918), p. 1150; BSES, AR 1918–19, *QPP*, Vol. 2 (1919–1920), p. 1286; BSES, AR 1926–27, *QPP*, Vol. 2 (1927), p. 725; Easterby, *The Cultivation of Sugar Cane in Queensland*, pp. 14–15 & 18.

20 CSR, *Green Manuring and Trash Conservation. Circular to Canegrowers* (Sydney, 1926), pp. 7–8; E.W. Knox to Goondi Mill Mang., 23 April 1924, CSRR, 142/1528, No. 315 of 1924; E.W. Knox to Victoria Mill Mang., 9 April 1924, CSRR, 142/1574, p. 231.

21 H. Freeman, 'Agricultural work in Queensland and New South Wales', *Proc. ISSCT*, Honolulu, 1924, p. 27; Broadwater Mill Cane Insp., AR 1927, p. 7, CSRR, Z/109, Box 460.

22 CSR, *Green Manuring and Trash Conservation. Circular to Canegrowers*, pp. 2–5.

23 E.W. Knox to Macknade Mill Mang., 8 September 1926, CSRR, 142/1530, Letter No. 671; K. MacLean, 'Report on Trash Conservation and Green Manuring, Macknade Mill', 5 January 1934, CSRR, NS126/129, pp. 53–54.

24 H. Easterby, 'Cane cultivation in Queensland', *Proc. ISSCT*, 1930, pp. 254–258; Ev. H. Easterby, in 'Gunn Royal Commission Evidence, 1930', pp. 711–712, CSRR, Z/109, Box 46.

25 Broadwater Mill Cane Insp., AR 1932, pp. 11, 12 & 21 & Broadwater Mill Cane Insp., AR 1933, pp. 9–10, in CSRR, Z/109, Box 460; Ev. A. Watson, in 'Gunn Royal Commission Evidence, 1930', p. 842, CSRR, Z/109, Box 47.

26 W. McPherson, 'Agricultural Field Report, Broadwater Mill, 1936', p. 3, CSRR, Z/109, Box 496; for tractor numbers in NSW in 1939 see Bureau of Census and Statistics, NSW, *Statistical Register. Rural Industries and Settlement and Meteorology*, Cat. No. 7101.1, 1954.

27 Extract from *The Canegrower Weekly*, 3 May 1934, in AER, 165/196, File No. 53 C, NBAC; W. Gibson, 'Horse and tractor power in farm practice', *Proc. QSSCT*, 1932, pp. 53–54; K. Maclean, 'Report on Agricultural Methods at Goondi Mill, 1933', Letter No. 114 N (b), p. 69, CSRR, 142/1528; Macknade Mill Cane Insp., AR 1939, pp. 15–16, CSRR, Z/109, Box 472; *ASJ*, 10 December 1937, advt. for McCormick-Deering F-30 tractors.

28 Macknade Mill Cane Insp., AR 1935, p. 19, CSRR, Z/109, Box 472; K. Maclean, 'Report on Agricultural Methods at Goondi Mill, 1933', Letter No. 114 N (b), p. 69, CSRR, 142/1528; The Agricultural Staff, 'Caneland cultivation in Queensland', *Proc. ISSCT*, Brisbane, 1935, p. 73; H. Kerr and A. Bell, *The Queensland Cane Growers' Handbook* (Brisbane, 1939), p. 57.

29 The Agricultural Staff, 'Caneland cultivation in Queensland', pp. 75 & 78; *ASJ*, 6 July 1933, p. 191; A. Gibson, 'Trials with the stubble shaver in the Burdekin district', *CGQB*, Vol. 1, no. 2 (1 October 1933), p. 34.

30 Manning, *In their Own Hands*, p. 207; Rolleston, *The Defiance*, p. 79; C. Story, 'Cutter planters in the Mackay district', *CGQB*, Vol. 12, no. 4 (April 1949), pp. 183–186.

31 *ASJ*, 9 March 1933, p. 631; NSW Department of Agriculture, *The Farmers' Handbook*, 5th edition (Sydney, 1933), pp. 622–624.

32 H. Kerr, 'Molasses as fertiliser', *Proc. ISSCT*, 1932, Bulletin No. 56; *ASJ*, 4 September 1934, p. 321; N. King, 'Soils investigations in the sugar-cane producing areas of Bundaberg', *The Qld J. of Agricultural Science*, Vol. 6, no. 3 (September 1949), p.172; BSES, AR 1937–38, pp. 12–13.

33 G. Bates, 'Some notes on the soils of the North Queensland sugar areas', *Proc. QSSCT*, 1934, pp. 259–263; N. King, 'Soils of the Bundaberg area', *Proc. QSSCT*, 1935, pp. 29–32; N. King, 'Cane soils of North Queensland', *QAJ*, Vol. 47, Part 4 (April 1937), pp. 375–380.

34 W. McPherson, Broadwater Mill, 'Agricultural Report for 1933', p. 4, Z/109, Box 496; for examples of agricultural circulars see CSR, Agricultural Circular, Herbert River District, 1935, Z/109, Box 370 & *ASJ*, 9 April 1931, pp. 41–43, 8 June 1935, p. 163 & 10 September 1937, p. 385; for these fertiliser trials in the 1930s see reports in CSRR, Z/109, Box 386; for securing legume seed see CSR, Cane Report 1931, pp. 91–92, CSRR, Z/109, Box 387.

35 Kerr and Bell, *The Queensland Cane Growers' Handbook*, pp. 25–26; CSR, Cane Report 1933, p. 67, CSRR, Z/109, Box 386.

36 Broadwater Mill Cane Insp., AR, 1936, p. 9, CSRR, Z/109, Box 461; W. McPherson, 'Agricultural Report for Broadwater Mill area, 1937', p. 2 & W. McPherson, 'Agricultural Report for Broadwater Mill area, 1938', p. 2, CSRR, Z/109, Box 496; CSR, Cane Report 1936, p. 65, CSRR, Z/109, Box 387.

37 E. Smith and W. Dixon, 'Report on Pioneer and Inkerman Mills: General, 30 September 1931', p. 2, CSRR, Z/109, Box 520; H. Kerr, 'Some agricultural practices of the Mackay district', *Proc. QSSCT*, 1940, p. 22; Kerr, Mungomery and Hughes, *Manual of Cane-Growing*, 1953 edition, pp. 48–49.

38 Goondi Mill Cane Insp., AR 1933, p. 14 & Goondi Mill Cane Insp., AR 1939, p. 11, CSRR, Z/109, Box 465; Ev. H.W. Kerr, in 'Royal Commission into Sugar Peaks and Cognate Matters Evidence, 1939', p. 1003, CSRR, Z/109, Box 47.

39 *ASJ*, 5 November 1931, p. 412, 8 June 1934, p. 149, 9 May 1935, p. 109 & 8 June 1935, p. 166; *Year Book of Australia*, 1946–1947, p. 965.

40 Ev. H.W. Kerr, in 'Royal Commission into Sugar Peaks and Cognate Matters Evidence, 1939', p. 1003, CSRR, Z/109, Box 47; BSES, AR 1938–39, p. 11.

41 BSES, AR 1940–41, p. 5; *ASYB*, 1940–41, p. 292; *ASYB*, 1944–45, pp. 243–244.

42 BSES, AR 1940–41, p. 5; BSES, AR 1941–42, p. 3; *ASYB*, 1940–41, p. 293; *ASJ*, 12 August 1941, p. 205 & 10 January 1942, p. 375.

43 C. Story, 'Cutter planters in the Mackay district', *CGQB*, Vol. 12, no. 4 (April 1949), p. 180; K. Gard, 'Report on Pests, Diseases and General Field Practices, Macknade Mill area, 1940', p. 9, CSRR, Z/395, Box 63, Letter 303N.

44 Harwood Mill Cane Insp., AR 1941, p. 9, CSRR, Z/109, Box 470; Goondi Mill Cane Insp., AR 1943, p. 22 & Goondi Mill Cane Insp., AR 1944, p. 14, CSRR, Z/109,

Box 465; for green manuring see CSR, Cane Report 1945, p. 59, CSRR, Z/109, Box 387 & Ev. of D. Robins, President of the NSWCGA. in 'Tariff Board Inquiry Evidence, 1944', p. 3, CSRR, Z/109, Box 52.

45 Macknade Mill Cane Insp., AR 1942, p. 11, CSRR, Z/109, Box 472; Ev. of D. Robins, President of the NSWCGA, in 'Tariff Board Inquiry Evidence, 1944', p. 3, CSRR, Z/109, Box 52.

46 Based upon figures in 'Papers Relating to the Royal Commission to Investigate Certain Aspects of the Sugar Industry, 1942/1943', in RS1236/1/270, QSA.

47 Hambledon Mill evidence, Q. 20, Babinda Mill evidence, Q. 20 & Mulgrave Mill evidence, Q. 20, in Folder 146/17, Cairns Sitting, in 'Papers Relating to the Royal Commission to Investigate Certain Aspects of the Sugar Industry, 1942/1943', RS1236/1/270, QSA; Harwood Cane Insp., AR 1943, p. 8, CSRR, Z/109, Box 470; Ev. of D. Robins, President of NSWCGA in 'Tariff Board Inquiry Evidence, 1944', p. 3, in CSRR, Z/109, Box 52.

48 BSES, AR 1943–44, pp. 1–4; BSES, AR 1944–45, pp. 1–4.

49 BSES, AR 1945–46, p. 7.

50 R. Jackson, *Salute to the Pioneers. Biographies of Pioneers of the Home Hill District* (Home Hill, 1993), p. 61; E. Brady, *The Land of the Sun* (London, 1924), p.80; for descriptions of forest clearing see A. Mackay, *Sugar Cane in Australia* (Sydney, 1883), pp. 25–26, A. Bicknell, *Travel and Adventures in Northern Queensland* (London, 1895), p. 42, J. Devanny, *Bird of Paradise* (Sydney, 1945), pp. 41–42 & King, Mungomery and Hughes, *Manual of Cane-Growing*, 1953 edition, p. 60.

51 *ASYB*, 1954, p. 99; King, Mungomery & Hughes, *Manual of Cane-Growing*, 1953 edition, pp. 59–61; Rolleston, *The Defiance*, p. 115; for use of explosives see Hudson, *By the Banyan*, p. 120 & *Herbert River Express* (Ingham), 11 September 1952, p. 2.

52 *ASYB*, 1970, pp. 29 & 38; L. Tilley, 'An efficient stone picker', *CGQB*, Vol. 34, no. 1 (July 1970), p. 36; J. Usher, 'New land clearing devices', *CGQB*, Vol. 29, no. 1 (July 1965), pp. 12–14.

53 King, Mungomery and Hughes, *Manual of Cane-Growing* revised 1965 edition, p. 41.

54 Harwood Mill Cane Insp., AR 1947, p. 7, CSRR, Z/109, Box 470; Clarence River Branch, NSWCGA, AR 1947–48, in Minute Book, Clarence River Branch, NSWCGA, March 1936 to Feb. 1951; *The Producers' Review*, October 1958, p. 71; CSR, *The Preparation of Land for Sugar Cane in NSW* (Sydney, 1958), p. 6; for tractor figures see Australian Bureau of Statistics, *Machinery on Rural Holdings in New South Wales*, Cat. 7103.1, 1958/59.

55 Papers Relating to the Royal Commission to Investigate Certain Aspects of the Sugar Industry, 1942/1943, RS1236/1/270, QSA; *ASJ*, 16 February 1953, p. 771.

56 Hambledon Mill Cane Inspector, AR 1957, p. 12, CSRR, Z/109, Box 468; *ASYB*, 1952, p. 184; *ASYB*, 1957, p. 98; *ASJ*, 15 September 1948, p. 363; *The Producers' Review*, 15 March 1956, p. 111 & 15 August 1956, p. 63.

57 *The Producers' Review*, 15 August 1956, p. 63; Macknade Mill Cane Insp., AR 1955, p. 12, CSRR, Z/109, Box 473.

58 Based upon figures in Bureau of Census and Statistics, Brisbane, *Tractors on Rural Holdings, Queensland*, 1962/63 & 1968/69.

59 C. Toohey, 'New fertilizing machines in the Mackay district', *CGQB*, Vol. 17, no. 3 (January 1954), pp. 87–88; S. Skinner, 'Mechanical fertilizing', *CGQB*, vol. 11, no. 1 (July 1947), pp. 40–42; BSES, AR 1947–48, p. 26.

60 King, Mungomery & Hughes, *Manual of Cane-Growing*, revised 1965 edition, pp. 44–45; *The Producers' Review*, 15 January 1954, p. 42; BSES, AR 1951–52, p. 44; *ASYB*, 1954, p. 27.

61 *The Producers' Review*, July 1958, p. 49 & April 1967, p. 20.

62 I. Freshwater, 'Growers' revolutionise planting techniques', *CGQB*, Vol. 35, no. 3 (January 1972), p. 76; *ASJ*, July 1969, pp. 202–203.

63 Freshwater, 'Growers revolutionise planting techniques', p. 76; *ASJ*, June 1971, pp. 142–143, October 1971, p. 332 & September 1972, pp. 300–303; Anon, 'Labour-saving planter', *CGQB*, Vol. 35, no. 4 (April 1972), p. 124.

64 *ASJ*, April 1972, p. 59 & August 1973, p. 241–251; O. Myatt, 'A "New Look" cane planting system', *CGQB*, Vol. 36, no. 1 (July 1972), pp. 11–12; *ASYB*, 1974, p. 179–180; CSR, AR Technical Field Work 1974–75, p. 204, CSRR, Z/109, Box 393.

65 Anon, 'Tests show billet planter sound but can be improved', *CGQB*, Vol. 41, no. 3 (January 1978), p. 86; CSR, AR Technical Field Work, 1974–75, p. 205, CSRR, Z/109, Box 393.

66 BSES, AR 1949–50, p. 17; BSES, AR 1954–55, p. 24; C. Toohey and A. Matthews, 'Chemical weedicides- Mackay district', *CGQB*, Vol.18, no. 1 (July 1954), pp. 19–20.

67 BSES, AR 1957–58, pp. 28–29; *ASJ*, January 1958, p. 801; C. Rehbein, 'A boom spray outfit', *CGQB*, Vol. 19, no. 1 (July 1955), pp. 14–15; I. Linedale, 'A boom sprayer for weeds in mature cane', *CGQB*, Vol. 33, no. 4 (April 1967), pp. 138–139; H. Young, 'Weed control', *CGQB*, Vol. 25, no. 4 (April 1962), pp. 32–33.

68 BSES, AR 1960–61, p. 21; BSES, AR 1962–63, p. 26; A. Linedale, 'Fighting the common reed', *CGQB*, Vol. 38, no. 2 (October 1974), pp. 46–47; O. Myatt, 'Vine weeds in the Herbert River district', *CGQB*, Vol. 34, no. 4 (April 1971), pp. 121–122.

69 BSES, AR 1967–68, p. 24; BSES, AR 1968–69, p. 24; E. Pembroke, 'The control of weeds in the Mackay area', *CGQB*, Vol. 33, no. 3 (January 1970), pp. 96–97; *ASJ*, June 1969, p. 176; D. Neal and L. Barnes, *Cardwell Shire Handbook* (Brisbane, 1976), pp. 7–10.

70 BSES, AR 1946–47, p. 9; BSES, AR 1949–50, p. 10; *ASYB*, 1949, p. 121; *ASYB*, 1953, p. 105.

71 BSES, AR 1952–53, p. 10; N. King, *Soils investigation in the sugar cane producing areas of Bundaberg*. BSES Technical Communication No. 3, 1949; L. Vallance, 'The history of sugar soils investigation and agricultural research', p. 36.

72 BSES, AR 1948–49, p. 16; AR BSES 1949–50, p. 11; BSES, AR 1953–54, p. 19; BSES, AR 1956–57, p. 25.

73 BSES, AR 1971–72, p. 27; BSES, AR 1974–75, p. 19; L. Chapman, M. Haysom and C. Chardon, 'Checking the fertility of Queensland's sugar land', *Proc. ASSCT*, 1981, pp. 325–333; O. Sturgess, 'Soil fertility monitoring a basis for cane nutrition research', *CGQB*, Vol. 45, no.1 (July 1981), p. 3.

74 For these trials see papers in CSRR, Z/109, Box 387 & CSR, Agricultural Circular for New South Wales, 1963, p. 4, CSRR, Z/109, Box 370.

75 CSR, Cane Report 1957, p. 53, CSRR, Z/109, Box 388; CSR, Cane Report 1960, p. 51, CSRR, Z/109, Box 389; R. Farquhar, 'The interpretation and use of tissue analysis within a fertiliser advisory service for sugar cane in North Queensland', *Proc. ISSCT*, 1965, pp. 227–236; B. Schroeder *et al.*, 'Leaf analysis – what does it offer the Australian sugar industry', *Proc. ASSCT*, 1999, p. 127.

76 The Australian Molasses Pool comprised all the Australian sugar mills. CSR acted as administrator of the pool, which maintained bulk molasses tanks at the bulk sugar

terminals. This organisation was superseded by the Australian Molasses Trading Pty. Ltd. in 1993 to act as the sugar industry's exporter of bulk molasses.

77 CSR, Cane Report 1951, p. 88 & CSR, Cane Report 1953, p. 84, CSRR, Z/109, Box 388.

78 BSES AR 1957–58, p. 11; King, Mungomery and Hughes, *Manual of Cane-Growing*, revised 1965 edition, pp. 77 & 79; L. Vallance, 'Speaking of nitrogen', *CGQB*, Vol. 27, no. 2 (October 1963), p. 64.

79 BSES, AR 1960–61, pp. 9–10; BSES, AR 1961–62, p. 8; *ASYB*, 1963, p. 139; *ASJ*, 28 August 1961, p. 369 & 30 April 1962, p. 93; J. Pulsford, 'Historical inputs of fertiliser nutrients on to agricultural lands of coastal Queensland', in D. Yellowlees (ed.), *Land Use Patterns and Nutrient Loading of the Great Barrier Reef Region* (Townsville, 1991), p. 48.

80 *ASYB*, 1961, p. 29; *ASYB*, 1963, p. 139; C. Chardon, 'Bulk fertiliser is here', *CGQB*, Vol. 34, no. 3 (January 1971), pp. 103–104; *ASJ*, September 1972, p. 289.

81 BSES, AR 1955–56, p. 10; BSES, AR 1957–58, p. 11; *The Producers' Review*, 15 December 1957, p. 91; Harwood Mill Cane Insp., AR 1965, p. 13 & Harwood Mill Cane Insp., AR 1972, p. 22, CSR, Z/109, Box 471.

82 Calculated from figures in Bureau of Census and Statistics, NSW, *Statistical Register. Rural Industries and Settlement and Meteorology*, various issues, and Australian Bureau of Statistics, *Agricultural Sector: Fertilisers and Other Improvements*, Cat. No. 7104.1., various issues.

83 *ASJ*, 19 April 1956, p. 82, 14 March 1958, p. 909 & 15 August 1959, p. 393; *ASYB*, 1961, p. 57; CSR Mill Insp., memo: NSW Mill Inspection, August 1961, CSRR, Z/109, Box 538; for a description of the Cairns superphosphate plant see *The Producers' Review*, 15 May 1957, p. 38.

84 N. King, 'Rising costs of cane production', *CGQB*, Vol. 28, no. 3 (January 1965), pp. 4–5; BSES, 'Notes on nitrogen fertilising from cane growing areas', *CGQB*, Vol. 34, no. 2 (1970), pp. 45–48.

85 *ASYB*, 1960, p. 61; BSES, AR 1970–71, p. 21; BSES, AR 1956–57, pp. 21–22.

86 R. Moller, 'Phosphate requirements of new soils', *CGQB*, Vol. 28, no. 2 (October 1964), p. 42; J. Anderson, 'Fertilisers for new land in the Isis area', *CGQB*, Vol. 28, no. 2 (October 1964), p. 51.

87 Victoria Mill Cane Insp., AR 1946, p. 16, CSRR, Z/109. Box 474; Hambledon Mill Cane Insp., AR 1946, p. 14, CSRR, Z/109, Box 468; CSR, Agricultural Circular for Broadwater Mill, 1953, p. 4, CSRR, Z/109, Box 370.

88 BSES, AR 1945–46, p. 24; King, Mungomery & Hughes, *Manual of Cane-Growing*, 1953 edition, pp. 50–57; King, Mungomery Hughes, *Manual of Cane-Growing*, revised 1965 edition, pp. 128–129.

89 BSES, AR 1950–51, p. 11; BSES, AR 1951–52, p. 11; G. Christie, 'Legume seed for the sugar industry', *CGQB*, Vol. 24, no. 4 (April 1961), pp. 130–131; CSR, Cane Report 1961, Part 2, p. 60, CSRR, Z/109, Box 389; King, Mungomery & Hughes, *Manual of Cane-Growing*, revised 1965 edition, pp. 122–127; CSR, *The preparation of land for sugar cane* (Sydney, 1958), pp. 7–9.

90 King, Mungomery & Hughes, *Manual of Cane-Growing*, revised 1965 edition, p. 109; Qld. Dept. of Mines, AR 1947, *QPP*, Vol. 2 (1948–89), p. 101; J. Wright, *The Coral Battleground* (Melbourne, 1977), pp. 7–14.

91 CSR, Cane Report 1949, p. 89, CSRR, Z/109, Box 387; BSES, AR 1953–54, p. 10; King, Mungomery and Hughes, *Manual of Cane-Growing*, revised 1965 edition,

pp. 129–130; C. McAleese, 'Bagasse surplus may be useful', *CGQB*, Vol. 23, no. 4 (April 1970), pp. 118–119.

92 King, Mungomery & Hughes, *Manual of Cane-Growing*, revised 1965 edition, pp. 88–89; BSES, 1949–50, p.11; BSES, AR 1955–56, p.10; BSES, AR 1956–57, p. 22.

93 D. Ridge, A. Hurney and M. Hayson, 'Some aspects of calcium and magnesium nutrition in North Queensland', *Proc. ASSCT*, 1980, p. 55.

94 J. Reghenzani, 'A survey of the nutritional status of North Queensland sugarcane soils with particular references to zinc', *Proc. ASSCT*, 1993, p. 298.

95 *ASJ*, July 1974, p. 187; *ASYB*, 1977, p. 118; *ASYB*, 1979, p. 58; *ASYB*, 1983, p. 269.

96 Burrows and Morton, *The Canecutters*, pp. 186–187; *ASYB* 1970, pp. 27–28; *ASJ*, August 1966, pp. 313–315 & December 1966, pp. 615–617.

97 *ASJ*, August 1976, p. 247–248; *ASYB*, 1978, pp. 221–225; L. Brennan *et al.*, 'Improving sugarcane harvesting procedures', in B. Keating and J. Wilson (eds.), *Intensive Sugarcane Production: Meeting the Challenges Beyond 2000* (Wallingford, UK, 1997), p. 329.

98 See for example R. Hyde and L. Teske, 'Surface management systems in the Central District canelands, Queensland', *Proc. ASSCT*, 1987, pp.71–73; A. Wood, 'Management of crop residues following green cane harvesting of sugar cane in Queensland', *Soil Tillage Research*, Vol. 20 (1991), pp. 69–85; L. Tilley, 'Trash blankets prove successful in the Central District', *BSES Bulletin* No. 52, October 1995, p. 14.

99 J. Hardman *et al.*, 'Agronomic and economic aspects of various farming systems for sugarcane in the Bundaberg district', *Proc. ASSCT*, 1985, p. 148; Tilley, 'Trash blankets prove successful in the Central District', p. 14; A. Wood, 'Green cane trash management in the Herbert Valley. Preliminary results and research priorities', *Proc. ASSCT*, 1986, pp. 85–94.

100 R. Hyde and L. Teske, 'Surface management systems in the Central District canelands, Queensland', p. 42; T. Wilcox, A. Garside and M. Braunack, 'The sugarcane cropping system', in M. Hogarth and P. Allsopp (eds.), *Manual of Cane Growing* (Brisbane, 2000), pp. 137–138.

101 N. Smith *et al.*, 'Green cane harvesting – a review with particular reference to the Mulgrave Mill area', *Proc. ASSCT*, 1984, pp. 25–26 & J. Hardman, L. Tilley and T. Glanville, 'Agronomic and economic aspects of various farming systems for sugar cane in the Bundaberg District', *Proc. ASSCT*, 1985, pp. 149–153.

102 A. Garside *et al.*, 'The yield plateau in the Australian sugar industry: 1970–1990', in B. Keating and J. Wilson (eds.), *Intensive Sugarcane Production. Meeting the Challenges Beyond 2000* (Wallingford, UK, 1997), p. 111; J. Hardman *et al.*, 'Agronomic and economic aspects of various farming systems for sugarcane in the Bundaberg District', p. 148; AR BSES 1982–83, p. 27; AR BSES 1984–85, p. 5; G. McMahon, P. Lawrence and T. O'Grady, 'Weed control in sugarcane', in M. Hogarth and P. Allsopp (eds.), *Manual of Cane Growing* (Brisbane, 200), pp. 244–246 & 253–254.

103 CSR, AR Technical Field Work 1973–74, p. 159, CSR, AR Technical Field Work 1974–75, p. 207 & CSR, AR Technical Field Work 1975–76, pp. 110 & 115, CSRR, Z/109, Box 393; *ASYB*, 1994, p. 60; *The Producers' Review*, November 1975, pp. 34–35; *AC*, January 1982, pp. 16–17.

104 D. Burgess, 'High clearance tractor conversion saves money', *BSES Bulletin*, No. 51, July 1995, pp. 22–23; McMahon, Lawrence and O'Grady, 'Weed control in sugarcane', pp. 254–255.

105 R. Dick and A. Hurney, 'Technological change in sugar cane agriculture', *Proc. ASSCT*, 1986, pp. 69–75; Hyde and Teske, 'Surface management systems in the Central District canelands, Queensland', p.42; Anon, 'Ingham growers modify machines for trash retention', *BSES Bulletin*, Jan. 1984, No. 5, p. 14; G. McMahon and P. English, 'Reduced tillage popular in the central district', *BSES Bulletin*, Oct. 1984, No. 8, pp. 8–9.

106 Anon, 'Tests show billet planter sound but can be improved', *CGQB*, Vol. 41, no. 3 (January 1978), pp. 85–86; *AC*, April 1979, p. 51, July 1980, pp. 75–76 & June 1982, p. 68.

107 *AC*, January 1987, p. 22.

108 L. Chapman, 'Constraints to production in ratoon crops', *Proc. ASSCT*, 1988, p. 189.

109 Hyde and Teske, 'Surface management systems in the Central District canelands', p. 42; Anon, 'Legumes provide major benefits for cane crops', *CGQB*, Vol. 45, no. 1 (July 1981), pp. 32–33; A. Garside and M. Bell, 'Fallow legumes in the Australian sugar industry: review of recent research findings and implications for the sugarcane cropping system', *Proc. ASSCT*, 2001, p. 231; Garside *et al.*, 'Yield plateau in the Australian sugar industry', p. 107.

110 L. Chapman, 'Estimating sugar yield responses from N, P and K fertilisers in Queensland', *Proc. ASSCT*, 1982, pp. 147–153; B. Schroeder, A. Wood and G. Kingston, 'Re-evaluation of the basis for fertiliser recommendations in the Australian sugar industry', *Proc. ASSCT*, 1998, pp. 240–242.

111 Anon, 'Sugarcane nutrition review', *BSES Bulletin*, No. 10, April 1985, pp. 4–19; L. Chapman, 'Fertiliser management in Australia', *Proc. ASSCT*, 1994, p. 83–84; D. Calcino, *Australian sugarcane nutrition manual* (Brisbane, 1995).

112 A. Wood *et al.*, 'Opportunities for improving nutrient management in the Australian sugar industry', in B. Keating and J. Wilson (eds.), *Intensive Sugarcane Production. Meeting the Challenges Beyond 2000* (Wallingford, 1997), pp. 250 & 252–253; Schroeder, Wood and Kingston, 'Re-evaluation of the basis for fertiliser recommendations in the Australian sugar industry', pp. 242–243.

113 L. Chapman, 'Rationalising fertiliser practices', *Proc. ASSCT*, 1985, p. 156; A. Wood, 'Soil degradation and management under intensive sugarcane cultivation in North Queensland', *Soil Use and Management*, Vol. 1, no. 4 (December 1985), p. 120.

114 L. Chapman, 'Rationalising fertiliser practices', *Proc. ASSCT*, 1985, pp. 155–156; M. Smith, *Recent Production Trends in the Queensland Sugar Industry* (Brisbane, 1991), p. 20; Garside *et al.*, 'Yield plateau in the Australian sugar industry', p. 112.

115 P. Bloesch, G. Rayment and J. Pulsford, 'Regional total phosphorous budgets for sugar production in Queensland', *Proc. ASSCT*, 1997, p. 216.

116 *ASYB*, 1979, p. 230; *AC*, July 1982, p. 22; CANEGROWERS, *Cane Growers' Information Handbook 1995/96* (Brisbane, 1996), p. 62.

117 G. Kuhn, 'Fertiliser products for coastal Queensland', in D. Yellowlees (ed.), *Land Use Patterns and Nutrient Loading of the Great Barrier Reef Region* (Townsville, 1990), pp. 32–38.

118 P. Prammanee, A. Wood and P. Saffinga, 'Nitrogen loss from urea applied to sugarcane crop residues', *Proc. ASCT*, 1988, p. 119; J. Freney *et al.*, 'Ammonia loss following urea addition to sugar cane trash blankets', *Proc. ASSCT*, 1994, p. 114; G. Kuhn, 'Fertiliser products for coastal Queensland', p. 37.

119 See for example P. Prammanee, A. Wood and P. Saffinga, 'Nitrogen loss from urea applied to sugarcane crop residues', *Proc. ASCT*, 1988, pp. 119–124 & D. Calcino and P. Makepeace, 'Fertiliser placement on green cane trash blanketed ratoons in North Queensland', *Proc. ASSCT*, 1988, pp. 125–130.

120 J. Freney *et al.*, 'Ammonia loss following urea addition to sugar cane trash blankets', *Proc. ASSCT*, 1994, pp. 114–121.

121 D. Ridge, A. Hurney and M. Haysom, 'Some nutritional aspects of calcium and magnesium nutrition in North Queensland', *Proc. ASSCT*, 1980, p. 55; Garside *et al.*, 'The yield plateau in the Australian sugar industry: 1970–1990', pp. 112–113; D. Calcino, G. Kingston and M. Haysom, 'Nutrition of the plant', in M. Hogarth and P. Alsopp (eds.), *Manual of Cane-Growing* (Brisbane, 2000), p. 181.

122 *ASYB*, 1988, p. 34; L. Chapman, 'A review of sulphur nutrition of sugarcane in Queensland', *Proc. ASSCT*, 1985, pp. 113 & 119; Calcino, Kingston and Haysom, 'Nutrition of the plant', p. 179; J. Reghenzani, 'A survey of the nutritional status of North Queensland sugarcane soils with particular references to zinc', *Proc. ASSCT*, 1993, p. 298–304.

123 G. Kingston, 'Soil salinity a hazard to productivity in southern areas', *BSES Bulletin,* No. 12, October 1985, pp. 14–17; G. Ham, A. Cox and G. McMahon, 'A review of sodic soils research in the Queensland sugar industry', *Proc. ASSCT*, 1997, p. 200.

124 D. Burgess, 'Innisfail growers use minimum tillage planting', *BSES Bulletin* No. 13, January 1986, pp. 20–21; F. Chapman, 'Controlled traffic zones increase yields', *BSES Bulletin* No. 52, October 1993, pp. 10–11.

125 Based upon details in CANEGROWERS, *Cane Growers' Information Handbook,* 1994/95 (Brisbane, 1995), pp. 52–56.

CHAPTER 12
Managing Water

In March 1915, John Drysdale, after surveying fields of dying or dead cane on the Inkerman Estate farms, near Home Hill, decided not to operate the newly erected Inkerman Mill as not enough cane would be available to keep the factory fully supplied. The failure of the cane crops on the southern side of the Burdekin River had resulted from the inability of the Home Hill district canegrowers to convince the Queensland government (Denham; Liberal) of the need for a communal irrigation scheme, as viable crops of cane could only be produced in the Lower Burdekin district with irrigation. After the disaster of the 1915 season, John Drysdale offered to provide the finance so that the Home Hill district canegrowers could purchase their own irrigation plants, like the canegrowers on the northern side of the Burdekin River. Most Home Hill district canegrowers, however, were content to continue lobbying the Queensland government for a communal irrigation scheme. Eventually, their persistence was rewarded with an announcement in August 1916 that the new Ryan Labor government in Queensland would build the long awaited irrigation scheme.[1]

The Inkerman Irrigation Scheme incorporated most of the lands that had been opened to settlement as part of the Inkerman Estate between 1911 and 1914. Central to the Inkerman Irrigation Scheme was a coal-fired powerhouse that supplied the electricity to power pumps which raised water from concrete wells, instead of iron spears that were used by the canegrowers on the northern side of the Burdekin River. The Queensland Treasury optimistically predicted that the scheme would take two years to complete, but the project was beset with shortages of material, delays in the arrival of materials and the construction engineers found the initial layout of the scheme unsuitable. Six years after construction commenced, Edward Theodore, Premier of Queensland, officially opened the scheme in May 1922, although it was not fully operational until late 1924. Plagued by problems from the outset, the Inkerman Irrigation Scheme continued to attract controversy during the 1920s. The Home Hill district canegrowers complained about water charges, poorly constructed wells, inadequate supplies of water to the farms and the lack of electricity from the powerhouse. By June 1930 the accumulated debts

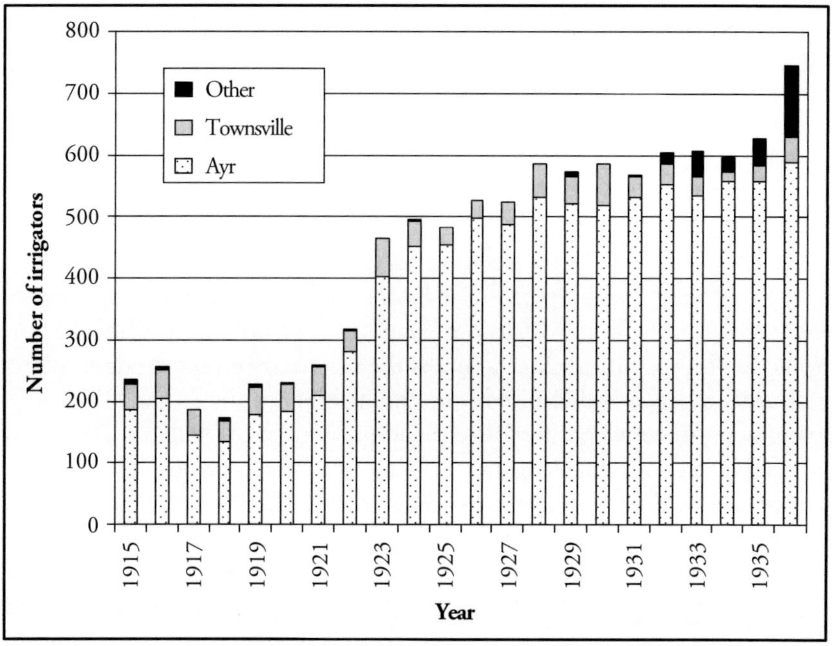

Notes: (1) The Townsville Petty Sessions District contained the Haughton River district.
(2) 'Other' includes figures for Mackay, Proserpine and Bundaberg Petty Sessions Districts.

Figure 12.1. The number of irrigators in the Queensland Petty Sessions Districts where sugar cane was irrigated, 1915–1936.

Source: Based upon figures in ARs of the Queensland Registrar-General on Agricultural Production.

of the Inkerman Irrigation Scheme amounted to £228 939. John Drysdale was correct in his assertion that economically run individual pumping units would have provided canegrowers in the Home Hill district with irrigation in half the time and at considerably less cost than the communal scheme.[2]

The number of canegrowers irrigating and the area under irrigation slowly increased throughout the Lower Burdekin district (i.e. Ayr Petty Sessions District) as the Inkerman Irrigation Scheme became operational (Figures 12.1 and 12.2). However, this increase was also assisted by two new developments which made irrigation more affordable for canegrowers. Pumping plants throughout the Lower Burdekin district had been traditionally run by costly suction gas or steam-driven pumps. During the late 1910s and early 1920s, tractors had been acquired by many Lower Burdekin canegrowers, and they provided an alternate and cheaper source of power to run pumping plants. The introduction of windmills in 1920 also lessened the

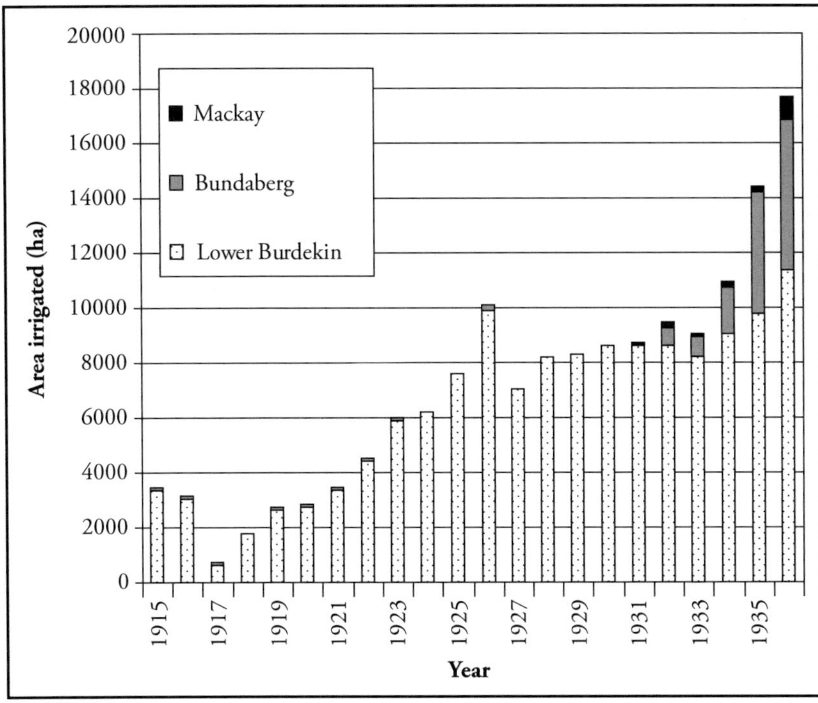

Notes: (1) Lower Burdekin contains figures for Ayr and Townsville Petty Sessions Districts.
(2) Mackay includes figures from the Proserpine Petty Sessions District.

Figure 12.2. Area irrigated (ha) in the Queensland Petty Sessions Districts where sugar cane was irrigated, 1915–1936.

Source: Based upon figures in ARs of the Queensland Registrar-General on Agricultural Production.

cost of pumping water onto the fields.[3] However, these two developments were not used by many other Queensland canegrowers to facilitate irrigation. Most irrigated sugar cane in Queensland during the late 1910s and 1920s remained concentrated in the Lower Burdekin and Haughton River districts (i.e. Ayr and Townsville Petty Sessions Districts), with less than 100 ha of cane being irrigated in the Bundaberg district during this period (Figure 12.2). Here Gibson & Howes Ltd. used the substantial irrigation plant established on Bingera plantation in the early 1900s (see Chapter 7) to continue applying water to their crops as required, but the cost of coal deterred constant pumping from the Burnett River. A few canegrowers in the Goodwood locality, south of Bundaberg, had also commenced irrigating, using the spear system pioneered in the Lower Burdekin district to tap underground water supplies.[4]

The apparent abundance of water available for irrigation in the Lower Burdekin district made some canegrowers recklessly extravagant in their use of the resource. In 1916, A. Gibson, a BSES field assistant, reported that some Lower Burdekin cane paddocks received as many as seven waterings, although the average was around four. Dr H. (Bill) Kerr, the BSES's soil chemist, visited the Lower Burdekin district in 1928, and afterwards described the farmers' irrigation methods as 'crude and wasteful'. He observed that the continued flooding of the land had caused packing of the surface soil, which cracked badly on drying and baked to a hard crust under the action of the sun and wind. Kerr concluded that some of the district's soils were so over-watered that their structure had or was in the process of being completely destroyed. Ernest Smith and W. Dixon, two CSR officers, reported upon the agricultural practices of the Lower Burdekin canegrowers in 1931. Their observations were part of a lengthy report which was used by CSR to assess the desirability of purchasing Pioneer and Inkerman Mills. They were told that the canegrowers applied water to the paddocks six to eight times in a normal growing season. In their report, Smith and Dixon mentioned seeing water being 'wastefully used' and they concluded that the farmers had 'no knowledge' of correct irrigation procedures. BSES field assistants advised Lower Burdekin canegrowers to reduce the volume of water applied to the paddocks and to shorten the water furrows.[5]

Lack of water did not trouble the New South Wales canegrowers during the late 1910s. By 1915, eight groups of landholders in northern New South Wales, including some canegrowers, had formed drainage unions or drainage trusts, taking advantage of the special conditions provided by the New South Wales *Water and Drainage Act* of 1902 and *Water Act* of 1912 (Figure 12.3). Under this legislation, landholders paid an annual levy to repay funds advanced by the New South Wales government to conduct drainage work in a particular gazetted area. Construction work was normally carried out by the New South Wales Department of Public Works, but maintenance of the drains became the responsibility of the landholders. However, these drainage unions and trusts were often short of funds to complete annual maintenance work. The Chinderah Swamp Drainage Trust's main drain, for example, was described in 1918 as being 'silted up very badly and in a rather bad condition with weed growth'. Moreover, the drainage unions and trusts had in some instances built channels or drains that were not large enough to remove the volume of water that troubled the landholders. The Secretary of the Robb Swamp Drainage Trust, for instance, complained in 1922 that their drains were 'insufficient in size' for draining all the land that was suitable for cane growing.[6]

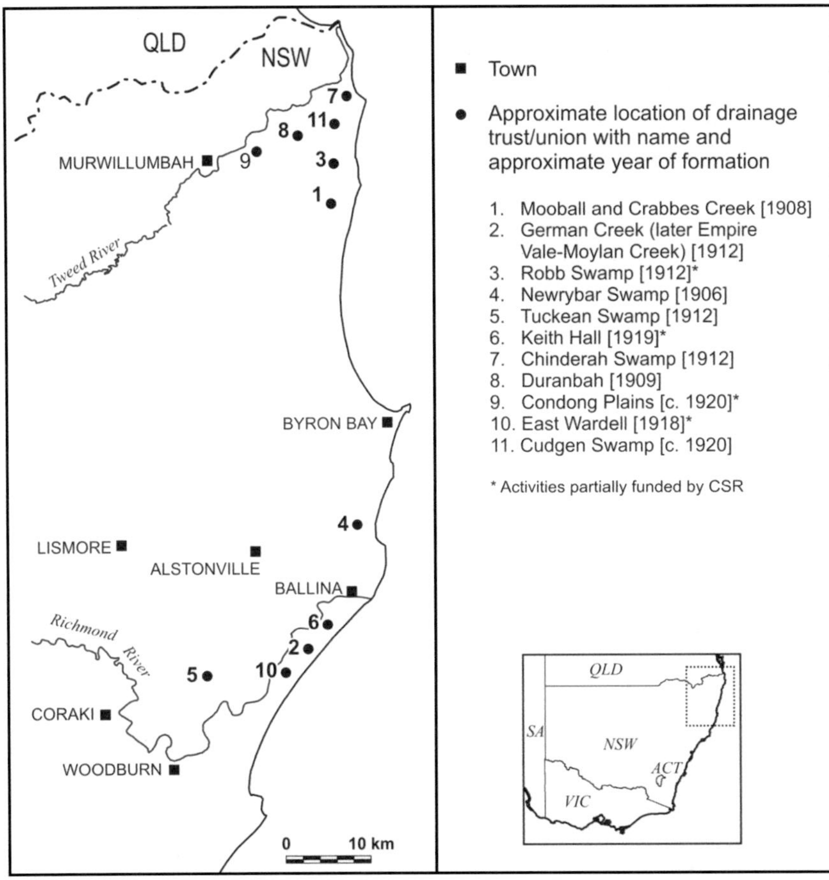

Figure 12.3. The location of drainage trusts/unions formed in northern New South Wales, 1900–1920.

Source: Based upon drainage union/trust files held by State Records Authority of New South Wales and CSR correspondence. File numbers available on request.

Immediately after World War 1, CSR realised that the survival of the New South Wales sugar industry depended upon expanding the area under cane and obtaining bigger crops from existing farmers. Sugar cane favours well drained and frost-free land, but it was originally in short supply throughout coastal northern New South Wales. Yet it could be created through reclaiming land from the many swamps that existed along the frost-free lower reaches of the Richmond, Tweed and Clarence Rivers. Better drainage would achieve this goal and lift the productivity per hectare of existing canegrowers. To achieve its goal,

CSR engaged in three main activities during the late 1910s and 1920s. Firstly, CSR drained the Cudgen plantation lands (Tweed River district) which it had acquired in 1912, demonstrating how to build effective drainage channels. Secondly, the firm advanced funds at 5% interest to some of the existing drainage unions, enabling improvements to be made to their network of drains. Thirdly, CSR encouraged the establishment of new drainage unions amongst groups of canegrowers by promising them the funds to construct the drains.[7] The latter two activities were particularly championed by CSR's mill manager at Broadwater Mill, Robert Donnan. In 1918, he commented that there was 'magnificent land lying idle for mere lack of drainage'. Donnan worked assiduously to convince the Richmond River district canegrowers of the benefits of growing sugar cane in paddocks that were not waterlogged for months at a time.[8]

Implementation of the above strategies commenced in early 1917 when CSR began building drains throughout the former Cudgen plantation lands. Part of this work involved constructing new channels that were joined to the main drains which had been built as part of the Chinderah and Robb Swamp Drainage Schemes (Plate 12.1). CSR even advanced £700 to the Robb Swamp Drainage Trust, thereby facilitating the enlargement of its main drain to cope with the increased in-flow of water. Eventually, CSR took over the operation of the Robb Swamp Drainage Scheme in 1923, repaying the debt to the New South Wales government and then ensuring the regular maintenance of the main drains. Following the effective drainage of the former Cudgen plantation lands, CSR created five new farms which were sold to interested settlers during 1918. CSR could then demonstrate to the existing Tweed River district canegrowers that good drainage improved the land for farming. This strategy achieved results, for a group of canegrowers on the Condong Plains agreed to form an unincorporated communal drainage scheme. By 1923, they had used the funds advanced by CSR to build several new channels that allowed the drainage of approximately 800 ha of farmland.[9] In the Richmond River district, CSR had encouraged the formation of the East Wardell Drainage Scheme (1918) and the Keith Hall Drainage Union (1919), and partially funded the construction of the drainage channels during the late 1910s and early 1920s.[10] CSR then built several major drains for the Richmond River district canegrowers between 1921 and 1923. The conditions associated with the building of these drains are unclear, although the District Surveyor at Grafton remarked in the late 1920s that CSR's drainage schemes throughout the Richmond River district had been carried out under such liberal terms that 'no financial embarrassment was incurred by

Plate 12.1. Excavating a new drain that formed part of the extension to the Robb Swamp Drainage Scheme, c. 1918.

Source: CSRR, 171/396 (b). Image reproduced courtesy of CSR Ltd.

the settlers'. Numbered one to sixteen and mentioned in the correspondence, the exact location of these drains could not be determined as no map survives in the extant records.[11]

CSR's focus on improving the drainage network in the Tweed and Richmond River districts during the late 1910s and early 1920s was not repeated in its North Queensland mill supply areas (i.e. Hambledon, Goondi, Macknade and Victoria), where the canegrowers were also troubled by waterlogged soils and regular flooding of their farms. Edward W. Knox had noted in 1921 that many cane growing areas in the Herbert River district would profit from properly organised drainage schemes. Yet CSR made no attempt to promote drainage schemes in the Herbert River district during the 1920s, although there was no legislative provision in Queensland before 1926 that would have allowed Queensland canegrowers to create incorporated drainage syndicates. After 1926, Queensland canegrowers could petition the Queensland Irrigation and Water Supply Commission (IWSC) to form drainage areas under Sections 18–19 of the *Water Act* of 1926. Under this arrangement, drainage boards could be established to administer the operation of the drainage works. Any scheme under this legislation attracted a 10% government subsidy and the unsubscribed part of the cost of building the drains were met by the Boards taking out loans. The method of rating each landholder benefiting from the drainage works to meet the annual repayment of loans plus annual maintenance costs was fixed by the Boards. However, interest in the provisions of this legislation amongst the state's canegrowers was non-existent during the late 1920s.[12]

Communal drainage schemes, either unincorporated voluntary arrangements or more formal, incorporated syndicates, may not have been popular amongst groups of Queensland canegrowers between 1915 and 1930, but some individual Queensland canegrowers did attempt to improve their on-farm drainage. Mr G. Wood, a canegrower supplying Victoria Mill, was observed in 1923 to have laid pipes on his property to drain away excess water. A more unusual method was adopted in 1928 by an unnamed Deeral canegrower, who had constructed 5.5 km of drains on his property just south of Cairns. Palm-tree trunks – some 12 m in length and exceedingly hard – had been used as the base for these drains. By the late 1920s, many cane farms along the Maroochy River were also observed to be crossed by drains, although 'leaky floodgates, wrong levels and drains being too narrow' often spoiled the efficiency of the drains. Despite these individual efforts, E. J. Ferguson Wood, a BSES assistant entomologist, claimed in 1928 that efficient drainage on cane growing properties was 'a feature seldom met with in Queensland'. He believed that the state's canegrowers had neglected this area for far too long.

The spread of irrigation and drainage works, 1930 to 1945

Canegrowers around Bundaberg failed in their repeated attempts during the early 1930s to convince the IWSC to construct a communal irrigation scheme using water pumped from the Elliott River, south of Bundaberg.[14] Instead, towards the end of 1932 – a year when only 430 mm of rain fell at Bundaberg – the Queensland government (Forgan Smith; Labor) approved a scheme proposed by Gibson & Howes Ltd. for enlarging the amount of cane irrigated in the South Kolan district, south-west of Bundaberg. Government support for the scheme was conditional on Gibson & Howes Ltd. funding half the cost of the work associated with the project and extending irrigation to the local canegrowers. Aided by a £5 000 grant from the Queensland government, Gibson & Howes Ltd. built a low level weir across the Burnett River at a spot known as The Falls, below their pumping station, to stop salt water intrusion upstream and to impound enough water for irrigation purposes. The firm also made extensive improvements to its aged pumping plant and constructed new irrigation piping, enabling the greater irrigation of the extensive Bingera plantation and many adjacent farms. Elsewhere in the Bundaberg district, the dry season of 1932 prompted the Fairymead Sugar Co. Ltd. and the Millaquin Sugar Co. Ltd. – firms which still engaged in plantation agriculture – to install new irrigation plants which pumped from the subterranean water supplies on their properties.[15] As a result of these developments, the area of cane irrigated in the Bundaberg district rose from under 1 000 ha in 1932 to almost 5 500 ha in 1936 (Figure 12.2). Other Bundaberg district canegrowers followed the plantation owners lead, so that approximately 200 irrigation units operated throughout the Bundaberg district by the mid-1940s, the majority pumping from the underground aquifers.[16]

Irrigation was also adopted by other Queensland canegrowers during the 1930s. In the Mackay district, the Australian Estates and Mortgage Co. Ltd. began irrigating on its Palms Estate in 1932, but decided in the following year that the cost of irrigation precluded its extension beyond 60 ha. Other Mackay and Proserpine canegrowers were more certain about the benefits of irrigation and were watering almost 700 ha of sugar cane by 1936 (Figure 12.2). Six farmers in the Tully district, Queensland's wettest locality, also showed faith in the potential of irrigation, when they installed tractor-driven pumps on their properties in 1935, claiming that they would use them during the dry spells over autumn to reduce the check to the growth of cane brought about by diminished supplies of moisture during the crucial growing season. Three canegrowers supplying Macknade Mill also established tractor-driven pumps in 1936, relying on water sourced from wells built on their farms. Senator Thomas Crawford's sprinkler irrigation system became quite a local curiosity when it commenced operating on his Mossman cane farm in 1938.[17] Those canegrowers who adopted irrigation for the first time during the mid-1930s boosted the amount of the state's cane crop irrigated from less than 8% before 1934 to just over 12% in 1936 (Table 12.1).

As canegrowers in different Queensland sugar-producing regions experimented with irrigation to sustain their cane crops, the Lower Burdekin irrigators were encountering some difficulties. The McCormack Labor government had been replaced with the Moore Country Party administration in 1929. The new government announced its intention to privatise the Inkerman Irrigation Scheme. After negotiations, the Moore government in early 1932 accepted the proposal from the Home Hill canegrowers to take over the debt-laden scheme. From 1 July 1932, the Inkerman Irrigation Board assumed the management of the scheme and the farmers became responsible for maintaining their own pumps, motors and wells.[18] A more serious problem was the dwindling underground water supply throughout the Burdekin River delta. By the early 1930s, the underground water levels were almost two metres lower than the levels of the mid-1920s, and saline intrusion contaminated the underground water supplies which could not be used on the crops of cane. The Ayr Shire Council

Table 12.1. The percentage of the annual sugar cane crop irrigated in Queensland, 1915–1936.*

Year	%	Year	%	Year	%
1916	4.7	1923	6.7	1930	7.2
1917	0.9	1924	6.0	1931	6.8
1918	2.8	1925	6.9	1932	7.8
1919	4.5	1926	9.4	1933	6.9
1920	4.4	1927	6.3	1934	8.7
1921	4.7	1928	7.1	1935	11.3
1922	5.5	1929	7.0	1936	12.9

Notes: * Irrigation figures were only collected until 1936.

Source: Based upon figures in ARs of the Queensland Registrar-General on Agricultural Production.

decided to seek approval to divert floodwaters into Plantation Creek from the Burdekin River. A loan to undertake the construction work was sought from the Queensland government, in preference to establishing a water board to facilitate the task. After the IWSC approved the scheme in 1935, the tunnels linking the Burdekin River to Plantation Creek were built in time for the floods of early 1936 to begin recharging the underground supplies. The huge flood of 1940 completely recharged the underground aquifers.[19]

The effective disposal of excess amounts of water occupied the attention of some New South Wales canegrowers during the 1930s. An upsurge in the building of mole drains (i.e. sub-surface channels for drainage formed by pulling a solid object through the soil) occurred throughout the Tweed River and Richmond River districts over the decade, facilitated by CSR's willingness to lend farmers an implement that created the mole drains. Other canegrowers built subsidiary drains on their properties, given they now had effective larger local outlets to receive the water from these smaller channels. CSR officers supervised teams of men who cleaned, maintained and improved the larger channels of the various drainage schemes in the Tweed River and Richmond River districts, thereby ensuring their effective operation during the 1930s.[20]

CSR's success in convincing the New South Wales canegrowers to consider drainage finally prompted the firm to try and duplicate these achievements throughout its North Queensland mill supply areas. The strategy of promoting the benefits of better drainage amongst its North Queensland canegrowers yielded some positive outcomes. In 1934, the organisation noted that several large drains had recently been built through the former Innisfail plantation lands, near Innisfail. By 1938, the Hambledon Mill technical field officer reported that many local farms were now well-drained, with some canegrowers even installing sub-surface pipe drainage.[21] Most activity, however, occurred in the Herbert River. In 1930, CSR concluded that the drainage network had become useless because so many existing drains had been destroyed by careless ploughing and others clogged with debris through years of neglect. The firm commenced assisting its Herbert River district canegrowers by providing a surveying service that took field levels and organising communal drainage activities. The canegrowers in the Foresthome locality, for example, agreed in 1931 to a levy of £10 each, and used the funds to clear the debris blocking the local creeks so that floodwaters would drain away quicker. Seven canegrowers in the Upper Fairford locality worked together to build a large drain across their properties in 1935, thereby reducing waterlogging on approximately 40 ha. In 1937, eleven canegrowers supplying Macknade Mill cleared Ripple Creek of the debris obstructing the flow of water. In addition, they used tractor scoops to build a large drain that emptied into Ripple Creek. By 1941, the Macknade Mill technical field officer observed that these canegrowers were now constructing better subsidiary drains now as they could connect their drains to an effective outlet.[22]

The implementation of more extensive drainage works by local governments in Queensland became possible following the introduction of the *Local Government Act* of 1936. Under Section 47 of this legislation, shire councils could receive 25% of the funds needed to build drains as a subsidy from the Queensland government and were permitted to borrow the rest of the capital through State Treasury. The borrowed monies were recovered after the payment of a special levy by those landholders who benefited from the drains. Cardwell Shire Council bec ame the first Queensland local authority with sugar cane growing lands to take advantage of this arrangement. In 1937, it secured a loan to commence an agricultural drainage scheme in the Tully River district to afford 'relief to farmers in low-lying lands'. Such an initiative by the Cardwell Shire Council was the beginning of its long involvement with local drainage schemes, although this association is not surprising as this locality receives the highest rainfall in Australia, and the local canegrowers suffered reduced yields because of waterlogged soils and regular flooding of their fields.[23] However, no other Queensland shire council with sugar cane cultivation within its boundaries followed the Cardwell Shire Council's example during the late 1930s and early 1940s. Canegrowers located in districts where CSR did not operate a sugar mill were left very much to their own devices when it came to agricultural drainage.

Australian canegrowers in the pre-World War II period also made their first tentative steps towards improving drainage by manipulating floodwaters, although with unexpected consequences. In 1919, the Richmond River district canegrowers, for instance, obtained £500 from the New South Wales government (Holman; Nationalist) to remove debris from the local creeks. Four gangs comprising forty to fifty men were employed in de-snagging, widening and deepening smaller watercourses that flowed into the Richmond River, thereby improving the run-off from nearby farms. In addition, riparian vegetation was often cleared along the banks of watercourses in the mistaken belief that removing the vegetation would assist in the speedier removal of floodwaters from the paddocks.[24] The loss of riparian vegetation, however, meant that the river banks began eroding. In 1926, a BSES extension officer observed that the erosion of river banks in the Babinda district had been brought about by the 'injudicious practice of removing the trees and undergrowth from the banks of speedy-running and winding streams'. Severe river bank erosion had also become prevalent along the lower reaches of the Herbert and Burdekin Rivers during the

1930s. The IWSC focused its remedial efforts upon the Burdekin River. By 1939, it had attempted to arrest river bank erosion by building a 24.4 m experimental timber groyne along a section of right river bank on the lower reaches of the Burdekin River, near Home Hill. Unsuccessful funding applications to the Queensland government (Forgan Smith; Labor) from the Hinchinbrook Shire Council between 1938 and 1940 meant that the lower reaches of the Herbert River were left unprotected against flooding.[25]

Major flooding of the Burdekin River in April 1940 recharged the underground water supplies, but also severely damaged the river banks, created new watercourses through cane farms, deposited sand and debris on some farms and in other localities removed considerable amounts of topsoil.[26] The investigation into the impact of the Burdekin River floods in 1940 concluded that the predisposing cause of the damage was the absence of riparian vegetation and the cultivation of cane extending to the river banks. In response to this natural disaster, the Queensland government enacted the *River Improvement Trust Act* in October 1940, enabling the formation of a trust to raise funds to repair the damage caused by the floods and to begin planning against future erosion. By the time the Burdekin River Improvement Trust held its first meeting in May 1941, the Forgan Smith government in Queensland had authorised repairs to the damaged river banks. Timber bulkheads, which were later protected by stone, had been built in front of the eroded sections of the river banks and shrubs had been planted to try and stabilise parts of the river banks. Following an application to the Queensland government by the Hinchinbrook Shire Council, Queensland's second river trust was gazetted in January 1942. The Herbert River Improvement Trust originally covered a comparatively short reach of the lower Herbert River around the Halifax and Macknade localities, but following flood damage to the river banks in 1946, it was extended to include the riverside localities in the Gairloch and Ripple Creek areas. Bank protection works focussed upon an area known as the Halifax Washaway, although they were hindered by a lack of funds.[27]

First concerns about soil erosion, 1930 to 1944

Soil erosion is the loosening and removal of soil from its previous resting place, through the agency of wind and water. It often occurs when farmers cultivate inappropriate locations, mostly sloping land. The steepness or gradient of the land has a very direct influence on the degree of erosion present. Land with a slope greater than 10% is very difficult to cultivate. Such land should be used only for grazing or forestry. Yet cane growing in Queensland during the 1930s was being conducted on slopes over 10%. Farmers employing incorrect farming techniques can also cause soil erosion. Contour tillage was unknown in Australian sugar-producing districts before 1940, claimed Dr H. (Bill) Kerr, the BSES Director during the late 1930s. The long straight drills went up and down hills with no thought given to how excess water might form channels seeking lower levels. Land was often left fallow during summer months and fields with young plant or ratoon crops suffered from heavy rainfall due to limited mulching with trash; scouring and small rills can occur on land with even a slight slope. In 1939, Charles Young told the Royal Commission on Sugar Peaks and other Cognate Matters that Fairymead Sugar Co. Ltd. had been forced to abandon cultivation on 105 ha because of 'soil erosion whilst it was fallow'. A similar picture was painted by CSR's Hambledon Mill technical field officer, when he reported that the 'hillside fields were scoured to cultivation depth and fallow fields had large washes taken out', a result of the summer rains in January 1969.[28]

By the mid-1930s, evidence of soil erosion on cane growing lands was observable in the Finch Hatton district, near Mackay, for some Italian cane-growers had stone pitched the heads of gullies on their farms to stop them from getting any larger. In 1938, J. L. Tardent, a local forester, estimated that 800 ha in the Isis district once grew sugar cane but 'are absolutely derelict land now or carry very poor crops'. He concluded that the damage from soil erosion had now reached an 'alarming total' and was a result of 80% of the area planted in the district being located on slopes between 6 and 20%.[29] Interestingly, Edward W. Knox hinted at this problem much earlier. In 1899, he visited the firm's sugar mill near Childers. As part of this visit, he toured the district, inspecting the farms of those settlers who grew sugar cane. In his report to CSR's Board of Directors, Knox expressed concern about what he had witnessed, writing, 'already some drawbacks are showing themselves. Many of the hills are very steep and in course of time trouble will be caused through the soil being washed into the gullies in heavy rains'.[30]

Dr H. (Bill) Kerr had become aware of the growing soil erosion hazard on some of Queensland's cane growing lands during the 1930s. However, he faced several challenges in trying to devise effective methods that would reduce the problem. In some parts of the world, soil erosion was slowed by alternating row crops like sugar cane with legumes/and or grasses so that the erosion under the row crops was counteracted by lower rates under other crops. A variation on this method was strip cropping which combined row crops and protection-effective crops in alternating strips aligned on the contour. Erosion was limited to the row crops and the soil lost around the row crops was trapped in the next strip downslope. However, because of the land assignment system, Queensland canegrowers suffering soil erosion could not alternate sugar cane with other row or tree crops every few years, nor easily engage in strip cropping. They had to grow sugar cane each year. Moreover, even if no legislative restriction upon the cultivation of other crops existed, there was the difficulty of finding a profitable alternative crop to grow on these farms. Rubber, rice, cotton, tea, coffee and bananas had all been tried in Queensland's coastal districts during the late nineteenth century, but all had proved unprofitable, except in the Innisfail district where a small banana industry co-existed alongside sugar cane cultivation. Mulching using the trash to protect the soil was an option, but the trash was increasingly being burnt in some North Queensland mill supply areas (e.g. Macknade; Mourilyan) to curtail outbreaks of Weil's disease and to kill the weevil borers that inhabited the crop residues, thereby reducing the damage they caused to the stalks of cane (see Chapter 14 for more details). In addition, cane growing could not be easily stopped on sloping land in Queensland, as cane farmers were required to cultivate such assignments each year under the land assignment system. Thus, the agricultural scientists were required to develop erosion control methods suited to such steep gradients.[31]

Kerr's initial advice to Queensland's canegrowers on soil erosion occurred in 1936, when he suggested a combination of protective methods. Firstly, he urged the state's canegrowers to adopt better soil management practices on their properties. The absorptive capacity of the soil could be improved by sub-drainage. Providing channels through which the water could pass was achievable by deep ploughing, subsoiling or growing deep-rooted crops (e.g. lucerne) on the land when it was not cultivated with sugar cane. Secondly, mechanical measures should be adopted. The rows of cane needed to follow the contour or run parallel to the slope, avoiding the long straight drills that went up and down hills. When this method proved inadequate or unsuitable, the farmers were urged to build contour banks or terraces and grassed waterways on their properties. Fields could be crossed by broad, shallow waterways which were flanked on the downhill by mounds or ridges of earth following the contours. The contour banks diverted run-off water to larger outlet channels before it had time to obtain sufficient velocity to erode the soil from the cultivated areas between the terraces. Finally, the field should be protected by a green manure crop or a trash blanket during the fallowing of the land.[32]

Kerr's advice was not supported by the results from any field experiments under Queensland conditions and the outbreak of World War II curtailed any further investigations into this matter. Assessing if Queensland canegrowers suffering soil erosion adopted any of Kerr's methods during the late 1930s and early 1940s has proved impossible due to the silence of the historical records on the topic. Timber and stone dams built across gullies to check the loss of soil from cultivated fields by causing silting of channels, however, were observed in the Isis and Innisfail districts during the early 1940s, so some canegrowers were trying to reduce the impact of soil erosion on their properties. Moreover, the Australian sugar industry's concern about soil erosion did not vanish. In March 1945, the QCGC asked John Curtin, Prime Minister of Australia, to authorise a survey of the extent of soil erosion in Queensland's sugar-producing districts. The Commonwealth government acceded to this request. During the latter months of 1945, officers from the Division of Soils, Council for Scientific and Industrial Research, and the BSES, toured half the sugar-producing regions of Queensland, the northern localities being the focus of their investigations. They identified serious field erosion in all areas visited, except the Herbert River district, with the Isis and South Johnstone districts being the worst affected. This investigation also found that slopes with angles of 25% in the Mackay district were being cultivated and the resulting erosion was 'very far advanced'.[33] The BSES faced a major post-war challenge: how to slow this erosion and how to convince the state's canegrowers to adopt the proposed ameliorative measures?

Watered and drained fields: beginnings of an irrigation and drainage mosaic, 1945 to 1970

Concern about drainage in the New South Wales sugar-producing districts had dwindled during World War II and its immediate aftermath. By 1948, CSR's Condong Mill Cane Inspector, for example, observed that the drains throughout the Tweed River district were in a 'deplorable condition' and that many canegrowers were content to 'let the water find its own way from the fields'. The number of drainage unions in the Richmond River district had been reduced from eight to five by 1951, so the drains in fewer areas were being effectively maintained.[34] Part of the problem was a general shortage of field labourers to undertake the necessary cultivation tasks, but

CSR admitted in 1956 that the firm had ceased emphasising the importance of good drainage. However, the Tweed River district suffered severe flooding in February 1954. The floodwaters caused the loss of 200 ha of cane and the paddocks were waterlogged for several weeks afterwards. This situation prompted CSR to once again promote the need for good drainage. In 1956, CSR instructed its technical field officers to give a high priority to drainage matters and that the success of local improvements must be conveyed to all CSR's suppliers so that the 'inertia and prejudice of certain canegrowers must break down'. Loans were provided to the local drainage unions for the cleaning and repair of existing drains and the construction of new drainage channels, and CSR's technical field officers commenced taking field levels for interested canegrowers so they could build more effective drains. A small booklet published by CSR in 1957 showed canegrowers how to improve the drainage on their properties.[35]

CSR's promotional activities prompted a flurry of drainage-related activity in the New South Wales sugar-producing districts during the late 1950s and early 1960s. Existing drainage unions showed more interest in their network of drains. The Condong Drainage Union, for example, used a £5 000 loan from CSR to hire a mechanical drag-line dredge to clean the drains it maintained throughout the Condong Plains locality. Nineteen landholders in the Duck Creek area (Richmond River district) formed a new drainage union in 1959, borrowed £10 000 from CSR, and used the funds to rehabilitate some old drains clogged with mangroves and silt and to build new drains and floodgates so that their farms were no longer inundated for several months after heavy rainfall. In 1959, nine canegrowers who subscribed to the Tygalgah Drainage Union (Tweed River district) pooled their resources and raised £360 which was spent deepening and widening a drain so that their farms were free of floodwaters in two to three days. An unidentified number of canegrowers farming on Harwood Island in the Clarence River district formed the Harwood Island No. 1 and No. 2 Basins Drainage Union in 1960. Funds were raised to enlarge and floodgate Narhoo Creek and to construct new drains and deepen the existing drains emptying into Narhoo Creek.[36] By 1961, some canegrowers in the Tweed River and Clarence River districts had even commenced lowering the water table levels on their properties by using low lift pumps to discharge the excess water into main drains. CSR advertised that canegrowers paying attention to drainage had increased the yields of cane per acre by as much as 16%.[37]

Interest in drainage had also dwindled in Queensland's sugar-producing regions during World War II. However, as mentioned in the previous chapter,

Queensland canegrowers acquired bulldozers immediately after the conclusion of the war, making the task of drain building much easier. Twelve canegrowers on the south bank of the Maroochy River, for example, formed the South Maroochy Drainage Scheme in 1947. Explosives and bulldozers obtained from Army Disposals were used to create drainage channels through the swampy soils. Over the next decade, almost 550 ha of land was reclaimed and protected from further inundation by the building of floodgates on the main channel and levee banks along the river. The canegrowers benefiting from the drains paid annual levies, providing the funds necessary to maintain the drains. Contributing canegrowers then built a variety of both surface and sub-surface drains on their properties, enabling the excess water to be chanelled into the main drains.[38] In North Queensland, the Macknade Mill Cane Inspector reported that many canegrowers throughout the Herbert River district during the early 1950s had 'bulldozer work done on their farms with the object of filling in depressions in fields and deepening existing drains'. Further south in the Mackay district, three community drainage schemes in the Te Kowai, Peri and Wallingford localities were established during 1955. Bulldozers constructed drains through swampy areas and the displaced soil was used to raise the height of low-lying parts of paddocks to eliminate localised flooding.[39] Towards the end of the 1950s, small groups of canegrowers in the Cairns district had also formed two voluntary drainage schemes. Natural waterways in the Green Hill and Smithfield localities had been cleared of debris to improve the escape of water after heavy rainfall.[40]

Lack of water, however, troubled the Bundaberg district canegrowers during the late 1940s. They again failed to convince the Queensland government (Hanlon; Labor) to build a communal irrigation scheme in the Bundaberg district, and were forced to continue relying on individual pumping plants to supply irrigation water from underground supplies. Additional viable irrigation water was discovered during 1953 in the Hummock area, near Bundaberg, and it was from these early bores that this locality grew into a major area of groundwater use.[41] By the mid-1950s, approximately 5 000 ha of sugar cane were being irrigated from underground supplies in the Bundaberg district, which was part of the Maryborough Statistical Division (Figure 12.4). Other canegrowers in the Mackay and Proserpine districts were also prompted by the 1946 drought to begin irrigating their cane. About 2 500 ha of cane was irrigated in Mackay Statistical Division during the late 1940s and early 1950s (Figure 12.4). Some canegrowers used the spear system pioneered in the Lower Burdekin district; others pumped water from wells or bores. A few canegrowers around Cairns were also reported as using sprinkler

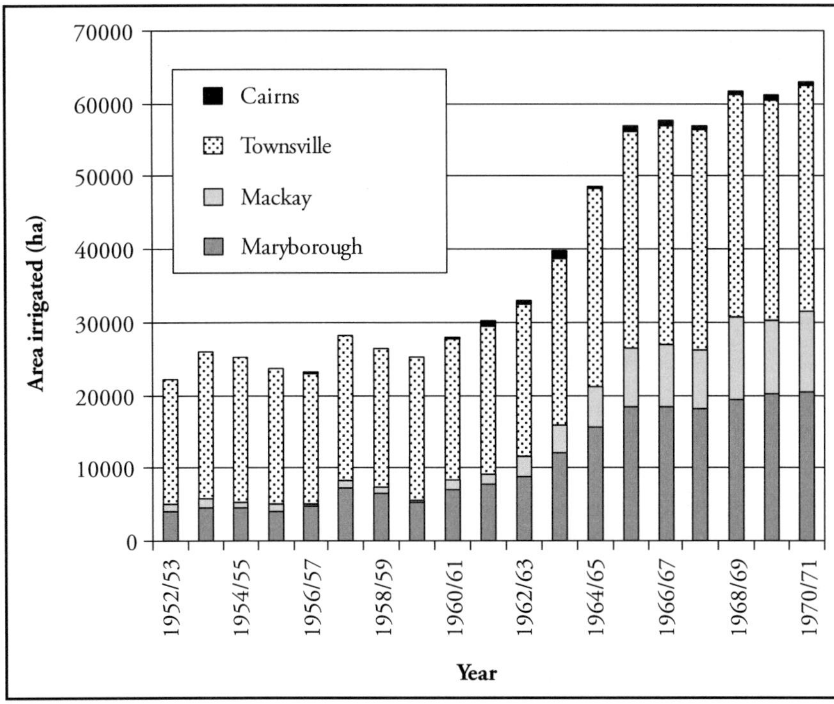

Notes: (1) Irrigation statistics were collected by Statistical Division only.
(2) The names of the Statistical Divisions are those used at the time of the data collection.

Figure 12. 4. Area of sugar cane irrigated (ha) in Queensland's Statistical Divisions, 1952/1953–1970/1971.

Source: Based upon figures in the *Queensland Year Book*, various issues.

irrigation in the late 1940s.[42] Consequently, by the mid-1950s, around 12% of the state's sugar cane was being irrigated, much the same level achieved in the mid-1930s (c.f. Figure 12.5 and Table 12.1). Most of this irrigated sugar cane continued to be found throughout the Lower Burdekin district.

As the Lower Burdekin canegrowers embarked on the mid-1950s expansion, their increased demand for irrigation water led to the removal of so much underground water that the risk of massive salt water intrusion emerged, threatening the viability of sugar cane cultivation in some of the areas supplying Kalamia and Inkerman Mills. Measures to alleviate this threat included artificial recharge from the Burdekin River via pits, together with the building of some low earth weirs on tidal inlets to restrict local seawater inflow.

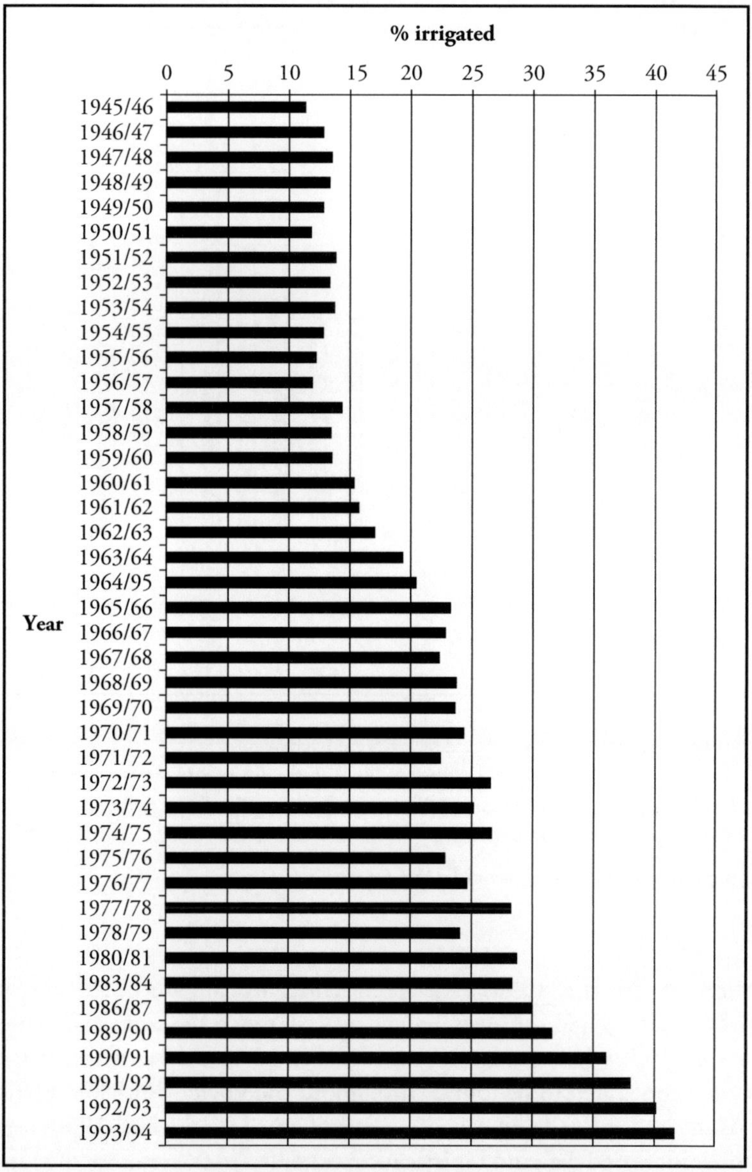

* Irrigation data was not collected during World War II and collected irregularly after 1980/81.

Figure 12. 5. The percentage of the annual sugar cane crop irrigated in Queensland, 1945/1946–1992/1993.

Source: Based upon figures in the *Queensland Year Book,* various issues.

Subsequent to a request from the Ayr Shire Council, the IWSC commenced a program in 1959 to determine the feasibility of artificial replenishment of the underground water in the light of supply and future demand from the canegrowers. The program included widespread drilling to establish water levels and availability, a detailed survey for pumping station sites and assessing the technical and economic feasibility of supplementing the underground supplies. As the officials conducted their investigations, rainfall throughout the Lower Burdekin district each year between 1960 and 1963 was below 1 100 mm (i.e. minimum amount required to practice non-irrigated cane cultivation), with only 625 mm falling in 1961. So much underground water was used in 1963, the IWSC concluded that maintaining development at the 1963 level (let alone any future expansion) under similar climatic conditions would result in a worsening of both water level and quality in the delta.[43]

The IWSC's final report on the water resources of the Burdekin River delta was issued in October 1964 towards the end of another dry year – 655 mm only had fallen between January and September. Its authors advised that 87 000 to 125 000 megalitres of the underground supplies in the Burdekin River delta needed to be replenished annually to avoid future water shortages and extensive salt water intrusion. To achieve this goal, the IWSC proposed the progressive construction of pumping stations to lift water into natural and artificial channels on both sides of the river. Proposals for the constitution of a Water Area covering the entire Burdekin River delta and a Water Board to operate and maintain the works were advertised in November 1964. Many of the district's canegrowers, fearing restrictions on their water use and increased charges for water, objected to the proposal. The Queensland government (Nicklin; Coalition of Country and Liberal Parties) reacted to the canegrowers' strong opposition by announcing in February 1965 that it would not proceed with the project. Subsequently, a majority of the landholders on the north side of the Burdekin River withdrew their objections, subject to the modification of the scheme to provide for pumping direct from the Burdekin River into both Sheepstation and Plantation Creeks and damming of tidal estuaries. As a result, the North Burdekin Water Board was constituted on 13 May 1965 and embraced the areas supplying Pioneer and Kalamia Mills (Figure 12.6). The Inkerman Mill Suppliers' Committee then began discussions with the IWSC about a recharge scheme for the south side of the Burdekin River. Following investigations into the feasibility of the scheme, the Queensland government on 31 May 1966 approved the formation of the South Burdekin Water Board,

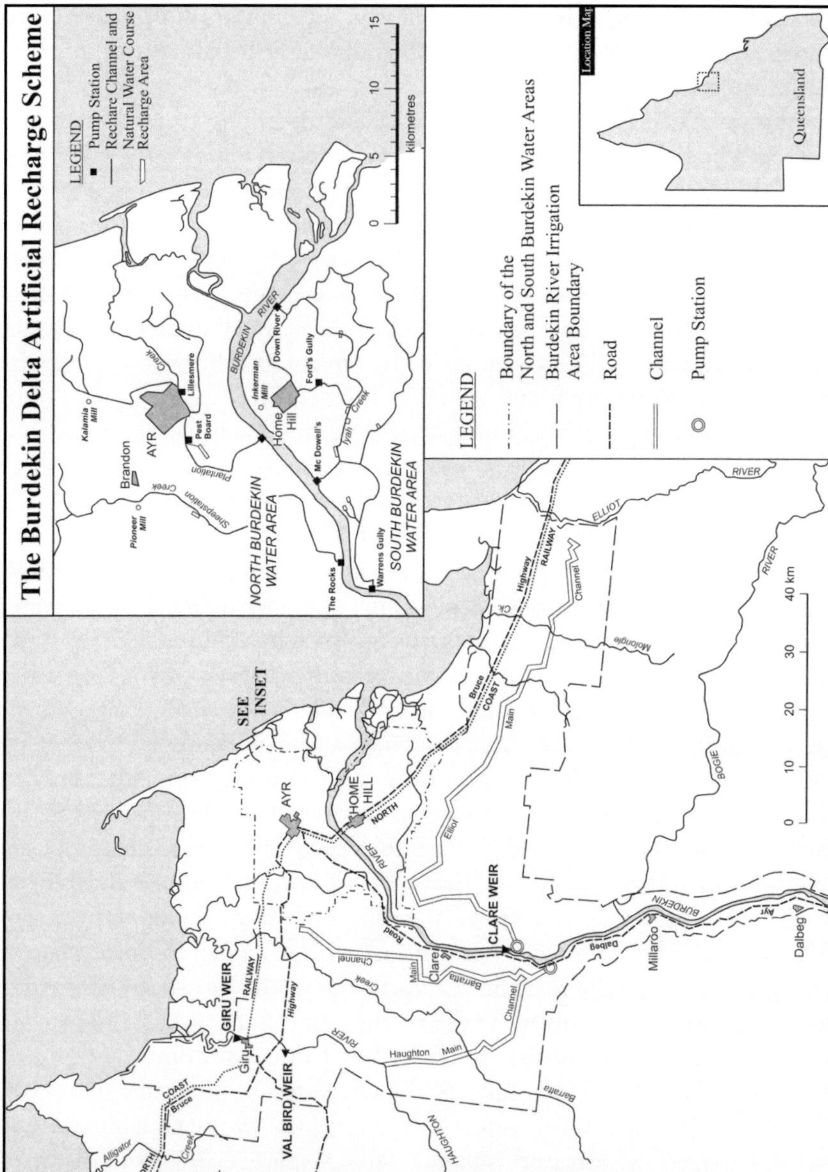

Figure 12. 6. The Burdekin River Irrigation Area and Burdekin Delta Artificial Recharge Scheme, 1990.

Source: Based upon a map in the Queensland Water Resources Commission, AR 1983–84. Reproduced courtesy of The State of Queensland, Department of Environment and Resource Management.

incorporating the Inkerman Mill supply area. In both instances, canegrowers were levied per tonne of cane to raise the necessary finance to commence and maintain operations.[44]

To achieve their goal of consistently recharging the underground aquifers, the North and South Burdekin Water Boards since their inception have built and maintained a dozen pumping stations which deliver water into artificial channels and natural watercourses that crisscross the Burdekin River delta (Figure 12.6). By 1973, the North Burdekin Water Board was responsible for 70 km of channels; this figure had grown to 120 km by 1984. To stop salt water intrusion into the watercourses, some tidal estuaries have been dammed. Pumping of water into the channels on the north side of the Burdekin River commenced during late 1965; on the south side pumping began in early 1967. By 1994/1995, the North Burdekin Water Board's facilities pumped almost 200 000 megalitres into the recharge system, compared to under 60 000 megalitres for much of the 1970s. After 1988, controlled releases of water from the Burdekin Falls Dam (see discussion later in the chapter) ensured a steady flow of water through the lower reaches of the Burdekin River, permitting this upsurge in pumping. Overall, the scheme has brought benefits to the district's farms, especially those located in the near vicinity of the replenishment channels, as spear water levels have been maintained.[45]

Drought conditions during the late 1950s and early 1960s also troubled the canegrowers throughout the Wide Bay-Burnett region. Seven of the nine years between 1960 and 1969 recorded less rainfall than the long-term mean of 1 423 mm, making rain-fed sugar cane cultivation very marginal. To enable the production of adequate crops of sugar cane, the underground water basin continued to be tapped. In the Fairymead Mill supply area, for example, the number of bores sunk into the underground supplies increased from 12 in 1953 to 312 in 1965. Some pumping restrictions were enforced, but salt water intrusion had occurred by 1969, leading to blocks of irrigated cane in the Avoca-Millbank locality being affected by saline water.[46] During the 1960s, the Bundaberg district canegrowers continued to press successive Queensland governments about the establishment of an irrigation scheme in the Wide Bay-Burnett region. Officers from the IWSC and Queensland Department of Primary Industries conducted investigations during the 1960s into the feasibility of a large scheme designed to irrigate almost all the existing sugar-producing lands in the region, using water from two major storages – one each on the Kolan and Burnett Rivers – several barrages and weirs on the two rivers and from the aquifers. Eventually, their report was presented to the Queensland government (Bjelke-Petersen; Coalition of Country and Liberal Parties) in March 1969 and recommended the irrigation scheme be constructed in two stages. Stage 1, estimated to cost $47.1 million, provided for storages and reticulation works to supply the existing cane assignments with irrigation water. Stage 2 proposed supplying water for about 13 300 ha of additional land that could be assigned for cane growing.[47]

Instead of the prevarication usually displayed by politicians, the Bjelke-Petersen government decided in April 1969 to proceed with Stage 1 of the proposed Bundaberg Irrigation Scheme, but requested Commonwealth government financial assistance for the entire capital cost. In 1970, the Commonwealth government (Gorton; Coalition of Liberal and Country Parties) agreed to provide $12.8 million towards the project. This amount, together with the $8.3 million pledge by the Queensland government in September 1969, financed the construction of Phase 1 of Stage 1. Phase 1 works included the building of a major dam at Monduran on the Kolan River, a pumping station at the dam, a 40 km diversion channel to deliver the water to the Burnett River and a tidal barrage each on the Burnett and Kolan Rivers. The funds were a far smaller commitment than asked for by the local canegrowers, but it was a start. The Monduran Dam, later re-named the Fred Haig Dam in honour of the Queensland Water Commissioner between 1955 and 1974, began filling with water in 1974.[48] Construction of the remainder of Phase 1 had been undertaken by 1980. During the 1980s, the concrete pipes and open channels associated with bringing the water to the canegrowers of the Woongarra and Gooburrum localities were completed. The last part of the Bundaberg Irrigation Scheme was finalised in the early 1990s when the Isis district received its surface irrigation water (Figure 12.7).

In the Queensland sugar-producing districts not affected by drought during the 1960s, three factors combined to ensure drainage assumed a much higher profile. Firstly, as discussed in Chapter 10, the industry embarked upon a major expansion, but some land available for this expansion needed drainage before it could be easily cultivated. Secondly, successive Queensland governments became more interested in the issue of drainage, and willingly funded drainage schemes. The IWSC was instructed to become more involved in the design and promotion of communal drainage schemes throughout the state, instead of focussing on irrigation works. Thirdly, many canegrowers were forced to address the issue of waterlogged paddocks following the widespread introduction of mechanical harvesting in the 1960s. Poorly drained paddocks with depressions and wet patches

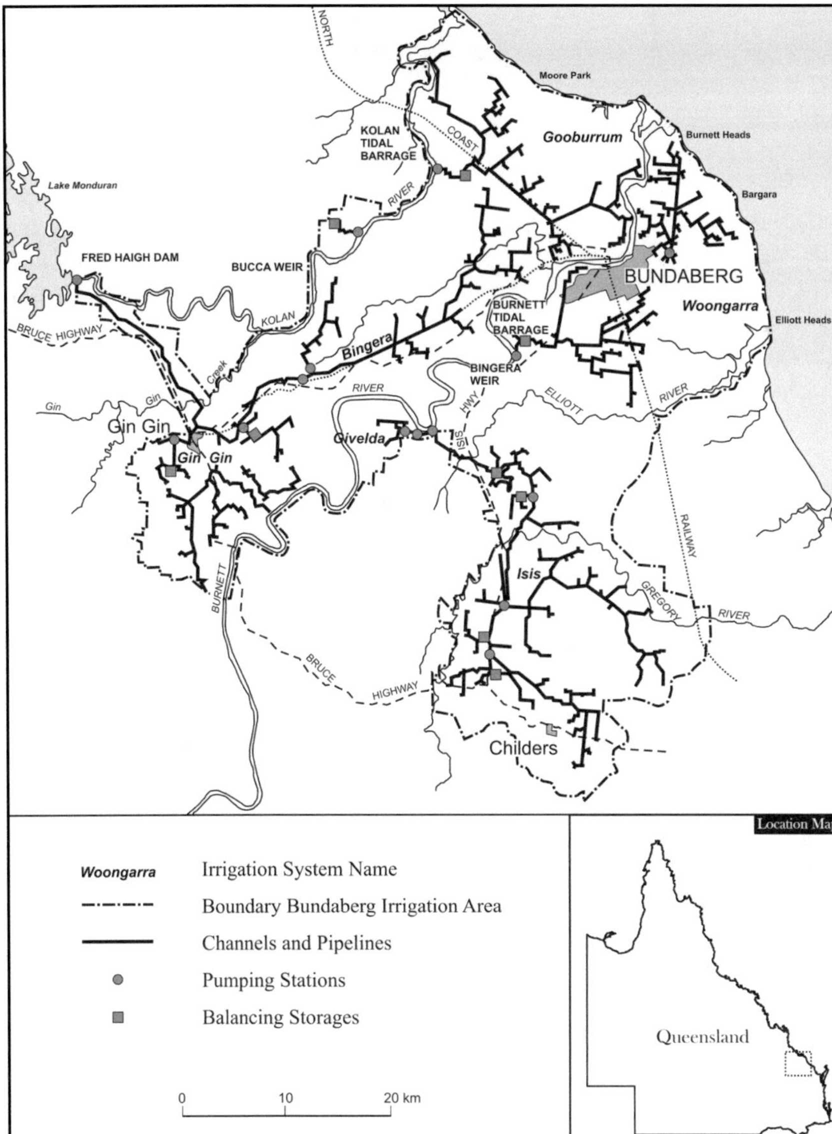

Figure 12.7. Bundaberg Irrigation Scheme, 1992.

Source: Based upon a map in Queensland Water Resources Commission, AR 1991–92. Reproduced courtesy of The State of Queensland, Department of Environment and Resource Management.

Plate 12.2. A Toft hydraulic loader fitted with a clam-shell type grab being used to create a farm-based drain in the Innisfail district, c. 1960.

Source: CSRR, Z/109, Box 461. Image reproduction courtesy of CSR Ltd.

were places where mechanical cane harvesters could bog and trying to harvest such paddocks often led to their damage.

Individual canegrowers could undertake drainage activities more easily during the 1960s following the development of a new piece of machinery. Toft Bros. of Bundaberg, local machinery inventors, had perfected the design of a hydraulic mechanical cane loader during the late 1950s, enabling the mechanisation of cane loading. Many canegrowers purchased this machinery, but they quickly found it could be used for other tasks. By the mid-1960s, some North Queensland canegrowers had fitted scoops or buckets to Toft mechanical cane loaders (Plate 12.2), and were using the equipment to build new drains or clean existing drains.[49] In CSR mill supply areas, technical field officers offered assistance with on-farm field surveys to determine the correct location of these new drains and supplied mole drainage equipment to improve sub-surface drainage. Indeed, between 1960 and 1964, CSR's Technical Field Department reported that canegrowers supplying their northern mills had built approximately 160 km of new channels. Farmer interest in improving drainage was so great that the Goondi Mill technical field officer estimated that 50% of his time during 1967 was devoted to drainage matters.[50] In addition,

Plate 12. 3. A levelling frame being used to remove small depressions in a paddock so as to improve field drainage.

Source: CSR, Agricultural Circular, Queensland, 1965, p. 6, CSRR, Z/109, Box 370. Image reproduction courtesy of CSR Ltd.

canegrowers levelled their land by filling in hollows that contributed to pondage areas and by eliminating all ridges that blocked the flow of water. During the 1950s and 1960s, this task was completed by soil or land levellers – flat, box-shaped equipment pulled by tractors (Plate 12.3). Later, this machinery was replaced by conventional land planes and graders, and then by laser-controlled scrapers and graders during the 1980s.[51]

Other Queensland canegrowers during the 1960s decided that cooperation with their neighbours was needed to maintain and/or expand local drainage networks. Unincorporated, communal drainage syndicates were formed in several of the sugar-producing districts. One operated in the Maroochy district, where four canegrowers developed a single drainage network in 1969 to serve their properties. A concrete floodgate was built on a tidal creek to stop the ingress of salt water during high tides and a pumping plant was established to dispose of excess water during times of flooding. Seven kilometres of newly constructed farm drains then emptied into this creek. Cooperation amongst canegrowers was particularly strong in the Innisfail district, where twelve communal drainage schemes had been formed by 1969 (Figure 12.8). Construction work

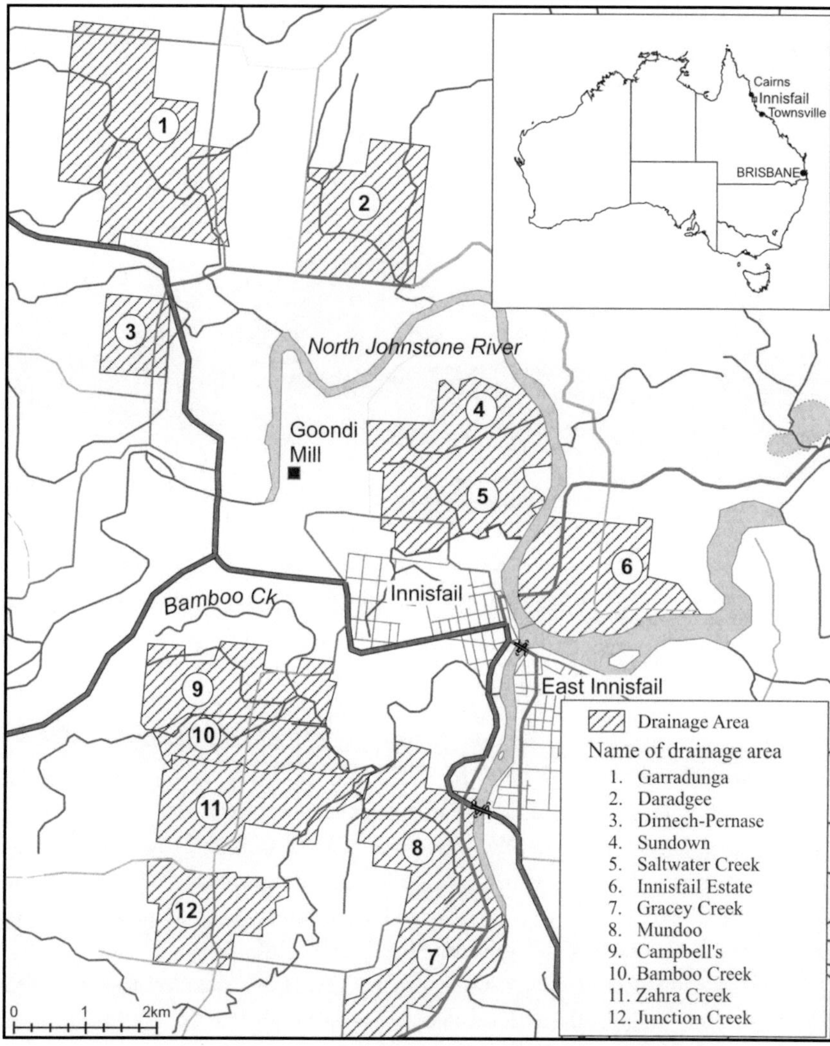

Figure 12.8. Voluntary drainage schemes formed in the Innisfail district by 1969.

Source: Map to accompany W. Maclean's, 'Drainage Report 1969', Report 31 N, Goondi Mill, CSRR Z/395, Box 25. Image reproduction courtesy of CSR Ltd.

associated with these schemes ranged from deepening existing creeks, diverting creeks, building new channels and installing floodgates on drains joining the network of tidal creeks associated with the Johnstone River.

CSR promoted such cooperation amongst its canegrowers and sometimes funded the construction work. The Mundoo Swamp Drainage Scheme is a good example. In 1960, three canegrowers borrowed £4 000 from CSR. During the early 1960s, they built a floodgate and a 100 m drain of 1 m diameter concrete pipes, thereby draining Mundoo Swamp by connecting it to the South Johnstone River and reclaiming 160 ha of land.[52] CSR's efforts in the Herbert River also led to the formation of nine voluntary drainage schemes by 1970, involving at least 183 canegrowers and contributing to improved drainage over almost 10 000 ha. Canegrowers combined to fund the clearing of debris blocking many local creeks, the building of new main drains and the deepening of existing watercourses. The most elaborate of these schemes began in 1964 when CSR commissioned the engineering firm Gutteridge Haskins and Davey Pty. Ltd. to design works that would improve the drainage of the Ripple Creek locality by extending the length of Ripple Creek for several kilometres. Further investigations were conducted by the IWSC. Eventually, the initial work associated with the scheme was undertaken during 1966–1967, and involved the construction of two new channels totaling 14.6 km in length, the removal of approximately 91 000 m³ of earth, stone pitching of erosion sites along Ripple Creek and the removal of blockages along the watercourse.[53]

Other canegrowers preferred a more structured approach to drain construction. They turned to different sources for funds and created more formal associations. Two groups of canegrowers near Mourilyan petitioned the IWSC in the early 1960s to establish drainage areas under the *Water Act* of 1926–1968. After investigations by the IWSC, the Cowley and Silkwood Drainage Areas were gazetted in 1969 (Figure 12.9). The Boards of Directors assumed the responsibility for existing drains and commenced constructing additional new drains.[54] Two Shire-based drainage schemes were also formed in the 1960s. A group of canegrowers in the Syndicate area on the north bank of the Tully River, just south-west of Tully, petitioned the Cardwell Shire Council in 1964 to establish a network of drains to facilitate the expansion of cane growing throughout this poorly drained locality (Figure 12.9). The Cardwell Shire Council secured funding under the provisions of Section 47 of the *Local Government Act* of 1936–1964, and the building of these drains occurred over three stages between 1967 and 1981, cost $225 000, and provided improved drainage throughout 1 067 ha of cane growing lands.[55] In Southern Queensland, the Woongoolba and Area Drainage Committee commenced agitating in 1962 for a drainage scheme to reduce flooding in the Woongoolba locality, which was found in the Rocky Point Mill supply area (Beenleigh district). By 1964, the Albert Shire Council had acceded to the Committee's request. Stage 1 of the Woongoolba Drainage Scheme was built over the next three years and consisted of a series of

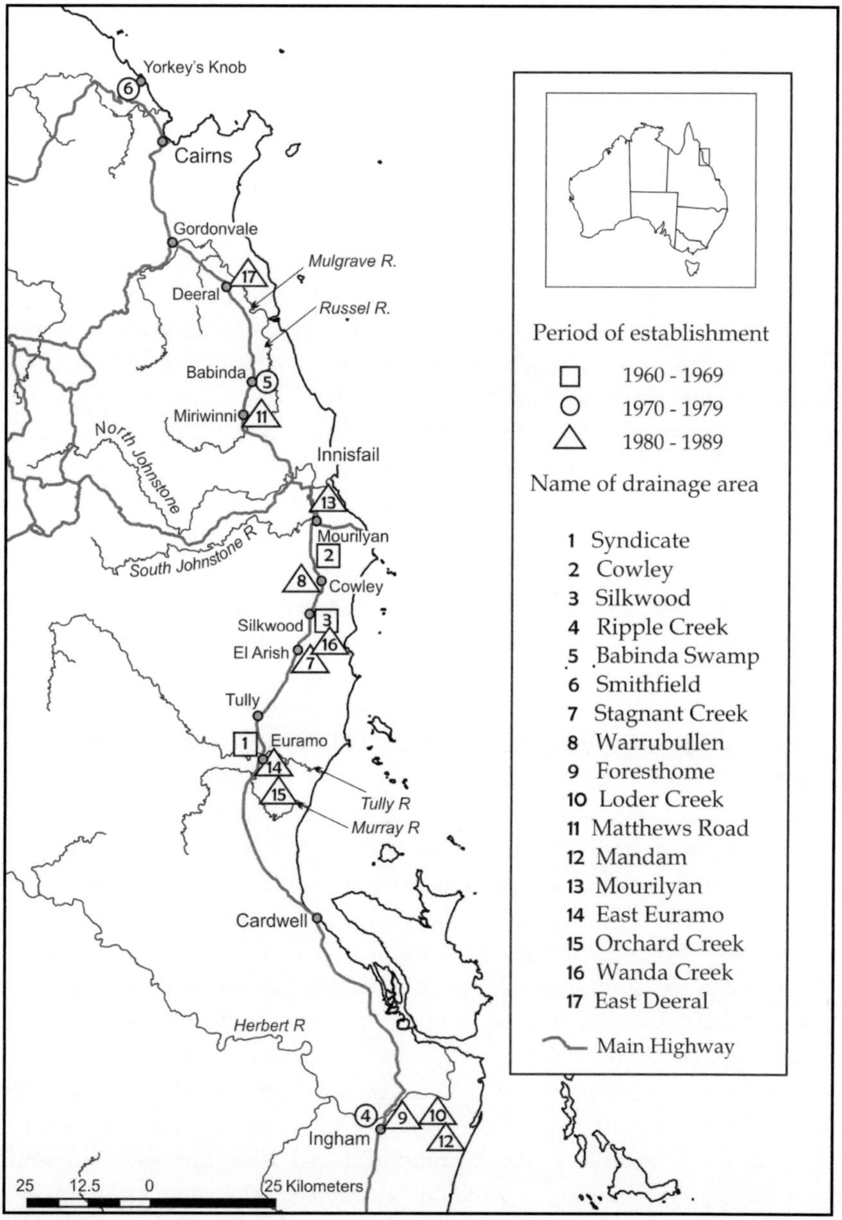

Figure 12.9. The location and period of establishment of the Far North Queensland drainage schemes, 1960–1990.

Source: Based upon gazettal notices in the *Queensland Government Gazette*, various issues.

levee banks and floodgated outlets in four locations to stop high tides spreading sea water over paddocks. The next stage, completed in 1977, included the construction of 25 km of channels, some having widths of 28 m and depths of up to 2 m. Funds totaling almost $100 000 for the scheme were provided by the Queensland government (Bjelke-Petersen; Coalition of National and Liberal Parties), with additional contributions coming from W.H. Heck & Sons Pty. Ltd., the owners of the Rocky Point Mill, and by way of a levy on the canegrowers benefiting from the drains.[56]

Communal drainage activities in New South Wales, as in Queensland, attempted to rid specific properties of excess water. A catchment-wide approach to flood mitigation and drainage required greater government involvement. Following the 1954 and 1956 floods in northern New South Wales, the Richmond River Valley Flood Mitigation Committee and the Tweed River Flood Mitigation Committee conducted investigations into ways to provide some relief against the damage and losses caused by flooding in the Richmond River downstream of Lismore and the Tweed River downstream of Murwillumbah. Both committees concluded that the capacity of the existing drainage channels was too small for the quick removal of flood rains of over 50 mm per day and that many channels were poorly maintained. The complete exclusion of the floodwaters was necessary to reduce the damage to the crops of sugar cane and this could only be achieved by the construction of low levee banks, deeper drains and floodgates.[57] In response to these reports, the New South Wales government (Cahill; Labor) approved the formation of authorities specifically constituted to oversee flood mitigation and drainage works. The Clarence River and Richmond River County Councils were constituted in 1959, with member local authority areas delegating their responsibility for floodplain management to the relevant county council. The Tweed Shire Council decided to maintain responsibility for flood mitigation in the Tweed Shire.

During the early 1960s, the northern New South Wales flood mitigation authorities commenced their program of construction. By 1963, the Clarence River County Council had spent approximately £25 000 in the Harwood Mill supply area. Debris was removed from existing channels and new main drains and floodgates were built on Harwood, Lower Palmer, Woodford and Goodwood Islands. Several large drainage canals, including the Kilgin No. 1 canal and Keith Hall No. 1 canal, and new floodgates on the Sneesby's Lane Drain, had also been built in the Broadwater Mill supply area. However, with limited finance available, construction of flood mitigation and drainage works in northern New South Wales was slow, and some landholders wondered if the original program of proposed works would ever be completed.

To expand the pace of construction, greater financial assistance was sought in 1963 by the New South Wales flood mitigation authorities from both the New South Wales and Commonwealth governments. The Commonwealth government (Menzies; Coalition of Liberal and Country Parties) accepted that flood mitigation works had appreciable national significance and agreed to match New South Wales government expenditure up to a maximum of $5.5 million within a six year period from 1 July 1963. In 1966, with the program of construction underway, the New South Wales flood mitigation authorities sought additional Commonwealth funding. The Commonwealth responded by raising the maximum amount available in loans to $8 million. At the termination of the grant from the Commonwealth in June 1969, the three flood mitigation authorities dealing with lands devoted to growing sugar cane had completed their original program of works. The funding for these projects came from Commonwealth, New South Wales and local governments, in proportions of 40%, 40% and 20%, respectively.[58]

The extra funds obtained by the three northern New South Wales flood mitigation authorities permitted the completion of further drainage works. Cropped land was protected against minor to moderate floods, by floodgating open creeks and drains and the construction of levee banks, and drains were constructed to remove floodwaters after the levee banks had been overtopped. By 1969, the Tweed Shire Council had spent $1.093 million building levee banks to reduce flooding in the Tygalgah, North Tumbulgum and Chinderah localities and deepened, cleaned and added floodgates to the main drainage channels throughout the Tweed Shire. The Richmond River County Council had spent $3.5 million on flood mitigation works by 1970, including the building of floodgates on all the outlets draining into the lower Richmond River and the construction of several levee banks and canals which specifically benefited the cane growing lands in the Wardell, Pimlico, Empire Vale and Keith Hall localities.[59]

CSR considered that the new, government-sponsored drainage and flood mitigation works in northern New South Wales during the 1960s had a most beneficial effect on the canegrowers' attitude towards improving their farm drains. Landholders could now connect subsidiary drains to deep, well maintained and floodgated main channels and creeks. This development coincided with the appearance of Toft mechanical cane loaders which could be modified by the addition of bucket dredges. This machinery reduced the manual labour associated with building and maintaining these on-farm drains. They were first used during the non-harvest season in the Tweed River and Richmond River districts in 1959 and the Harwood Mill supply area in 1960. The increased presence of these machines during the 1960s, together with the

Table 12.2. The length of new drains created and the length of existing drains cleared (metres) in New South Wales mill supply areas, 1959–1964.

	Harwood		Broadwater		Condong	
Year	Old drains cleared	New drains created	Old drains cleared	New drains created	Old drains cleared	New drains created
1959	3 359	3 359	52 304	2 313.5	1 408	*
1960	1 408.2	1 428	79 301.2	*	3 218.7	*
1961	*	*	65 561.3	*	12 774.3	20 157
1962	*	3 178.5	96 823.1	*	*	*
1963	7 845.6	11 406	*	*	264 458	54 014.2
1964	23 335.7	23 617.4	53 108	*	241 404	87 549.2
TOTAL		42 988.9		23 313.5		161 720.4

* Details not mentioned in records.

Source: Based upon figures in CSR, General Reports on Cane, Part 2, 1959 to 1965, in CSRR, Z/109, Boxes 389 and 390.

introduction of weedicides such as Dowpon and 2, 4-D, encouraged many New South Wales canegrowers and the drainage unions to undertake new drainage work and/or clean existing drains regularly during the early 1960s. They were assisted by CSR's technical field officers, who undertook survey work for the canegrowers so they could locate their drains in the correct positions.[60] CSR's figures on this topic between 1959 and 1964, although possibly incomplete, suggest that approximately 228 022 m (or 228 km) of new drains were built in the New South Wales sugar-producing districts, with many of these new drains being built in the Condong Mill supply area (Table 12.2). The introduction of mechanical cane harvesting during the early 1970s also encouraged New South Wales canegrowers to develop effective on-farm drainage systems, including mole drains instead of open drains.[61]

Flood damage to farms and stream bank erosion also troubled North Queensland canegrowers during the period 1945 to 1970, although the more coordinated approach to the problem that emerged in northern New South Wales was not duplicated in Queensland. Severe stream bank erosion along Cattle Creek occurred during the 1946 floods in the Mackay district, leading to the formation of the Upper Pioneer River Improvement Trust in 1948. This organisation completed extensive creek bank repairs after the 1958 and 1959 floods led to the loss of more cane growing land in the Finch Hatton locality.

Queensland's three River Improvement Trusts located in sugar-producing districts were not called upon to undertake much repair work throughout the early 1960s due to reduced rainfall and flooding activity. However, large amounts of rain fell in parts of North Queensland during early 1967, and the severe flooding in the Herbert River district that followed caused substantial damage to both the river banks and adjacent farms. During 1969 and 1970, the Herbert River Improvement Trust conducted an extensive program of bank stabilisation works and improvements to the carrying capacity of various stream channels by undertaking tree clearing from channel islands in the Herbert River and several kilometres of the Stone River, a tributary of the Herbert River.[62]

Growing concern about soil erosion, 1945 to 1975

In September 1945, a group of Isis district canegrowers met with Norman King and L.G. Vallance of the BSES and Walton G. Wells, the cotton specialist with the Queensland Department of Agriculture and Stock. The canegrowers agreed to support experiments into ways to reduce soil erosion throughout the district. During the wet season of 1946–47, L.G. Vallance conducted a soil erosion control experiment in the Isis district. The trial was placed on a slope having a maximum gradient of 16%. Part of the paddock was terraced with contour banks, and all the run-off was diverted by waterways into one main outlet. A considerable amount of soil movement occurred and the diversion waterways became almost completely silted up, but the contour banks had stopped the formation of deep gullies that were noticeable in nearby fields that were unprotected by contour banks.[63] As this experiment progressed, Percival Yeomans – an Australian mining engineer who owned a small property near Sydney – was developing his Keyline planting method. At the heart of this method is the Keyline, a very specific contour line that occurs in all valley and ridge topography. This contour line delineates the transition areas above which all contour tillage must proceed up the slope and below which all contour tillage must proceed down the slope. Terraces and waterways are established to channel run-off water into a series of farm dams, thereby drought-proofing properties. To improve soil infiltration, the ground is contour ripped parallel to the terraces.[64]

Armed with the details gained from the earliest experiments in the Isis district and the approach adopted by Yeomans, the BSES scientists designed further trials during the early 1950s in order to refine the advice they provided to canegrowers. Contour banks were built on additional cane growing paddocks in the Isis district, using the Keyline planting method, although elements of the approach such as contour ripping and on-farm dams were omitted. These trials determined the ideal distance for the spacing of contour terraces on sloping land. The local canegrowers had expressed dissatisfaction with placing contour banks every 30 m apart, as they did not deal adequately with the large amount of run-off. By the late 1950s, the BSES scientists had concluded that contour banks placed between 20 and 25 m apart achieved the most reduction in the velocity of the water flowing across the cane paddocks. In addition, the contour banks had restricted the amount of water flowing across the fields, thereby increasing water penetration and improving the growth of the cane adjacent to the contour banks.[65] These efforts were also supported by other organisations. From the early 1950s onwards, CSR conducted its own trials in the Innisfail district to determine the effectiveness of contour banks in slowing soil loss. Results from these trials convinced CSR that contour banks effectively reduced soil loss, so the firm urged its canegrowers who cultivated sloping land in the Innisfail district to build contour banks and cover fallow paddocks with green manure crops. The Queensland Department of Agriculture and Stock (later the Department of Primary Industries) also turned its attention to soil loss in the sugar cane growing lands. Soil conservation officers were appointed at Bundaberg and Mackay in 1955 and 1967, respectively.[66]

Equipped with the knowledge gained from these investigations, extension officers from CSR, the BSES and the Queensland Department of Primary Industries promoted soil conservation measures throughout Australia's sugar-producing districts over the next two decades. To slow the flow of water across their properties, Australian canegrowers were urged to adopt one or more of the following measures: contour tillage (later known as contour row direction); contour bank construction across their properties; encircling their fields on sloping land with grassed waterways or ditches to divert water away from paddocks; and the building of stabilisation structures across gullies to prevent their extension. Australian canegrowers were also urged to minimise the duration of the bare fallow during the summer season of high rainfall by planting a summer cover crop.[67]

Some Australian canegrowers tried implementing the advice provided by the agricultural extension officers. By the late 1950s, contour banks had been established on a few farms in the Isis district, and sporadic use of contour ploughing and stone walls to slow the loss of soil were observed in the South Johnstone district. A handful of canegrowers in the Hambledon Mill supply area had constructed contour banks and diversion drains during the mid 1960s, aiming to slow soil erosion on their properties. By the early 1970s, soil conservation measures were employed by seventy-two farmers in the Pioneer Shire which encompassed the heart of the cane growing area in the Mackay district. These measures protected about 480 ha of cane growing land. CSR also reported that some canegrowers in the Clarence River district during the early 1970s were building drains to divert hillside water off their cane fields and where fields had been planted up and down the slope, frequent cross drains had been constructed to break the momentum of the down hill flow of water and to divert it across the slope. Several canegrowers in the Rous and Alstonsville localities had also received assistance from the New South Wales Soil Conservation Service to build contour banks and diversion banks on their properties.[68]

Those Australian canegrowers adopting soil conservation measures between 1945 and 1975, however, turned out to be a minority of landholders. Instead, many canegrowers had an ambivalent attitude towards the issue. Evidence of this view is apparent in a report by CSR's Goondi Mill technical field officer. He complained in July 1967 that the local canegrowers had 'shown little interest in erosion control measures'. Similarly, his counterpart for the Hambledon Mill supply area complained in March 1971 that the summer rains had caused extensive soil erosion, but the local canegrowers 'on the whole do not regard this very seriously'. This limited success in convincing Queensland's canegrowers to take more care of their soil by better water management meant the level of soil erosion in some Australian sugar-producing districts grew steadily worse during the 1950s and 1960s. By the early 1970s, CSR reported 'very serious soil losses' from the Daradgee and Garradunga localities north of Innisfail. Farming land in parts of the Mackay region was also being subjected to significant soil erosion.[69]

The Isis district, however, suffered the severest soil erosion in any of the sugar-producing districts in Queensland. Yet ironically only twenty-three canegrowers or 9% of the district's canegrowers used soil conservation measures in 1970. The Isis Land Use Committee noted forlornly that even if the number of canegrowers adopting soil conservation measures doubled from six per year it would still be about twenty years before an acceptable level of soil conservation would be applied over the whole district.[70] Eventually, the Isis and Gin Gin districts were declared areas of soil erosion hazard by 1975, resulting in the Queensland government agreeing to subsidise landholders 50% of the cost of any soil conservation works undertaken on their properties in accordance with approved plans. Moreover, the Queensland government (Bjelke-Petersen; Coalition of Liberal and Country Parties) had commenced assisting seventy canegrowers farming very sloping land move their assignments to more level farms created from vacant Crown land or areas surrendered from local State forests.[71]

Removing water, impounding water: irrigation and drainage schemes come to fruition, 1970 to 1995

Drainage and flood mitigation assumed greater importance for many Queensland canegrowers during the early 1970s due to a succession of years when heavy coastal rain fell during the summer months. Major flooding occurred in the Burdekin, Proserpine, Herbert, Tully, Haughton and Murray Rivers resulting in damage to river banks and adjacent properties. These flood events led to the formation of the following River Improvement Trusts: Haughton River (1972); Mulgrave Shire (1972); Johnstone River (1973); Proserpine River (1974); and Cardwell Shire (1975). During the late 1970s, all the river improvement trusts operating in Queensland's sugar-producing districts obtained some funds from the Queensland and Commonwealth governments under Natural Disaster Relief Arrangements to build levee banks, protect eroding river banks with rocks, repair breakaways and improve stream flow by removing vegetation and sand islands from stream beds.[72]

Major flood events in North Queensland during early 1981, 1986 and 1991 caused more river bank erosion along sections of the Haughton, Mulgrave, Tully, Johnstone and Herbert Rivers, and the local river improvement trusts tried to restore this flood damage.[73] However, instead of just reacting to episodic flood damage, the Mulgrave Shire River Improvement Trust decided in 1975 to try and mitigate the flooding of existing cane farms. The engineering firm Gutteridge, Haskins and Davey Pty. Ltd. was commissioned to formulate an integrated plan of work to assure better flood plain management.[74] Other river improvement trusts adopted a similar approach during the 1980s and early 1990s, trying to link flood mitigation and drainage together so as to reduce damage to river banks and adjacent cane growing lands.

The very wet seasons during the summer months of the early 1970s had also encouraged groups of Queensland canegrowers to maintain an interest in drainage. By 1980, eleven drainage areas operated in Queensland. With the exception of the South Maroochy Drainage Area, all were located in North Queensland (Figure 12.9). In addition, individual Queensland canegrowers also improved their on-farm drainage during the 1970s. Some Mossman district canegrowers were observed to be making 'very extensive use' of drainage equipment and land forming and land planing machinery to improve farm drainage. CSR continued assisting its northern suppliers by taking levels for drainage purposes. Herbert River canegrowers were singled out as particularly taking an active interest in eliminating depressions on their farms, by levelling their paddocks with levellers, land planes, graders or mobile scoops. Even BSES extension officers had become involved in this process, surveying properties to obtain the correct levels in order to improve drainage in the districts between Mossman and Innisfail.[75] Canegrowers created more rectangular fields suited to mechanical cane harvesters and reclaimed land to permit an expansion in the area under cane. Creek realignment and deepening of watercourses also occurred in North Queensland sugar-producing districts during the 1970s (Figure 12.10).

Despite their efforts, North Queensland canegrowers recognized that productivity on substantial areas of cropped land was still being affected by inadequate drainage and flood inundation. Between 1970 and 1974, Productivity Committees were formed in eight North Queensland mill supply areas to consider increasing local productivity by improved drainage and flood mitigation. In September 1974, the Northern District Sugar Industry Productivity Group was formed to co-ordinate the activities of the eight Productivity Committees. Drainage and flood mitigation assumed even greater importance when the Queensland government approved an expansion of mill peaks in 1974 and 1975. Small parcels of waterlogged country were often the only land left available for expansion. Thus, the Northern District Sugar Industry Productivity Committee sought funds from the Bjelke-Petersen government in Queensland to undertake a detailed study into ways of substantially overcoming the problems of poor drainage and flooding. The organisation claimed that the annual financial losses of $12-20 million for sugar prices between $150 and $250 per tonne were caused by flooding and impeded drainage during the early 1970s. Subsequently, a study was undertaken by the IWSC using the Consulting Engineers Gutteridge, Haskins and Davey Pty. Ltd. Two reports were produced: an assessment of local drainage problems; and an overview of flood problems and possible flood mitigation options. The reports' author realised that in some localities flood mitigation measures were an essential prerequisite before drainage networks could be fully effective.[76]

The Flood Mitigation Study recommended that several flood mitigation proposals could be implemented, including the construction of alternate outlets for the overflows of the Tully and Murray Rivers at an estimated cost of $3 million and the building of floodgates and levees on the lower Herbert River at a cost of approximately $5 million. Storage dams on the Mulgrave and Herbert Rivers would also provide appreciable flood mitigation in these areas. The Drainage Study reached the conclusion

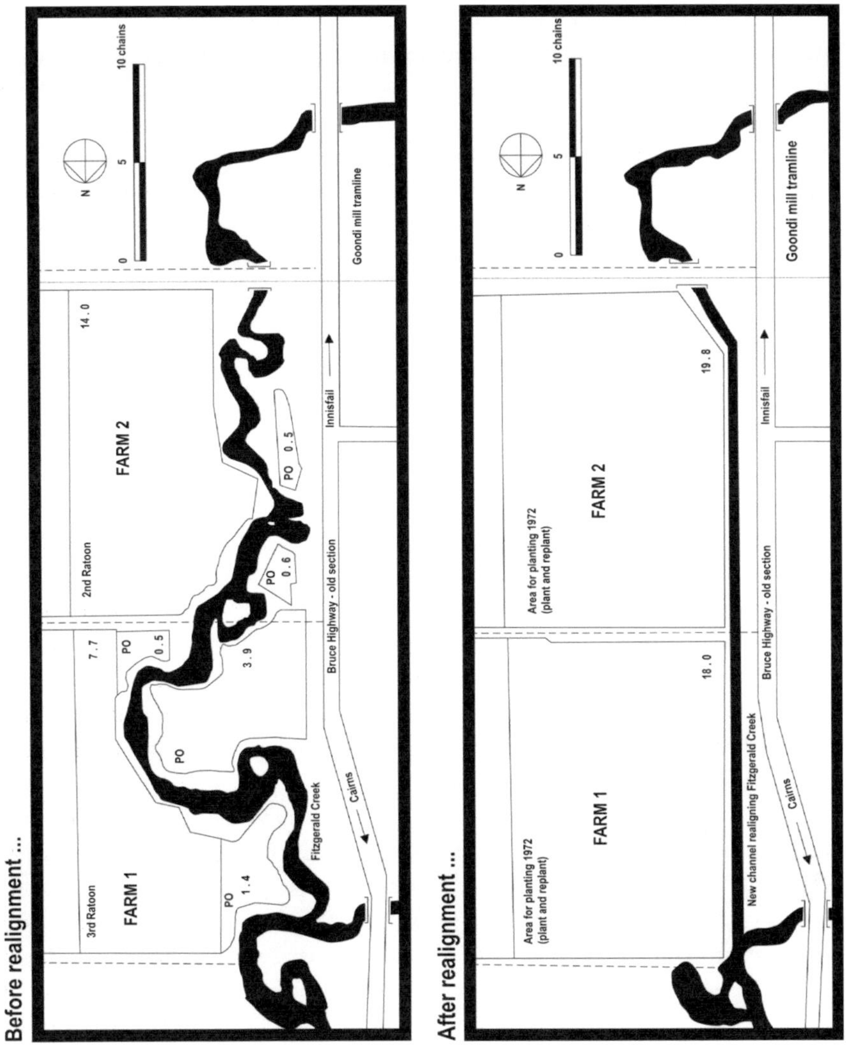

Figure 12.10. Plans of cane farms in the Innisfail district showing how the realignment of Fitzgerald Creek permitted the creation of more regular field layouts. The creek was also deepened to provide faster drainage, making flooding less likely.

Source: CSR, Agricultural Circular, Queensland, 1972, pp. 10–11, CSRR, Z/109, Box 370. Image reproduction courtesy of CSR Ltd.

that some North Queensland canegrowers had taken steps to improve their on-farm drainage. However, in many instances the drains were ineffective, failing to connect to efficient outlets which provided a suitable escape of water to natural streams. In some mill supply areas (e.g. Mossman; South Johnstone), farmers had shown little interest in on-farm drainage. Throughout the region, at least seventy-five private, communal drainage schemes were operational, but many of the drains were undersized and needed channels of greater capacity. Most drains in these schemes had been constructed with no engineering design, having been sized on a trial and error basis. Drain maintenance was ad hoc, depending upon the degree of cooperation between canegrowers and the efforts of various individuals. The overall effectiveness of the drainage networks throughout the sugar-producing localities in North Queensland was assessed as being fair to poor. The Goondi Mill supply area near Innisfail was identified as being the most effectively drained area, while the Victoria Mill supply area near Ingham was assessed as having the least effective drainage network due to a 'lack of interest by farmers'.[77]

Subsequent to the North Queensland sugar industry flood and drainage studies, the IWSC funded comprehensive flood mitigation investigations in the Hinchinbrook, Cardwell and Johnstone Shires on a 50:50 basis with the respective Councils. No assistance was forthcoming from the Commonwealth government for these investigations, although flood mitigation works were being constructed using Commonwealth funds in the sugar-producing districts of New South Wales during the 1970s (see discussion later). These reports had been prepared by the mid-1980s and contained suggestions on how to integrate individual farm and communal drainage schemes with the overall drainage of the rivers to reduce the flooding troubling the canegrowers in three North Queensland sugar-producing districts. The Hinchinbrook Shire Council had even developed a rural drainage scheme by 1983, providing a coordinated plan for drainage development throughout its local area.[78] Funds to implement the recommended solutions, however, proved difficult to secure, although support for specific projects came from grants as part of the Queensland Sugar Industry Infrastructure Package of 1993.[79]

Lack of funds did not slow flood mitigation works in northern New South Wales during the 1970s. The New South Wales government (Askin; Liberal) had replaced the flood mitigation legislation of 1964 with the *New South Wales Grant (Flood Mitigation)* Act of 1974, enabling continuous Commonwealth financial assistance to the flood mitigation authorities. After 1978, this arrangement was altered and the New South Wales government had to make annual submissions to the Commonwealth government to receive flood mitigation funding under the latter's *National Water Resources Assistance Act* of 1978. During the 1970s, the northern New South Wales flood mitigation authorities were granted additional funding that allowed extra drainage work to be completed.

The new drainage works involved a range of activities. In the Richmond River district, the Bagotville Barrage was completed in 1971 and protected the Tuckean Swamp floodplain from inundation by flooding from the Richmond River and tidal in-flows of saline water. Grazing lands, once periodically inundated, were placed under sugar cane. Levee banks were constructed in the Teven locality (north-west of Ballina) to reduce flooding to once every four years and the Pimlico-Emigrant Creek drains were further extended during 1972 so that no areas in the Pimlico locality would be inundated for more than three days. By 1982, the Richmond River County Council was responsible for 16 km of main channels, 112 floodgates and 15 km of levee banks (Figure 12.11). This new flood mitigation work ensured that cane growing areas were free of flood waters within seven days of the flood peak and the flood waters had receded to normal after about fourteen days. In general a tractor could be used on the land within a month.[80] The Clarence River County Council used its additional funds to build new levee banks near Maclean and at Goodwood and Harwood Islands, and constructed new channels in the Palmers Channel, Palmers Island and Harwood Island areas, allowing canegrowers to improve their on-farm drains. Normally, the cane growing areas of the lower reaches of the Clarence River were flooded once or twice a year, and in some areas floods could occur due simply to tidal influences. After the completion of the flood mitigation works during the 1970s, flooding was reduced to once every three of four years. Funds were also secured by the Tweed River Council for further bank stabilisation along the Tweed River and channel improvements and outlet works for Cudgen and Mooball Creeks.[81]

Individually, New South Wales canegrowers were particularly active during the 1970s and early 1980s, redesigning fields and farm layouts to improve drainage even further. Existing minor watercourses were filled and new drains installed using the latest drain digging equipment fitted with laser levellers. Fields were enlarged and became more rectangular in shape. The location of water furrows had to be altered so as to not interfere with the access for or the operation of mechanical cane harvesters. Instead of situating water furrows between two normally spaced rows of cane, at least 3 m had to be left between the cane rows where water furrows were located. Land planes were also deployed to eradicate depressions in the paddocks and to level a paddock's surface prior to planting. Approximately fifty land levellers were used throughout the Broadwater Mill supply area by the early 1980s.

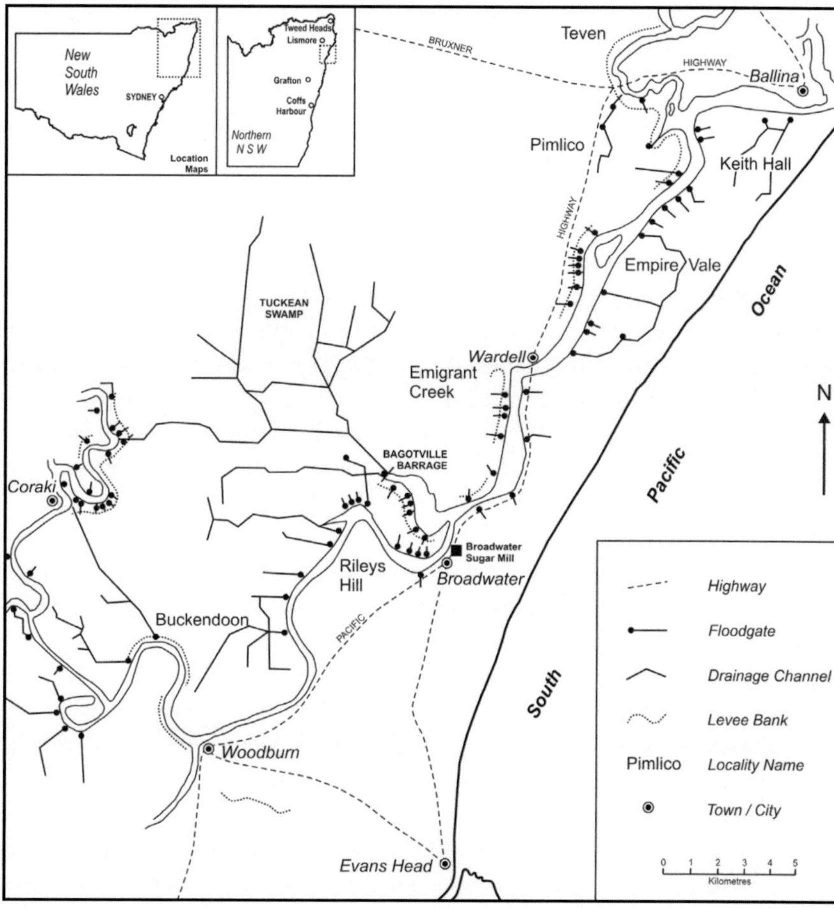

Figure 12.11. Richmond River district drainage improvements, 1981.

Source: Based upon details in Sinclair, Knight and Partners, *Richmond Valley: summary report* (Sydney, 1981), various pages.

Floodgating of most outlets to the Clarence, Tweed and Richmond Rivers encouraged the use of larger capacity pumps, enabling excess water to be discharged to the rivers twenty-four hours a day. During the 1960s, the impeller pumps had 25–30 cm diameter pumps, but by the early 1980s larger capacity, 60 cm, 75 cm and 90 cm diameter impeller pumps were being installed by canegrowers. An increasing number of New South Wales canegrowers also displayed a preference for portable or mobile pumps that gave them more flexibility in their drainage activities. A few communal pumping schemes also emerged, joining the already existing drainage unions that still maintained some of the drains. A group of farmers in the Richmond River district, for example, had formed the East Wardelll Pumping Scheme by 1982, installing and managing three large capacity pumping installations in the Wardell locality.[82]

Drainage remained a high priority for many Queensland canegrowers during the 1980s. For some, it was a matter of fine tuning their practices. Fifty Herbert River canegrowers, for example, complemented their open surface drains by installing sub-surface drainage pipes. Investigations by CSR's technical field officers found that the yields on these properties had increased by an average of 13 tonnes per hectare. A large amount of underground drainage work was undertaken by Mossman canegrowers during 1980, and had proved to be most successful in removing excess water. Other canegrowers sought to cultivate waterlogged land so as to participate in the industry expansion approved by the Queensland government during the early 1980s. Landholders in the Sippy Downs area in the Maroochy district, for instance, were permitted to grow cane as a result of this expansion. During the early 1980s, they were observed to be installing open drains on their properties to enable the successful cultivation of sugar cane on approximately 2 000 ha of podzolic soils.[83] In contrast, use of the free drainage service provided by staff at Condong Mill in the Tweed River district during the early 1980s had been disappointing. Tony Hayes, the Condong Mill technical field officer, claimed that there was 'still plenty of room for improvement' in the drainage network throughout the district.[84]

North Queensland canegrowers continued to support communal drainage schemes during the 1980s, raising funds that were spent on various activities, not just the building new drains. The Ripple Creek Drainage Board, for instance, erected floodgates at the mouth of Ripple Creek in 1984 to prevent floodwaters from the Herbert River backing up the creek and causing minor flooding. Nearby, the Mandam Drainage Board funded the clearing of mangrove re-growth in Mandam Creek during the 1980s, ensuring that the flow of water from drains entering this watercourse was not impeded by a build up of debris. Moreover, ten new communal drainage schemes were formed in North Queensland during the 1980s (Figure 12.9). As a result of the formation of these new drainage schemes, the majority of the cane growing lands in the Mourilyan Mill supply area, incorporating the localities around the townships of El Arish, Silkwood, Cowley and Mourilyan, were comprehensively covered by drainage schemes. By 1990, most cane-producing lands in the Herbert River district were also covered by a drainage scheme. In reality, however, some Herbert River Schemes (e.g. Foresthome,

Mandam, Loder Creek) had functioned since the 1960s or early 1970s, operating as voluntary associations. CSR claimed that the lengthy delays in the formal gazettal of these schemes occurred because of the reluctance of some canegrowers to become involved in semi-government bodies, fearing a loss of control over drainage works.[85]

Drainage works disturb potential acid sulfate soils, thereby leading to the emergence of acid hazards. Acid sulfate soils (ASS) is the common name given to sediments deposited under estuarine conditions (i.e. close to sea level), and which contain significant amounts of iron sulfides, particularly pyrites (FeS). These soils are formed when seawater or sulfate-rich water mixes with land sediments containing iron oxides. ASS remain relatively harmless if undisturbed and are termed potential ASS. However, if the sediments are disturbed, as occurs when drains are constructed, and exposed to air, the pyrite is oxidised and sulfuric acid is generated. Large volumes of sulfuric acid may then be stored in the soil or groundwater, and displaced into drains and natural watercourses by rainfall. Floodgated drains may act as acid reservoirs, releasing acid into coastal streams gradually or in large flushes following heavy rainfall. Moreover, as the rate of acid formation exceeds the neutralising capacity of the parent material, actual ASS are formed, and soil pH may fall below 4. Oxidation and subsequent reactions produce iron and aluminum which may become soluble in toxic quantities, with their precipitates affecting water quality and local aquatic wildlife. Thus, debris and soil excavated when drains are cleaned, deepened or widened may be highly toxic.[86]

Drainage measures had probably disturbed some ASS in North Queensland by the early 1980s. Drain clearing throughout the Tully district and the Fishery Falls, Deeral and El Arish localities during the mid-1980s created acidified drain spoil that was placed on adjacent fallow fields. The subsequent death of the crops of cane planted in the affected areas and poor growth of cane in adjacent areas was blamed on this material.[87] A large fish kill in the Tweed River in 1987, after rain followed the driest period in eight years, was attributed to acid releases into the watercourses. The Tweed River district canegrowers were strongly condemned by the local fishing and oyster industries after the 1987 incident. Some Tweed River district canegrowers commenced using additional amounts of dolomite and lime on their fields, attempting to lessen the problem caused by ASS. In response to the emerging acid threat, the New South Wales sugar industry, in conjunction with state government authorities, developed a set of guidelines in 1994 to minimise the off-farm effects of drain construction and maintenance. These guidelines were circulated to all New South

Wales canegrowers and were supported by a series articles in canegrower newsletters.[88] Additional small areas (mostly <0.5 ha) where sugar cane failed to flourish because of ASS had been identified in Queensland's sugar-producing districts by the mid-1990s, but the problem did not appear to be as severe as in northern New South Wales. Nevertheless, mapping areas of ASS risk had commenced in Queensland's coastal regions by 1995. This preliminary mapping had identified that farmers supplying the Moreton and Rocky Point Mills were the most vulnerable, with three quarters of the area assigned for growing sugar cane being considered as having a potential high acid sulphate risk.[89]

Canegrowers in other eastern Australian sugar-producing districts during the 1970s and early 1980s were more concerned about adding water to their fields of sugar cane and bringing more land under irrigation. The total area of sugar cane irrigated in Queensland rose from just under 60 000 ha in 1971/1972 to just over 100 000 ha in 1982/1983 (Figure 12.12). Regionally, the area of irrigated sugar cane in the Wide Bay-Burnett Statistical Division almost doubled between 1971/1972 and 1983/1984. Part of this increase occurred because water from the Bundaberg Irrigation Scheme finally became available to canegrowers. Some Maryborough district canegrowers had also commenced irrigating their crops during the 1960s and 1970s by obtaining permits to pump water from the Mary River and its tributaries.[90] The largest increase, however, occurred in the Mackay Statistical Division, following a trebling in the amount of irrigated cane over this period (Figure 12.12). A proportion of this growth occurred in the Proserpine district, where the canegrowers used bores or spears sunk into the bed of the Proserpine River to extract water from the alluvium.

At Mackay, drought conditions of the early 1960s had also prompted greater use of supplementary irrigation, mostly sourced from underground supplies. The number of irrigated farms in this district in December 1963 was 205 and this figure had increased to 426 by early 1969. This trend encouraged some canegrowers in the North Eton Mill supply area to begin agitating for a communal irrigation scheme centered on Sandy Creek and using water pumped from the Pioneer River.[91] Investigations by the IWSC culminated with the release of a report in January 1971 recommending the building of a major water conservation scheme based on diverting water from the upper Pioneer River into a dam on upper Sandy Creek and the distribution of this water to 9 300 ha of land spread over 150 cane farms. Funds for the Eton Irrigation Scheme were provided by both the Queensland (Bjelke-Petersen; Coalition of Country and Liberal Parties) and Commonwealth (Whitlam; Labor) governments and work began on what became known as the Kinchant Dam in 1974. A weir was also constructed at Mirani in 1986, enabling excess flows in the

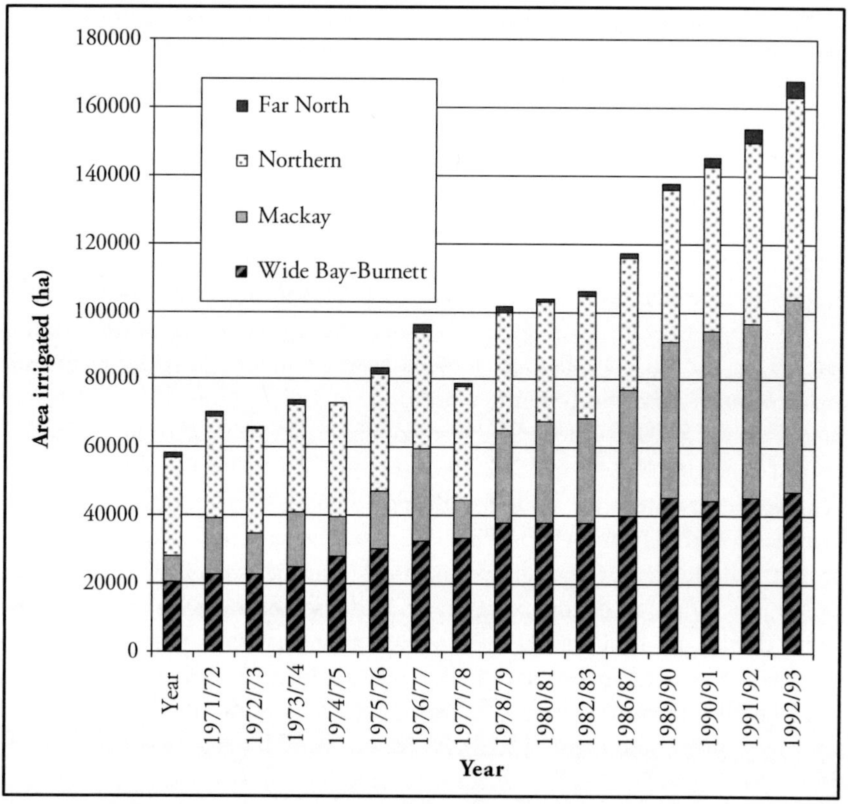

Notes: (1) Irrigation statistics cease to be collected each year after 1978/1979. (2) Names of Statistical Divisions as of early 1990s.

Figure 12.12. Area (ha) of sugar cane irrigated in Queensland's Statistical Divisions, 1971/1972–1993/1994.

Source: Based upon figures in the *Queensland Year Book*, various issues.

Pioneer River to be harvested and diverted to the Kinchant Dam. Canegrowers supplying North Eton Mill were the first to benefit in 1979 and subsequent work expanded the distribution network to include farms in the Racecourse Mill supply area.[92] As a consequence of these developments, the percentage of the total area of sugar cane irrigated in Queensland slowly rose, reaching almost 30 % of the season's crop in 1983/1984 (Figure 12.5).

Delivery of irrigation waters to the cane crops changed in some Queensland sugar-producing districts during the 1960s and 1970s. Flood or furrow irrigation was still the most commonly adopted method of distributing water in the early 1960s, especially in the Lower Burdekin

Table 12.3. Area (ha) irrigated by different methods in selected Queensland shires where sugar cane was the dominant crop in 1962/1963, 1972/1973 and 1978/1979.

Region/Shire	1962/1963		1972/1973		1978/1979	
	Furrows and flood	Spray	Furrows and flood	Spray	Furrows and flood	Spray
Far Northern						
Mulgrave	6	514.7	87	1 026	281	1 001
Hinchinbrook	31.2	114.2	1	983	0	350
Lower Burdekin						
Thuringowa	1 639.8	274.2	1 065	509	1 817	500
Ayr	22 016.2	144.9	24 326	972	34 137	1 364
Mackay						
Proserpine	97.6	906.4	232	2 160	534	1027
Pioneer	565.8	466.1	259	8 621	937	5 222
Mirani	197.6	588.9	63	3 368	370	2 016
Sarina	0	43.4	*	*	20	435
Bundaberg						
Gooburrum	3 310.8	1 977.6	1 280	8 009	2 165	10 858
Isis	42.5	249.0	0	1 854	0	3 504
Woongarra	2 850.8	1 067.6	1 698	5 280	6 545	3 066
Kolan	*	*	129	1 557	283	2 692
TOTAL	30 758.3	6 347	29 140	34 339	47 089	32 035
% of total	82.9	17.1	45.9	54.0	59.5	40.5

*Figures not provided

Source: *ASYB*, 1965, p. 95 and Australian Bureau of Statistics, *Irrigation on Rural Holdings*, Queensland, Cat. No. 7403.3, various issues.

and Bundaberg districts (Table 12.3). This irrigation method required the availability of a large volume of water. Overhead or spray irrigation using hand shift sprinklers and aluminum piping became more common during the late 1950s (see Plate 12.4). It was best suited to areas with a limited water supply and allowed sugar cane to be irrigated on lands unsuited to furrow

Plate 12.4. Hand shift sprinklers being used to irrigate young cane on a farm in the Mackay district, 1963.

Source: Image reproduction courtesy of the National Library of Australia. Image no. nla.pic- vn3294957.

irrigation because of soil texture and/or topography. By 1962/1963, sprinklers delivered water to 17.1% of the total area irrigated in Queensland shires where sugar cane was the dominant crop and had become particularly favoured by canegrowers in the Far Northern, Mackay and Bundaberg Regions. The popularity of overhead irrigation grew quickly over the next decade, with half the area irrigated in Queensland shires where sugar cane was the dominant crop receiving water from sprays in 1972/1973. Six years later, however, spray irrigation had lost its popularity. (Table 12.3). Some canegrowers had abandoned or reduced the use of overhead irrigation due to its disadvantages. Wind changes play havoc with the delivery of the water and shifting portable spraylines in advanced cane was a very arduous and labour-intensive task. A very marked swing to travelling irrigators also occurred in the Bundaberg Region during the 1970s, as they saved on labour costs which had increased considerably during the decade. By the early 1980s, the BSES reported that portable spraylines were now generally only used in young plant and ratoon crops and furrow or travelling irrigation systems had become more popular for more advanced cane.[93]

By the early 1970s, the reasons for irrigation in the different sugar-producing districts had moved beyond just keeping the crop of cane alive during a time of drought to a more sophisticated understanding of the relationship between soil moisture and productivity. Bundaberg district canegrowers regarded irrigation as a supplement to rainfall and favoured infrequent heavy irrigations, rather than frequent light irrigation. Most Mackay canegrowers strived to maximise production through irrigation rather than using irrigation to stabilise production. Lower Burdekin canegrowers irrigated to prevent the crop suffering little more than minimum stress at any time, aiming to provide adequate moisture for the crop at all stages of growth through to harvest. Even as harvest approached, a balance tried to be struck between the necessity of having the soil sufficiently dry for harvest and the maintenance of the crop in good condition.[94]

Some Queensland canegrowers had also realised that the water supply was not limited. Traditionally, Queensland canegrowers who irrigated delivered the water to their cane using open, earthen head ditches (i.e. main irrigation ditch at the edge of a field of cane). Losses of water were high through evaporation or soakage, and the drains had to be broken down and then rebuilt each time a new crop of cane was planted. However, from the late 1950s onwards, both underground and surface pipes of various materials replaced earth ditches for the distribution of water around farms, while gated pipes of aluminum, galvanized iron, rigid PVC, or flexible polythene were used to control the flow of water into the furrows (Plate 12.5). These new methods of water delivery reduced water loss and eliminated the labour-intensive task of creating and maintaining open, head ditches. The supervisory cost of monitoring water flows could be eliminated, allowing irrigation to be conducted during the night without any of the frequent inspections necessary when using head ditches. Corrosion problems were a drawback with the use of galvanised pipes, but the gated pipe system had none of the disadvantages of open, head ditches.[95] Despite these advances, the BSES in 1970 was highly critical of the irrigation practices of the state's canegrowers. In the Bundaberg district, the BSES complained that 'very limited reliable data' was available regarding the amount of water applied to the crops of cane and 'very few Lower Burdekin farmers had any idea of the amount of water applied'.

During the 1960s and 1970s, some Australian canegrowers using irrigation also began sculpturing paddocks, seeking to obtain suitable and uniform gradients to permit more efficient use of water. Hills and

Plate 12.5. Plastic irrigation tubing on a Queensland cane farm, c. 1970. This method of delivering irrigation water was adopted on some Queensland cane farms during the late 1950s.

Source: CGQB, Vol. 33, no. 3 (January 1970), p. 78. Image reproduction courtesy of BSES Limited.

hollows within fields were eliminated by the use of a blade or a scoop attached to a tractor. After grading, scooping or filling, the surface was then smoothed with a land plane. In other instances, canegrowers embarked upon more ambitious levelling schemes when a suitable gradient for furrow irrigation could not be produced by levelling the topsoil. All topsoil from the area to be levelled was first bulldozed into piles on the side of the fields. The sub-soil was then graded into a predetermined slope, before the topsoil was bulldozed back and graded to a uniform depth. Such activities in the mid-1960s required high capital outlays of £100 to £150 per acre, but the costs were recouped through better yields due to the increased efficiency of the irrigation and the generally favourable returns from sugar during this period. Moreover, they had to be undertaken by specialist earthmoving contractors using heavy bulldozers, scoops and graders. By the early 1980s, laser technology had brought greater precision to this levelling work and it subsequently became common practice on farms as canegrowers sought greater accuracy and more efficient use of flood irrigation.[96]

Other Queensland canegrowers during the late 1970s and 1980s were more concerned about gaining access to a securer supply of water for irrigation. One such group of canegrowers farmed in the Haughton River district, where they relied extensively on underground water supplies to irrigate their crops of cane. In 1936, the Haughton River canegrowers failed to convince the IWSC to build a weir on the Haughton River, above the township of Giru, to conserve water for irrigation and to stop salt water intrusion into the underground aquifers. After low rainfall in the early 1960s, the Haughton River district canegrowers resurrected the weir idea. Findings from investigations by the IWSC, released in 1971, showed that the water supply in the district's aquifer system was insufficient to meet the requirements for full irrigation of the existing assigned areas every second year. After a four-year delay, the Queensland government (Bjelke-Petersen; Coalition of Country and Liberal Parties) approved the building of two weirs on the Haughton River, although the Giru Weir and Val Bird Weir were not constructed until 1977 and 1981, respectively.[97]

Canegrowers around Maryborough had also commenced lobbying for some irrigation scheme in their district, following reduced rainfall during much of the 1960s. Investigations by the IWSC led to the development of a proposal that could irrigate double the area cropped with sugar cane in the Maryborough district. In late 1977, the Queensland government (Bjelke-Petersen; Coalition of National and Liberal Parties) approved the Lower Mary River Irrigation Scheme. Construction began the following year and was completed by 1987. At the heart of the scheme was the construction of tidal barrages on the Mary River and Tinana Creek to impound water and prevent salt water intrusion, and the building of pumping stations and distribution channels to convey water to the cane farms. At the same time, the

Proserpine district canegrowers were agitating for a dam that would reduce flooding of vulnerable areas along the Proserpine River and supply the cane-growers with additional water for irrigation. In late 1984, the Queensland (Bjelke-Petersen; Coalition of National and Liberal Parties) and Commonwealth (Hawke; Labor) governments agreed to construct a dam on the Proserpine River. Built between 1986 and 1990, the Peter Faust Dam enabled the creation of the Proserpine River Water Supply Scheme which supplied water for the irrigation of sugar cane in the Proserpine district.[98]

Accompanying the above developments was a major expansion in the Burdekin River Irrigation Area. Initially, irrigated agriculture using water pumped from the Burdekin River and distributed through 40 km of concrete channels and pipelines to tobacco farms commenced in the Clare district during the early 1950s. This development was Stage 1 of a supposedly much larger irrigation project that envisaged the building of the Burdekin Falls Dam and several weirs with the goal of supplying water to 780 new farms.[99] The IWSC then constructed the Gorge Weir above Clare in 1953 and the Blue Valley Weir below the junction of the Burdekin and Bowen Rivers in 1963, impounding water to enable the tobacco scheme to proceed. The impetus to complete the scheme then faltered until the idea was resurrected in the early 1970s. The first practical outcome was the construction of the Clare Weir between 1978 and 1980, thus increasing the storage capacity along the Burdekin River. In 1979, the Bjelke Petersen government in Queensland also announced its intention to construct the Burdekin Falls Dam and an extensive distribution network. The main channels would run from the Burdekin River above the Clare Weir to the Haughton River, and on the eastern side from the Clare Weir to the Elliott River. Lower down, the Barratta Main Channel would deliver water to landholders situated between the Haughton River and the Burdekin River delta lands around Ayr (Figure 12.6). In August 1980, the Commonwealth government (Fraser; Coalition of Liberal and National Parties) announced it would provide $84 million for the scheme; the Queensland government then contributed $126 million towards the project. The Burdekin Falls Dam was completed by mid-1987 and floodwaters filled the dam in March 1988. Water then commenced being delivered through the expanding network of channels to the canegrowers, including those new farmers who had acquired irrigation farms at auction.[100]

As a result of the above developments, the area of sugar cane irrigated in Queensland increased by 62 000 ha between 1986/1987 and 1993/1994, the last available figures for the topic (Figure 12.12). Regionally, the biggest increase of almost 26 000 ha in the amount of irrigated sugar cane occurred in the Mackay Statistical Division. Much of this growth came from the large upsurge in the area irrigated throughout the Eton Irrigation Area. An additional 23 000 ha of sugar cane were irrigated in the Northern Statistical Division in this period, a direct result of a doubling of the area irrigated in the Burdekin River Irrigation Area. An extra 9 500 ha of sugar cane were irrigated between 1986/1987 and 1993/1994 in the Wide Bay-Burnett Statistical Division. Some of this increase was a consequence of new farmers in the Lower Mary Irrigation Area securing access to water for irrigation and a continued growth in the area irrigated in the Bundaberg Irrigation Area. Almost 5 000 ha of cane were irrigated in the Far North Statistical Division in 1993/1994, as some canegrowers, mostly around Cairns, used supplementary irrigation to improve their yields (Figure 12.12). Consequently, nearly 42% (168 132 ha) of Queensland's sugar cane crop was irrigated in 1993/1994 (Figure 12.5).

During the 1980s and early 1990s, flood or furrow irrigation continued to be the most widely used irrigation method, especially in the Lower Burdekin, Mackay and Proserpine districts. Overhead irrigation was favoured by canegrowers around Cairns and in the Bundaberg and Isis districts. New equipment used in overhead irrigation such as water cannons, lateral or linear move irrigators and centre pivot irrigators appeared in the fields of sugar cane throughout Queensland during the 1980s.[101] Canegrowers in Southern Queensland also developed an interest in trickle or drip irrigation during the early 1990s. In 1993, nearly 600 ha of sugar cane were being watered by this method in Queensland. Despite the high initial cost, trickle irrigation had the advantage of using approximately one third less water than some sprinkler systems, especially water winches. By 1994, a trickle irrigators group had been formed in Bundaberg to facilitate informal discussions and technology transfer. Bundaberg district canegrowers using trickle irrigation reported producing more sugar per megalitre of water, and had significantly reduced their irrigation labour requirements and power use for pumping.[102]

Other Queensland canegrowers adopted more refined irrigation methods during the late 1980s and early 1990s. Furrow irrigation is often the least effective in terms of water use efficiency. Irrigation water applied by flooding can be lost when water passes below the crop root zone or from water that leaves the paddock and ends up in the drainage system (i.e. tail water). Long crop row lengths, sometimes one kilometre or longer on new farms in the Burdekin River Irrigation Area, were one factor which contributed to tail water on farms. The long rows make it difficult to determine when to shut irrigation water off. To reduce this problem, many canegrowers in the Burdekin River

Irrigation Area had established water recycle pits on their farms. The excess irrigation water was collected from farm drains and stored in small on-farm dams. The water was then pumped back into the main irrigation channels.[103] In addition, other canegrowers had reduced their water use by up to 40% by changing from a broad-based U-shaped furrow to a narrow-based V-shaped furrow. By 1995, a few Lower Burdekin canegrowers had even introduced irrigation scheduling on their farms based upon readings from evaporation minipans (i.e. containers calibrated to a particular soil type to enable measurement of water loss). This approach reduced water use by delivering exact amounts of moisture to the crop as it was required and avoided the traditional subjective approach to irrigation scheduling based upon visual signs of moisture stress.[104]

Environmental impacts associated with irrigated agriculture include water loss through run-off or leaching, irrigation-induced salinity, raising or lowering watertables and salt water intrusion. Evidence of over-exploitation of underground water, particularly in delta and coastal aquifers, had become apparent in the Lower Burdekin, Mackay and Bundaberg districts by the early 1990s. The growing vulnerability of salt water intrusion meant that land and water managers in the districts where cane crops were extensively irrigated had commenced trying to evaluate and manage the groundwater yields to ensure their sustainable use. Groundwater pumping restrictions had been introduced in the Mackay district and Bundaberg Groundwater Area in 1993 and 1994, respectively, in an attempt to slow salt water intrusion into the underground aquifers.[105] Small areas of irrigation-induced salinity or secondary salinity were also reported emerging in the Lower Burdekin, Mackay and Bundaberg districts during the 1980s. Remedial actions involved sourcing alternative suitable water supplies, the use of gypsum in specific problem situations and heavy applications of mill ash to improve the drainage from the salt-affected area. The emergence of these two irrigated-related environmental problems was a reminder that water supplies in Queensland's sugar-producing districts needed to be managed more carefully, especially considering the rapid escalation in the area of sugar cane being irrigated during the early 1990s.[106]

Slowing soil erosion, 1975 to 1995

By the mid-1970s, the worsening soil erosion problem in Queensland's sugar-producing lands prompted the Queensland Department of Primary Industries to commence investigations into quantifying just how much soil was being lost from sloping paddocks cultivated with sugar cane.

The first of these studies conducted in the Mackay district during the wet seasons of 1976/1977 and 1977/1978 concluded that losses ranging from 42–227 tonnes/ha per year were occurring on paddocks with slopes of between 6 and 8%. These amounts were extremely high, as maximum losses of 12.5 tonnes/ha per year were considered acceptable. In 1980, a soil conservation unit was also established at the Queensland Department of Primary Industries' South Johnstone Research Station, near Innisfail, in order to tackle the loss of soil in North Queensland sugar-producing districts. Officers from this unit investigated the magnitude of soil loss on cane paddocks under conventional cultivation in North Queensland. Average annual losses were calculated to be 150 tonnes/ha, with a range of annual measurements being 70–500 tonnes/ha.[107] Actual figures, however, only confirmed the obvious: many of Queensland's canegrowers were making no or limited attempts to reduce soil erosion on their properties.

During the late 1970s and early 1980s, Australian canegrowers were being provided with increasingly more refined advice on ways to reduce soil erosion. The suitability of land for cane growing in Queensland was being determined via land resource surveys. Moreover, official confirmation of the assigned areas during the expansion of cane growing to the Julatten district (near Mareeba) in the early 1980s was made conditional on satisfactory soil conservation measures being established from the outset.[108] In addition, canegrowers were urged to complete plough out-replant operations or the replanting of fallow land early to ensure an adequate crop cover by the rainy season. Contour banks, diversion waterways and contour row direction continued to be promoted. However, the old layouts with sharp corners and many short rows were being superseded by layouts that contained slighter curves, parallel banks and fewer short rows.

By the early 1980s, BSES officers reported that some canegrowers farming sloping land had constructed head ditches to prevent excess water flowing onto their fields, but except in a few instances, further control measures had not been implemented. Similar observations were made by the northern soil conservation officers with the Queensland Department of Primary Industries. They noted that 'little adoption of the more traditional soil conservation methods had been achieved' in North Queensland. Throughout the Innisfail district, for example, only 1 300 ha had been treated with soil conservation measures by 1986. This amount was a tiny fraction of the 23 000 ha with a slope above 3% requiring contouring.[109] A similar situation existed in the Mackay district, where just less than 15% of the 38 700 ha requiring some form of protection with soil conservation works had been treated by 1985.

Adoption of soil conservation methods amongst the canegrowers throughout the Maryborough, Isis and Bundaberg districts, however, had improved by the mid-1980s, with the local soil conservation officer estimating that a third of the existing cultivated land in 1987 was protected by soil conservation measures. Despite these changes, the Queensland Department of Primary Industries expressed concern in 1983 that the efforts by the State's canegrowers to reduce soil erosion were not quick enough. Soil conservation measures did not match the rate of development of erosion prone land in Queensland's sugar-producing districts, let alone make inroads into the 100 000 ha of existing land cropped with cane that had untreated erosion problems. In the absence of any further expansion in cane growing, Queensland's canegrowers would take forty to fifty years to treat all the eroding land. The Queensland sugar industry could not rely on additional staff from the Department of Primary Industries to hasten the implementation of these measures.[110]

Various reasons can explain the slow acceptance by some Queensland canegrowers of the facts about soil erosion. Firstly, acceptance that soil loss needed to be reduced required a canegrower to make two confessions: that a problem existed; and that this problem existed because of poor agricultural practices adopted currently or in the past. W. Sloan of the BSES claimed in 1947 that many canegrowers were still not fully 'cognisant of the dangers of soil erosion' or that they did not understand that a problem existed.[111] However, contrasting views existed. In 1946, C.K. Simpson, the Goondi Mill technical field officer, noted that the local canegrowers knew about the dangers of soil erosion, but were 'apt to consider soil erosion was something that nothing can be done about'. A.R. Taylor of the South Johnstone Cane Pest and Disease Control Board claimed in 1957 that Queensland's canegrowers had heard about the need for soil conservation measures, but the 'subject seldom progresses beyond the discussion stage with the average Queensland cane farmer'. A survey of sixty canegrowers at Innisfail in 1982 confirmed that they recognised erosion as a problem on their farms, but it was rated as a lesser concern than rising production costs and declining sugar prices.[112] This fragmentary evidence suggests that some Queensland canegrowers knew that soil loss was a problem, but were slow to respond for other reasons.

A second factor explaining the slow uptake of soil conservation measures by Queensland's canegrowers was that the loss of topsoil by soil erosion, except in the most extreme instances, had not transferred into lost productivity. Improvements in plant breeding had provided the state's canegrowers with sucrose-rich cane varieties suited to each district and the threat of diseases and pests were reduced by the BSES control programs (see Chapters 13 to 15).

Greater fertiliser use reduced soil deterioration. Moreover, in some sugar-producing districts, the eroding soils had sufficient depth to permit repeated turning up of a fresh layer for cultivation. However, the canegrowers in the Mackay district by the early 1970s were reduced to using subsoil, following the removal of so much topsoil.[113] A third reason advanced for the slow uptake of soil conservation measures was the perceived disadvantages of contour tillage and terraces. Canegrowers, including those surveyed at Innisfail in 1982, claimed that a great number of short rows almost always occurred between the contour banks at frequent intervals along their length. Such short rows were difficult to cultivate and became less acceptable following the introduction of mechanical harvesting in the 1960s. Field layouts had to be designed for heavy mechanical cane harvesters requiring well-defined turning points. In addition, contour banks absorbed an appreciable amount of scarce tillable land on their assignments, they became places where weeds and grasses accumulated, they silted up every few years and difficulties arose using travelling irrigators on non-straight layouts.[114]

The final reason suggested as an explanation for the slowness in implementing soil conservation measures in Queensland's sugar-producing districts was that a lack of equipment and staff with expertise slowed the building of contour banks. In 1959, R. Moller, a BSES agronomist, claimed that the construction of soil conservation measures in the Isis district was hindered because earth-moving equipment was not readily available for hire. Even when earth-moving equipment became more readily available, the Queensland Department of Primary Industries admitted in 1983 that canegrowers wishing to implement soil conservation measures were hindered by the following factors: lack of suitable topographic information about their properties; the ratoon system which inhibited implementation across an entire property as part of the farm is always under a crop; and a restricted implementation period (i.e. April to September). Under these constraints, soil conservation officers in sugar-producing districts had only been able to plan and survey soil conservation measures on approximately 250 ha per year. In contrast, rates of 7 000 ha per year had been achieved in the extensive grain cropping lands of central Queensland.[115]

In 1982, the Queensland Department of Primary Industries and the BSES commenced trials to determine the effectiveness of trash retention as a method of reducing soil erosion. These studies concluded that average annual soil loss over the period 1982–1987 was 5 tonnes/ha under a green cane harvest with 100% of the trash retained as a blanket cover. If no trash blanketing was practiced, the soil losses grew to 15 tonnes/ha. Therefore,

Table 12.4. The percentage of green cane harvesting in Queensland sugar-producing regions, 1985–1994.

Region	1985	1987	1989	1990	1992	1994
Northern*	21	45	68	73	79	88
Lower Burdekin	0	0	0	5	4	3
Central	0	0	11	12	12	27
South	0	0	18	15	17	24

* Sugar-producing districts north of Townsville include Mossman, Cairns, Innisfail, Tully and Herbert River.

Source: P. Griggs, 'Soil erosion, scientists and the development of conservation tillage techniques in the Queensland Sugar Industry', *Environment and History*, Vol. 12 (2006), p. 258.

trash blanketing or trash blanketing and no tillage, provided a substantial soil conservation measure regardless of soil type. These results were confirmed when additional studies into the magnitude of soil loss from sugar cane growing on sloping land under conventional cultivation and various no-tillage operations were completed by the Queensland Department of Primary Industries during the early 1990s.[116] As a result of these trials, canegrowers now had a soil conservation measure that did not solely involve contour tillage and/or building contour banks and ditches. Moreover, they were not required to build any structures: the mechanical cane harvesters supplied the necessary residues that provided an effective blanket against soil loss. Ease of implementation was a very attractive feature of this new soil conservation method.

The successful implementation of trash retention and minimum tillage techniques by these pioneering canegrowers and improvements in the green cane capability of mechanical cane harvesters accelerated the adoption of green cane harvesting. Furthermore, the slump in world sugar prices in the early 1980s encouraged more canegrowers to reduce their cultivation costs, achievable by adopting trash blanketing and reduced tillage. Hence, green harvesting was used on half the entire Queensland crop by 1996, although its adoption was initially greater in cane growing districts north of Townsville (Table 12.4). Moreover, an estimated 80% of the sloping land (> 2%) in Far North Queensland by 1991 was being farmed with trash blanketing and zero tillage methods, thereby reducing soil erosion. Trash retention and minimum tillage with its soil erosion reduction characteristics had occurred much faster than the implementation of traditional soil conservation measures.

Moreover, some North Queensland canegrowers had even commenced building artificial wetlands on their properties to trap eroded material before it reached local watercourses.[117]

Despite the implementation of trash blanketing, the Australian Conservation Foundation still expressed concerns in the mid-1990s about the vulnerability of soil to erosion in Australian sugar-producing districts. The Mackay district was singled out for special mention, as soil conservation planning and structural works were needed on 41 000 ha, but as of June 1992 only approximately 9 000 ha (or 22%) had been treated. Moreover, freshly cultivated soils in all sugar-producing districts remained at a high risk, especially if heavy rainfall occurred in spring or early summer before the soil had settled. The risk of erosion on newly cultivated lands, fallow lands and headlands – places not covered by trash – remained high, particularly in the sugar-producing districts south of Townsville, where the adoption of green cane harvesting had been much slower during the early 1990s. The thick trash blanket posed difficulties for irrigation in the Lower Burdekin district, while the ratooning of cane under the colder conditions of Southern Queensland presented significant problems. Nevertheless, more Australian canegrowers had finally commenced tacking the soil erosion problem by 1995, but there was still some way to go before the problem would be considered fully under control.[118]

Drainage and irrigation works and soil conservation measures by their very nature lead to landscape change. The natural flow of watercourses in the Lower Burdekin, Mackay, Proserpine, Bundaberg and Maryborough districts have been interrupted by the building of dams and weirs. The water impounded by these structures was delivered via artificial channels and pipes to the canegrowers who used it to irrigate their fields of sugar cane. The sugar-producing lands of northern New South Wales and of the Maroochy, Herbert River, Tully and Innisfail districts in Queensland have lost most of their freshwater wetlands which have been drained in order to maximise sugar production in these areas. These districts are also full of straightened watercourses, often floodgated, instead of the sinuous patterns associated with undisturbed drainage networks. Today, these districts are also crisscrossed by channels, where once none existed. Indeed, so many new channels have been built in the Tully district that the engineering firm Ian Drummond and Associates described the Tully-Murray Rivers system in 1993 as one now characterised 'by an extensive pattern of artificial drains'. A similar sentiment was expressed by the authors of a report on Queensland wetlands in 1999 when they claimed that so many artificial channels had been built throughout the Herbert River district that drains existed in place of natural watercourses.[119] The presence of these channels is perhaps the most visible legacy of the post World War II efforts by Australian canegrowers to manage water on their properties.

Notes

1 Griggs, 'Plantation to small farm', pp. 347–350.

2 Griggs, 'Plantation to small farm', pp. 350–354; Kerr, *Black Snow and Liquid Gold*, pp. 155–157.

3 BSES, AR 1920–21, *QPP,* Vol. 1 (1922), p. 1035; BSES, AR 1923–24, *QPP*, Vol. 2 (1925), pp. 710–711; *ASJ*, 5 September 1924, p. 381.

4 Kerr, *Southern Sugar Saga*, p. 103; B. Anderson, *The History of Irrigation in the Bundaberg Area* (Brisbane, 1978), p. 58.

5 A. Gibson, 'Sugar prospects in the north: Lower Burdekin district', *QAJ*, Vol. 6 (July 1916), p. 25; BSES, AR 1927–28, *QPP*, Vol. 2 (1929), pp. 869–870; E. Smith and W. Dixon, 'Report on Pioneer and Inkerman Mills: General', pp. 2–3, CSRR, Z/109, Box 520; H. Kerr and N. Cassidy, 'The soils and irrigation waters of the Burdekin delta', *QAJ*, Vol. 37, no. 2 (February 1932), pp. 126–127.

6 Sec. Robb Swamp Drainage Trust, Murwillumbah, to H. Osborn, NSW Public Works Dept., Lismore, 19 June 1922, In-letter 497 of 1922, 17/1281, File 24/319, AONSW; Condong Mill Mang. to E.W. Knox, 15 February 1918, CSRR, 142/1220, Letter No. 746 of 1918, p. 2.

7 For an example see P. O'Connor, Empire Vale, to G. Nesbitt MLA, Lismore, 28 August 1918, In-letter D658 of 1918, 17/1281, File 25/158, AONSW.

8 Broadwater Mill Mang. to E.W. Knox, 5 April 1918, CSRR, 142/881, p. 638; Broadwater Mill Mang. to E.W. Knox, 29 April 1919, CSRR 142/882, p. 220; Smith, *Sweet Beginnings*, p. 179.

9 Condong Mill Mang. to E.W. Knox, 8 September 1917, In-letter No. 712 of 1917, CSRR, 142/1220; E.W. Knox to Condong Mill Mang., 6 March 1917, CSRR 142/1172, p. 10; E.W. Knox to Condong Mill Mang., 4 May 1918, CSRR, 142/1172, p. 377; E.W. Knox to Condong Mill Mang., 26 October 1922, CSRR, 142/1175, Letter No. 83, pp. 133–134.

10 Broadwater Mill Mang. to E.W. Knox, 27 July 1918, CSRR, 142/881, p. 737; Broadwater Mill Mang. to E.W. Knox, 1 May 1920, CSRR, 142/882, pp. 548–549; Broadwater Mill Mang. to E.W. Knox, 3 May 1921, CSRR, 142/883, pp. 173–176.

11 Broadwater Mill Mang. to E.W. Knox, 17 February 1922, CSRR, 142/882, pp. 440–448; Broadwater Mill Mang. to E.W. Knox, 19 January 1927, CSRR, 142/885, p. 664.

12 CSR, Cane Report 1921, p. 56, CSRR, Z/109, Box 386.

13 E.W. Knox to Victoria Mill Mang., 5 April 1923, CSRR, 142/1573, Letter No. 104, p. 274; E. Wood, 'Cane pests and diseases', *QAJ*, Vol. 29, no. 1 (1 January 1928), p. 6; E. Wood, 'Gumming disease', *QAJ*, Vol. 29, no. 2 (1 March 1928), p. 189; *QAJ*, Vol. 30, no. 9 (1 September 1928), pp. 194–195.

14 Anderson, *The History of Irrigation in the Bundaberg Area*, pp. 46–47; *ASJ*, 3 April 1930, p. 40, 6 October 1932, p. 362–363, 6 August 1933, pp. 272–273 & 4 December 1934, p. 495.

15 Kerr, *Southern Sugar Saga*, p. 104; *ASJ*, 3 November 1932, p. 395 & 7 March 1935, p. 655; N. King, 'Soils investigations in the sugar-cane producing areas of Bundaberg', *The Qld J. of Agricultural Science*, Vol. 6, no. 3 (September 1949), p. 143.

16 *ASJ*, 10 October 1935, p. 401; King, 'Soils investigations in the sugar-cane producing areas of Bundaberg', p. 143.

17 Extracts from 'Notes of Mr Colman's visit to Queensland, June 1933', p. 371, AER, 165/196, File No. 53 C, NBAC; *ASJ*, 8 December 1932, p. 438, 7 February 1935, p. 596 & 10 November 1939, p. 481; Macknade Mill Cane Insp., AR 1936, pp. 20–21, CSR, Z/109, Box 472.

18 Kerr, *Black Snow and Liquid Gold*, pp. 157–158; *ASJ*, 4 February 1932, p. 621.

19 *ASJ*, 4 May 1933, p. 75; Kerr, *Black Snow and Liquid Gold*, pp. 240–241.

20 Broadwater Mill Cane Insp., AR 1930, p. 20 & Broadwater Mill Cane Insp., AR 1931, p. 12, CSRR, Z/109, Box 460; P. Goldfinch to Broadwater Mill Mang., 29 July 1933, p. 87, CSRR, 142/855, p. 87; CSR, Cane Report 1937, p. 72 & CSR, Cane Report 1939, p. 65, CSRR, Z/109, Box 387.

21 L. Maynard, 'The development of drainage in the Broadwater Mill cane area', *Proc. ASSCT*, 1982, pp. 84–85; K. MacLean, Report on the Agricultural Methods at Goondi Mill, 20 February 1934, Letter 114N (b), CSRR, 142/1428; CSR, Cane Report 1938, p. 81, CSRR, Z/109, Box 387.

22 *ASJ*, 13 January 1931, p. 603; CSR, Herbert River Agricultural Circular 1930, Doc. N 101, CSRR, Z/395, Box 2; CSR, Cane Report 1935, p. 65, CSRR, Z/109, Box 387; CSR, Cane Report 1937, p. 72, CSRR, Z/109, Box 387; K. Gard, 'Report on Pests, Diseases and General Field Problems at Macknade Mill, 1941 season', p. 5, CSRR, Z/395, Box 63, Report 326 N.

23 IWSC, AR 1936–37, *QPP*, Vol. 2 (1937), p. 554; IWSC, AR 1937–38, *QPP*, Vol. 2 (1938), p. 510.

24 Broadwater Mill Mang. to E.W. Knox,, 22 February 1919, CSRR, 142/882, p. 162; A. Arthington *et al.*, 'Potential impact of sugarcane production on riparian and freshwater environments', in B. Keating and J. Wilson (eds.), *Intensive Sugarcane Production. Meeting the Challenges Beyond 2000* (Wallingford, 1997), p. 404; S. Levett and P. Price, *Managing Riparian Lands in the Sugar Industry. A Guide to Principles and Practices* (Brisbane, 2001), p. 10.

25 *QAJ*, Vol. 26, no. 4 (1 October 1926), p. 287; IWSC, AR 1938–39, *QPP*, Vol. 2 (1939), p. 571; Herbert River Improvement Trust, *Herbert River Flood Management: initial submission for Australian government financial assistance* (Ingham, 1975), p. 25.

26 Kerr, *Black Snow and Liquid Gold*, 210–211; H. Kerr, 'The recent Burdekin Flood and its lesson', *QAJ*, Vol. 54, no. 8 (August 1940), pp. 111–112.

27 Office of the Qld. Co-ordinator General of Public Works. Minute – re: cyclone and flood damage in the Burdekin Area, 9 August 1940, in Chief Sec.'s Batch No. 409c, SRS1043/1, Box 721, Item 2165, QSA; M. Medley, *Presenting the Burdekin River Improvement Trust* (Ayr, 1989), pp. 5–11; Herbert River Improvement Trust, *Herbert River Flood Management: initial submission for Australian government financial assistance* (Ingham, 1975), p. 26.

28 Kerr, 'Soil erosion', p. 33; Ev. C. Young, p. 97, in 'Royal Commission on Sugar Peaks Scheme and Cognate Matters Evidence', CSRR, Z/109, Box 48; D. Quinn, 'Agricultural Report, Hambledon Mill area, 1969', p. 3, CSRR, Z/395, Box 21, Report No. 455 N.

29 Kerr, *Top Mill in the Valley*, p. 66; J. Tardent, 1938, 'Soil erosion in Queensland', p. 12, Queensland Forestry Files, SRS 5213/1, Box 12, Item 65: Soil Erosion, 1935–1966, QSA.

30 J. Tardent, 1938, 'Soil erosion in Queensland', p. 12, QSA; E.W. Knox, 'Notes on a visit to the Isis district, 21 July 1899', pp. 1–2, in 'Memoranda and Reports for the Board of Directors 1887–1899', CSRR, 142/2753.

31 N. King, 'Trash conservation', *CGQB*, Vol. 1, no. 4 (April 1934), pp. 127–128; H. Kerr and A. Bell, *The Queensland Agricultural and Pastoral Handbook. Vol. 4: Sugar Cane and its Culture* (Brisbane, 1939), p. 114; Kerr, 'Some agricultural problems of the Mackay district', p. 222; R. Moller, 'Soil conservation in the Childers area', *CGQB*, Vol. 21, no. 3 (January 1958), p. 93.

32 Kerr, 'Soil erosion', pp. 28–33.

33 C. Stephens, *The Nature and Incidence of Soil Erosion on the Sugar Cane Fields of Queensland.* Divisional Report No. 20 of 1945, Division of Soils, Council for Scientific and Industrial Research (Canberra, 1945), p. 3; C. Simpson, Technical Field Officer, Goondi Mill, to C. Rothe, CSR Gen. Mang., 9 October 1946, Letter No. 255 N, CSRR N/126/131.

34 Harwood Mill Cane Insp., AR 1945, p.5, CSRR, Z/109, Box 470; Circular to Condong Cane Contractors, 14 May 1948, CSRR, Z/109, Box 463; NSW Dept. of Public Works, *Report of the Richmond River Valley Flood Mitigation Committee* (Sydney, 1954), p. 23.

35 CSR Mill Insp., Memo: Harwood Mill Inspection, to CSR Gen. Mang., 24 August 1956, p. 2, CSRR, Z/109, Box 538; CSR, *Newsletter* No. 43, December 1956, p. 7, CSRR, Z/109, Box 371; A.G. Lowndes, *The Drainage of Sugar Canelands in New South Wales* (Sydney, 1957); *ASJ*, 15 May 1954, p. 129; CSR Mill Insp., Memo: NSW Mill Inspection, to CSR Gen. Mang., 6 April 1959, p. 1, CSRR, Z/109, Box 538.

36 CSR, *Newsletter* No. 43, December 1956, p. 7, CSRR, Z/109, Box 371; CSR Mill Insp., Memo: NSW Mill Inspection, to CSR Gen. Mang., 6 April 1959, p. 1, CSRR, Z/109, Box 538; CSR, Agricultural Circular, Condong Mill area, 1959, pp. 8–9, CSRR, Z/109, Box 370; Correspondence about Harwood Island No. 1 and No. 2 Basins Drainage Union, in File 17/1265, D59/394, AONSW.

37 CSR, *Newsletter* No. 43, December 1956, p. 7, CSRR, Z/109, Box 371; CSR, Cane Report 1959, p. 64, CSRR, Z/109, Box 389; CSR Mill Insp., Memo: NSW Mill Inspection, to CSR Gen. Mang. 4 September 1961, p. 6, CSRR, Z/109, Box 538; CSR, Condong Mill Cane Insp., AR 1962, p. 11, CSRR, Z/109, Box 463.

38 N. McD. Smith, 'The South Maroochy Swamp Drainage Scheme', *Proc. QSSCT*, 1949, pp. 149–153; N. McD. Smith, 'Land drainage in the Moreton Area', *CGQB*, Vol. 12, no. 3 (1 January 1949), pp. 110–114.

39 Transcripts of Evidence before the Queensland Royal Commission on the Sugar Industry 1942–43, A/6401 & A/6402, QSA; BSES, AR 1955–56, p. 45; Macknade Mill Cane Insp., AR 1951, p. 9 & Macknade Mill Cane Insp., AR 1954, p. 11, CSRR, Z/109, Box 473; Goondi Mill Cane Insp., AR 1957, p. 20, CSRR, Z/109, Box 464.

40 Macknade Mill Cane Insp., AR 1957, p. 11a, CSRR, Z/109, Box 473; Hambledon Mill Cane Insp., AR 1957, p. 12, CSRR, Z/109, Box 468.

41 Anderson, *The History of Irrigation in the Bundaberg Area*, pp. 82–85 & 88–89; *ASJ*, 15 May 1952, p. 186.

42 CSR, Cane Report 1946, p. 67, CSRR, Z/109, Box 387; Rolleston, *The Defiance*, p. 118.

43 J. Wright and I. Freshwater, 'Underground water replenishment in the Burdekin district', *CGQB*, Vol. 31, no. 4 (April 1968), pp. 138–139; IWSC, *Progress Report on the Water Resources of the Burdekin Delta* (Brisbane, 1964), pp. 9–10; Kerr, *Black Snow and Liquid Gold*, p. 241.

44 IWSC, *Report on Works for Replenishment of Underground Water Supplies, Burdekin Delta: south side* (Brisbane, 1965), pp. 6–8; Kerr, *Black Snow and Liquid Gold*, pp. 241–242.

45 North Burdekin Water Board, ARs 1969 to 1995, JOL; Wright and Freshwater, 'Underground water replenishment in the Burdekin district', p. 142; South Burdekin Water Board, *Brief Resume on the Replenishment of the Burdekin Delta Aquifer* (Home Hill, 1973), pp. 3–8; Kerr, *Black Snow and Liquid Gold*, pp. 242–243.

46 Nolan, *Bundaberg. History and People*, p. 226; Anderson, *The History of Irrigation in the Bundaberg Area*, pp. 139–141.

47 Anderson, *The History of Irrigation in the Bundaberg Area*, pp. 91–118 & 141–151; IWSC, AR 1966–67, p. 13; Australian Bureau of Agricultural Economics, *Bundaberg Irrigation Scheme, Queensland* (Canberra, 1975), pp. 4–5.

48 Australian Bureau of Agricultural Economics, *Bundaberg Irrigation Scheme, Queensland*, pp. 5–7; Kerr, *Southern Sugar Saga*, p. 144.

49 Goondi Mill Cane Inspector, AR 1963, p. 20, CSRR, Z/109, Box 464; N. Maclean, Technical Field Officer, Goondi Mill, Memo to CSR General Mang., 10 May 1967, re: drainage report 1966, p. 8, CSRR, N126/132, No. 577N.

50 Calculated from figures provided in CSR's Annual Reports of Technical Field Work, 1960–1965, CSRR, Z/109, Boxes 390–392; see also N. Maclean, Technical Field Officer, Goondi Mill, Memo to CSR General. Mang., 12 August 1968, re: drainage report 1967, pp. 1 & 8, CSRR, N126/132, No. 608N; CSR, Cane Report 1963, p. 75, CSRR, Z/109, Box 390; CSR, AR Technical Field Work 1967–68, pp. 51–52, CSRR, Z/109, Box 392; & CSR, AR Technical Field Work 1968–69, p. 46, CSRR, Z/109, Box 392.

51 For laser leveling see *BSES Bulletin* No. 46, April 1994, p. 22.

52 C. Toohey, 'Syndicated farm drainage a positive answer', *The Producers' Review*, Vol. 59, no. 2 (February 1969), p. 23; N. Maclean, Technical Field Officer, Goondi Mill, Memo. to CSR Gen. Mang., 28 December 1959, re: drainage, p. 2, CSRR, N126/131, No. 487N; Goondi Mill Cane Insp., AR 1963, p. 20, CSRR, Z/109, Box 464.

53 CSR, Agricultural Circular: Queensland, 1967, pp. 12–15 & CSR, Agricultural Circular: Queensland, 1971, pp. 6–8, CSRR, Z/109, Box 370; CSR, AR Technical Field Work 1968–69, pp. 44–45, CSRR, Z/109, Box 392; J. Sneesby, Drainage Officer, 'Drainage of Herbert River District 1969–1970', CSRR, Z/395, Box 8, Report No. 66N.

54 *Queensland Government Gazette*, Vol. 230 (12 April 1969), pp. 1241–1242; *Queensland Government Gazette*, Vol. 232 (2 November 1969), pp. 1250–1251.

55 *ASJ*, July 1985, p. 157; R. Cargnello, S. Dibella and T. Fuelling, 'Drainage in the Tully Region', in R. Smith and A. Rixon (eds.), *Rural Drainage in Northern Australia: Proceedings of a Symposium held at Darling Downs Institute of Advanced Education* (Toowoomba, 1982), p. 325.

56 C. Toohey, 'The Woongoolba and Area Drainage Scheme', *Proc. QSSCT*, 1966, pp. 87–92; J. Turnbull, 'Rocky Point Drainage Scheme', *ASYB*, 1966, pp. 25–26;

G. Holz, *Rocky Point: a sugar caneland suitability study* (Brisbane, 1979), pp. 3–4; *ASJ*, July 1984, p. 159.

57 For more details see Tweed River Valley Flood Mitigation Committee, *Report of the Tweed River Valley Flood Mitigation Committee* (Sydney, 1957); NSW Dept. of Public Works, *Report of the Richmond River Valley Flood Mitigation Committee* (Sydney, 1954); & NSW Dept. of Public Works, *Report of the Richmond River Valley Flood Mitigation Committee* (Sydney, 1958).

58 Harwood Mill Cane Insp., AR 1963, p. 7, CSRR, Z/109, Box 471; CSR, Agricultural Circular, New South Wales 1963, pp. 14–15, CSRR, Z/109, Box 370; NSW Dept. of Public Works, *Application for Commonwealth financial assistance for flood mitigation* (Sydney, 1969), pp. 1–2; CSR, Cane Report 1962, p. 88, CSRR, Z/109, Box 389.

59 *ASYB,* 1971, p. 30; Tweed Shire Council Case for Subsidy for Additional Flood Mitigation Works in Tweed Shire, March 1969, pp. 1–3, as an attachment in NSW Dept. of Public Works, *Application for Commonwealth financial assistance for flood mitigation* (Sydney, 1969).

60 For the use of Toft mechanical loaders see CSR, Cane Report 1959: Part II, p. 65 & CSR, Cane Report 1960, p.62, in CSRR, Z/109, Box 389; for an example of CSR's technical field officers engaged in survey work see CSR, AR Technical Field Work, 1964–65, p. 70, CSRR, Z/109, Box 392.

61 CSR, Agricultural Circular for New South Wales, 1968, p. 8, CSRR, Z/109, Box 370; CSR, AR Technical Field Work, 1964–65, p 71 & CSR, AR Technical Field Work, 1969–1970, p. 6, in CSRR, Z/109, Box 392.

62 Kerr, *Top Mill in the Valley*, p. 57; IWSC, AR 1967–68, p. 10; IWSC, AR 1968–69, p. 12; Herbert River Improvement Trust, *Herbert River Flood Management: initial submission for Australian government financial assistance* (Ingham, 1975), pp. 31–32.

63 J. Kerr, *Only Room For One. A History of Sugar in the Isis District* (Childers, 1996), p. 133; L. Vallance, 'A soil erosion experiment in the Isis district', *CGQB,* Vol. 10, no. 3 (January 1947), pp. 121–126.

64 For an abbreviated account of this method see J. Holmes, *Soil Erosion in Australia and New Zealand* (Sydney, 1946), pp. 138–139.

65 Kerr, *Only Room For One*, p. 134; N. Smith, 'Combatting the soil erosion problem in the Childers area', *CGQB*, Vol. 18, no. 3 (January 1955), p. 103; R. Moller, 'Soil conservation in the Childers area', *CGQB*, Vol. 21, no. 3 (January 1958), pp. 94–95.

66 For this advice see CSR Ltd., Agricultural Circulars for the Goondi Mill area, 1951 to 1970, CSRR, N 126/131; N. Dawson, R. Berndt and B. Venz, 'Land use planning: Queensland canelands', *Proc. ASSCT*, 1983, p. 50.

67 For examples of this advice see A. Taylor, 'Attempts at soil conservation on South Johnstone farms', *Proc. QSSCT*, 1957, pp. 149–153; BSES, AR 1957–1958, p. 28; Moller, 'Soil conservation in the Childers area', pp. 93–96; J. Rosser, 'Mackay growers use contour banks to control erosion', *CGQB*, Vol. 25, no. 1 (January 1961), pp. 13–16 & 21–23; & CSR, Agricultural Circular, Goondi Mill area, 1956, CSRR, N126/131.

68 C. Filet, *Pioneer Shire Handbook* (Brisbane, 1971), p. 4–2; A. Taylor, 'Attempts at soil conservation on South Johnstone farms', *Proc. QSSCT*, 1957, pp. 150–152; Kerr, *Only Room For One*, p. 134; Moller, 'Soil conservation in the Childers area', p. 93; CSR, AR Technical Field Work 1973–74, pp. 159–160 & CSR, AR Technical Field Work

1975–76, p. 109, in CSRR, Z/109, Box 393; A. Hayes, Hambledon Mill area agricultural report 1966, p. 3, CSRR, Z/395, Box 21, Report 421 N.

69 W. Maclean, Agricultural Report, Goondi Mill area, 1966, p. 2, CSRR, Z/395, Box 25, Report No. 580N; D. Quinn, Agricultural report, Hambledon Mill area, 1971, p. 2, CSRR, Z/395, Box 21, Report 468N; D. Parsons, Agricultural report, Goondi Mill area, 1972, p. 2, CSRR, Z/395, Box 25A, Report 94 N.

70 Isis Land Use Committee, *Report on a Land Use Study of the Isis District* (Brisbane, 1970), p. 7–3.

71 H. Pink, 'Land and water management on cane farms', *Proc. QSSCT*, 1975, p. 88; Isis Land Use Committee, *Report on a Land Use Study of the Isis District*, pp. 7–3 & 7–4; Kerr, *Only Room For One,* pp. 136–137; M. Capelin and B. Prove, 'Soil conservation problems of the humid tropics of North Queensland', *Proc. ASSCT*, 1983, p. 88.

72 E. Copley, 'River improvement trusts – an asset to the industry', *CGQB*, Vol. 39, no. 1 (July 1975), p. 33; Ian Drummond & Associates, *Tully and Murray Rivers Management Strategy* (Wangaratta, 1993), Appendix 6: List of Repair Works; CSR, AR Technical Field Work 1973–74, p. 161, CSRR, Z/109, Box 393; *The Producers' Review*, September 1975, p. 70; Qld. Dept. of Primary Industries, *Progress Report on Proserpine River Irrigation Project* (Brisbane, 1983), pp. 2–7.

73 Ian Drummond and Associates, *February 1991 Flood, Proposal for Damage Restoration* (Wangaratta, 1991); Qld Water Resources Commission, AR 1980–81, p. 40; Qld. Water Resources Commission, AR 1985–86, p. 41; see also ARs of the Cardwell Shire River Improvement Trust, 1987/88 to 1994/95 and ARs of the Burdekin River Improvement Trust, 1986/87 to 1995/96, JOL.

74 Gutteridge, Haskins and Davey Pty. Ltd., *Russell River project. Flood Mitigation Study – Stage 1 Report* (Brisbane?, 1975), pp. 1–7.

75 *ASJ*, July 1976, p. 221; CSR, AR Technical Field Work 1970–71, p. iv & CSR, AR Technical Field Work 1974–75, p. 190, in CSRR, Z/109, Box 393; C. Chardon and A. Rudd, 'The problem of inadequate drainage in the Lower Herbert', *Proc. QSSCT*, 1978, 161; BSES, AR 1974–75, p. 21.

76 D. Beattie, 'Drainage and flood mitigation problems in the North Queensland sugar region', in James Cook University of North Queensland. Department of Civil and Systems Engineering and Water Research Foundation of Australia. North Queensland Regional Committee (eds.), *Water Management and Drainage in North Queensland* (Townsville, 1977), pp. 6–7.

77 Gutteridge, Haskins and Davey Pty. Ltd., *North Queensland sugar industry flood mitigation study: an overview*, (Brisbane, 1977), pp. 2–4; Gutteridge, Haskins and Davey Pty. Ltd., *North Queensland sugar industry drainage study* (Brisbane, 1976), pp. 37, 55–56 & 59–62.

78 *ASJ*, July 1985, p. 158; D. Horsley, A. Wood and R. Stewart, 'Drainage of the Herbert valley canelands: an overview', *Proc. ASSCT*, 1982, pp. 69–70; Ullman and Nolan Pty. Ltd., *Cardwell Shire Flood Management Study Report* (Cardwell, 1983); Cameron McNamara, *Herbert River Flood Management* Study (Ingham, 1980); *ASYB*, 1983, p. 152; IWSC, AR 1983–84, p. 42.

79 See for example Tully Sugar Ltd. and CANEGROWERS, Tully, *Water Management and Cane Transportation for Productivity Enhancement and Increased Production for the Tully and Murray River Valley* (Tully, 1993); Qld. Dept. of Primary Industries, Water

Resources and CANEGROWERS, Mulgrave, *A Water Management Strategy for Improved Sugar Production in the Russell and Mulgrave River Valleys* (Townsville?, 1993).

80 M. Lee, *Pimlico area drainage report* (Lismore?, 1972), pp. 1–4; E. Smith and J. Baldwin, *A Landuse History of Tuckean Swamp*. Tuckean Swamp Study Technical Report No. 6 (Lismore?, 1997), p. 69; Richmond River Inter-departmental Committee, *Richmond River Flood Problems* (Sydney, 1982), pp. 16 & 32; D. Smith *et al.*, *Flood Damage in the Richmond River Valley, New South Wales* (Canberra, 1979), p. 92.

81 Soros-Longworth and McKenzie Pty. Ltd., *The Tweed Valley: summary report* (Sydney, 1980), p. 9; T. Stewart, *The Goodwood Island Proposed Levee* (Grafton, 1972), p. 1; T. Stewart, *Harwood Island Levee: engineering investigation report* (Grafton, 1974); CSR, *Agricultural Circular New South Wales 1973*, pp. 22–23, CSRR, Z/109, Box 370; CSR, AR Technical Fieldwork 1975–76, p. 92, CSRR, Z/109, Box 393; Soros-Longworth and McKenzie Pty. Ltd., *New South Wales Coastal Rivers Flood Plain Management Studies. Clarence Valley Main Report* (Sydney, 1980), p. 5–5.

82 CSR, Agricultural Circular: New South Wales, 1970, pp. 5–6; CSR, Agricultural Circular: New South Wales, 1972, p. 6; Broadwater Mill Cane Insp., AR 1973, p. 33, CSRR, Z/109, Box 478; E. Maynard, 'The development of drainage in the Broadwater Mill cane area', *Proc. ASSCT*, 1982, p. 86; F. Barrow, 'Flood mitigation and drainage of the cane growing areas of the Richmond River and Sandy Creek Flooding', in Richmond River Interdepartmental Committee, *Richmond River Flood Problems. Part C* (Sydney, 1982), p. 194.

83 G. Kingston, 'A review of drainage requirements of canelands in Southern Queensland', p. 72; BSES, AR 1982–83, p. 27; D. Horsley *et al.*, 'Sub-surface drainage in the Herbert valley canelands', in R. Smith and A. Rixon (eds.), *Rural Drainage in Northern Australia: proceedings of a symposium held at Darling Downs Institute of Advanced Education* (Toowoomba, 1982), pp. 109–113; *ASJ*, March 1981, p. 585.

84 *ASYB*, 1985, p. 179.

85 Horsley, Wood and Stewart, 'Drainage of the Herbert valley canelands: an overview', p. 68; *AC*, December 1984, p. 18; see Ripple Creek Drainage Scheme, ARs 1986/1987 to 1989/1990 and Mandam Drainage Scheme, ARs 1987/1988 & 1988/1989, JOL.

86 Environmental Protection Agency, Queensland, *State of the Environment Report, Queensland, 1999* (Brisbane, 1999), p. 5–14.

87 Anon, 'Acid sulfate subsoils cause severe problems', *BSES Bulletin*, No. 15 (July 1986, pp. 4–5.

88 R. Beattie, 'Achieving self-regulation for drain maintenance in the NSW sugar industry', *Proc. ASSCT*, 2001, pp. 51–53; Minutes of Meeting of the Tweed River Branch of the NSWCGA, 5 November 1991, in Volume of Minutes 26 October 1989 to 5 November 1991, Tweed River Branch of the NSWCGA, Murwillumbah.

89 A. Arthington *et al.*, 'Potential impact of sugarcane production in riparian and freshwater environments', in B. Keating and J. Wilson (eds.), *Intensive Sugarcane Production: Meeting the Challenge Beyond 2000* (Wallingford, 1997), p. 412; P. Muller & A. Coutts, *Acid sulfate soils and estuarine water quality of the Mackay district* (Brisbane, 2005); C. Smith and C. Ahern, *Report on the Acid Sulfate Soils Investigation, Ingham and Tully Drainage Schemes* (Brisbane, 1998); B. Powell and C. Ahern, 'Risk assessment of acid sulfate soils in southern Queensland canelands', in J. Wilson *et al.* (eds.), *Sugarcane: Research Towards Efficient and Sustainable Production* (Brisbane, 1999), p. 32.

90 Kerr, *Sugar at Maryborough*, pp. 231–232.

91 Kerr, *Top Mill in the Valley*, p. 59; Rolleston, *The Defiance*, pp. 118–119; C. Story, 'Supplementary irrigation in the Central District', *Proc. QSSCT*, 1964, p. 112; I. Cribb, *Proserpine Shire Handbook* (Brisbane, 1973), pp. 3–1 & 3–2; M. Haysom, 'The evaluation of irrigation waters in the Central District', *Proc. QSSCT*, 1970, p. 78.

92 Kerr, *A Century of Sugar*, pp. 199–200; IWSC, AR 1969–70, p. 15.

93 G. Bieske, G. Kingston and K. Pharr, 'The future of irrigated sugar cane production in Bundaberg', in Water Research Foundation of Australia, *Water Resources of the Burnett Region, 1967 to 1983* (Brisbane, c. 1983), p. 143; C. Story, 'Supplementary irrigation in the Central District', *Proc. QSSCT*, 1964, p. 112; K. Leverington, 'Irrigation practices in the Queensland sugar industry', *Proc. ISSCT*, 1971, p. 902; R. Moller and L. Chapman, 'Irrigation systems in the Queensland sugar industry', *Proc. QSSCT*, 1975, p. 52; Anon, 'Methods of irrigating', *CGQB*, Vol. 44, no. 3 (January 1981), p. 83.

94 Moller and Chapman, 'Irrigation systems in the Queensland sugar industry', pp. 52–54.

95 BSES, AR 1956–57, p. 8; Kerr, *Southern Sugar Saga*, p. 142; I. Freshwater, 'Developments in the methods of distribution of flood irrigation water', *CGQB*, Vol. 33, no. 3 (January 1970), pp. 76–78; Leverington, 'Irrigation practices in the Queensland sugar industry', pp. 899–905.

96 J. Wright, 'A unique method of land levelling', *CGQB,* Vol. 28, no. 3 (January 1965), pp. 121–122; H. Boyle, 'Land leveling on the Lower Burdekin', *CGQB*, Vol. 37. no. 2 (October 1973), pp. 69–71; J. Reimers, 'Farm levels at Bundaberg', *CGQB*, Vol. 39, no. 1 (July 1975), pp. 26–27; *AC*, October 1982, p. 28 & 18 September 1995, p. 27; R. Kane, 'Drainage development – Fairymead Plantation', in Institution of Engineers, Australia. National Committee on Agricultural Engineering (ed.), *Conference on Agricultural Engineering, 1984* (Canberra, 1984), pp. 4–8.

97 IWSC, AR 1936–37, *QPP*, Vol. 2 (1937), p. 553; IWSC, AR 1964–65, p. 7; IWSC, AR 1976–77, p. 23; IWSC, *Report on Improvement to Water Supplies Haughton River (Giru)* (Brisbane, 1971), pp. 4–6 & 33–34; Kerr, *Black Snow and Liquid Gold*, p. 300.

98 Kerr, *Sugar at Maryborough*, pp. 234–236; *ASJ*, November 1977, pp. 422–423; IWSC, *Lower Mary River Irrigation Project* (Brisbane, 1977), passim; J. Kerr, *Crystal Century. The Proserpine Sugar Industry* (Proserpine, 1997), pp. 162–165.

99 J. Powell, *Plains of Promise. Rivers of Destiny. Water Management and the Development of Queensland 1824–1990* (Brisbane, 1991), pp. 242–244.

100 Kerr, *Black Snow and Liquid Gold*, pp. 297–300.

101 Anon, 'Centre pivot irrigation at North Eton', *CGQB*, Vol. 45, no. 4 (April 1982), pp. 110–111; *AC*, November 1983, p. 49; *ASYB*, 1985, p. 172; R. Kane, 'The development of a lateral move irrigator in sugar cane production', *Proc. ASSCT*, 1985, pp. 91–94.

102 BSES, AR 1993–94, p. 22; T. Blatch, 'Bundaberg growers increase productivity with trickle irrigation', *BSES Bulletin*, No. 47, July 1994, pp. 14–15; *AC*, 16 November 1992, p. 15.

103 G. McMahon, 'Recycling efforts held reduce water loss', *BSES Bulletin* No. 43, July 1993, pp. 14–15; *AC* 6 February 1995, p. 13.

104 S. Raine, 'Efficient furrow irrigation a high priority', *BSES Bulletin* No. 50, April 1995, p. 7; E. Shannon, 'Burdekin growers take their hats off to evaporation mini-pans', *BSES Bulletin*, No. 48, October 1994, p. 17; G. Ham, P. McGuire and G. Kingston, 'Irrigation of Sugar Cane' in M. Hogarth and P. Allsopp (eds.), *Manual of Cane Growing* (Brisbane, 2000), p. 200.

105 S, Murphy and R. Sorensen, 'Saltwater intrusion in the Mackay coastal plains aquifer', *Proc. ASSCT*, 2001, p. 70; L. Tilley, C. Sarich and T. Capp, *Irrigation Best Practice for the Mackay Region*. BSES Technical Report TE96005, August 1996, p. 2; Environmental Protection Agency, Queensland, *State of the Environment. Queensland 1999* (Brisbane, 1999), p. 4.61.

106 G. Kingston, 'Soil salinity a hazard to productivity in southern areas', *BSES Bulletin*, No. 12 (October 1985), p. 14; P. McGuire, *Irrigation of Sugar Cane* (Brisbane, 1991), p. 34; G. Ham and L. Chapman, 'A review of salinity related drainage problems in cane in the central region', *Proc. ASSCT*, 1982, pp. 78–79.

107 M. Sallaway, 'Soil erosion studies in the Mackay district', *Proc. ASSCT*, 1979, p. 130; M. Capelin and B. Prove, 'Soil conservation problems of the humid tropics of North Queensland', *Proc. ASSCT*, 1983, pp. 87–93; B. Prove and W. Hicks, 'Soil and nutrient movements from rural lands of North Queensland', in D. Yellowlees (ed.), *Land Use Patterns and Nutrient Loading of the Great Barrier Reef* (Townsville, 1991), p. 68.

108 Anon, 'Soil conservation in the Julatten area', *CGQB*, Vol. 45, no. 4 (April 1982), p. 123; for land use studies see Anon, *Moreton Region Non-urban Land Suitability Study*. Division of Land Utilisation, Qld. Dept. of Primary Industries Technical Bulletin No. 11 (Brisbane, 1974); M. Capelin, *Moreton Mill Area – A Sugar Caneland Suitability Study*. Division of Land Utilisation, Qld. Dept. of Primary Industries Technical Bulletin No. 37 (Brisbane, 1979); G. Holz, *Rocky Point – A Sugar Caneland Suitability Study*. Division of Land Utilisation, Qld. Dept. of Primary Industries Technical Bulletin No. 38 (Brisbane, 1979); & G. Holz and P. Shields, *Mackay Sugar Caneland Suitability Study*, 2 vols. (Brisbane, 1985).

109 A. Matthews and P. Makepeace, 'A new slant on soil erosion control', *CGQB*, Vol. 45, no. 2 (October 1981), p. 43; T. Wilson, A. Wissemann and G. Dwyer, *Report on the Soil Conservation Practices and Related Attitudes of Innisfail Canegrowers*. Project Report No. Q082015 (Brisbane, 1982), p. 5; B. Prove, P. Truong and D. Evans, 'Strategies for controlling caneland erosion in the Wet Tropical Coast of Queensland', *Proc. ASSCT*, 1986, p. 78.

110 *AC*, October 1987, p. 26; *ASJ*, February 1985, p. 460; N. Dawson *et al.*, 'Land use planning – Queensland canelands', *Proc. ASSCT*, 1983, p. 50.

111 W. Sloan, 'Some aspects of the problem of soil erosion control in Queensland cane fields', *CGQB*, Vol. 10, no. 4 (April 1947), p. 157.

112 Wilson, Wissemann and Dwyer, *Report on the Soil Conservation Practices and Related Attitudes of Innisfail Canegrowers*, p. 11; A. Taylor, 'Attempts at soil conservation on South Johnstone farms', *Proc. QSSCT*, 1957, p. 14; C. Simpson, Technical Field Officer, Goondi Mill, to CSR Gen. Mang., 27 November 1946, Letter No. 258, re: soil erosion, CSRR, N126/131.

113 J. Veurman, 'Soil conservation in the central cane district', *QAJ*, Vol. 103, no. 6 (November 1977), p. 582.

114 N. Smith, 'Combatting the soil erosion problem in the Childers area', *CGQB*, Vol. 18, no. 3 (January 1955), p. 104; J. Wright, 'Is soil erosion a problem on the farm?', *CGQB*, Vol. 33, no. 4 (April 1970), pp. 134–136; E. Sypkens, 'Are short rows necessary in contoured cane fields', *CGQB*, Vol. 23, no. 4 (April 1970), p. 137–140; Veurman, 'Soil conservation in the central cane district', p. 582; Wilson, Wissemann and Dwyer, *Report on the Soil Conservation Practices and Related Attitudes of Innisfail Canegrowers*, p. 9.

115 R. Moller, 'Soil conservation in the Isis area', *Proc. QSSCT*, 1959, p. 90; Dawson *et al.*, 'Land use planning – Queensland canelands', p. 50.

116 BSES, 'A review of results of trails with trash management for soil conservation', *Proc. ASSCT,* 1984, pp 101–106; Prove, Truong and Evans, 'Strategies for controlling cane-land erosion in the Wet Tropical Coast of Queensland' pp. 78–79; Prove and Hicks, 'Soil and nutrient movements from rural lands of North Queensland', pp. 68–69.

117 Prove and Hicks, 'Soil and nutrient movements from rural lands of North Queensland', p. 70; A. Harris, 'Sweet Waters', *Habitat Australia,* June 1995, pp. 10–11.

118 I. Christiansen, 'Environmental Management', in M. Hogarth and P. Allsopp (eds.), *Manual of Cane Growing* (Brisbane, 2000), p. 45; R. Quabba, 'Achieving ESD as a cane grower', in P. Hale *et al.* (eds.), *Management of Sustainable Ecosystems* (Brisbane, 2000), p. 160; L. Geno, *Keeping it sweet: sustainability and sugar cane* (Melbourne, 1996), pp. 45 & 79.

119 J. Blackmann *et al., Characteristics of Important Wetlands in Queensland* (Brisbane, 1999), p. 362; Ian Drummond & Associates, *Tully and Murray Rivers Management Strategy.* Report to the Cardwell Shire River Improvement Trust (Cardwell?, 1993), p. 28.

CHAPTER 13

Breeding and Improving Cane Varieties

In November 1912, the BSES took delivery of 158 new cane varieties from New Guinea. They had been collected in the mountainous country on the southern side of the Owen Stanley Range near Sogeri (inland from Port Moresby) by Thomas H. Wells, an Isis district canegrower. He had been commissioned by the BSES to supply new genetic material for Queensland's canegrowers. Identifying and testing the potential commercial value of these new cane varieties then occurred over the next five years at the Mackay and Bundaberg BSES stations. Eventually, 119 cane varieties were given N.G. numbers following on from No. 66, the last cane variety brought back from New Guinea by Tryon in 1896. Some of these cane varieties developed gumming disease and so were discarded immediately. Only fifteen cane varieties were sufficiently promising to be planted in a yield trial in 1918, and only six of these emerged as potential commercial cane varieties. However, there was a limited distribution of these varieties to the state's canegrowers, and none became popular because their sucrose content did not compare favourably with Badila, the main cane variety grown by North Queensland canegrowers during the late 1910s and 1920s.[1]

Varietal improvement can also occur through breeding plants that have desired characteristics, such as high sucrose content or resistance to a particular disease. Harry Easterby, the BSES's General Superintendent during the late 1910s, was well aware of the cane variety improvement programs and disease investigations in other sugar-producing countries. The BSES lagged far behind Mauritius, Barbados, Demerara and Java, for example, having failed to breed any new cane varieties at its Queensland stations since the inception of the organisation in 1900.[2] Easterby decided in 1917 to rectify this shortcoming, after gaining the approval of Ernest Scriven, the Under Secretary for the Queensland Department of Agriculture and Stock and the BSES Director. The organisation's scientific reputation was at stake, but also a sense of urgency prevailed due to the return of outbreaks of gumming disease in the Herbert River district in the late 1910s (see Chapter 15). The BSES needed to find new, gumming disease resistant cane varieties or the industry's future could again be under threat, like in the mid-1890s. Producing new, hybrid cane varieties before 1945 took between ten and fifteen years, so as a stop gap measure the BSES continued importing hybrid cane varieties into Queensland, mostly from India, Mauritius, Java and Hawaii, seeking to determine if any could replace those varieties succumbing to diseases.[3]

Breeding new cane varieties requires one essential element–fertile seed which is found in matured tassels or arrows of the cane plants during the flowering season which occurs in North Queensland throughout May and June. In 1917, the BSES did not possess an experiment station north of Mackay. Attempts to gather fertile seed at the Mackay BSES station had failed because the cane did not arrow freely and many varieties failed to produce mature flowers. Easterby decided that the cane variety improvement program needed to be sited in North Queensland where the environmental conditions were more conducive to profuse arrowing. This goal was achieved by the BSES establishing a new station near the town of South Johnstone in North Queensland. Its first officer-in-charge was Peter McWalters, a returned soldier with 'considerable experience in chemistry and agriculture'.[4]

At the South Johnstone BSES Station, McWalters assembled the cane varieties imported by the BSES or gathered from the Mackay and Bundaberg BSES stations and the Kairi State Farm, near Atherton. Breeding new cane varieties had commenced at the South Johnstone BSES station in 1919, with McWalters following Easterby's instructions which were based upon the experience of James Clark who had successfully bred H.Q. 426 (Clark's Seedling) at CSR's Hambledon Cane Nursery in the mid-1900s. He was to gather the fuzz or seed from the arrows of cane once it became 'fluffy'. This material was then planted in boxes filled with sterilised soil. Promising seedlings which grew from the germination of this seed material were then nurtured, propagated and eventually transplanted into field plots where they were assessed for sucrose content, yield per hectare, disease susceptibility and growing habits.[5]

During the first few years of research at the South Johnstone BSES station, seedlings were obtained from field crosses and self-pollination. However, initial results from the seedling raising program were disappointing, so the BSES decided to develop a better understanding of the fertility of sugar cane flowers. W. Cottrell-Dormer, one of the BSES plant pathologists, visited the South Johnstone BSES Station in mid-1924 and investigated the processes of pollination and fertilisation of the cane flower. Eighteen months after this visit, the breeding program was re-organised at the South Johnstone BSES Station, following the introduction of the Hawaiian sulfurous method. Under this procedure, arrows of the cane variety selected as the male parents were stood in a dilute solution of sulfuric acid to maintain them during pollination. The solution kept the stalk and arrow alive for several weeks, permitting the normal shedding of pollen to continue.[6] Such an arrangement allowed for the development of controlled cross-pollination. The cane plant selected as the female parent was left growing in the field. The chosen male parent with its arrows, standing in a container filled with the special solution of sulfuric acid, was then carried into the field and placed near the female plant, just before the flowers of the female cane plant commenced opening. The stalks of cane of the male variety were tapped lightly each morning to facilitate the shedding of pollen. The cane breeders hoped that these new procedures would produce more fertile seed and give them better control over the pollination process.

Over the first decade of investigations at the South Johnstone BSES station, the field staff raised approximately 15 000 seedlings, although less than a third proceeded to the next stage of testing. About a dozen showed enough promise to be subjected to further analytical examination. In 1928, the BSES selected one hybrid cane variety– S.J. 4 – for farm trial plots. Originating in the batch of seedlings raised in 1922, S.J. 4's parentage was a cross between the Badila cane variety and an unknown cane variety. It had moderate sucrose content, produced particularly strong ratoon crops and withstood cane grub attacks, but was very susceptible to gumming disease.[7] Its popularity during the early 1930s was confined mostly to the Cairns district where the canegrowers suffered particularly bad cane grub attacks. However, S.J. 4's susceptibility to gumming disease led to its abandonment after 1940, following outbreaks of gumming disease throughout the Cairns district in the late 1930s. These early investigations also led to the BSES releasing S.J. 2 and S.J. 16 for commercial use during the 1930s, although these two cane varieties never gained much favour amongst North Queensland canegrowers.[8]

As the BSES scientists commenced their program to breed new cane varieties, the state's canegrowers relied heavily on the cane varieties that had emerged as favourites in the mid-1910s. Before 1934, BSES figures on the amount of a particular cane variety growing in each Queensland sugar-producing district were collected only intermittently, so details about the popularity of particular cane varieties during the 1920s had to be reconstructed from these limited BSES figures, descriptions of what cane varieties grew in each district or CSR's figures for its mill supply areas.

North of Cardwell, Badila remained the favourite cane variety, accounting for 70–90% of the cane harvested annually for Mulgrave, Hambledon and

Goondi Mills during the 1920s. Other cane varieties grown in smaller amounts included D. 1135 and H.Q. 426. Badila was less popular in the Herbert River district, contributing only 50% of the cane harvested during the 1920s. The gumming disease resistant H.Q. 409, Q. 813, Korpi and Oramboo cane varieties made up the remainder of the crops harvested in the Herbert River district. In the Lower Burdekin district, Badila and B. 208 had assumed greater importance during the 1920s, although the Goru cane varieties were still grown by a small percentage of canegrowers.[9] A more complicated varietal pattern existed in the Mackay district throughout the 1920s. Malagache and D. 1135, two cane varieties with low sucrose content, had virtually disappeared by 1929, being replaced by more sucrose-rich cane varieties such as Badila, M. 1900, E.K. 28 and Q. 813. In contrast, 80% of the cane crushed in the Bundaberg district in 1920 was D. 1135. Following gumming disease outbreaks throughout the Bundaberg district in 1921, the popularity of D. 1135 waned rapidly, due to its susceptibility to this disease. By 1931, only 30% of the cane harvested in the Bundaberg district was D. 1135. Its place had been taken by more gumming disease resistant cane varieties such as Q. 813 and Uba.[10]

CSR resumes breeding cane varieties in Australia

A growing concern about disease outbreaks in all its mill supply areas, over-reliance on the Badila variety in its northern mill supply areas and a lack of sucrose-rich varieties prompted CSR in May 1914 to send Messrs A. Carne and C. Baker to New Guinea, searching for new cane varieties. This expedition was CSR's second attempt at gathering new cane varieties in New Guinea. Two of the firm's officers had gathered 125 cane varieties from the D'Entrecasteaux Islands and islands of the Louisiade Archipelago in 1908, but the material went to Rarawai in Fiji, and none survived to commercial cropping. In 1914, CSR's officers concentrated their searches in a different locality – the elevated country in the Hydrographers Range area located to the west of Buna, on the north-east coast. The expedition returned to Queensland in September 1914 with ninety-six new cane varieties which were tested at CSR's experimental stations in Australia and Fiji.[11] Three cane varieties collected on the Buna sorties were eventually used commercially in the sugar-producing districts north of Mackay and in New South Wales during the late 1920s and early 1930s. They were the moderately sucrose-rich and very gumming-resistant Korpi (14 N.G. 124), Oramboo (14 N.G. 190) and Nanemo (14 N.G. 241). The other cane varieties gathered during the Buna expedition were used extensively as parents in seedling work. Korpi, for example, featured in the parentage of Trojan, Eros and Vesta, well-known CSR-bred cane varieties that were widely grown during the 1940s and 1950s.[12]

Against a backdrop of worsening gumming disease outbreaks in the Herbert River and Richmond River districts, and after seven years of testing the 14 N.G. series of cane varieties with limited success, CSR decided in 1921 to sponsor the acquisition of more New Guinea cane varieties. Mr E. Chinnery supplied CSR with forty new cane varieties collected on the southern side of Mount Obree in the Owen Stanley Range, while Mr. A. Joubert collected eighteen new cane varieties for CSR during a visit to New Ireland and Tabar Island, to the north of New Ireland. However, there was no guarantee that the 21 N.G. series of cane varieties would yield any useful varieties – as it turned out none of the 21 N.G. series of cane varieties reached commercial status, although one of them, 21 N.G. 22, was used later as a parent in CSR's breeding program. CSR also decided to resume breeding cane varieties in Australia, but this time on a more elaborate scale than during the early 1900s. Moreover, David North, CSR's plant pathologist, would have input into the process, particularly ensuring that any new hybrid cane varieties were resistant to the main diseases troubling the canegrowers, especially gumming disease (see Chapter 15).

As cane rarely arrows in New South Wales, the seed has to be produced elsewhere. Most cross-pollination work was carried out at CSR's Macknade Experiment Station, under the supervision of Kenneth Gard, one of CSR's chemists who had been David North's assistant at the Broadwater Experiment Station during 1922 and 1923.[13] Some cane seed was retained and used to establish seedlings which were then tested under North Queensland environmental conditions for their suitability as possible commercial cane varieties. In 1922, cane seed from the Macknade Experiment Station was planted at the Broadwater Experiment Station. Eight seedlings were raised – the 22 B.W. series – thus demonstrating that cane seed could be successfully transferred from the tropics and germinated in the sub-tropics. A further trial in 1923 of raising seedlings at the Broadwater Experiment Station proved unsuccessful, so CSR transferred seedling germination operations to Sydney where the process was placed under the supervision of the firm's chemical staff. In 1924, the S.N. series was commenced using cane seed from the Macknade Experiment Station and CSR's plant breeding facility at Rarawai, Fiji. At two years old, cuttings of each seedling were forwarded to each of the three New South Wales sugar mills for further testing. This method of transfer proved cumbersome and was replaced in 1928 by a new arrangement of sending three to four month old seedlings in 8 cm paper pots to CSR's New South Wales mills for further testing in isolation plots, before going to farm-based selection trials. From 1930 onwards, the New South Wales Department of Agriculture also assisted CSR by raising seedlings at its Grafton Experiment Farm, chiefly from cane seed imported from overseas cane-breeding stations. Cuttings of the most promising cane varieties were selected by CSR for further testing.[14]

Overseas cane varieties, especially those displaying any resistance to gumming disease and frost, continued to be imported by CSR during the late 1920s and 1930s. CSR even arranged to receive a duplicate set of the most promising new cane varieties assembled by Dr E. W. Brandes during his 1927 collecting expedition along the Fly and Strickland Rivers in western New Guinea. Initially, CSR's imported cane varieties spent some time in an insect-proof quarantine house built at Bellevue Hill in 1925, before they were transferred to the isolation plots around Sydney. This approach ensured that imported canes were grown for at least one generation in some locality far removed from the nearest sugar-producing centre so that any pest or disease outbreaks did not infect the local industry. CSR's dual system of quarantine operated until the last isolation plot around Sydney was abandoned in 1933, with all subsequent imported cane varieties being raised only in the quarantine house.[15]

Once CSR's breeding program was in full swing, anywhere between 500 and 5 000 seedlings were raised annually, with yearly fluctuations dependent upon weather conditions which impacted upon cane arrowing. Seedling raising, however, was suspended at Sydney in 1926, 1927 and 1931. A large number of seedlings had built up and this pause in proceedings allowed them to be culled, and the more promising seedlings were selected and distributed to CSR's sugar mills. Overall, 30 700 seedlings were raised at the Macknade Experiment Station between 1925 and 1941; another 56 200 seedlings were grown at Sydney and Grafton.[16]

Initially, the cane breeders at the Macknade Experiment Station focused upon crossing sucrose-rich and gumming disease resistant parents such as Badila, Korpi, Oramboo, Innis, H.Q. 409 and Q. 813, with the aim of producing new cane varieties that displayed both characteristics. Crosses using the sucrose-rich P.O.J. 2878, the Javanese wonder cane, commenced in 1927, after CSR imported this variety for its cane variety improvement program. The ability to withstand low temperatures also received special attention when blood lines and parents were being chosen to produce cane seed for the New South Wales mill supply areas which suffered from frost. Following outbreaks of Fiji disease in the New South Wales sugar-producing districts during the early 1930s, the cross-pollination work switched its focus to breeding Fiji disease resistant cane varieties. Crossings were first achieved by either natural pollination or by the Java method of tying arrows of the pollen-bearing male parent to the arrows of the female parent plant which was still growing in the field. In 1928, the Hawaiian innovation of preserving the arrow-bearing stalks in weak sulfuric acid was adopted, and during part of 1930 a plant breeder from the Experiment Station of the Hawaiian Sugar Planters' Association spent several months at the Macknade Experiment Station, assisting with the cross-pollination work.[17]

Germinating seedlings was just the start of CSR's breeding program. At three to four months old, the new seedlings were planted 1.5 m apart in field plots at the Macknade Experiment Station. They were interspersed amongst a standard variety to every ten seedlings. The first selection occurred when the seedlings were thirteen to fourteen months old following a visual comparison with the standard variety in respect to freedom of disease and to physical characteristics such as vigorous growth, number of stalks per stool and thickness of stalk. Between 10 and 20% of the seedlings grown in any one year were selected for further testing. Those selected were planted in first selection beds containing frequent replications of standard varieties for further comparison. Sucrose content was then determined from juice samples, and details gathered on growing habits, especially ratooning vigour. Promising seedlings were then retained for second selection trials, when they were tested throughout CSR's other mill supply areas. Those seedlings showing sufficient promise after second selection trials were sent to variety plot trials on cane farms where they competed with standard varieties under normal farming conditions and practices. Those seedlings judged worthy of commercial status after this process were planted in experimental variety blocks and a few were eventually approved for general planting. In addition, the BSES received cuttings from those seedlings which were likely to become standard commercial varieties so they could be included in its testing program.

An extra procedure was followed at CSR's New South Wales sugar mills. Promising seedlings were subjected to gumming-resistance field trials. Those that survived these trials were split into two groups: one lot was allowed to grow for one more year; the other group was permitted to grow for two years. An assessment of their growth and disease-susceptibility continued to be made against standard cane varieties. Eventually, two groups of seedlings existed: the purely two-year old types; and the two-year old types that would make enough growth under favourable conditions to be harvested after one year. The next selection was made after testing for sucrose content and yield. The survivors then followed the same testing process undertaken elsewhere.

The production of suitable biennial cane varieties was emphasised during the first decade of breeding cane varieties in New South Wales; from 1934 onwards CSR also experimented with creating annual-type cane varieties.[18]

Even with its considerable financial resources, CSR's post-1920 cane variety improvement program yielded no commercial results during its first fourteen years in operation. The great majority of seedlings raised before 1928 were discarded because of their susceptibility to gumming disease.[19] Suppliers of CSR's North Queensland sugar mills, as mentioned above, continued relying heavily on Badila during the 1920s and early 1930s. In New South Wales, fresh epidemics of gumming disease during the 1920s forced canegrowers to abandon the gumming disease-susceptible Mahona cane variety and switch to the gumming resistant Badila, Malabar and H.Q. 5 cane varieties, although the latter two varieties were only moderately sucrose-rich (Figure 13.1). Other New South Wales canegrowers persisted with cultivating the frost-resistant and sucrose-rich N.G. 16.[20]

CSR, however, considered the use of Badila a stop-gap measure as it was not frost-tolerant and unable to be cultivated along the upper reaches of the Richmond and Clarence Rivers. Moreover, Badila had succumbed to gumming disease in the Tweed River district by 1929, so its usefulness in New South Wales was becoming compromised.[21] Consequently, Badila, Malabar and N.G. 16 were replaced during the early 1930s, mostly by the gumming disease resistant and sucrose-rich Q. 813 and P.O.J. varieties. During this period, CSR also convinced some of its canegrowers to adopt the gumming resistant Korpi, Oramboo and Nanemo, the three cane varieties from the 14 N.G. series, although less than 10% of the cane harvested during this period comprised these cane varieties (Figure 13.1). Thus, most cane varieties used in New South Wales during the late 1920s and 1930s had either been bred by the Queensland Acclimatisation Society or imported from New Guinea and Java (Figure 13.2).

Re-organising Queensland's cane variety improvement program, 1929–1946

By the late 1920s, the BSES realised that its cane variety improvement program was not yielding rapid results. In 1928, the BSES attempted to improve the genetic material being used at the South Johnstone BSES Station by importing a selection of the P.O.J. varieties, including P.O.J. 2878 which had given such pronounced success in Java.[22] A year later the BSES expanded its cane variety improvement

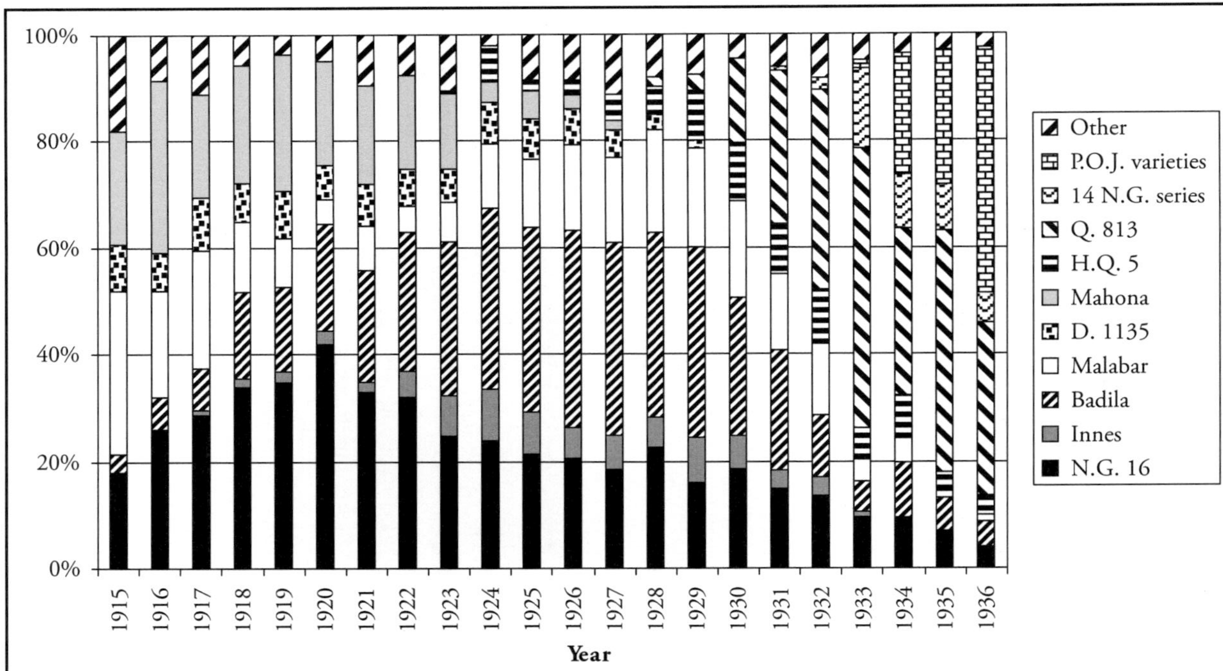

Notes: (1) 14 N.G. series = Korpi, Oramboo and Nanemo cane varieties. (2) P.O.J. varieties = P.O.J. 2878 & P.O.J. 2883.

Figure 13.1. The percentage of harvested cane in New South Wales classified according to variety, 1915–1936.

Source: Calculated from varietal census figures in CSR, Cane Reports, 1915 to 1936 seasons, in CSRR, Z/109, Boxes 378 to 386.

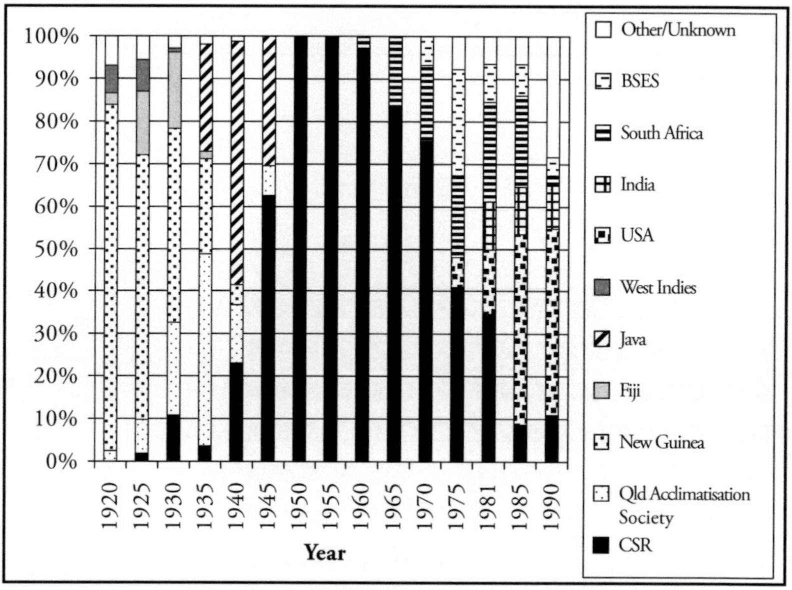

Notes: Figures from 1920 to 1975 are based upon percentage of the crop harvested. From 1981 onwards the figures are for percentage of the crop planted annually.

Figure 13.2. The origin of cane varieties in New South Wales, 1920–1990.

Source: Calculated from varietal census figures in CSR, Cane Reports, 1920 to 1975 seasons, CSRR, Z/109, Boxes 378 to 391 and the *ASYB,* various issues.

activities by trying to raise seedlings at the Mackay and Bundaberg stations, using seed sourced from South Johnstone. These initial experiments failed, the cold weather not permitting the germination of the seed. Successful seedling germinations occurred in 1930, however, following the erection of heating frames at the Macky and Bundaberg BSES stations. In 1932, this expanded program of seedling propagation was placed under the direct supervision of the plant breeder, E.J. Barke, who had replaced McWalters after his retirement in 1927. The BSES hoped procedures would become more standardised under Barke's guidance, thereby improving the outcomes from the cane variety improvement program that was showing no signs of great success.[23] The main concern, however, was that limited cross-pollinations occurred at South Johnstone, for the frequent incidence of wet weather at arrowing time had caused too much shedding of pollen. To overcome this difficulty, plantings of parental cane were made in 1930 on a plot where arrowing had been observed to be unusually heavy at Freshwater, to the north of Cairns. This departure proved very successful and the arrow fuzz obtained from the Freshwater plot yielded a much higher proportion of viable seed than the South Johnstone crossings. Increasingly, cross-pollinations were carried out at Freshwater, so that when the BSES decided in 1934 to transfer the entire cane variety improvement program to the Meringa BSES station, most cross-pollinations were being undertaken at Freshwater.[24]

The troubled BSES's cane variety improvement program was further reorganised in 1935 following the formation of a Committee on Seedling Propagation. Consisting of Messrs E. Barke (Meringa), D.L. McBryde (Mackay) and N. King (Bundaberg), under the chairmanship of Arthur Bell, the Assistant BSES Director and officer in charge of the Division of Entomology and Pathology, this committee concluded that all future cross-pollinations would only occur in the Cairns district. C. Graham Hughes, the BSES plant pathologist, however, assumed control over the breeding operations at Meringa in 1937 and directed proceedings until 1947. The other stations continued receiving dried arrow fuzz from Meringa. The Committee on Seedling Propagation issued instructions that carefully detailed the standarised methods for raising seedlings and selecting seedlings for further trials and the types of records to be kept during each trial.[25]

This administrative reorganisation was also accompanied by the provision of more appropriate buildings and equipment to enable the completion of better investigations. In 1935, a new quarantine house of fine bronze-gauze and glass construction on a concrete base was built in the Government Domain, inner Brisbane (i.e. today's site of the Queensland University of Technology), eliminating the previous system of assessing imported cane varieties by growing them in isolation plots. Glasshouses were built at the Meringa (1935), Bundaberg (1936) and Mackay (1938) BSES Stations, with each glasshouse being equipped with a hot-water boiler, heating pipes and provision for maintaining a high humidity without excessive temperatures on hot days.[26]

The new glasshouses enabled the BSES to raise approximately 22 000 seedlings annually between 1935 and 1941. Cross-pollinations and seedling raising ceased due to staff shortages at Meringa in 1942 and seedling raising stopped at Meringa in 1942 and at Bundaberg in 1942 and 1943. Normal cane variety improvement activities resumed at all BSES stations in 1944.[27] In 1937, Arthur Bell thought the most pressing requirements now demanded of new Queensland cane varieties in the northern and central districts was resistance to beetle borer damage and strong rooting systems which enabled them to withstand cane grub attack (see Chapter 14). Thus, crosses with several different varieties of *S. spontaneum* at various stages of 'nobilisation' were made during the late 1930s and early 1940s; the cane breeders hoped their progenies would display the strong rooting system and vigorous growth habit of the parents. Hardness of rind was also desirable in cane varieties grown North Queensland as a means of keeping the beetle borer in check. To achieve this characteristic in seedlings, crosses were made between Badila and four clones of *S. robustum* which had particularly hard rind. Imported Turkmenistan cold-resistant hybrids were also crossed with many different cane varieties, with a view to producing new cane varieties which would be resistant to the mild frosts of Southern Queensland.[28]

Cross-pollinations at Meringa during the late 1930s and early 1940s were completed in what became known as the 'crossing paddock', an area of ground lightly timbered with regrowth forest. Experiments showed that the shade afforded by the trees did not have any harmful effects on seed setting. Each combination was normally kept separate from the others in its own drum of solution and with the arrows tied to a timber stand. Between 100 and 150 cross-pollinations were completed annually at Meringa during the early 1940s (Figure 13.3). Investigations were also conducted into identifying the best method to conduct the cross-pollinations. Fabric lanterns – a light wooden framework covered with thin cotton voile to protect crosses from stray pollination – had been introduced into the breeding

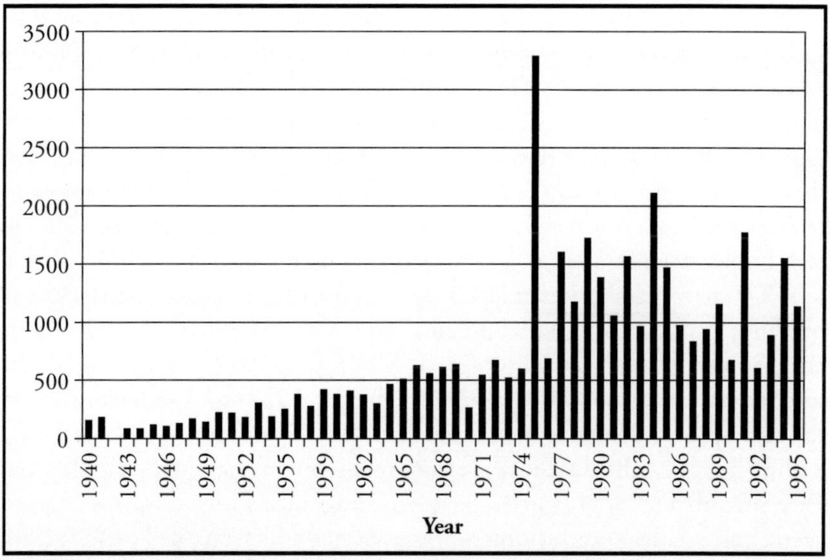

Figure 13.3. The number of cross-pollinations completed annually at the Meringa BSES station, 1940–1995.

Source: Based upon figures in BSES, ARs 1939–1940 to 1995–1996.

program in 1939 (see Plate 13.1). Trials concluded that the complete solution method, in which both male and female arrows were combined together while maintained in solution, allowed more cross-pollinations to be completed and led to the production of more viable seed than the open pollination approach in which a detached arrow from the male parent was stood beside an arrowing female parent growing either in the field or in a drum filled with soil. Experiments showed that the most suitable solution was a mixture of .01% sulfur dioxide and .01% phosphoric acid. Arrowing of early arrowing cane varieties such as C.P. 29/116 or P.O.J. 2725 was delayed by using the lopping method – removing leaves on half the stalks of any particular cane variety during the growing season. Soaking cane stalks in different chemicals, however, failed to stimulate arrowing in non-arrowing cane varieties. Investigations between 1938 and 1941 also showed that arrow fuzz could be dried, placed in cold storage for twelve months and then used to successfully raise seedlings.[29]

Procedures for selecting promising seedlings were also altered after Arthur Bell assumed responsibility for the cane variety improvement program. Sketchy details on earlier selection practices indicated that the process

Plate 13.1 Voile-covered crossing lanterns in use at the Meringa BSES station, c. 1953. Note the stalks of cane standing in tins containing the solutions of water and sulfuric acid.

Source: N. King, R. Mungomery and C. Hughes, *Manual of Cane-Growing* (Sydney, 1953), p. 236.

commenced when about 2% of seedlings were selected on the basis of a visual inspection for vigorous growth. Brix was then determined using a hand refractometer (i.e. a field device for the measurement of the total sugars present in a solution of mixed sugars) and any seedling with a low brix was discarded immediately. The remaining seedlings were then planted in 1-row, 10-sett plots of first selection seedlings. Those seedlings chosen for further testing were planted in 1-row, 10-sett plots of second stage seedlings and four such plots of third year seedlings. After disease resistance trials, the most promising seedlings were picked for yield selection trials and then finally farm yield trials. However, the BSES scientists in the mid-1930s had become concerned about the desirability of such small samples and from 1938 onwards planted out 40-sett plots (4 rows x 10 setts) in the second and third years. Similar sized plots of standard varieties were evenly distributed throughout the field so that every seedling contacted a standard and its progress could be assessed compared to a standard. Further modifications occurred in 1940 when Arthur Bell decided that selections from the yield observation trials would be made from ratoons, owing to the increasing importance of ratooning as a determining factor.[30]

At the commencement of the reorganisation of the BSES cane variety improvement program in 1935, almost 80% of the cane varieties crushed

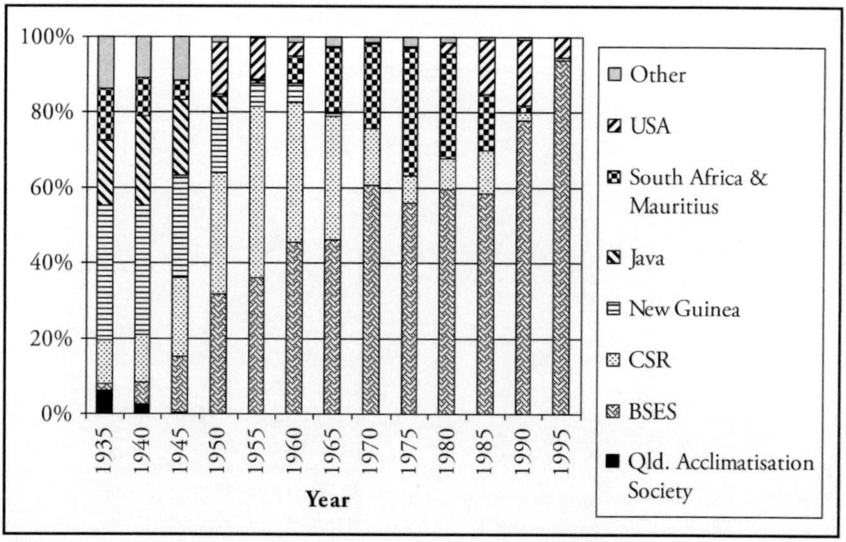

Figure 13.4. The origin of cane varieties in Queensland, 1935–1995.

Source: Calculated from varietal census figures in the BSES annual reports and the *ASYB*, various issues.

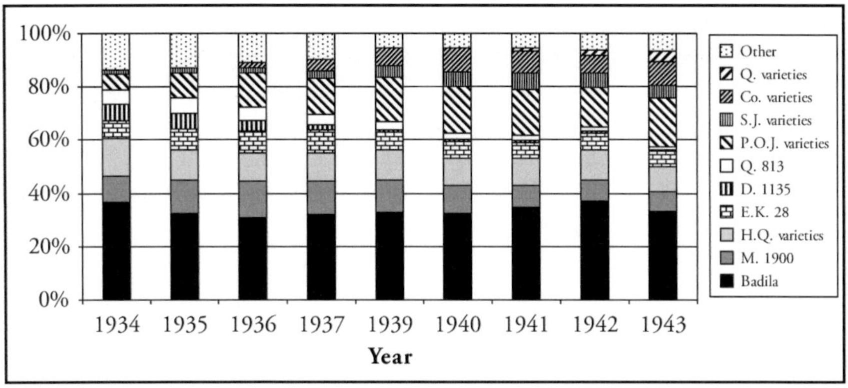

Notes: (1) No varietal census conducted in 1938. (2) Q. varieties = Q. 2, Q. 10, Q. 20 & Q. 25. (3) Co. varieties = Co. 210, Co. 213 & Co. 290. (4) S.J. varieties = S.J. 2 & S.J. 4. (5) P.O.J. varieties = P.O.J. 2878, P.O.J. 2714, P.O.J. 213 & P.O.J. 234. (6) H.Q. varieties = H.Q. 285, H.Q. 409 & H.Q. 426.

Figure 13.5. The percentage of harvested cane in Queensland classified according to variety, 1934–1943.

Source: Based upon the varietal census figures in the BSES, ARs, 1934–35 to 1944–45.

in Queensland had been imported. This remained the situation in 1940 (Figure 13.4). In particular, Queensland canegrowers relied heavily on overseas-sourced cane varieties such as Badila (from New Guinea in 1896), the P.O.J. varieties and E.K. 28 (sourced from Java), M. 1900 (secured from Mauritius) and the Co. varieties (obtained from India). The only locally bred cane varieties of note were H.Q. 426 and Q. 813 which were developed thirty years earlier by CSR and the Queensland Acclimatisation Society, respectively (Figure 13.5). Together these varieties had reduced the outbreaks of gumming disease and ensured the industry's survival.

This lack of BSES success in breeding new cane varieties had not gone unnoticed. A staff member from Fairymead plantation on a visit to CSR's Broadwater Experiment Station in April 1939 told David North that the BSES's work in supplying improved cane varieties to the Bundaberg district canegrowers was considered to be 'inadequate'. A year later, Norman King of the BSES, claimed he was aware of canegrower dissatisfaction with the progress being made by the BSES in releasing new cane varieties for commercial use, but when all aspects of the difficult work were considered it could hardly have been expected to have borne results any sooner.[31] However, this situation was about to alter. Two BSES-bred varieties – Q. 2 and Q. 20 – appeared in the Queensland varietal census for 1940. Over the next two years, they were joined by Q. 10 and Q. 25. This development, even though these Q. varieties made up less than 5% of the cane harvested in 1943 (Figure 13.5), signalled a turnaround in the fortunes of the BSES cane variety improvement program. Moreover, the cross-pollinations completed by C. Graham Hughes during the late 1930s and early 1940s, eventually contributed to the development of the Q. 50, Q. 57, Q. 58 and Q. 59 cane varieties which assumed some importance during the 1950s.

Australian-bred cane varieties commence their ascendancy, 1937–1959

In 1937, a subtle change occurred in the cane varieties being cultivated in New South Wales. Comus, the name given to the seedling numbered 28 S.N. 947, commenced being grown by the suppliers of Broadwater and Condong Mills. A cross between the Oramboo and Q. 813 cane varieties, Comus was an annual cane variety resistant to gumming disease and frost tolerant, making it ideal for New South Wales.[32] Even though Comus never made up more than 10% of the cane harvested in New South Wales during the late 1930s and 1940s (Figure 13.6), its cultivation heralded at long last success for CSR's cane variety improvement program. Comus was followed by new CSR-bred cane varieties such as Jason and Atlas – the two cane varieties of the 30 S.N. Series – and Solon, Akbar, Pollux and Trojan – the four cane varieties making up the 33 S.N. Series. Jason displayed a particularly high degree of Fiji disease-resistance and it was prominent amongst the varieties of the 30 S.N. and 33 S.N. series that assumed greater importance during the early 1940s, slowly displacing Q. 813 and the P.O.J. varieties which succumbed to Fiji disease.

The most significant of the 1933 seedlings, however, was a variety that became known as Vesta (33 S.N. 115). Bred from crossing P.O.J. 2878 with another Macknade seedling (31 M.Q. 228), Vesta grew quickly, displayed resistance to gumming and Fiji diseases, and was moderately sucrose-rich.[33] After its release in 1944, Vesta was readily adopted by New South Wales canegrowers. By 1950, it made up 60% of all the cane harvested in New South Wales (Figure 13.6). Vesta's popularity only began waning after 1957, its place being taken by Pindar – another cane variety bred at the Macknade Experiment Station – and two cane varieties of the 40 S.N. series (i.e. 40 S.N. 5819 and 40 S.N. 746).

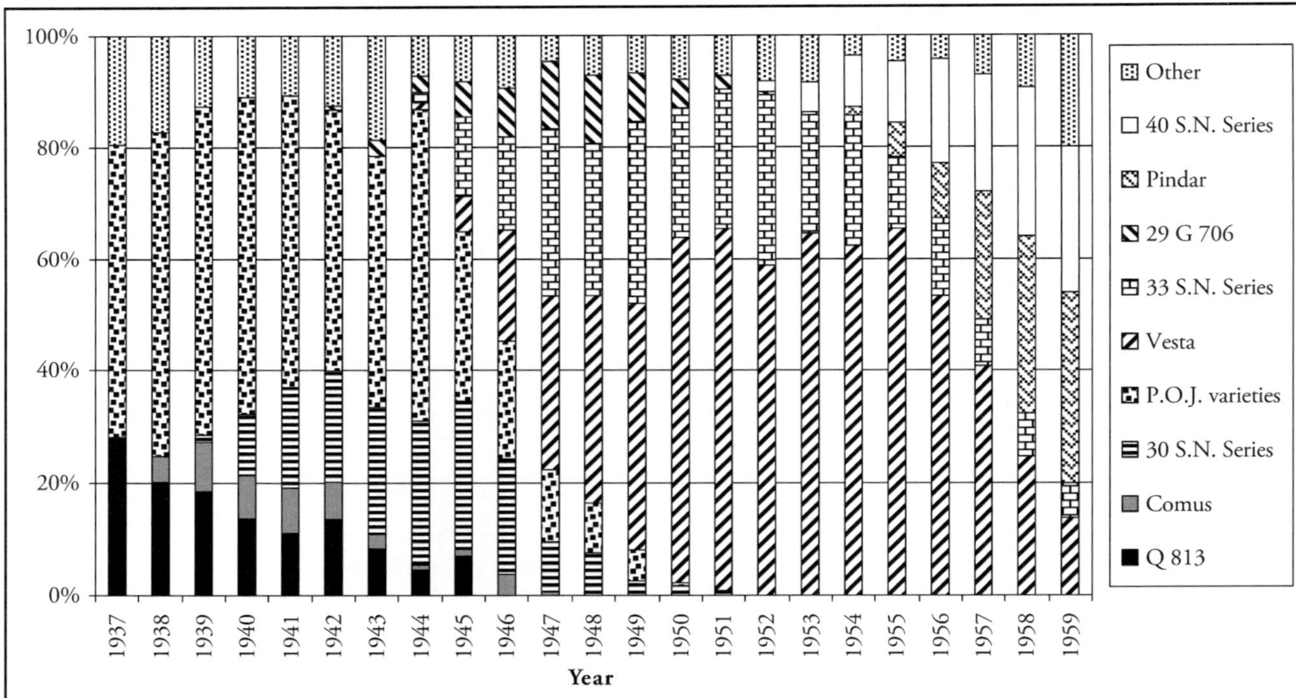

Notes: (1) 30 S.N. Series = 30 S.N. 361, 30 S.N. 673 & 30 S.N. 874. (2) 33 S.N. Series = 33 S.N. 1145, 33 S.N. 1166, 33 S.N. 1873 & 30 S.N. 1882. (3) 40 S.N. Series = 40 S.N. 746, 40 S.N. 1133 & 40 S.N. 5819.

Figure 13.6. The percentage of harvested cane in New South Wales classified according to variety, 1937–1959.

Source: Calculated from varietal census figures in CSR, Cane Reports, 1937 to 1959 seasons, CSRR, Z/109, Boxes 387 to 389.

In Queensland, Juno and Cato, two CSR-bred cane varieties developed at the Macknade Experiment Station, were first adopted by North Queensland canegrowers in 1940. Neither became a major Queensland cane variety, unlike Pindar and Trojan, two other CSR-bred cane varieties that originated from the batch of seedlings grown at the Macknade Experiment Station in 1933. Trojan was a cross between Co. 270 and another seedling that had the Korpi cane variety as one of its parents. It produced heavy crops, displayed good germination and ratooning ability and had very high sucrose content (Plate 13.2). Trojan first appeared in the Queensland varietal census of 1944 and was quickly adopted by the canegrowers of the Lower Burdekin district and in the mill supply areas north of Townsville. At the height of its popularity in 1952, almost a quarter of the cane harvested in Queensland was Trojan (Figure 13.7). Pindar was listed initially in the Queensland varietal census of 1949. Its parents included Co. 270, P.O.J. 2878, Badila and Q. 813, so Pindar was a gumming disease resistant cane variety that germinated and grew quickly, displayed excellent ratooning

Plate 13.2. A first ratoon crop of Badila (left) and Trojan (right), showing the superior vigour of the latter CSR-bred cane variety.

Source: J. Wright, 'Badila in the Burdekin', *CGQB*, Vol. 31, October 1967, p. 72. Image reproduction courtesy of BSES Limited.

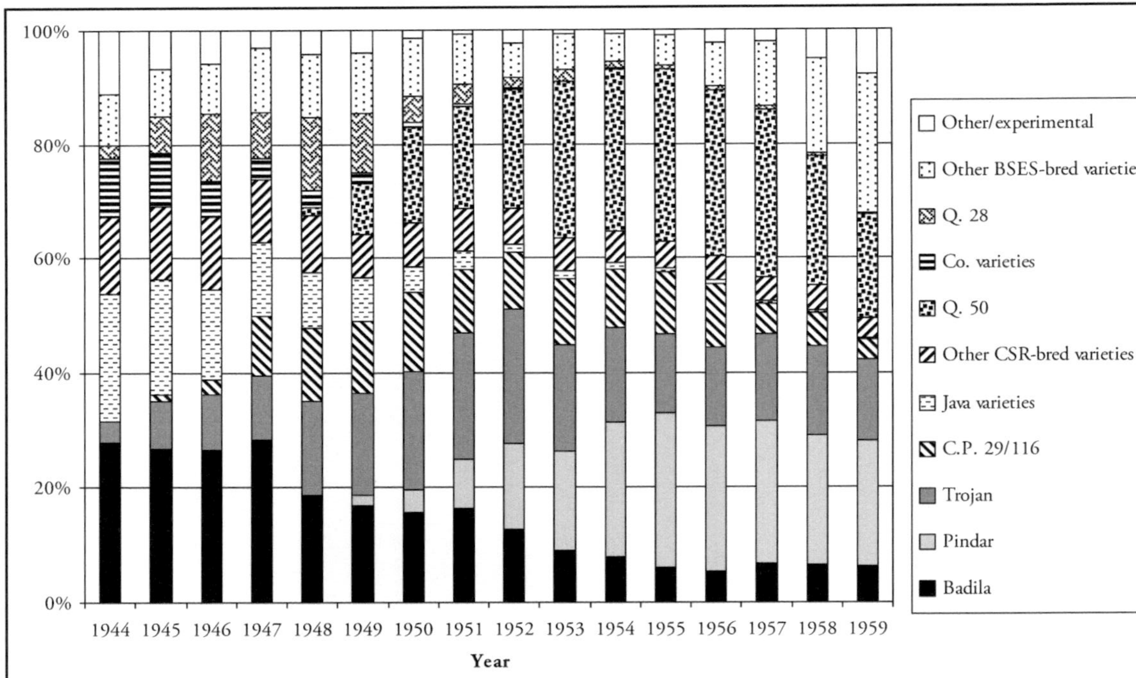

Notes: (1) Other BSES-bred varieties = Mostly S.J. 2, S.J. 4, S.J. 16, Q. 25, Q. 44, Q. 45, Q. 47, Q. 49, Q. 57 & Q. 58. (2) Other CSR-bred varieties = Mostly H.Q. 426, H.Q. 409, Pompey, Comus, Eros, Cato & Ragnar. (3) Java varieties = E.K. 28, P.O.J. 2878 & P.O.J. 2725. (4) Co. varieties = Co. 290 & Co. 301.

Figure 13.7. The percentage of harvested cane in Queensland classified according to variety, 1944–1959.

Source: Based upon figures in the ARs, BSES 1944/45 to 1959/60.

capabilities and produced medium to good sucrose content.[34] Pindar's ascendancy during the late 1940s led to the loss of Trojan's dominance. By 1955, a quarter of the Queensland crop was Pindar. H.Q. 426 had also waned in importance during the late 1940s, but other cane varieties developed by CSR such as Comus, Cato, Eros, Pompey and Ragnar were grown by North Queensland canegrowers during the late 1940s and 1950s, although together these cane varieties never exceeded 10% of the cane harvested in Queensland (Figure 13.7).

During the late 1940s and early 1950s, BSES-bred cane varieties such as S.J. 2, S.J. 4, S.J. 16, Q. 25, Q. 44, Q. 45 and Q. 49, were adopted by some Queensland canegrowers, although none ever reached more than 5% of the total Queensland crop. In contrast, Q. 28 and Q. 50 gained some significance during this period (Figure 13.7). Q. 28 was raised at the Mackay BSES station in 1935 from a cross between Co. 290 and Q. 1098, a variety bred by the Queensland Acclimatisation Society. Sucrose-rich, a good and sure germinator that displayed vigorous growth and good ratooning ability, Q. 28 was resistant to gumming, red rot, downy mildew and Fiji diseases. Grown overwhelmingly in the Mackay region, Q. 28 made up a third to half the cane harvested in this region between 1946 and 1949 and almost 15% of the cane harvested in Queensland during 1948. Yet its susceptibility to ratoon stunting disease, which was first identified in this cane variety during the summer of 1944–1945, led to Q. 28's abandonment throughout the early 1950s (see Chapter 15). On the other hand, Q. 50 – a seedling raised at the Mackay BSES Station in 1938 from a cross between P.O.J. 2725 and Co. 290 – assumed some importance during the early 1950s. At the peak of its popularity in 1955, 30% of the cane harvested in Queensland was Q. 50 (Figure 13.7). It was grown in the sugar-producing districts south of Proserpine, but especially throughout the Mackay district where it accounted for between 65 and 85% of the cane harvested over the period 1950 to 1958. Q. 50's popularity stemmed from its excellent ratooning ability, high sucrose content and tolerance of dry conditions and frost, which made it ideal for the Bundaberg district and the drier parts of the Mackay district that had no access to irrigation.

The BSES also introduced C.P. 29–116 into Queensland in 1935. This hybrid cane variety from Canal Point, Florida, was resistant to frost and downy mildew and Fiji diseases, able to withstand the long spells of dry weather often encountered in Southern Queensland, and exhibited vigorous growth and good ratooning ability, but was only moderately sucrose-rich. Yet its characteristics made this cane variety very suitable for the Southern Queensland, where it was widely grown during the 1950s.[35]

These developments ensured a major change in the origin of the cane varieties in Queensland. New Guinea's importance as a source of cane varieties had dwindled to less than 10% by 1955, a consequence of canegrowers growing Trojan and Pindar in place of Badila. Cane varieties from Java, especially E.K. 28 and P.O. 2878, were phased out during the 1940s, so that less than 5% of the cane varieties harvested in Queensland in 1950 had originated in Java. The demise of Q. 813 meant the defunct Queensland Acclimatisation Society was no longer a source of cane varieties in Australia, its pioneering work breeding new cane varieties during the early 1900s finally being no longer of any use. M. 1900, the favourite of canegrowers in the Mackay and Bundaberg districts for almost twenty years, had finally been superseded by newer cane varieties, so Mauritius disappeared as a source for cane varieties in Queensland after 1950. Instead, CSR- and BSES-bred cane varieties assumed rapid importance, boosting the amount of Australian-produced cane varieties grown in Queensland from 35% in 1945 to 82% in 1960 (Figure 13. 4).

As Queensland canegrowers commenced adopting the earliest Q. cane varieties, the BSES cane variety improvement program underwent further administrative changes. A BSES reorganisation at the end of 1947 disbanded the Committee on Seedling Propagation and plant breeding was given the status of a separate division. C. Graham Hughes surrendered his control over breeding new cane varieties at the Meringa BSES Station, his place being taken in November 1947 by James Hardie Buzacott (1902–1984), who was appointed Senior Plant Breeder in charge of the crossing work at Meringa and the seedling selection activities at the other stations and substations. Since 1930, Buzacott had spent much of his career at Meringa devising methods to control the cane grub and weevil borer and he was responsible for the early testing of the insecticide BHC against the cane grubs in the late 1940s (see Chapter 14). Affectionately known as 'Buz', he remained in charge of plant breeding until becoming the BSES's Chief Administration Officer in 1966 before retiring in 1970.[36] Buzacott was joined at Meringa in March 1953 by a senior geneticist, Dr James C. Skinner, who had completed postgraduate studies in plant breeding at Manchester University. Skinner designed and conducted many of the important BSES cane variety improvement experiments at Meringa. The findings from these investigations led to the development of the BSES procedures for breeding new cane varieties. His vital role in this endeavour needs more recognition.[37]

Seedling production and testing of potential new cane varieties expanded to new localities during the late 1940s and 1950s. In 1947, a site was acquired for the Burdekin BSES station near Brandon. Seedlings commenced being raised at the Burdekin BSES station in 1950. An additional 5.2 ha of land was purchased at the Meringa BSES Station in 1959, enabling an enlargement of the research work into the breeding of new cane varieties. By 1956, new lantern frameworks built from iron had replaced the old wooden frames, enabling eighty-six lanterns to be deployed during the cross-pollination season. Four years later, 170 lanterns were used at Meringa. Consequently, cross-pollinations at Meringa doubled from around 150 in the late 1940s to approximately 300 during the late 1950s (Figure 13.3). Moreover, substations functioned at Babinda, Innisfail, Tully, Ingham, Childers and Nambour by 1960, together with quarantine introduction plots at Beerwah and Millaroo and introduction plots in all mill supply areas. Every sugar-producing district in Queensland now had a better opportunity to share in the selection of cane varieties suited to its particular environmental conditions. The new facilities also enabled seedling production to almost treble from the levels reached in the early 1940s, so that approximately 75 000 seedlings were planted annually by 1960.[38]

The aim of the BSES plant breeding program during the late 1940s and early 1950s shifted from just producing disease- and pest-resistant cane varieties to improving the overall productivity of the state's crops of sugar cane. The Queensland sugar cane crop is harvested on an annual basis, so any cane planted late in the year (i.e. August) could be only ten months old when it was harvested the following year. Queensland cane-growers were paid according to the cane's sucrose content, based on individual analysis. Consequently, the cane breeders decided in the late 1940s that it was highly desirable to produce early maturing cane varieties that displayed good sucrose content at the beginning of the crushing season. The second goal was also linked to cane harvesting and the associated high labour costs of employing unionised European workers to undertake the task in Australia. Higher wage rates had to be paid for cutting lodged cane, so the cane breeders aimed to produce cane varieties that remained erect under most growing conditions (i.e. anti-lodging). Erectness and lodging-resistance also became more important characteristics with the adoption of mechanical harvesting during the 1960s, as the earliest mechanical cane harvesters did not function effectively in very lodged and tangled crops of cane. Other less pressing needs, according to Buzacott, were new high-yielding varieties to replace the present popular cane varieties that would undoubtedly 'run out' in a few years and the development of better cane varieties resistant to frost and diseases such as red rot, leaf scald and chlorotic streak.[39]

To achieve its goals, the BSES decided in 1950 that the cane variety improvement program needed an infusion of new genetic material. New Guinea was again selected as the source for this material. From mid-April to early June 1951, Jim Buzacott, C. Graham Hughes, and Dr John Warner of the Experiment Station of the Hawaiian Sugar Planters' Association, spent several weeks in New Guinea, collecting new cane varieties in the central highlands between Mendi, Mt Hagen and Goroka, along the Markam River valley and along the coast to the east of Lae. The object of the expedition was twofold. Firstly, 'noble' cane varieties (i.e. clones of *S. officinarum*) were sought, as they might be valuable for direct commercial production, as Badila had been previously. Secondly, outstanding 'noble' and wild cane varieties were collected, especially those varieties that had early maturing qualities or which could confer such characteristics on their progeny. Buzacott and Hughes returned to Queensland with an extensive collection of 165 cane varieties, covering the species *S. officinarum*, *S. spontaneum* and *S. robustum* (wild canes). Three quarters of these cane varieties eventually became established at a quarantine plot at Redlands, near Brisbane, where they were assessed for signs of diseases which could affect the Australian sugar industry.[40]

Further new breeding material was acquired a year later, when cane fuzz from thirteen different crosses was received from Hawaii. These crosses had been made at the Experiment Station of the Hawaiian Sugar Planters' Association using parental material which was expected to yield progeny suited to Queensland conditions.[41] Additional new genetic material was obtained in 1957, when the BSES acted as the quarantine agent for an expedition to New Guinea headed by Dr John Warner and Carl Grassl of the United States Department of Agriculture. Of the 206 cane varieties that survived the quarantine treatment and establishment on Magnetic Island, just under three-quarters were 'noble' canes (*S. officinarum*).[42]

Cross-pollinations at the Meringa BSES Station continued using the Hawaiian method of crossing in a mixture of water and chemicals, the constitution of which changed several times, but in 1959 consisted of a water solution containing 150 p.p.m. of sulfur dioxide and 85 p.p.m. of phosphoric acid. However, the former technique of leaving most of the leaves on the flowering stalks, cutting off a section of stalk and changing the solutions every second day was discarded in favour of the less laborious method of stripping all the leaves, except the flag leave, from the stalks, changing the solution twice per week and cutting off a section of stalk only once during the period of cross-pollination. During the late 1940s and early 1950s, some cross-pollinations occurred under lanterns while the remainder were maintained and dispersed at intervals in open forest to reduce pollen contamination. Before 1956, the lanterns were made of voile, a light open-mesh, cotton material designed to keep out pollen but to allow the circulation of air. However, Skinner found after extensive tests in 1956 and 1957 that fairly serious contamination occurred from pollen both in the open crosses and in the voile lanterns. Consequently, more closely woven materials were used for the lanterns and open crossing had been discarded by 1959, so that contamination was reduced to a negligible figure.

Artificial induction of arrowing was attempted using chemicals (e.g. gibberellic acid) and the manipulation of the amount of daylight in a small make-shift photoperiod room, but with limited success – only one cane variety was induced to flower out of season. Buzacott concluded that these processes were too time consuming to be effective and they were not pursued, especially as ample flowers from the most required cane varieties were available during the crossing seasons. Synchronisation of flowering between *S. spontaneum* clones which did not normally flower in season with the 'noble' and near 'noble' cane varieties was achieved, however, by the use of artificial light, maleic hydrazide sprays and lopping of foliage.[43]

Improvements were also made to the BSES seedling production and assessment techniques during the 1950s. A reduction in seedling losses in the glasshouses occurred after the standardisation of the composts used for seedling raising. Better glass-house temperature control was instigated, ensuring appropriate and uniform temperatures during the germination period. Experiments under Queensland conditions were conducted from 1953 onwards to determine the merits of either the Hawaiian system of bunch planting (i.e. five to fifteen seedlings) or single planting of seedlings, and showed that both methods had merits. By 1960, bunch planting was being used mostly at Meringa, particularly for the selection of new parent canes as it enabled a rapid elimination of those seedlings unlikely to flower.

The selection of seedlings for further testing initially relied upon some arbitrary decisions. These were largely eliminated by the development of a grading system to evaluate cane varieties at all stages of selection. The scale finally adopted was one (very poor) to twenty (excellent), with the standard cane variety being equal to ten.[44] Following theoretical studies of the influence of heredity and environment on the performance of the sugar cane seedlings, experiments in 1954 showed that a higher selection rate than the 2% previously used was desirable in the early stages of selection to overcome the possibility of discarding superior clones which may not have expressed their full potential in highly variable original seedlings. After trials to assess the reliability of the results from different combinations of selection systems, the BSES by 1960 had settled upon an arrangement whereby the best 2% of seedlings were selected and planted in 30-sett plots and another 10–20% more original seedlings were planted in a new 4-sett plot stage. A tenth of the best seedlings were then selected from the 4-sett stage for additional testing in the 30-sett stage. Replicated trials were also introduced earlier. From 1955 onwards, the last two stages of selection at the experiment stations were replaced by a single stage using two or three replicated yield observation trials. This enabled the testing of about thirty times more potential new varieties than previously.[45]

After a burst of releasing ten new approved Q. cane varieties in the early 1940s, the BSES cane variety improvement program had limited success in the early 1950s (Table 13.1). The turnaround commenced in 1954 with the release of Q. 57 and over the next five years, eleven new Q. cane varieties were approved for cultivation by the state's canegrowers. Moreover, this batch of new Q. cane varieties contained three important cane varieties – Q. 57, Q. 66, and Q. 68. The former was bred at Meringa in 1945 from a cross between Q.27 and Q. 31. It matured early in the harvest season and possessed the ability to produce high sugar content, even under low sunlight conditions. This feature contributed to Q. 57 becoming a major cane variety in North Queensland which sometimes suffers lengthy rainy periods. Q. 66 produced very erect crops that

Table 13.1. New Q. cane varieties and their year of approval, 1940–1959.

Year	New approved cane varieties	Year	New approved cane varieties
1940 and earlier	Q. 2; Q. 20	1950	Q. 51
1941	Q. 10; Q. 25	1951	Q. 55
1942	Q. 28	1952	Nil
1943	Q. 44	1953	Nil
1944	Q. 45	1954	Q. 57
1945	Q. 47	1955	Q. 58
1946	Q. 48; Q. 49	1956	Q. 56; Q. 59
1947	Q. 50	1957	Q. 61; Q. 64
1948	Nil	1958	Q. 63; Q. 65; Q. 66; Q. 70
1949	Nil	1959	Q. 67; Q. 68

Source: Based upon details in BSES, AR s 1939–40 to 1944–45 and J. Buzacott, 'Silver anniversary of cane breeding at Meringa', *CGQB*, Vol. 23, no. 2 (October 1959), p. 41.

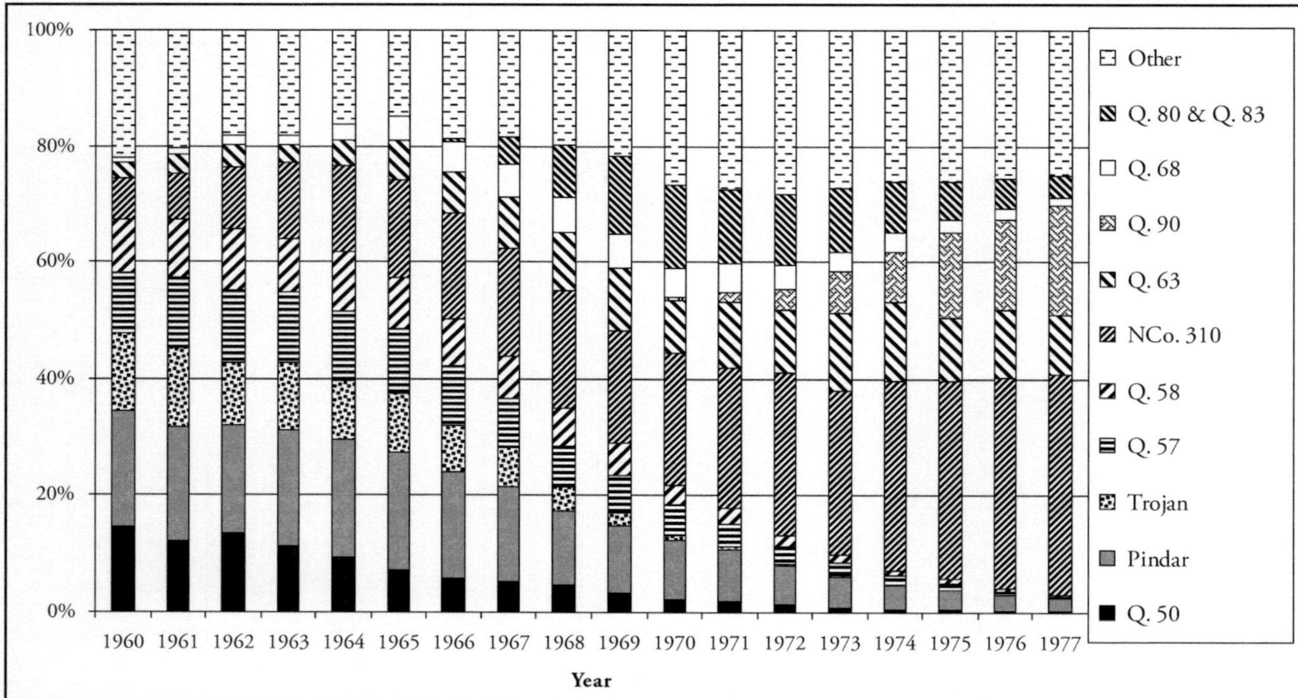

Note: (1) The 'Other' category contained numberous Q. varieties, each mostly having less than 3% of the amount of cane harvested annually.

Figure 13.8. The percentage of harvested cane in Queensland classified according to variety, 1960–1977.

Source: Based upon figures from the annual varietal census published in the *ASYB*.

resisted lodging under high-wind and flood conditions. In the early 1960s, Q. 66 was distinguished as the 'most lodging resistant variety grown in Queensland'. These features had been bred into this new cane variety through the use of genetic material from *S. robustum*, a wild cane variety which resisted lodging during flooding and while growing in waterlogged soils throughout its natural habitat. However, Q. 66 never gained much popularity amongst Queensland canegrowers, being grown mostly in North Queensland where it made up less than 2% of the cane harvested during the early 1960s.[46] Q. 68 also produced very erect and heavy crops, and had good early sugar content that was maintained throughout the crushing season. It was favoured by Southern Queensland canegrowers, making up around 5% of the cane harvested in Queensland during the early 1960s (Figure 13.8). Eventually, New South Wales canegrowers also considered Q. 68 worthy of cultivation, the variety making up a quarter of the cane harvested in that state during the mid-1970s (Figure 13.9).

Running parallel to the BSES cane variety improvement program was CSR's continued research into breeding new cane varieties. Initially, the outbreak of World War II did not curtail crossing work, but staff shortages eventually led to the cessation of seedling raising at Macknade and Sydney between 1942 and 1944. CSR-sponsored cane importations ceased and the Sydney quarantine house remained empty during the war years. After this hiatus, seedling raising resumed at Macknade in 1945, but not at Sydney as CSR decided to shift its New South Wales cane variety improvement program to the Broadwater Experiment Station. A new glasshouse was built at the Broadwater Experiment Station, but the seed continued to be supplied from cross-pollinations undertaken at Macknade. All future seedlings raised at Broadwater were given the distinguishing letters B.N.[47] The extent of plant breeding activities also expanded after World War II, as making more crosses and growing greater number of seedlings increased the chances of finding new cane varieties. At Broadwater, twenty-five crosses were undertaken in 1947; this number had quadrupled by 1959. Seventy crosses took place annually at Macknade in the late 1940s; this figure had risen to 100 by 1959. Approximately 7 000 seedlings were planted each year in New South Wales during the late 1940s; this amount had quadrupled by the late 1950s. A similar trend occurred at Macknade. Part of this upsurge in the number of seedlings planted resulted from the adoption of bunch planting after 1957.[48]

During the late 1940s and early 1950s, CSR's cane breeders made extensive use of the bloodlines from Co. 270, Co. 280 and Uba Marot – a natural hybrid variety from Mauritius – in their cross-pollination activities, because each cane variety had *S. spontaneum*, a wild cane variety, as one of its

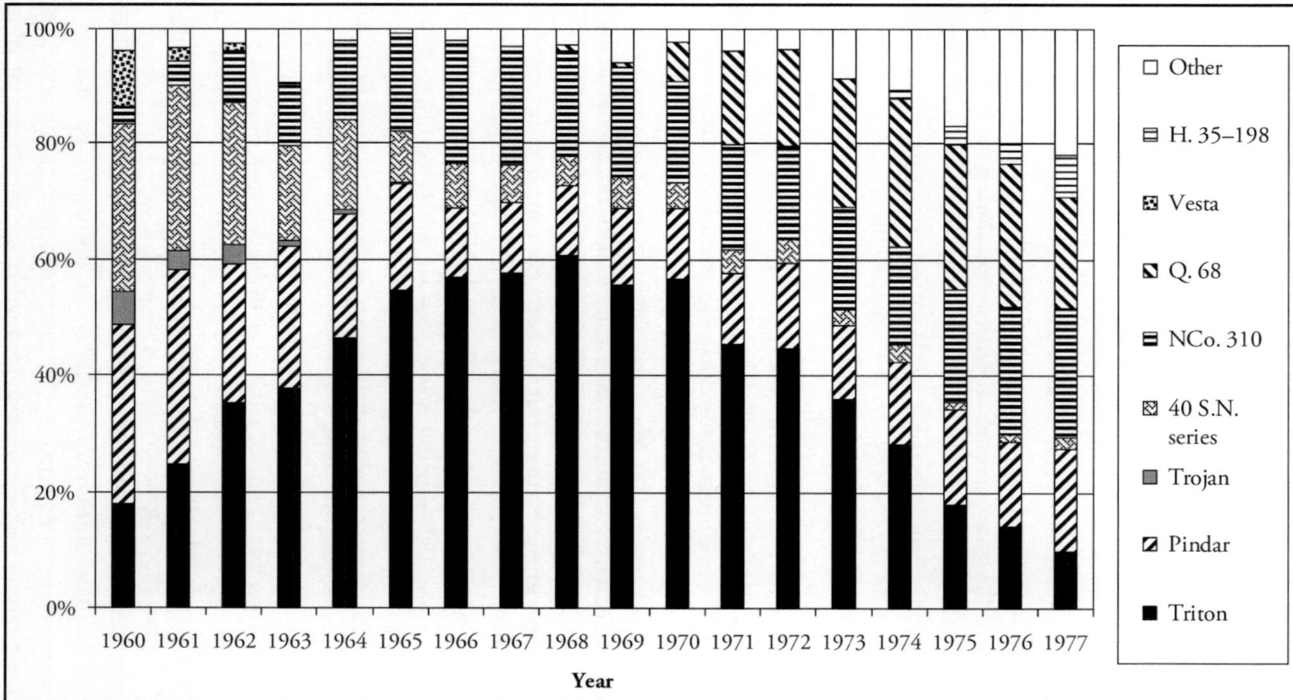

Notes: (1) 40 S.N. Series = 40 S.N. 746, 40 S.N. 1133 & 40 S.N. 5819.
(2) Other = After 1974 mostly B49–119, B52– 405, H49–104, Hebe and Hermes.

Figure 13.9. The percentage of harvested cane in New South Wales classified according to variety, 1960–1977.

Source: Calculated from varietal census figures in CSR, Cane Reports 1960 to 1977 seasons, CSRR, Z/109, Boxes 389 to 391.

parents. CSR's cane breeders aimed to create new hybrid cultivars with the characteristics of the wild cane varieties such as vigorous growth, heavy stooling and tolerance of harsh conditions, together with high, early season sucrose content. In addition, CSR's cane breeders after 1945 tried unsuccessfully to breed frost-resistant cane varieties by using the blood line of the cold resistant Turkmenistan cane varieties, although similar trials in Queensland and Louisiana during the 1940s and 1950s had also met with negative results. New genetic material for the breeding program was constantly being sourced from hybrid cane varieties imported from the BSES, Taiwan and the United States Department of Agriculture and cane arrow fuzz supplied by the Experiment Station, Hawaiian Sugar Planters' Association. All this new material spent time in CSR's Sydney quarantine house before being dispatched to either the Macknade or Broadwater Experiment Stations.[49]

Results from CSR's cane variety improvement program during the late 1940s and 1950, however, were not as significant as previously. In Queensland, minor cane varieties such as Ragnar, Sirius, Vidar, Castor, Luna and Cadmus – bred mostly from crosses at Macknade during the 1940s – were trialed by some CSR suppliers in the sugar-producing districts north of Townsville, but they never individually assumed more than 5% of the cane harvested in these regions. A little more success was achieved in New South Wales. Two cane varieties of the 40 S.N. series (i.e. 40 S.N. 5819 and 40 S.N. 746) made up a quarter of the cane harvested in the late 1950s (see Figure 13.6). However, the most promising cane variety developed during this period was Triton (47 B.N. 1843), one of the seedlings raised at the Broadwater Experiment Station in 1947. By 1959, Triton made up a third of the cane planted in the Harwood Mill supply area, and its popularity was increasing due to the cane variety's outstanding characteristics which included high sucrose content, good germination qualities, frost-resistance and resistance to the disease red rot. Trials indicated that Triton produced half a ton of sugar per acre (1.2 tonne per hectare) more than Vesta, the standard variety in New South Wales during the 1950s.[50]

BSES leads the way in breeding cane varieties, 1960–1977

During the early 1960s, the BSES cane variety improvement program each year consisted of undertaking approximately 400 cross-pollinations and germinating between 60 000 and 70 000 original seedlings. A few extra plant breeding staff at Meringa, the use of an additional forty-eight crossing lanterns and more streamlined procedures enabled a modest expansion in cane variety improvement activities during the late 1960s. By the early 1970s, between 500 and 600 cross-pollinations were completed each year (Figure 13.3), and the number of original seedlings raised annually had risen nearer to 90 000. Between 100 and 150 potential new cane varieties were tested annually in farm-based trials. Further increases in the number of cross-pollinations occurred from 1974 onwards following the adoption of the factorial polycross technique. This method involved several female varieties being crossed simultaneously with one male in a single cross pollination lantern. The same group of female varieties was then crossed with different male varieties in different lanterns, to give an m x n factorial mating system.[51] This method led to 3 286 cross-pollinations being made in 1975 – a record that had not been surpassed even by 1995 (Figure 13.3). Crossing work continued to be completed entirely at Meringa, and the seedlings were raised at the Burdekin, Mackay and Bundaberg BSES stations. Additional intellectual rigour for the breeding program came in August 1964, when the BSES employed Douglas Mac Hogarth, who had a background in statistical genetics. Hogarth commenced gathering information on the inheritance of characters such as yield and sucrose content and the environmental effects on varietal performance. In 1973 he received a PhD for his research on estimating genetic variances in populations of sugar cane. [52]

Attempts to improve the efficiency of the BSES's cross-pollination techniques continued throughout the 1960s. Following several experiments during the late 1950s on chemical combinations to determine if they induced better rates of cross-pollination, the BSES decided in 1960 that its crosses would be made in a new mixture consisting of 100 p.p.m. sulfur dioxide, 50 p.p.m. phosphoric acid, 25 p.p.m. sulfuric acid and 25 p.p.m. nitric acid. The crossing solution was further increased in strength by 50% in 1961, after the discovery in Hawaii that stronger solutions produced enhanced cross pollination rates.[53] The new technique developed in Mauritius of enclosing crossing solutions in plastic liners to slow sulfur dioxide loss was adopted in 1970, thereby reducing the addition of sulfur dioxide to the containers holding the crossing solutions to once a week. This change also involved the substitution of plastic containers for the painted iron containers.

Research into cross-pollination techniques continued during the 1970s. An experiment in 1971 confirmed that the standard technique of using lanterns closed at the base all the time completely avoided contamination by foreign pollen. Trials during 1973 showed that the frequency of topping-up the crossing solution could be reduced from three times per week to twice or even once a week without harmful effects. A switch to bulk supplies of crossing solutions also minimised labour requirements in this phase of the cane variety improvement program. Marcotting of the arrowed stalks selected for crossing had commenced by 1968, as the process kept them alive longer, although this procedure was eventually abandoned, the BSES cane breeders concluding they gained no real benefit from this approach. From 1969 onwards, the BSES cane breeders deliberately used longer stalks for cross-pollination, thereby substantially increasing the survival rate of arrows. Such a practice, however, meant that the crossing lanterns needed to be elevated higher to accommodate the longer stalks and this led to the development of the system whereby the crossing lanterns were raised or lowered on pulleys.[54]

Research on methods of selecting seedlings for future testing continued during the 1960s and early 1970s. The long-term experiments comparing single planting and bunch planting selection systems, commenced in 1956, were finalised in 1968. The Hawaiian bunch planting system appeared to be less efficient than single planting, although the results were not completely conclusive. An improved bunch planting system developed during these experiments involved increased selection rates (16%) and planting of selected seedlings in 4-sett instead of 2-sett plots. This system was slightly superior to single planting, but not sufficient to justify a change from single planting to bunch planting entirely in Queensland.[55] Another significant change to the research program during the 1960s was the development of a sampling method for measuring the weight of cane varieties in replicated farm trials. With the rapid adoption of mechanical harvesting, the previous system of weighing entire plots cut by hand could no longer be used (see Plate 13.3). Instead, the cane breeders weighed either a 45- or 60-stalk samples to calculate the weight per stalk. The result from this procedure was then multiplied by the number of stalks per plot. In the 1967 season, thirty-five replicated cane variety trials were harvested throughout Queensland, thirteen by sampling and twenty-two by weighing the entire plot. By 1973, the number of cane variety trials was sixty-six, and sixty-five were processed by the sampling method. This arrangement was further modified in 1973 and 1974, following the introduction of a bin-mounted weighing machine which required only one person to collect, weigh and record the amount of cane harvested mechanically from the experimental plots.[56]

The BSES crossing program during the 1960s and early 1970s con tinued trying to incorporate the desirable characteristics of wild canes (*S. spontaneum; S. robustum*) into new, hybrid varieties. By 1964, all the New

Plate 13.3. Hand weighing stalk samples during a field trial in Southern Queensland, c. 1959.
Source: BSES, AR 1958–59, p. 53. Image reproduction courtesy of BSES Limited.

Guinea cane varieties collected in 1951 had been discounted as potential commercial varieties, but their potential as possible parents in the breeding program was still being assessed. Crossing with a wider range of wild canes intensified after the completion of a new climate or photo-period room at Meringa in early 1966. This structure – used to induce flowering in wild cane varieties – contained an air conditioning unit, artificial lighting and fine water sprays to increase humidity.[57] Specific goals of the breeding program during the 1960s included eliminating seedlings (and therefore future potential varieties) that displayed a susceptibility to leaf scald disease and the development of new anti-lodging varieties, particularly one suited to the Lower Burdekin district. Breeding potential future varieties that displayed Fiji disease resistance became a priority in the mid-1970s, following the severe outbreaks of Fiji disease in Southern Queensland during the 1970s (see Chapter 15). Many experimental crosses were made between parents displaying a reasonable to good resistance to Fiji disease, although the benefits of this work were not available until the 1980s.[58]

In July 1960, Jim Buzacott wrote that the BSES cane variety improvement program had 'gradually gathered impetus and new varieties are now rolling off the assembly line at a steady rate'.[59] Over the 1960s, nineteen

Table 13.2. New Q. varieties and their year of approval, 1960–1977.

Year	New approved varieties	Year	New approved varieties
1960	Q. 69; Q. 71; Q. 72; Q. 73	1969	Q. 87
1961	Q. 63	1970	Q. 89; Q. 90; Q. 91; Q. 92
1962	Q. 62; Q. 75	1971	Q. 93
1963	Q. 76; Q. 74	1972	Q. 94; Q. 95; Q. 97
1964	Q. 77; Q. 78	1973	Nil
1965	Q. 79; Q. 80; Q. 82	1974	Q. 96; Q. 98; Q. 99; Q. 100; Q. 101
1966	Q. 83	1975	Nil
1967	Q. 86	1976	Nil
1968	Q. 84; Q. 85	1977	Q. 102

Source: Based upon details in BSES, ARs 1959–1960 to 1977–1978.

new BSES-bred Q. varieties were approved for use by the state's canegrowers. This momentum was maintained, with fourteen new BSES-bred Q. varieties being released between 1970 and 1977 (Table 13.2). Some varieties such as Q. 75, Q. 76 and Q. 78 only ever made up 1 or 2% of the cane harvested in Queensland in any one year. In contrast, Q. 63 and Q. 90 became quite popular varieties during the late 1960s and 1970s (Figure 13.8). The former was a cross between Trojan and C.P. 29–116 and had been selected by the technicians at Mackay. A very rapid and good germinator, Q. 63 had the capacity to produce heavy crops, with very high sugar content being achieved early and maintained throughout the crushing season. Between 1968 and 1977, Q. 63 made up 10% of the cane harvested in Queensland. It was especially favoured by the Mackay and Lower Burdekin canegrowers – Q. 63 made up two-thirds of the cane grown in the Lower Burdekin district during the mid-1970s. Moreover, its popularity persisted into the late 1980s (Figure 13.11). Q. 90, approved for commercial use in 1970, rapidly gained favour amongst canegrowers north of Townsville, replacing Pindar and Q. 83. Exceptionally sucrose-rich, results from numerous farm trials showed that Q. 90 yielded about a third more sugar per ha than Pindar.[60] By 1977, Q. 90 made up almost 20% of the cane harvested in Queensland

(Figure 13.8), although its popularity waned rapidly during the mid-1980s after it succumbed to the fungal disease common rust (*Puccinia melanocephala* H. & P. Sydow).

Expansion, but not success, characterised CSR's plant breeding program during the 1960s and 1970s. In November 1959, Dr Keith Sutherland was appointed CSR's new Director of Research. With a background in engineering and industrial experience with CSIRO, Sutherland convinced the company's directors of the need to undertake more basic research on the sugar cane plant, especially in order to understand the environmental factors influencing the production of sucrose. Consequently, CSR announced in June 1960 that it would spend £300 000 on building a new plant research centre in Brisbane. The Centre would house a phytotron – a series of laboratories of special design containing rooms and greenhouses for growing sugar cane under conditions where the temperature, wind, humidity and light were strictly controlled. At the time of its construction, CSR's phytotron was the only facility in the world devoted solely to studying the physiology of the sugar cane plant. Named the David North Plant Research Centre in honour of CSR's first plant pathologist, it was to be the centrepiece of CSR's plan to rejuvenate and expand its lagging cane variety improvement program.[61] In addition, a new and enlarged glasshouse was erected at the Macknade Experiment Station in 1962 to permit the production of extra seedlings, as the area available for experimental plantings at Macknade had been doubled from 11 ha in 1959 to 22 ha by 1961. Seedling raising also re-commenced at the Hambledon Cane Nursery in 1961. Further improvements in 1969 involved enlarging the glasshouse at the Broadwater Experiment Station, thereby doubling the space available for seedling raising. The glasshouse was also fitted with controlled air flow, humidifying sprays and heating lamps to improve the plant breeders' control over the environmental conditions under which to raise the seedlings.[62]

CSR's decision to expand its cane variety improvement program coincided with a major change of personnel at Macknade. Kenneth Gard, CSR's chief cane breeder since the mid-1920s, retired in 1959. He was replaced by Brian Roach, who had university qualifications in agriculture from the University of Sydney.[63] Roach supervised the increase in the annual number of cross-pollinations from around 100 during the early 1960s to 200 by the early 1970s. The total number of seedlings raised by CSR in the early 1960s stood at around 45 000 annually; this figure had risen to approximately 75 000 in the early 1970s. CSR's cross-pollination techniques had changed little during the 1940s and 1950s, but arrow preservation experiments in 1960 showed that it would be unwise to depart from the standard method of changing the acid solution and trimming the stalks every two days. Further experiments on cross-pollination methods during 1961 and 1962, however, did lead to some changes such as the use of de-ionised water in the preserving solution, plastic crossing lanterns and marcotting (or layering) of arrowed stalks. Atomized spraying of suspensions of pollen and macerated anthers from poor flowering *S. spontaneum* cane varieties onto arrows from 'noble' varieties were successful in achieving crosses. Covered ripening frames were erected at the Macknade Experiment Station in 1963 to improve seed setting which was adversely affected by rainy periods.[64]

CSR's cross-pollination activities during the 1960s and 1970s were guided by specific company goals. The first of these goals emerged from a meeting at Macknade in 1962 between Joe Daniels (CSR Fiji), Dr Kenneth Glasziou, Manager of the David North Plant Research Centre, Dr Barry Latter of CSIRO's Plant Industry Division, and CSR staff, including Brian Roach. Those attending this meeting recognised that the world's commercial hybrid canes were derived from a very narrow genetic base. Projects to determine whether *S. spontaneum* could be selected on the basis of its cane yields and sugar content for hybridising with *S. officinarum* were proposed to improve the genetic base of cane variety improvement in Australia (i.e. introgression breeding or the crossing of closely related plant relatives). A considerable proportion of time and equipment during the early 1960s was then devoted to making crosses associated with the *S. spontaneum* research activities at the phytotron. Following outbreaks of Fiji disease throughout CSR's New South Wales mill supply areas in 1969 and severe injury to the cane crops from frost during the winters of 1968 and 1969, CSR decided that two new research programs should commence in 1970. Cold tolerant varieties identified at the David North Plant Research Centre by artificial freeze treatments were to be evaluated in field trials for their potential usefulness. Other cane varieties known to possess resistance to Fiji disease were to be assessed for their frost tolerance and general commercial usefulness. CSR was seeking a particularly frost-resistant and sucrose-rich early maturing cane variety that could be harvested after one year. Crosses between other cane varieties displaying varying degrees of Fiji disease resistance aimed to study more fully the inheritance of this characteristic.[65]

CSR's cane variety improvement program during the 1950s, 1960s and early 1970s produced several new minor commercial varieties such as Hera, Woden, Midas, Sidon, Damon, Apollo and Cassius. The most successful of these varieties was Cassius, which accounted for between 5 and 6% of the cane harvested in the sugar-producing districts north of Townsville during the early 1980s.[66] In New South Wales, CSR-bred varieties such as Triton and Pindar, developed in the 1940s, gained some popularity during the 1960s and early 1970s, but were then quickly replaced by Q. 68, a sucrose-rich and high yielding BSES-bred cane variety and the imported South African hybrid cane variety, NCo. 310 (Figure 13.9). Overall, CSR's cane variety improvement program failed to repeat the success achieved in the 1930s and 1940s, despite the increased resources devoted to it by the company. By 1975, only 40% of the cane harvested in New South Wales came from a CSR-bred cane variety, half the level reached a decade earlier (Figure 13.2). A similar trend occurred in Queensland, where Pindar and Trojan were replaced by new BSES-bred varieties such as Q. 90 or the imported NCo. 310. In 1975, only 7% of the cane harvested in Queensland came from a CSR-bred cane variety, compared to 32% in 1965 (Figure 13.4).

This reversal in fortunes of the two Australian cane variety improvement programs requires some explanation. Three reasons can be advanced to explain the turnaround in the outcomes from the cane variety improvement activities of CSR and the BSES. Firstly, the overall size of the BSES cane variety improvement program had become much larger than CSR's endeavours. In the early 1960s, the BSES undertook approximately 400 cross-pollinations – CSR managed around 100 cross-pollinations. This disparity had grown considerably by 1977, when CSR made 200 cross-pollinations annually compared to the BSES's 1 000 cross-pollinations. The BSES also consistently raised around 25 000 more original seedlings each year than CSR. The chances of creating a new and useful cane variety were more enhanced in the BSES program given its sheer size. Secondly, the main BSES cane breeders were university educated and included geneticists, giving them a considerable advantage over the smaller CSR program which never employed a trained geneticist. Thirdly, the CSR cane variety improvement program concentrated on introgression breeding and its activities became distracted by investigations into the physiology of the sugar cane plant – a result of the demands from the CSR scientists working at the David North Plant Research Centre – instead of the practical focus of finding new commercial cane varieties. In contrast, the BSES's plant breeding program, although underpinned by strong scientific research, remained focused upon improving arrowing, the survival of arrows and seedlings and the rates of cross-pollination – activities that assisted in the creation of new cane varieties.

Re-organising the Australian cane variety improvement programs, 1978–1995

In 1978, cane variety improvement arrangements changed in Australia, a result of CSR selling its New South Wales sugar mills to the NSW Sugar Milling Coop. However, the NSW Sugar Milling Coop. was aware of the need for ongoing scientific input into the breeding of cane varieties suited to the environmental conditions of northern New South Wales. Shortly after CSR's withdrawal from New South Wales, the BSES was asked to become involved in a program of breeding new cane varieties suited to New South Wales conditions, but the request was declined. Consequently, the NSW Sugar Milling Coop. decided to maintain the cane variety improvement arrangements instituted by CSR, using cane seed supplied by CSR from its Macknade Experiment Station.[67]

Throughout the 1980s, 30 000 to 40 000 seedlings were raised annually in northern New South Wales from the seed delivered by CSR. Each year approximately 100 cross-pollinations were completed.[68] New cane varieties continued to be sourced from Florida and Louisiana because the temperate climate in these regions was similar to that of northern New South Wales, and their canegrowers, to save costs, grew cane in a two-year cycle, like in New South Wales. Imported frost-tolerant C.P. varieties from the United States gained considerable importance in New South Wales during the 1980s and early 1990s. Some were given local names like Dart and Delta; others continued to be known by their original names such as C.P. 57–526 and C.P. 44–101 (Figure 13.10). Fiji disease resistance testing of promising varieties assumed greater importance in the mid-1980s, following the resumption of Fiji disease outbreaks throughout the Richmond River district in 1983. Links were forged with outside agencies when Sugar Research and Development Corporation (SRDC)-funded projects commenced in 1988 under the guidance of BSES senior research officer: plant breeding, Dr Mac Hogarth. Approximately 100 potential varieties and crosses were planted across a range of sites in Southern Queensland and northern New South Wales to determine whether selection in Southern Queensland could be adequate for New South Wales. In addition, scientists from the CSIRO Division of Tropical Crops and Pastures cooperated with Peter Nielsen, one of NSW Sugar Milling Coop.'s agricultural extension officers, in a series of experiments during 1991–1992 to identify the physiological basis of variation in frost resistance in a diverse range of sugar cane genotypes with a view to developing new, high yielding frost resistant varieties.[69]

After a decade of maintaining its own cane variety improvement program, a total review of the NSW Sugar Milling Coop.'s plant breeding activities was carried out by CSR personnel in late 1987. A major change adopted from 1988 onwards required that all the seedlings, presently shared by the three New South Wales mill supply areas, would be grown at Broadwater. A common selection program would then be developed for the canegrowers supplying Harwood and Broadwater Mills, using common varieties and pooling information for selection purposes. A separate program to develop cane varieties that matured in twelve months would be maintained at Condong Mill. Yields were to be determined by machine harvesting and weighing.[70] Despite these changes, the NSW Sugar Milling Coop. expressed concern in January 1991 about the performance of the new cane varieties from the breeding program relying upon seed supplied by CSR. Imported Co. and CP. cane varieties and NCo. 310 remained dominant in New South Wales, not locally-bred varieties (Figure 13.10). Negotiations commenced with CSR and CSIRO about a joint program for sugar cane plant breeding, but a satisfactory arrangement could not be reached. Eventually, an agreement was reached with the BSES which offered its technical services on a contract basis. The BSES assumed control of the variety testing program for New South Wales as of 1 July 1992, and implemented a selection program that aimed to identify suitable clones for one-year and two-year growth. Early results were promising. By 1995, several BSES-bred varieties such as Q. 124, Q. 135, Q. 155 and Q. 159 had been added to the approved lists of cane varieties for the New South Wales mills. Five years later, BSES-bred varieties made up almost 40% of the cane harvested in New South Wales.[71] This figure was less than 5% in 1990 (Figure 13.2).

New directions and reorganisation also characterised CSR's cane variety improvement program during the 1980s and early 1990s. As mentioned earlier, a key activity of the CSR cane variety improvement program during the 1980s was the supply of seed to the NSW Sugar Milling Coop. CSR also focused upon developing new cultivars for the Herbert River district, but this goal was expanded to include the Lower Burdekin district, following CSR's acquisition of Kalamia and Invicta Mills in 1975 and Pioneer and Inkerman Mills in 1987. Breeding new cane varieties suited to the Innisfail and Cairns districts ceased to be a priority following the closure of CSR's Goondi and Hambledon Mills in 1987 and 1991, respectively. Other investigations during the 1980s involved developing a repeatable test for screening sugarcane varieties and breeding material for waterlogging tolerance, genotype x environment effects on selection for rust resistance in cultivars and continued research into the use of wild canes in crosses with 'noble' canes or

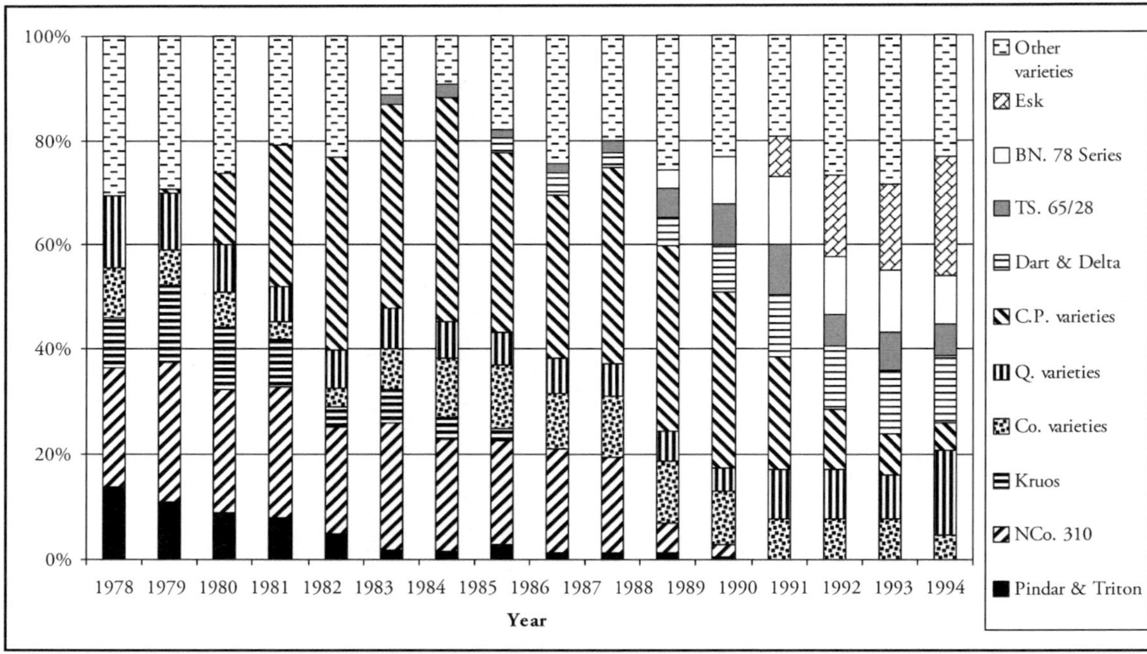

Notes: (1) No data available for 1980. (2) Co. varieties = Co. 740 & Co. 954. (3) C.P. varieties = C.P. 44–101 & CP 57–526. (4) Q. varieties = Mostly Q. 68 before 1993; then Q. 124 & Q. 135. (5) BN Series = Mostly BN 78–8031, BN 78–8032.

Figure 13.10. The percentage of area planted in New South Wales classified according to variety, 1978–1994

Source: Based upon figures in the *ASYB*, various issues, and annual reports of the Tweed River, Clarence River and Richmond River Branches of the New South Wales Cane Growers' Association.

commercial hybrids.[72] New minor commercial varieties that emerged from CSR's breeding work in the 1970s and early 1980s included Leda, Pelorus, Orpheus, Helius and Demos – all grown in CSR's Far North Queensland mill supply areas – and Eos which was released in the Lower Burdekin district in 1994. However, under 2% of the cane harvested in the sugar-producing districts north of Townsville during the early 1990s were new CSR-bred cane varieties – Herbert River district canegrowers preferred the BSES bred cane varieties, particularly Q. 124.[73]

By the late 1980s, CSR realised that its Queensland and New South Wales breeding programs had not been particularly successful in producing new cultivars that were adopted by many canegrowers. CSR's managers questioned the benefits of maintaining a separate cane variety improvement program and its activities were scrutinised in 1989 by staff from CSIRO's Division of Tropical Crops and Pastures. A year later an external review of the breeding program was conducted by scientists with plant breeding and crop science experience. Following these reviews, changes in the selection program for the Herbert River district were initiated in 1990 and 1991. Greater emphasis was placed on identifying high sucrose content during the early stages of selection by adopting family selection. Cane yields were estimated by mechanical harvesting and weighing, instead of manually obtaining samples of stalks at maturity. Evaluation of genetic material occurred across a number of sites throughout all stages of selection, as some genotypes may have been adapted to specific sites or environmental niches, providing opportunities to develop cultivars adapted to one or more niches. The Lower Burdekin district breeding program also moved to estimating yield from weighing harvested crops and the adoption of family selection in the early stages of evaluating seedlings. In addition, greater cooperation with the BSES occurred, with the release of Q. 164 taking place in 1995 after CSR conducted its final-stage selection trials.[74]

Throughout these changes, the BSES operated the largest and most sophisticated Australian cane variety improvement program, consisting of around fifteen staff. Between 1 000 and 1 500 bi-parental and factorial polycross cross-pollinations were completed each year at the Meringa BSES station, although 2 108 crossings were managed in 1984 – the second highest on record (Figure 13.3). Details on the number of seedlings planted annually by the BSES ceased to be published in 1989, but between 40 000 and 80 000 single original seedlings were planted every year during the 1980s. To cater for this expanded program of cane variety improvement, the BSES built two large steel-framed sheds in 1979 to house the 180 cross-pollination lanterns; another two cross pollination sheds were erected in 1981. A neighbouring cane farm was purchased in 1981 and the extra space devoted to new cultivar selection trials. Dr James Skinner guided all of these activities – as he had done since 1966 – until the end of 1988. He remained associated with the BSES, retiring on 6 July 1990. Following an administrative re-organisation of the BSES in early 1989, cane variety improvement became part of the Cane Improvement Group, headed by Dr Mac Hogarth.[75]

During the 1980s and early 1990s, the major objective of the BSES cane variety improvement program continued to be increasing the amount of sugar recovered per hectare. Subprograms within the cane variety improvement program, however, were devised to develop specific characteristics within the new cane varieties. Screening for resistance to Fiji disease using bunch planting that allowed more seedlings to be screened commenced in the 1970s and continued throughout the 1980s.[76] One of the newer sub-programs focused upon developing cultivars with high, early season sucrose content. Industry expansions during the 1980s put strains on milling capacity in some districts and research identified the potential to breed for high sucrose content in the May-June period as a way of extending the harvesting season-length.[77]

Following outbreaks of common rust (*Puccina melanocephala* H. & P. Sydow) in Queensland's crops of cane between 1978 and 1980, another sub-program was established to develop resistant clones. Dr Nils Berding at the Meringa BSES station and his colleagues developed a rapid test using potted plants to screen for rust resistance. Consequently, the degree of susceptibility of the cultivars under selection rapidly decreased and susceptibility to common rust had become a minor selection problem by the mid-1990s.[78] In 1981, the discovery of the soil-borne pathogen *Pachymetra chaunorhiza* Croft and Dick which attacks the root system of sugar cane and causes yield losses up to 40% in susceptible varieties led to investigations into the possibility of breeding resistant clones. Seedlings commenced being screened for resistance to the pathogen, but the screening procedure was expensive and time consuming, requiring the testing of clones in pots on temperature controlled benches. A compromise was reached whereby crosses between susceptible parents were avoided and all new cultivars in the advanced stages of selection received testing for resistance to the pathogen. More resistant cultivars had become available by the mid-1990s.[79]

Sparse, unreliable flowering and poor pollen fertility had hindered sugar cane improvement in Australia. During the 1980s, the BSES cane breeders tried improving the availability of arrows. Investigations in 1979 and 1980 discovered that sugar cane flowered much better at the Walkamin Research Station of the Queensland Department of Primary Industries, 35 km inland of Meringa. The

Queensland Department of Primary Industries then agreed to maintain a BSES sub-station at Walkamin, supplying arrows to the BSES cane variety improvement program. The construction of a new photoperiod house at the Meringa BSES station in early 1986 enabled the production of additional arrowed stalks. Flowers were obtained from some cane varieties which had never flowered naturally in Australia, while flowering intensity and pollen fertility improved in other cultivars which showed limited natural flowering. Irrigation of clones selected for breeding also improved the flowering intensity of some clones, but did not induce flowering of naturally non-flowering clones.[80]

Improved information management by the BSES during the late 1970s and 1980s allowed larger breeding populations to be maintained and used more efficiently. A crossing chart showing all proven male and female parents in all combinations had been developed manually during the early 1970s. In 1976, the BSES decided to extend the chart to include all parents in the collection, whether proven or not. Such a task would have been very difficult to prepare manually, so a system was developed to produce the chart using a computer. Refinements to the system occurred over the next two years, so that daily crossing charts printed by a computer were used during 1979 to decide upon the crosses made throughout the breeding season. Time saved in preparing the crossing program released staff for other field work associated with cross-pollination and allowed more crosses to be completed. After 1979, the computerised system was expanded to summarise the vast amount of agronomic, disease and breeding performance data available on each clone and it used the information to predict the likely performance of a cross between two parents. Following the introduction of computerised breeding records, the population of breeding clones increased from 600 to 2 000 by the mid-1990s, but professional input was reduced, although parental assessment and cross prediction had become more objective and reliable.[81]

Reducing the length of the BSES selection process also became a priority in the early 1980s and by the early 1990s the time taken to release a new variety had been reduced from fifteen to twelve years. This reduction had come through omitting one year of testing in both stages 3 and 4 of the selection process. In stage 3, potential new varieties started being tested at two locations in the plant crop rather than at one location as investigations found that genotype x location effects were far more important than genotype x crop year effects. The efficiency of the system also allowed more clones to be assessed. Throughout new stage 3 selection trials in the Lower Burdekin district during the late 1980s, for example, the number of new cultivars assessed rose from 600 to 3 000 per year. From 1984 onwards, greater emphasis was also placed on replicated farm trials, using substations on farms. Canegrowers agreed to accept more responsibility for the final testing of newly released cane varieties on their farms.[82] Backlogs of samples for sucrose determination in the juice laboratory had also slowed the selection process. Near infrared spectroscopy was evaluated in the late 1980s and early 1990s. This technology permitted the rapid determination of sugar and fibre content in fibrated cane stalks (i.e. cane stalks that had been shredded into threads) or sugar content in juice. It allowed larger numbers of samples to be processed more rapidly and permitted an improved assessment of sugar and fibre content during earlier stages of selection. From 1996 onwards, this method was used routinely for all clonal improvement trials processed at the Meringa BSES Station.[83]

The BSES cane variety improvement program produced a steady output of new Q. varieties in most years between 1978 and 1995. Sixty new Q. varieties were released to canegrowers during the 1980s and early 1990s (Table 13.3). Many of these Q. varieties never accounted for more than 5% of the cane harvested annually in Queensland. Q. 96, Q. 117 and Q. 124, however, did assume some prominence during the 1980s and early 1990s (Figure 13.11). The former had been approved for use in 1974. A vigorous cropper with exceptionally high sucrose content, Q. 96 was susceptible to leaf scald disease, and this feature may have slowed its initial acceptance amongst canegrowers.[84] Yet by 1981, Q. 96 accounted for 10% of the cane harvested in Queensland. It became particularly popular in the Lower Burdekin district and to a lesser extent the Mackay district. Moreover, Q. 96's usefulness was long lasting – a tenth of the crop harvested in Queensland in 1995 was still this variety (Figure 13.11). Q. 117 was released to canegrowers in 1981 and became a favourite variety of canegrowers in the sugar-producing districts north of Townsville, taking the place of Triton and the rust disease-susceptible Q. 90. Four years later, Q. 124 became available, and although moderately susceptible to Fiji disease, it was high in sucrose and produced strong ratoon crops. A decade later, Q. 124 accounted for almost 30% of the cane harvested in Queensland. It was the most favoured variety in the central district – 70% of the cane harvested in this region in 1995 was Q. 124 – and increasing numbers of Herbert River district canegrowers also showed a preference for Q. 124. The BSES, however, expressed concern about the Queensland sugar industry becoming too dependent on this one cane variety, although the organisation acknowledged that Q. 124 had contributed significantly to increased productivity in the Mackay and Herbert River districts. [85]

In the mid-1970s, 55 % of the cane harvested in Queensland came from BSES-bred varieties (Figure 13.4). The Fiji-disease susceptible South African

Table 13.3. New Q. varieties and their year of approval, 1978–1995

Year	New approved varieties	Year	New approved varieties
1978	Q. 103	1987	Q. 138; Q. 139; Q. 140; Q. 141; Q. 142; Q. 143
1979	Q. 104; Q. 105; Q. 106; Q. 107	1988	Q. 144; Q. 145
1980	Q. 108; Q. 109; Q. 110; Q. 111; Q.112; Q. 113; Q. 114; Q. 115	1989	Q. 146; Q. 147
1981	Q. 116; Q. 117; Q. 118	1990	Q. 149
1982	Nil	1991	Q. 150; Q. 151
1983	Q. 119; Q. 120; Q. 121	1992	Q. 152; Q. 153
1984	Q. 123	1993	Q. 154; Q. 155; Q. 156; Q. 157; Q. 158; Q. 159; Q. 160
1985	Q. 122; Q. 124; Q. 125; Q. 126; Q. 127; Q. 128; Q. 129; Q. 130; Q. 131; Q. 132	1994	Q. 161; Q. 162
1986	Q. 133; Q. 134; Q. 135; Q. 136	1995	Q. 163; Q. 164

Source: Based upon details in BSES, ARs 1978–79 to 1995–96.

variety NCo. 310 was the most prominent non-Australian cane variety. Following Fiji disease outbreaks in Southern Queensland during the late 1970s and Mackay district in the early 1980s, NCo. 310 was replaced in Southern Queensland by Q. 87 and Q. 110 and two American varieties – C.P. 44–101 and H. 56–752 – and in the Mackay district by H. 56–752. Such a reliance on American varieties, however, prompted the BSES cane breeders in the late 1980s to focus upon selecting locally-bred varieties that would replace them. In Southern Queensland, Q. 141, Q. 146 and Q. 151 took the place of the American varieties; Q. 124 and Q. 136 fulfilled this role in the Mackay district.[86] This trend, together with the absence of any new, widely adopted CSR-bred cane varieties, meant that BSES-bred cane varieties assumed even greater importance during the 1980s and early 1990s. In 1995, BSES-bred cane varieties accounted for 94% of the cane harvested in Queensland (Figure 13.4).

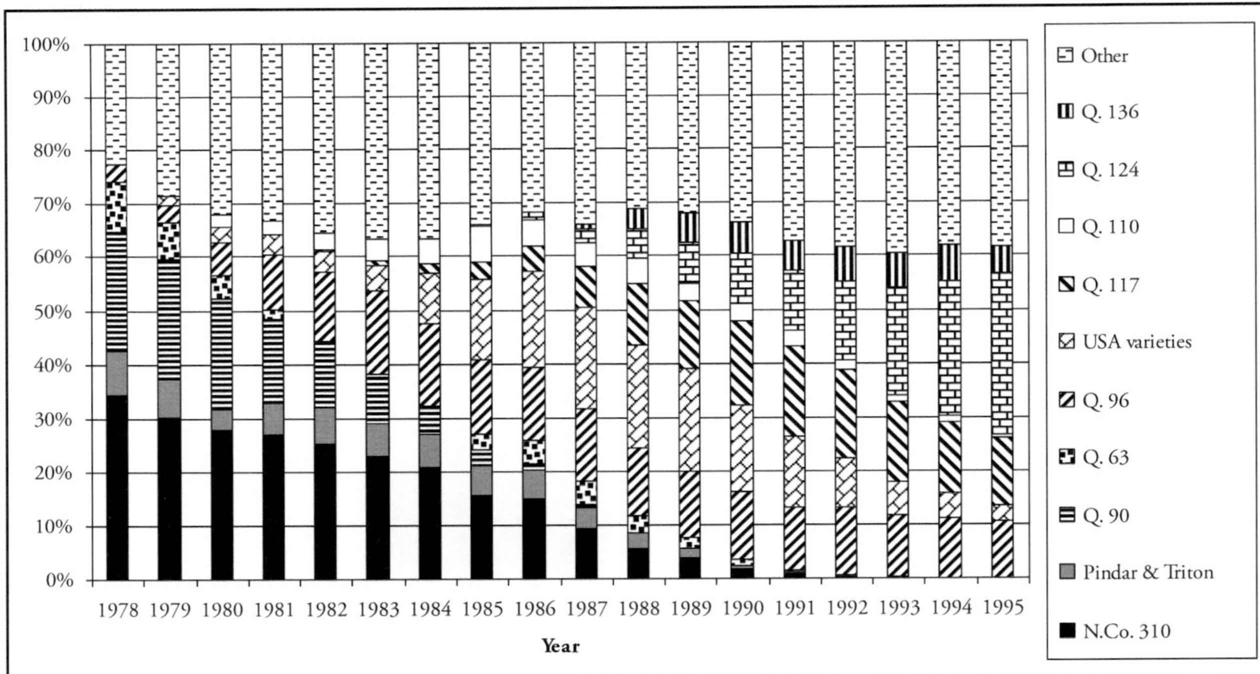

Notes: (1) USA varieties = H. 56–752 & C.P. 44–101.
(2) The 'Other' category contained numerous Q. varieties and CSR-bred varieties, each mostly having less than 3% of the amount of cane harvested annually.

Figure 13.11. The percentage of harvested cane in Queensland classified according to variety, 1978–1995.

Source: Based upon figures from the annual varietal census published yearly in the *ASYB*.

Benefits of the Australian cane variety improvement programs

Assessing the benefits of Australia's cane variety improvement programs since 1915 is not straight forward. Productivity improvements as measured by tonnes of cane per ha or tonnes of sugar per ha have been linked to the introduction of new cane varieties. A BSES study by Dr Nils Berding and Dr James Skinner in 1987, for example, showed that when Pindar was replaced by the new cultivar Q. 90, productivity increased by 31%. However, when common rust began attacking the cane crops in 1978, Q. 90 suffered a loss of productivity of 25% as measured against resistant clones in a range of trials. Q. 113, a rust-resistant cultivar, produced only 95% of Q. 90 in the absence of common rust, but was clearly more productive than Q. 90 in the presence of rust. BSES scientists M. Cox and P. Hansen analysed productivity data for the period 1966 to 1994 from the sugar mills in the Bundaberg and Mackay districts. They concluded that about two-thirds of the increase in tonnes of sugar/ha at Bundaberg was due to new varieties such as Q. 87, Q.110, Q. 146, C.P. 44–101 and H. 56–752, while Q. 124 was responsible for 40–50% of the increase in the Mackay district.[87] Yet yield increases are also influenced by improved plant nutrition and drainage, access to irrigation and a reduction in the deleterious impacts of cane diseases and pests. These factors also need to be considered when examining the productivity figures in Figures 13.12 and 13.13.

Productivity data for New South Wales during the period 1915–1919 to 1930–1934 was static (Figures 13.12 and 13.13). These years coincided with gumming disease outbreaks, poor drainage and the first wave of attacks from African black beetles (*Heteronychus arator*), which will be discussed in Chapter 14. A similar yield plateau was apparent in Queensland and can be partially attributed to outbreaks of gumming, Fiji and downy mildew diseases (see Chapter 15), poor drainage, limited irrigation and the negative impacts of attacks by cane grubs, weevil borers and rats (see Chapter 14). Productivity figures for Queensland and to a lesser extent New South Wales then showed a general upward trend until the period 1970–1974. This increase occurred at the same time as the introduction of new cane varieties such as Vesta, Pindar and Triton in New South Wales and Trojan, Pindar, Q. 50, Q. 57, Q. 58 and NCo. 310 in Queensland. Clearly, this improvement was not all related to new cane varieties, as it also coincided with the introduction of insecticides which reduced the impact of insect pests, especially the cane grub (see Chapter 14), increased

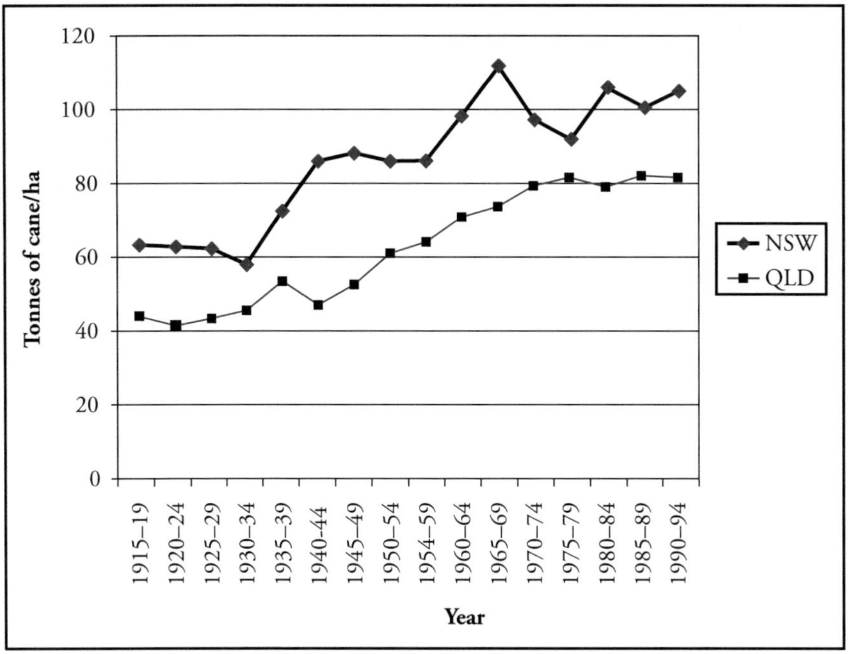

Figure 13.12. Average tonnes of cane per hectare in New South Wales and Queensland for five year periods, 1915/1919–1990/1994.

Source: Calculated from yield figures in the *Year Book of Australia*, various issues.

fertiliser use, and improved drainage and access to irrigation water. However, Dr Mac Hogarth calculated that improved yields (tonnes/ha) of 1.9% per annum occurred in Queensland from 1948 to 1975, and that plant breeding was likely to have contributed half of this increase.[88]

Yields in Queensland and New South Wales plateau in the 1970s and 1980s, despite at least eighty new cane varieties being released by Australian cane breeders. On the basis of productivity increases of 1.0% per year estimated by the cane breeders, this plateau should have been a productivity increase, perhaps in the order of 15–20%. Measured yield improvements in cane variety evaluation and research plots had not been transferred to the commercial cane fields. Industry leaders in the early 1990s suggested that the farming system or soil-related factors may have imposed a limitation on potential productivity increases through genetic improvement. A Sugar Yield Decline Joint Venture (SYDJV) involving the Sugar Research and Development Corporation, BSES, the CSIRO Division of Soils and Queensland Department of Primary Industries was established in 1993 to research the

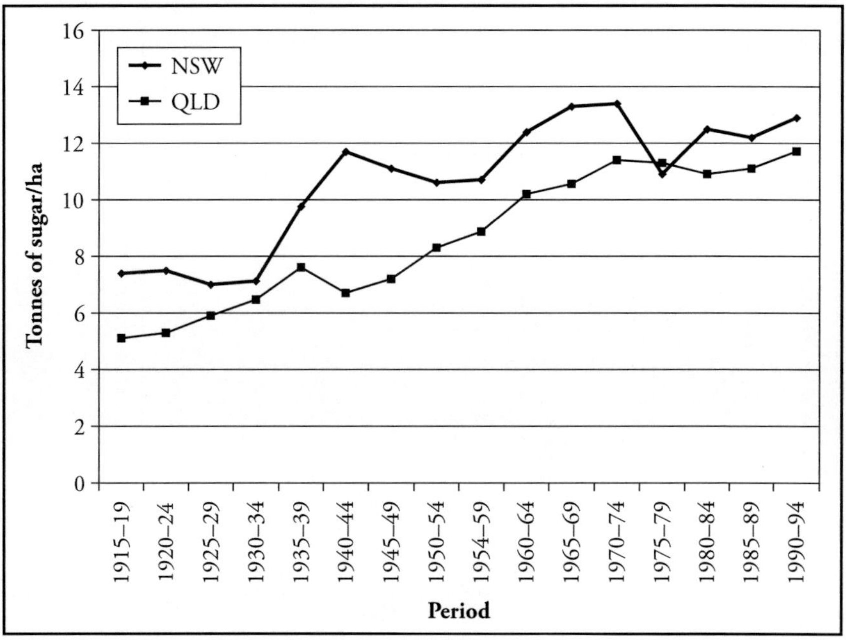

Figure 13.13. Average tonnes of sugar per hectare in New South Wales and Queensland for five year periods, 1915/1919–1990/1994.

Source: As in Figure 13.13

causes of this productivity plateau. When the joint venture commenced, the organisations involved believed that the apparent diminishing ability of cane growing soils to produce sugar was related to the direct effects of long-term sugar cane monoculture and heavy machinery use. Initial results from early SYDJV investigations into paired old and new land sites suggested that crop growth, as measured by stalk development, was reduced on old land. In general old land was more acidic, had lower amounts of zinc and copper, lower infiltration and water holding capacity and more root pathogens. Interestingly, tonnes of sugar per ha increased slightly again in Queensland between 1990 and 1994 (Figure 13. 13), possibly because of cane cultivation spreading onto better quality soils and/or irrigated areas in the Burdekin River Irrigation Area, the Murray River area (near Tully) and Atherton Tableland and the benefits of green cane trash blanketing.[89]

Dr Alan Garside and his BSES colleagues concluded that gauging the impact of new varieties *per se* on the productivity plateau was difficult as genetic improvement in a stable environment could not be measured. The impact of Fiji disease, common rust and *Pachymetra* root rot created important diversions in the realisation of genetic gain and had resulted in potentially less productive but more resistant cane varieties being grown during the 1970s and 1980s. However, it is probable that the yield plateau since 1970 may well have been a yield decline but for the influence of the many new cane varieties released by the plant breeders. Cane variety improvement had once again proved to be vital to the Australian sugar industry's survival and prosperity, as it had been in the 1920s and 1930s when the canegrowers were struggling against the threat from gumming disease. However, Dr Alan Garside and his BSES colleagues posed a very important question: how long can a continually degrading system be maintained through genetic manipulation?[90] In the mid-1990s, Australia's canegrowers were just starting to grapple with that issue.

Notes

1 *ASJ*, 9 January 1913, pp. 752–753; BSES, AR 1914–1915, *QPP* Vol. 2 (1915–1916), p. 1134; J. Buzacott and C. Hughes, 'The 1951 cane collecting expedition to New Guinea', *CGQB*, Vol. 25, no. 2 (October 1951), pp. 39–40.

2 P. Cramer, 'Sugar-cane breeding in Java', *Economic Botany*, Vol. 6 (1952), pp. 143–1950; J. Galloway, 'Botany in the service of Empire: The Barbados cane-breeding program and the revival of the Caribbean sugar industry, 1880s–1930s', *Annals of the Association of American Geographers*, Vol. 86, no. 4 (1996), pp. 682–706.

3 H. Easterby, 'Varieties of Sugar Cane in Queensland'. *BSES General Series Bulletin No. 2* (revised 2nd edition, 1926), pp. 12–14.

4 Easterby, *The Queensland Sugar Industry*, p. 209; BSES, AR 1921–22, *QPP*, Vol. 2 (1923), p. 281.

5 BSES, AR 1922–1923, *QPP* Vol. 2 (1924), p. 339; H. Easterby, 'Varieties of Sugar Cane in Queensland', p. 15.

6 BSES, AR 1927–28, *QPP*, Vol. 2 (1929), pp. 922–923; J. Buzacott, 'A short history of Bureau cane breeding', in BSES, *Fifty Years of Scientific Progress* (Brisbane, 1950), pp. 62–63.

7 BSES, AR 1926–27, *QPP*, Vol. 2 (1927), p. 797–798; BSES, AR 1927–28, *QPP*, Vol. 2 (1929), pp. 922–923; King, Mungomery and Hughes, *Manual of Cane-Growing*, 1953 edition, p. 253.

8 BSES, AR 1939–40, p. 16; A. Bell, 'The resistance of the seedling S.J. 4 to grub damage in North Queensland', *CGQB*, Vol. 2, no. 1 (July 1934), pp. 9–10; G. Bates, 'Varietal trials in North Queensland', *CGQB*, Vol. 2, no. 2 (October 1934), p. 40; 'Percentage of leading varieties crushed at each CSR mill', CSR, Cane Reports 1930 to 1935, CSRR, Z/109, Boxes 385 & 386.

9 See varietal statistics in Mulgrave Central Mill Co. Ltd., ARs, 1924–1930, ML; H. Easterby, 'The Cultivation of Sugar Cane in Queensland'. *BSES General Series Bulletin No. 3,*

(revised 2nd edition, 1924), pp. 40–42; 'Percentage of leading varieties crushed at each CSR mill' in CSR, Cane Reports 1920 to 1935, CSRR, Z/109, Boxes 385 & 386.

10 A. Bell, 'Variation of varietal constitution of the cane crops of the Mackay and Bundaberg Districts', *Proc. QSSCT,* 1936, pp. 133–136.

11 Buzacott and Hughes, 'The 1951 cane collecting expedition to New Guinea', pp. 39–41; J. Daniels, J. Ashley and C. Daniels, 'The 1914 sugarcane collecting expedition to New Guinea', *Sugarcane Breeders' Newsletter*, No. 43 (1980), pp. 24–38.

12 E. Ferguson Wood, 'Cane varieties grown in Queensland. Their resistance to disease', *QAJ*, Vol. 29, no. 4 (1 April 1928), p. 264 and 'Percentage of leading varieties crushed at each CSR mill' in CSR, Cane Reports 1925 to 1935, CSRR, Z/109, Boxes 385 & 386.

13 *ASJ,* 17 February 1958, p. 835.

14 D. North to W. Rothe, CSR Gen. Mang., 6 December 1923, Letter No. 120D, p. 2, CSRR, Z/109, Box 239; *ASJ*, 5 June 1930, p. 160; D. North, 'Sugarcane improvement work in New South Wales', *Proc. ISSCT*, 1938, pp. 80–81.

15 CSR, Cane Report 1927, p. 17; CSR, Cane Report 1930, pp. 40–41; & CSR, Cane Report 1931, p. 30, all in CSRR, Z/109, Box 386.

16 CSR, Cane Report 1926, pp. 18–19 & CSR, Cane Report 1927, p. 19, CSRR, Z/109, Box 386; K. Gard, 'Twenty-five years of sugar-cane breeding at Macknade', *Proc. ISSCT*, 1950, pp. 63–64 & 69.

17 CSR, Cane Report 1928, pp. 23–24; CSR, Cane Report 1930, p. 35; & CSR, Cane Report 1932, p. 32, all in CSRR, Z/109, Box 386; Gard, 'Twenty-five years of sugar-cane breeding at Macknade', pp. 63–64 & 69.

18 Gard, 'Twenty-five years of sugar-cane breeding at Macknade', *Proc. ISSCT*, 1950, pp. 64–65; H. Clarke, 'The methods of selecting seedlings practiced by the Colonial Sugar Refining Company', *Proc. ISSCT,* 1935, pp. 341–342; D. North, 'Sugarcane improvement work in New South Wales', *Proc. ISSCT*, 1938, pp. 80–81.

19 D. North, 'Obtaining resistant cane varieties for the control of gumming disease', *Proc. ISSCT*, 1932, Bulletin No. 30, p. 2.

20 New South Wales Department of Agriculture, *The Farmers' Handbook*, 5th edition (Sydney, 1933), pp. 624–625; A. Haywood, 'The culture of sugar cane in New South Wales: the question of varieties', *Agricultural Gazette of NSW*, Vol. 32, no. 3 (March 1921), pp. 181–182.

21 D. North to H. Dixon, CSR Gen. Mang, 9 December 1927, Letter No. 199D, p. 7, CSRR, Z/109, Box 240; D. North to P. Goldfinch, CSR Gen. Mang., 13 December 1929, Letter No. 261D, p.1, CSRR, Z/109, Box 240.

22 BSES, AR 1928–29, *QPP,* Vol. 2 1929, p. 944; G. Bates, 'Varietal trials in North Queensland', *CGQB,* Vol. 2, no. 2 (October 1934), p. 40; P.O.J. 2878 was produced in Java in 1921 and subsequently exported worldwide. By 1929, 95% of the crop harvested in Java was this one variety. See C. Edgerton, *Sugarcane and Its Diseases* (Baton Rouge, 1955) p. 46.

23 Buzacott, 'A short history of cane breeding', p. 63; BSES, AR 1929–30, *QPP*, Vol. 2 (1930), pp. 921, 924 & 935; BSES, AR 1931–32, *QPP*, Vol. 2 (1932), p. 577.

24 BSES, AR 1930–31, *QPP,* Vol. 2 (1931), p. 684; Buzacott, 'A short history of cane breeding', p. 63; A. Bell, 'The breeding of new varieties of sugar-cane', *CGQB*, Vol. 3, no. 4 (1936), pp. 135–137.

25 BSES, AR 1935–36, p. 11; BSES, AR 1936–37, p. 32.

26 Buzacott, 'A short history of cane breeding', p. 64; *ASYB*, 1944–45, p. 279; N. King, 'Seedling raising at Bundaberg and some notes on Q. 25', *CGQB*, Vol. 8, no. 2 (July 1940), p. 29; C. Hughes, 'The importation of sugar cane into Queensland', *Proc. ISSCT*, 1959, p. 1030.

27 BSES, AR 1942–43, p. 16; BSES, AR 1943–44, p. 19.

28 BSES, AR 1937–38, p. 36; BSES, AR 1944–45, p. 16; C. Hughes, 'The production and testing of sugar-cane seedlings', *The Qld. J. of Agricultural Science*, Vol. 6, no. 1 (March 1949), pp. 19–21.

29 BSES, AR 1938–39, pp. 40–44; BSES, AR 1940–41, p.16–18; BSES, AR 1941–42, p. 10; Hughes, 'The production and testing of sugar-cane seedlings', pp. 24–27; Buzacott, 'A short history of cane breeding', pp. 64–65.

30 BSES, AR 1936–37, p. 33; BSES , AR 1937–38, p. 37; BSES, AR 1938–39, p. 41; BSES, AR 1939–40, p. 14; BSES, AR 1940–41, p. 17; Hughes, 'The production and testing of sugar-cane seedlings', pp. 47–49.

31 D. North to P. Goldfinch, CSR Gen. Mang., 6 April 1939, Letter No. 534D, pp. 1–2, CSRR, Z/109. Box 242; King, 'Seedling raising at Bundaberg and some notes on Q. 25', p. 29.

32 King, Mungomery and Hughes, *Manual of Cane-Growing*, 1953 edition, p. 246.

33 King, Mungomery and Hughes, *Manual of Cane-Growing*, 1953 edition, p. 254.

34 Anon, 'Trojan', CSR, *Newsletter* No. 6, May 1946, p. 5, CSRR, Z/109, Box 371; King, Mungomery and Hughes, *Manual of Cane-Growing*, 1953 edition, pp. 248 & 254.

35 C. Hughes, 'The variety Q. 28', *CGQB*, Vol. 18, no. 2 (October 1954), pp. 46–47; King, Mungomery and Hughes, *Manual of Cane-Growing*, 1953 edition, pp. 244, 250 & 252; Kerr, *Southern Sugar Saga*, p. 123; C. Toohey, 'The displacement of C.P. 26–116', *CGQB*, Vol. 26, no. 2 (October 1962), pp. 51–52.

36 Anon, 'Mr J.H. (Jim) Buzacott', *CGQB*, Vol. 34, 1 (July 1970), p. 4.

37 BSES, AR 1952–53, p. 29.

38 BSES, AR 1955–56, pp. 56–57; BSES, AR 1960–61, p. 44; N. King, 'New seedling sub-station at Innisfail', *CGQB*, Vol. 19, no. 3 (January 1956), pp. 86–87.

39 BSES, AR 1948–49, p. 31; BSES, AR 1949–50, p. 32; BSES, AR 1959–60, p. 61; J. Buzacott, 'Recent trends in the production of cane seedlings in Queensland by the Bureau of Sugar Experiment Stations', *Proc. ISSCT*, 1950, p. 94; J. Buzacott, 'The parental performance of some sugarcane varieties in Queensland', *Proc. ISSCT*, 1953, pp. 490–491; N. King, 'Cane varieties as a solution to a harvesting problem', *Proc. QSSCT*, 1955, pp. 137–140.

40 J. Buzacott and C.G. Hughes, 'Cane collecting in New Guinea in 1951', *Proc. ISSCT*, 1953, pp. 521–528.; BSES, AR 1950–51, pp. 36–37.

41 BSES, AR 1951–52, p. 33.

42 C. Hughes, 'New Guinea once more in the news', *CGQB*, Vol. 20, no. 4 (April 1957), pp. 130–131; B. Egan, 'Quarantine measures for the 1957 N.G. varieties', *CGQB*, Vol. 22, no. 4 (April 1959), pp. 112–113.

43 J. Buzacott, 'Methods of cane breeding used by the Queensland Bureau of Sugar Experiment Stations', *Proc. ISSCT*, 1959, pp. 690–693; BSES, AR 1950–51, p. 33; BSES, AR 1952–53, pp. 29–30; BSES , AR 1953–54, pp. 49–50; BSES, AR 1955–56, p. 58–59.

44 BSES, AR 1954–55, pp. 50–51; J. Skinner, 'Grading varieties for selection', *Proc. ISSCT*, 1965, pp. 938–939.

45 J. Buzacott, 'Progress in the development of new varieties', *CGQB*, Vol. 24, no. 1 (July 1960), pp. 7–8; BSES, *75 Years of Scientific Progress. Bureau of Sugar Experiment Stations 1900–1975* (Brisbane, 1975), pp. 36–37; BSES, AR 1956–57, pp. 59–60; BSES, AR 1959–60, p. 60; J. Skinner, 'Sugarcane breeding – in retrospect and prospect', *Proc. ASSCT*, 1989, p. 55.

46 Buzacott, 'Methods of cane breeding used by the Queensland Bureau of Sugar Experiment Stations', p. 692; BSES, AR 1958–59, p. 8; King, Mungomery and Hughes, *Manual of Cane-Growing*, revised edition, 1965, pp. 244–245.

47 Gard, 'Twenty-five years of sugar-cane breeding at Macknade', p. 68; CSR, Cane Report 1945, p. 14, CSRR, Z/109, Box 387.

48 For figures on numbers of cross-pollinations and seedlings raised see CSR, Cane Reports 1946 to 1959, in CSRR, Z/109, Boxes 387 to 389.

49 CSR, Cane Report 1949, p. 21, CSRR, Z/109, Box 387; CSR, Cane Report 1952, pp. 33 & 39 & CSR, Cane Report 1957, p. 12, CSRR, Z/109, Box 388.

50 *ASJ*, June 1960, p. 20; 'Table V: Percentage of leading varieties crushed at each CSR mill, 1959', CSR, Cane Report 1959, CSRR, Z/109, Box 389.

51 BSES, AR 1965–66, p. 41; BSES, AR 1964–65, p. 39; BSES, AR 1974–75, pp. 44–45.

52 BSES, AR 1963–64, p. 49; for Hogarth's early work see D. M. Hogarth, *Sugar cane selection experiments. 6. Factors influencing selection in original seedlings*. BSES Technical Communications, 1971, no. 1 and D. M. Hogarth, 'Methods of selection and estimation of genetic variances in populations of sugar cane', PhD thesis, University of Queensland, 1973.

53 BSES, AR 1959–60, pp. 56–57; BSES, AR 1960–61, p. 45.

54 BSES, *75 Years of Scientific Progress. Bureau of Sugar Experiment Stations* 1900–1975, p. 39; BSES, AR 1967–68, p. 41; BSES, AR 1969–70, p. 34; BSES, AR 1970–71, p. 34; BSES, AR 1971–72, p. 55; BSES, AR 1973–74, p. 42; BSES, AR 1978–79, p. 37.

55 J. Skinner, 'Selection in sugarcane: a review', *Proc. ISSCT*, 1977, p. 151; BSES, AR 1964–65, p. 39; BSES, AR 1966–67, p. 38; BSES, AR 1967–68, pp. 42 & 44.

56 BSES, *75 Years of Scientific Progress. Bureau of Sugar Experiment Stations 1900–1975*, p. 39; BSES, AR 1963–64, p. 47; BSES, AR 1966–67, p. 38.

57 J. Buzacott, 'Where are the New Guinea canes?' *CGQB*, Vol. 27, no. 4 (April 1964), p. 136; BSES, *75 Years of Scientific Progress. Bureau of Sugar Experiment Stations 1900–1975*, p. 37; BSES, AR 1965–66, p. 44.

58 BSES, AR 1961–62, pp. 40–41; BSES, AR 1977–78, pp. 35–36.

59 J. Buzacott, 'Progress in the development of new varieties', *CGQB*, Vol. 24, no. 1 (July 1960), p. 7.

60 King, Mungomery and Hughes, *Manual of Cane-Growing*, revised 1965 edition, p. 243; BSES, AR 1975–76, p. 17.

61 D. Miller, 'Sutherland, Keith Leonard (1916–1980)', *ADB*, Vol. 16 (2002), pp. 347–348; CSR, *Newsletter* No. 59, June 1960, p. 3; *ASYB*, 1961, pp. 127–128; *ASYB*, 1963, p. 109.

62 CSR, *Newsletter* No. 68, December 1961, p. 25; CSR, Cane Report 1963. Part II, p. i, CSRR, Z/109, Box 390; CSR, Agricultural Circular: New South Wales, 1969, pp. 8–9.

63 G. Shannon, 'Brian Roach – A lifetime of research', *Australian Sugarcane*, Vol. 8, no. 1 (April–May 2004), pp. 19.

64 CSR, Cane Report 1962, Part II, p. i & CSR, Cane Report 1963. Part II, p. i, CSRR, Z/109, Box 390.

65 Shannon, 'Brian Roach – A lifetime of research', pp. 19–20; N. Berding and B. Roach, 'Germplasm collection, maintenance and use', in D. Heinz (ed.), *Sugarcane Improvement through Breeding* (Amsterdam, 1987), p. 184; CSR, AR Technical Field Work 1969–70, p.i, CSRR, Z/109, Box 392; R. Sloan and R. Farquhar, 'The effects of frosting on varieties of sugar cane with commercial potential in New South Wales', *Proc. QSSCT*, 1978, pp. 293–296.

66 See the varietal census for Queensland during the late 1970s and 1980s, published annually in the *ASYB*.

67 Briody, *Secrets of Success*, pp. 95–96.

68 For cane variety breeding statistics see the ARs of the NSW Sugar Milling Coop. Ltd., held by the Richmond River Branch of the NSWCGA, Wardell, NSW.

69 Briody, *Secrets of Success*, pp. 98–100; BSES, AR 1993–94, p. 11; K. Weaich, M. Ludlow and P. Nielsen, 'Identification of trails and germplasm to improve sugarcane resistance to frost damage', *Proc. ASSCT*, 1993, p. 256–260.

70 Clarence River Branch of the NSWCGA, AR 1987–88, p. 10 & Clarence River Branch of the NSWCGA, AR 1989–90, p. 9.

71 Briody, *Secrets of Success*, pp. 102–104; BSES, AR 1992–93, p. 10; NSW Sugar Milling Coop. Ltd., AR 1994–95, p. 13; NSW Sugar Milling Coop. Ltd., AR 1995–96, p. 12.

72 P. Jackson *et al.*, 'The CSR sugarcane breeding program – future directions and strategies', *Proc. ASSCT*, 1992, pp. 123–129; B. Roach and R. Mullins, 'Testing for waterlogging tolerance', *Proc. ASSCT*, 1985, pp. 95–101; W. Symington, 'Commercial potential of Macknade nobilisations for yield, sugar content and stress tolerance', *Proc. ASSCT*, 1989, pp. 48–53; W. Symington, 'Genotype x environment effects on selection for rust resistance in sugarcane', *Proc. ASSCT*, 1987, pp. 99–103.

73 BSES, AR 1994–95, p. 8; see also the varietal censuses for Queensland during the late 1980s and early 1990s, published annually in the *ASYB*.

74 Jackson *et al.*, 'The CSR sugarcane breeding program – future directions and strategies', p. 123; BSES, AR 1995–96, p. 37.

75 BSES, AR 1978–79, p. 36; BSES, AR 1981–82, p. 61.

76 BSES, AR 1980–81, p. 39; D. Hogarth *et. al.*, 'Quantitative inheritance of Fiji disease resistance in sugarcane', *Field Crops Research*, Vol. 34 (1993), pp. 175–176.

77 BSES, AR 1989–90, p. 8; M. Cox, D. Hogarth and R. Mullins, 'Clonal evaluation of early sugar content', *Proc. ASSCT*, 1990, pp. 90–98; M. Cox, D. Hogarth and P. Hansen, 'Breeding for high early season sugar content in a sugarcane (*Saccharum spp.* hybrids) improvement program', *Australian J. of Agricultural Research*, Vol. 45 (1994), pp. 1569–1575; M. Cox, M. Hogarth and G. Smith, 'Cane breeding and improvement', in M. Hogarth and P. Allsopp (eds.), *Manual of Cane Growing* (Brisbane, 2000), p. 102.

78 BSES, AR 1980–81, p. 43; BSES, AR 1983–84, p. 43; BSES, AR 1984–85, p. 9; D. Hogarth, M. Cox and J. Bull, 'Sugarcane improvement: past achievements and future prospects', in M. S. Kang (ed.), *Crop Improvement for the 21st century* (Trivandram, India, 1997), p. 33.

79 BSES, AR 1988–89, p. 24; BSES, AR 1989–90, p. 18; Hogarth, Cox and Bull, 'Sugarcane improvement: past achievements and future prospects', p. 34; B. Croft and R. Magery, 'A review of research into *Pachymetra* root rot, an important new fungal disease of sugarcane', *Proc. ISSCT*, 1989, p. 691.

80 BSES, AR 1980–81, p. 41; BSES, AR 1985–86, p. 9; D. Hogarth and R. Mullins, 'Changes in the BSES plant improvement program', *Proc. ISSCT*, 1989, p. 958.

81 BSES, AR 1975–76, p. 41; BSES, AR 1978–79, p. 36; Hogarth and Mullins, 'Changes in the BSES plant improvement program', p. 957; D. Hogarth and J. Skinner, 'Computerisation of cane breeding records', *Proc. ISSCT*, 1986, pp. 478–491.

82 Cox and Hogarth, 'Progress and changes in the south Queensland variety selection program' p. 253; Hogarth and Mullins, 'Changes in the BSES plant improvement program', p. 960; BSES, AR 1984–85, pp. 4 & 8.

83 BSES, AR 1987–88, p. 11; BSES, AR 1989–90, p. 9; N. Berding and J. Bull, 'Australian sugarcane improvement: current concerns', *Proc. ASSCT*, 1997, p. 93.

84 BSES, AR 1978–79, p. 13.

85 BSES, AR 1981–82, p. 15; BSES, AR 1984–85, p. 8; BSES, AR 1992–93, p. 9; BSES, AR 1995–96, p. 37.

86 BSES, AR 1989–90, p. 8; see also the Queensland varietal censuses during the late 1980s and early 1990s, published annually in the *ASYB*.

87 M. Cox and P. Hansen, 'Productivity trends in southern and central regions and the impact of new varieties', *Proc. ASSCT*, 1995, p. 1–7; Hogarth, Cox and Bull, 'Sugarcane improvement: past achievements and future prospects', p. 35.

88 D. Hogarth, 'New varieties lift sugar production', *The Producers' Review*, Vol. 66, no. 10 (1976), pp. 21–22.

89 Garside *et al.*, 'The yield plateau in the Australian sugar industry: 1970–1990', pp. 113 & 115.

90 Garside *et al.*, 'The yield plateau in the Australian sugar industry: 1970–1990', pp. 113 & 117.

CHAPTER 14
Combatting Pests

In May 1914, CSR's Head Office arranged for a curious delivery to its technical field staff at Goondi Mill – a cage full of tachinid flies (*Ceromasia sphenophori* Vill) from its Fijian plantations. The tachinid fly preys upon the beetle borer or sugar cane weevil borer, sometimes called the New Guinea borer (*Rhabdoscelus obscurus* Boisduval). By the mid-1910s it had become established in most sugar-producing districts from Mackay northwards following its probable introduction with the New Guinea cane varieties brought to Queensland during the mid-1890s, despite precautions being taken to avoid bringing any insect pests into the colony. This insect is a pest because the young larvae or borer grubs eat the soft internal tissues of the cane stalks, frequently making tunnels 45 cm or more in length, which are filled with waste material resembling sawdust. Weevil borers are strongly attracted to injured and broken cane which is caused by disease, especially red stripe (top rot), attacks from rats, flood and cyclone damage or lodging of the crop. Pre-World War II cane crops in North Queensland were vulnerable to weevil borer attacks because they were often lodged, especially in cane grub affected areas, and damaged by rats. The succulent cane varieties with soft rinds such as H.Q. 426, Badila and Goru which were widely grown in North Queensland by the early 1910s also provided ideal canes in which the borer could breed in large numbers.[1]

The tachinid flies were released at Daradgee, just north of Innisfail. CSR hoped they would prey on the weevil borer grubs, thereby reducing their depredations which were troubling the Goondi Mill suppliers. As an interim measure until the flies became sufficiently established to make an impact, CSR encouraged its canegrowers to trap the weevil borers, although this activity never became very popular.[2] This first attempt at establishing a population of tachinid flies in the Innisfail district failed, and CSR's technical field officers concluded that the fly could not survive in Australia. However, Edmund Jarvis, one of the BSES entomologists, discovered in 1918 that the tachinid flies flourished in the Mossman district. Jarvis suggested that the tachinid flies were the offspring of those flies that had escaped from the breeding cages established in 1910 by Mr F. Muir, the entomologist of the Hawaiian Sugar Planters' Association, who had brought the flies to a breeding depot at

Mossman. Here they were propagated in large numbers and transported to Fiji and Hawaii in an attempt to reduce the damage being done to the sugar cane by the weevil borer in those two countries. Jarvis realised that CSR's failure to establish a population of tachinid flies in the Innisfail district was not a result of the flies being unsuited to Australian conditions.[3]

Initially, Jarvis caught tachinid flies at Mossman and released them into other northern sugar-producing districts – several colonies of the parasitic fly had been established in the Babinda district by 1920. From 1920 onwards, Jarvis also bred the tachinid flies at the Meringa BSES station, then the main site of the BSES's entomological program. Over the next decade he supplied them to interested North Queensland canegrowers who tried to establish populations of the flies on their properties. Jarvis urged canegrowers to set traps for the weevil borers or to collect them from by inspecting the leaves on the cane plants, although his advice appears to have been unheeded.[4] CSR also commenced breeding the tachinid fly at its Macknade Experiment Station in 1922, after seeking advice from the Hawaiian entomologist F. Muir, who spent three months at Macknade in 1921. Yet CSR's officers struggled during the 1920s to successfully establish large numbers of the fly in the Innisfail and Herbert River districts. Moreover, CSR's Macknade Mill technical field officer complained in 1922 that their efforts were hindered by 'the apathy and ignorance of the farmer in regard to the borer'. He continued encountering canegrowers planting borer-infected setts, thereby allowing populations of borers to survive throughout the district. Nevertheless, damage from the weevil borer had been reduced in North Queensland mill supply areas by 1930 as a result of the liberation of the tachinid flies, but not eliminated entirely.[5] Figures collected for CSR's mill supply areas during the late 1920s showed that 5–10% of the stalks crushed at the sugar mills had been damaged by weevil borer.[6]

Weevil borer depredations may have been reduced by 1930, but damage to the crops of sugar cane from cane grubs remained a major concern during the late 1910s and 1920s. CSR, for example, estimated that its four North Queensland mill supply areas had lost approximately 63 000 tonnes of sugar cane to grub attack between 1915 and 1919 – not far short of an average season's crushing for one sugar mill. At least another 50 600 tonnes of cane were lost over the 1920s, with half of this loss occurring in the Goondi Mill supply area where cane grub attacks were particularly severe during the decade. Reports from BSES entomologists after visits to the Mackay, Bundaberg and Isis districts in 1925 described how entire fields of cane had been 'eaten out' by the cane grubs, which were responsible for 'large losses'.[7]

Such losses in the late 1910s stemmed from a combination of North Queensland canegrowers loosing interest in controlling cane grub attacks and two dry seasons (i.e. 1915 and 1919) which reduced beetle numbers naturally. The hand collection of cane beetles and cane grubs – used so widely during the late 1900s – had waned by 1920. Disagreement over its effectiveness between BSES entomologists Edmund Jarvis – a consistent supporter of the efficacy of hand collection – and Dr James Illingworth, who believed drought played a more crucial role in controlling beetle numbers, no doubt contributed to the ambivalence amongst the local pest destruction committees towards funding such activities.[8] Soil fumigation with carbon bisulphide, another control measure which had gained a small following in the early 1910s, was according to Edmund Jarvis still not widely used in 1921. All the reasons outlined in Chapter 8 for canegrowers disliking soil fumigation still persisted. Jarvis observed in 1921 that many canegrowers had not achieved positive results from using carbon bisulphide as they persisted in applying the chemical at the wrong time (i.e. during the wet season) and/or had used incorrect dosages.

Edmund Jarvis, who had worked as Henry Tryon's assistant since 1908, assumed the position of Acting Entomologist at the BSES's building in Gordonvale in September 1914. Jarvis shifted the focus of the entomological research from understanding the cane beetle's life-cycle to the 'very urgent question of remedial measures', an area neglected by his predecessor Alexandre Girault. Jarvis initiated both laboratory and field experiments, testing the insecticidal effect of various substances on the cane beetles and their grubs and the impact of these chemicals on the young and growing cane plant. Substances tested included creolin, borax, cyanide of potassium, copper and lead arsenic, barium chloride, creosote and paradichlorbenzene. All substances tested, except barium chloride, killed the cane beetles and grubs. None interfered with the growth of the cane plant. Borax proved very effective but was far too expensive for widespread application. Paradichlorbenzene was also a particularly efficient insecticide, as its odour remained in the ground up to three weeks after the original crystals had decayed, causing the cane grubs to avoid any treated area. The chemical was not known to be poisonous or flammable and did not dissolve in water, meaning it would be ideal for application during the wet summer months. However, using this substance would be expensive, costing farmers 30s to treat one acre (0.45 ha), and more than one treatment might have been needed during the growing season. Moreover, the chemical could not be immediately procured due to the commencement of World War I.

Dr James Illingworth, an American entomologist with a PhD from Cornell University, took control of the entomological investigations at the newly established Meringa BSES Station on 1 June 1917. His initial contract was for three years, but it was extended in mid-1920 for another twelve months. Illingworth appeared an ideal choice for the position of BSES entomologist. He had gained practical experience assisting orchard owners combat insect pests and had been studying the insect pests affecting the Hawaiian sugar industry. Illingworth commenced more extensive field experiments into ways of controlling the cane grubs. Over the next three years, 4 ha trial plots at Meringa were subjected to tests to determine if lime, commercial fertilisers, arsenic mixed with bagasse or creosote eased the impact of the cane grub attacks. Similar experiments, but over entire paddocks, were conducted at Green Hill Estate, a nearby CSR-owned property that suffered greatly from grub attacks.

By late 1920, Illingworth had concluded that none of the substances tested were effective as soil fumigants. His other investigations showed that cane grubs preferred feeding upon soil laden with organic matter in preference to the roots of growing cane and that cultivated soil deterred the female beetles from laying their eggs. He urged the state's canegrowers to practice trash conservation in preference to burning the post-harvest residues and to engage in more thorough cultivation of their paddocks in spring and early summer – eggs and young grubs were harmed by ploughing even to the shallow depth of a few centimeters. However, upon his departure for the USA in May 1921, Illingworth had not provided Queensland's canegrowers a cheap and effective method for controlling cane grubs.

Jarvis' and Illingworth's investigations were probably being monitored by CSR, although the organisation's annual seasonal reports on cane production contain no mention of the subject. CSR had not pursued any independent research into reducing cane grub damage since the mid-1890s, although as mentioned above, the firm had assisted the BSES with field experiments on its Green Hill Estate during the late 1910s. CSR's annual report on cane production for 1922–1923, however, suddenly mentions that soil fumigation experiments were being conducted by Hunter Freeman, CSR's manager at Green Hill Estate. Freeman had joined CSR as a junior chemist in 1912. After military service during World War I, he gained additional practical experience as a chemist at CSR's Goondi and Broadwater Mills. Therefore, he was ideally suited to pursuing research testing the effectiveness of particular chemicals.[9]

The stimulus for CSR undertaking its own research again could not be determined, but it may have been the firm's frustration with the BSES's inability to develop an effective method to reduce cane grub damage. Freeman was testing substances such as carbon bisulphide and calcium cyanide which were being investigated by Jarvis who had resumed control of the BSES entomological program following Illingworth's departure, but he was also reported as using a mixture of carbon bisulphide and paradichlorbenzene.[10] This mixture emerged as the most effective combination to reduce cane grub damage and was eventually promoted by the BSES as the preferred method of cane grub destruction (see Plate 14.1). CSR continued to conduct sporadic soil fumigation tests throughout its North Queensland mill supply areas during the late 1920s and 1930s, but often the experiments involved applying or testing discoveries made by the BSES entomologists (e.g. mechanical application of carbon bisulphide).[11]

Hunter Freeman began fumigating regularly with mixtures of carbon bisulphide and paradichlorbenzene at CSR's Green Hill Estate after 1923. CSR's promotion of soil fumigation led to an upsurge in the practice amongst the suppliers of its Goondi and Childers Mills during the late 1920s.[12]

Plate 14.1. A comparison of an experimental plot injected with paradichlorbenzene (on left) and an untreated plot (on right), six months after treatment. Cane stools in the untreated plot have lodged owing to grub injuries to their root system.

Source: E. Jarvis, *Notes on Queensland Cane Insects and their Control.* Queensland Bureau of Sugar Experiment Stations, Division of Entomology, Bulletin No. 19, 1926, p. 38. Image reproduction courtesy of BSES Limited.

Other canegrowers used different substances. See Chin, a canegrower in the Green Hill locality, near Cairns, reputedly reduced grub attacks on his crops during 1921 by burying moth-balls (i.e. naphthalene) in the paddocks. His approach, however, appeared unique.[13] The application of different types of arsenic attracted a following amongst the canegrowers in the Herbert River and Lower Burdekin districts during the late 1910s and 1920s, even though Illingworth's investigations showed that it was not an effective deterrent. Eventually, the BSES warned Queensland canegrowers in 1930 that persisting with applying arsenic to cane grub infested paddocks was futile and could only 'result in disappointment and financial loss'.[14] Overall, however, soil fumigation with any chemical, especially carbon bisulphide, remained very unpopular amongst the state's canegrowers during the 1920s.

Collecting beetles regained some popularity during the 1920s. The considerable crop losses caused by cane grub attacks in the Herbert River district during 1920 and 1921 prompted a resumption of hand collection in that district during 1922. Moreover, it persisted over the rest of the decade. The Burdekin Canegrowers' Association also funded beetle collection during the early 1920s. Farmers in the Mackay district were observed over the summer of 1927–1928 to be 'judiciously collecting by hand large numbers of cane beetles from the feeding trees'. Canegrowers in the Cairns district, however, were not convinced that hand collection reduced grub attacks and still had not resumed the practice in 1927. Edmund Jarvis lamented that they had made little attempt of late year 'to prepare for or combat any possible future attacks from this notorious beetle pest'.[15]

Funding pest control activities in Queensland

In 1915, funds for the control of pests affecting Queensland canegrowers came from BSES levy funds, although the ASPA and United Cane Growers' Association had been lobbying successive Queensland governments for additional funds. In late 1915, the new Ryan Labor government in Queensland agreed to amend the legislation controlling the collection of BSES levies by providing for an additional compulsory levy for pest control, new entomology research facilities at Meringa and other cane grub investigations. This suggestion was not welcomed by Queensland's canegrowers and the amending legislation was blocked by the Queensland Legislative Council. Amendments proposed by the Legislative Council were rejected by the Legislative Assembly, thereby depriving Queensland's canegrowers of additional pest control funds.[16]

Eight years then lapsed before another attempt was made to expand the funding for pest control activities in Queensland's sugar-producing districts. This time the Queensland sugar industry and the Queensland government (Theodore; Labor) cooperated to devise the legislative changes. The 1923 amendments to *The Sugar Experiment Stations Act* of 1900 provided for the creation of statutory cane pest control boards in specified sugar-producing districts, with members elected by canegrowers and millowners every two years. These boards – funded by a compulsory levy of not more than 3d per ton of cane delivered to the sugar mills – were required to suppress and prevent the damages from cane pests. Interestingly, the mill-owners now agreed that pest control on cane farms was equally as important for them as for the canegrowers.[17]

Interest in forming the new cane pest control boards was initially slow. Only seven compulsory boards had been set up by 1930: Plane Creek, Mackay, Lower Burdekin, Invicta (i.e. Ingham Line suppliers), Tully, South Johnstone and Mossman. Pest control activities were also undertaken by another twenty-three voluntary organisations. By 1938, the number of cane pest control boards had doubled, while the voluntary ones had declined to seven. Annual conferences of the pest control boards commenced in 1935, enabling representatives from each organisation to share ideas about pest control, develop policies about pest control (e.g. paying for hand collection of beetles) and to hear papers from BSES officers on research into pest control methods. These cane pest control boards and the cane disease control boards established in 1938 were amalgamated in 1941 to create Cane Pest and Disease Control Boards. Several voluntary pest control boards, however, persisted until the formation of statutory disease and control boards in their mill supply areas. Their names and year of closure are as follows: Mulgrave (1942); Babinda (1952); Proserpine (1953); Hambledon (1960); and Mourilyan (1975). Goondi Voluntary Pest Control Board ceased to exist in 1987, after the closure of Goondi Mill. By the early 1990s, the Macknade Voluntary Pest Control Board was the last organisation of its kind in existence in Queensland, having operated for almost 100 years with CSR support. It finally ceased to exist following an amalgamation with the Victoria Cane Pest and Disease Control Board in mid-1993 to form the Herbert Cane Protection and Productivity Board.[18]

Controlling rats, weevil borers, wireworms and African black beetles, 1930–1939

Rats had periodically troubled the Australian sugar industry during the nineteenth and early twentieth centuries. Several Queensland sugar plantation owners appearing before the Royal Commission into the Queensland Sugar Industry in 1888 outlined how their crops of sugar cane had been attacked by rats. Poisoning the rodents was the preferred control measure. Plagues of rats troubled North Queensland canegrowers between 1905 and 1911, and coincided with several years of high rainfall.[19] A variety of control measures were implemented during these outbreaks in the 1900s and early 1910s. Shareholders in the Mossman Central Sugar Mill Co. Ltd. decided in 1907 that the company should supply rat poison free to the mill's suppliers. Phosphorous paste was distributed in large quantities, but individual efforts were inadequate. In 1911, the company employed gangs of labourers who distributed baits laced with strychnine and burnt the vegetation near the cane paddocks, thereby depriving the rats of natural food and cover. A further concerted campaign occurred in 1913, when rat-infested localities were systematically poisoned as the cane was cut. Poisoning was also CSR's preferred strategy in the Herbert River district over the period 1904–1906. In contrast, management at Mourilyan Mill offered a 1d bounty for every rodent killed – nearly 8 000 rats were killed in the Mourilyan Mill supply area during 1905.[20]

Several species of introduced and native rats are responsible for damaging Australian crops of sugar cane. The introduced house rat or black rat (*Rattus rattus*) is often present in cane fields, but also nests in trees and buildings where the native rats are not normally found. The house rat is found in greatest numbers in cane fields near townships or farm buildings. The canefield rat (*Rattus sordidus* Gould) and pale field rat (*Rattus tunneyi*) have a coastal distribution down the eastern coast of Australia to northern New South Wales. Grassland melomys (*Melomys burtoni* Ramsay) and the fawn-footed melomys or large khaki rat (*Melomys cervinipes* Gould) are also found in or near cane fields. These rat species favour refuges such as grassy areas, forested areas with grassy understoreys or fringes, swamps, sedgelands and drains. Therefore, cane fields provide rats with major habitat requirements of friable soil in which to make their extensive burrows and good cover against their natural predators (e.g. owls and hawks). Rat numbers generally increase following good seasons and are affected by periods of drought. Stalks of cane chewed by rats are vulnerable to weevil borer and bacterial and fungal degradation which lowers sugar content.[21]

Rat injuries to the sugar cane crops persisted in Queensland throughout the late 1910s and 1920s. CSR, for instance, calculated that about 10 200 tonnes of cane were lost through rat injury in its four North Queensland mill supply areas during 1921, an amount almost equivalent to the damage done over the four-year period 1915–1918. By the early 1930s, however, rat damage to the cane had escalated in North Queensland, especially the Herbert River district where CSR claimed that the rodents had reached 'plague proportions'.[22] Consequently, during 1932 and 1933, trapping and poisoning attempted to control the rodents in the Herbert River district – one Macknade canegrower in 1933 caught 3 200 rats in break-back traps over a four week period. CSR's Head Office, alarmed at the growing damage to the sugar cane crops during 1933, arranged for 6 500 break-back traps and 500 cages to be dispatched from Sydney to Macknade Mill in an effort to expedite control measures. However, Kenneth Gard, CSR's technical field officer for the Macknade Mill supply area, noted that the rat damage had not waned, despite the firm having distributed in the vicinity of 30 to 35 tonnes of poisoned baits and traps to the local farmers between 1930 and 1933. He concluded that the local canegrowers were not taking the menace seriously. Baits were not laid at all or laid carelessly and the destruction of natural cover and breeding places remained inadequate.[23]

Some North Queensland canecutters went on strike during August and October 1934 over demands to burn paddocks of cane so as to rid them of rats, which had been linked to the recent deaths of several canecutters from Weil's disease (see Chapter 16 for full details). Not surprisingly, campaigns against the rats intensified throughout North Queensland during late 1934. In the Herbert River district, CSR co-coordinated the distribution of poison baits and burning campaigns in June, August and December. During the year, 11 000 break-back traps and 500 cage traps were issued to the local canegrowers; an estimated 277 000 rats were caught. The Mulgrave Central Mill Co. Ltd. purchased two tonnes of poisoned wheat in mid-November 1934 and distributed it to cane-growers throughout the Mulgrave Mill supply area. In the same month, the Goondi Voluntary Pest Control Board commenced laying poison baits, and urged its members to 'clean up all places harbouring rats, such as dirty head-lands, creeks and gullies'. By the end of November 1934, rat poisoning had also commenced throughout the other mill supply areas in the Innisfail district (i.e. South Johnstone and Mourilyan).[24]

Government action against the rodents began in November 1934. Under the direction of Sir Raphael Cilento, the Queensland Department of Public Health commenced a large-scale anti-rat campaign in the Herbert River district. Estimated to cost £12 000, the funds were raised by a combination of government grant, a levy of 3d per ton of sugar cane harvested, paid by both the mill-owner and canegrowers, and a rate of an equivalent amount on the ratepayers of the Hinchinbrook Shire. Eleven gangs of men each laid poison baits and used flame throwers to burn long grass found in road, rail and tramway easements. By the end of 1935, approximately three million baits had been distributed throughout the Herbert River district, with a very high percentage of them being nibbled or taken. Farmers were also instructed to burn any rat harbourages on their farms, and inspectors made sure that the instructions had been completed. Yet CSR claimed that much of the campaign's initial effort was misdirected. Its technical field officers argued that the poisoned baits consisting of a phosphorous mixture spread on bread were too laborious to prepare, expensive and had to be used soon after preparation. Moreover, the distribution was made during the wet season when the rats were not numerous in the cane fields; persistent rain also created unattractive baits.[25]

A combination of an exceptionally dry summer of 1934–1935 and the anti-rat campaigns conducted by various organisations during 1934 and 1935 brought the rat numbers under control in all North Queensland mill supply areas. Levels of rat-damaged cane fell. Throughout the Lower Burdekin district, for instance, where rat damage was observed to be severe in 1934, the BSES reported that 'practically no damage' occurred during 1935. Moreover, rat poisoning campaigns continued in many North Queensland mill supply areas during the late 1930s, keeping the amount of rat damage at reduced levels.[26] This turnaround compared to the early 1930s can be illustrated by considering the figures collected as part of the annual surveys of the cane crushed at the two Herbert River sugar mills during the late 1930s. From 1937 to 1940, around 10% of the stalks delivered to Macknade Mill had been rat-damaged, a third of the levels reached during the early 1930s. About 4 100 tonnes of cane annually were lost to rat damage in the Victoria Mill supply area over the period 1936–1939, substantially less that the 33 000 tonnes of cane lost during the period 1932–1935.[27]

Weevil borers posed less of a problem than rats during the early 1930s, but their presence in North Queensland still caused concern. A noticeable upsurge of borer damage occurred throughout the mill supply areas north of Townsville during 1933 and 1934 and was attributed to the serious rat damage allowing easy oviposition by the borer. Greater borer activity also occurred in the Mackay district over the period 1934–1936, although it was often confined to the borer-susceptible Badila cane variety.[28] Borer numbers were kept under some control during the early 1930s by the tachinid fly, bred by Edmund Jarvis in two large cages at Meringa, and distributed to interested canegrowers. Propagating supplies of the parasite under artificial conditions, however, could not meet farmer demand, so a new strategy was implemented in 1933. Badly bored stalks of cane in localities known to contain the flies were cut open and the puparia of the parasitic flies removed and taken to Meringa. After hatching, they were liberated on borer-infested cane farms where no flies had been found previously. An intensive tachinid fly liberation campaign was conducted during 1934 to re-establish the fly in areas where its numbers had been depleted – 10 000 flies were bred and distributed by the entomological staff in North Queensland.[29]

Doubts about the effectiveness of relying on the tachinid fly to control the weevil borer had started emerging during the mid-1930s. Following the creation of the Division of Entomology and Pathology after a reorganisation of the BSES in 1934, the entomologist Reginald Mungomery was transferred from Bundaberg to Meringa. He assumed control over entomological investigations following the retirement of Edmund Jarvis. Mungomery and Jim Buzacott, the entomological assistant, commenced a critical examination of the relationship between the weevil borer and its parasite. Infestation and parasitism were examined statistically and large-scale experimental liberations of the flies were made to asses their survival rates at different times of the year. Results from these investigations showed that the yearly harvesting of the crops of cane and the consequent scarcity of the host larvae in positions where they could be readily parasitised meant that the fly population often collapsed after each harvest and never caught up with the natural increase of the borer pest. Liberations during late spring and early summer were of little value as the natural increase in fly numbers was also checked by heavy wet season rains. The entomologists realised that fly numbers needed to be replenished throughout the sugar-producing districts every few seasons, an unappealing prospect. In addition, the experiments showed that parasitism from the tachinid flies often peaked during September and October, not throughout autumn prior to the harvest when most benefit would accrue.[30]

Varietal tests conducted by Buzacott for his Masters of Science thesis showed that rind-hard varieties like S.J. 4, Q. 2 and Q. 10 were less susceptible to borer damage than Badila. However, Q. 2 had only been released to canegrowers in 1937, Q. 10 was still being tested for commercial release and S.J. 4 displayed susceptibility to gumming disease so its widespread use could not be countenanced.[31] Further BSES investigations during the late 1930s found that trashing contributed to a significant reduction of the population of weevil borers, as it removed their favoured sheltering positions. Yet its use as a control measure was regarded as economically impracticable given the high pre-World War II cost of labour and the likely field worker resistance to undertaking such an unpleasant task. Burning post-harvest crop residues reduced beetle borer numbers, but pre-harvest burning coupled with careful selection of borer-free planting material emerged as the most effective control method. Pre-harvest burning destroyed the living weevil borers, preventing oviposition in any surviving post-harvest residues or neighbouring, unharvested fields. However, the practice was frowned upon by mill-owners. Thus, Arthur Bell believed in 1937 that the borer problem would be ultimately solved by the substitution of resistant cane varieties, although he acknowledged that the canegrowers would not readily abandon the sucrose-rich Badila for inferior cane varieties just to reduce weevil borer attacks.[32]

In the absence of a truly effective control measure, North Queensland canegrowers were lucky that weevil borer damage was very benign during the period 1936–1939. The Innisfail district was the worst affected locality, with CSR estimating that a fifth to a quarter of the stalks of cane crushed at its Goondi Mill had been damaged by this insect pest.[33] Perhaps a combination of the tachinid flies, the reduction in rat numbers which eliminated stalk damage, making them less attractive to the egg-laying female weevil borer, and some post-harvest trash burning and localised pre-harvest burning had managed to control the numbers of weevil borers. Yet at the outbreak of World War II, the weevil borer pest, although in check, had not been eliminated.

Wireworms were first mentioned as being responsible for losses in Australian cane fields during 1891. In some fields of sugar cane along the Tweed River, a distance of 35–40 m had to be replanted for a second time following the poor germination of many setts. During the early part of the twentieth century, sett failures continued to be ascribed to many causes, including wireworms. By the early 1920s, however, wireworms had definitely emerged as an insect pest of major importance in the Mackay district, when large areas of poorly drained clay-bottomed soils were brought under cultivation following the expansion of the sugar industry.[34] Heavy damage from these insect pests was reported from the Mackay and Proserpine districts during the 1930s.[35]

Wireworms are the immature stages or larvae of what are known as click beetles – these insects if they land on their backs lay motionless for a time until suddenly there is a click and they pop up into the air, eventually landing on their feet and then scurrying away. The larva grows to 20 mm long and has a firm, slightly flattened shiny cream-white segmented body with a hard yellow-brownish head. Several species of wireworms are found in Australian cane fields. The major incidence of this pest, *Agryphus variabillis* Candèze, occurs in the wetter, poorly drained soils in the Proserpine, Mackay and Herbert River districts, and in poorly drained fields north of Mossman. *Heteroderes* spp. and *Conderus* spp. wireworms are common in southern sugar-producing districts as far south as Maryborough. Wireworms bore into the eyes of germinating setts or into the growing point of young shoots. Symptoms of damage are poor or patchy germination. Ratoon crops do not suffer from direct wireworm attack.[36]

No serious attempts had been made to investigate wireworm control before W. McDougall, an Assistant BSES entomologist at Mackay, began his inquiries during 1931. He bred the wireworms so as to understand their life history and habits. McDougall tested many substances as potential insecticides, but concluded that wireworms were notoriously immune to poisons and no parasites or predators of any economical value were identified. Fertilisers or lime were of little use in their control and there was no correlation between the percentage of organic matter in the soil and the occurrence of wireworm damage. Hand collection of the grubs proved futile – very few larvae were brought to the surface during ploughing. Some relief could be obtained by late plantings (i.e. August–September), as very little feeding takes place towards the end of the last larval stage, and/or very thorough drainage during the wet season immediately prior to planting in areas known to be badly affected by wireworms.[37] Throughout the late 1930s, the BSES issued pre-seasonal warnings to the Mackay canegrowers regarding the probable intensity of wireworm attack that year, urging those farmers with low-lying and ill-drained fields to delay planting until at least late September. However, Arthur Bell noted that these warnings had not served to reduce the losses in the Mackay district to any great extent because of the 'general dislike of late planting' by the canegrowers. On the eve of World War II, this insect pest remained just as troublesome as the weevil borer and cane grub.[38]

Injuries to the New South Wales crops of sugar cane were rarely discussed in CSR's annual cane production reports during the 1920s, except for mention of damage from black beetles in 1919 and 1923. During 1929, 1930 and 1931, black beetles were more numerous and widespread damage occurred in the New South Wales sugar-producing districts, especially along the Richmond River. Some black beetles damaging the cane crops turned out to be the African black beetle (*Heteronychus arator*), accidentally introduced into New South Wales sometime in the early part of the twentieth century. Later investigations showed that adult African black beetles chewed ragged deep cavities in young shoots, causing 'dead hearts', and could also chew the eyes of germinating setts. Generally, African black beetles are a problem in plant cane but can also damage ratoon crops. They are associated with *Paspalum* spp. of grasses which grew in abundance in northern New South Wales as pastures for dairy cattle.[39]

Damage from the black beetles became severe enough for some New South Wales canegrowers to try hand collection during 1930 in an attempt to reduce their losses, but this practice was abandoned because the farmers found it too tedious and costly. Throughout 1931, David North investigated control methods, although he admitted that the possibility of finding an effective control measure without prolonged and exhaustive investigations was 'remote'. Fumigating the soil with carbon bisulphide failed to kill the beetles and bright lights did not attract them so that approach held little promise as a practical means of control. Dipping the setts in lead arsenate (a 10% suspension in water), however, protected them from black beetle attack and had negligible impact upon the growth of the young cane plant. This control method was promoted by CSR from 1932 onwards, and mention of black beetle damage to the New South Wales crops of sugar cane dwindled in CSR's annual cane production reports and had ceased by 1939.[40]

Cane grubs and the introduction of the cane toad

During the late 1920s and early 1930s, the BSES entomologists continued their numerous experiments on ways to reduce the numbers of cane grubs and cane beetles. At the Meringa BSES Station, Edmund Jarvis unsuccessfully attempted to develop an effective fumigating machine to replace hand injectors. His investigations, however, showed that spraying native fig trees (*Ficus* spp.) with arsenicals effectively killed the beetles, but the practical application of this finding was never pursued. Jim Buzacott, his assistant, also tested a range of different chemicals (e.g. ethylene dichloride, propylene chloride and propylene oxide), but concluded that their killing power was inferior to that of mixtures of carbon bisulphide and paradichlorbenzene. Assistant entomologist A. N. Burns conducted investigations at Mackay, and concluded that the soil fumigation with any substance was not effective due to the heavy nature of the local soils. Reginald Mungomery's work at Bundaberg explored different avenues. His inquiries into the phototropism of the cane beetles revealed that female beetles were not attracted to any particular coloured light, so that avenue as a control measure could not be pursued. A foray into biological control during 1931–1932 using the imported Canadian fly parasite *Microphthalma michiganenis* was also unsuccessful. The colonies of this fly failed to become established throughout the Isis district. Mungomery's experiments with a modified rotary hoe were more successful, with the rotor blades killing most grubs to a depth of 20 cm during the process of pulverising the soil.[40]

Mungomery assumed control of entomological investigations at Meringa in early 1935. His colleague Arthur Bell had been the Queensland government's representative at the Fourth Conference of the International Society of Sugar Cane Technologists held in San Juan, Puerto Rico, 1932. At this conference, Bell most likely heard about the success achieved by the Puerto Rican sugar industry in controlling their cane beetles by using the giant American toad (*Bufo marinus*), and probably shared this information with Mungomery upon his return to Queensland. Sometime during the early months of 1935, Mungomery must have decided to copy not only his Puerto Rican, but Hawaiian and Philippino colleagues who had brought the cane toads to Hawaii and the Philippines in 1932 and 1934, respectively, to control the native beetles damaging the local sugar cane crops. Mungomery never left any pronouncements about his reasoning, but it is reasonable to assume that he may have been frustrated with the pace of investigations under Jarvis' direction, especially the cautious testing of the impact of control measures. He wrote, however, that the BSES had been 'impressed with the possibilities of such an animal operating against our indigenous cane beetles'.[42]

Mungomery travelled to Hawaii in early June 1935, returning with 102 cane toads by the end of the month. They were initially quarantined in a special pond at Meringa, and bred immediately. During the latter part of 1935, the cane toads were released in the Cairns and Innisfail districts. However, the Commonwealth Director-General of Health prohibited the release of any further cane toads in December 1935, claiming that they 'might eradicate insects of economic value'. Immediate representations by sugar industry leaders to the Commonwealth government were initially successful in getting permission to liberate more cane toads in the original areas of release, but not in new areas. This ban was rescinded in September 1936 and the cane toads were released throughout the other Queensland sugar-producing districts as far south as Bundaberg.[43]

All the cautious testing characterising the previous investigations into cane grub control methods was completely forgotten when it came to the cane toad. No evidence could be found of any pre-release testing by the BSES entomologists to determine if the toads even ate the local cane beetles. The first systematic examination of the stomach contents of an unidentified number of cane toads appears to have been completed during the period December 1938 to April 1939. Results looked encouraging. Some cane toads had gorged themselves on *Lepidiota frenchi* Blackburn and others had consumed weevil borers. Additional analyses of the stomach contents of cane toads were repeated a year later. However, the BSES reported two alarming observations: first, grub damage had increased in northern cane growing areas where the cane toad population was high; and second, the number of beetles destroyed by the cane toads was relatively small when compared to the nightly swarms. Moreover, from these observations the BSES concluded that the 'number of toads present in the field was insufficient to effect a material reduction in the pest'.[44] No further mention of the effectiveness of the cane toad appears in the annual BSES reports. The BSES entomologists had made a terrible blunder. Subsequently, cane toads have become one of northern Australia's major pests.

Throughout the 1930s, canegrowers in Queensland adopted various control measures against the cane grubs and cane beetles. Collecting the beetles and grubs persisted, especially in the Mackay district and in Southern Queensland where Reginald Mungomery observed in 1934 that small boys still followed behind the ploughs collecting any grubs exposed during the cultivation of the fields. Widespread vegetation destruction, thereby depriving the cane beetles of their feeding trees, was tried in the mid-1930s, when a group of canegrowers near Mourilyan cleared entirely 450 ha of the nearby Basilisk Range, mostly by firing the slopes. The BSES reported in 1940 that previously infested farms had remained free of the pest since the clearing of this range.[45] Soil fumigation persisted during the 1930s and became more beneficial as the entomologists delivered better information about the correct application of the fumigants (i.e. size of dosages; timing; spacing between dosages). The recommendation appearing consistently in many publications was that the most effective fumigant was a mixture of carbon bisulphide and paradichlorbenzene (two parts to one). Moreover, canegrowers did not have to fumigate consistently, but only when soil inspections determined that the presence of grubs of the correct species (i.e. not the frenchi cane beetle) were sufficiently numerous to justify the expense.[46] In July 1938, soil fumigation became the preferred control method when delegates at the annual conference of the Queensland Cane Pest Control Boards resolved that the payment for collecting cane beetles was 'uneconomic' and that it should be replaced by soil fumigation. Two years later the BSES reported that most mill supply areas had adopted 'soil fumigation as their method of front-line attack on the grub pest'.[47]

Wartime losses – interruptions to pest control activities

Soil fumigation against cane grubs in Queensland did not receive an immediate setback following the commencement of World War II, but fumigant shortages had emerged by 1941. The BSES entomologists anticipated these possible fumigant shortages and tested a range of organic dilutants (e.g. benzol; vegetable and mineral turpentine; sinapis oil) in field trials during 1940 and 1941 to see if they would increase the toxicity of reduced amounts of carbon bisulphide. None of the dilutants improved the numbers of cane grubs killed, but this avenue of research showed that under suitable soil conditions (i.e. particularly high soil temperatures) it was possible to halve the standard dosage of carbon disulphide, although the practical application of this finding would have been difficult. Eventually, however, labour shortages and disruptions to the supply of the fumigants contributed to a reduction in the area fumigated in 1942 and 1943. Mackay and Lower Burdekin canegrowers during the summer of 1942–1943 re-commenced collecting cane beetles in an attempt to reduce the severity of cane grub attacks. Normal soil fumigant activities resumed during 1944 following an easing of labour shortages and improved fumigant supplies, but were hindered in 1945 due to sodden paddocks and severe lodging of the cane following heavy summer rains. Only 400 ha were fumigated, the lowest amount since 1938. Collecting beetles and grubs occurred throughout the Mackay district over the summer of 1944–1945 as an alternative control measure.[48]

Despite the wartime conditions, soil fumigation became a more palatable control method during the 1940s following the development of the Blundell knapsack fumigant injector and Blundell machine injector in 1939 and 1944, respectively. The former had a much simplified injecting mechanism, which eliminated the tedious and frequent interruptions with choked Dank's injectors. In addition, it had the advantage of being more comfortable to use compared to the Dank's injectors, as the fumigant mixture was housed in containers on the operator's back instead of filling a container where the entire weight of the fumigant was taken by the hands and arms of the operator (cf. Plates 14.2 and 14.3). The latter sped up fumigation, allowing up to 1.2 ha daily to be treated, compared to 0.2–0.25 ha with the Blundell knapsack injector and 0.1 ha with other hand injectors. Moreover, to assist canegrowers, some Cane Pest Control Boards subsidised fumigation costs to the extent of 50% and supplied injectors free of charge. Such support halved the cost of fumigation which stood at £3–£4 per ha in the late 1930s.[49]

Plate 14.2. A farmer demonstrating the use of the Dank's injector to fumigate soil.
Source: *ASJ*, 11 June 1940, p. x.

Wartime exigencies assisted in the control of the weevil borer, ensuring damage from its depredations were minimal. Scarcity of experienced canecutters and the need for a lessened labour force to cope with the harvesting of much the same tonnages led to faster cutting operations. This goal was accomplished by pre-harvest burning to rid the crops of trash and weeds which slowed harvesting. These fires killed the weevil borers. North Queensland canegrowers also claimed that the growing numbers of cane toads were responsible for keeping this insect pest in check. Their investigations of the stomach contents from the cane toads showed that the amphibians were eating the beetle borers.[50]

Other insect pests were not a major trouble to Australian canegrowers during the early 1940s, except for yellow-winged locusts, (*Gastrimargus musicus* Fabr.) which assumed plague proportions towards the conclusion of the war. Flights of grasshoppers had troubled canegrowers previously during late 1914 and early 1915, but their numbers had been relatively scarce since that year.[51] However, in the early part of 1944, swarms of locusts visited the coastal areas of Queensland, but were the heaviest around Mackay – older residents

Plate 14.3. The Blundell knapsack fumigant injector.

Source: W. McDougall, 'Fumigation for the control of cane grubs', *CGQB*, Vol. 8, no. 1 (1940), p. 27. Image reproduction courtesy of BSES Limited.

claimed it was the worst visitation since 1896. No district-wide control measures were instituted, but attempts at destroying individual swarms were made with arsenical sprays and baits and the burning of grassed areas where the locusts congregated at night. The insects were reputed to have destroyed 20 000 tonnes of cane in the North Eton Mill supply area alone.[52]

A reduction in wartime shooting activities following the confiscation of weapons and ammunition shortages, however, contributed to an upsurge in the numbers of swamp hens (*Porphyrio melanotus*) – known as red bills or coots – and feral pigs in some North Queensland mill supply areas. At the war's conclusion, these pests were reported to being doing considerable damage to the crops adjoining swampy areas. Scarcity of proper fencing wire and wire netting had also hindered control activities against feral pigs in at least the Tully and Cairns districts. A poisoning campaign against the red bills had commenced in the Goondi Mill supply area during 1945, as CSR sought to reduce the damage from this bird.[53] Wallabies troubled the Lower Burdekin canegrowers by the later stages of the war. The Lower Burdekin Cane Pest and Disease Control Board responded by introducing beagle hounds into the district in 1944. These dogs proved effective in either destroying the marsupials or keeping them on the move.[54]

Increased damage by rats was also reported throughout the mill supply areas north of Townsville and Mackay district by 1942. Owing to labour and fuel shortages, many canegrowers found keeping farm lands clean difficult, thereby providing ideal conditions for the build-up of rat populations. Shortages of rodenticides also hindered this control measure. By 1944, a scarcity of fencing wire was reported to be preventing canegrowers from pasturing stock on gullies and waste lands, thereby destroying the sanctuaries of these rodents. Despite the disruption to normal control operations, rats never assumed plague proportions during the wartime years, probably because of the drier conditions that existed in North Queensland during parts of 1943, 1944 and 1945.[55]

The insecticide era begins, 1946–1959

Crop destruction from cane grubs throughout the North Queensland sugar-producing districts during early 1946 was the 'worst in twenty years', according to the BSES. Losses attributable to the cane grubs between Tully and Mossman were estimated to be at least 106 000 tonnes of cane. Soil fumigation with carbon bisulphide or carbon bisulphide-paradichlorbenzene mixtures continued to be the main control method, although farmers supplying Mulgrave Mill again collected beetles in an attempt to reduce the damage to their crops.[56] As the North Queensland canegrowers struggled to control the impact of the attacks from the grubs of the greyback cane beetle (*Dermolepida albohirtum* Waterhouse; hereafter referred to as greyback cane beetle), CSR and the BSES were testing the effectiveness of two new chemicals against the grubs: dichloro-diphenyl trichloroethane (DDT) and benzene hexachloride (BHC). The former had been synthesised in 1874, although its insecticidal properties were not discovered until 1939. The latter had been synthesised in 1825 by passing chlorine gas through liquid benzol, producing several isomers (forms) of BHC. Discarded as useless, its effectiveness as an insecticide became clear during 1942. BHC, currently known as hexachlorocyclohexane (HCH), exists in eight isomers. Imperial Chemical Industries (ICI) began marketing BHC during the late 1940s under the commercial name 'Gammexane', now identified as Technical HCH. This product contained the α-, β-, γ- and δ-BHC isomers in the approximate proportion of 70–75%, 5–7%, 13–15% and 6–8%, respectively, although virtually all of the insecticidal properties resided in the gamma isomer. As most historical documents refer to BHC or 'Gammexane', these terms will be used throughout the remainder of the narrative.[57]

CSR's experiments with DDT in North Queensland during 1945 and 1946 revealed that the chemical failed to deter cane grub attacks and further trials with it ceased. BSES investigations at Meringa, under the control of Jim Buzacott, also concluded that DDT offered very little possibility as an effective control agent. However, trials with BHC, although constrained due to the limited availability of the product, showed considerable promise. Buzacott's investigations using a 10% BHC dust – at that stage the only product being produced by the manufacturers – revealed that spraying the insecticide to the foliage of beetle feeding trees killed beetles at a similar rate to that obtained with a lead arsenate spray. Small exploratory field trials applying 'Gammexane' to the soil in furrows of a plant crop produced outstanding results. Arbitrary dosages of 18–72 kg per ha caused the death of 85% of the grubs of the greyback cane beetle in the field trial sites. Moreover, the insecticide rendered the soil toxic to the young cane grubs for several months, and its application did not involve the unpleasant conditions associated with the use of carbon bisulphide. CSR's field trials also showed that the insecticide reduced cane grub numbers, but applying the chemical over the setts at planting adversely affected germination.[58]

During the late 1940s, further tests with the insecticide were carried out by the BSES entomologist George Wilson, Buzacott's replacement following his transfer to the plant breeding section in 1947. Wilson also tested a stronger product as the manufacturers during 1948 had commenced releasing a special formulation designed specifically for the sugar industry – 20% BHC mixed with rock phosphate dust. His investigations concluded that BHC did not have any toxic effect on cane growth and the most effective and economical dosage was an application of 14 kg of 20% BHC per ha which would confer protection for three years. The most satisfactory method of application was to deliver the dust on each side of half-open drills after the young plant cane had become adequately established.[59] Moreover, trials demonstrated that the insecticide was effective against the grubs of other species of cane beetles (e.g. *Lepidiota frenchi* Blackburn) and wireworms.[60] CSR also conducted trials during the late 1940s and early 1950s on the effectiveness of BHC applied under different conditions and at varying rates, duplicating much of the BSES research and reaching many of the same conclusions as the BSES scientists.[61]

BHC was a very different substance to carbon bisulphide and required canegrowers to change their thinking about cane grub control. Carbon bisulphide was used as a remedial insecticide, being applied only to fields actually infested with cane grubs. Its application was tedious and difficult, so the chemical could not be used over large areas. In contrast, BHC was relatively easy to apply over large areas. However, it was a preventative insecticide applied to fields liable to become infested, but before the actual infestation occurred. Therefore, Australian canegrowers were required to make some forecast regarding the yearly risk of infestation. Farmers who had lived on their properties for many years had gained this experience, often through suffering losses. The newcomers who joined the sugar industry during the early 1950s relied on advice from their local Cane Pest and Disease Control Boards. BSES experiments showed that applications of BHC made just prior to the beetle flights in late spring or summer obtained the best results.[62]

BHC dust was usually deposited in narrow bands along each side of the planting furrow, being incorporated into the soil by means of small tines on the distributing implement. Surface or very deep applications of the dust proved largely ineffective; applications with fertiliser led to incomplete control over cane grubs as the dust was buried too deeply, but this arrangement proved effective against wireworms.[63] Initially, the BHC was delivered into the furrows by fertiliser distributors. These machines gave satisfactory delivery with the larger quantity of the more dilute dust, but they were incapable of delivering accurately the smaller quantity of the 20% product. During the late 1940s and 1950s, local machine makers designed and manufactured modified fertiliser distributors that handled the finer product, ensuring a more even distribution of the insecticide throughout the soil.[64]

As soon as BHC became commercially available in 1947, North Queensland canegrowers commenced using the insecticide against greyback cane beetle grubs, rapidly abandoning soil fumigation with carbon bisulphide. A shortage of BHC in 1949, however, witnessed infested crops being fumigated again with carbon bisulphide in an attempt to salvage some harvestable cane. Hand collections of beetles also occurred throughout the Cairns and Lower Burdekin districts during 1947 and 1948, respectively.[65] Such control methods then ceased, replaced entirely by the widespread application of BHC. The area treated with BHC in Queensland increased rapidly from 2 800 ha in 1948 to 24 450 ha in 1954. Most canegrowers throughout Far North Queensland commenced using the insecticide – CSR reported that 85% of the cane harvested for its Goondi and Hambledon Mills in the early 1950s had been treated with BHC. Indeed, so much BHC was being purchased that the BSES concluded some North Queensland canegrowers were wasting money by using excessive amounts of the product which gained them no extra control benefits.[66]

BHC-fertiliser mixtures proved to be very effective against wireworms, prompting the BSES to declare in 1950 that this insect pest had ceased to be of major concern in securing satisfactory germinations of the setts.[67] Severe and widespread attacks by greyback cane beetle grubs also eased during the mid-1950s after the widespread use of BHC, although sporadic losses continued because infestations occurred occasionally on farms not usually attacked, and thus not treated with BHC. After observing these trends, the BSES entomologist Reginald Mungomery claimed in late 1954 that the greyback cane beetle grub pest would not 'reassert itself to any marked extent'.[68] Such optimism, however, was shattered during late 1956 and early 1957. Unexpectedly large numbers of cane beetles emerged throughout the Lower Burdekin district and grub infestations increased considerably in the Mackay district; older residents at Carmila, south of Mackay, reported the heaviest beetle flights for thirty years. Losses, although not as great as those of the mid-1940s, stood at approximately 22 500 tonnes of cane. Lower Burdekin canegrowers who had previously only suffered light infestations of cane grubs abandoned an unorthodox BHC dosage in favour of the conventional BHC formulation and greater overall use of BHC occurred immediately throughout the Lower Burdekin and Mackay districts. This response ensured that losses from cane grubs throughout Queensland during 1958 and 1959 then fell to less than 1 000 tonnes each year.[69]

Severe African black beetle damage to newly planted cane was widespread in the Clarence River district during 1946. The traditional control method of soaking setts in a lead arsenate solution was used, but other canegrowers obtained relief by spraying the growing crops with a 0.1% solution of DDT. Damage from the black beetles declined over the next two years, especially as large numbers of them were destroyed during the flooding of the northern New South Wales rivers in June 1948. Over these two years, the New South Wales canegrowers must have become acquainted with BHC, for during the 1949 season they were reported to be applying 10% BHC dust to fields at the rate of 3–6 kg per ha in order to reduce the numbers of black beetles. Some canegrowers applied the dust in the drills at planting; others used a BHC-fertiliser mixture. The regular application of BHC at planting then reduced the extent of black beetle infestation in the New South Wales sugar-producing districts, although their impact does not disappear entirely. CSR reported 'appreciable' black beetle damage to the young sugar cane crops throughout the Condong Mill supply area in 1959.[70]

Insecticides, however, had no impact upon the rats and wallabies, the two main vertebrate pests which continued troubling Australian canegrowers and their depredations had to be controlled by more conventional methods. The amount of rat-damaged cane stalks during the late 1940s and early 1950s remained low, as yearly poisoning campaigns combined with dry seasonal conditions in some years (e.g. 1947 and 1951) kept their numbers under control.[71] However, following a succession of favourable seasons in the mid-1950s, a marked increase in rat-injured cane occurred over the period 1955 to 1957 – losses amounted to 61 000, 41 000 and 71 000 tonnes of cane, respectively. Crops in North Queensland and along the Kolan, Burnett and Maroochy Rivers were most affected. A return of dry spring weather in 1957 and again in 1958, combined with widespread poisoning by canegrowers, reduced annual losses in 1958 and 1959 to around 21 500 tonnes each year.[72] Wallabies attacked young fields of cane, mostly in the Lower Burdekin and Mackay districts, especially in the drier years of 1947, 1951 and 1957. Shortages of wire netting immediately after World War II made their control difficult, but additional beagle hounds were bred and made available to new owners. Other Cane Pest and Disease Control Boards increased the amount of bonus paid per scalp, attempting to provide greater inducement for the destruction of these marsupials.[73]

New pests and old menaces: soldier flies, cane grubs and rats, 1960–1979

During the late 1950s and early 1960s, soldier fly infestations caused many poor ratoon crops in the Bundaberg district. This insect was not a new pest, having been recorded as damaging cane in the Mackay and Bundaberg districts during the mid-1920s. However, soldier fly caused more losses in Queensland cane crops than rats or cane grubs by the mid-1960s, increasing the overall amount of cane damaged each year (Figure 14.1). The soldier fly, wrote BSES scientists Norman King, Reginald Mungomery and C. Graham Hughes, had gone from being a minor pest to one of the worst pests of sugar cane in Southern Queensland, being responsible for 'vast areas of indifferent and poor ratoons and even complete ratoon failures'.[74]

Sugar cane soldier fly, *Inopus rubriceps* Macquart, occurs around Innisfail, in the Macknade locality (Herbert River district) and more extensively in mill supply areas from Proserpine south to Harwood in New South Wales.

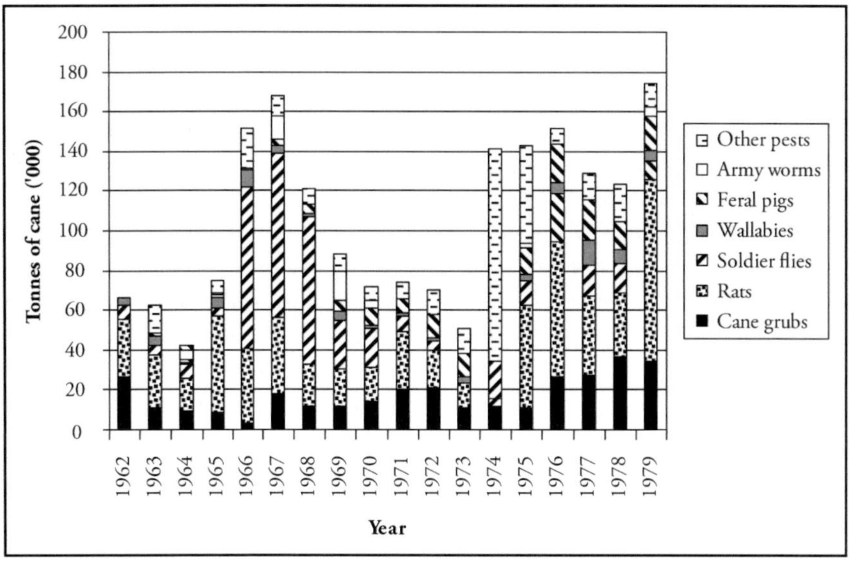

Note: Losses for 1974 were not recorded in any detail. Rat damage, for example, was not listed as a separate category.

Figure 14. 1. Estimated annual losses (tonnes of cane) to the Queensland sugar crop from pests, 1960–1979.

Source: Based upon figures in BSES, ARs 1960–61 to 1979–80.

The yellow soldier fly, *Inopus flavus* James, occurs throughout the Lower Burdekin, Proserpine and Mackay districts. Soldier fly larvae or maggots feed on roots, reducing germination if present in sufficient numbers at planting, but mostly their presence causes poor ratoon crops as the underground buds fail to germinate after harvest. The upsurge in soldier fly numbers during the 1950s and 1960s was linked to various developments. In pre-war years, soldier fly damage was usually noticeable only after the occurrence of two successive wet autumn-winter periods which allowed the numbers of this pest to increase appreciably. The introduction of irrigation provided these conditions permanently. Less cultivation exposing the soldier fly larvae to surface predators and a longer crop cycle were also considered contributory factors. However, BSES entomologists in the mid-1960s discovered that a soldier fly build up occurred in plots treated with low rates of dieldrin – a chlorinated hydrocarbon with a molecular structure based on naphthalene – and lindane (i.e. gamma isomer of BHC). Later studies showed that low concentrations of dieldrin and lindane killed insect predators, thereby allowing soldier fly numbers to increase. Therefore, increased soldier fly numbers in Queensland probably coincided with the introduction of organochlorine insecticides.[75]

Investigations into methods of controlling soldier fly depredations commenced in 1955 and finished a decade later. This work concluded that the most reliable treatment was a pre-planting, broadcast dressing of BHC applied at a rate of 1.5 kg of gamma isomer per ha. This dosage could be achieved by applying 56 kg of 20% BHC dust or 28 kg of 40% BHC dust per ha. Dieldrin applied as a spray at the rate of 1.2 kg per ha was equally effective. Yet lindane when used either as an emulsion or as a dispersible powder, and aldrin – an organochlorine insecticide which is oxidised in the insect to form dieldrin – applied as an emulsion, failed to control the soldier flies. These materials were not sufficiently persistent in the soil to kill the soldier fly maggots beyond the earliest stages of the plant crop.[76]

Farm applications of lindane against soldier flies commenced at Bundaberg in 1961 and at Mackay in 1962, but this insecticide was replaced with BHC in 1963. Dieldrin usage against soldier fly stepped up appreciably after 1966, and by the early 1970s most canegrowers in the Bundaberg and Mackay districts used dieldrin to control soldier fly, following experiences of poor control with BHC.[77] Canegrowers throughout soldier fly-infested areas in Queensland were also advised to grow vigorously ratooning cane varieties such as NCo. 310, Q. 58 and C.P. 29–116 in preference to other cane varieties that were more susceptible to maggot damage.[78] Following the introduction of these control measures and the treatment progressively

Plate 14.4. Ratoon crop failure attributed to soldier fly in the Innisfail district, c. 1968.
Source: BSES, AR 1967–68, p. 57. Image reproduction courtesy of BSES Limited.

of more land subjected to infestation, damage from soldier flies in Queensland receded in the early 1970s, but increased again in the late 1970s as cane farms in the Isis, Maryborough and Beenleigh districts succumbed to soldier fly infestation. Monitoring of the incidence of soldier fly in the Mackay district also showed that the area infested had doubled between 1976 and 1977.[79] In addition, a small outbreak of soldier fly infestation occurred throughout the Clarence River district in New South Wales during the early 1970s.[80] Cane losses attributed to soldier flies in New South Wales over the 1970s, however, were quite small, and in Queensland the damage was considerably less than the levels reached in the mid-1960s (Figure 14.1).

By the early 1960s, control measures for cane grubs had been well proven and canegrowers liable to infestations customarily applied an insecticide to protect their crops. Yet cane grub attacks during the 1960s followed the negligent attitude towards cane grub control in some localities or when greyback cane beetle grubs began causing losses in hitherto uninfested localities throughout the Herbert River and Mackay districts in 1962 and 1966, respectively, and when the grub of the Childers cane beetle (*Antitrogus parvulus* Britton) started damaging cane throughout the Bundaberg district in 1963. These outbreaks were limited to one or two years because BHC was applied on the most affected farms and no further expansion of the pest occurred.[81] In addition, during the period 1967–1971, *Lepidiota frenchi* Blackburn and *Lepidiota negatoria* Blackburn became more troublesome in the Burdekin and Mackay districts, while increased attacks from a hitherto minor species, *Antitrogus consanguineus* Blackburn, occurred in the Bundaberg district as new farms became affected each year. BHC still had not been widely used in the Bundaberg and Mackay districts – only an estimated 10% of the total area under sugar cane in those two districts in 1972 had been affected by cane grubs. Therefore, canegrowers were advised to watch ratoon crops for the first signs of grub attack. If grubs were detected, all new plantings should be treated with BHC.[82] Despite the increased use of BHC or other insecticides in Queensland, losses from cane grubs increased significantly during the late 1970s (Figure 14.1). Severe cane grub damage was also reported in New South Wales cane growing districts throughout the late 1970s.[83]

BHC remained the most widely used insecticide against cane grubs during the 1960s and 1970s, especially amongst Southern Queensland canegrowers who could only use BHC, as other insecticides were not effective against the grub of the Childers, Bundaberg and frenchi cane beetle (i.e. *Antitrogus parvulus* Britton; *Lepidiota crinite* Brenske; *Lepidiota frenchi* Blackburn). Moreover, the application of aldrin, heptachlor or dieldrin in Southern Queensland sugar-producing districts was shown to exacerbate the grub problem as they killed the natural predators of the grubs. Yet the increased cost of BHC dust and its application during the early 1960s induced some North Queensland canegrowers who knew beyond doubt that their cane was being infested with only greyback cane beetle grubs to experiment with different insecticides for the control of cane grubs. They applied lindane, aldrin, heptachlor or dieldrin either broadcast as a dust or as a spray emulsion, although use of a water spray to deliver the insecticide was concentrated in the Lower Burdekin district.[84] Between 1959 and 1968, lindane and aldrin were quite popular, but heptachlor had taken their place by 1970. However, these insecticides were applied to less than 10% of the area treated against cane grubs in Queensland during the 1960s; figures on lindane and aldrin use for the 1970s were not published by the BSES so ascertaining if this trend persisted could not be determined.[85]

Other insect pests troubled Queensland canegrowers episodically during the 1960s and 1970s. Army worm caterpillars, for instance, were quite a problem in the mid-1960s – the largest loss of 15 000 tonnes occurred in 1969 (Figure 14.1). Army worm caterpillars are the progeny of various moth species that occur in each Queensland sugar-producing district. Infestations by army worms, if particularly extensive, can lead to severe defoliation of the cane plant and successive defoliations cause considerable crop losses. This insect pest's depredations were brought under control by applications of insecticides, chiefly DDT.[86] The spur-throated locust, *Austracris guttulosa* Walker, began causing small losses in the early 1970s, especially in the Mackay district. Dieldrin applied through boomsprays or misting equipment was used as a control measure. However, locust infestations were widespread in all districts from Mackay northwards during 1975, and caused a total loss of 42 000 tonnes of cane. Extensive aerial spraying helped bring the outbreak under control.[87] Two species of cicada, *Cicadetta crucifera* Distant and *Parnkalla muelleri* Distant, were first recorded as minor cane pests during the early 1960s, but they destroyed several thousand tonnes of ratoon cane in 1974, 1978 and 1979. Such losses resulted from stunted ratoon crops caused by the cicada nymphs feeding on the roots of ratoon stools. The upsurge in cicada-related losses during the late 1970s prompted the BSES to warn in 1979 that the potential for cicadas to develop into a serious cane pest in Queensland remained a distinct possibility, especially as the only control measure remained the ploughout of affected blocks – no chemical or other means had been devised for eliminating cicadas from ratoons.[88]

Rats occasionally troubled New South Wales canegrowers during the 1960s and 1970s.[89] In contrast, annual losses attributable to rats could be quite significant in Queensland sugar-producing districts, especially throughout North Queensland. During years when weather conditions were favourable in Queensland (e.g. 1965, 1976, 1979), anywhere between 50 000 and 90 000 tonnes of cane were injured by rats (Figure 14.1). Hand distribution of baits remained the most effective control measure against rats, but it was a time consuming and tedious task. CSR complained that 'grower apathy' towards bait laying was pronounced in its Macknade Mill supply area and much of the loss from rats occurred because canegrowers delayed the distribution of baits.

To speed up the process, experiments with aerial baiting commenced in the Tully Mill supply area during 1964. Three years later, the Macknade Voluntary Pest Destruction Board also conducted aerial rat-baiting trials using wheat baits smeared with thallium sulphate and raw linseed oil. These investigations showed that the aerial distribution of rat baits was physically and economically practicable over a wide area. Its advantages included all fields were baited at the same time, baits were placed in all fields, especially fields full of lodged cane which could not be easily hand-baited, and canegrowers were relieved of the unpleasant task of walking through the cane fields to lay the baits.

Moreover, aerial baiting brought about an appreciable degree of rat control in the trial areas.[90] Distribution of rat baits by a fixed wing aircraft across an entire mill supply area then occurred in May 1968, when rat baits were applied to approximately 8 100 ha in the Macknade Mill supply area. Baiting by means of an aircraft then became the standard practice in the Herbert River district and was used in the Proserpine district during 1971; in other affected localities baits continued to be laid by hand or occasionally by air.[91] Yet the BSES concluded that rats, even when baits were distributed by aircraft, remained the most difficult pest to eliminate, for the control methods only reduced damage by 30–50% in heavily affected areas and they had to be undertaken annually.[92]

Various other vertebrate pests continued damaging the sugar cane crops during the 1960s and 1970s. Wallabies proved to be not particularly bothersome over these two decades, except in the drought years of 1966 and 1977 when they caused considerable losses (Figure 14.1). In districts where they were a major nuisance (e.g. Lower Burdekin; Cairns; Mackay), some canegrowers applied blood and bone fertiliser to the young shoots at the rate of approximately 20 kg per ha and this practice proved to be an effective repellant. At Mackay, electric fences excluded wallabies from young cane.[93] Feral pig damage to cane crops in Queensland grew slowly during the 1960s, but rose alarmingly in 1972 (Figure 14.1). By 1975, feral pig damage was being reported in the Lower Burdekin and Mackay districts, where previously feral pigs had been of little importance. Sporadic damage to cane crops from feral pigs also occurred in New South Wales during the late 1970s.[94] More widespread use of control measures, chiefly wire netting, electric fences and hunting throughout the 1970s, prevented even more damage from occurring.[95] Cockatoos also caused occasional damage, except in 1979 when 6 000 tonnes of cane was lost in the Tully and South Johnstone districts from the attacks by flocks of these birds.[96]

New approaches to pest management, 1980–1995

During the mid-1970s, BSES scientists became aware of poor growth problems in young plant and ratoon crops of sugar cane. They decided to commence further investigations – some had been undertaken in the late 1950s – into the effect of parasitic nematodes on yields and ways to control any impact. Nematodes are microscopic, worm-like animals that break down dead organic matter in the soil. Crop damage occurs when the population density of a pathogenic species is large enough to curtail root development in growing conditions that are not ideal for the cane crop.[97] The BSES commenced trials during the late 1970s and early 1980s to investigate if commercial organophosphate and carbamate nematocides and insecticides such as Temik® (active substance aldicard), Nemacur® (active substance fenamiphos) and Mocap® (active substance ethoprophos) reduced nematode numbers, thereby having a beneficial impact upon the growing crops. Results showed that applications of nematocides improved yields, in some instances by 50%.[98] In 1980, the first attempt at estimating losses from nematodes in Queensland cane crops was conducted over approximately 6 600 ha of coastal sandy soils in the Bundaberg district. The results indicated that these pests were responsible for an estimated loss of 85 000 tonnes of cane. The application of nematocides from 1979 onwards, particularly ethoprophos and aldicard, reduced the annual nematode losses occurring mostly in Southern Queensland.[99] By the mid-1990s, nematode losses were estimated at only 7 000 tonnes of cane annually (Figure 14.2).

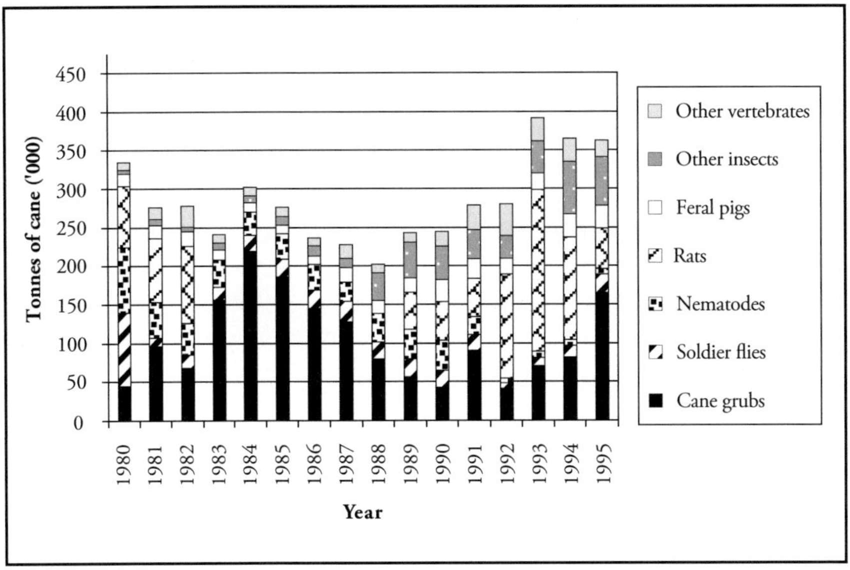

Note: Figures on rat damage from 1983 to 1988 were given as tonnes of sugar not tonnes of cane.

Figure 14. 2. Estimated annual losses (tonnes of cane) to the Queensland sugar crop from pests, 1980–1995.

Source: Based upon figures in the BSES, AR 1980–81 to 1995–96.

Losses from nematodes following the introduction of control measures, however, were dwarfed by the growing damage done by cane grubs in both New South Wales and Queensland. Most grub-related injuries to the crops of sugar cane in New South Wales happened in the Tweed River and Richmond River districts, especially on farms with red-volcanic or peat soils.[100] In Queensland, record levels of damage from cane grubs occurred during the mid-1980s (Figure 14.2), following a vast upsurge in the incidence of the pest, mainly *Lepidota consobrina* Girault, *Lepidota frenchi* Blackburn and the various *Antitrogus* species of cane beetle. Increased grub numbers were also associated with the growing failure of broadcast applications of BHC to control cane grubs, especially the larvae of the Childers cane beetle (*Antitrogus parvulus* Britton) and *Antitrogus consanguineus* Blackburn at Bundaberg. The situation was further compounded in the Bundaberg district by the rapid escalation to major pest status of *Lepidiota picticollis* Lea, a previously rare species with known natural tolerance to BHC. In addition, poor control of cane grubs had been achieved following the incorrect application of BHC in widely separated bands: cane grubs feeding between the bands were not killed.[101] Alternative control measures were becoming a priority.

Escalating cane grub resistance to BHC had also been accompanied by a growing global disquiet about the environmental impact of organochlorine insecticides – the publication of Rachel Carson's book *Silent Spring* in 1962 was a significant trigger. Brian Hitchcock, a BSES entomologist, together with the precursors of the agricultural chemical supplier Crop Care Australia, had responded to these concerns by developing the alternative insecticide suSCon®Blue, based upon the organophosphate chlorpyrifos, and using controlled release technology (i.e. releasing a bioactive substance at a sufficient rate over a period of time in order to achieve a desired effect on a target species). Other commercial organophosphate or carbamate insecticides available by 1980 included Mocap®10G (active substance ethoprophos), Dasanit® 10G (active substance fensulfothion) and Counter®10G (active substance terbufos).[102] These new insecticides were tested by the BSES and NSW Sugar Milling Coop. during trials in Queensland and New South Wales throughout the late 1970s and early 1980s. Ethoprophos offered some protection against cane grub attacks, but the insecticide's adoption would be limited because of its short persistence and the need for sufficient soil moisture at the time of use. Applications of chlorpyrifos reduced cane grub numbers and improved yields, so it looked a promising alternative to BHC.[103]

During 1984, some Queensland canegrowers, especially in the Bundaberg district, began using ethoprophos and fensulfothion, in preference to BHC.

Consolidated Fertilisers Ltd. also marketed a controlled release formulation of chlorpyrifos, which was applied to about 600 ha in Queensland.[104] Figures on the percentage of the area treated in Queensland with a particular insecticide applied to control cane grubs were collected between 1985 and 1992 (Table 14.1). The figures showed that BHC quickly lost favour, as canegrowers switched to suSCon®Blue, ethoprophos and fensulfothion. This trend was accelerated after September 1987, when BHC was declared a prohibited chemical in Queensland, following the discovery of unacceptable residual concentrations of DDT in Australian beef imports to the United States of America.[105] By 1992, the use of ethoprophos and fensulfothion against cane grubs had all but vanished in Queensland, the state's canegrowers preferring to apply suSCon®Blue which had proved to be more effective in reducing the cane grub attacks (Table 14.1).

The new insecticides, however, were not always effective. A marked increase in the tonnage of cane damaged by cane grubs occurred during the early 1990s (Figures 14.2) and was blamed upon the apparent failures of suSCon®Blue to control the grub of the Childers cane beetle (*Antitrogus parvulus* Britton) throughout the Isis district in 1989 and the greyback cane beetle grub in the Lower Burdekin district in 1992, 1993 and 1994 and the Innisfail district in 1995. Additional BSES investigations were then conducted into factors

Table 14. 1. The percentage of the area treated in Queensland with particular insecticides to control cane grubs, 1985–1992.

Year	BHC	Heptachlor	Ethoprophos (Mocap) and fensulfothion (Dasanit)	suSCon® Blue	Total area treated (ha)
1985	63	12	12	13	*
1986	35	*	17	37	*
1987	20	*	15	58	29 360
1988	10	*	14	76	33 640
1989	0	0	14	86	31 950
1990	0	0	9	91	33 620
1991	0	0	12	88	28 200
1992	0	0	4	96	23 540

* Not mentioned

Source: Based upon figures in BSES, ARs 1985–1986 to 1992–1993.

limiting the product's performance under different conditions. Studies showed that some species of cane beetles (e.g. *Lepidiota noxia* Britton, *Lepidiota caudata* Blackburn and *Lepidiota picticollis* Lea) appeared to be inherently less susceptible to chlorpyrifos, so suSCon®Blue was probably unsuitable for controlling these species and different control measures needed to be implemented. Moreover, placement of the chlorpyrifos granules was critical for achieving control, as the insecticide's impact was concentrated near the granules. Investigations showed that optimum placement varied with target species. Granules positioned 15–20 cm below the soil surface gave better control of greyback cane beetle grubs than did granules positioned 20–30 cm deep. The Childers cane beetle grub ((*Antitrogus parvulus* Britton) was better controlled by granules 20 cm deep. Moreover, suSCon®Blue must be covered with at least 50 mm of compacted soil immediately after application or losses of up to 90% of the active ingredient can occur within two months.[106]

The withdrawal of BHC and then the failure of the new products highlighted the need for alternative control measures against the cane grubs. BSES entomologists began reconsidering methods investigated during the pre-World War II period. Throughout 1989 and 1990, various substances (e.g. eugenol, citral, citronella, valeric acid, eucalyptus oil) were tested as beetle attractants, although none of the compounds proved sufficiently attractive to continue this line of inquiry. Attempts to control greyback cane beetle grub numbers by trapping adult beetles at lights near Hone Hill – reminiscent of Edmund Jarvis' early work into control measures – failed to prevent severe cane grub damage to 300 ha in this area in 1993 and 1994. Interest in pathogens as microbial insecticides led the scientists to reconsider the work of Henry Tryon, who discovered in the mid-1900s that *Metarhizium anisopliae* Metchnikoff, a fungal pathogen, was fatal to cane grubs (see Chapter Eight). Field trials were commenced during 1991 to test the effectiveness of *M. anisopliae* in controlling populations of one-year old cane grubs for the species *Antitrogus consanguineus* Blackburn. These investigations, conducted in conjunction with Incitec Fertilisers Ltd. – the main Queensland fertiliser supplier – and CSIRO showed that the application of the spore suspensions into infested ratoon crops did reduce cane grub populations. However, the delay of one to two months before significant mortality occurred was a major drawback. Screening cane grubs for diseases revealed other pathogens with the potential for suppressing cane grub numbers. By 1995, however, the development of a *M. anisopliae* formulation and other formulations were still in their early stages.[107]

Cane grubs were not the only insect pest developing resistance to organochlorine pesticides during the early 1908s. In mid-1982, some South Kalkie canegrowers near Bundaberg reported problems with wireworm control following the application of the aldrin-fertiliser mixtures. BSES entomologist Brian Hitchcock reported in October 1982 that a similar problem had emerged on one farm in the Mackay district. Wireworms, it was surmised, after many years of aldrin usage had become resistant to the insecticide.[108] A spray of chlorpyrifos as Lorsban® over the setts at 1.5 litres of product per ha replaced the use of aldrin from 1983 onwards. By 1988, applying the controlled-release granules of chlorpyrifos in suSCon®Blue at 21 kg of the product per ha achieved wireworm control equivalent to that reached with liquid chlorpyrifos. By 1990, wireworm damage to Queensland' sugar cane crops appeared to have been sufficiently brought under control through the use of insecticides containing chlorpyrifos, for it no longer rated a mention in the BSES annual reports during the early 1990s.[109]

Organochlorine pesticides had not failed against all insect pests troubling the Australian sugar industry during the mid-1980s. Dieldrin was still applied against soldier flies, heptachlor retained its efficacy against funnel ants and applications of BHC controlled the attacks of African black beetles. However, the phasing out of organochlorine pesticides in Australia during the mid-1980s meant that dieldrin, heptachlor and aldrin all became prohibited substances after 1987, although restricted usage of dieldrin for soldier fly control and heptachlor for funnel ant was permitted in some sugar-producing districts until 1991 and 1994, respectively.[110] Insecticide trials against African black beetles in the early 1990s showed that terbufos and chlorpyrifos (Lorsban®) reduced damage from African black beetles, so the gains in yields warranted the expense associated with the application of the insecticide.[111] In contrast, trials with a variety of insecticides (e.g. chlorpyrifos; aldicard; carbofuran) during the late 1980s and early 1990s proved mostly ineffective against soldier fly. Investigations, however, revealed that soldier fly losses could be managed by cultivation methods. Long, bare fallows caused heavy population reductions in infected fields and severe ploughout methods using rotary hoes destroyed stools and injured larvae. A rating system was also developed during the early 1990s to determine the tolerance of many commercial cane varieties to soldier fly. The cane varieties C.P. 44–101, Q. 144, Q. 146, Q. 151 and Q. 154 were shown to be the most tolerant; H. 56–752 emerged as the most susceptible to soldier fly attack. This system allowed canegrowers to make informed decisions about cane variety selection in soldier fly prone areas.[112]

As dieldrin use against soldier flies was phased out, losses from the pest during the late 1980s remained consistent at approximately 23 000 tonnes of cane annually, costing the Queensland sugar industry around $3–5 million each year. After a contraction in soldier fly losses during 1992 and 1993, the amount of damage inflicted by these insect pests again increased during 1994 and 1995, with a trend towards increased damage in the Mackay district (Figure 14.2). Moreover, a worrying sign had emerged – soldier fly damage had been found for the first time in the Lower Burdekin district during 1994.[113] Thus, soldier fly remained a major concern for Australian canegrowers in 1995, especially as they had no registered insecticide that could be used to reduce its damage to the crops of sugar cane. In contrast, even though new infestations of African black beetles were recorded in the Maryborough district in 1990, the extent of losses during the early 1990s were being kept under control by the application of terbufos and chlorpyrifos. Yet canegrowers throughout the Bundaberg and Isis districts in 1995 were faced with the prospect of this exotic insect pest moving even further northwards and infesting their districts.[114]

A forgotten insect pest re-surfaced as a major menace in North Queensland during the early 1990s. Mulgrave and Mourilyan Mills each lost an estimated $1 million in sugar from weevil borer attacks in both 1993 and 1994. Weevil borer damage was then assessed in affected northern mill supply areas during 1995. Half the cane examined in the Mourilyan Mill supply area had suffered damage from weevil borers, a third of the cane delivered to the Mulgrave, Babinda and South Johnstone Mills had been affected by weevil borers and isolated losses occurred in the Herbert River and Mackay districts. Increased damage from this insect pest had occurred following an escalation in weevil borer numbers, a direct consequence of reduced pre-harvest burning after the introduction of green cane harvesting and trash blanketing. By 1995, the BSES entomologists were considering alternative control strategies in order to retain trash blanketing in North Queensland as no insecticides were registered for control of the weevil borer. In addition, the BSES plant breeding program had initiated a sub-program to breed weevil borer-resistant cane varieties.[115]

Feral pigs and rats remained the two most significant vertebrate pests for the Australian sugar industry between 1980 and 1995. The tonnage of cane damaged by feral pigs in Queensland escalated during the 1980s, reaching 30 000 tonnes in both 1994 and 1995, double the amount lost in 1980 (Figure 14.2). The cost of feral pig damage reached over $1 million each year during the early 1990s. Rat damage in the New South Wales cane growing areas during the early 1980s was described as 'high', for a lack of flooding had not reduced their numbers.[116] In Queensland, rats caused damage to approximately 87 000 tonnes of cane each year over the period 1980–1982. Figures on the tonnage of cane damaged by rats in Queensland were then not collected during the mid-1980s, as the BSES decided that rat injury caused a loss of sugar. However, when the figures were again collected in the early 1990s, the amount of rat damaged cane each year was still substantial (Figure 14.2). Presumably wallabies, cockatoos and red bills also caused some damage during the 1980s and early 1990s, but they were not mentioned as separate categories in the BSES annual reports, which increasingly contained fewer details on pest losses.

Reducing the numbers of feral pigs affecting Queensland's sugar cane farms continued to be achieved through trapping and shooting organised by the Cane Pest and Disease Control Boards (known as Cane Protection and Productivity Boards after 1991). During the mid-1990s, an organised feral pig trapping program managed by the Land Protection Branch of the Queensland Department of Lands also aimed to protect sugar cane crops bordering the rainforests of the Wet Tropics World Heritage Area in North Queensland.[117] Rat numbers continued to be managed by the aerial distribution of thallium sulphate baits. However, the BSES commented in 1980 that even with the use of 'massive amounts of baits', damage in years of high rat incidence could only be reduced by 30–50%. Moreover, growing concerns about the use of thallium sulphate baits meant this option was not feasible in the long-term. Therefore, new BSES investigations during the 1980s into rodent control methods showed that reducing nearby non-crop habitats available as refuges and slowing the immigration of rats to the cane fields could reduce rodent injury to the sugar cane. Suitable habitats near cane fields could be made unsuitable by mowing grassy headlands or replacing grassed areas with closed-canopy forest by revegetating stream banks and gullies.[118]

Thallium sulphate was eventually banned nationally in 1992. A new rodenticide, Brodifacoum (Klerat®), was provisionally registered in 1991 for use in sugar cane against canefield rats, subject to further investigations into the potential for secondary poisoning of other animals or birds. However, it could only be applied to fields at risk from rodents and not from aircraft, so the baits needed to be distributed when the cane was short, before rat populations were large and damage was evident. Therefore, rodent control in the mid-1990s had become more complex than the previous strategy of aerial baiting each year. Canegrowers at risk from rodent damage had to plan to control the pests during the summer months by reducing weed infestations and applying baits if model predictions based on rainfall, potential weed growth and numbers of rodents trapped during early growth of the crops suggested high numbers of rodents throughout the following year.[119]

By the early 1990s, 350 000 tonnes of sugar cane was being lost annually to pests in Queensland (Figure 14.2). For some canegrowers, these record losses were an alarming sign that they were loosing the battle against the animals damaging their crops. On the other hand, some Australian canegrowers had realised that single-strategy controls, so favoured by the industry since World War II, based on replacing one insecticide or rodenticide with another would be unviable in the long term. A limited range of alternative pesticides and rodenticides, potential resistance to these pesticides and public concern about the pest control strategies used by the Australian sugar industry contributed to this realisation. Moreover, many Australian canegrowers who had made the switch to minimum tillage techniques were also demanding more environmentally friendly pest control strategies that included greater responsible use of registered pesticides and more benign substitutes for increasingly costly pesticides. Instead of attempting to eliminate or directly control the pests, Australian canegrowers had commenced redesigning the production system, improving management strategies and introducing integrated pest management strategies. By 1995, the Australian sugar industry's approach to pest management had entered a period of transition.[120]

Notes

1 R. Veitch, *The Cane Beetle Borer in Australia.* CSR Agricultural Circular No. 3 (Sydney, 1917), pp. 4 & 8; R. Mungomery, 'The rise and fall of the sugarcane weevil borer pest in Queensland', *Proc. ISSCT*, 1953, p. 587; *ASJ*, 5 November 1914, p. 550.

2 *ASJ,* 6 August 1914, p. 374; W. Farquhar, CSR's Mill Insp., to Goondi Mill Mang., 19 November 1914, p. 2, CSRR, N 126/150; Mungomery, 'The rise and fall of the sugarcane weevil borer pest in Queensland', p. 587.

3 J. Buzacott, 'The tachinid fly', *CGQB*, Vol. 1 (October 1933), p. 57; BSES, AR 1917–1918, *QPP*, Vol. 1 (1918), pp. 1123 & 1149.

4 Buzacott, 'The tachinid fly', p. 57; E. Jarvis, 'Cane pest combat and control', *QAJ*, Vol. 28, no. 8 (August 1927), p. 109; BSES, AR 1919–1920, *QPP*, Vol. 2 (1920), p. 202; BSES, AR 1924–25, *QPP*, Vol. 2 (1926), p. 703.

5 K. Gard, 'Notes on tachinid fly breeding at Macknade, 19 June 1922', pp. 2–3 & K. Gard, 'Notes on tachinid fly breeding at Macknade, 4 August 1926', pp. 2–4, in CSRR, Z/395, Box 1; E. Jarvis, 'Present position of entomology with regard to the control of our chief insect pests of sugar cane', *Proc. QSSCT*, 1930, p. 145.

6 CSR, Cane Report 1930, p. 50, CSRR, Z/109, Box 386.

7 Calculated from estimated annual losses in CSR's annual cane reports, 1915 to 1929, in CSRR, Z/109, Boxes 377 & 386; BSES, AR 1925–26, *QPP*, Vol. 2 (1927), p. 661.

8 For these views see BSES, AR 1924–25, *QPP*, Vol. 2 (1926), p. 701.

9 Staff Register, Officers Service Series, 1904–1958: A–G, entry for Hunter Freeman, CSRR, Z/109, Box 555.

10 CSR, Cane Report 1922–1923, p. 23, CSRR, Z/109, Box 386.

11 For an example of this advice see R.W. Mungomery, 'The control of insect pests of sugar cane', *CGQB*, Vol. 2, no. 1 (July 1934), p. 4; for these experiments see CSR, Cane Report 1926, p. 21 & CSR, Cane Report 1931, p. 56, in CSRR, Z/109, Box 386; & CSR, Cane Report 1939, p. 57, CSRR, Z/109, Box 387.

12 *ASJ*, 6 May 1909, p. 29, 3 February 1910, pp. 412–413 & 3 March 1910, p. 481; CSR, Cane Report 1925, pp. 38–39; CSR, Cane Report 1926, p. 21; CSR, Cane Report 1928, p. 32 – all located in CSRR, Z/109, Box 386.

13 E. Jarvis, *Notes on Queensland Cane-Insects and Their Control.* BSES, Division of Entomology, Bulletin No. 17, 1923, p. 79–80.

14 BSES, AR 1929–30, *QPP* Vol. 2 (1930), p. 945.

15 BSES, AR 1921–22, *QPP*, Vol. 2 (1923), pp. 278 & 279; BSES, AR 1927–28, *QPP*, Vol. 2 (1929), p. 881; BSES, AR 1926–27, *QPP*, Vol. 2 (1927), p. 739.

16 B. Egan, 'Origins and development of the sugarcane pest and disease control board system in Queensland, 1890–1920', *Proc. ASSCT*, 2002, p. 277.

17 Egan, 'Origins and development of the sugarcane pest and disease control board system in Queensland, 1890–1920', p. 277.

18 Egan, 'Origins and development of the sugarcane pest and disease control board system in Queensland, 1890–1920', p. 278; Easterby, *The Queensland Sugar Industry*, p. 223; BSES, AR 1935–36, p. 19.

19 *QVP*, Vol. 4 (1889), p. 294 (Ev. W. Paget, Nindaroo plantation, Q. 6810) and p. 388 (Ev. W. Robertson, Habana plantation); *Q*, 8 October 1910, p. 39; *Cairns Morning Post*, 22 July 1907, p. 2 & 4 December 1908, p. 5.

20 *ASJ*, 9 September 1909, p. 333 & 6 April 1911, p. 45; Union Bank, Report on Half Yearly Accounts, Qld. Division, 28 February 1906, p. 53, U/220/1, UBR, ANZA; Cultivation Overseer, Macknade Mill, 'Report on the Work Done Towards the Extermination of the Rat Pest on Macknade Cane Areas, 6 March 1906', p. 358, CSRR, 142/2766; Kerr, *Northern Outpost*, pp. 95–96.

21 P. Allsopp, P. Samson and K. Chandler, 'Pest Management', in M. Hogarth and P. Allsopp (eds.), *Manual of Cane Growing* (Brisbane, 2000), pp. 331–332; King, Mungomery and Hughes, *Manual of Cane-Growing*, 1953 edition, pp. 339–341.

22 CSR, Cane Report 1921, pp. 48–49, CSRR, Z/109, Box 386; O. Stevenson, 'Rat campaign at Macknade and Victoria Mills, 7 December 1922', pp. 8–10, CSRR, Z/395, Box 1; BSES, AR 1921–22, *QPP*, Vol. 2 (1923), p. 278; CSR, Cane Report 1933, pp. 47– 48, CSRR, Z/109, Box 386.

23 K. Gard, 'The rat pest in cane areas', *Proc. ISSCT*, 1935, p. 595; *ASJ*, 10 April 1934, p. 46; CSR, Cane Report 1933, pp. 47–49, CSRR Z/109, Box 386.

24 CSR, Cane Report 1934, pp. 40– 41, CSRR, Z/109, Box 387; *ASJ*, 4 December 1934, pp. 477– 479.

25 CSR, Cane Report 1934, pp. 41–42, CSRR, Z/109, Box 387; *ASJ*, 9 July 1935, p. 189 & 12 December 1935, p. 511.

26 BSES, AR 1934–35, p. 829; BSES, AR 1935–36, p. 22; BSES, AR 1938–39, p. 50; *ASJ*, 10 August 1937, pp. 305–306, 10 September 1937, p. 385, 11 February 1938, p. 669 & 9 June 1938, p. 179.

27 McDougall, 'An investigation of the rat pest problem in Queensland cane fields', p. 39.

28 BSES, AR 1932–33, *QPP*, Vol. 2 (1933), pp. 818–819; BSES, AR 1933–34, *QPP*, Vol. 2 (1935), p. 788; BSES, AR 1935–36, p. 21; W. McDougall, 'Cane borers in the Mackay district', *CGQB*, Vol. 2, no. 1 (July 1934), p. 21.

29 E. Jarvis, 'Entomological hints to canegrowers', *QAJ*, Vol. 37, part 3 (March 1932), p. 155; BSES, AR 1930–31, *QPP*, Vol. 2 (1931), p. 713; BSES, AR 1931–32, *QPP*, Vol. 2 (1932), p. 614; A. Bell, 'Sanctuaries for the tachinid fly parasite of the beetle borer', *CGQB*, Vol. 3, no. 1 (July 1935), p. 19.

30 BSES, AR 1933–34, *QPP*, Vol. 2 (1935), pp. 788–789; BSES, AR 1934–35, p. 827–828.

31 R. Mungomery, 'The superiority of S.J. 4 over Badila in regard to beetle borer attack', *CGQB*, Vol. 2, no. 2 (October 1934), p. 49; J. Buzacott, 'Relationship between hardness of sugar cane and varietal resistance to beetle borer (*Rhabdocnemis obscura* Biosd)'. MSc. thesis, Uni. of Qld., 1938.

32 BSES, AR 1937–38, pp. 44–45; BSES, AR 1938–39, pp. 49–50; King, Mungonery and Hughes, *Manual of Cane-Growing*, 1953 edition, p. 336; A. Bell, 'Recent advances in cane borer investigations in North Queensland', *CGQB*, Vol. 4, no. 4 (April 1937), p. 151–156.

33 *ASJ*, 10 August 1937, p. 305; CSR, Cane Report 1940, p. 57, CSRR, Z/109, Box 387.

34 A. Koebele, 'Sugar-cane insects in New South Wales', *Insect Life*, Vol. 4 (1891–92), pp. 388–389; Qld. Entomologist, AR 1910–11, *QPP*, Vol. 2 (1911–12), p. 863; R. Mungomery, 'A review of sugar cane entomological investigations', in BSES, *Fifty Years of Scientific Progress* (Brisbane, 1950), p. 46; W. McDougall, *The Wireworm Pest and its Control in Central Queensland Sugar-Cane Fields*. BSES Division of Entomology, Bulletin No. 22, 1935, p. 1.

35 BSES, AR 1931–32, *QPP*, Vol. 2 (1932), p. 619; W. McDougall, 'The wireworm pest in the Mackay and Proserpine Cane Fields', *CGQB*, Vol. 1, no. 1 (July 1933), p. 47–48.

36 Allsopp, Samson and Chandler, 'Pest Management', pp. 312–313.

37 McDougall, 'The wireworm pest in the Mackay and Proserpine Cane Fields', p. 48; W. McDougall, 'Wireworms (*Elateride*) and their control in Queensland sugar-cane fields', *Proc. ISSCT*, 1935, p. 584–585.

38 BSES, AR 1936–37, p. 37; BSES, AR 1937–38, p. 42; BSES, AR 1939–40, p. 17.

39 D. North, Report on the Black Beetle Pest, 4 March 1937, Letter No. 475D (b), pp. 1–2, CSRR, Z/109, Box 241; P. Samson and P. Downs, 'Efficacy of Terbufos against African black beetle', *Proc. ASSCT*, 1994, p. 159; Allsopp, Samson and Chandler, 'Pest Management', p. 318.

40 D. North to P. Goldfinch, CSR Gen. Mang., 18 September 1931, Letter No. 334 D; D. North to P. Goldfinch, CSR Gen. Mang., 1 December 1931, Letter No. 340D, p. 3; D. North, Report on the Black Beetle Pest, 4 March 1937, Letter No. 475 D (b), pp. 3–5, all in CSRR, Z/109, Box 241; D. North to P. Goldfinch, CSR Gen. Mang., 10 March 1933, Letter No. 375 D, pp. 3–7, CSRR, Z/109, Box 241.

41 For more details see P. Griggs, 'Entomology in the service of the state: Queensland scientists and the campaign against cane beetles, 1895–1950', *Historical Records of Australian Science*, Vol. 16, no. 1 (2005), pp. 15–16.

42 M. Tyler, *Australian Frogs. A Natural History* (Sydney, 1994), pp. 105–106; R. Mungomery, 'The giant American toad (*Bufo marinus*)', *CGQB*, Vol. 3, no. 1 (July 1935), p. 21.

43 R. Mungomery, 'A short note on the breeding of *Bufo marinus* in captivity', *Proc. ISSCT*, 1935, pp. 589–590; *ASJ*, 12 December 1935, p. 485 & 10 October 1936, p. 421.

44 BSES, AR 1938–39, pp. 48–49; BSES, AR 1939– 40, p. 17.

45 R. Mungomery, 'The control of insect pests of sugar cane', *CGQB*, Vol. 2, no. 1 (July 1934), p. 2; E. Fox, 'Experimental attempts to control grub damage: clearing of beetle feeding trees on the Basilisk Range, Mourilyan, 1933–1935', *Proc. QSSCT*, 1939, pp. 61–64; BSES, AR 1937–38, p. 42; BSES, AR 1938–39, p. 48; BSES, AR 1939– 40, p. 16.

46 H.G. Knust, 'Control of cane grubs in North Queensland by fumigation', *CGQB*, Vol. 2, no. 1 (July 1934), pp. 16–18; CSR, Agricultural Circular: Hambledon Mill, 1936 (reproduced in *ASJ*, 11 June 1936, p. 147); R. W. Mungomery and J. Buzacott, 'Control of the "Greyback Cane Beetle" (*Lepidoderma albohirtum* Waterh.) in North Queensland', *Proc. ISSCT*, 1935, p. 457.

47 *ASJ*, 12 July 1938, p. 240; BSES, AR 1937–38, p. 42; BSES, AR 1939– 40, p. 16.

48 BSES, AR 1940–41, pp. 19–20; BSES, AR 1942–43, p. 18; BSES, AR 1943–44, p. 21; BSES, AR 1944–45, p. 20.

49 J. Buzacott, 'The Blundell knapsack injector', *CGQB*, Vol. 9, no. 1 (July 1941), pp. 1–8; H.G. Knust, 'Control of cane grubs in North Queensland by fumigation,' *CGQB*, Vol. 2, no. 1 (July 1934), p. 18; W. McDougall, 'Fumigation for the control of cane grubs,' *CGQB*, Vol. 8, no. 1 (July 1940), pp. 27–28; J. Buzacott, 'The Blundell mechanical fumigant injector', *CGQB*, Vol. 10, no. 1 (July 1946), pp. 10–11; *ASJ*, 10 August 1935, p. 271.

50 BSES, AR 1940–41, p. 20; BSES, AR 1942–43, p. 18; BSES, AR 1944–45, p. 20; Mungomery, 'A review of sugar cane entomological investigations', p. 46; Mungomery, 'The rise and fall of the sugarcane weevil borer pest in Queensland', p. 592.

51 For more details see CSR, Chemical Report on Manufacture of Raw Sugar, 1914, p. 56, CSRR, Z/109, Box 377; *ASJ*, 4 February 1915, p. 809 & 11 March 1915, p. 890.

52 BSES, AR 1943–44, pp. 22–23; BSES, AR 1944–45, p. 21; Rolleston, *The Defiance*, p. 129.

53 Hambledon Mill Cane Insp., AR 1943, pp. 10–11, CSRR, Z/109, Box 468; Goondi Mill Cane Insp., AR 1945, p. 13, CSRR, Z/109, Box 465; Hudson, *By the Banyan*, p. 261.

54 F. Reading, 'Wallaby Control', *CGQB*, Vol. 10, no. 4 (April 1947), p. 171.

55 BSES, AR 1942–43, p. 18; BSES, AR 1943–44, p. 23; BSES, AR 1944–45, p. 21.

56 BSES, AR 1945–46, p. 32; P. Volp, 'Notes on beetle collections in the Mulgrave area, 1946–47 Season', *CGQB*, Vol. 11, no. 2 (October 1947), p. 77.

57 R.W. Mungomery, 'The use of "Gammexane" in controlling "White Grubs" in Queensland cane fields', *CGQB*, Vol. 12, no. 1 (July 1948), p. 21; BSES, AR 1971–72, p. 43; J. Cavanagh, 'Organochlorine insecticide usage in the sugar industry of the Herbert and Burdekin River regions: chemical, biological and risk assessments'. PhD thesis, James Cook Uni., 2000, pp. 14–15.

58 BSES, AR 1945–46, p. 33; BSES, AR 1946–47, pp. 36–37; CSR, Cane Report 1945, pp. 39–40 & CSR, Cane Report 1946, pp. 46–48, in CSRR, Z/109, Box 387.

59 BSES, AR 1947–48, pp. 32–33; BSES, AR 1948–49, pp. 37–38; R. Mungomery, 'The use of benzene hexachloride in controlling "White Grubs" in Queensland cane fields', *Proc. QSSCT*, 1948, pp. 35–42.

60 BSES, AR 1946–47, p. 38; G. Wilson, 'Frenchi grub control in North Queensland by benzene hexachloride', *Proc. ISSCT*, Brisbane, 1950, pp. 414–416; R. Mungomery,

'The status of benzene hexachloride in the control of some Queensland sugar cane pests', *Proc. ISSCT*, 1950, p. 404.

61 For these trials see CSR, Cane Reports 1947 to 1952, in CSRR, Z/109, Boxes 387 & 388.

62 King, Mungomery and Hughes, *Manual of Cane-Growing*, 1953 edition, pp. 311–312; C. Morton, *By Strong Arms. The Mulgrave Central Mill Company Limited* (Gordonvale, 1995), p. 139.

63 BSES, AR 1948–49, p. 39; BSES, AR 1949–50, p. 41; CSR, Cane Report 1950, p. 64, CSRR, Z/109, Box 388.

64 King, Mungomery and Hughes, *Manual of Cane-Growing*, 1953 edition, pp. 313–314; Mungomery, 'The status of benzene hexachloride in the control of some Queensland sugar cane pests', pp. 406–407; CSR, Cane Report 1947, p. 60, CSRR, Z/109, Box 387.

65 BSES, AR 1946–47, p. 37; BSES, AR 1947–48, p. 32; BSES, AR 1948–49, p. 37; CSR, Cane Report 1949, p. 57, CSRR, Z/109, Box 387.

66 BSES, AR 1952–53, p. 40; CSR, Cane Reports for 1950 to 1954, various pages, in CSRR, Z/109, Box 388.

67 BSES, AR 1949–50, p. 41; Mungomery, 'The status of benzene hexachloride in the control of some Queensland sugar cane pests', *Proc. ISSCT*, 1950, p. 409.

68 BSES, AR 1952–53, p. 38; BSES, AR 1953–54, pp. 61–62.

69 BSES, AR 1956–57, p. 70; BSES, AR 1957–58, p. 79; BSES, AR 1958–59, p. 73.

70 CSR, Cane Report 1946, p. 60; CSR, Cane Report 1948, p. 63, CSR, Cane Report 1949, p. 64, all in CSRR, Z/109, Box 387; CSR, Cane Report 1959, p. ii, CSRR, Z/109, Box 389.

71 BSES, AR 1948–49, p. 39.

72 BSES, AR 1955–56, p. 78; BSES, AR 1956–57, p. 77; BSES, AR 1957–58, p. 86; BSES, AR 1958–59, p. 80; BSES, AR 1959–60, p. 78.

73 Hudson, *By the Banyan*, p. 261; BSES, AR 1946–47, p. 39; BSES, AR 1947–48, p. 35; BSES, AR 1948–49, p. 39; BSES, AR 1954–54, p. 66; Reading, 'Wallaby Control', p. 170–171.

74 B. Hitchcock, 'The soldier fly pest', *Proc. QSSCT*, 1970, p. 207; King, Mungomery and Hughes, *Manual of Cane-Growing*, 1965 edition, pp. 325–326.

75 Allsopp, Samson and Chandler, 'Pest Management', pp. 310–311; P. Samson, 'Management of soldier flies in sugarcane', *Proc. ASSCT*, 1995, pp. 75–76.

76 BSES, AR 1961–62, pp. 50–51; BSES, AR 1962–63, p. 51; R. Moller, 'Soldier fly-past, present and future', *CGQB*, Vol. 32, no. 1 (July 1968), p. 4; King, Mungomery and Hughes, *Manual of Cane-Growing*, 1965 edition, pp. 326.

77 BSES, AR 1972–73, p. 12; Moller, 'Soldier fly-past, present and future', p. 4; Hitchcock, 'The soldier fly pest', p. 210.

78 King, Mungomery and Hughes, *Manual of Cane-Growing*, 1965 edition, p. 328.

79 BSES, AR 1976–77, p. 9; BSES, AR 1979–80, p. 9; J. Wright, 'Alarming increase in soldier fly damage at Maryborough', *CGQB*, Vol. 40, no. 4 (April 1977), pp. 126–127.

80 CSR, Agricultural Circular NSW, 1970, pp. 1–2; CSR, Report on Technical Field Work, 1969–70, p. vi, CSRR, Z/109, Box 392.

81 BSES, AR 1960–61, p. 54; BSES, AR 1961–62, p. 56; BSES, AR 1962–63, p. 50; BSES, AR 1965–66, p. 51.

82 BSES, AR 1966–67, p. 47; BSES, AR 1971–72, p. 10; BSES, AR 1972–73, p. 13; BSES, AR 1975–76, p. 10; BSES, AR 1976–77, pp. 8 & 10; BSES, AR 1977–78, p. 9;

B. Hitchcock, 'The changing status of pests in Queensland cane fields', *Proc. QSSCT*, 1974, p. 111.

83 NSW Sugar Milling Coop., AR 1978–79, p. 8; NSW Sugar Milling Coop., AR 1979–80, p. 8; NSW Sugar Milling Coop., AR 1980–81, p. 13.

84 G. Wilson, 'Selecting an insecticide for cane grub control', *CGQB*, Vol. 30, no. 1 (July 1966), pp. 12–15; R. Mungomery, 'Modern trends in grub control in the Burdekin district', *CGQB*, Vol. 25, no. 2 (October 1961), pp. 40–41; BSES, AR 1960–61, p. 55; BSES, AR 1961–62, p. 56; BSES, AR 1962–63, pp. 58–59; King, Mungomery and Hughes, *Manual of Cane-Growing*, revised 1965 edition, p. 308.

85 Calculated from figures listed in BSES, ARs 1959–60 to 1969–70.

86 Allsopp, Samson and Chandler, 'Pest Management', pp. 313–314; King, Mungomery and Hughes, *Manual of Cane-Growing*, revised 1965 edition, pp. 331–333; BSES, AR 1964–65, p. 52; BSES, AR 1967–68, p. 60.

87 BSES, AR 1975–76, p. 12; P. Amiet, 'Locusts in the Mackay district, 1971–75', *Proc. QSSCT*, 1976, pp. 63–65; D. Williams, 'Locust control in the Inkerman Mill area, 1974–75', *Proc. QSSCT*, 1976, pp. 67–69.

88 G. Wilson, 'Cicadas as pests of sugarcane in Queensland', *Proc. ISSCT*, 1968, pp. 1410–1414; BSES, AR 1962–63, p. 63; BSES, AR 1972–73, p. 14; BSES, AR 1974–75, p. 10; BSES, AR 1978–79, p. 9.

89 CSR, Cane Report 1962, p. iv, CSRR, Z/109, Box 389; CSR, Report on Technical Field Work 1967–68, p. iv, CSRR, Z/109, Box 390; NSW Sugar Milling Coop., AR 1978–79, p. 8; NSW Sugar Milling Coop., AR 1980–81, p. 13.

90 CSR Cane Report 1963, p. iv, Box 390; Hudson, *By the Banyan*, p. 263; BSES, AR 1966–67, p. 52.

91 BSES, AR 1967–68, p. 62; BSES, AR 1968–69, p. 47; BSES, AR 1969–70, p. 46; BSES, AR 1970–71, p. 47; BSES, AR 1971–72, p. 13.

92 BSES, AR 1972–73, p. 12.

93 BSES, AR 1960–61, p. 62; BSES, AR 1961–62, p. 62; BSES, AR 1978–79, p. 9.

94 BSES, AR 1972–73, pp. 15–16; BSES, AR 1975–76, p. 13; NSW Sugar Milling Coop., AR 1978–79, p. 8.

95 BSES, AR 1975–76, p. 13; BSES, AR 1976–77, p. 10; BSES, AR 1978–79, p. 9; R. Kerkwyk, 'Pigs as a problem to sugar cane', *CGQB*, Vol. 37, no. 1 (July 1973), pp. 8–9.

96 BSES, AR 1979–80, p. 10.

97 B. Croft, R. Magarey and P. Whittle, 'Disease Management', in M. Hogarth and P. Allsopp (eds.), *Manual of Cane Growing* (Brisbane, 2000), p. 284.

98 BSES, AR 1975–76, p. 34; BSES, AR 1976–77, p. 29; BSES, AR 1977–78, pp. 29–30; BSES, AR 1978–79, p. 28; BSES, AR 1980–81, pp. 29–30.

99 BSES, AR 1979–80, p. 8; BSES, AR 1980–81, p. 10; BSES, AR 1982–83, p. 18; BSES, AR 1987–88, p. 25.

100 NSW Sugar Milling Coop., AR 1981–82, p. 12; NSW Sugar Milling Coop., AR 1982–83, p. 12; NSW Sugar Milling Coop., AR 1983–84, p. 12.

101 BSES, AR 1981–82, p. 9; BSES, AR 1982–83, p. 15; BSES, AR 1983–84, p. 18; BSES, AR 1985–86, p. 24; R. Bull, '*Lepidiota picticollis* Lea, an increasing cane grub pest problem in the Bundaberg district', *Proc. ASSCT*, 1986, pp. 141–147.

102 P. Allsopp, 'The research process – canegrub management in the Australian industry', *Proc. ASSCT*, 2001, p. 11.

103 L. Geno, *Keeping it sweet: sustainability and sugar cane* (Melbourne, 1996), p. 63; BSES, AR 1983–84, pp. 18–19; BSES, AR 1984–85, p. 18; NSW Sugar Milling Coop., AR 1982–83, p. 12; NSW Sugar Milling Coop., 1983–84, p. 12; B. Hitchcock, K. Chandler and B. Stickley, 'Controlled release pesticides for soil insect control in sugar cane', *Proc. ASSCT*, 1984, pp. 87–94.

104 BSES, AR 1984–85, p. 18.

105 Cavanagh, 'Organochlorine insecticide usage in the sugar industry of the Herbert and Burdekin River regions', p. 140.

106 *ASYB*, 1995, p. 57; BSES, AR 1991–92, p. 16; BSES, AR 1992–93, p. 18; L. Robertson *et al.*, 'Integrated management of canegrubs in Australia: current situation and future research directions', *Aust. J. of Agricultural Research*, Vol. 46 (1995), pp. 3–4.

107 BSES, AR 1989–90, p. 15; BSES, AR 1990–91, p. 13; BSES, AR 1992–93, p. 17; P. Allsopp *et al.*, 'Management of canegrubs: future directions', *Proc. ASSCT*, 1995, p. 93.

108 BSES, AR 1982–83, p. 19; Anon, 'Severe crop damage even after aldrin damage', *BSES Bulletin* No. 1 (Janauray 1983), p. 11.

109 BSES, AR 1983–84, p. 20; BSES, AR 1985–86, p. 27; BSES, AR 1986–87, p. 29; BSES, AR 1987–88, p. 26; Allsopp, Samson and Chandler, 'Pest Management', p. 313.

110 Cavanagh, 'Organochlorine insecticide usage in the sugar industry of the Herbert and Burdekin River regions', p. 15.

111 P. Samson and P. Downs, 'Efficacy of Terbufos against African black beetle', *Proc. ASSCT*, 1994, pp. 159–162; BSES, AR 1994–95, p. 23.

112 BSES, AR 1989–90, p. 16; BSES, AR 1990–91, p. 12; BSES, AR 1991–92, p. 16; BSES, AR 1992–93, p. 17; BSES, AR 1993–94, p. 16; P. Allsopp and R. Bull, 'Cultural methods for soldier fly control', *BSES Bulletin* No. 20 (October 1987), pp. 20–21; P. Samson, 'Management of soldier flies in sugarcane', *Proc. ASSCT*, 1995, pp. 77–78.

113 BSES, AR 1994–95, p. 20; Samson, 'Management of soldier flies in sugarcane', p. 76.

114 BSES, AR 1990–91, p. 12; BSES, AR 1992–93, p. 19; BSES, AR 1993–94, p. 17; *AC* 16 May 1994, p. 33.

115 BSES, AR 1995–96, p. 18; L. Robertson and D. Webster, 'Strategies for managing cane weevil borer', *Proc. ASSCT*, 1995, pp. 83–87.

116 Clarence River Branch of NSWCGA, AR 1982–83, p. 11 & Clarence River Branch of NSWCGA, AR 1984–85, p. 11.

117 Rural Lands Protection Board, Queensland, AR 1993–94, p. 22.

118 BSES, AR 1980–81, p. 10; BSES, AR 1989–90, p. 17; BSES, AR 1991–92, p. 16; BSES, AR 1993–94, p. 17.

119 P. Allsopp and J. Manners, 'Novel approaches for managing pests and diseases in sugarcane', in B. Keating and J. Wilson (eds.), *Intensive Sugarcane Production: Meeting the Challenges Beyond 2000* (Wallingford, UK, 1997), p. 177.

120 P. Allsopp and J. Manners, 'Novel approaches for managing pests and diseases in sugarcane', pp. 174 & 180.

Chapter 15

Defeating Cane Diseases

In 1908, CSR decided that the firm needed a plant pathologist to assist in controlling the widespread outbreaks of Fiji disease that were threatening the very existence of the Fijian sugar industry. CSR selected a twenty-six year old chemist, David Shepherd North (1882–1967), as the staff member who would fulfill this role. North joined CSR as a junior laboratory assistant in 1900 and had subsequently gained experience in the laboratories of several CSR sugar mills and its Pyrmont Refinery. During the latter part of 1908 he was given 'special training' in plant pathology at Sydney University and by the Linnaean Society of New South Wales, although just what he learnt was never spelt out in the documents mentioning his training. In January 1909, North was transferred to Fiji where he devised a scheme to control Fiji disease, thereby saving the Fijian sugar industry and CSR's Fijian interests from almost certain extinction.[1]

From 1910 onwards, North also investigated the type of cane diseases occurring throughout CSR's Australian mill supply areas. These inquiries were interrupted following his enlistment in 1916. North resumed his employment with CSR in 1918, but he was transferred from Fiji to the Broadwater Experiment Station in 1920. Here he established the organisation's pathology laboratory and cane variety improvement program, and commenced training CSR's technical field officers who would undertake experimental work at each cane nursery attached to CSR's sugar mills.[2] He retired from CSR in 1941. His contributions to the organisation's plant pathology program were recognised when CSR's Plant Research Centre in Brisbane was named after him in 1960.[3]

During his earliest Australian investigations, North identified the symptoms of the disease that became known as leaf scald. He argued, however, that the disease was probably prevalent in the North Queensland sugar-producing districts at the turn of the century.[4] Leaf stripe disease, subsequently given the name downy mildew, was found to be present in all CSR's North Queensland mill supply areas.[5] Fiji disease was present in New South Wales during the early 1910s, and some seedlings at CSR's New South Wales cane nurseries displayed symptoms of the disease during 1910, but their subsequent destruction stopped the disease's spread to any of CSR's Queensland mill supply areas.[6] Yellow stripe disease, subsequently named mosaic, was present in the Herbert River, Isis and

Tweed River districts. Gumming disease was mentioned for the first time in 1915, when it affected the H.Q. 426 cane variety in the Herbert River district.[7]

Details about the causal agents, symptoms and means of transmission of the diseases identified in the Australian sugar-producing districts before 1955 are shown in Table 15.1. Of all the cane diseases, Fiji disease is one of the easiest to recognise, with no special training being required; even colour blindness is no drawback in its detection. Fiji disease has one distinctive characteristic which is never found associated with any other cane disease – the presence of smallish pale green to yellowish galls or lumps on the under surface of affected leaves. Soon after the appearance of the galls, the leaves develop a stiff appearance and the cane top becomes deformed, giving the impression of having been bitten off by an animal.[8] Stools of cane become severely stunted (Plate 15.1). Gumming is also relatively easy to identify. Freshly cut stalks

Plate 15.1. The main symptoms of Fiji disease: stunted stools (bottom left) displaying leaf galls (shown in the inset).

Source: *CGQB*, Vol. 33, no. 3 (1 January 1970), p. 79. Image reproduction courtesy of BSES Limited.

Table 15.1. Characteristics of the main diseases affecting the Australian sugar industry, 1920–1955.

Disease	Causal agent or organism	Scientific name of causal agent or organism*	Agents of transmission	Symptoms
Gumming	bacterium	*Xanthomonas axonopodis* pv.*vasculorum* (Cobb 1894) Vauterin *et al.* 1995	PM; RW; MD	Exudation of yellowish gum from freshly cut stalks; yellowish streaks on leaves
Leaf scald	Bacterium	*Xanthomonas albilineans* (Ashby 1929), Dowson 1943	PM; MD	Leaf streaks varying from narrow white pencil lines to broader pale yellow, with withering and in-curling of leaf extremities, giving the 'scalded look'; also leaf chlorosis, sideshooting and sudden stool death
Red Stripe or Top Rot	Bacterium	*Acidovorax avenae* subsp. *avenae* (Manns 1909) Willems *et al.*, 1992	RW; S	Dark red stripes on leaves; death of tops, rooting of top internodes & foul smell
Ratoon stunting	Bacterium	*Leifsonia xyli* subsp. *xyli* (Davis *et al.,* 1984) Evtushenko *et. al.* 2000	PM; MD	Marked reduction in yields; reddish-yellow discolouration of vascular bundles in the lower part of the nodal region
Chlorotic streak	Virus?	Causal agent unknown	PM; FW; S	Yellowish to creamy white streaks with wavy irregular margins on leaves
Mosaic	Virus	*Sugarcane mosaic virus* (SCMV), of the genus *Potyvirus*	PM; I	Yellowish-green blotches on leaves causing mottled pattern
Fiji	Virus	*Fiji disease virus* (FDV) of the genus *Fijivirus*	PM; I	Small, whitish-yellow galls on under surface of leaves; cane top becomes deformed with short stiff leaves; stunted ratoon crops
Pineapple	Fungus	*Ceratocystis paradoxa* (Dade) C. Moreau	S	Failure of sett to germinate; odour of fermenting fruits develops in infected setts
Yellow spot	Fungus	*Mycovellosiella koepkei* (W. Krüger) Deighton	W	Yellow spotting of leaves; partial defoliation of plants
Downy mildew	Fungus	*Peronosclerospora sacchari* (T.Miyake) C.G. Shaw	PM; W	White down on lower surface of leaves, occurring as well-marked, longitudinal streaks which later assume a rusty appearance as the leaves mature
Red Rot	Fungus	*Glomerella tucumanensis* (Speg.) Arx and E. Mull.	PM; RW; W; S?	Initially, stalks display red discolouration of rind; later the affected tissues turn dull red, interrupted by occasional whitish patches

Notes: * Modern scientific names are used; PM = Planting material; I = Insects; FW = Flood water; RW = Wind-blown rain water; W = Wind-borne spores; MD = Mechanical devices such as cane knives or cutter planters; S = Soil

Source: P. Griggs, 'Defeating cane diseases: plant pathologists and the development of disease control strategies in the Australian sugar industry, 1920–1955', *Historical Records of Australian Science*, Vol. 18, no. 1 (May 2007), p. 45.

exuded a yellowish, slimy gum that filled the cavities emerging from the death of stalk cells. Yellow spot, downy mildew and leaf scald are harder to recognise, but symptoms of both diseases appear on the leaves of the cane plant. With the exception of gumming disease, Australian canegrowers during the late 1910s and early 1920s had no information on the causal agent for each of these diseases or how they were transmitted. A bacterium identified in 1894 caused gumming disease, and it was often associated with poorly drained fields, so CSR's technical field staff and the first BSES plant pathologists wondered if it was a water-borne disease.[9]

During the late 1910s and early 1920s, CSR's technical field staff took over the monitoring of the extent of disease outbreaks in their mill supply areas, no doubt after some training in disease identification by David North. Leaf scald, downy mildew, mosaic and Fiji diseases continued troubling CSR's canegrowers. The latter disease was particularly prevalent throughout the Clarence River district, despite efforts to eradicate its presence.[10] The most disturbing trend, however, was the escalating spread of gumming disease throughout the Herbert River district and its appearance in the Richmond River district during the period 1918–1920. A touch of xenophobia crept into CSR's views on the matter, with the firm blaming the outbreaks of the disease in the Herbert River district upon the Italian canegrowers and their 'careless' cultivation methods.[11] These gumming disease outbreaks then worsened during the early 1920s, becoming more widespread in the Richmond River district, and emerging throughout the Clarence River, Tweed River and Isis districts.[12] Eventually, the control measures implemented by CSR reduced their severity (see discussion later in chapter).

In response to the gumming disease crisis, David North decided in April 1926 that eliminating the re-occurrence of this disease in Australia could only be solved by research into understanding how it spread throughout the fields of sugar cane. Moreover, a sense of urgency prevailed, because Badila, the main cane variety throughout the Tweed River district, was showing signs of susceptibility to gumming and its replacement had not been clearly identified. North's inquiries during late 1926 and early 1927 concluded that insects did not transmit gumming disease, but that it was spread by means of a combination of wind and rain during storms. The bacteria gain entry to the leaves of neighbouring uninfected plants, mainly through small wounds caused by the leaves coming into violent contact with one another under the action of the wind. The bacteria are readily killed by exposure to direct sunlight or drying so infection could only occur during wet weather, thereby confirming observations that wet conditions always contributed to the spread of the disease.[13]

North's discovery, although solving a plant pathology mystery, did not contribute much to controlling the outbreaks of gumming disease. His research now turned to identifying gumming disease resistant, but sucrose-rich cane varieties that would be suitable for New South Wales. Gumming disease resistance trials were commenced at the Broadwater Experiment Station in 1927.[14] Results from these trials were first reported in 1928, when North advised CSR's General Manager that from the fifty-four cane varieties tested, only six had showed themselves to be highly resistant to gumming and sucrose-rich. The most promising was Q. 813, and he was encouraging canegrowers to plant this cane variety in preference to gumming-susceptible cane varieties.[15]

The first BSES plant pathologists

The BSES had been much slower than CSR to appreciate the need for a plant pathologist amongst its staff. However, after Harry Easterby assumed the BSES directorship, he decided in 1921 that the organisation needed 'the services of a pathologist specialising in sugar-cane diseases'. Efforts throughout Australia and then abroad to secure a suitably qualified plant pathologist to head a pathology section met without success. Easterby, therefore, concluded that the BSES would have to train its own specialised scientists. A recommendation was made to William Gillies, the Queensland Minister for Agriculture and Stock, that three-year scholarships be awarded to selected graduates in the disciplines of pathology, soils and agriculture and mill technology. As an interim measure, the BSES appointed a series of cadet assistant plant pathologists who worked twelve months for the BSES, and then returned to studies at The University of Queensland. Easterby hoped that on completion of their studies, the newly qualified graduates would return to the BSES and staff its pathology division. During the mid-1920s, Messrs W. (Bill) Cottrell-Dormer, L. Kelly, George Wilson and E. J. Ferguson Wood participated in this program.[16] Each returned to work for the BSES, with Cottrell-Dormer perhaps being the most remembered, undertaking some of the earliest Queensland investigations into Fiji, red stripe and leaf scald diseases.[17]

The BSES cadet assistant plant pathologists conducted the first comprehensive survey of the extent of cane diseases in Queensland. They travelled to each of the state's sugar-producing districts, sometimes many times, between August 1924 and October 1927. Interestingly, their investigations were aided by CSR, for Cottrell-Dormer and Kelly spent several weeks at the Broadwater Experiment

Station during 1925. Here they studied cane disease identification under North's guidance. This formal co-operation between CSR and BSES marked a sudden thawing in the hitherto frosty relationship between both organisations. Co-operation between CSR and the BSES had been virtually non-existent, with CSR stridently opposing the formation of the BSES and refusing its first staff access to the firm's mills and cane nurseries (see Chapter 8). Just why this new arrangement was permitted by CSR managers could not be determined from the surviving archival records. Perhaps CSR recognised that Easterby was trying to recruit more skilled scientists to the BSES and that the firm's desire to control disease outbreaks would be better served by greater co-operation.

The subsequent reports and papers by the BSES cadet assistant plant pathologists outlined the geographic distribution of the main cane diseases they identified and which cane varieties were most susceptible to particular diseases. These reports contained five alarming findings. Firstly, incidences of gumming disease were found in most Queensland sugar-producing districts, but they were particularly prevalent in the Maroochy, Bundaberg and Isis districts. Secondly, Fiji disease was present in the Bundaberg, Maryborough and Beenleigh districts, and it was very destructive to D. 1135, the main cane variety favoured by Southern Queensland canegrowers. The infection was traceable to the introduction of infected cane from New South Wales. Thirdly, the very sucrose-rich H.Q. 426 (Clark's Seedling) and B. 208 cane varieties were susceptible to gumming, leaf scald, red rot and mosaic diseases. These cane varieties, grown mostly in North Queensland, needed to be phased out. Fourthly, Badila, the main cane variety grown in districts north of Mackay, appeared not to be susceptible to any of the main diseases except red stripe/top rot. Yet northern canegrowers were very vulnerable, given their heavy reliance on this one particular cane variety. Moreover, Badila was highly susceptible to gumming disease when grown in Southern Queensland, so it was not a possible replacement for D. 1135. Fifthly, downy mildew was present in several Queensland sugar-producing districts and Cottrell-Dormer remarked prophetically that this disease promised to be 'the most difficult to control without discarding valuable, though susceptible, varieties'.[18]

As the BSES cadet assistant plant pathologists conducted their preliminary disease investigations, Arthur Bell (1899–1958) became the recipient of the Queensland sugar industry's Travelling Research Scholarship in plant pathology. Bell left Australia in October 1924 and initially completed a Master of Science degree at the University of California (1925–1926) and a post-graduate Diploma of Membership from the Imperial College at London University (1927). On his return to Australia, Bell assumed control of the

BSES's Division of Pathology, after a restructuring of the organisation's administrative units in 1929. Eventually, the BSES's pest and disease control work was combined in the Division of Entomology and Pathology under Bell's leadership in 1934. He continued to head this unit after becoming Assistant BSES Director in 1935.[19] Bell produced publications about dwarf, Fiji and gumming diseases, but he is most remembered for producing *A Key to the Field Identification of Sugar Cane Diseases*, published in 1929. This publication contained detailed illustrations, photographs and descriptions of the sugar cane diseases found throughout the world.[20] Bell eventually became BSES Director in 1945, before taking up the post of Under-Secretary for the Queensland Department of Agriculture and Stock in 1947.[21]

Upon his return from overseas, Arthur Bell assessed the information gathered by the cadet assistant plant pathologists and planned an appropriate research strategy to tackle the most pressing cane disease problems in Queensland.[22] During the late 1920s, the BSES plant pathologists at their newly constructed laboratories in Brisbane commenced the following investigations: determining the morphology and physiology of the organisms causing gumming and leaf scald diseases; how gumming disease was spread; the identification of cane varieties; and classifying the degree of resistance of particular cane varieties to specific diseases. Field trials were also organised throughout the state to test the resistance of various cane varieties to different diseases. In 1928, locally bred (e.g. Q. 813; S.J. 3; S.J. 4) and imported cane varieties from Java (e.g. P.O.J. 2714 and P.O.J. 2878) and the Coimbatore Experiment Station in India (e.g. Co. 210, Co. 213, Co. 227 & Co. 290) were assessed at the Bundaberg BSES Station to determine their susceptibility to gumming disease. At Maryborough, twenty-five different cane varieties were assessed for susceptibility to Fiji disease, while the vulnerability of imported Indian and Javanese cane varieties to mosaic was considered at Mackay. Further north, the standard cane varieties (e.g. Badila; Goru) were tested at the South Johnstone BSES Station in the spring of 1930 for their susceptibility to leaf scald disease. These investigations were the first properly planned trials into the susceptibility of different cane varieties to diseases in Queensland.[23]

By 1935, some positive outcomes from these early investigations were being announced by BSES. The systematic description of the morphological characteristics, history, distribution and disease resistance of at least 130 cane varieties in Queensland had been completed. For the first time in the history of sugar production in Queensland, both canegrowers and scientists had a reliable key to assist them identify any particular cane variety. Southern Queensland canegrowers were advised that the cane varieties Co. 290, P.O.J. 2878, P.O.J. 2725 and to a lesser extent P.O.J. 234, displayed the most resistance to gumming disease.[24] In contrast, the investigations into leaf scald at the South Johnstone BSES Station had yielded little useful information. The trials were shifted to the Meringa BSES Station in 1932 and eventually abandoned in 1934. The main conclusions reached were that any clone containing the cane variety N.G. 34 as a parent produced an unduly large number of susceptible progeny and Bell observed 'disconcerting experiences with the masking of symptoms'. Bell advised North that he considered leaf scald to be a 'most complex disease'.[25]

During the early 1930s, the BSES plant pathologists also commenced new avenues of inquiries. Several gumming disease resistant cane varieties were identified as being susceptible to Fiji disease (e.g. Uba; P.O.J. 2878). This finding was unwelcome, as the sucrose-rich P.O.J. 2878 produced large numbers of stalks, thereby making it an ideal cane variety to cultivate. As such, Australian canegrowers were loathe to abandon its cultivation. This discovery, however, led to investigations into the conditions governing the transmission of Fiji disease under natural conditions, and the establishment of trials to identify which cane varieties were resistant to the disease. In 1933, trials were also commenced to determine family resistance to gumming and leaf-scald diseases, in order to ascertain whether any particular crosses were likely to yield progeny which would include a high percentage of individuals resistant to either or both these diseases. Downy mildew resistance trials commenced throughout the Mackay and Lower Burdekin districts during 1932 and 1933, respectively, in an effort to determine which cane varieties were most susceptible to these diseases.[26]

As the BSES plant pathologists conducted their investigations during the late 1920s and early 1930s, they also identified new diseases and clarified the etiology of existing diseases. Chlorotic streak was recognised as a distinct disease in 1929. Symptoms included yellowish to creamy white streaks with wavy irregular margins on the leaves. Preliminary studies commenced into determining the extent of yield losses associated with the disease, the effectiveness of hot water treatment on eliminating infection in setts, and whether its transmission occurred via knife infection or insects. In 1930, Cottrell-Dormer continued inquiries, commenced in 1926 but not finalised, to determine if the disease top rot found in Queensland was the same as the red stripe disease which had been identified in overseas sugar-producing countries during the mid-1920s. Dwarf disease which produced stunted abnormal shoots was recognised for the first time throughout the Mackay district in 1930 and investigations sought to identify its causal agent, means of transmission and most susceptible cane varieties.[27]

The BSES investigations during the late 1920s yielded useful results quickly. By 1932, Cottrell-Dormer had concluded that red stripe and top rot

*Plate 15.2. Leafhoppers (*Perkinsiella saccharicida, *Kirk) on cane leaves. These insects carry the virus which causes Fiji disease.*

Source: CSR, *Newsletter* No. 114, October 1969, p. 11, CSRR, Z/109, Box 372. Image reproduction courtesy of CSR Ltd.

were different manifestations of the same disease. In 1933, BSES entomologist Reginald Mungomery, in collaboration with Arthur Bell, discovered that the sugar cane leafhopper (*Perkinsiella saccharicida* Kirk) was the insect vector of Fiji disease (see Plate 15.2). Five years later, the scientists had identified that the cane varieties Jason, Co. 290, P.O.J. 213 and Q. 813 were the most resistant to Fiji disease. By 1935, the plant pathologists had also shown that B. 208 and all the P.O.J. cane varieties were the most susceptible to downy mildew and that they should not be grown in the Lower Burdekin, Mackay and Bundaberg districts where the disease was very prevalent. Family resistance trials concluded that the P.O.J. 2725 and P.O.J. 2878 cane varieties conferred high resistance against gumming disease on a very high percentage of their progeny. Yet no crosses showing a high degree of resistance to both gumming and leaf scald had been identified.[28]

Earliest disease control measures, 1915–1934

Arthur Bell during his Presidential Address to the Australian Institute of Agricultural Science in 1949 commented that disease control in sugar cane presented some unique challenges. Sugar cane is a tall bulky plant grown in large fields. Consequently, the usual control techniques of spraying and dusting the crops were not realistic options in any campaign to reduce specific disease outbreaks. Sugar cane is not propagated commercially from seed, but from setts. Therefore, ordinary seed treatments were out of the question. Moreover, cane crops re-sprout or ratoon after harvest, so systemic diseases persist in subsequent crops. Finally, sugar cane was grown in non-contiguous districts along almost 2 500 km of eastern Australian coastline, encompassing various climatic regions, soil types and other agricultural pursuits that could influence the incidence of diseases (e.g. maize cultivation; improved pastures for dairying). CSR, for example, found that controlling cane disease outbreaks throughout the Richmond River district during the 1920s required encouraging local dairy farmers to grow only disease-resistant cane varieties in the fodder cane plots for their dairy cattle. CSR supplied the local Primary Producers' Union with suitable cane varieties for distribution to local dairy farmers.[29]

Combating the earliest disease outbreaks troubling the Australian sugar industry fell into two broad categories: legislative and direct methods. The legislative approach included the following measures: regulations restricting the indiscriminate importation of cane varieties from overseas and the uncontrolled movement of cane varieties between sugar-producing districts; the gazettal of lists of approved cane varieties that could be cultivated in each sugar-producing district; and harvest or plough-out orders requiring canegrowers to harvest or destroy diseased fields of cane. Direct methods included the following activities: supplying canegrowers with disease-resistant cane varieties and disease-free setts; roguing or removing diseased stools or 'rogues' from the field; chemical control or heat treatment methods to destroy the pathogen causing the disease; and altered cultivation methods. The introduction of these measures required the cooperation of canegrowers and mill-owners, and their implementation was often underpinned by the research and experimentation of the plant pathologists. Therefore, the adoption of one particular method or combination of methods depended very much upon the knowledge about the disease that had been assembled by the plant pathologists, and the seriousness and extent of the disease outbreak.

Indiscriminate movement of infected cane was one of the first issues tackled in the control of cane diseases. Alerted by David North about the risks associated with the uncontrolled movement of cane varieties and planting material, CSR enlisted the assistance of its canegrowers to combat the spread of gumming disease by establishing a quarantine zone in the Clarence River district in 1925. This move was described by CSR as 'an entirely new departure in the control of cane diseases in Australia'. Under the New South Wales *Plant Diseases Act* of 1924, a local quarantine district was created covering the farms throughout the lower reaches of the

Clarence River. Sugar cane plants could not be moved out of this area. A district quarantine covering all Clarence River cane growing areas was also created and no cane material from outside areas could be moved into the district quarantine zone, except under special permit. Later, district quarantine zones were established along the Richmond and Tweed Rivers, and two local quarantine areas covering the farms affected by gumming disease were proclaimed in the Tweed River district.[30] These measures were part of wider efforts implemented in New South Wales to eliminate gumming disease (see discussion later in the chapter).

In Queensland, no official controls existed over the movement of cane between sugar-producing districts during the 1910s and early 1920s. David North was scathing of such laxity, particularly complaining about the 'indiscriminate and haphazard distribution' of cane varieties by the BSES before 1920 as being a 'serious menace' to the viability of the cane supply for CSR's mills. North believed that the BSES had too few safeguards in place to stop the spread of diseases at their experiment stations, which were in his opinion not staffed by competent entomologists and pathologists.[31] However, the discovery of Fiji disease in Queensland finally forced the BSES to approach quarantine more seriously. In October 1926, the Queensland government (McCormack; Labor) prohibited the introduction of any sugar cane from New South Wales into Queensland and the movement of any cane material out of the Beenleigh, Maroochy and Maryborough districts because of Fiji disease outbreaks. Eventually, a proclamation was issued in Queensland under the *Diseases in Plants Act* of 1929, prohibiting the movement of setts for planting purposes from or into the sugar cane quarantine districts, except after permission had been granted by an Inspector under the legislation. In addition, the sugar-producing belt in Queensland was split into nine quarantine districts which coincided with the main sugar-producing regions (e.g. Mackay; Bundaberg; Maryborough), although this arrangement was amended in 1938 when the Far Northern district (i.e. Cardwell northwards) was split into two quarantine areas.[32]

These earliest quarantine measures were further assisted by amendments to the *Regulation of Sugar Cane Prices Act* of 1934. The new measures required the BSES Director to supply each Local Cane Prices Board with a list containing approved cane varieties. All cane varieties not approved would only be crushed upon payment of a penalty not exceeding 10s per ton of cane. The scientists hoped this regulation would gradually eliminate disease-susceptible cane varieties, reduce the incentive to remove cane varieties from one district to another, remove any dangerous sources of infection and assist in the breeding of disease-resistant cane varieties. Moreover, by reducing the number of approved cane varieties to less than ten per mill supply area, BSES officials could more easily recognise non-approved varieties. However, these restrictions could not be adequately enforced, as the system of field inspections was inadequate, with some farmers never getting around to digging out diseased stools. Other Queensland canegrowers displayed 'bower-bird instincts', maintaining private collections of many cane varieties on their properties, including some which often turned out to be the source of infection for a particular disease.[33]

Legislative measures worked effectively only if supported by the canegrowers, a point stressed by cadet assistant plant pathologist Ferguson Wood who stated in 1927 that BSES could not hope to 'control the disease situation without the co-operation of the farmers'. Gaining this cooperation was a daunting task, given that the Australian sugar industry consisted of approximately 8 000 canegrowers during the 1920s. Moreover, at the beginning of the campaign to reduce cane disease outbreaks in 1920, David North was very pessimistic about the capabilities of Australian canegrowers, observing they were:

> quite incompetent to carry out recommendations for controlling diseases. None of them are able to recognize such well known diseases as Fiji, Yellow Stripe or gumming in their early stages, so they are unable to select healthy setts.[34]

From 1920 onwards, CSR assisted its canegrowers by holding farm demonstrations about how to obtain disease-free setts. Within a few years, North aimed to ensure what canegrowers did now 'with the assistance and constant prompting of the technical staff, they should do more or less automatically'. By 1927, the BSES had also commenced regular field days, where specimens of diseased cane were displayed so that Queensland canegrowers could familiarise themselves with the symptoms of particular cane diseases. Photographs and preserved specimens, illustrating the symptoms of the main diseases, were also shown at the annual local Agricultural Shows.[35]

Insisting that canegrowers use disease-free setts was one way to combat disease outbreaks. Planting whole sticks of cane into the drills and then chopping them through with a cane knife was discouraged as it prevented the detection of gumming disease. This practice waned with the adoption of cutter planters, but disease-free stalks of cane still had to be sourced for the machines. That task could be a source of infection, for David North discovered in 1919 that healthy setts were infected with gumming disease via infected cane knives. North advised

CSR's general manager that the firm's field officers should encourage the cane-cutters to soak their cane knives frequently in solutions of sheep dip to kill the bacterium causing gumming disease. This advice was further emphasised by the BSES during the late 1920s, when its plant pathologists also advised the state's canecutters to clean their cane knives by soaking them in kerosene tins of boiling water or a carbolic solution. The effectiveness or adoption of this measure was hard to judge, but CSR did claim that 'knife sterilization' was one important measure that contributed to the eradication of gumming disease in the Clarence River district during the late 1920s.[36]

Sourcing large numbers of disease-free setts in localities affected by disease outbreaks required coordinated local efforts. In New South Wales, CSR took the lead, with its field officers during the early 1920s identifying disease-free farms from which planting material could be obtained. In Queensland, district surveys carried out during 1928 and 1929 located farms free of gumming and Fiji diseases. Isolation plots, operated by the local Canegrowers' Association under BSES supervision, provided disease-free planting material. By 1930, such plots had been established in the Cairns, Tully and Bundaberg districts. In addition, the BSES acquired a site on Stradbroke Island (near Brisbane) in 1930 and established an isolated cane nursery which supplied disease-free setts and acted as a place where cane varieties released from quarantine could be propagated until sufficient quantities of seedlings were grown for distribution.[37] However, this strategy was not always suitable; for example leaf scald was difficult to control using this method because of the frequent 'masking' of symptoms. Plants could outwardly appear healthy and show no signs of the disease, yet when they were used as planting material, diseased fields resulted.[38]

Several examples can illustrate how disease-free planting material assisted in bringing specific pre-1935 disease outbreaks under control. During the gumming disease outbreaks in the Herbert River district throughout the late 1910s and early 1920s, CSR arranged for its canegrowers to be supplied with 2 200 tonnes of disease-free planting material. This action contributed significantly to the reduction of gumming disease in this district. During the late 1920s, 6 000 tonnes of disease-free planting material was supplied to canegrowers throughout the Bundaberg district in an attempt to reduce the incidence of gumming disease until resistant cane varieties became available. Farms free of Fiji disease were identified in the Beenleigh and Maryborough districts in the late 1920s, and used to supply non-infected setts. This strategy assisted in keeping Fiji disease in check in Southern Queensland until resistant cane varieties could be identified.[39]

Disease outbreaks could also be reduced by eliminating susceptible cane varieties and planting those varieties more resistant to particular diseases. Many examples can illustrate the success of this strategy between 1915 and 1935. Downy mildew was brought under control in the Herbert River and Isis districts during the mid-1910s by abandoning the B. 208 cane variety because of its susceptibility to the disease.[40] Leaf scald and gumming diseases, prevalent in New South Wales during the early 1920s, were reduced in severity by discarding the Mahona and D. 1135 cane varieties. Their place was taken initially by the N.G. 16 and Malabar cane varieties which were more disease-resistant, but less sucrose-rich. By the early 1930s, these two cane varieties had lost popularity in favour of Q. 813 which was both sucrose-rich and resistant to gumming and Fiji diseases.[41] At Bundaberg, the gumming disease susceptible cane varieties D. 1135 and M. 1900 were initially replaced during the 1920s by Q. 813, Uba, Co. 210 and Co. 213, so that by 1932 they represented 51% of the crop. These new cane varieties, although gumming disease resistant, had certain disadvantages. The Co. varieties were not very sucrose-rich and Q. 813 developed poor root systems, making it susceptible to lodging. Uba was seriously disliked by mill-owners, its high fibre content requiring additional crushing power to extract the juice. Moreover, the juice was difficult to clarify, slowing work at the vacuum pans. Therefore, these cane varieties were quickly replaced by the gumming disease resistant P.O.J. varieties during the 1930s. In the Herbert River district, the gumming disease outbreaks throughout the late 1910s and early 1920s were eliminated by replacing the gumming disease susceptible cane variety H.Q. 426 with resistant cane varieties such as Korpi, Oramboo, H.Q. 409 and Q. 813.[42]

In 1935, Arthur Bell reflected upon the efforts to control cane diseases in Queensland over the previous seven years. He claimed that gumming was the State's most serious cane disease problem at the creation of the Division of Pathology within the BSES in 1928. Present throughout the entire State, gumming disease threatened the very existence of sugar production in Southern Queensland. Yet control measures had almost eradicated the disease in 1935, except for a recent outbreak in the Mulgrave Mill supply area. On the other hand, Fiji disease was posing a major threat to the viability of cane growing in Southern Queensland. Downey mildew was an emerging threat, especially in the Mackay and Lower Burdekin districts.[43] In 1923, David North also believed that the New South Wales sugar industry was on the 'verge of a catastrophe' because of gumming disease, yet a decade later CSR announced the 'eclipse' of the disease. CSR's efforts had also yielded positive results in the control of other diseases, for leaf scald had nearly been eliminated and New South Wales losses from Fiji disease had been reduced considerably by 1930. Downy mildew was of little concern and mosaic only troubled canegrowers throughout the Clarence River district where large amounts of maize was grown near the cane crops.[44]

New plant pathologists; new disease discoveries, 1935–1964

In the mid-1930s Arthur Bell faced a staffing shortage. One by one the first cadet assistant plant pathologists had sought other employment, so none remained with the BSES in 1934.[45] Rebuilding the BSES plant pathology group commenced in 1934, when agricultural scientists Cecil Graham Hughes (1913–2002) joined the BSES as a plant pathologist, eventually serving as the organisation's Chief Plant Pathologist from 1966 until his retirement in 1977. From 1937 to 1947, Hughes was diverted from pathology research for two to three months each year, as he supervised the cane variety improvement program at the Meringa BSES Station. Nevertheless, he made an important disease-related contribution in 1939, identifying that a range of grasses, sweet corn and sweet sorghum could be infected by gumming disease and act as alternative hosts for the disease. During the late 1940s and early 1950s, Hughes developed mercurial control methods against pineapple disease and worked with David Steindl on some aspects of ratoon stunting disease.[46]

A year after Hughes became a BSES staff member, Bell appointed another graduate in Agricultural Science from Sydney University. David Steindl (1912–1981) joined the BSES as an assistant plant pathologist. He was promoted to Senior Pathologist in 1955 and retired in 1977. Steindl is best remembered for his pioneering work into the identification of ratoon stunting disease and the development of hot water treatments to eliminate this disease in cane planting material. Steindl spent some time in late 1952 at the University of California (Berkeley), studying virus isolation and detection techniques in an effort to improve the methods of recognition of ratoon stunting disease. He also conducted investigations into leaf scald, downy mildew and bacterial mottle diseases and assisted in the development of targeted roguing of diseased stools in the control of Fiji and downy mildew diseases.[47] Both Hughes and Steindl were jointly awarded the Medal of the Australian Institute of Agricultural Science in 1959 in recognition of their work on sugar cane diseases, especially the investigations into ratoon stunting disease.[48]

The early success of the BSES plant pathological investigations were not repeated during the late 1930s. Experiments to transmit dwarf disease by mechanical means or insects were unsuccessful and extensive field investigations failed to find any grasses or weeds that displayed symptoms of the disease. Attempts to control the spread of the disease by treating the setts with hot water also proved futile. In 1937, the BSES plant pathologists were still very unclear about the origins or methods to stop the spread of this uniquely Australian cane disease, except to recommend the elimination of susceptible cane varieties which had been identified as P.O.J. 2714, H.Q. 426 and Malagache. Similarly, attempts to understand the etiology of chlorotic streak disease proved futile. By 1938, laboratory investigations had failed to produce evidence of a visible causal organism. Experiments to spread the disease by insects or mechanically via cane knives had also proved unsuccessful, but the disease was found to spread most rapidly in paddocks of cane prone to flooding. Uba, D.1135 and Co. 290 had been identified as the cane varieties most resistant to the disease. Breeding trials to create a cane variety incorporating resistance to the disease had commenced.[49]

BSES research into cane diseases slowed with the outbreak of World War II and ceased altogether during 1942. Efforts were concentrated upon maintaining the Bureau's advisory staff. Arthur Bell noted with some relief that operating with such a skeleton staff was possible because downy mildew, gumming and Fiji diseases had been brought under control or eliminated in most sugar-producing districts (see discussion later in the chapter).[50] Bell, however, had foreshadowed during the war years that understanding chlorotic streak disease would be the focus of BSES plant pathological investigations immediately after the war.[51]

Experiments to determine the time of year when chlorotic streak was transmitted and varietal resistance to the disease commenced in several Queensland sugar-producing districts during 1946. Unsuccessful attempts to transmit the disease by different insects commonly found in cane fields were also conducted during 1947 and 1948. Interest in the disease did not wane, despite the difficulty in understanding its etiology, and extensive additional investigations into its causal organism, method of spread and the effectiveness of hot water treatment on the disease's subsequent development continued during the 1950s and early 1960s. Some of these experiments were conducted by Brian Egan and Owen Sturgess, who had joined the BSES in 1947 and 1956, respectively.[52]

Other BSES investigations immediately after the war included trials to identify varietal resistance to leaf scald. Commenced in 1945, these leaf scald resistance trials were still being conducted during the early 1960s. Several test plots to gauge the effectiveness of the pre-planting treatment of setts with various mercurial preparations against pineapple disease also began in the Home Hill district during 1945 and were expanded to the Mackay district in 1949. In addition, investigations were carried out into a serious new problem in the Mackay district between 1945 and 1950. Initially called 'Q. 28 disease', this cane disease is now known world-wide as ratoon stunting. The chief symptoms of this disease are a stunting of the cane, especially in the ratoons, and a marked reduction in yields, particularly in susceptible cane varieties (see Plate 15.3). Experiments during the late 1940s and early 1950s sought to identify the most susceptible cane varieties and looked for a causal agent, methods of transmission and possible control measures.[53]

Some findings from these investigations proved very useful to cane-growers. The discovery that ratoon stunting disease could be transmitted mechanically by cane knives and the blades in cutter planters highlighted a need to sterilise cane knives and planting machinery. This sterilisation could be achieved by pouring boiled water onto the knives of the cutter-planters and soaking the cane knives in boiling water. Additional sterilisation could be obtained from washing the cane knives in methylated spirits or by lighting the spirits on the blade. Eventually an ammonium compound marketed under the trade name 'Mirrol' became the standard knife sterilant. Initial experiments with Q. 28 showed that setts immersed in water at 54°C for one

Plate 15.3. The impact of ratoon stunting disease on the Q. 28 cane variety (left of photograph). Healthy Q. 28 appears on the right.

Source: N. King, R. Mungomery and C. Hughes, *Manual of Cane-Growing* (Sydney, 1953), p. 271.

hour or 52°C for one and a half hours did not kill the setts and yielded plants which could be guaranteed to be free from the disease. However, applications of these treatments to other cane varieties showed that they were not as tolerant to hot water as Q. 28 and that a treatment at 50°C for two to three hours was not detrimental to germination.[54] Other investigations demonstrated that dipping setts into tanks containing solutions of organic mercurial fungicides would protect them from the fungus that caused pineapple disease. A preparation of methoxy ethyl mercury chloride, marketed in Australia under the name 'Aretan', proved the most effective. Results from the leaf scald trials confirmed that the cane varieties most susceptible to the disease were Q. 44, Q. 63, Q. 64 and Trojan.

The chlorotic streak trials, however, produced mixed results. The scientists were unable to identify the causal organism or transmit the disease by insects, although investigations revealed that ratoon crops grown from setts planted late in the season displayed a slight decrease in the incidence of the disease. Nevertheless, the BSES was certain by the early 1960s that the disease was spread by using diseased stalks for planting material and transmission from diseased to healthy plants occurred through the root system, generally under wet conditions. The transmissibility of the pathogen through the soil and the infectivity of soil leachates from contaminated soils had also been conclusively shown. The recommended method for controlling chlorotic streak was improved farm drainage.[55]

BSES disease-resistance trials had been conducted in different parts of Queensland on isolated farms well away from commercial plantings of sugar cane or at the Government Domain, in inner Brisbane (i.e. today's site of the Queensland University of Technology). Apart from the risk of spreading disease into commercial crops of cane, supervision of the trials proved difficult. Moreover, the small disease plot at the Government Domain could no longer cope with the expanded disease investigation program. Consequently, the BSES decided in the late 1940s to concentrate its disease-resistance trials nearer to Brisbane. They were shifted to The University of Queensland's farm at Moggill, a western suburb of Brisbane, following the BSES leasing part of the property during 1949. The BSES had planned that the majority of these trials would proceed uninterrupted there for a number of years, although the site proved to be not ideal as it suffered from frost. However, following an expansion of investigational work by The University of Queensland, the BSES was advised in 1950 that its lease at Moggill would not be renewed. During late 1951, the BSES then acquired a neglected market garden at Eight Mile Plains on the southern side of Brisbane.

This property became known as the Pathology Farm and the centre of the organisation's pathological investigations. Initially, the property was not equipped with a glasshouse, but one was erected during 1965/1966 when the BSES lost its inner Brisbane glasshouse facilities. Eventually, the BSES's pathology laboratory at Head Office was transferred in 1969 to the Pathology Farm, where it remained until suburban expansion encroached upon the property during the 1980s and 1990s. By 1998, the BSES had moved its pathological investigations to a new farm at Woodford, approximately 60 km north of Brisbane.[56]

Research by the BSES pathologists throughout the 1950s also led to the positive identification of three new cane diseases in Queensland: yellow spot; bacterial mottle and striate mosaic. Yellow spot disease is caused by the fungus *Cercospora koepkei* Krüger. Its most obvious symptom is yellow spotting on the leaf surface. In 1908, Henry Tryon had recorded a *Cercospora* causing a conspicuous leaf freckle on sugar cane in North Queensland, so it was possible that this was C. *koepkei*. However, yellow spot disease was positively identified in 1950, although the BSES plant pathologists thought that it had been present for a number of years prior to this identification. Infection with yellow spot caused the early death of leaves, resulting in a heavy, sudden drop in ccs and a cessation of growth for infected plants. The disease clearly caused a considerable loss of tonnage and the lack of cover allows a profuse growth of winter weeds with a consequent increase in harvesting problems.[57]

Symptoms of the disease known as bacterial mottle were first detected by BSES officers between 1955 and 1957 in the Haughton River and Mackay districts, and subsequently its presence was recorded in all mill supply areas between Mossman and Sarina. Its occurrence, however, had most likely been recorded by CSR's technical field officers in the Herbert River district some time earlier. The poorly developed root systems of diseased plants had prompted CSR's field technical staff to assign the disease the name 'root disease', but its identification was not reported in any published paper and no specific casual organism had been identified as causing the disease. Plants affected by bacterial mottle initially display parallel, creamy-white regular stripes on the leaves. Infected setts give rise to shoots with a whitish mottling. The disease can cause death of the leaves and shoots, and diseased stalks frequently produce numerous side shoots at the base. Bacterial mottle was soon associated with areas prone to submergence during heavy flooding.

The specific symptoms of a disease that was subsequently named striate mosaic were noticed in 1956, although its formal identification was not completed at the BSES Pathology Farm until 1960. The disease which only affected the Q. 57 and Pindar cane varieties throughout the Lower Burdekin district could be identified by a marked striping effect on the leaves or even a yellowing to the entire top. Infection caused severe stunting and ratoon failure. Infested patches did not lead to new outbreaks nearby, except when diseased setts were planted into new areas.[58]

Investigations to devise control measures for these new diseases yielded mixed results. Bacterial mottle proved relatively easy to control. BSES plant pathologists quickly identified that Q. 57, Q. 66, Q. 73 and Trojan were the most susceptible cane varieties, so the use of resistant cane varieties, the selection of healthy planting material and better drainage reduced the occurrence of this disease. Striate mosaic could not be eliminated through hot water treatment of the setts, so control was only achieved by using disease-free plants and resistant cane varieties.[59] Yellow spot proved harder to control. Investigations showed that Trojan, Q. 42 and Q. 28 were the most susceptible cane varieties; Badila, Pindar, Vesta, Q. 44, Q. 47 and Q. 813 were the most resistant cane varieties, but their cultivation was on the wane. Trials showed that weekly dustings of a fine copper oxychloride dust would control yellow spot disease. Numerous dustings, however, would be impracticable in normal commercial fields, on account of both the expense and the inaccessibility of the fields during the wet season when the disease developed most rapidly. Some attempts at control by dusting only after periods of high humidity and rain were not successful. Moreover, such a method would be difficult to implement owing to the variability in weather, even in adjacent localities. By the late 1950s, the only effective way to control yellow spot disease was to breed new disease-resistant cane varieties and this task had only just commenced.[60]

Running parallel to the BSES plant pathology program were CSR's investigations in cane diseases and their control. The upsurge of Fiji disease in the New South Wales sugar-producing districts during the early 1930s prompted David North to investigate this disease. Between 1930 and 1932, numerous attempts to transmit the disease by inoculating different parts of the plant failed, and trials using insects commonly found around cane fields returned negative results. North also supervised trials each year throughout the 1930s to determine which cane varieties were resistant to Fiji disease under New South Wales conditions. These trials confirmed that the cane varieties Jason, Mahona, Q. 813 and Co. 290 were moderately to highly resistant to Fiji disease.[61] In addition, North continued trials to determine the resistance of new cane varieties to gumming, leaf scald and mosaic diseases. He argued that although gumming had now practically disappeared from Australia it was important to identify as many cane varieties as possible that were resistant to gumming in order to 'be well fortified against any threatened recurrence of this disease'. Experiments between 1935 and 1938 also confirmed earlier BSES work that chlorotic streak was not a mild form of leaf scald, that they were definitely two distinct diseases, and that the former could be cured via a short hot water treatment.[62]

The expanding BSES program of research into cane diseases was not duplicated by CSR after 1945. Following North's retirement in 1941, CSR did not replace him immediately. No explanation for CSR's slowness in replacing North could be uncovered in the firm's archival records. The company may have faced difficulties in recruiting a suitable person due to World War II. On the other hand, CSR may have been satisfied that the new university-trained BSES scientists were satisfactorily advancing the understanding of the etiology of Australian sugar cane diseases. CSR, however, did not entirely abandon investigations into cane diseases. Supervision of the research work at the Broadwater Experiment Station passed to Esmond G. Baber, the firm's assistant plant pathologist. He continued assessing the susceptibility of new seedlings to Fiji, mosaic, leaf scald and gumming diseases. The most significant finding was that the sucrose-rich cane variety known as Trojan displayed considerable resistance to gumming and Fiji diseases. Subsequently, it was released to canegrowers during the late 1940s and early 1950s.[63]

Eventually, CSR appointed Perry E. Robinson as its new plant pathologist in 1950.[64] Robinson continued supervising CSR's annual trials to test the susceptibility of potential new cane varieties to disease, especially Fiji and gumming diseases, although eventually the gumming resistance activities were discontinued after 1963, following the disappearance of the bacterium from the trial areas. Investigations during the mid-1950s at the Macknade Experiment Station showed that losses in the Trojan cane variety from chlorotic streak disease could be as high as 10 tonnes/ha. This finding prompted further investigations which led Robinson to conclude that chlorotic streak disease was not transmitted by aerial insects, but most likely in the soil by the medium of soil water. Such a conclusion, together with the results from the BSES investigations, led CSR to emphasise the importance of good on-farm and district drainage during the 1960s (see Chapter 12). Robinson also concluded that the treatment of cane affected by chlorotic streak disease with Gibberellic acid caused an appreciable increase in growth compared to untreated stools, but the high price of this hormone meant that it was unlikely to become a serious control measure against the disease. A series of trials were commenced in 1957 to determine the yield losses from ratoon stunting disease, so that the results could be compared with the BSES findings on the topic. Trials involving the aerial spraying of different fungicides to test their effectiveness against yellow spot were also conducted during 1957, but proved ineffective in controlling outbreaks of the disease and were not continued. A survey of parasitic nematodes in the soil in CSR's northern mill supply areas commenced during 1960. None of this research work contained any major discoveries that benefitted the entire Australian sugar industry, unlike CSR's previous advances during the 1920s and 1930s. CSR's plant pathological investigations during the 1950s and early 1960s, like their cane variety improvement activities, appeared to lack direction and often duplicated BSES activities. The firm, which had once led the way in scientific investigations associated with the growing of sugar cane and disease-control methods, surrendered its dominance to the BSES.[65]

Expanded control measures, 1935–1965

In 1937, the BSES Advisory Board, comprising canegrower, mill-owner and government representatives, agreed it was necessary to radically overhaul the existing *Sugar Experiment Stations Act* of 1900 to 1934. This recommendation was forwarded to Frank Bulcock, the Queensland Minister for Agriculture and Stock. Bulcock had both an academic and a practical background in agricultural science and was prepared to reform legislation to assist the state's rural industries. He agreed to the legislative changes.

The *Sugar Experiment Stations Act Amendment Bill* of 1938, largely drafted by Arthur Bell, incorporated the relevant sections of *The Diseases in Plants Act*, tailored to sugar industry requirements, the quarantine regulations described earlier and the annual gazettal of approved and non-approved cane varieties. This new legislation also provided for the decentralisation of disease control work by creating local Cane Disease Control Boards, composed of representatives from the canegrowers and millowners, and financed by compulsory levies. These new Cane Disease Control Boards were charged with the responsibility of bringing disease outbreaks under control in their local mill supply areas. They employed field staff who maintained constant field inspections, and when necessary, teams of employees to dig out diseased plants. The new legislation received formal assent on 27 October 1938 and the first local Cane Disease Control Boards were formed the following year. In 1941, these organisations were amalgamated with the Pest Control Boards to form single Cane Pest and Disease Control Boards.[66] C. Graham Hughes claimed in 1950 that these legislative measures ensured that Queensland had the most comprehensive disease control programme of any other sugar-producing country.

Moreover, this arrangement remained unique to Australia until late 1981, when the South African sugar industry introduced Pest and Disease Control Committees based upon the Queensland system.[67]

Incidences of disease continued to be controlled by the substitution of susceptible cane varieties by more disease-resistant cane varieties. The gumming disease outbreaks during the late 1930s in the Mulgrave Mill supply area were stopped by replacing the gumming disease susceptible cane varieties S.J. 4 and H.Q. 426 with resistant cane varieties such as Badila, H.Q. 409 and Q. 813. Downy mildew infections throughout the Lower Burdekin district between 1934 and 1944 were eliminated by a prohibition on the cultivation of the susceptible B. 208 cane variety. This same disease was also brought under control in the Mackay district during the late 1930s by a ban on the planting of the susceptible P.O.J. 2878 cane variety. At Bundaberg, the rapid replacement during the early 1940s of P.O.J. 2878 by C.P. 29/116 and Q. 49 helped reduced downy mildew outbreaks. In 1948, P.O.J. 2878 was removed from the approved cane variety list for the Bundaberg district.[68] A reduction in the occurrence of Fiji disease throughout the Maryborough district during the late 1930s and early 1940s occurred after the widely grown, but susceptible D. 1135 and M. 1900 cane varieties, were replaced with the Fiji disease resistant Co. 290. By 1942, it represented 42% of the crop crushed at the Maryborough Sugar Factory. Yet its importance had been halved by 1948, as canegrowers switched to P.O.J. 2878 which was re-introduced into the district in 1944 following the reduction in the incidence of Fiji disease. Dwarf disease in the Mackay district was eliminated during the 1930s by the substitution of the susceptible P.O.J. 2714 and P.O.J. 2878 cane varieties with Q. 813.[69]

Outbreaks of leaf scald disease in North Queensland after 1950 were also minimised by the cultivation of new cane varieties. Leaf scald infections peaked in North Queensland during 1947. From then onwards, the leaf scald susceptible cane varieties of S.J. 4, S.J. 16 and Trojan were replaced by more leaf scald resistant cane varieties such as Pindar, Triton, Q. 50, Q. 57, Q. 68, Q. 70, Q. 75 and Q. 82. By 1969, 95% of the cane crushed in the mill supply areas north of Townsville came from cane varieties that displayed moderate to high resistance to leaf scald disease. Consequently, incidences of leaf scald in the sugar-producing localities north of Townsville petered out during the 1950s and 1960s, but the disease still caused trouble in mill supply areas south of Townsville (see discussion later in chapter).[70]

Controlling pineapple disease required an entirely new approach – applying a fungicide to setts. This practice was first used in the Lower Burdekin and Bundaberg districts during the late 1940s, and then spread to other

Australian sugar-producing districts, becoming a standard agricultural practice by 1960.[71] Moreover, Australian canegrowers displayed considerable ingenuity when making dipping equipment. Devices included small wire baskets holding half a bag of setts dipped into a small tank of solution (see Plate 15.4), double-row planter boxes full of setts or a large installation equipped with a mobile crane and a 22 700 litres (6 000 gallons) tank at Fairymead plantation. These devices were slowly superseded during the 1950s, following the development of spray attachments for cutter planters. A series of sprays were fitted to the chute, covering the sett with a mercurial solution as it slid down from the knife. Other cutter planters were modified so that the setts could be immersed entirely in the mercurial dip, before dropping them into the furrows.[72]

Another disease that could be eliminated in setts was chlorotic streak, if the planting material was treated in hot water. However, C. Graham Hughes noted in 1950 that this method was not widely used, even in areas where the disease caused serious economic losses. Canegrowers found this approach expensive and time consuming, but more importantly they lacked centralised facilities

Plate 15.4. Hand dipping of setts against pineapple disease. The setts in small wire baskets were placed into boxes containing the solution of fungicide.

Source: N. King, R. Mungomery and C. Hughes, *Manual of Cane-Growing* (Sydney, 1953), p. 279.

in which to treat their setts.[73] This situation changed when the campaign to eliminate ratoon stunting disease commenced during the early 1950s. The plant pathologists realised that canegrowers could not complete this task individually because of the need for accurate temperature control in the heating facilities. Moreover, it had to be completed on a massive scale. The Mackay district alone, for example, required 45 000 to 55 000 tonnes of planting material in the 1953 season. Therefore, hot water treatment of planting material in Queensland became a matter undertaken by each Cane Pest and Disease Control Board working in conjunction with the local sugar mills. This service began in 1952, when the Ayr Cane Pest and Disease Control Board set up the first hot water treatment plant. In New South Wales, CSR constructed the hot water treatment units at each of its sugar mills. By 1967, twenty-five such units existed throughout Australia's sugar-producing districts. These units had all been custom-built by the individual Cane Pest and Disease Control Boards or CSR, and steam from the sugar mills was used for heating in the majority of the treatment plants. Bundles of whole stalk cane or metal baskets filled with setts were immersed in tanks fitted with a circulation pump that circulated the entire tankful of boiling water at least six times an hour, thereby ensuring that every sett received the required treatment (see Plate 15.5). Consequently, efforts to reduce chlorotic streak and ratoon stunting diseases occurred simultaneously. Improved on-farm and district drainage in North Queensland, outlined in Chapter 12, also assisted in the control of chlorotic streak disease.[74]

Roguing or the removal of diseased stool or rogues from fields infected with particular diseases was also used effectively as a control measure, allowing the retention of susceptible cane varieties while more resistant cane varieties were developed by the plant breeders. It was employed to control localised outbreaks of several diseases, including downy mildew throughout the Mackay and Bundaberg districts, but was applied extensively in various Southern Queensland mill supply areas to control Fiji disease in localities not too badly affected by the disease between 1935 and 1955. Teams of men employed by the local Cane Pest and Disease Control Boards worked throughout the Southern Queensland sugar-producing districts destroying diseased stools. By the mid-1950s, these campaigns had virtually eliminated Fiji disease in Southern Queensland.[75] Rouging, however, proved to be ineffective in the control of leaf scald, with experiments throughout the Mulgrave Mill supply area in the late 1940s showing that apparently disease-free setts sourced from fields that had been rogued still produced new crops infected with the disease. Canegrowers were advised that in no circumstances should any setts from infected farms be used as plant material.[76]

Plate 15.5. Baskets of setts being loaded into a hot water treatment tank at Isis Mill, near Childers, c. 1957.

Source: BSES, AR 1956–57, p. 50. Image reproduction courtesy of BSES Limited.

Harvest and plough-out orders tackled the issue of disease outbreaks by roguing on a field basis rather than stool by stool. They were issued primarily to control infection by disease, but also for the destruction of plantings made illegally, such as those using non-approved cane varieties or cane varieties brought across a quarantine boundary. Plough-out orders were also issued on nearby crops that threatened sugar cane. Downy mildew, for example, will infect maize crops and can be easily transmitted from that crop back to sugar cane. The destruction of maize crops was necessary to protect the sugar cane crops.

This approach had been used by CSR (although not called harvest or plough-out orders) throughout the Cudgen locality in the Tweed River district during the gumming disease outbreaks of the mid-1920s. CSR convinced canegrowers to destroy entire fields of sugar cane badly infected with gumming. To compensate these farmers for their losses, local canegrowers contributed pro rata to their cane areas in order to raise funds for the compensation and CSR contributed a like amount.[77] In Queensland, harvest or plough-out orders were first made possible by the *Sugar Experiment Stations Act Amendment Act* of 1938.[78] They were used widely by the BSES to combat outbreaks of downy mildew throughout the Mackay and Bundaberg districts during the late 1930s and early 1940s. This approach, combined with roguing of lightly infected fields and the prohibition on planting of the highly susceptible P.O.J. 2878, led to the successful eradication of downy mildew at Mackay. Subsequently, P.O.J. 2878 was able to be reinstated on the approved list of cane varieties to be cultivated in 1944.[79]

The return of Fiji disease, 1965–1978

Fiji disease appeared to have been brought under control in Australian sugar-producing districts by the late 1950s. However, it reappeared in the Tweed River district during 1962. By 1969, half the area cultivated with cane in the Tweed River district was infected with the disease. Isolated incidences of Fiji disease then occurred on the farms supplying Harwood, Moreton and Rocky Point Mills during the mid-1960s. The small outbreaks in the latter two localities were brought quickly under control by roguing and ploughing out the diseased blocks.[80]

Fiji disease made its next appearance in the Gooburrum locality near Bundaberg during 1969. The outbreak was blamed on either Fiji disease persisting undetected in some tolerant commercial variety or infected setts being sourced illegally from farms where the disease persisted, most likely the Tweed River or Beenleigh districts. The disease then spread quickly throughout the entire Bundaberg district during the early 1970s, a consequence of an upsurge in the population of leafhoppers, the vector of the disease, over the late 1960s, and the widespread use of susceptible cane varieties such as NCo. 310, Q. 70 and Q. 71. Moreover, Fiji disease was also again present throughout the farms supplying Moreton and Rocky Point Mills in 1970; nearly half the farms in the Rocky Point Mill supply area were known to be infected by 1975. The disease first appeared in the Isis and Maryborough districts in 1971 and 1976, respectively. Eventually a small outbreak occurred on some farms supplying the Plane Creek and Racecourse Mills (Mackay district) during the early 1980s.[81]

Incidences of Fiji disease peaked in the Tweed River district during the mid-1970s, when approximately 70% of the area under cane was infected.

Nearly 75 million stools showed signs of Fiji disease at the height of the disease outbreak in the Bundaberg district throughout 1979, compared to 0.25 million infected stools in 1973. Elsewhere, the disease reached its major extent in 1985, when 95% of the assignments in the Maryborough Sugar Factory supply area displayed evidence of the disease, while an estimated 3 million stools were infected in the Rocky Point Mill supply area, compared to 200 000 in 1978.[82]

The susceptible cane variety NCo. 310 was the common link in each Fiji disease outbreak. Yet this cane variety could not be immediately abandoned. During the 1970s, it made up 60–80% of the crop in Southern Queensland and 20–40% of the cane grown in the Tweed River district. Consequently, a variety of interim control measures were implemented to slow the spread of the disease. In the Tweed River district between 1965 and 1967, inspections were conducted by Pest Board survey teams who marked stools with diesel and the weed killer Gramoxone, hoping that the farmers would destroy the marked stools. After 1967, farmers were asked to undertake their own inspections. After consultation with the New South Wales Department of Agriculture, all known fodder cane plots for dairy cattle were inspected and plough-out orders were issued when infected fodder cane plots were encountered, thereby attempting to prevent such plots acting as a form of infection for commercial crops.[83] In Queensland, a prohibition was introduced on the movement of cane plants between mill supply areas and districts in Southern Queensland. Employees of the Bundaberg Cane Pest and Disease Control Board commenced inspection and roguing during the early 1970s, the staff of three inspectors being expanded to forty-five in January 1973. In addition, the Fairymead Sugar Co. Ltd. employed ten men to inspect its plantations. By 1974, however, farm inspections and roguing were abandoned because the task had become so great. Many Bundaberg canegrowers also voluntarily used roguing, but their numbers decreased after 1974, following the cessation of serious attempts at mass roguing. In the Maryborough Sugar Factory supply area where the outbreaks were not as extensive, Pest Board staff rogued all diseased stools during inspections. In addition, the susceptible Q. 70 and Q. 71 cane varieties, except under permit, were removed from the approved lists of cane varieties for the Maryborough Sugar Factory and Fairymead, Bingera, Millaquin and Qunaba Mills.[84]

Until NCo. 310 could be replaced, Cane Pest and Disease Control Board personnel supplied non-infected NCo. 310 setts. In New South Wales, all planting material in the Condong Mill supply area from 1965 onwards had to be certified as being free of disease (i.e. clean seed program). Isolated NCo. 310 plots were established during the early 1970s and sprayed with insecticide to reduce leafhopper numbers, but not all the setts sourced from these plots were disease-free, illustrating the difficulty in containing the disease.[85] At Bundaberg, the task initially proved a challenge for members of the Bundaberg Cane Pest and Disease Board, who discovered it was quite difficult to detect Fiji disease in well-grown crops of NCo. 310 and that the disease could appear anytime in this cane variety throughout the plant's life.

In order to secure healthy NCo. 310 plants for canegrowers in the Bundaberg district, a ban was introduced on the use of planting material from within a 400 m radius of a block of cane known to be infected with Fiji disease; this approach was also used throughout the Maryborough district during the late 1970s. The Fiji disease-free fringe localities of Bucca and Elliott Heads became the major source of cane plants, but these had to be completely inspected beforehand by Cane Pest and Disease Control Board personnel who found the task very time consuming. In July 1972, the Bundaberg Approved Plant Sources Scheme was inaugurated. Two isolated mother plots were established at Carmila, south of Mackay, and Toweran near Bundaberg. The Fairymead Sugar Co. Ltd. established its own isolation plots at Baffle Creek and in pasture at Miara. During 1973, an additional six secondary increase plots were established throughout the Bundaberg district. They successfully provided approximately 25 000 tonnes of setts to farmers in 1974, their first full year in operation. By 1976, farmers supplying Rocky Point and Isis Mills also had access to approved plant schemes or clean seed programs.[86]

As mentioned above, CSR throughout the 1950s had been undertaking Australian trials to identify cane varieties resistant to Fiji disease. Similar work had also occurred in Fiji, but the field trials were so unsatisfactory that the firm's scientists developed the insectary technique for screening cane varieties for resistance to Fiji disease. Seedling canes were grown in pots and exposed to infective leafhoppers in greenhouses, instead of being tested in the field for Fiji disease susceptibility. As the Fiji disease outbreaks worsened in New South Wales, CSR adopted this new approach in Australia during the mid-1960s (the exact date could not be identified).[87] The firm's plant breeding program was modified to ensure that potential cane varieties not resistant to Fiji disease were rejected as early as possible in the testing cycle. Infective leafhoppers used in the tests were bred in the Condong and Broadwater insectaries, new facilities built specially for the Fiji disease eradication program. From 1962 onwards, experiments were also commenced at the Broadwater Experiment Station to test the effectiveness and practicability of releasing large numbers of leafhoppers sterilised by irradiation amongst the normal field population of leafhoppers. Since no offspring will result from the union involving a sterile male, it was hoped that the total population of leafhoppers would be reduced. This work showed that a sterile male release program was biologically feasible, but the logistics and monitoring costs were excessive. Widespread aerial chemical control of the insects was also possible, but expensive and not all the leafhoppers were eliminated, so it would have been an ongoing control measure.[88]

Because of CSR's experience, the BSES investigated the reliability of the insectary method for screening cane varieties compared to field trials. Norman King, the BSES director who was well experienced with Fiji disease from his years in charge at the Bundaberg BSES station when the disease troubled the local canegrowers during the late 1940s, allocated extra staff to the Fiji disease resistance testing program in the early 1970s. Part of a special government grant for new projects in 1972 was devoted to constructing a glasshouse and insectary at the Bundaberg BSES station; the remainder of the funds financed the construction of a second glasshouse at the BSES Pathology Farm for additional Fiji disease studies.

The BSES research using the insectary method commenced in 1974 and focused upon quickly screening a large number of seedlings through the adoption of bunch planting and identifying replacement varieties for NCo. 310. Any promising resistant seedling was accelerated through the testing program and all parent cane and new seedlings were tested for resistance at an early stage.[89] Even though 70–80% of the clones in the BSES cane variety selection program were susceptible to Fiji disease at the outbreak of the epidemic, the BSES scientists by 1977 had identified that the Q. 87 and CP. 44–101 cane varieties were relatively resistant to the disease and could be used as a possible replacement for NCo. 310. By 1980, Q. 103, Q. 108, Q. 109, Q. 110 and Q. 111 had been identified as cane varieties resistant to Fiji disease, so the farmers now had several cane varieties to use instead of NCo. 310. Moreover, results from a trial using the insectary method were available in around 100 days or less, compared with two to three years required for field trials conducted at the Pathology Farm, so it was a speedier research technique.[90]

By the late 1970s, interim control measures such as roguing and clean seed programs had not stabilised the outbreaks of Fiji disease in the Bundaberg district. However, a more effective control measure began in Southern Queensland during 1978 when Q. 87 and CP. 44–101 – two newly approved cane varieties resistant to Fiji disease – made up almost 80% of the plantings on Bundaberg cane farms. In 1980, Bundaberg canegrowers were also permitted to use Q. 103, Q. 108, Q. 109, Q. 110 and Q. 111 – each of these cane varieties was resistant to Fiji disease. By 1989, the BSES reported that Fiji disease in the Bundaberg district was 'now of relatively minor importance'. In New South Wales, many farmers in the Tweed River and Richmond River districts during the late 1970s switched to using various Fiji disease resistant CP. and Co. cane varieties (see Figure 13.11), thereby bringing Fiji disease outbreaks under control, albeit temporarily (see discussion later in chapter). Eliminating the use of NCo. 310 took a little longer in the Mackay district, where the last of this cane variety was harvested in 1990, its place being taken by the Fiji disease resistant cane varieties Q. 124, Q. 135 and Q. 136.[91]

Elsewhere in Australia's sugar-producing districts, various well known cane diseases troubled the canegrowers during the late 1960s and 1970s. Red stripe (top-rot) outbreaks throughout the Lower Burdekin district in 1966 and 1967 led to the abandonment of the susceptible cane variety Q. 67. A severe outbreak of red rot between Mossman and Tully during 1968 affected canegrowers cultivating the new cane variety Q. 83. The year had been especially dry which favoured a build up of the disease, it being most prevalent in areas affected by moisture stress. The outbreak was brought under control following drought breaking rains in mid-January 1969.[92] Leaf scald infections in the high sucrose, but very susceptible Q. 63 cane variety commenced at both Mackay and Bundaberg in 1960. Proportionally, the outbreak of this disease was much more serious at Bundaberg than Mackay. A quick stop to commercial plantings of Q. 63 at Bundaberg led to the apparent disappearance of the disease from this district. At Mackay, the losses were confined to a few farms and the destruction of the diseased blocks and rigid control over planting material eliminated the disease. Consequently, some Mackay canegrowers persisted in using Q. 63, so that almost 13% of the cane crushed in the Central District in 1970 was this cane variety. Not surprisingly, lead scald outbreaks re-occurred in the Mackay district during 1971 and 1972, although the infections were more widespread than the early 1960s. Future crop protection against leaf scald was achieved by the large-scale ploughout of heavily diseased Q. 63 fields in the Mackay district during 1973 and the removal of Q. 63 from the list of approved cane varieties for the Mackay mill supply areas. Late in 1976, leaf scald was also found for the first time in Q. 63 in the Lower Burdekin district. The removal of this cane variety from the approved list of cane varieties at the end of 1978 brought this outbreak under control.[93]

Yellow spot also periodically affected canegrowers north of Townsville throughout the late 1960s, but the disease became particularly widespread during 1970 and 1971 in susceptible cane varieties such as Pindar, Triton, Q. 83 and Q. 91. This outbreak caused losses of approximately 30 000 tonnes of sugar in the mill supply areas north of Townsville. A further severe epidemic of the disease swept through the cane fields north of Townsville during the summer months of 1973, 1974 and 1977. Control measures were confined to reducing the planting of susceptible cane varieties.[94]

Ratoon stunting also became more troublesome throughout Australian sugar-producing districts during the late 1960s. The upsurge in the incidence of this disease occurred because of a combination of factors. In New South Wales, CSR noted a decline in the number of canegrowers using hot water treatment. Its field officers also complained that:

> many growers have also not yet realised the importance of sterilising mechanical harvesters and other farm machinery when changing fields. Stricter farm hygiene is needed to prevent re-infection with ratoon stunting disease.[95]

Similarly, the initial enthusiasm for hot water treatment of planting material in Queensland waned due to the belief that treated cane would remain healthy for a long time. In addition, massive re-infection occurred following the widespread failure to sterilise base cutters on mechanical cane harvesters.

Reducing the incidence of ratoon stunting disease in Queensland became a BSES priority during 1969. BSES-sponsored extension tours of farms affected with ratoon stunting disease were conducted in ten North Queensland mill supply areas during 1969 and 1970; over 1 000 canegrowers attended. Some direct effects such as increased demands for plant source inspections and bookings for heat treatment of planting material became apparent over the next few years. Yet a substantial reduction occurred in the tonnage of cane subjected to hot water treatment between 1974 and 1978, compared to the previous five years, again partly due to a waning of interest by canegrowers, although the reduction was offset by the increased tonnage sold from special plots planted with treated cane and maintained by the local Cane Pest and Disease Control Boards. By the late 1970s, approximately only 25% of Queensland canegrowers used hot water treatment each year, while an average of 35% treated and/or brought plant cane from special plots.[96]

One consequence of the efforts by the BSES and Cane Pest and Disease Control Boards to control disease outbreaks during the late 1960s and 1970s was a greater emphasis on providing and maintaining healthy planting material. ‘Significant advances were made during the 1970s in improving the overall quality of cane used for plants as far as freedom from pests and diseases was concerned', wrote BSES plant pathologist Brian Egan in 1980.[97] He suggested that pre-planting inspections were probably made on less than 50% of assignments throughout the 1960s, although no figures were available to support this claim. The situation improved considerably during the early 1970s, when almost two-thirds of all assignments in Queensland had their setts inspected prior to planting. This figure had risen to 92% by 1978. Such an improvement followed the introduction of an Approved Plant Sources Scheme in the Bundaberg district and Compulsory Plant Source Inspection Schemes for canegrowers supplying the Maryborough Sugar Factory and Moreton, Rocky Point, Kalamia, Inkerman and Pioneer Mills. These two schemes had been implemented in an attempt to reduce the incidence of Fiji and leaf scald diseases (see discussion earlier). In the Innisfail and Herbert River districts, few planting material inspections had been done prior to 1970, and the increase during the 1970s had been achieved by an increased commitment from the local Cane Pest and Disease Control Boards to overcome considerable canegrower apathy. However, almost a quarter of assignments in the Innisfail and Tully districts in 1978 still received no inspections.[98]

Combatting new and established diseases, 1979–1995

As the Fiji disease crisis commenced, cane on one farm in the Babinda Mill supply area during 1967 displayed a soft, flaccid rot of the larger roots. This symptom was associated with patchy poor growth, stools lodging, small root systems concentrating near the surface and ratoon failures due to the mechanical cane harvester pulling out the stools. In 1972, a second affected farm was discovered and more affected farms were located during the mid-1970s. The first report of the widespread presence of what had been tentatively called 'Babinda root rot' occurred throughout the Babinda district in 1978. 'Babinda root rot' and its associated problems were widespread in the Babinda Mill supply area during 1979 and 1980. Initial investigations into the cause of the problem focused upon identifying if nematodes were responsible for the symptoms, but these earliest studies were inconclusive. The possibility of a nematode/fungi complex causing the problem was investigated, but it proved impossible to isolate the pathogenic fungus. Soil salinity was also ruled out as a primary causal agent. By 1980, the BSES had excluded nutritional and other soil effects as the cause of the problem.[99]

Discovery of 'root rot' symptoms in the Mourilyan Mill supply area during 1980 alarmed the BSES. A far greater research effort was needed and the highest priority was accorded to the problem by the Sugar Experiment Stations Board. A special coordinated research program was established in August 1980 involving pathologists, entomologists and BSES extension officers. This research constituted the major activity of five BSES staff members in North Queensland and was a significant but varying activity for at least six other BSES staff. Some research was also carried out at the BSES Pathology Farm, while divisional heads and others senior staff at Head Office spent a considerable amount of time in planning and coordinating the program which focused upon identifying the biological agents causing the disease. In June 1982, the Sugar Experiment Stations Board invited a panel of specialist scientists to review and evaluate the research and to recommend improvements or changes in emphasis for the investigations. The Review Committee agreed that pathogenic fungi were likely to be the cause of 'root rot' and recommended that a major research effort be continued to test this hypothesis.

Part of these earliest investigations also involved gauging the extent of the disease. This work commenced in 1981 when a survey of young plant cane in mill supply areas from Mossman to Ingham, but excluding Babinda, revealed that approximately 12% of the blocks in the Tully, Innisfail and Cairns districts showed some rotten roots, while a further 11% had some dead roots. Further field surveys during 1984 and early 1985 were undertaken on major cane varieties to determine the distribution of the pathogen in the mill supply areas south of Tully. These surveys were based on the observation of rotted roots on excavated stools from commercial fields and the confirmation of the causal agent by observation under a microscope. They confirmed that 'root rot' was widespread in the Herbert River and Mackay districts, was present on only one Lower Burdekin farm, but was not located in Southern Queensland. However, its apparent absence in Southern Queensland sugar-producing districts proved illusionary, as 'root rot' was detected in a few sites throughout the Bundaberg district during 1987.[100]

The new disease that the BSES discovered was assigned the term 'Northern Poor Root Syndrome' in 1980, although the term initially covered the entire range of problems associated with poor or restricted root growth in North Queensland mill supply areas. Symptoms included flaccid primary roots, lesions on primary, secondary and tertiary roots and reduced fine root growth. By the late-1980s, the prefix 'Northern' had been dropped when field surveys revealed the disease to be present in some Southern Queensland mill supply areas. Moreover, by then the cause of the disease had been identified by BSES pathologists Barry Croft and Robert Magarey, who had first isolated an unidentified fungus from diseased roots in 1981. However, this fungus was unknown previously. It was eventually described as *Pachymetra chaunorhiza* Croft and Dick in 1989, the scientists creating a new genus to contain the fungus. Since then the disease has been known as *Pachymetra* root rot and subsequent investigations have shown that it could reduce the yield of susceptible cane varieties by up to 40% at soil spore levels of 50 sp/g and above.[101]

BSES research during the mid-1980s into methods that would control *Pachymetra* root rot failed to provide canegrowers with a quick solution to the problem. Tests on the effectiveness of fungicides against the disease were unable to identify a suitable fungicide that could be used for commercial control of the disease and no cultural treatments or nutritional amendments were found to significantly reduce the level of *Pachymetra* root rot in crops of sugar cane grown in affected soils. In the absence of any other control measure, cultivar resistance became the basis of the BSES control strategy for the disease. Screening of parent cultivars for resistance commenced and crosses between susceptible varieties were avoided. Consequently, the percentage of new cultivars resistant to *Pachymetra* root rot in the North Queensland mill supply areas rose from 21% in 1979 to 58% by 1987. Further field surveys of the extent of the disease between 1991 and 1994 revealed that it was reaching near critical levels in some mill supply areas (e.g. Fairymead; Victoria), while Mossman and Mackay districts had low levels of *Pachymetra* root rot. As a result of these findings, the BSES plant breeders were asked to place additional emphasis on selecting resistant cane varieties, especially for the Bundaberg district.[102]

As the symptoms of root rot first appeared in the Babinda Mill supply area, another new cane disease was discovered in North Queensland. Common rust or *Puccinia melanocephala* H. & P. Sydow was identified in the Cairns district for the first time in October 1978, although a subsequent survey indicated that the earliest known occurrence was in late September 1978 in the Mourilyan Mill supply area. Reports of rusty-coloured patches in cane crops were received from all mill supply areas from Tully to Mossman by mid-November 1978. Common rust was identified in the Herbert River district in late December 1978, in the Lower Burdekin, Mackay and Southern Queensland districts in March or April 1979 and in New South Wales by June 1979. Therefore, all sugar-producing districts in eastern Australia had been colonised by this disease within six months of its initial discovery in North Queensland. This outbreak was the first record of this rust species in the Australian/South Pacific region, the nearest countries previously known to have this rust being India and Taiwan. Spore dispersal by wind was considered to be the most likely method of entry into Australia.[103]

The earliest symptoms of common rust on the leaves are small, elongated yellowish spots which are visible on both leaf surfaces. These spots increase in size, eventually turning red-brown in colour. When common rust is severe, numerous lesions occur on individual leaves giving them an overall brown or rusty appearance. Affected crops assume a distinctive rusty to brown colour, depending upon cane variety, as disease intensity builds up. Consequently, infected blocks of cane stand out dramatically from a distance. The number of live leaves per plant affected by common rust can be seriously reduced, while in very susceptible cane varieties there may be fewer stalks per stool and a reduced diameter and length of stalk. Common rust is most severe in the young stages of plant and ratoon crops (i.e. less than six months growth), and it generally declines with increasing crop age. Crops most affected by common rust were those suffering from some moisture stress. In highly susceptible cane varieties, yield losses up to 45% were reported.[104]

In response to the common rust outbreaks, the BSES had commenced screening seedlings for common rust susceptibility in 1980, trying to identify new common rust resistant cane varieties that could take the place of common rust susceptible cane varieties such as Q. 90 (Townsville northwards), Q. 97 and Q. 117 (Lower Burdekin), Q. 142 (Mackay) and Q. 87 and Q. 108 (Southern Queensland). Young seedlings were sprayed with common rust spores and maintained in a humid atmosphere overnight. After two to three weeks growth in a glasshouse, the plants were rated for rust intensity. Investigations showed that there was only one physiologic race of common rust present throughout the pathogen population in Queensland, so this finding made selecting rust-resistant cane varieties a little easier. CSR also commenced screening its seedlings for common rust resistance at the Macknade Experiment Station. By the late 1980s, the adoption of common rust resistant cane varieties had become the main control method. Moderate to heavy common rust disease outbreaks, however, continued to affect susceptible cane varieties during the late 1980s. Farmers cultivating common rust susceptible cane varieties were advised to reduce moisture stress in their crops.[105]

During the 1980s, already identified diseases continued to trouble Queensland's canegrowers. Fiji disease remained a problem in the Mackay district and Rocky Point Mill supply area until the late 1980s. Moderate to heavy yellow spot disease outbreaks in the summers of 1983–1984 and 1988–1989 coincided with heavy wet seasons in North Queensland. These years were also particularly bad for incidences of chlorotic streak, which overall remained at relatively low levels for much of the decade due to the drier conditions. Outbreaks of striate mosaic/Trojan mottle occurred throughout the Lower Burdekin district during the 1980s. To control the spread of this disease, the Inkerman Cane Pest and Disease Control Board adopted a policy of restricting the use of planting material from farms with a history of the disease. Mosaic affected the susceptible variety Q. 95 in the Isis district during the mid-1980s, but a control program of plough-outs and restrictions on plant sources brought the outbreaks under control.[106]

Ratoon stunting disease, however, remained the main concern of the BSES and Cane Pest and Disease Control Boards. During the early 1980s, ten Cane Pest and Disease Control Boards purchased phase contrast microscopes to assist their supervisors diagnose the presence of ratoon stunting disease in crops of sugar cane. A BSES survey of hot water treatment tanks over 1983 found that 'several needed some attention to improve their performance', although the tonnages being treated by hot water were falling as the Cane Pest and Disease Control Boards had developed clean seed plots for the supply of disease-free setts. In 1989, a BSES survey of the ratoon stunting disease situation in Queensland showed that some mill supply areas suffered severe losses to this disease. Estimates of the area infected ranged from 13 000–20 000 ha which could have caused losses of $3–5 million annually. The failure of canegrowers to establish disease-free plant sources on their farms and to adopt adequate farm hygiene contributed to the build up of ratoon stunting disease throughout the state.[107]

In Queensland, controlling cane disease outbreaks had been the responsibility of the local Cane Pest and Disease Control Boards since the early 1940s. Minimal government control existed over these Boards and technical knowledge on pest and disease control was variable, as the expertise resided with the BSES. Many Boards sought BSES advice, but other Boards resented any interference in their affairs. A proposal for greater BSES involvement in the Boards was raised in 1970 but rejected by sections of the Queensland sugar industry. However, upsurges in outbreaks of Fiji and leaf scald diseases in the Mackay and Bundaberg districts highlighted the need for greater BSES technical input into the activities of these Boards. A second request for BSES representation on these Boards in 1972 was agreed to by the main Queensland sugar industry organisations. An amendment to the *Sugar Experiment Stations Act* in 1973 permitted this arrangement and a new BSES staff position known as Coordinator of Cane Pest and Disease Control Boards was created in January 1974. The holder of this position became a member of all the state's Cane Pest and Disease Control Boards.

The local BSES extension officer attended these meetings if the Coordinator was unable to attend.[108]

Brian Egan, the BSES pathologist who is best remembered for his research work during the mid-1960s into identifying the bacterium which caused sour storage rot of chopped cane (see Chapter 16 for a full discussion), assumed the role of Coordinator of Cane Pest and Disease Control Boards. Under BSES guidance, the technical competence of board members and staff increased. By the early 1980s, twenty-two of these Boards existed in Queensland and they employed a total of fifty-eight people who maintained the approved seed plots and conducted inspections of planting material. The future role of these Boards began to be reviewed in the mid-1980s, when the Bjelke-Petersen government in Queensland established a Sugar Industry Review Program to recommend industry-wide changes. A recommendation that the Boards be absorbed by the BSES was withdrawn when both parties objected. Eventually, the Queensland sugar industry agreed that the Boards should undertake local productivity functions as well as pest and disease control. Pest and disease control powers were brought under the *Plant Protection Act* of 1989 and were to be exercised by the Queensland Department of Primary Industries and the BSES. When the *Sugar Industry Act* of 1991 replaced much of the previous legislation concerning the Queensland sugar industry, the Cane Pest and Disease Control Boards became Cane Protection and Productivity Boards, with the authority to undertake productivity functions, although not obliged to do so. They remained statutory bodies with the right to statutory levies.[109]

The newly established Cane Protection and Productivity Boards in Queensland were confronted with various disease outbreaks during the early 1990s, but no new diseases troubled the canegrowers. Chlorotic streak was particularly widespread in North Queensland mill supply areas following widespread flooding during the early 1990s, but became less significant by 1995 (Table 15.2). Ratoon stunting remained a significant disease of sugar cane in Queensland during the early 1990s. Hot water treatment had dwindled in importance as a control measure, its place taken by the supply of disease-free plant material from approved seed plots maintained by the Cane Protection and Productivity Boards. Yellow spot and common rust still proved troublesome. Mosaic disease affected farms mostly in the Bundaberg and Isis districts and proved to be particularly difficult to bring under control. Incidences of Fiji and leaf scald diseases had been reduced to very low levels, while pineapple disease, red rot and striate mosaic remained minor diseases, much as they had been in previous decades.[110] Overall, the amount of the harvested crop affected by some disease, excluding *Pachymetra* root rot, fell during the early 1990s (Table 15.2).

Table 15.2. Estimated area (ha) affected by major sugar cane diseases in Queensland, 1991–1995.

Disease	1991	1992	1993	1994	1995
Chlorotic streak	31 000	26 300	19 200	13 150	8 700
Fiji disease	> 3 000	> 2 000	> 2 000	1 300	300
Leaf scald	130	90	20	0	0
Mosaic	3 300	3 000	4 400	3 650	3 300
Pineapple disease	902	170	130	80	130
Ratoon stunting disease	> 16 500	> 15 000	> 15 000	15 000	10 500
Red rot	600	3	3	300	150
Rust	16 000	16 100	16 800	17 400	17 500
Striate mosaic	140	170	89	33	24
Yellow spot	8 800	4 150	2 200	7 300	5 900
TOTALS	80 372 (24.6%)*	66 983 (20.8%)	46 343 (14.3%)	58 213 (16%)	46 504 (13.2%)

* The figure in brackets is a percentage of the total harvested crop.
Source: Based upon figures in the BSES, ARs 1991/92 to 1995/96.

In New South Wales, ratoon stunting and chlorotic streak disease troubled New South Wales canegrowers throughout the 1980s, especially during wetter years such as the summer of 1983–1984. Yet the major crop protection issue facing New South Wales sugar industry during this period was Fiji disease, a consequence of the sucrose-rich cane variety NCo. 310 being cultivated for as long as possible in the Harwood Mill supply area, despite its susceptibility to Fiji disease. By the early 1980s, most resources of the Harwood Cane Pest and Disease Control Board were devoted towards stopping the spread of Fiji disease by undertaking annual inspections and roguing. However, the problem escalated in 1983 when Fiji disease was found on three farms in the Richmond River district. This outbreak was contained, but the disease reappeared in the Richmond River district during 1989, infecting the newly released and frost-resistant cane variety H. 59–1966.

With Fiji disease spreading, the NSW Sugar Milling Coop. decided to replace Fiji-susceptible cane varieties, especially NCo. 310. Planting of

NCo.310 in the Harwood Mill supply area was prohibited in 1992 and the last of this cane variety was accepted for milling from farms up-river of Maclean during 1994 and from farms downstream of Maclean throughout 1995. Annual roguing campaigns against Fiji disease continued during the early 1990s in the Harwood Mill supply area in an attempt to stop the disease from infecting the replacement varieties.[111] During this transition phase, the NSW Sugar Milling Coop. also decided that it was important to enhance production in other cane varieties by reducing the incidence of ratoon stunting and chlorotic streak diseases. It assumed responsibility for maintaining and upgrading the hot water treatment facilities at each sugar mill, and commenced an extension program to promote the propagation of disease-free setts and improved hygiene with farm machinery. In addition, the three New South Wales Cane Protection and Productivity Boards during the early 1990s established Approved Seed Plots to supply the canegrowers with disease-free setts from the major cane varieties.[112]

At the beginning of this chapter, the discussion showed that the Australian sugar industry before World War II was affected by numerous cane diseases (e.g. gumming; leaf scald; Fiji disease). Yet university-trained plant pathologists employed by CSR and the BSES undertook research into the causes of these diseases, and developed a range of effective control measures which reduced their impact. Australian canegrowers were mostly compliant with the control strategies devised by the plant pathologists and industry officials; evidence of widespread resistance was not apparent. The legacy of this compliance was improved productivity which underpinned the great expansion in sugar output and exports after World War II and the total elimination of gumming and downy mildew diseases and the near eradication of leaf scald and Fiji diseases. Australia had moved from being a country known as a 'hot bed' of cane diseases to one where its plant pathologists led the way in identifying new cane diseases (e.g. ratoon stunting *Pachymetra* root rot; dwarf) and developing methods to reduce their impacts.

Notes

1 CSR, Staff Book, p. 307, CSRR, Z/109, Box 307; C. Hughes and P. Robinson, 'Fiji disease', in J. Martin, E. Abbott and C. Hughes (eds.), *Sugar-Cane Diseases of the World*, Vol. 1 (Amsterdam, 1961), pp. 390–391.

2 D. North, Memo to Gen. Mang., re: resignation, 10 April 1941, Report No. 580 D, CSRR, Z/109, Box 242; E. Baber, 'CSR Co. Agricultural History', pp. 27–30, unpubl. 59 pp. typescript dated March 1955, CSRR, Z/303, File A 6.0, Folder 2, Doc. 46.

3 *ASJ*, June 1960, pp. 217–218.

4 North, *Leaf Scald. A Bacterial Disease of Sugar Cane*, pp. 11–12; BSES, AR 1922–23, *QPP*, Vol. 2 (1924), p. 291.

5 N. King, 'Can downy-mildew be eradicated in Queensland', *CGQB*, Vol. 12, no. 1 (July 1948), p. 10; C. Leece, *Downy-Mildew Disease of Sugar Cane and Other Grasses.* BSES Technical Communications 1941, p. 1; D. North, 'Leaf stripe disease', Report No. 20, 20 June 1910, CSRR, Z/303, Box N239; D. North, 'Present Status of Cane Diseases and Pests at CSR's Australian Mills, 20 October 1914', p. 19, Report No. 45D, CSRR, Z/303, Box N239.

6 D. North, 'Fiji disease at CSR's Australian mills, 31 July 1911', p. 1, Report No 11 D, CSRR, Z/303, Box N239; D. North, 'Report on cane diseases at CSR's Australian mills, 1915', p. 2, Report No 61 D, CSRR, Z/303, Box N239.

7 D. North, 'Notes on cane diseases at CSR's Australian mills, 4 October 1912', Report No. 25D, CSRR, Z/303, Box N239; D. North, 'Present status of cane diseases and pests at CSR's Australian mills, 20 October 1914', p. 24, Report No. 45D, CSRR, Z/303, Box N239; CSR, Cane Report 1915, p. 35, CSRR, Z/109, Box 378.

8 W. Cottrell-Dormer, 'Report on the Fiji disease situation in Queensland', *QAJ*, Vol. 28, no. 2 (1 August 1927), p. 128.

9 For these views see CSR, Cane Report 1922–23, p. 27 & CSR, Cane Report 1924, p. 52, CSRR, Z/109, Box 386.

10 CSR, Cane Report 1916, pp. 35–36; CSR, Cane Report 1918, p. 45; CSR, Cane Report 1919, p. 34; CSR, Cane Report 1920, pp. 43–44 & 46–47; CSR, Cane Report 1921, pp. 51–52; CSR, Cane Report 1922–23, pp. 29–30; CSR, Cane Report 1924, p. 54; CSR, Cane Report 1925, p. 48, all in CSRR, Z/109, Box 386.

11 CSR, Cane Report 1918, p. 45 & CSR, Cane Report 1920, p. 43, both in CSRR, Z/109, Box 386.

12 CSR, Cane Report 1921, p. 50; CSR, Cane Report 1922–23, p. 28; CSR, Cane Report 1924, pp. 45–52; CSR, Cane Report 1925, pp. 44–47, all in CSRR, Z/109, Box 386.

13 CSR, Cane Report 1922–23, p. 28 & CSR, Cane Report 1926, p. 33, both in CSRR, Z/109, Box 386; D. North, 'The dissemination of gumming disease', *ASJ*, August 1927, pp. 285 & 292.

14 D. North, 'Obtaining resistant cane varieties for the control of gumming disease', *Proc. ISSCT*, 1932, Bulletin No. 30, pp. 1–3; R. Antoine, *Observations on Sugar Cane Diseases in Australia and Fiji*, Bulletin No. 5, Mauritius Sugar Industry Research Institute, 1955, pp. 35–36.

15 CSR, Cane Report 1927, pp. 27–28 & CSR, Cane Report 1928, pp. 28 & 46, both in CSRR, Z/109, Box 386; BSES, AR 1929–30, *QPP*, Vol. 2 (1930), p. 939.

16 Easterby, *The Queensland Sugar Industry*, pp. 211 & 214; Skerman, Fisher and Lloyd, *Guiding Queensland Agriculture*, p. 128; BSES, AR 1925–26, *QPP*, Vol. 2 (1927), p. 643; for BSES staffing details during the 1930s see BSES, *Fifty Years of Scientific Progress. A Historical Review of the Half Century Since the Foundation of the BSES* (Brisbane, c. 1950), pp. 19–20 & BSES, AR 1933–34, *QPP*, Vol. 2 (1935), p. 732.

17 For a selection of his papers see W. Cottrell-Dormer, *Red-stripe disease of sugar-cane in Queensland.* BSES Division of Pathology, Bulletin No. 3 (1932), pp. 25–59; W. Cottrell-Dormer, 'The estimation of losses due to red stripe disease in Queensland', *Proc. ISSCT*, Bulletin No. 43, 1932; A. Bell and W. Cottrell-Dormer, 'An improved

method for the isolation of the leaf-scald organism', *Proc. ISSCT*, Bulletin No. 72, 1932; W. Cottrell-Dormer, 'Report on the Fiji disease situation in Queensland', *QAJ*, Vol. 28, no. 2 (1 August 1927), pp. 125–135; William Cottrell-Dormer, 'Red rot in sugar-cane', in *ASJ*, 6 February 1925, pp. 734–737.

18 BSES, AR 1924–25, *QPP*, Vol. 2 (1926), pp. 705–707; BSES, AR 1925–26, *QPP* Vol. 2 (1927), pp. 662–665; BSES, AR 1926–27, *QPP*, Vol. 2 (1927), p. 752; E. Ferguson Wood, 'Fiji disease in the Maryborough district', *QAJ* Vol. 27 (May 1927), pp. 388–393; W. Cottrell-Dormer, 'Report on the Fiji disease situation in Queensland', *QAJ*, Vol. 28 (August 1927), pp. 125–135.

19 Skerman, Fisher and Lloyd, *Guiding Queensland Agriculture*, pp. 128 & 199; details about Bell's university studies were supplied by the Archives Section, University of Queensland.

20 A. Bell, *Dwarf Disease of Sugar-Cane*. BSES Farm Bulletin No. 8 (1933); A. Bell and R. Mungomery, *Fiji Disease of Sugar-Cane and its Transmission*. BSES Division of Pathology Bulletin No. 4, 1933; A. Bell, 'Fiji disease in the southern districts' *CGQB*, Vol. 3, no. 1 (July 1935), pp. 1–4; A. Bell, *A Key for the Field Identification of Sugar Cane Diseases*. BSES Division of Pathology Bulletin No. 2, 1929, p. 7.

21 Easterby, *The Queensland Sugar Industry*, p. 218; Skerman, Fisher and Lloyd, *Guiding Queensland Agriculture*, pp. 131 & 199.

22 A. Bell, 'Twenty years of disease control', *J. of the Australian Institute of Agricultural Science*, Vol. 15, no. 1 (March 1949), p. 3.

23 BSES, AR 1927–28, *QPP*, Vol. 2 (1929), p. 873; BSES, AR 1928–29, *QPP* Vol. 2 (1929), p. 970; BSES, AR 1929–30, *QPP*, Vol. 2 (1930), pp. 938–940; C. G. Hughes, 'Sugar cane pathology and disease control in Australia', in BSES, *Fifty Years of Scientific Progress. A Historical Review of the Half Century Since the Foundation of the BSES* (Brisbane, c. 1950), p. 53.

24 BSES, AR 1929–30, *QPP*, Vol. 2 (1930), p. 938; BSES, AR 1930–31, *QPP*, Vol. 2 (1931), pp. 707–708; BSES, AR 1933–34, *QPP*, Vol. 2 (1935), p. 774.

25 BSES, AR 1932–33, *QPP*, Vol. 2 (1930), p. 813; BSES, AR 1933–34, *QPP*, Vol. 2 (1935) p. 776; A. Bell, Brisbane, to D. North, Broadwater, 27 May 1933, Letter No. 380D, CSRR, Z/109, Box 241.

26 BSES, AR 1932–33, *QPP*, Vol. 2 (1933), pp. 808–813; Hughes, 'Sugar cane pathology and disease control in Australia', p. 54.

27 W. Cottrell-Dormer, 'Notes and observations on the red-streak associated with Queensland top rot disease', *QAJ*, Vol. 25, no. 5 (May 1926), pp. 406–414; W. Cottrell-Dormer, 'Red-Stripe disease of sugar cane in Queensland. Part 1', *QAJ*, Vol. 37, no. 1 (January 1932), pp. 23–40; W. Cottrell-Dormer, 'Red-Stripe disease of sugar cane in Queensland. Part 2', *QAJ*, Vol. 37, no. 2 (January 1932), pp.98–114; BSES, AR 1932–33, *QPP*, Vol. 2 (1933), pp. 809– 810; King, Mungomery and Hughes, *Manual of Cane-Growing*, 1953 edition, p. 276.

28 R. Mungomery and A. Bell, *Fiji Disease of Sugar Cane and its Transmission*. BSES Division of Pathology. Bulletin No. 4 (1933), pp. 20–21; BSES, AR 1932–33, *QPP* Vol. 2 (1933), pp. 809–810 & 813; BSES, AR 1934–35, *QPP*, Vol. 2 (1936), p. 820; BSES, AR 1936–37, p. 38; BSES, AR 1938–39, p. 54.

29 Bell, 'Twenty years of disease control', p. 3; D. North to Gen. Mang., CSR, Sydney, 13 October 1933, Letter 387 D, CSRR, Z/109, Box 241; D. North, *The Control of Sugar Cane Diseases*. CSR, Agricultural Report No. 6 (Sydney, 1923), p. 34.

30 CSR, Cane Report 1925, pp. 45–46, CSRR, Z/109, Box 386; Minute Book, Clarence River Branch of NSWCGA, 1925–1930, pp. 18 & 66.

31 Easterby, *The Queensland Sugar Industry*, p. 214; D. North, 'Report on Queensland Government Bureau of Sugar Experiment Stations, 1915', p. 2, Report No. 63 D & D. North, 'Memo to Gen. Mang. re: Cane Varieties, 22 December 1920', p. 2, Report No. 80 D, CSRR, Z/303, Box N239.

32 BSES, AR 1925–26, *QPP*, vol. 2 (1927), p. 665; BSES, AR 1926–27, *QPP*, Vol. 2 (1927), p. 747; BSES, AR 1929–30, *QPP*, Vol. 2 (1930), p. 936.

33 BSES, AR 1933–34, *QPP*, Vol. 2 (1935), p. 770; Bell, 'Twenty years of disease control', pp. 4–5.

34 BSES, AR 1926–27, *QPP*, Vol. 2 (1927), p. 747; D. North to E. W. Knox, Gen. Mang., CSR, 6 January 1920, No. 69D of 1920, CSRR, Z/109, Box 239.

35 CSR, Cane Report 1920, p. 43, CSRR, Z/109, Box 386; BSES, AR 1926–27, *QPP*, Vol. 2 (1927), p. 749; BSES, AR 1929–30, *QPP*, Vol. 2 (1930), p. 936.

36 CSR, Cane Report 1919, p. 34 & CSR, Cane Report 1928, p. 34, both in CSRR, Z/109, Box 386; BSES, AR 1921–22, *QPP*, Vol. 2 (1923), p. 278; BSES, AR 1926–27, *QPP*, Vol. 2 (1927), p. 748; BSES, AR 1927–28, *QPP*, Vol. 2 (1929), p. 873; D. North to CSR Gen. Mang., 6 January 1920, pp. 1–4, Letter No. 69 D, CSRR, Z/109, Box 239.

37 BSES, AR 1929–30, *QPP*, Vol. 2 (1930), pp. 936–937.

38 P. Volp, 'Some effects of leaf-scald disease in the Mulgrave area', *Proc. QSSCT*, 1948, p. 171.

39 CSR, Cane Report 1924, p. 50 & CSR, Cane Report 1926, p. 31, both in CSRR, Z/109, Box 386; BSES, AR 1927–28, *QPP*, Vol. 2 (1929), p. 10; Hughes, 'Sugar cane pathology and disease control in Australia', p. 53.

40 D. North, 'Present status of cane diseases and pests at CSR's Australian Mills, 20 October 1914', p. 24, Report No. 45D, CSRR, Z/303, Box N239; D. North, 'Cane Diseases at Childers, 22 February 1915', Report No. 51D, CSRR, Z/303, Box N239; D. North, 'Cane Diseases at Australian Mills, 27 October 1915', pp. 6–7, CSRR, Z/303, Box 239.

41 B. Egan, 'The decline of leaf scald as a major disease in northern Queensland', *Proc. QSSCT*, 1971, p. 157; D. North, *Leaf-Scald. A Bacterial Disease of Sugar Cane*. CSR, Agricultural Report No. 8 (Sydney, 1926), pp. 13 & 15, CSRR, Z/109, Box 369; CSR, Cane Report 1930, pp. 22–23 & CSR, Cane Report 1932, p. 13, both in CSRR, Z109, Box 386.

42 CSR, Cane Report 1922–1923, p. 28, CSRR, Z109, Box 386; N. King and G. Christie, 'Gumming resistant canes in southern Queensland', *CGQB*, Vol. 4, no. 1 (July 1936), p. 19; N. Bennett, 'Report on Uba cane milling and payment, 22 September 1927' in RSI, 14844/1/21, Box 3, BSES Field Assistants' Reports, QSA; compare figures for Victoria and Macknade Mills (Herbert River district) in the 'Table of Leading Varieties crushed at each CSR Mill', for 1923 and 1928, in CSR Cane Report 1922–23, p. 12 & CSR, Cane Report 1928, p. 13, both in CSRR, Z/109, Box 386.

43 BSES, AR 1935–36, p. 23; A. Bell, 'Fiji disease in the southern districts', *CGQB*, Vol. 3, no 1 (July 1935), p. 1; for the gumming disease outbreaks of 1934–36 see CSR, Cane Report 1934, pp. 51–52, CSR, Cane Report 1935, p. 52 & CSR, Cane Report 1936, p. 45, all in CSRR, Z/109, Box 387.

44 D. North, 'Memo to Gen. Mang. re: gumming disease, 4 February 1923', p. 3, Report No. 127 D, CSRR Z/109, Box 239; CSR, Cane Report 1927, p. 28, CSR, Cane Report 1928, pp. 41–44 & CSR, Cane Report 1933, p. 50, all in CSRR, Z/109, Box 386.

45 BSES, AR 1933–34, *QPP*, Vol. 2 (1935), p. 731.

46 BSES, AR 1933–34, *QPP*, Vol. 2 (1935), p. 732; C.G. Hughes, E. Abbott and C. Wismer (eds.), *Sugar-Cane Diseases of the World*, 2 vols. (Amsterdam, 1961 and 1964); C. Hughes, 'Treatment with fungicides as a help to better strikes', *CGQB*, Vol. 12. no. 2 (October 1948), pp. 54–58; C. Hughes, *Alternative hosts of* B. vasculorum, *the causal agent of gumming disease*. BSES Technical Communications, No. 3 (1939), pp. 35–64.

47 BSES, AR 1934–35, *QPP*, Vol. 2 (1936), p. 815; BSES, AR 1952–53, p. 37; BSES, AR 1954–55, p. 2; for a brief biography see O. Sturgess, 'Eminent pathologist dies', *CGQB*, Vol. 45, no. 3 (January 1982), p. 71; for his earlier works see D. Steindl, 'Q. 28 disease', *CGQB*, Vol. 12, no. 4 (1949), pp. 191–193; D. Steindl, 'Ratoon stunting disease', *Proc. ISSCT*, 1950, pp. 456–465; & N. King & D. Steindl, 'The relationship between varietal yield deterioration and ratoon stunting disease', *CGQB*, Vol. 47, no. 2 (October 1953), pp. 64–73.

48 Anon, 'The Australian Medal of Agricultural Science, 1959', *J. of the Australian Institute of Agricultural Science*, June 1959, pp. 160–161.

49 BSES, AR 1933–34, *QPP*, Vol. 2 (1935), pp. 770–773 & 777; BSES, AR 1934–35, *QPP*, Vol. 2 (1936), p. 822; BSES, AR 1936–37, p. 38; BSES, AR 1937–38, pp. 47–48; BSES, AR 1938–39, pp. 54–57; A. Bell, 'The present status of dwarf disease', *Proc. QSSCT*, 1936, p. 130.

50 BSES, AR 1941–42, p. 12; BSES, AR 1943–44, p. 24; BSES, AR 1944–45, pp. 21–22.

51 BSES, AR 1942–43, p. 19.

52 BSES, AR 1946–47, pp. 42–43; BSES, AR 1947–48, pp. 36–37; BSES, AR 1948–49, pp. 40–41; BSES, AR 1952–53, p. 45; BSES, AR 1955–56, p. 85; BSES, AR 1958–59, pp. 84–85; BSES, AR 1959–60, pp. 80–82; BSES, AR 1960–61, pp. 69–71; for a selection of published work on chlorotic streak research see B. Egan, 'Results from some chlorotic streak trials in North Queensland', *Proc. QSSCT*, Vol. 29 (1962), pp. 121–125; B. Egan, 'Host range and possible sources of resistance to chlorotic streak disease', *Proc. ISSCT*, 1965, pp. 1055–1059; & O. Sturgess, 'An understanding of chlorotic streak disease', *Proc. ISSCT*, 1965, pp. 1044–1045.

53 BSES, AR 1945–46, pp. 37–38; BSES, AR 1947–48, p. 39; BSES, AR 1948–49, pp. 41–42 & 44–45; BSES; AR 1949–50, pp. 42 & 44; BSES, 1954–55, pp. 72–73; BSES, AR 1955–56, pp. 81–83; BSES, AR 1956–57, pp. 79–80; BSES, AR 1963–64, p. 79.

54 Steindl, 'Ratoon stunting disease', pp. 462–463; BSES, *75 Years of Scientific Progress*, pp. 27; BSES, AR 1948–49, pp. 41–42; BSES, AR 1954–55, pp. 73–74; BSES, AR 1956–57, p. 80.

55 Hughes, 'Treatment with fungicides as a help to better strike', pp. 56–58; BSES, AR 1947–48, pp. 36–37; BSES, AR 1948–49, pp. 40–41& 44–45; BSES, AR 1949–50, p. 44; BSES, AR 1950–51, p. 45; BSES, AR 1951–52, p. 44; BSES, AR 1953–54, p.73; BSES, AR 1956–57, p. 85; BSES, *75 Years of Scientific Progress*, pp. 27–28; B. Egan, 'Chlorotic streak', in C. Ricaud *et al.* (eds.), *Diseases of Sugar Cane. Major Diseases* (Amsterdam, 1989), p. 250.

56 BSES, AR 1949–50, p. 38; BSES, AR 1951–52, p. 39; BSES, AR 1965–66, p. 62; BSES, AR 1969–70, p. 51; C. Hughes, 'The Bureau pathology farm', *CGQB*, Vol. 17, no. 2 (October 1953), pp. 57–61.

57 BSES, AR 1950–51, p. 44; C. Hughes and G. Ocfemia, 'Yellow Spot Disease', in International Society of Sugar-Cane Technologists. Pathology Section, *Sugar-Cane Diseases*

of the World. Vol. 1 (Amsterdam, 1961), pp. 357–360; King, Mungomery and Hughes, *Manual of Cane-Growing*, 1965 revised edition, p. 281.

58 King, Mungomery and Hughes, *Manual of Cane-Growing*, 1965 edition, pp. 274–275 & 281–282; BSES, AR 1956–57, p. 82; BSES, AR 1960–61, p. 77.

59 King, Mungomery and Hughes, *Manual of Cane-Growing*, 1965 revised edition, pp. 275 & 282.

60 Hughes and Ocfemia, 'Yellow Spot Disease', p. 365; BSES, AR 1950–51, p. 44; BSES, AR 1951–52, p. 45; BSES, AR 1958–59, p. 86; B. Egan, 'A preliminary report on the control of yellow spot disease by spraying with copper oxychloride', *Proc. QSSCT*, 1960, pp. 145–148.

61 CSR, Cane Report 1932, p. 54, CSRR, Z/109, Box 386; D. North, 'Fiji disease and varieties', *Proc. ISSCT*, 1935, pp. 498–505; D. North to CSR Gen. Mang., 10 June 1938, Letter No. 516 D (a), CSRR, Z/109, Box 242.

62 CSR, Cane Report 1933, p. 57, CSRR, Z/109, Box 386; CSR, Cane Report 1935, p. 52, CSRR, Z/109, Box 387; D. North to CSR Gen. Mang., 10 June 1938, Letter No. 515D & D. North to CSR Gen. Mang., 19 April 1940, Letter No. 561D, both in CSRR, Z/109, Box 242.

63 For these annual trials see the section on 'Cane diseases' in the annual cane reports produced by CSR, 1945 to 1950, in CSRR, Z/109, Boxes 387; for Esmond Baber's career with CSR see 'Staff Register Officers Service Series 2, 1916–1958', CSRR, Z/109, Box 307.

64 For Perry Robinson's career with CSR see 'CSR Staff Register', CSRR, Z/109, Box 560.

65 For details on CSR's plant pathological investigations during the 1950s and early 1960s see 'Pathology Memoranda, 1955 to 1962', in CSRR, Z/395, Box 57 & CSR, AR Cane Season 1954, p. 64; CSR, AR Cane Season 1956, p. 36; CSR, AR Cane Season 1957, pp. 32, 38 & 39; CSR, AR Cane Season 1959, pp. 38 & 43; & CSR, AR Cane Season 1964, Part II, p. iv, all in CSRR, Z/109, Boxes 388 and 389.

66 Skerman, Fisher and Lloyd, *Guiding Queensland Agriculture*, p. 112; BSES, AR 1938–39, p. 45; B. Egan, 'Origins and development of the sugarcane pest and disease control board system in Queensland, 1890–2000', *Proc. ASSCT*, 2002, pp. 277–278.

67 C. Hughes, 'The control of cane diseases in Queensland', *Proc. ISSCT*, 1950, p. 466; BSES, AR 1981–82, p. 65.

68 BSES, AR 1935–36, p. 24; BSES, AR 1938–39, p. 52; BSES, AR 1940–41, p. 21; BSES, AR 1947–48, p. 39; J. Buzacott, 'Varietal changes in the Burdekin district', *Proc. QSSCT*, 1952, p. 157; A. Bell, 'S.J. 4 in the Hambledon area', *CGQB*, Vol. 7, no. 4 (April 1940), p. 186.

69 N. King, 'Fiji disease in the Maryborough district', *Proc. QSSCT*, 1948, pp. 167–168; A. Bell, 'Dwarf Disease in P.O.J. 2878 at Mackay', *QAJ*, Vol. 50, no. 1 (July 1938), p. 86.

70 B. Egan, 'The decline of leaf scald as a major disease in northern Queensland', *Proc. QSSCT*, 1971, pp. 157–160.

71 BSES, AR 1950–51, p. 45; BSES, AR 1951–52, p. 44; CSR, AR Cane Season 1958: Part 2, p. 39 & CSR, AR Cane Season 1961: Part 2, p. 41, both in CSRR, Z/109, Box 389; S. Skinner, 'Mercurial treatment of cane setts in the Innisfail and Tully areas', *CGQB*, Vol. 18, no. 4 (April 1955), pp. 137–138.

72 BSES, AR 1949–50, p. 44; King, Mungomery & Hughes, *Manual of Cane-Growing*, 1953 edition, pp. 280–281; CSR, Cane Report 1952, p. 77, CSRR, Z/109, Box 389.

73 Hughes, 'The control of cane diseases in Queensland', p. 471.

74 R. Mungomery, 'Clean cane for combating ratoon stunting disease', *CGQB*, Vol. 17, no. 3 (January 1954), p. 92; S. Greenaway, 'Notes on the hot water treatment of cane setts in the Mackay district, 1953', *Proc. QSSCT*, 1954, pp. 201–202; C. Hughes and D. Steindl, 'Some further developments in the study of ratoon stunting disease in Queensland', *Proc. ISSCT*, Vol. 1, 1956, pp. 1015–1017; BSES, AR 1966–67, p. 56; *ASYB*, 1966, p. 39.

75 Kerr, *Southern Sugar Saga*, pp. 137–138; *ASYB*, 1991, p. 33; B. Egan, 'The fall and rise of Fiji disease in southern Queensland', *Proc. QSSCT*, 1976, p. 74.

76 P. Volp, 'The ineffectiveness of roguing leaf scald infected fields for use as a source of planting material', *Proc. QSSCT*, 1947, p. 101.

77 CSR, Cane Report 1924, p. 48, CSRR Z/109, Box 386.

78 King, Mungomery and Hughes, *Manual of Cane-Growing*, 1953 edition, p. 263.

79 BSES, AR 1939–40, p. 18; BSES, AR 1940–41, p. 20; A. Bell, 'Can downy-mildew disease be eradicated in Queensland', *CGQB*, Vol. 12, no. 1 (July 1948), p. 11.

80 *ASYB*, 1966, p. 95; BSES, AR 1964–65, p. 63; BSES, AR 1966–67, p. 56; BSES, AR 1967–68, p. 68; A. Hayes, 'Control of Fiji disease in sugar cane – a review of 15 years experience in New South Wales', *Proc. QSSCT*, 1977, p. 49.

81 BSES, AR 1970–71, p. 48; BSES, AR 1975–76, p. 15; C. Toohey and P. Nielsen, 'Fiji disease at Bundaberg', *Proc. QSSCT*, 1972, pp. 192–194; Anon, 'Fiji disease builds up in Central District', *BSES Bulletin*, No. 2 (April 1983), pp. 21–22.

82 Hayes, 'Control of Fiji disease in sugar cane – a review of 15 years experience in New South Wales', p. 49; BSES, AR 1980–81, pp. 13 & 15; BSES, AR 1985–86, p. 22; *ASYB*, 1985, p. 178.

83 Hayes, 'Control of Fiji disease in sugar cane – a review of 15 years experience in New South Wales', pp. 50–53; CSR, AR Technical Field Work 1972–73, p. 8 & CSR, AR Technical Field Work 1973–74, p. 8, both in CSRR, Z/109, Box 393.

84 BSES, AR 1969–70, p. 50; Toohey and Nielsen, 'Fiji disease at Bundaberg', p. 195; C. Hughes, 'Fiji disease – behind the scenes at Bundaberg', *Proc. QSSCT*, 1972, p. 198; B. Eagan and T. Fraser, 'The development of the Fiji disease epidemic at Bundaberg', *Proc. QSSCT*, 1977, pp. 43–44; J. Wright, 'Resurgence of Fiji disease in southern Queensland', *CGQB*, Vol. 43, no. 4 (April 1980), p. 105.

85 Hayes, 'Control of Fiji disease in sugar cane – a review of 15 years experience in New South Wales', p. 50; CSR, AR Technical Field Work 1971–72, p. iv, CSRR, Z/109, Box 393.

86 Eagan and Fraser, 'The development of the Fiji disease epidemic at Bundaberg', p. 43; Kerr, *Southern Sugar Saga*, p. 138; BSES, AR 1971–72, p. 17; B. Egan, 'The Bundaberg Approved Plant Sources Scheme', *Proc. QSSCT*, 1977, pp. 55–59.

87 D. Steindl, C. Hughes and D. Hogarth, 'Developments in Fiji field-trial design', *Proc. QSSCT*, 1977, p. 75.

88 CSR, *Newsletter* No. 114, October 1969, pp. 9–13, CSRR, Z/109, Box 372; CSR, AR Cane Season 1962, Part II, p. v, CSRR, Z/109, Box 389; Hayes, 'Control of Fiji disease in sugar cane – a review of 15 years experience in New South Wales', pp. 52–53.

89 BSES, AR 1974–75, p. 13; J. Powell, 'The glasshouse complex at Bundaberg', *CGQB*, Vol. 37, no. 4 (April 1974), pp. 142–144; Steindl, Hughes & Hogarth, 'Developments in Fiji field-trial design', p. 75.

90 D. Hogarth *et al.*, 'Quantitative inheritance of Fiji disease resistance in sugarcane', *Field Crops Research*, Vol. 34 (1993), p. 176; BSES, AR 1976–77, p. 4; BSES, AR

1979–80, p. 10; P. Ledger and C. Ryan, 'Screening of sugar-cane varieties for resistance to Fiji disease in Queensland – the insectary glasshouse method', *Proc. QSSCT*, 1977, p. 81.

91 BSES, AR 1977–78, p. 12; AR, BSES 1989–90, p. 19; *ASYB*, 1991, p. 33.

92 BSES, AR 1968–69, p. 57; BSES, AR 1969–70, p. 49; B. Egan, 'Red rot in North Queensland', *Proc. QSSCT*, 1969, pp. 73–74.

93 BSES, *75 Years of Scientific Progress*, p. 28; BSES, AR 1963–64, p. 69; BSES, AR 1965–66, p. 61; BSES, AR 1973–74, p. 14; BSES, AR 1977–78, p. 12.

94 B. Egan, 'Yellow spot attacks again', *CGQB*, Vol. 38, no. 1 (July 1974), pp. 7–9; BSES, AR 1969–70, p. 49; BSES, AR 1973–74, p. 16; AR BSES 1976–77, p. 12.

95 CSR, AR Technical Field Work 1971–72, p. iv, CSRR, Z/109, Box 393.

96 B. Egan, 'R.S.D. in North Queensland', *Proc. QSSCT*, 1970, pp. 221–224; BSES, AR 1967–68, p. 68; BSES, AR 1969–70, p. 48; E. Pembroke, 'Be Wise-Sterilise', *CGQB*, Vol. 37, no. 1 (July 1973), p. 27; B. Egan, 'Methods used to improve the health and status of plant sources during the 1970s', *Proc. QSSCT*, 1980, pp. 83–84.

97 B. Egan, 'Methods used to improve the health status of plant sources during the 1970s', *Proc. ASSCT*, 1980, p. 86.

98 Egan, 'Methods used to improve the health status of plant sources during the 1970s', p. 84.

99 B. Egan *et al.*, 'A review of the Northern Poor Root Syndrome of sugarcane in North Queensland', *Proc. ASSCT*, 1984, pp. 2–4.

100 BSES, AR 1980–81, pp 4 & 15; BSES, AR 1981–82, p. 47; BSES, AR 1986–87, p. 21; Egan *et al.*, 'A review of the Northern Poor Root Syndrome of sugarcane in North Queensland', pp. 4–5; R. Magarey, P. Taylor and C. Ryan, 'Distribution of root rot fungus involved in poor root syndrome in cane fields from Ingham to Rocky Point', *Proc. ASSCT*, 1987, pp. 105–107; R. Magarey *et al.*, 'The severity of *Pachymetra* root rot in Queensland cane fields', *Proc. ASSCT*, 1997, p. 67.

101 BSES, AR 1983–84, p. 11; B. Croft and R. Magarey, 'A review of research into *Pachymetra* root rot, an important new fungal disease of sugar cane', *Proc. ISSCT*, Vol. 2, 1989, pp. 687–688; Magarey *et al.*, 'The severity of *Pachymetra* root rot in Queensland cane fields', p. 67.

102 BSES, AR 1987–88, p. 20; B. Croft and R. Magarey, 'A review of research into *Pachymetra* root rot, an important new fungal disease of sugar cane', *Proc. ISSCT*, Vol. 2, 1989, pp. 690–692; Magarey *et al.*, 'The severity of *Pachymetra* root rot in Queensland cane fields', pp. 70–71.

103 BSES, AR 1978–79, pp. 10 & 12; BSES, AR 1981–82, p. 12; C. Ryan and P. Ledger, 'Studies and observations on rust disease in Queensland', *Proc. ASSCT*, 1980, p. 71.

104 BSES, AR 1983–84, p. 14; C. Ryan and B. Egan, 'Rust', in C. Ricaud *et al.* (eds.), *Diseases of Sugarcane. Major Diseases* (Amsterdam, 1989), p. 192.

105 BSES, AR 1982–83, p. 13; BSES, AR 1985–86, p. 23; BSES, AR 1988–89, p. 26; BSES, AR 1989–90, p. 19; Ryan and Ledger, 'Studies and observations on rust disease in Queensland', p. 72; W. Symington, 'Selection of varieties for rust resistance in the Herbert River area', *Proc. ASSCT*, 1981, pp. 287–291.

106 BSES, AR 1981–82, pp. 12–13; BSES, AR 1983–84, p. 16; BSES, AR 1984–85, p. 16; BSES, AR 1985–86, pp. 23–24; BSES, AR 1987–88, pp. 23 & 25; BSES, AR 1989–90, pp. 19 & 25.

107 BSES, AR 1981–82, p. 13; BSES, AR 1982–83, pp. 14–15; BSES AR 1983–84, p. 15; BSES, AR 1984–85, p. 16; BSES, AR 1988–89, p. 25; B. Croft, 'Integrated pest management of ratoon stunting disease of sugarcane in Australia', in J. Wilson *et al.* (eds.), *Sugarcane: Research Towards Efficient and Sustainable Production* (Brisbane, 1996), p. 232.

108 Egan, 'Origins and development of the sugarcane Pest and Disease Control Board system in Queensland', p. 278.

109 Egan, 'Origins and development of the sugarcane Pest and Disease Control Board system in Queensland', p. 279; *ASYB*, 1982, pp. 22–24.

110 BSES, AR 1991–92, pp. 16–17; Croft, 'Integrated pest management of ratoon stunting disease of sugarcane in Australia', p. 232.

111 Briody, *Sunshine Sugar*, pp. 98–99; Clarence River Branch, NSWCGA, AR 1985–86, p. 11; Clarence River Branch, NSWCGA, AR 1992–93, pp. 10 & 13; Clarence River Branch, NSWCGA, AR 1993–94, p. 12.

112 Briody, *Sunshine Sugar*, pp. 98–100; Clarence River Branch, NSWCGA, AR 1992–93, p. 13; NSW Sugar Milling Coop., AR 1994–95, p. 13; NSW Sugar Milling Coop., AR 1995–96, p. 11.

CHAPTER 16
Harvesting and Transport of Cane

In 1915, cane cutting by gangs of Europeans or the butty gang system – a phrase introduced into the sugar industry literature by the economic historian Ralph Shlomowitz – had been in existence for a decade. The team was paid on a collective piece-rate basis (shillings per ton of cane cut), with all canecutters sharing equally in the joint earnings of the team according to the number of days each man had worked during the crushing season. Most teams consisted of seven to ten men, and included a full-time cook who shared equally with the other members of the team in the proceeds. The canegrower retained a percentage of the team's earnings (i.e. retention money) until the contract was completed. If a canecutter departed the team before the end of the contract, the retention money was in some instances given to other team members or kept by the canegrower. The retention money was between 10 and 20% of the earnings, and it gave the canecutters an incentive to see the harvesting season through by holding the gang together. The historian Clive Morton, however, has stated that retention money was actually little incentive for canecutters to honour their contracts and that in fact some canecutters took their contracts lightly.[1]

'By the beginning of the 1920s, the procedures for the recruitment and employment of canecutters were becoming fixed', observed the historian Denis Conomos. The ganger or team leader recruited about seven others who were prepared to cut cane with him and one other to be the cook. The Sugar Industry Award stated that only men could cook for the gangs, but the historian Diane Menghetti claims that women, including the wives of gangers or canegrowers, regularly cooked for the gangs. Some of the ganger's responsibilities included completing time sheets, organising the gang's harvesting schedule and acting as a gang's spokesperson. The ganger also negotiated with the canegrower or Cane Inspector about the special rates to be paid to the gang for cutting lodged or weed-infested cane, or cutting in paddocks littered with stones, and when the cane should be burnt. Maintaining harmony in the field and barracks and letting a man know he was not pulling his weight were also other ganger responsibilities.[2]

Before the commencement of the season, which corresponded to the day the mill would start crushing the cane, the gangers assembled at a local sugar mill for a meeting with the cane inspector. The cane inspector had divided the farms in the mill supply area into groups of three or four depending upon the size of the mill supply area. The name of each ganger would be placed in a tin and as each group of farms was read out, a ganger's name would be taken from the tin by the cane inspector and the cutting for that group of farms would be given to that gang. Some entrepreneurial gangers, however, approached the canegrowers before the commencement of the crushing season, trying to secure a contract for the team. Canegrowers may have known the quality of local gangers, but they had no say in the membership of the gang and its potential to complete the task, unless they had previously employed the gang. If the ganger was successful in convincing a canegrower to offer a contract, the cane inspector would be informed and the need to draw lots would be avoided.[3]

The new season commenced with the sign-on day, when the canegrower and his prospective gang met to formalise their contract. In North Queensland this event usually occurred in June, although it could happen in late May if the season was to be unusually long because of large crops or possible difficulties over the crushing season, like the shortage of canecutters during World War II. In Southern Queensland, sign-on occurred in July or even early August if the crop to be harvested was quite small. On sign-on days, canecutters gathered outside the local cinema, the town hall, a sugar mill or the local office of the QCGA, waiting for their turn to enter the building and to sign-on for the season.

After sign-on, the canecutters would take up residence in the cane barracks. The form, size and quality of accommodation provided by the barrack had first been determined by *The Shearers and Sugar Workers' Accommodation Acts*, 1905–1906. The newly elected Ryan Labor government in Queensland repealed this legislation in 1915, replacing it with *The Workers' Accommodation Act*. Basically, the new regulations increased the amount of air space allowed for each worker from 360 to 480 ft^3 (10.1 m^3 to 13.6 m^3), and required the building used for accommodation to be provided with separate dining areas, bathrooms, an adequate supply of water and verandahs in order to protect the rooms from heat and enable the men to sit outside and when nights were hot, to sleep outside. Inspectors were to examine the barracks and report upon their condition at least once a year. Further amendments to the legislation occurred in 1921, 1937, 1946 and 1952, requiring employers to make several changes to the cane barracks, including the provision of stoves of various sizes according to the number of employees, the screening of all kitchens and dining rooms against flies and the screening of all tanks and other vessels for the storage of water against access by birds. Bill Scott, a canecutter during the 1940s, wrote about his experiences of living in cane barracks. He observed that 'Some farmers were pretty liberal with a few extras thrown in .... Others were skinflints who provided exactly what the Act specified, of the cheapest and poorest quality'.[4]

Cane barracks were 'quaint little structures' wrote the historian Bianka Balazategui. Located near river banks or tucked away 'under the shadows of mountains', they came to be constructed of different material over time: wood, brick, concrete blocks, fibrolite and most commonly, galvanized iron. Sometimes they were raised above ground level on short stumps, like many domestic buildings in North Queensland; in other instances they were barely raised above ground level. Floors were made from wood or concrete. Each structure consisted of several rooms, usually four or five for sleeping, and at one end a room served as a kitchen. Each room held two men, the upper limit set by the legislation. A verandah ran the full length of the building. Dennis Formiatti in his study of the architecture of cane barracks in Far North Queensland found several different barrack types. One arrangement shown in Figure 16.1 (A) was what he termed a 'linear single barracks' – designed to accommodate one gang of four to twelve men. Other cane barracks had more of a square shape with verandahs on both sides (i.e. linear, double banked rooms) or were L-shaped as in Figure 16.1 (B). Larger cane barracks could accommodate two gangs of men. Over time the cane barracks became the iconic feature of the cultural landscape associated with Australian sugar production, the symbol of the determination by British-Australians that their sugar industry should be dominated by 'whites' only.[5]

The daily and weekly routine of the canecutters during the crushing season has been outlined in some detail by the historians Diane Menghetti and Bianka Balazategui, so it will not be repeated here in any depth. After an early breakfast during weekdays, sometimes before dawn, the gang cycled or walked to the cane paddock being harvested. Canecutters could start the day by loading or cutting, but there had to be two loading shifts daily in each mill supply area as the sugar mills could not supply enough rail trucks to allow all farms to load simultaneously. Loading was the most arduous task in cane cutting. Some cutters preferred to load in the cool of the morning; others favoured the afternoon when the cane was drier and would not slip as it was being carried. Morning loading took two to three hours and once it had been completed, the gang would break for their morning tea-break. After this break the gang would resume its activities, which ended at around 11.00 am or 11.30 am, when the cutters stopped for lunch. After lunch, the men rested for about an hour before going back to the cane at half past one or two o'clock, after avoiding

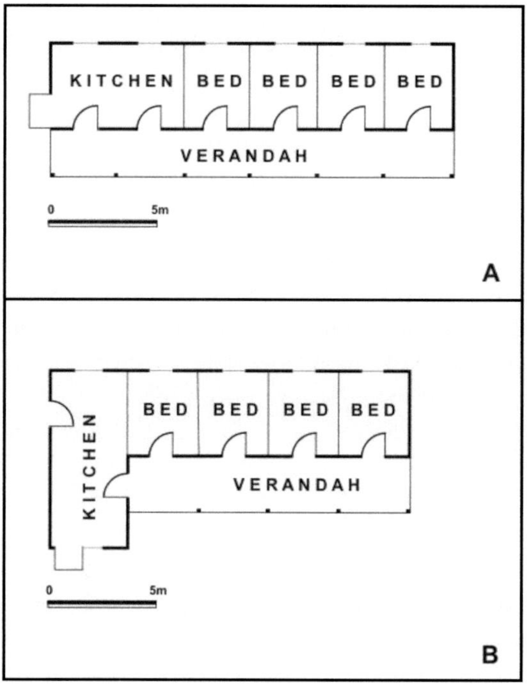

Figure 16.1. (A): Plan of linear cane barracks (single banked rooms); (B): Plan of L-shaped cane barracks (single banked rooms).

Source: D. Formiatti, 'Sugar cane barracks of Far North Queensland'. B. Architecture Thesis, Uni. of Queensland, 1984, pp. 45 & 48.

Plate 16.1. Canecutters in the Tully district laying portable rail track, c. 1930.

Source: Image no. 32434, John Oxley Library. Image reproduction courtesy of the State Library of Queensland, Brisbane.

the midday heat. More cane could be cut or the afternoon loading would be completed. In North Queensland sugar-producing districts, portable rail track was used extensively, and it was laid by the gang before loading; in the evening if loading was to occur the next morning (Plate 16.1). It was the second most arduous task in cane cutting. Burning the cane, if permission had been granted for it to be burnt (see discussion later in chapter), was mostly done on Sunday, Tuesday and Thursday evenings. Canecutters were supposed to work only between 6 am and 6 pm, and strict policing of work outside these hours was attempted by organisers from the Australian Workers' Union (AWU), but some gangs would start earlier or even work at night – it was difficult to police outlying farms effectively. After the early mornings and toil during the week, many canecutters rested over the weekend, cleaned the barracks, washed clothes and visited the nearby towns, especially their hotels.[6]

Industrial unrest and the British preference movement

During the late 1910s, the number of field hands, cutters and carters required to handle the Australian sugar cane crop stood at around 10 000 persons – this figure excluded the labour provided by the farmers and their families. Sufficient and suitable labour was now available to cut the crops of cane. The canecutters mostly came from other parts of Queensland, or other Australian states, and generally did not remain in the sugar-producing areas after the close of the crushing season.[7] Wages and working conditions for these canecutters (and mill workers) were determined by industrial awards. Since the 1911 sugar workers strike, tension between canecutters, mill workers, mill-owners and farmers had eased. Wages for canecutters had been increased under the Macnaughton Award in 1914, but this action had not led to industrial unrest. However, the 1916 crushing season was disrupted by further strike action, following the handing down of the infamous Dickson Award.

Prompted by the Commonwealth government's increase in the price of raw sugar after its takeover of the marketing of sugar in Australia in 1915 as a wartime measure to stop profiteering, and the perceived improved financial situation of the canegrowers and mill-owners, the AWU on 26 February 1916 filed a claim in the Queensland Industrial Court requesting an increase in the wages for sugar mill and sugar field labourers. Between 3 May and 4 July, evidence was heard at Cairns, Townsville, Mackay and Bundaberg before Acting Judge Dickson. The Queensland government (Ryan; Labor) asked for no more than a 5% increase in wages and requested that the forty-eight hour week be unchanged. This request was opposed by the mill-owners who asked for the Macnaughton Award of 1914 to be left unchanged. In late August, Acting Judge Dickson, after gathering hundreds of pages of evidence on the wages, accommodation conditions and profits made by some mill-owners, granted a 50% increase in wages to all sugar industry employees, set out a schedule of food to be provided to sugar industry workers, and ordered that lights, stretchers and mattresses should be supplied in the accommodation facilities provided for workers. The minimum wage for a forty-eight hour week rose from 39s and 37s in the northern and southern regions, respectively, to 58s and 55s, respectively.[8]

Upon hearing of the Dickson Award, Queensland's canegrowers claimed they could not afford to pay such wages and provide the extravagant scale of provisions. They refused to harvest their cane, and stood down the canecutters. The State's mill-owners supported their actions. The majority of sugar mills south of Townsville ceased crushing by way of protest, and several remained closed for two months, while three did not even commence crushing. As a result of the Dickson Award strike, the streets of Bundaberg 'thronged with unemployed men discharged from the plantations and farms', wrote the correspondent for the *Australian Sugar Journal.*[9] Premier Ryan sought to diffuse the situation quickly by asking the Commonwealth government (Hughes; Labor) to increase the price of a ton of raw sugar so that the mill-owners and canegrowers could afford to pay the increased wages and the increased prices for sugar cane required under the cane price-fixing legislation introduced by the Queensland government (see Chapter 17). This strategy failed, but many Queensland canegrowers realised during late September that any additional delay in the season's cane harvest would be deleterious. Cane harvesting re-commenced and the state's sugar mills resumed crushing, although the industry remained in a state of turmoil with Queensland canegrowers criticising Premier Ryan's inability to solve the crisis.[10]

After protracted negotiations, Prime Minister Hughes announced on 8 March 1917 that the price of a ton of raw sugar would be increased from £18 to £21 (see Chapter 18), subject to the condition that sugar mill and sugar field workers would not seek higher wages or changed working arrangements during the crushing season. Two months later, the ASPA asked the newly constituted Queensland Court of Industrial Arbitration to replace the Dickson Award with a new award. This matter was heard before Justice Thomas McCawley, the former Queensland Under-Secretary for Justice and Crown Solicitor. The ASPA, the UCGA and AWU had met and agreed that

Justice McCawley would use the same evidence that had been before Acting Judge Dickson as a basis for his judgment. On 1 June 1917 a new industrial award was gazetted, reducing the wages by 4d and 1s 2d per day for mill and field workers in northern and southern regions, respectively, but maintaining the other conditions. The AWU agreed to accept the award on the condition that it would be reviewed in twelve months. Edward Theodore, the Queensland Treasurer, advised the Commonwealth government that the labour question had now been resolved. Prime Minister Hughes again asked Ryan to accept his conditions, but the Queensland Premier refused the request. Some Queensland sugar producers urged Hughes to adopt a harsher attitude to Ryan – CSR was implicated in this pressure. On the other hand, some canegrowers and mill-owners in Queensland were not supportive of such a conflict, wishing to avoid another disrupted crushing season. Senator Thomas Crawford, still president of the ASPA, and various Queensland Chambers of Commerce began pressuring Hughes to reach an agreement. Prime Minister Hughes relented on 18 June 1917, agreeing to the increase in the price of a ton of raw sugar, subject to no conditions. Mill-owners and canegrowers could now cover their additional wage costs.[11]

The majority of workers affected by the Dickson Award strike were European, the industry supposedly being purged of its non-European canecutters. Yet the AWU noted in 1919 that the higher wages paid during the 1910s for cane cutting and field work on cane farms had drawn some non-European workers from other occupations into the sugar industry. The number of non-Europeans employed in the Australian sugar industry was again increasing. However, under an award made on 27 June 1919 by Justice Thomas McCawley of the Queensland Court of Industrial Arbitration, no non-European labour was allowed to be employed or continue in employment at cane cutting, while non-European labour was prohibited on any cane farm where more than 30. 4 ha were planted with cane. However, the award permitted non-European labour to be used in cane cutting on farms owned by non-Europeans, if the workers were of the same nationality as the farm owner and as long as the farms had less than 30. 4 ha under cane. Justice McCawley also observed that there had been an upsurge of southern European canecutters during the war period, but the canegrowers' associations should encourage British-Australian canegrowers to give preference to British canecutters (i.e. Australians of British stock were regarded as British citizens) now that normal conditions had returned. Such sentiments were again expressed a year later by Justice Allan Macnaughton in another judgment of the Queensland Court of Industrial Arbitration. He noted that a number of prosecutions for working excessive hours had been brought against non-European workers during the previous season in the Innisfail district.[12] The seeds of the British preference movement had been sown.

During the late 1910s and early 1920s, the AWU maintained its campaign against the employment of any non-European workers in the sugar industry. William Storrie, the AWU organiser at Ayr, for example, complained about the twenty Japanese men working at Inkerman Mill or as navvies on its tramlines. Considerable concern was expressed about the canegrowers at Cudgen in northern New South Wales employing a gang of Pacific Islanders who had legally avoided deportation. The historian Kay Saunders has suggested that some Queensland canegrowers were unwilling to cooperate with the AWU and the local Inspectors of Factories and Shops who had the responsibility of detecting illegal labour in the sugar cane fields, for they had made secret agreements with non-Europeans which enabled them to work at lower wages than those prescribed by the sugar industry awards.[13]

The presence of a few non-European workers, however, became a minor issue as the tension intensified between the unions, canegrowers and canecutters over the increasing number of 'foreign' or non-British workers harvesting cane. The workers at South Johnstone Mill advised the local canegrowers in July 1925 that 75% of the gangs during the coming crushing season had to be British if they were to handle the harvest. Within a few days the AWU branch at Cairns had declared that the 75% rule would also apply at Babinda Mill. Local canegrowers opposed such a demand, and after negotiations failed to resolve the matter it was referred to the Court of Industrial Arbitration in Brisbane. Justice William Webb, the presiding magistrate, ruled that first preference for the 1925 crushing season throughout Queensland should be given to men who had worked as canecutters in the same area the previous season, and second preference to men, irrespective of nationality, who held valid AWU membership tickets. Such a ruling did not modify the ratio of British to non-British canecutters, but new southern European immigrants would find it harder to gain employment as canecutters.[14]

Webb's decision did little to ease the concerns of the workforce at South Johnstone Mill. The men persisted in their pro-British stance, resolving to handle only British-cut cane until all Anglo-cutters in the district had been employed. Eventually they relented, agreeing to operate the sugar mill longer each day, processing enough surplus cane to accommodate the employment of eleven extra British gangs. Babinda Mill shut briefly during June 1925 over the British preference issue, but re-opened when the directors of the Babinda Cooperative Sugar Milling Association agreed to hire forty British canecutters and thirty British for work on the tramlines during the crushing season and promised that the canecutters for the next crushing season would be 75% British. Labour strife associated with hostility towards foreigners flared again at Mackay in August 1926 when mill workers refused to process Italian-grown cane. Negotiations resulted in the lifting of this ban on non-British harvested cane. A year later further industrial unrest occurred at South Johnstone Mill over the method of selecting labour. A strike of mill workers and canecutters lasted four months, and one British striker was killed by an unknown assailant thought to be Sicilian. The canegrowers tried keeping the sugar mill going and attempted to break the strike by recruiting Sicilian canecutters from the Tully district, but they were turned back by the strikers. The South Johnstone strike, one of the bitterest in the Australian sugar industry's history, was not entirely about British opposing foreigners, but rather British and non-British workers trying to protect their jobs against an attempt by the mill manager to reduce the factory's workforce, weed out the bad element in the workforce and employ only those who would carry out a fair day's work.[15]

By the late 1920s, opponents of non-British labour in the Australian sugar industry had grown impatient with Justice Webb's compromise to stabilise the proportion of British and non-British workers harvesting each season's crop. The pro-British lobby argued that the time had arrived to implement stronger measures aimed at ridding the Australian sugar industry of non-British workers. They pointed to the growing wave of complaints from southern Australians about the embargo on sugar imports causing higher domestic sugar prices and the unlikely support for its extension after 1930 (see Chapter 18 for full discussion), especially if the country's sugar industry was falling into the hands of foreigners. This opposition culminated in the formation of a British Preference Committee at Ayr in April 1930. This organisation sought to create a field labour force that was 90% British, and promoted British farm ownership and preferential patronage of British-owned businesses.

The four Lower Burdekin mill suppliers' committees failed to endorse such a radical scheme, but it quickly gained admirers elsewhere throughout the North Queensland sugar-producing districts, becoming known as the British Preference League. Ten branches of the organisation had been formed in North Queensland by February 1931. As the 1930 sign-on of canecutters commenced, representatives of the British Preference League urged farmers to hire more British canecutters. Their exhortations had some success – 76% of the canecutters in the Babinda Mill supply area were British, at South Johnstone it stood at 50%, and on the Lower Burdekin it reached 90%. As the sign-on proceeded, a mid-June conference of the ASPA and QCGC was held in Brisbane. This meeting agreed to an accord, referred to as the 'Gentlemen's

Agreement'. It specified that in sugar mill supply areas north of Townsville the numbers of British canecutters should remain at 75%, except in the Italian-dominated supply areas for Macknade, Victoria and Mourilyan Mills, where the meeting decided that a 'reasonable percentage' would be employed. South of Townsville, the British figure was to be increased to 90%. Overall, the industry should aspire to a labour force that was about 85% British. This arrangement was subsequently endorsed by the AWU.[16]

The implementation of the Gentlemen's Agreement during the early 1930s occurred without any major industrial unrest, although not without complaint. CSR's Hambledon Mill Cane Inspector in 1931 criticised the AWU's tactics, noting that quite a number of the best performing foreign gangs in the Hambledon Mill supply area had to be broken up deliberately to conform to the Gentlemen's Agreement. CSR, however, never took a public pro-alien stance, and even though its Herbert River sugar mills were not included in the 75% British-25% non-British workforce agreement, the firm cooperated with the local canegrowers' organisations to increase the number of British gangs in the Herbert River district and reduced the number of Italians employed in Victoria and Macknade Mills. Italian canegrowers and canecutters complained about the discriminatory objectives of the Gentlemen's Agreement and the racist demands of the British Preference League, and considered political alternatives, including support for the Communist Party of Queensland and the formation of the Association of Italians in the Sugar Producing Industry. These actions failed to overturn the Gentlemen's Agreement, but they did temporarily weaken the AWU's support base in the northern sugar-producing districts. Five years after the introduction of the Gentlemen's Agreement, the number of British canecutters in the sugar industry had increased by 47%, while the number of alien canecutters had increased by only 4%. Overall, a total of 704 new canecutter jobs had been added over the period to cope with the expanded cane crops, of which 632 went to British workers. This trend, however, had not really eroded the number of alien canecutters who remained an integral part of the labour force that was needed each year for the annual cane harvest.[17]

Improving and mechanising harvest procedures

The contentious issue of employing British or non-British canecutters had emerged because of the xenophobic attitudes prevailing in Australian society during the 1920s and 1930s, combined with the reliance of the Australian sugar industry on so many workers to harvest the crops of cane – this activity had still not been mechanised in 1930, unlike other field tasks such as ploughing, planting the cane or weeding. However, the transport of harvested cane to its final destination was made easier during the 1920s and 1930s by the introduction of motor lorries throughout the Australian sugar-producing districts. Canecutters did not have to lay portable rail track to get the loaded trucks to the railhead – they could load harvested cane directly onto these lorries and move it faster than by horse-drawn wagons or carts to either the sugar mill or loading spots. CSR's Condong Mill Cane Inspector estimated that one lorry could do the work of six horse teams. This change, together with the adoption of tractors for cultivation tasks, eliminated the need to maintain large numbers of horses.

Yet motor lorries contributed indirectly to bigger crops, thereby exacerbating the harvesting problem. Some canegrowers grew only limited quantities of sugar cane when its removal to the sugar mill was overwhelmingly by horse-drawn drays or wagons. However, they were now able to grow considerably more sugar cane and deliver it in half the time, a consequence of using motor lorries. The introduction of motor lorries throughout the Mt Bauple district, for example, meant it was possible to grow sugar cane in localities not previously cultivated because of a lack of transportation to the sugar mill. In addition, time saved by using motor lorries could be devoted to drainage and pest and weed control – all activities that improved the size of the crop – or extending the area under cane.[18]

Motor lorries may have improved the delivery phase of harvesting, but the cutting and loading of the cane proved more difficult to mechanise. Further efforts during the 1910s by mining engineer and Bundaberg canegrower Charles Hurrey, in partnership with wealthy Riverina grazier Ralph Falkiner, had produced several different mechanical cane harvesters, but they had never reached commercial production because these machines were costlier than canecutters and failed to handle lodged cane. Following Hurrey's death in 1919, Falkiner participated in a syndicate that funded the construction of a mechanical cane harvester designed by Hurrey's son James. Built around a 25 hp Cletrac caterpillar tractor, the machine was designed to cut one to two hectares a day and be operated by one man. The cut stalks of cane were topped by a guillotine, but trials revealed inadequacies in the cutting mechanism. Even though these investigations failed to produce a commercial mechanical cane harvester, they had stimulated Falkiner's interest in the problem and he was determined to succeed in the matter and share the potential rewards.[19]

After the failure of the Hurrey machines in 1921, Falkiner engaged Melbourne engineer Walter Charley to re-examine the problem from afresh and over the next decade he poured a huge amount of money and energy into the project, concentrating on overcoming the difficult problem of topping and trashing. By 1925, this effort had led to the development of the Falkiner whole stalk mechanical cane harvester (see Plate 16.2). Confident that the machine would gain industry acceptance, Falkiner set up the Queensland Cane Harvester Corporation Ltd. in December 1925 to manufacture the cane harvester and to contract harvest crops of cane. Several Queensland sugar milling companies, CSR and the ASPA invested in the company, but due to cash flow problems only four machines were ever built, and one was destroyed by fire while being railed to a trial site (perhaps sabotaged by angry canecutters who feared the loss of their livelihoods). Moreover, trials of the Falkiner cane harvester throughout both Queensland and New South Wales sugar-producing districts during late 1925, 1926 and 1927 proved disappointing – a canecutter at the demonstration of the machine in the Mulgrave Mill supply area in 1926 managed to manually cut the cane faster than the cane harvester, and later the same year the machine choked in a field of unburnt cane at Mackay. Australian interest in the machine waned, with the Falkiner cane harvester never cutting crops of cane commercially in Queensland. Yet the Falkiner cane harvester demonstrated that cane could be mechanically harvested, and Falkiner's and

Plate 16.2. The Falkiner mechanical cane harvester during a demonstration run, Pimpama Island district, 1925.

Source: Image reproduction courtesy of the Gold Coast City Council Local Studies Library, Image no. LS-LSP-CD101-IMG0102.

Charley's ideas were not forgotten, but resurrected by inventive canegrowers who designed mechanical cane harvesters during the late 1940s and 1950s.[20]

Running parallel to the Falkiner developments were other Australian investigations into the production of a mechanical cane harvester. William Miller and Francis Owen had patented a whole stalk mechanical cane harvester in 1926. Funds for modifications following field trials ran out in 1932, before the machine gained industry acceptance. Arthur Clifford Howard, the inventor of the rotary hoe which had gained popularity in Australian sugar-producing districts during the late 1920s, constructed his first whole stalk mechanical cane harvester in 1926. Three years later when Howard Cane Harvesters Ltd. was floated, the Fairymead Sugar Co. Ltd. invested capital in the venture. Further smaller scale models of Howard's whole stalk cane harvester were tested at Fairymead plantation during 1930 and 1931, but the emphasis suddenly shifted in 1932, when Howard and Martin 'Jim' Vichie decided to concentrate upon designing a chopper cane harvester. Early field trials were encouraging and work on the project continued until 1938, when this idea too was abandoned. Charles Young wrote that the machine was 'excessively heavy, had poor manoeuvrability and the many moving parts caused much mechanical trouble'.[21]

By 1939, Charles Young had also recruited Walter Charley, Falkiner's former chief engineer, who tried to devise another chopper cane harvester, but the project was abandoned because of the rapid deterioration of chopped cane, a problem encountered twenty-five years later. Being satisfied that chopper cane harvesters had too many disadvantages, the focus returned to whole stalk cane harvesters. In 1940, a single-row whole stalk machine built onto a Farmall tractor emerged from the workshops at Fairymead plantation. The capacity of this mechanical cane harvester ranged from 200 to 250 tonnes per day, and it dropped 160 kg bundles of harvested cane along every row – these were later picked up by imported jib-type grab loaders which were subsequently re-designed and mounted on a turntable to revolve through 360°. By the mid-1940s, the majority of the cane at Fairymead plantation, up to 800 tonnes daily, was being mechanically loaded and mechanically harvested.[22]

The vexed question of cane burning

Cane burning was a contentious issue before World War II. A trial on a farm in the Mulgrave Mill supply area during 1913 measured the weight loss and ccs deterioration of burnt and green cane left standing in the field for up to six days. These investigations concluded that burning the cane had a detrimental impact upon its quality. Hence, many mill-owners, especially CSR, were very opposed to cane burning. Other mill-owners, however, were more relaxed about the practice. Alex Draper, Chairman of Directors for the Mulgrave Central Mill Co. Ltd., suggested in 1921 that clean burnt cane was a better product than dirty green cut cane.[23] Yet the canecutters preferred to cut cane that had been burnt, for they avoided having to cope with the trash. Crops of cane full of weeds, lodged and tangled or affected by frost, which induced much side-shooting, were difficult to harvest and often burnt to make the canecutters' task a little easier. However, rules governed when canecutters could burn paddocks of cane. If a cane cutting gang wanted to burn off, it had to seek permission from the mill's cane inspector. If permission was granted, the canecutters were paid at the normal rate, but the canegrower would be penalised, usually 1s per ton. If permission was not granted, the canecutters could still decide, with the approval of the farmer, to burn the cane, in which event the farmer and gang were each penalised. Theoretically, the farmer supervised the amount of cane to be burnt, but not all farmers were conscientious in their attention to this practice. The Mulgrave Mill Cane Inspector, for instance, complained in 1926 that too many local farmers had been 'very lax in supervision when permission to burn has been granted. Some gangs are allowed to burn any quantity, even as much as a week's supply'.[24]

Large amounts of cane were being burnt in some mill supply areas during the late 1910s and 1920s, despite the penalties and mill-owners' opposition to the practice. A report on cane burning which was discussed at the ASPA conference in early 1913 mentioned that the practice, either by accident or design, had become 'common'. CSR reported in 1915 that cane burning was becoming 'an established practice' and in some mill supply areas it had reached 'unwarranted proportions'. Clive Morton in his history of Mulgrave Mill suggested that unpermitted burning of cane was so rife in the Mulgrave Mill supply area by 1917 that 'anarchy triumphed in the cane fields'. Indeed, figures on the amount of burnt cane received at Mulgrave Mill during the late 1920s show that in some years a third to half the crop had been burnt. Similar levels or even higher amounts of burnt cane were being received in the mill supply areas under the control of the Queensland Bureau of Central Sugar Mills. In 1927, A.P. Gibson, the BSES's Northern Field Assistant, reported that 'practically all the cane is burnt prior to harvesting' in the Mossman Mill supply area. Roy Harman, CSR's Chief Chemist (and later General Manager), addressed delegates attending the meeting of the International Society of Sugar Cane Technologists in Brisbane during August 1935. In his speech, Harman mentioned that about half the entire Australian cane crop was now burnt.[25]

Less cane burning occurred in CSR's mill supply areas, probably because of the tighter control the firm exercised over its suppliers and gangs of canecutters. Its Childers Mill supply area, however, was the exception, as the occurrence of drought and frost regularly affected the crops of cane, so they were intentionally burnt to assist with their harvest. CSR thought that too many farmers were content to submit to the payment deduction imposed for burnt cane, and that too many canecutters 'simply demand the firing of a field before they enter it. To them, the loss to grower and miller is of secondary consideration'. By 1919, CSR's officials reported that keeping control over burning had become very difficult due to other mill-owners allowing its widespread use, and that some harvested cane in its mill supply areas had been 'maliciously burnt'. Moreover, CSR's cane inspectors found that burning, either approved or malicious, always escalated during the latter weeks of the crushing season as the canecutters sought to finish harvesting the crops, especially in prolonged seasons such as 1924 and 1928.[26]

The amount of burning in CSR's Herbert River mill supply areas altered in the mid-1930s. The catalyst for this change originated during 1933 when canecutters fell ill throughout the crushing season with serious fevers, intense headaches and extreme mental depression; in the more severe cases jaundice emerged, accompanied by black vomit containing blood, bile and blood in the urine, bleeding from the gums and acute mania. Forty cases had been treated by the end of the year, and four patients had died from their illness. Dr Gordon Morrisey, the local medical practitioner, diagnosed Weil's disease. Laboratory and epidemiological investigations during early 1934 demonstrated the relationship between the virus *Leptospira icterohaemorrhagiae* – the causal agent for Weil's disease – and rats in the cane fields. The virus, spread by the rats urinating on the wet ground and on the stalks of cane, invaded the canecutters through abrasions suffered while handling the sharp cane trash. This trash could be eliminated by burning the crops of sugar cane, which would also partially sterilise the ground and kill some rats. Such a practice was opposed by the mill-owners. Employers of the canecutters argued that they had acquired the disease while walking through or standing in contaminated water as they worked barefoot. Canegrowers and mill-owners proposed that rather than burning the cane, the workers needed to be educated on the benefits of wearing boots.[27] Rat control which had been solely an agricultural problem now also loomed as a public health and industrial issue.

More canecutters died in the Herbert River district during July and early August 1934, leading to a communist-led unofficial strike of local canecutters, mostly Italian-Australians, on 20 August 1934. They demanded all cane be burnt before harvesting and that the wage for cutting burnt cane be the same as for cutting green cane. Employers contested these claims, but the Queensland Industrial Court ruled in favour of a general order to burn all cane in the Herbert River district, although this order stipulated that canecutters would receive 1s per ton less than the rate for harvesting green cane. Attempts by CSR, the QCGC and the ASPA to overturn the ruling a month later failed. The focus of the dispute then shifted to the Innisfail district, where two canecutters died from Weil's disease during October 1934, prompting their colleagues to strike until the Queensland Industrial Court issued an order to burn all cane. Again the QCGC and CSR opposed the AWU's application for an order to burn, claiming that the financial loss from burning was 'tremendous' and that the rats could have come from the town. Ronald Muir, the general secretary of the QCGC, accused the union of exploiting the situation to achieve the same rate of pay for harvesting either burnt or green cane, an arrangement long desired by the canecutters. The Queensland Industrial Court was not swayed by the employers' arguments and granted the union's application to burn all cane in the Innisfail district.[28]

As mentioned in Chapter 14, a major campaign during late 1934 and early 1935 sought to destroy the rats throughout the North Queensland sugar-producing districts. In May 1935, Dr Raphael Cilento, Queensland's new Director General of Health and Medical Services, proposed that before the crushing season commenced, cane inspectors, industrial inspectors or health inspectors should be given the authority to decide which cane paddocks should be burnt. The scheme was supported by AWU officials. In June 1935, the Queensland Industrial Court ratified Cilento's scheme by ruling that cane could only be burnt if the health inspector felt it was warranted, where a cane inspector directed it in writing, or where a canegrower agreed in writing, and that an order to burn would depend upon the extent of rat-damage. Towards the end of June 1935, the AWU in conjunction with the ASPA and QCGC, obtained a burning order for all cane in the Herbert River district, following the hospitalisation of twenty canecutters with Weil's disease.

The lack of a general burning order for all North Queensland's mill supply areas angered the canecutters in the Mourilyan Mill supply area. They went on strike on 4 August 1935. In defiance of union officials, and spurred on by local communists, canecutters in the areas served by the Tully, Goondi and South Johnstone Mills agreed to support their Mourilyan colleagues. By the end of August 1935, 2 000 canecutters were refusing to work. 'With four mills out of production and the farmers immobilized, some 4 000 people were directly affected by the stoppage', writes the historian Diane Menghetti. Moreover, strike action was being threatened in the Hambledon Mill supply area. Negotiations led by the AWU resolved the issue and the canecutters' strike in all mill supply areas ended by 8 October 1935. Yet two months later, as the 1935 crushing season drew to a close, the strike's goals had clearly failed. The Queensland Industrial Court had made only one ruling that permitted all the cane in the Herbert River district to be burnt. However, CSR was not impressed by the action of the Queensland Industrial Court, its Herbert River technical field officers complaining that wholesale burning had been very difficult to control, leading to cane being 'not merely scorched, but blistered and even charred'.[29]

Cases of Weil's disease did not vanish immediately during the late 1930s as the rat numbers were brought under control. Eight deaths were attributed to the disease between 1936–1937 and 1942–1943, and 88 cases of the disease were reported between 1936–1937 and 1938–1939, but those affected were assessed as having much milder infections than previously. Canecutter unrest and general community alarm in the North Queensland sugar-producing districts subsided, but as a precaution the AWU in July 1936 successfully applied for an order to permit the burning of all cane in the Herbert River district. Further general burning orders for all cane on the farms in the Johnstone and Hinchinbrook Shires where the epidemic was most severe in 1933–1935 were not sought by the AWU during the period 1937 to 1939. Inspectors appointed under the direction of the Queensland Department of Health and armed with the power to order the burning of rat-infested paddocks continued to be stationed in the northern sugar-producing areas. Burning paddocks of cane that posed a health risk to canecutters continued following their orders. Yet by the outbreak of World War II, levels of burning had fallen away again in at least CSR's northern mill supply areas.[30]

World War II, its aftermath and the harvesting crisis in Australian cane fields, 1939–1959

From the outset of hostilities in World War II, there was a fear in the Australian sugar industry about possible labour shortages during the cane harvesting season. 'In the best of times', writes William Douglass, 'cane cutting was arduous, seasonal and hence highly compensated'. Many men were not physically suited to such work, and under war conditions the ranks of existing workers were depleted by enlistments of the most able-bodied. Other industries offered good wages, the opportunity for overtime, and permanent employment closer to the capital cities, instead of isolated regions in Queensland, which soon came under threat from Japanese invasion as the war progressed. Moreover, Italian canecutters, both naturalised and un-naturalised, unlike during World War I, became suspected enemy aliens and subjected to restrictions, even internment. Despite growing anti-Italian sentiment, the 1940 and 1941 crushing seasons were completed without incident, but a review of the 1941 crushing season in January 1942 by the president of the Queensland chapter of the AWU painted a bleak picture. Attempts to avoid using Italian canecutters on the farms supplying Hambledon and Babinda Mills resulted in their replacement with over-sized British gangs of fourteen to twenty-five men, although not all these men completed the season. One hundred inexperienced British canecutters had to be sacked because of poor performance, their places being taken by Torres Strait Islanders, who had previously been excluded from the industry. The AWU suggested that the anti-Italian stand had created unnecessary racial tensions during a time of national crisis.[31]

As the commencement of the 1942 crushing season approached, mill officials and sugar industry leaders expressed grave concern about the labour situation. Further enlistments, additional internments, and the fear of Japanese invasion had reduced the population in North Queensland even further. Representatives of the QCGC conferred with the Minister for the Army, indicating that even when all avenues had been explored, they did not think it would be possible to obtain sufficient canecutters unless there were some organized military releases. In early June 1942, almost 800 canecutters were released from the militia to assist with the forthcoming harvest. Yet after the sign-on, the situation was quite clear: not enough cutters were available to harvest the crops of cane, even with a lengthier season. North Queensland mill supply areas faced an estimated shortage of at least 1 300 canecutters; this figure was revised upwards to 1 500 in mid-July, following the non-arrival of some militiamen. In the Herbert River district alone there was a shortage of 150 canecutters. Fewer than half the canecutters were experienced, and most were under eighteen or over forty years of age. After the commencement of crushing in the Bundaberg district, only approximately half the estimated number of required canecutters was available.[32] To alleviate these shortages, the canegrowers again used non-European labour. Eighty-three Aborigines from Cherbourg were employed in the Bundaberg district, mostly by the Fairymead Sugar Co. Ltd. One hundred Aborigines from Palm Island assisted with the harvest in the Macknade Mill supply area. However, even with the assistance of Aboriginal labour and militiamen, many farmers and their families were forced to cut the cane to keep the sugar mills supplied, thereby neglecting planting and cultivation tasks. At the conclusion of the 1942 season, the Queensland sugar mills had managed only 82% of their best capacity.[33]

As a result of the severe disruption to the crushing 1942 season, the Queensland government appointed The Hon. Mr. Justice Alan Mansfield as Chairman of a Royal Commission into how to manage the next crushing season. After examining 137 witnesses from each Queensland sugar-producing district, Mansfield delivered his report to Frank Cooper, Premier of Queensland, on 8 February 1943. The Commission recommended that the sugar industry be allocated sufficient harvesting labour to enable all sugar mills to crush at a weekly rate equal to 75% of their maximum average seasonal weekly crushing rate. Representatives of canegrowers, millers and the AWU should form a central committee in Brisbane to advise manpower authorities on how to achieve an equitable distribution of labour to all mill supply areas. This committee should be assisted by locally formed harvesting committees for each mill supply area, established for the purpose of controlling the size of harvesting gangs and their movement within the district. The Queensland government accepted these recommendations, and immediately approached the Commonwealth government with a request for sufficient labour to conduct the 1943 harvest.[34]

As part of its investigations, the Royal Commissioners inspected the mechanical loading and cutting machines operating in the Bundaberg district. At Fairymead plantation, bulk harvesting of heavy, fallen cane was dealt with by the Hawaiian method or rake harvesting method. The cane was first burnt and then a special machine costing over £500 was pulled backwards and forwards across the fields by a rope-winding tractor or engine powerful enough to shift two or three tonnes. The cane was stacked along the tramline, ready for loading onto rail trucks by jib-type grab loaders mounted on a turntable (see Plate 16.3). This type of loader required three men – one operating the mobile loader, one on the ground placing the grab, and one to guide the bundles of cane onto the rail trucks. Tops were not severed from the stalks in the field, but the cane was freed of tops, dirt and trash in a cane washing unit or cane cleaner at the sugar mill. More upright crops of cane were mechanically harvested by a machine mounted on a tractor. Such cutting machines were operated by two men. The cut cane, after being gathered into convenient heaps, was subsequently loaded onto rail trucks by a portable

Plate 16.3. Jib-type grab loaders mounted on turntables being used to fill rail trucks, Fairymead plantation, 1948.

Source: CSRR, 171/437. Image reproduction courtesy of CSR Ltd.

crane mounted on a tractor base. Again, the cane was cleaned in the mill's cane washing unit.

Mechanical loading using the machine developed by Joseph Toft Jnr., son of a Bundaberg canegrower Joseph Toft Snr., was also viewed in operation at Bundaberg. Toft's mechanical cane loader, constructed largely from motor-car parts, was mounted on a lorry or tractor base and consisted of a wooden swing boom and grab driven by a Model T Ford engine (see Plate 16.4). The machine was designed particularly for burnt crops so that the grab did not pick up a great deal of trash. Very little effort was required to load seven tonnes within thirty minutes. Normally, this job requiring six men working very hard could be completed by three men.[35]

The Mansfield Commission thought the system used at Fairymead was unsuitable for wider application, as the canegrower was also the mill-owner and willing to incur considerable capital expenditure in both sugar mill and field to make the system work. The mechanical cane harvester impressed the commissioners, but they saved their highest praise for the Toft mechanical cane loaders. The relief afforded by mechanising the loading had a twofold

Plate 16.4. The earliest type of Toft loader showing the loading of harvested cane onto a motor lorry, 1940.

Source: *ASJ*, 12 September 1940, p. 327.

effect of allowing the canecutters more actual cutting time and also permitting the employment of men less physically fit for harvesting as the heavy lifting with loading was eliminated. Such an arrangement could be applied in any sugar-producing district. The Commissioners suggested that the development of these machines should be encouraged. In the meantime, the Commission recommended that the Sugar Industry Award be amended to allow for contract cutting alone where mechanical cane loaders were used.[36]

For the remainder of the war years, Queensland's sugar mills were troubled by shortages of canecutters, even operating under the Mansfield Royal Commission's recommendations that they crush a weekly rate equal to 75% of their maximum average seasonal weekly crushing rate. Farmer-cut cane continued being delivered to the sugar mills, much to the chagrin of the ASPA, the organisation warning that such arrangements jeopardised the cultivation of young cane crops for the following year. In the Herbert River district, some assistance was provided by teams of Aborigines from Palm Island, while the lack of canecutters in the Fairymead Mill supply area was so acute that if the three mechanical cane harvesters had not been present, the mill would not have had enough cane to even manage two shifts daily. The crushing rate set by the Directorate of Manpower was reached in few Australian sugar mills.[37]

During 1942 and 1943, some Australian canegrowers also encountered considerable difficulties transporting their harvested cane to the sugar mills. Military traffic was given priority in the allocation of locomotives and rolling stock. In 1942, only a quarter of the railway wagons normally available throughout the crushing season were available to the sugar industry. Congestion on the single rail track between Brisbane and Cairns made cane traffic a nuisance – it was often sidetracked so as not to delay the long-distance military trains. Consequently, cane trains often ran hours late, thereby affecting the progress of the crushing at the sugar mills which had to cope with erratic cane deliveries. As a result, more harvested cane was moved by motor lorries than previously, the war starting the long process of diverting cane away from the government-owned railways onto the roads. Such arrangements, however, added to the cost of harvesting. Normal transport costs to a railway siding in the Mt Bauple Mill supply area, for example, was around 4s per ton, but transport costs rose to 7s per ton during 1943 when the canegrowers had to use motor lorries to shift their cane to the sugar mill. Petrol shortages and rationing also affected the use of motor lorries, delaying harvested cane deliveries to the sugar mills. In addition, harvested cane was often moved to the nearest sugar mill, even if the farm was not assigned to that factory. Such practices also commenced the long process of re-assessing the distribution of assignments to each Queensland sugar mill so that more logical transport arrangements emerged in the sugar-producing districts.[38]

New cane cutting styles developed during the wartime period. The traditional practice involved cutting one or two sticks at a time and then slashing the top off while the sticks were held. 'Stooling' or 'bangtailing' replaced this traditional method. Canecutters endeavoured to slice whole stools in an armful, pivoting the bundle on a hip and dropping it to the ground; it was topped later. Previously, the leafy tops were removed before the stalks were laid to the ground. Cane knife blades were specially bent and fitted with longer handles, reducing the need to bend over so far during the cutting process. Some canecutters' output per day increased markedly using the new procedures, but it meant an alarming fall in the quality of harvested cane being delivered to the sugar mills. In 1942, Jim Pollard, the Chief Chemist at Farleigh Mill, reported trash, roots, tops and soil at more than twice the level of the previous five year average. Moreover, the new styles could only be performed satisfactorily in burnt cane. Thus, most cane delivered to the sugar mills after 1941 had been burnt, even in New South Wales where the least amount of burning had been used before the war. [39]

‘Conditions in the industry are still far from normal, which is the natural aftermath of the war, and this is clearly reflected in the lack of canecutters’, remarked Jim McFadden, Chairman of Directors of the South Johnstone Cooperative Sugar Milling Association, when he addressed the annual shareholders’ meeting in May 1947. McFadden and other Australian canegrowers had just survived the 1946 crushing season, but the cessation of hostilities failed to ease the gap between the supply of labour and the numbers required to harvest the cane crops, a consequence of the general full employment situation in Australia and the attractiveness of less arduous occupations. A shortage of canecutters delayed cane deliveries so much during 1946 that the North Queensland sugar mills were again forced to work only two shifts. The labour supply failed to improve during the 1947 crushing season – before the war about 9 000 canecutters were required to maintain all the Australian sugar mills with a full supply of sugar cane, but during the 1947 season the maximum number of professional canecutters available in any one week stood at 5 500. Hence, the sugar mills crushed cane at only 70–80% of their normal capacity, and only by virtue of the fact that about one-sixth of the crop was harvested by the farmers.[40] The situation in North Queensland during the 1946 and 1947 crushing seasons would have been even worse, except for the employment of non-European labour. Canegrowers in the Mulgrave Mill supply area relied on gangs of Torres Strait Islanders; Aborigines from Yarrabah assisted with the harvest in the Hambledon Mill supply area; fifty Aborigines, mostly from Palm Island or Cherbourg, cut cane in the Herbert River district.[41]

The shortage of canecutters during the 1945–1947 crushing seasons caused alarm amongst sugar industry leaders, especially as the industry was being asked to absorb ex-servicemen (see Chapter 10). Moreover, the industry could not commence even a modest expansion if it continued having difficulties at harvest time. In anticipation of another labour shortage during the 1948 crushing season, the Australian sugar industry during late 1947 embarked upon an initiative with the Commonwealth authorities to try and ease this shortage. Australia had agreed to accept 12 000 displaced persons annually from the European refugee camps. In Australia, these displaced persons would be known variously, and often disparagingly, as Balts (i.e. Lithuanians, Estonians or Latvians), DPs (displaced persons) or Reffos (refugees). They were admitted under a ‘Certificate of Exemption’ which compelled them, under threat of deportation, to remain in approved occupations for two years. Australia’s canegrowers requested 2 200 Balts for the 1948 harvest season, but Arthur Caldwell, the Minister for Immigration, announced that the sugar industry would be supplied with 1 000 Balts.

This plan to secure displaced persons was the first formal mass-recruitment program for the Australian cane fields since the Spanish immigration scheme of 1907, yet it was fraught with difficulties. The potential canecutters would have to be carefully selected, as only healthy, young single men were wanted for this occupation. The status of these immigrants during the slack season posed challenges. How were they to secure slack season employment, and would they return for the 1949 harvest if they did find work beyond the sugar-producing districts during the non-crushing period? Most would have no knowledge of English and lack personal contacts in their adopted land, unlike previous Italian immigrants who became canecutters. Presumably none would have any idea about cane cutting.[42]

As the commencement of the 1948 crushing season approached, migrant cane cutting gangs arriving by train at railway stations along 2 000 km of the sugar belt were 'welcomed by committees of civic dignitaries, canegrowers, AWU officials and Commonwealth Employment Service (CSE) officials', writes Geoff Burrows and Clive Morton in their book *The Canecutters*. By November 1948, about 900 displaced persons, mostly Balts and Poles, had eased the expected labour shortage of 1948 – professional canecutters reached a peak of 6 278 during the 1948 season compared to 5 500 in 1947.[43] Gangs of entirely displaced persons and mixed gangs of displaced persons and Australians were then common over the next few years, accounting for about one-sixth to one-fifth of the harvest workforce. They created a very favourable impression upon the canegrowers, especially in North Queensland, with the historian Bianka Balanzategui observing that the Australian sugar industry lauded them 'as saviours of the industry, coming to the rescue at a critical time of acute labour shortage'. Yet their numbers had peaked by 1950, as the European refugee camps rapidly emptied and the Australian sugar industry competed for these workers with the factories of southern Australia. Probably no more than 100 first time displaced persons were sent to Queensland for the 1951 harvest. The Balts, noted Burrows and Morton, gave the sugar industry 'a respite from labour problems, but it was to be brief'.[44]

At the QCGC's annual meeting at Mackay in March 1955, delegates heard that the organisation had taken steps to ease the possible labour shortage during the 1955 season, following the return of labour shortages during 1954. Principal amongst these measures was the recruitment of specially selected Italian migrant labour.[45] 'Cheering passengers were greeted by Italian songs from the Cairns passenger terminal when the S.S. *Flaminia* berthed on 27 May 1955', noted Burrows and Morton in their history of canecutters in Australia. Jubilation quickly turned into wretched misery for some of the Italian migrant canecutters, as unseasonal rain hampered the harvest. By August 1955, one-fifth of *Flaminia's* contingent had vanished south, despite migration officials urging the newcomers to remain in North Queensland. A further 700 single men arrived aboard S.S. *Toscanella* on 13 July 1955, but they had been allotted to the cane fields of Southern Queensland and New South Wales. Few of these immigrants absconded, as they encountered fine and cooler weather during the harvest season, making their task easier.[46] Yet the special selection program which recruited Italians for the Australian cane fields, despite its initial problems, was successful enough to be repeated in 1956, 1957 and 1959, although the number of Italians recruited declined each year. This reduced demand for Italian canecutters emerged after Spanish-Australian canegrowers had recruited several hundred Spaniards following the signing of an immigration agreement with Spain. Immigrant Yugoslavs and Finns were also making their way to the cane fields in search of work.[47]

By 1960, Italian immigration to North Queensland was being stymied by Italian government officials who tried to retain the best workers in Italy where a major economic revival was underway. Australian sugar industry officials were increasingly displeased with the quality of arriving Italian workers. However, the demand for immigrant canecutters was also waning. Extensive mechanical loading of harvested cane had now been implemented and mechanical harvesting was commencing (see the discussion in the next section). Manual canecutters thrown out of work by canegrowers who were early adopters of mechanised loading and harvesting meant there were sufficient canecutters to harvest the annual crops during the early 1960s. CSR ceased recruiting Aborigines from Palm Island in 1960 as enough workers were now available for the annual harvest in the Herbert River district. In 1963, when the Herbert River Canegrowers' Executive polled its membership about sending a recruiter to Italy or Spain, the proposal was rejected overwhelmingly. The days of the manual canecutter, especially immigrant cutters, was numbered.[48]

Mechanical loading and harvesting, 1945 to 1974

As the Australian sugar industry sought to secure sufficient labour for the yearly harvests during the 1940s and 1950s, investigations commenced into mechanising the harvesting procedures, thereby reducing the industry's reliance on so many workers. In December 1943, the QCGC formed a Mechanical Harvesting sub-committee which was allocated a budget of £1 000 and charged with the task of evaluating the potential of current machines for wider use. The QCGC, however, was wary of investing canegrowers' funds in speculative mechanical cane harvester development after the disappointments of the Falkiner machine, but the Mechanical Harvesting sub-committee in 1944 provided £200 to assist P. F. Tierney, an Innisfail canegrower, complete his whole stalk cane harvester. A demonstration of Tierney's cane harvester on his farm in December 1944 was reported as being a 'splendid success', but the machine appears to have never progressed beyond the experimental stage. The Mechanical Harvesting sub-committee also decided that the mechanical cane harvester built by Joe Toft Snr. and his sons showed the most promise and authorised the building of one of the machines and a Toft mechanical loader to work in conjunction with the mechanical cane harvester.[49]

During 1945, local demonstrations of the newly constructed Toft mechanical cane harvester and a Toft mechanical cane loader were held at Bundaberg and Mackay and throughout the Far North Queensland sugar-producing districts. The Herbert River Canegrowers' Executive purchased two Toft mechanical cane loaders for local use; the Hambledon Mill Suppliers' Committee purchased both pieces of machinery. However, the canegrowers who had used the machines expressed disappointment in their performance at the conclusion of the 1945 crushing season. The mechanical cane harvester was 'not a success' and the mechanical loaders, although useful, were 'slower than good man labour'. AWU district organiser George Lendich, according to Bill Kerr and Ken Blyth in their history of mechanical cane harvesting in Australia, told reporters that 'hand cutters need not worry too much about their jobs if that was the best the machines could do'.[50]

Aware of the industry's labour difficulties, the QCGC then levied its members one fifth of a penny per ton of cane for a special fund to encourage and assist the development of mechanisation. Ronald Muir, the QCGC's general secretary who had worked in his early years promoting the Queensland Cane Harvester Corporation Ltd., went overseas between November 1945 and January 1946 with Stan Toft of Bundaberg – one of the sons of Joe Toft Snr. – and E. (Roly) Behne, then of the BSES (later Managing Director of Pioneer Sugar Mills Ltd.), to study the most recent developments in mechanisation of the cultivation and harvesting of sugar cane. They returned recommending that development should concentrate on mechanical loading instead of harvesting, even though mechanical cane harvesters at Fairymead plantation were cutting over 400 tonnes in an eight hour day. To promote more mechanisation, the QCGC and the ASPA formed the Mechanical Harvesting Committee in 1947. Over the next seventeen years, members of this committee travelled throughout Australia's sugar-producing districts, inspecting various types of machines and providing limited financial support to inventors who had developed machines that showed commercial promise. Eventually, the Mechanical Harvesting Committee was disbanded in August 1964, once the mechanical harvesting of sugar cane in Queensland had become a foregone conclusion.[51]

Additional Toft mechanical cane loaders were purchased by a few canegrowers during the late 1940s. When worked effectively, Toft mechanical cane loaders did speed up the loading process. A report on their performance in the Richmond River district during 1951 showed that a gang of men working without a Toft loader averaged 5.5 tons per cutter daily and earned £3 3s 4d per cutter per day. Working with the mechanical cane loader, the gang averaged 7.6 tons per cutter daily and received £3 16s 10d per cutter per day.[52] Yet despite canecutter shortages during the late 1940s, Toft's original loading machine never gained much popularity amongst Australian canegrowers. The arrangement of a boom and grab operated by cables was cumbersome and always required at least two men to operate the loader, as a ground assistant had to position the load over the rail truck or motor lorry. Farmers wanted a loading machine that was lighter and more manoeuverable. In addition, the machine loading of hand-cut cane was still not economically attractive. Occupying about a third of a cane cutting gang's time, loading was the most exhausting part of their work. The schedule piecework rates for canecutters under the Sugar Industry Award were prescribed for the composite task of cutting, topping and loading, but the Queensland Industrial Court in 1943 provided a reduction of 1s 9d per ton in piecework rates when the canegrower used a mechanical loader and relieved the canecutters of the loading task. This reduction had not been enough to tempt many canegrowers into purchasing even the smallest of mechanical cane loaders which cost around £1 000 and those being operated were attributable solely to the desperate labour shortages.[53]

In early 1951, George Quaid Snr., a Mossman canegrower and member of the Mechanical Harvesting Committee, developed a hydraulically operated front-end cane loader consisting of a removable modified silage rake attachment mounted on a small Ferguson TE 20 tractor (see Plate 16.5). Quaid's invention had several advantages. Its low capital cost of under £1 000 made the machinery available to most canegrowers and the removable attachment meant that canegrowers could use it on existing tractors. A separate piece of machinery did not have to be acquired. In addition, the machine could be operated by one person. The operator formed a bundle of cane within the grab by driving, with an open grab, into the heaps of harvested stalks of cane. The loader operator did not require a ground assistant to position

Plate 16.5. A front-end cane loader being used in the Mackay district, 1959.

Source: Image reproduction courtesy of the National Library of Australia, Canberra. Image No. nla. pic-vn4590143.

the load over the tram trucks or motor lorries. Once loading had been completed, the loader operators could then attend to other duties such as hauling loaded or empty trucks to rail sidings. First demonstrated in August 1951, the Bundaberg machinery makers Wyper Bros. were building them by the end of the year. CSR immediately introduced front-end cane loaders into New South Wales during the 1952 season, for the firm sponsored their trial use on several farms throughout the Richmond River and Clarence River districts. Yet in 1954, two years after Quaid's invention and despite its advantages, only about forty front-end cane loaders were used in Queensland.[54] Less than 5% of the cane harvested in Queensland was mechanically loaded (Table 16.1).

The initial slow uptake of front-end cane loaders was still related to the unfavourable economics of the machine loading of hand-cut cane, but this situation was about to change. The Queensland Industrial Court during 1954 increased the reduction in piecework rates between the composite task of cutting, topping and loading and cutting and topping only by 3s per ton. Following further representations by the QCGC, the deduction for mechanically loaded cane was increased to 4s per ton in May 1955. Moreover, this deduction was indexed to future movements in the Award – it was increased to 5s in 1959. 'An economic basis for mechanical loading now existed', observed Burrows and Morton, 'creating the opportunity for this operation to be performed by contractors'.[55] Alternatively, farmers using front-end cane loaders during the harvest period could make substantial savings by paying cutters to harvest their cane only, thereby offsetting the cost of purchasing the machinery. Consequently, there was an upsurge in the adoption of front-end cane loaders – nearly 1 000 were used during 1956. By 1959, 2 100 of the 2 165 mechanical cane loaders in Queensland were front-end cane loaders. The amount of mechanically loaded cane in Queensland had quadrupled by 1960 and stood at nearly half the crop (Table 16.1).

Table 16.1. Mechanical cane harvesting and loading statistics for Queensland, 1954–1976.

Year	Mechanically harvested ('000 tons)	Mechanically loaded ('000 tons)	Year	Mechanically harvested ('000 tons)	Mechanically loaded ('000 tons)
1954	Negligible	247.0 (2.5)	1965	5 387.3 (39.1)	12 873.6 (93.5)
1955	Negligible	741.0 (8.6)	1966	7 651.0 (48.5)	15 099.2 (95.8)
1956	180 (2.0)	2 595.0 (28.9)	1967	9 351.9 (59.2)	15 581.4 (97.5)
1957	107 (1.2)	2 746.0 (30.7)	1968	12 673.6 (71.6)	17 447.6 (98.6)
1958	190 (1.9)	3 574.0 (36.0)	1969	12 682.9 (84.5)	14 759.0 (98.8)
1959	138 (1.6)	3 265.0 (38.7)	1970	15 141.0 (91.9)	16 407.6 (99.66)
1960	238.7 (2.7)	4 223.7 (47.8)	1971	17 861.9 (97.0)	18 341.4 (99.63)
1961	492.1 (5.4)	4 708.4 (51.4)	1972	17 866.6 (98.8)	18 078.8 (99.95)
1962	1 067.8 (8.7)	7 930.8 (64.5)	1973	18 201.1 (99.5)	18 274.7 (99.97)
1963	1 570.1 (13.4)	8 561.1 (73.3)	1974	19 383.8 (99.8)	19 420.7 (99.99)
1964	3 457.9 (23.8)	1 2 811.6 (88.1)	1975	21 058.4 (99.9)	21 068.9 (100.0)

Notes: (1) The figure in brackets refers to the percentage of the total. (2) The figures for 1954–1959 are definitely tons. It is unclear from the *ASYB* if the figures were tons or tonnes from 1960 onwards.

Source: J. See and H. Crouch, 'Mechanisation of sugar cane harvesting', *Personal Practice Bulletin*. Department of Labour and National Service, Australia, 1963, p. 46 & *ASYB*, 1978, p. 60.

Other mechanical cane loaders were also developed by innovative canegrowers after 1952. They were responding to the introduction of whole stalk mechanical cane harvesters which deposited different-sized bundles of cane in various directions on the ground. In 1954, Don Leighton, a Mulgrave canegrower, acted on a suggestion from Jack Skidmore, his neighbour, and replaced the dirt bucket on a Nuffield front-end loader with two pair of improvised claws. Subsequently, this machine was manufactured in a modified form in Brisbane and marketed in Australia as the 'Nuffleigh'. More robust than Quaid's machine, it could load up to 0.6 tonnes of cane. Toft Bros. also designed a one-person operated hydraulic grab-type cane loader in 1956. The first model was mounted on a Ford Blitz 3-ton truck. With its fully rotatable grab and considerable reach, the Toft hydraulic grab-type cane loaders could lift bundles of between 0.2–0.3 tonnes, loading approximately 25 tonnes per hour. The second, built at the request of the Mechanical Harvesting Committee, bundled as well as loaded, and was tractor mounted. However, the capital cost of these machines limited their use to circumstances where a considerable tonnage needed to be loaded during the crushing season. About 400 000 tonnes were loaded by other mechanical cane loaders in 1960, compared to 3.9 million tonnes by front-end cane loaders.[56]

The adoption of mechanical loading in Queensland by 1960 was quite uneven, with the proportion of the crop loaded mechanically in each region being as follows: north of Townsville (21%); Lower Burdekin (46%); Mackay-Proserpine (87%); and Southern Queensland (47%). Burrows and Morton argued that the front-end cane loader was ideally suited to the Mackay district, for here a different cane cutting sub-culture had evolved. Instead of the larger cane cutting gangs found in the sugar-producing districts north of Townsville, cutters worked solo, in pairs or threesomes and the farmer assisted with the loading. They used 'truck wagons' – only one rail truck sat on these wagons (Plate 16.5). These were loaded and hauled continuously over half a day, so it was to the farmer's advantage to help load and hurry the process, allowing more time for other farm work. In contrast, several rail trucks or a 'rake' sat on portable rail track in the paddocks of the sugar-producing districts north of Townsville. The cane cutting gang filled them towards the end of the day and the rakes were hauled as one unit to the permanent tramline. In addition, the mechanical cane loaders were operated in the districts south of Townsville consistently each day throughout the season. However, throughout the mill supply areas north of Townsville, except the Mossman Mill supply area, the canecutters continued relying on manual loading, with the mechanical cane loader being held in reserve, and brought into use as gang strengths declined, or when the harvesting became particularly difficult for the canecutters in fallen or damaged crops or towards the end of the season when the weather became much hotter.[57]

Mechanical cane loading was also adopted by New South Wales canegrowers. A quarter of the crop in the Clarence River district was loaded mechanically in 1955; two years later this figure had swelled to 97%. Initially, the Clarence River district canegrowers used front-end cane loaders, but Toft grab-type cane loaders became more popular during the early 1960s because of their better stability in boggy paddocks, and greater versatility. A fifth of the crop had been mechanically loaded in the Richmond River district by 1958, although the Richmond River canegrowers preferred the Toft grab-type cane loaders to front-end cane loaders.[58] Farmers in this district also saw that mechanical cane loaders were a positive insurance against labour shortages. The Richmond River Branch of the NSWCGA had decided by 1958 that 'the ultimate aim should be to organize the harvest so that the crops can be taken off by the use of machines and local labour. This has been practically achieved on the Clarence this year'. In 1960, almost 80% of the harvested cane in the Richmond River district had been loaded with mechanical cane loaders; this amount reached 99% the following year. Mechanical cane loading was slower to be adopted in the Tweed River district, with a quarter of the crop not being mechanically loaded until 1962; the following year, however, three-quarters of the Tweed River district crop was loaded mechanically. Again the farmers preferred the Toft grab-type cane loaders over the front-end cane loaders.[59]

Mechanical cane loaders reduced the physical standard of the labour force required for harvesting and meant that the industry was no longer dependent upon young men; older men could continue harvesting cane. They also enabled canecutters to increase their typical daily output to around 13 tonnes per man, almost double the amount managed by a canecutter before World War II. However, improved productivity came with a set of problems. Soil compaction emerged as a major concern. Canegrowers soon found that their fields were criss-crossed with concrete-like tracks where the loaders had operated. Mechanical cane loaders bogged easily during wet weather. Days might pass before loading operations would re-commence, leaving burnt cane to spoil in the field. Sometimes the canegrower could persuade the canecutters to load, but increasingly they shunned this arduous task, preferring to concentrate only on harvesting and topping.[60] Responsibly used, mechanical cane loaders gathered little dirt with the cane. Careless operators, however, not only collected the harvested stalks of cane, but stones, soil and trash, and this 'extraneous matter' then found its way into the crushing rollers.

Edward Ritchie Knox, CSR's Chairman of Directors, had warned the Australian sugar industry in 1949 about the risk of an upsurge in extraneous matter if mechanical cane loaders were adopted, citing the manufacturing difficulties experienced by the sugar millers in Louisiana and Hawaii, following the switch to mechanical cane loading in those two regions. Knox's warning proved correct. In 1956, despite some protests from the New South Wales canegrowers, CSR's new cane agreement for its New South Wales mill supply areas contained a clause that outlined a scale of penalties per ton of cane for various percentages of extraneous matter. CSR also tried discouraging the use of mechanical cane loaders in its mill supply areas, but by 1957 had only succeeded in getting the Hambledon Mill Suppliers' Committee to agree to this strategy. Yet the extraneous matter problem became so great in Queensland that the Central Board increased deductions to 3s in certain mill supply areas for the 1956 crushing season.[61]

CSR, however, decided that penalties could be paid by its suppliers, but the firm still had to process dirty cane and produce sugar that was acceptable for export. Therefore, CSR in 1957 obtained the agreement of the suppliers in all its mill supply areas to modify the local awards by including a bonus arrangement for cane which was cleaner than a set standard. In its Queensland mill supply areas, CSR paid a bonus of 15 cents per ton per 1.0% variation from above 97.5% cleanliness. Specially trained assessors evaluated every truck of cane by visual inspection, allocating it to one of the following classes: clean cane (> 97.5% clean); dirty cane (95–97% clean); very dirty cane (< 95% clean). Deductions of 30 cents and 60 cents were made per ton of dirty and very dirty cane, respectively.[62] Some CSR canegrowers received bonuses for several years, but following the introduction of mechanical cane harvesting in the mid-1960s, bonuses disappeared and penalties increased. Eventually, CSR's canegrowers in 1969 and 1970 sought to overturn this system. The Central Board in 1970 resolved the matter by introducing a scheme at CSR's mills, except Hambledon, whereby bonuses and penalties were paid for cane on the basis of a six year average for its cleanliness. Suppliers to Hambledon Mill were penalised or rewarded for cane above or below 97% cleanliness.[63]

As mechanical cane loaders were adopted by some Australian canegrowers during the 1940s and 1950s, investigations continued into the development of a viable mechanical cane harvester. By 1945, Mackay canegrower Larry Moloney had produced a tractor-mounted cane harvester, built mostly from wood due to wartime steel shortages. The ground level cutting knife was attached to a spring, enabling it to bounce over obstructions such as stones, while an operator controlled the topping device from a platform on the timber frame which was attached to the right hand side of a steel-wheeled Oliver Hart-Parr tractor. The machine harvested 20–25 tonnes per hour when carefully operated, or as much as two competent canecutters could cut in a day. Yet the harvester worked best in straight cane and the harvested cane was left in a windrowed condition, no attempt being made to bundle it. Hailed as being better than the Toft or Fairymead machines, the Moloney mechanical cane harvester, however, gained favour only in the Mackay district where most cane was loaded onto truck wagons or lorries, so a lack of bundles was not a major consideration.[64]

Jack Kinnear, a canegrower at Shinfield near Sarina, also produced a cane cutting machine during the late 1940s. It was developed as a tractor attachment rather than as a separate unit, thus sparing the farmer the expense of having to purchase an entirely new piece of machinery. Two shear knives set at an angle to each other cut the cane at ground level, and the cut stalks were guided into the machine in an upright position and allowed to fall towards the left. Beneath the cane harvester a set of tines bundled the fallen cane stalks. More manoeuverable than the Toft or Fairymead units, it would be cheap to mass-produce. Yet the main disadvantages of the Kinnear cane harvester were its need for two operators, and after cutting one row it was necessary to remove the cane from that row before the row next to it could be cut. Under sponsorship from the Mechanical Harvesting Committee, an improved version of the Kinnear cane harvester was manufactured by Bonel Bros. who operated an engineering workshop at Bundaberg during 1954, but it never reached commercial production. A few canegrowers in Southern Queensland had also attached cutting blades to tractors, pioneering an arrangement that became known as shear blade technology. By the early 1950s, Ray and Dixon Scott, Bundaberg canegrowers, had built what became known as the Scott cane harvester – an enlarged, removable shear blade attached to a very narrow gauge tractor accompanied by a two-pronged rake attachment that left the cane in bundles. Harvested cane was then topped by hand and loaded, either manually or by a grab-type mechanical loader. They were most popular in the mid-1950s, but only nine shear blade cane harvesters operated in Queensland during 1960.[65] However, all these machines, although assisting in the mechanical harvesting of cane on individual properties, never achieved industry-wide appeal.

Throughout the investigations into the development of a viable mechanical cane harvester for the entire Australian sugar industry, Fairymead plantation remained an enclave of successful mechanical cane harvesting. Immediately after World War II, Charles Young and Martin Vichie improved upon the design of their mechanical cane harvester. They developed two-row machines that cut the cane from two rows simultaneously, combining it into a single bin which was emptied between the rows in half-ton bundles. A two-row machine cutting 400 tonnes a day, supported by four mechanical loaders, involved fourteen operators and a ground crew. The average daily output per man was thirty tonnes, four times the rate reached by men cutting and loading manually. In 1948, the Fairymead Sugar Co. Ltd. patented their one- and two-row mechanical cane harvesters and they had commenced being manufactured under licence by Kilner Bros. of Brisbane. Canegrowers flocked to see a demonstration of the machine at Bundaberg in October 1948, but few farmers purchased them over the next decade.[66] CSR employees had also visited Fairymead plantation during 1948, but their findings were not enough to convince the firm to pursue the development of a mechanical cane harvester. Eventually, however, CSR ventured into mechanical cane harvester design during the late 1950s. Its machine – nicknamed 'The Butterfly' because of its unusual moth-like appearance – was scrapped in 1964 after field trials demonstrated that it worked reasonably well in erect cane, but any obstruction caused chokages.[67]

Norman King, the BSES Director wrote in 1951 that 'outmoded conservatism' was preventing the majority of Australian canegrowers from displaying an active interest in mechanical cane harvesters, although other reasons can explain why mechanical harvesting was so slow to be adopted elsewhere in Australian sugar-producing districts. Mechanical cane harvesters cannot do a good job in lodged or tangled cane, observed Charles Young in 1950. This type of cane was often found in the northern sugar-producing districts which were subjected to cyclones or floods, so some northern canegrowers shunned the mechanical cane harvesters because they did not cope with their particular environmental conditions. The earliest mechanical cane harvesters were also too large and cumbersome to manoeuver effectively on small family farms and they often broke down – small farmers could not afford a delayed harvest or the expense of machinery repairs. In addition, mechanical cane harvesters were expensive, costing at least £1 000 or more, so they had to be used effectively to recoup this capital outlay. Charles Young noted that the machines cut too much cane in eight hours for the average Queensland canegrower, and so could only be used to their full capacity if controlled by a contractor operating amongst a small group of farmers or on plantations.[68]

By the early 1950s, the contractor arrangement described by Charles Young had emerged in the sugar-producing districts of Queensland. One operated in the Farleigh Mill supply area between 1951 and 1957. Its mechanical cane harvester was a locally-designed Rasmussen machine that consisted of a wooden framework with topping blades operating on a pivoting beam balanced by a counterweight. The machine cut up to 12 000 tonnes from twelve farms during the season, usually cutting on six farms each day, but this extensive shifting from farm to farm caused mechanical difficulties. Another drawback was that harvested stalks of cane were dropped along the drills of cane instead of at right angles to the drills, making loading more difficult. Eventually, the Rasmussen machine was replaced by a Fairymead mechanical cane harvester.[69]

In 1957, Queensland's canegrowers used ten Fairymead machines and about forty-six miscellaneous mechanical cane harvesters, of which about half were shear blades.[70] Yet their effectiveness was still limited with only around 2% of the Queensland crop being harvested mechanically (Table 16.1). But this arrangement altered dramatically over the next five years, as the uptake of mechanical cane harvesters gained momentum, a result of a combination of factors. Firstly, farmers increasingly had to harvest more cane, a consequence of industry expansions, so the point had been reached where it was more economic for contractors or individual farmers to purchase the machines. Secondly, farmers had more capital to invest in new machinery, following a succession of high prices for sugar during the late 1950s and early 1960s. Thirdly, better designed models of mechanical cane harvesters had become available.

Two distinct types of mechanical cane harvester evolved by 1960: whole stalk and chopper harvesters. The former had found some early commercial acceptance because they could be easily used with front-end cane loaders. In 1962, nearly half the mechanical cane harvesters in Australia were whole stalk models (Table 16.2). One of the more popular whole stalk models during the early 1960s was the Venton, developed by the Plane Creek mill supplier Ray Venton in 1956. Designed around a tractor, two-power driven discs cut the cane, and carrying chains conveyed the stalks vertically up a conveyor for topping as they fell from an elevated position on to the bundler. Another very popular whole stalk model during the mid-1960s was the Crichton, named after its inventor, the Plane Creek Mill supplier William Crichton (Plate 16.6). By 1966, nearly 200 of these machines made up half the fleet of whole stalk mechanical cane harvesters operating in Queensland. Colin and Harold Toft also released a whole stalk mechanical cane harvester in 1961. Mounted on a Farmall M tractor, the single-purpose machine laid the harvested stalks of cane in large bundles parallel to the cane row, thus restricting loading to grab-type mechanical loaders. Its sale laid the foundation of the highly successful Toft Bros. Industries Ltd. (later Versatile Toft and then Austoft Industries Ltd.).[71]

Chopper cane harvesters cut burnt cane into short lengths or billets, removed the extraneous matter and discharged them into hoppers or bins which replaced the conventional cane trucks. These machines automated

Table 16.2. The percentage of the Queensland sugar cane crop harvested by different types of mechanical cane harvesters, 1960–1971.

Year	No. of chopper harvesters	% of crop harvested	No. of whole stalk harvesters	% of crop harvested	Total no. of harvesters
1960	*	*	*	*	56
1961	38	*	72	*	110
1962	91	6.0	70	2.6	161
1963	155	9.9	93	3.5	248
1964	354	19.0	129	5	483
1965	571	30.0	251	9	822
1966	665	32.4	403	16.1	1 070
1967	814	38.6	509	20.0	1 327
1968	1 050	50.7	572	20.9	1 627
1969	1 271	67.1	472	17.4	1 743
1970	1 478	78.0	442	14.0	1 920
1971	1 568	88.3	331	8.7	1 899

* Data not supplied.

Source: Assembled from figures in *ASJ*, March 1967, p. 787, Department of Labour and National Service, *Men and Machines in Sugar Cane Harvesting* (Melbourne, 1970), p. 20 & G. Willis, *Harvesting and Transport of Sugar Cane in Australia* (Townsville, 1972), p. 14.

Plate 16.6. A Crichton whole stalk mechanical cane harvester operating in the Moreton Mill supply area, August 1962. Note how the harvested stalks of cane form a bundle in the attached bin.

Source: Image reproduction courtesy of Sunshine Coast Libraries, Picture Ref. No. M670808.

both the harvesting and loading operations, thereby eliminating the need for mechanical loaders. Prominent in the development of these new models was J. K. (Ken) Gaunt, a dour Englishman who was the Superintendent of Engineering Research and Design at Massey-Harris Ltd., the Australian subsidiary of the Canadian firm Massey-Ferguson. Design work had commenced in 1956 on a small chopper cane harvester guided by three main principles: the machine should be capable of harvesting erect and lodged crops; the machine should be a true cane harvester, not a cutter, so it needed to deliver the crop loaded ready for transport to the sugar mills; and it should be a farmer's machine, easily purchased by canegrowers and preferably tractor-mounted. After extensive tests, the 1.5 tonne tractor-mounted MF 515 was first used by the Rossi Farming Co. at Aloomba, near Gordonvale, to supply chopped up cane to Mulgrave Mill in 1959. The MF 515 was commercially released the following year, becoming an instant success and revolutionising cane harvesting in Australia (Plate 16.7). Ninety-four MF 515s cut almost three quarters of a million tonnes of sugar cane in Queensland during 1962; their number had swelled to over 400 by 1965.[72] The MF 515, however, soon had strong competition. By 1962, Herbert River canegrower, Laurence Mizzi had used a grant from the Mechanical Harvesting Committee to develop a whole stalk cane harvester. Rights to manufacture the Don Mizzi cane harvester were assigned to Wyper Bros. of Bundaberg, but sales were initially slow. After the machines were redesigned in 1964 to incorporate a side chopper configuration they became hugely popular. Wide axles provided versatility in the handling of cane on fields of all types – they became standard on hillside farms where their stability was outstanding and the machines coped with wet and boggy fields. Toft Bros. Industries Ltd. had also commenced manufacturing chopper cane harvesters by the early 1970s.[73]

Just under 3% of the entire sugar cane crop in Queensland had been mechanically harvested in 1960. Fifty-six mechanical cane harvesters operated in the state – five Fairymead machines, eleven Massey Ferguson, eighteen Venton, nine shear blade and thirteen other miscellaneous models.[74] Two years later, nearly 10% of the Queensland sugar cane crop was harvested by 167 mechanical cane harvesters; this number of cane harvesters had grown to 483 during the 1964 crushing season. Almost half the state's sugar cane was mechanically harvested in 1966 – the days of the manual canecutter were clearly numbered. By 1971, 97% of the Queensland sugar cane crop was handled by mechanical cane harvesters (Table 16.1), with the entire crop being mechanically harvested in several mill supply areas for the first time in 1971, namely Goondi, Mourilyan, Macknade and Proserpine.

Plate 16.7. The Massey-Ferguson mechanical chopper cane harvester in operation, 1958.

Source: *ASYB*, 1959, p. 6. Image reproduction courtesy of Rural Press Queensland Pty. Ltd.

Moreover, mechanical cane harvester numbers had nearly quadrupled in the intervening seven years (Table 16.2). By 1976, the transition process was virtually complete, when 99.99% of the Queensland sugar cane crop was harvested by machines. In 1979, all sugar cane in Queensland was mechanically harvested and loaded after Denis Gravener and Allan Atkinson, the only two manual canecutters remaining in Queensland, completed their last season in the Moreton Mill supply area in 1978.[75]

Mechanical harvesting of cane in New South Wales occurred much later than in Queensland. Limited trials using three different models of mechanical cane harvesters (i.e. MF 515; Harvestall; Toft) were conducted in the Condong Mill supply area during 1965, with the MF 515 chopper cane harvester performing poorly. About 1 000 tonnes of cane were also cut by a Crichton whole stalk cane harvester in the Broadwater Mill supply area in 1965, but the machine kept bogging and only worked successfully on very erect, one-year old cane. Further restricted trials in each New South Wales sugar-producing district using chopper and whole stalk cane harvesters occurred during 1966, but the results were not promising: the machines' capabilities appeared to be limited to light, non-lodged cane and firm ground. Clearly, New South Wales conditions were quite different to those found in parts of Queensland. Many New South Wales canegrowers had two-year old crops, which were difficult to harvest because they were often sprawled, tangled and trash-bound, and poor drainage and winter rains meant the fields were sometimes waterlogged at the crucial harvest time. 'It will never happen in our heavy cane' was one catch cry. 'I'll never have one of those bloody things bogging up my paddock' was another typical reaction to the possibility of mechanical cane harvesters, claims local canegrower and author Graham Smith in his history of cane farming in the Richmond River district.[76] Both views, however, soon proved to be wrong.

By the mid-1960s, CSR handled mechanically harvested cane in its Queensland mill supply areas, so the firm considered machine harvesting in New South Wales. In 1968, CSR in conjunction with the New South Wales Mechanical Harvesting Committee (NSWMHC) – an organisation formed by the NSWCGA – decided to investigate the possibility of mechanical cane harvesting in New South Wales, taking into account its special environmental conditions. A pilot trial commenced in the Condong Mill supply area using a chopper cane harvester together with specially designed in-field transporters and cane cleaning equipment. The NSWCGA considered the trial results a failure and refused to fund further investigations during the 1969 season.

CSR, however, persevered with the chopper cane harvester and in-field transporters during 1969, demonstrating that the machines could remove sugar cane crops from the field and successfully place billets in the bins. Broadwater Mill canegrowers who watched the demonstrations of the mechanical cane harvester realised that the solution lay in using a tracked machine to overcome the soft ground, a modification being investigated by Toft Bros. The following year, just under 5% of the cane was mechanically harvested in the Broadwater Mill supply area (Table 16.3). Subsequently, the uptake of machine harvesting in New South Wales was rapid, following the introduction of cane harvesters on tracks and tracked in-field transporters, aided by simple arithmetic: cane

Table 16.3. The percentage of the sugar cane crop mechanically harvested in New South Wales mill supply areas, 1970–1975.

Year	Harwood	Broadwater	Condong
1970	0	4.7	0
1971	0	7.2	21.7
1972	0	18	42
1973	15	32	70
1974	56	74	89
1975	86	97	100

Source: Based upon figures in CSR, ARs on Cane. Part 1, 1970 to 1976, CSRR, Z/109, Boxes 390 & 391.

cut by hand in the Richmond River district in 1972 cost $3.27 per tonne, whereas machine harvested cane cost $2.35 per tonne, while in the Harwood Mill supply area during 1974–1975 men cost 50% more than machines. By 1975, the entire crop in the Condong Mill supply area was removed by chopper cane harvester; the following year no manual canecutters were employed in the Broadwater Mill supply area. Four gangs cut cane in the Clarence River district during 1975 – the last manual canecutters in New South Wales. The final load of hand cut cane was delivered to Harwood Mill on 5 January 1976, ending an era that had commenced a century earlier.[77]

As the amount of mechanical harvesting increased, the popularity of whole stalk cane harvesters declined. By 1971, just under a fifth of the mechanical cane harvesters in Queensland were whole stalk models and under 10% of the entire sugar cane crop had been harvested by these machines (Table 16.2). They remained most popular in Southern Queensland where farmer-owned machines provided low cost harvesting on the many small properties throughout the region, and their replacement with the more costly chopper cane harvesters was resisted by the canegrowers. In 1972, no new whole stalk cane harvesters were manufactured or sold in Australia, and those still operating were being phased out, eventually disappearing after the 1979 crushing season. Whole stalk cane harvesters lost their dominance because they were incapable of handling the badly lodged crops found in North Queensland, especially throughout the Lower Burdekin district. This failing became very apparent to the Queensland Harvesting Co. Pty. Ltd., formed in March 1966, with the aim of using Toft whole stalk cane harvesters to harvest Burdekin cane on contract. The investors – Pioneer Sugar Mills Ltd., the Haughton Sugar Co. Ltd. and Toft Bros. – lost heavily when the company was wound up after losing $154 312 in 1967, its machines struggling to remove the Lower Burdekin cane. Moreover, whole stalk models were doomed to obsolescence because they had the disadvantage of dropping the bundled cane on the ground, necessitating further mechanical loading.[78]

Chopper cane harvesters functioned very successfully in erect or semi-erect crops, severing and discarding the tops. The cut-up material delivered to the bins was clean and relatively free from extraneous matter if the pre-harvest burning of the crops had been satisfactory. Yet in lodged cane, the tops were dragged into the machine with the stalks, cut up and conveyed to the bins with the billets. This bin tainting was eventually overcome by the installation of air separation devices on the cane harvesters which reduced the amount of leaf and top material.[79] Cane harvested by chopper cane harvesters, however, suffered losses from what became known as sour storage rot of chopped cane. First noted in Queensland during the 1962 crushing season, the loss became acute when the chopped-up cane was stored in bins during the weekend shutdown of the sugar mills, and when the delay between harvesting and crushing extended to forty-eight hours or more.

Early investigations into the phenomenon were conducted by BSES pathologist Brian Egan at Mulgrave Mill. He found that this deterioration is due to invasion of the cut ends by the bacterium *Leuconostoc mesenteroide* Cienkowski. The organism enters the cut ends at the instant of cutting by the cane harvester's knives. A 3 m length of whole stalk cane had only two exposed ends, but cut into smaller sections it had more than a dozen ends which could be attacked by the bacterium. If shredded by blunt chopping knives it had even more of the interior surface exposed. *Leuconostoc* activity is assisted by high temperatures and humidity, so the cane deterioration becomes more acute in the hotter months of the year, during the second half of the crushing season. Apparent ccs losses could be as high as 1.31 units compared with whole stalk cane. Moreover, juice from cane contaminated by the bacterium failed to crystallise satisfactorily during processing, leading to a decline in both the quality and quantity of sugar produced. Marked improvements in the sharpness of the cutting blades and in cane harvester hygiene assisted in reducing the problem, but the main approach to avoiding the losses was to eliminate weekend stocks of chopped-up cane. The latter was achieved at Australian sugar mills by a reduction of the harvesting rate on Fridays, preferential crushing of chopped-up cane, and the introduction of Sunday work by chopper cane harvesters so that the sugar mills did not have to store sufficient cane over the weekend to allow a resumption of crushing on Monday. [80]

The initial cost of mechanical harvesting equipment was quite high in the early 1960s, with purchasers expected to outlay between £4 000 and £6 000 to acquire the cane harvester and any associated bins or trailers. Alternatives to sole ownership included group ownership by a number of canegrowers or the use of contractor harvesting. A single canegrower or groups of canegrowers cut 'on contract' for others as well as using their machines for their own crops. In 1962, the Australian Department of Labour and National Service assembled details on cane harvester ownership, finding that 67% of all cane harvesters were owned by individual farmers, 22% by groups, 10% by the sugar mills or grower organisations such as local mechanical harvesting committees, and only 1.0% by independent contractors.[81] Details obtained in further annual surveys made by the Department of Labour and National Service revealed that contractor ownership of machines had increased substantially to about 20% of all cane harvesters by 1964. L. G. Vallance, agricultural consultant to the ASPA, reported on mechanical harvesting for the 1968 and 1969 seasons. He found that about a quarter of the new machines had been purchased by independent contractors, thereby confirming that independent, non-farmer cane harvester contractors had become an established feature of the Australian sugar industry. Marked regional differences also existed between individual mill supply areas and between districts. In Far North Queensland, most machines were owned by groups; in the neighbouring Herbert River district, contractor-owned machines dominated (61%). Individual ownership was most marked in the Mackay district and Southern Queensland, where about 40% of the machines were owned by canegrowers who cut only their own cane.[82]

The switch to mechanical harvesting prompted changes in a range of farming practices. Field drainage, as mentioned in Chapter 12, became more important, and many fields had to be redesigned to accommodate mechanical cane harvesters – small and odd-shaped fields crossed by winding water-courses were eliminated. The width of headlands had to be redesigned to enable the machines to turn comfortably – about 6 m was required for most machines. Mechanical cane harvesters were prone to damage from stones, so ridding paddocks of stones became a greater priority; some farmers developed their own mechanical stone pickers to assist with this task. In the Garradunga locality near Innisfail, local canegrowers during the early 1970s used the stones removed from their paddocks to make retaining walls which aided in the reduction of soil erosion, although they did pose a hazard to tractors. Traditional, low-roofed implement sheds could not cope with mechanical cane harvesters. Many canegrowers during the 1960s either built new high-roofed implement sheds or modified existing sheds so that they incorporated a high-roofed section for the mechanical cane harvester.[83]

In 1961, just under 8 100 canecutters were employed in Australia's sugar-producing districts. Following the introduction of mechanical harvesting, this number had been halved by 1968. Parallel with this trend was an increasing demand for workers to operate the mechanical loading and harvesting equipment. Some of this demand was filled by seasonal workers, but there was a switch towards a larger participation of resident farm labour in the actual harvesting operations, and a higher retention in permanent farm employment of men skilled in operating harvesting equipment. Eventually, the annual influx of labour from southern Australian states disappeared during the late 1960s and early 1970s as younger local men – sometimes the sons of canegrowers – became the mechanical harvesting employees.[84] Consequently, the temporary accommodation provided for the canecutters by canegrowers lost its main purpose. Instead, many cane barracks began to be used by canegrowers for storage. Others were abandoned, and became derelict, being places where canegrowers left rusted farm machinery and other old equipment. A few cane barracks have been maintained and refurbished, providing accommodation for small families.[85]

During this period of transition, the AWU sought to protect the interests of its members, even though the organisation recognised the inevitability of the transition. In September 1955, the AWU successfully obtained the addition of three new classifications to the Sugar Industry Award: mechanical cutter driver, mechanical cutter driver's assistant and driver of a mechanical loader. When the ASPA and CSR sought a new Award for the mechanical harvesting of sugar cane in 1960, the application was vigorously opposed by the AWU. The employers argued that the high level of capital investment in the mechanical cane harvesters required them to be worked for the maximum possible time. They sought 'round-the-clock' harvesting. When the Queensland Industrial Court brought down its decision in June 1961, it acceded to the application by granting a new Award which specified that extra payment of 4s daily plus 3d an hour for working between the hours of 6 pm and 8 am.[86] Limited night time harvesting was conducted in the Mossman, Herbert River and Bundaberg districts, but eventually in October 1969, following escalating canegrower opposition to the practice, the Queensland Industrial Commission refused to allow its extension throughout all mill supply areas (see discussion later in chapter).

Transport arrangements for harvested cane, 1915–1975

Between 1915 and 1950, a 'typical' Australian sugar mill maintained 50–100 km of tramline, although Isis Mill with nearly 120 km of permanent tramline had one of the most extensive tramline systems in Australia. Each sugar mill operated five to seven steam locomotives that relied on wood or coal as fuel. They hauled from 1 000 to 2 000 rail trucks with stake ends, which sometimes held as little as one tonne of cane each. Harvested cane was delivered to this transport system by 100–200 canegrowers, divided up into harvesting groups of varying sizes. These farmers supplied the mill at an average rate of 10–20 tonnes daily, and the cane harvesting and its loading onto rail trucks was limited to daylight hours. The cane would be moved from the fields and assembled at various loading points on the tramline system. The locomotive then collected the rail trucks from the different assembly points and transported them to the mill yard for inspection, weighing and finally their load was emptied into the cane carrier. Empty rail trucks collected from the mill yard were deposited back at various assembly points near the fields, and the entire process started again. Normally the locomotives made two runs per day, collecting the cane and delivering the empty rail trucks for loading, over three 8-hour locomotive shifts. The entire basis of such a tramline network was that it collected harvested cane during a twelve-hour period, yet kept the sugar mill supplied on a twenty-four hour basis. Journeys by the locomotives, however, were not always smooth sailing. These trains were braked only by the engine brake and operations over the light tracks were a hazardous procedure. Smashes or derailments were common and caused by broken axles, loads shifting on the trucks or poor track. These smashes were known colloquially as 'gutzers', and it was a lucky and skilful locomotive driver who did not have at least one 'gutzer' a week. Gangs of navvies would then be dispatched to clear the track quickly so the locomotives could pass once more as soon as possible in order to keep the sugar mill supplied with cane.[87]

Three other methods of transporting cane to Australian sugar mills existed between 1915 and 1950. After 1920, some canegrowers used road transport to bring their harvested cane directly to the sugar mill. The Maryborough Sugar Factory, for example, received part of its cane by road delivery in 1937, after a dozen suppliers ceased punting cane to the factory's riverside carrier. It was one of only two Queensland sugar mills that did not have tramlines, the other being Rocky Point. Harvested cane was also transported over government-owned railways to at least eleven Australian sugar mills, namely Maryborough, Prosperpine, Farleigh, Mourilyan, South Johnstone, Racecourse, Cattle Creek, Plane Creek, Invicta, Mt Bauple and Condong (see Plate 16.8). All these sugar mills were located in Queensland, except Condong Mill, which continued receiving cane carried on the New South Wales government railway from the Crabbes Creek locality.[88] Water transport of harvested cane was also still undertaken in 1950. In Queensland, punts laden with cane destined for Moreton Mill continued being towed along the Maroochy River. The movement of cane by punts remained the major form of transport in New South Wales, although the wooden punts were no longer towed by paddle steamers, but diesel powered tugs and steel barges. Loading of cane onto the punts had also become easier during the twentieth century. By the outbreak of World War II, most New South Wales canegrowers using punts had their own loading derrick, so 'cane slipping' or carrying bundles of cane to the punts across wooden planks was in decline.[89]

By the early 1950s, the overall transport arrangements for harvested cane in Australia had altered little since 1915. In 1952, approximately 80% of Queensland's harvested cane was hauled to the sugar mills from the canegrowers' delivery points on a mill's tramline, about 15% by government railways and the remaining 5% by road or punts.[90] The main change, however, had been the escalation in the use of road vehicles to move harvested cane to delivery points on the sugar mill's tramline in some sugar-producing

Plate 16.8. Loaded rail trucks waiting for unloading at Invicta Mill, c. 1925. Rail trucks with stake ends are shown on the left and Queensland government railway wagons are shown on the right of the photograph. The sugar cane in the Queensland government wagons had been transported nearly 150 km.

Source: Image reproduction courtesy of the Burdekin Shire Library.

districts (see Table 16. 4). Between 60 and 80% of the cane in the Mossman, Lower Burdekin and Bundaberg districts and all or almost all the cane in the Maryborough, Proserpine and Mackay districts was now delivered to the assembly points by various types of road vehicles. In the Proserpine district, farmers made use of low slung dual truck trailers hauled by tractors to take two cane trucks to the field for direct loading. Pick-a-back cane trucks were favoured in the Mossman district. They consisted of two short rails attached to a table-top truck. A short length of rail track was slung beneath the truck for use in picking up trucks from anywhere in the field and running them off at the assembly point on the mill tramway.[91] In contrast, rail trucks pulled over portable line still dominated in the Herbert River and other Far North Queensland sugar-producing districts.

In the late 1940s, where it was necessary to trans-load cane at tramway sidings, it was still done by the chain and block method. At the community railway sidings, the old style circle boom horse type derricks shown in Plate 8.8 were still being used in some places. Loading bundles of cane averaging two tonnes using these methods was physically demanding and time consuming. During the early 1950s, groups of canegrowers loading at the same derrick combined resources and installed improved loading devices. Many railway lifts were converted to be operated by tractor, electric or diesel power. The most ingenious of these new style lifts was the Treacy lift, which was invented by a Mackay canegrower. In this device the lift was operated by driving the rear wheels of the loaded motor-truck onto two sets of rollers which were set in a

Table 16.4. The percentage of harvested sugar cane in Queensland delivered by either portable line or road vehicles to mill tramline assembly points, 1951.

Region	Portable line	Road vehicles	Region	Portable line	Road vehicles
Mossman	40	60	Proserpine	3	97
Cairns	91	9	Mackay	5	95
Innisfail	99	1	Bundaberg	40	60
Tully	95	5	Isis	63	37
Herbert River	98	2	Maryborough	Nil	100
Lower Burdekin	27	73	Moreton	78	22

Source: *ASJ,* 15 January 1954, p. 655.

metal framework at ground level. By using the truck's engine in second gear, the rollers were set in motion, and the power transmitted from the front roller axle by gearing and this operated a winding drum. By means of a hook, the wire slings on the cane load were picked up and the load was lifted from the vehicle. When elevated far enough, the motor truck was driven forward, its place taken by a rail truck and the load lowered onto the rail truck. The truck driver could load in less than five minutes with limited effort.[92]

During the sugar industry expansion of the 1950s and early 1960s, the length of tramlines operated by Australian sugar mills increased by 587 km (Table 16.5). Some of this expansion was related to the opening up of new areas to cane growing. During the early 1950s, CSR added another 40 km to its already extensive tramline network at Victoria Mill, enabling the cane from the Abergowrie Soldier Settlement area to be transported to the mill for crushing. In the Isis district, the Isis Central Sugar Mill Co. Ltd. built a 5 m tramline extension during 1956–1957, connecting the Isis Mill to the North Isis Soldier Settlement area, north of Cordalba.[93] Other Queensland mill-owners built new tramlines so as to avoid paying the freight charges, following a 212% increase in freight charges on government-owned railways between July 1947 and August 1953. Two examples will illustrate this trend. A 55 km extension costing about £750 000 was built from Farleigh Mill north to Pindi Pindi between 1959 and 1961, eliminating the need for the sugar cane to be railed to the sugar mill over the Queensland government's North Coast Railway. Just over £300 000 was spent on constructing a 23 km tramline extension south from Plane Creek Mill to Koumala in 1959–1960, again eliminating the need for the sugar cane to be delivered to the sugar mill by government railway. In New South Wales, CSR built a 5 km permanent tramline from Condong Mill into the Tumbulgum locality in 1952, eliminating the punting of harvested cane from this area along the Tweed River to the sugar mill. Between 1957 and 1959, this tramline was extended another 5 km to the Cudgen locality, leading to the dismantling of the Chinderah Wharf tramline and the punting of cane from the Cudgen locality was abandoned.[94]

The conveyance of sugar cane to the nearest sugar mill during wartime conditions proved popular and there seemed little reason why the benefits of reduced transport costs could not be extended to peace time, especially as the freight rates had been increased on the Queensland government-owned railways during the 1950s. Consequently, the Central Board commenced in the mid-1950s an extensive program of rezoning assigned land which eventually extended over twenty years. Assignments were re-allocated to the

Table 16.5. Total number of locomotives and rail trucks and length of track at Australian sugar mills, 1952–1978.

	1952	1959	1961	1966	1970	1974	1978
Locomotives							
Steam	170	137	98	63	*	Est. 20	12
Diesel	7	94	119	190	*	*	242
Total	177	231	217	253	Approx. 200	Approx. 200	252
Rail trucks							
Whole Stalk	30 810	43 306	* (most)	46 055	*	* (few)	*
Bulk-bins	0	0	* (few)	17 333	*	* (most)	*
Total	30 810	43 306	44 810	63 388	62 500	50 000	*
Length of track (km)	2 179	2 563	2 766	2 967	3 218	3 218	3 072

Source: I. Crellin, 'Australian sugar tramways. The challenge of the 1980s', *Light Railways,* No. 66 (October 1979), p. 7.

nearest sugar mill, making compensating adjustments to avoid disadvantaging either sugar mill, the changes aiming to reduce transport costs and replace government rail haulage by mill tramlines. 'Rezoning was easiest to achieve during periods of expansion', argued the historian John Kerr, as 'mills which lost in the rezoning could be compensated by the granting of new assignments to new growers. Also sugar mills were more prepared to make the necessary investment in tramlines while sugar prices were high'.[95]

Rezonings were mostly concentrated in the Herbert River, Lower Burdekin, Mackay and Bundaberg districts. One of the largest swaps involved the transfer of 3 236 ha of assignments in the Ingham Line area – the locality between Ingham and Rollingstone, just north of Townsville – from Invicta Mill to Victoria Mill. Railage of cane to Invicta Mill ceased and Invicta Mill was compensated for this loss by receiving assignments in the Lower Burdekin localities of Clare, Dalbeg and Millaroo.[96] This transfer commenced in 1958 and was completed during 1965. In the Mackay district, some rezonings of assignments during 1964 and 1965 centered around relocating farms from Pleystowe and Marian Mills to Cattle Creek Mill and Pleystowe Mill to Marian Mill. The aim was to stop a canegrower's cane being hauled beyond a closer mill's tramline so that it could be sent to a more distant sugar mill where it had been assigned historically. In the Bundaberg district, the assignment re-allocations were quite complicated, involving swaps between all the sugar mills to rationalise transport arrangements. These rezonings commenced in 1955 when assignments at Moorland and Gooburrum – Millaquin Mill supply areas north of the Burnett River – were transferred to Fairymead Mill which lost assignments at the Spring Hill locality to Qunaba Mill and the Goodwood locality to Isis Mill. More rezonings occurred during 1965, involving the transfer of assignments in the Splitter's Creek locality from Fairymead Mill to Bingera Mill; Fairymead gained assignments in the Waterview locality and cane from the Windemere locality was transferred from Millaquin Mill to Qunaba Mill.[97] As a consequence of these rezonings, the number of Queensland sugar mills in the mid-1960s still relying on Queensland Railways had fallen to a handful, namely Maryborough, Proserpine, Mourilyan, South Johnstone, Racecourse, Cattle Creek and Plane Creek.

Expanded tramline networks and larger volumes of cane to be moved meant almost eighty extra locomotives were added to the fleet of locomotives operated by Australian sugar mills between 1952 and 1966 (Table 16.5). However, the high cost of coal and labour in the 1950s prompted a strong swing towards diesel-powered locomotives. By 1961, the number of diesel locomotives outnumbered the number of steam locomotives. The advantages of these early diesels were soon recognised by the sugar mills. Through the elimination of watering and reduced refueling times, more time was available for haulage. They also offered economy in maintenance, required less staff and offered power in excess of what had been available with steam locomotives using light track. CSR concluded that a diesel locomotive did the work of at least one and a half steam locomotives of the same weight. In 1955, Lou Prince, the innovative manager at Mossman Mill, arranged for two-way radios to be fitted to the mill's new diesel locomotive and the cane inspector's vehicle. Mossman's innovation was soon copied at other Australian sugar mills, as their transport staff sought to more effectively control the movement of increased rail traffic brought about by the expansion in production during the 1950s and 1960s.[98]

The locomotives operating during the 1950s and early 1960s were designed basically like shunting locomotives and served the industry well while hauls were relatively short with frequent stops. By the late 1960s, this design was being questioned for cane field application by scientists from the Sugar Research Institute (SRI). In the early 1970s, the first diesel-hydraulic locomotives with 'lock-up' torque convertors appeared. This modification allowed the engine power to be utilised at the high speed end of the operating range, thereby allowing for long express hauls and the movement of heavier loads. At some Australian sugar mills, the introduction of a second radio-controlled locomotive towards the rear of the train enabled longer trains to be used, while two smaller locomotives coupled together at the front of the train increased haulage capacity at other sugar mills. However, longer trains and higher speeds could not be used effectively until better control had been achieved over a train of unbraked, loose-coupled rail trucks on light and often irregular track. During the early 1970s, SRI experiments with brakevans were carried out, promising the locomotive crews a better way of controlling the trains. Early brakevans were merely towed behind the locomotive and provided the driver with extra breaking power at the head-end. Getting the braking power at the tail-end of the train was solved by the introduction of radio-controlled brakevans. As a result of these developments, the mass of a typical Australian canefield locomotive increased from about 8–16 tonnes during the 1930s to about 24–30 tonnes in the late 1970s. Approximate engine power had increased from about 25 kW for a ten tonne locomotive used in the 1930s to about 350 kW for 26–32 tonne locomotives employed during the 1970s.[99]

To cater for the expanded sugar cane crops being harvested during the 1950s, an extra 14 000 rail trucks with stake ends for whole stalk cane were acquired by Australian sugar mills between 1952 and 1961 (Table 16.5). However, portable rail track and rail trucks with stake ends were suddenly made obsolete by the development of chopper cane harvesters. Yet chopper cane harvesters could not be introduced unless the mill-owners agreed to supply specially designed box-like cane bins to hold the billets. The sides and bottoms of the earliest cane bins were made of galvanized roofing iron, but this material was replaced with heavy wire mesh, enabling a better assessment of the amount of extraneous matter in a bin (i.e. stones, dirt and trash). The earliest cane bins of 2 tonnes capacity were fitted to the chassis of the cane trucks. At the rail siding delivery point these were run onto a low trailer which was hauled to the cane harvester by a tractor. CSR conducted experiments on cane bin sizes in the Herbert River district during 1963, even testing a bin that weighed 7 tonnes. Subsequently, cane bins of 3 tonnes and then 4 tonnes capacity were adopted and the chassis of the rail trucks were redesigned and strengthened to take the heavier loads. By the mid-1980s, cane bins of 4 tonne capacity were used on the tramlines of half Queensland's sugar mills; a quarter of the mills used 5, 5.5 or 6 tonne capacity bins; the other sugar mills, except Mossman, used 2.5, 3 or 3.5 tonne bins.[100]

Initially, mill-owners required canegrowers or groups of canegrowers planning to use chopper cane harvesters to apply for a supply of modified cane bins. By 1964, Proserpine, Racecourse, Mourilyan, Inkerman, Victoria, Mulgrave and Tully Mills, to list a few, had provided cane bins to their suppliers using chopper cane harvesters, although the sugar mills imposed a rental charge for the new bins based on tonnage, and usually for a period of three or four years. Most sugar mills charged 1s to 3s a ton, with the highest rental by a mill being 6s a ton charged by Inkerman Mill to Arthur Cannavan, the inventor and builder of the first chopper cane harvester used in the Inkerman Mill supply area in 1961. This higher charge reflected Pioneer Sugar Mills Ltd.'s view that the heavy and tangled crops of cane would defeat the early chopper cane harvesters and the sugar mill would be left with bins that were of no use.[101]

The issue of mill-owners charging rentals on cane bins came before the Central Board in June 1965, when it heard an appeal against the Mourilyan Mill Local Board Award which had been modified to include a provision requiring the mill-owner to supply cane bins free of charge to named canegrowers. The Central Board in a majority decision ruled that Mourilyan Mill should no longer charge rent on its cane bins. The Central Board members who supported the decision argued that the supply of cane bins had become a normal procedure of harvesting and delivery of sugar cane in the Mourilyan Mill supply area, so it had ceased to be a special service provided by the mill-owner and that the canegrowers no longer were required to contribute towards the capital cost of the bins or their maintenance as they remained the mill-owner's property.[102] Even though the decision applied to the Mourilyan Mill supply area only – a similar application from the Pleystowe Mill supply area was refused – the Central Board had set a precedent: cane bins to receive chopped-up cane were becoming the normal practice in mill supply areas.

Before 1964, most mill-owners had been prepared to modify their transport system to accommodate the introduction of chopped cane. Their attitudes towards this new form of harvested cane, however, had hardened by the commencement of the 1965 crushing season because too much chopped cane had been soured by the *Leuconostoc* bacteria. In August 1965, representatives of the mill-owners appeared before the Central Board and argued that there was a direct link between stale cane, mostly chopped-up cane, and the quality of raw sugar. The Central Board, by a majority decision, gave a Queensland sugar mill the right to impose penalties on burnt cane on a graduated scale according to the length of time between burning and delivery to the mill, and also the right to reject cane delivered more than five days after burning or which was reasonably regarded as unfit for the manufacture of raw sugar. Burnt cane delivered on or before the second day of burning was penalised 1s per ton; this amount rose to 8s per ton on the sixth day after burning. George Mocatta, the canegrowers' representative on the Central Board disagreed with the decision, arguing that the felon in this situation was not burning but delay, and that the answer did not lie in a stepped-up scale of monetary penalties, but improved mill transport arrangements.[103]

Some Queensland canegrowers grew increasingly disgruntled with the new penalties, the unwillingness of some mill-owners to supply cane bins and the continued rental charges on existing cane bins. A deputation representing local cane-grower organisations met John Row, the Queensland Minister for Primary Industries, in Ayr during April 1966 and requested that steps should be taken immediately to give the Central Board the authority to order mill-owners to supply cane bins to canegrowers for the transport of chopped-up cane from the farms to the sugar mills. Row was sympathetic to their plight. In November 1966, his amendment to the *Regulation of Sugar Cane Prices Act* of 1962 to 1965 gave the Central Board additional power to make orders that compelled mill-owners to accept mechanically harvested chopped cane and to supply the cane bins or other containers to receive this chopped-up cane. The Central Board then began hearing applications for cane bin supply, and in most instances, unless the sugar mill was severely hampered by a lack of finance because of its commitments to mill upgrades (see Chapter 17), a suitable arrangement was reached to facilitate the increased use of chopped-up cane. Canegrowers not supplied with cane bins could use whole stalk cane harvesters until they received an allocation of cane bins. CSR was the exception to this arrangement.[104]

During the late 1960s, CSR tried to limit the amount of chopped-up cane in its North Queensland mill supply areas to around 80% of the total harvest by refusing bin applications on the grounds that the deterioration of chopped-up cane affected the quality of sugar produced. Oddly, the Central Board sided with CSR, despite the growing evidence from other Queensland sugar mills (e.g. Mulgrave) that high quality raw sugar could be manufactured, even though increased amounts of chopped-up cane were processed following the implementation of careful transport scheduling (see discussion later in chapter). CSR nominated that the only condition under which it would allow 100% chopper harvesting would be an extension of the traditional daylight harvesting hours into a round-the-clock concept.[105] Night harvesting had been introduced in the Mossman district during 1961 and shown to be possible. A 'crush as you cut' system of twenty-four hour harvesting had also been introduced at Fairymead plantation.[106] During the 1968 crushing season, some CSR suppliers in the Herbert River and Innisfail districts succumbed to the firm's pressure and cane harvesters were equipped with floodlights, ground crew operated with miner's lights on foreheads and sidings were lit with floodlights. This trend, however, alarmed canegrowers elsewhere and the QCGC in early 1969 expressed its strong opposition to this development. An application by the ASPA in June 1969 to extend night harvesting to other mill supply areas, however, was rejected by the Queensland Industrial Commissioner in October 1969.[107]

CSR's Hambledon Mill suppliers had rejected night harvesting, claiming that it was dangerous to field operators and technical information from the QCGC showed that dew on cane cut at night depressed sugar content and that extraneous matter was increased due to the inferior visibility of the operator. Yet canegrowers without cane bin allocations supplying Hambledon and Mulgrave Mills continued pressuring CSR, but with no success. Eventually in mid-1972, seven Hambledon Mill suppliers refused to cut their cane mechanically by whole stalk cane harvesters. CSR thought the strike would last a week, but after three months they called a conference to resolve the matter. Chopper cane harvesters were diverted from harvesting groups which were ahead in their entitlements to cut the cane on the striking suppliers' properties. Subsequently, chopped cane comprised more than 80% of the total supply of cane to Hambledon Mill, and the raw sugar made was of a high standard. The strike was over and the Hon. Mr. Justice Ronald Matthews, Chairman of the Central Board, issued an order at the conclusion of the 1972 crushing season that Hambledon Mill would in 1973 accept a total harvest of chopped cane. During the 1973 crushing season, CSR succumbed to pressure from its suppliers and phased out night harvesting in its other North Queensland mill supply areas.[108]

As mentioned earlier, the use of a three or four tonne capacity cane bins hauled behind a tractor and filled at the side of a mechanical cane harvester had emerged as the main type of infield transport in many mill supply areas by the mid-1960s. This arrangement continued relying upon a mill's tramline network. However, Lou Prince, Mossman Mill's innovative manager, ordered specially designed ten-tonne bulk cane bins during 1967. They became known as canetainers (see Plate 16.9). Built 2 m across and 6 m long,

Plate 16.9. A 9–10 tonne canetainer loaded onto an infield trailer hauled by a tractor, Mossman district, c. 1970.

Source: *ASYB*, 1971, p. 16. Image reproduction courtesy of Rural Press Queensland Pty. Ltd.

the canetainers were filled with billets and were initially used for end tipping, feeding their load straight into the cane carrier. Portable stands were provided to support the bins and to assist in the transfer from field vehicles to semi-trailers for rapid delivery to the sugar mill.

The new canetainers revolutionised cane hauling in the Mossman Mill supply area. Mechanical cane harvesters had fewer halts to change bins. Triangular blocks of concrete were placed at assembly points as permanent stands for the bins to be transferred to and from field vehicles. Semi-trailers fitted with hydraulic transfer gear each brought two empties from the sugar mill, and returned with two laden canetainers. Strengthened bogie underframes were designed, enabling canetainers to be moved to the mill over its tramline, so rail haulage was not phased out entirely, although some branch lines were closed. By 1971, 56% of the harvested cane in the Mossman Mill supply area was moved by canetainers.[109]

A further improvement was made concurrently with the development of canetainers when it was found that they were unsuitable for use in the fields during wet conditions. Elevating side-tipping bins were first introduced throughout the Mossman Mill supply area in 1969, and then were adopted elsewhere during the 1970s, especially as they emerged as a partial solution to the wet-weather harvesting problem for they could be partly loaded to a weight that could be conveniently handled under the wet conditions (Plate 16.10). The first models had similar tyres to the roll-on-roll off trailers, but later models were fitted with large tractor and aircraft tyres that allowed lower ground pressures which improved performance in wet conditions. Compared

Plate 16.10. An elevating side-tipper field bin loading a canetainer, 1975. This in-field cane transport equipment was used to shuttle chopped cane from the cane harvester to canetainers at strategic assembly points.

Source: *ASYB,* 1976, p. 180. Image reproduction courtesy of Rural Press Queensland Pty. Ltd.

to traditional 'piggy-back' haulage equipment, they offered several advantages. The transfer of the billets to the sugar mill's cane bins or canetainers could be completed in a matter of seconds, and the haul-out driver did not have to leave the driving seat. Spillage was negligible. Another innovation associated with infield transport was the development of transloaders. They were permanent fixtures at a tramline siding. Multi-lift prime movers or other tipping trucks were used for the transport of the cane from the field to the transloaders.[110] At the siding the trucks tipped the cane into the transloaders which delivered it into the sugar mill's bins (Plate 16.11).

Canetainers moved on semi-trailers were adopted in two other Queensland mill supply areas, namely Plane Creek (1969) and Rocky Point (1975).[111] However, their use received a major boost following the commencement of mechanical harvesting in New South Wales, where canetainers became known as the Freighter Multi-lift system. First used in 1972, one semi-trailer or multi-lift prime mover began moving twelve canetainers in the

Plate 16.11. Harvested cane being transferred from a multi-lift prime mover to a transloader, which delivered the cane into cane bins destined for Bingera Mill, Bundaberg district, 1976.

Source: *ASYB,* 1976, p. 373. Image reproduction courtesy of Rural Press Queensland Pty. Ltd.

Condong Mill supply area. During 1974, this arrangement was adopted in the other two New South Wales mill supply areas, beginning the replacement of the small trucks with loads of 8–12 tonnes of cane that were being trialed as a possible form of transport. By 1976, 300 canetainers moved by twenty-four semi-trailers handled almost all the sugar cane harvested in New South Wales. CSR bought the canetainers which could hold 20 tonnes of cane, but the semi-trailers were owned by the two contractors who hauled the bins. Under the Multi-lift system, the chopper cane harvesters fed billets into elevating and side-lift in-field transporters which shuttled between the cane harvester and the canetainer. The canetainers sat on the ground at pick up sites located strategically throughout the cane growing areas. When full, they were winched onto the semi-trailers and then delivered to the sugar mills.[112]

The shift to chopped cane and the need to avoid spoilage forced the mill-owners to pay more attention to tramway operations. Prior to 1960, the running of locomotives could be more relaxed, with units stopping at many assembly points to drop off empty rail trucks or to collect loaded rail trucks. Such methods could not continue if the sugar mills were to control cane deterioration. In 1966, CSR introduced manually-produced traffic plans which determined pick-up and delivery schedules of all locomotives over a 24-hour period in an attempt to minimise delays between harvesting and crushing. To assist other Queensland mill-owners, the SRI recruited Rod Murry from The University of Queensland in 1965, and assigned him the task of investigating ways to improve the logistics of the state's cane transport system. Murry began his investigations at Tully Mill in 1966 and concluded that cane transport arrangements were primitive and focused upon demand which was estimated daily by mill traffic officers. Transport surveys were then conducted at eleven different Queensland sugar mills to determine bin turn-around time, the number of locomotives and bins in service and the amount of cane waiting in the mill yard to be processed. These surveys found that a rail truck was loaded about half or less of the time it was in use, actually used for transport for only about four hours every thirty hours and used for cane storage for about thirteen and a half hours every thirty hours with much of this time being spent at the delivery point, not the mill yard.[113]

From its studies, the SRI research team concluded that every Queensland sugar mill operated more transport equipment than was really needed for efficient operations. Contrary to the perceived wisdom that the firm lost money if the factory ran out of sugar cane, the cost of disruption to supply was not as high as the mill managers believed. The researchers worked out a desk system of scheduling cane trains, which involved the coordinated harvesting by geographic groupings of farmers, enabling the locomotives to operate fast, direct services to and from a compact supply area. Several years of research work were required to sort out the desk scheduling system using manual calculations, but as computer use became more widespread, the job became easier. By 1971, Tully Mill, for instance, had eliminated many of the delay times on tramlines so that 80% of its cane had been crushed within ten hours. In addition, the tramline network at some Queensland sugar mills began to alter as a result of transport scheduling. Some small sidings disappeared, with the focus shifting to fewer, but larger collection points.[114] Moreover, better scheduling reduced the demand for so many rail trucks, resulting in their number falling by about 13 000 between 1966 and 1974 (Table 16.5).

Larger volumes of cane to be moved and heavier locomotives travelling at higher speeds required more attention be given to tramway maintenance and improvement. Prior to 1960, tramlines at many Australian sugar mills were generally in poor condition. Little if any ballast was used and the short life of hardwood timber sleepers and light rail resulted in high maintenance costs. During the 1950s, the track provided for some of the new tramline extensions was heavier rail, often 40–45 lbs/yard weight. During the 1960s and 1970s, many Queensland sugar mills began re-railing their tramlines using heavier rail, in some instances laying rail that was 60–62 lbs/yard weight. Queensland Railways was the major source of replacement rail, although Tully Mill obtained used rail from New South Wales and Western Australia. Along with heavier track, better sleepers were introduced. Pre-stressed concrete sleepers commenced being used by sugar mills, namely Victoria and Macknade (by 1960), Tully (1965), South Johnstone (1967) and Proserpine (1973), to list a few.[115] Getting the best from improved track material required the use of proper ballast. During the 1960s, CSR led the way by improving the general standard of its mill tramlines by extensive ballasting with crushed metal. This innovation was soon introduced by other Queensland sugar mills. Mechanical tramway maintenance equipment such as ballast tampers and new hopper wagons for carrying crushed rock ballast also began to be acquired by some Queensland sugar mills. Ballast tampers particularly improved tramway maintenance as these machines could level and tamp 400 sleepers per hour, considerably increasing the results achieved by gangs manually undertaking the task. Programs of bridge and culvert strengthening were undertaken, and low-level wooden bridges were replaced by higher, steel bridges.[116]

By 1971, the geographer Geoffrey Willis found that 82% of harvested cane in Australia was still moved via tramline, although only nine out of the thirty-four sugar mills used this system exclusively for hauling their cane. Sugar mill tramlines had persisted despite suggestions that the narrow gauge tramlines would become obsolete. Personnel Administration, a firm of management consultants engaged by Tully Mill to conduct a major overhaul of its operations between 1958 and 1961, for instance, advised the directors of the Tully Cooperative Sugar Milling Association Ltd. to reduce maintenance on the mill's tramline as it was inevitable that all cane would one day be moved by road haulage.[117] Road transport was used by two-thirds of Australian sugar mills in 1971, although it accounted for the movement of only just on 13% of the harvested cane that year. Considerable variation existed in the level of road transport used at each sugar mill. The Maryborough Sugar Factory and Rocky Point Mill relied entirely on road transport, for as mentioned earlier, neither sugar mill had tramlines. Half the harvested cane delivered to Invicta Mill came by road; a third of the harvested cane received at Condong, Marian and Mossman Mills was moved by road. Interviews with mill officials conducted by Willis discovered that some mill-owners had persisted with the tramlines because of the disadvantages of road transport, despite its cheapness. These disadvantages included the more involved scheduling of cane transport using road haulage, especially as road contractors were outside the direct control of the sugar mill, the erection of expensive sidings for the transfer of road hauled cane and a 40 km limit on the transport of cane by road due to taxation arrangements – farmers living less than 40 km from a mill paid no road tax, while a farmer living 41 km away had to pay 3 cents per ton per km for the total 41 km.[118]

The introduction of chopper cane harvesters posed those remaining Queensland sugar mills still relying on the government railways a major problem – chopped cane deteriorates rapidly and the government railway service was unsuitable for a fast delivery of cane to the sugar mills. In addition, rising Queensland government railway freight charges during the 1950s and early 1960s prompted some mill-owners to investigate alternative transport methods. Pleystowe Mill during 1962 introduced a combined semi-trailer and tramline transportation system to handle a substantial portion of the harvested cane that used to be delivered by the government railway. During 1969, Proserpine Mill eliminated the transport of 85 000 tons of harvested cane by government railway after introducing semi-trailers which carried three, four-tonne bins to delivery points on the sugar mill's tramline. Canegrowers in the Maryborough Sugar Factory supply area who relied on Queensland Railways changed from loading cane at rail sidings to using bins which were moved to transfer stands situated on each farm. The cane was then picked up from these points by a semi-trailer and transferred to the sugar mill, where some of it was tipped directly onto the cane carrier, but the majority was emptied onto an elevator belt at a transfer station and loaded into 6-tonne containers in the mill yard for storage.[119]

As a consequence of the increased usage of road transport to move harvested cane, the volume of cane carried by Queensland Railways fell from 903 581 tonnes in 1959–1960 to 374 694 tonnes in 1971–1972, despite an overall expansion in the amount of cane harvested in Queensland. By 1971, only five Queensland sugar mills still relied on the government railways – Mourilyan, South Johnstone, Racecourse, Cattle Creek and Plane Creek. Each, with the exception of Plane Creek, received less than 15% of their harvested cane by government railways. Plane Creek still received almost a quarter of its cane by the government service, as it came from almost 100 km away and the alternative of road transport was too expensive because of the heavy road tax on cane transported distances in excess of 40 km. During the 1970s, the haulage of cane by Queensland Railways dwindled even further so that only Plane Creek Mill received consignments of harvested cane by this method in 1981.[120]

In 1971, 3% of harvested cane in Australia was still moved by river punt. The phasing out of the transport of cane along rivers in Australia had commenced with the building of a new traffic bridge across the Maroochy River in 1958. This structure eliminated the need for the punting of cane along the Maroochy River, as canegrowers adopted road transport to get their harvested cane to Moreton Mill. This form of cane transport then existed only in New South Wales. Movement of cane to the punts in New South Wales had become faster during the late 1940s and early 1950s following the adoption of trailers mounted on a pair of motor lorry wheels. One trailer pulled by a tractor carried up to 1.5–1.8 tonnes of cane, equivalent to two or three horse-drawn wagon loads. Loading of cane onto the punts also became quicker and easier during the late 1950s following the introduction of power winches as a means of transferring mechanically loaded cane from tractor-trailers to the punts.[121]

These developments, however, could not stop the inevitable demise of river transport. It began to vanish in 1962 when Condong Mill instituted a system of haulage by tandem-trailers (which persisted until all units were finally replaced by the Freighter Multi-Lift system 1981). CSR assisted the process by partially funding the erection of a road bridge across the Tweed River. Following the road bridge's opening in 1966, Condong Mill received all its cane by land transport during 1967. Road haulage of cane to Broadwater and Harwood Mills began replacing punting during the mid-1970s following the introduction of mechanical harvesting. Punt deliveries ceased on the Richmond River at the conclusion of the 1974 season and the final punt laden with cane made its journey along the Clarence River on 5 January 1976.[122]

Searching for cane harvester and transport efficiencies, 1975 to 1995

Mechanical cane harvester development since 1975 favoured the production of higher capacity machines designed to cut crops under a wide variety of conditions. In the mid-1980s, the Mizzi family, for example, had produced a massive two-row cane harvester that could cut both green and burnt cane at accelerated rates. During the 1980s, high flotation rubber tyres were added to the mechanical cane harvesters and transporters in the sugar-producing districts in an attempt to reduce soil compaction and deep rutting in paddocks that occurred when harvesting was undertaken during wet weather. In areas receiving very high rainfall, newer models of cane harvesters such as the

Austoft 7700 were equipped with half or full tracks.[123] By the early 1990s, cane harvesters could achieve pour rates (i.e. a measure of the machine's capacity in tonnes cut per hour) of 140 tonnes per hour in burnt cane. In contrast, pour rates were only 35–50 tonnes per hour in the early 1970s. Green cane pour rates, however, still ranged from 50 to 80% of burnt cane rates depending on crop size, as engineers struggled to develop machines that could undertake green cane harvesting at rates achieved in burnt cane. Overall, these innovations meant that a single high-capacity Austoft 7000 or Austoft 7700 tracked version could easily harvest 50 000 tonnes of cane a year.[124]

Accompanying these machinery developments was the slow decline in cane harvester numbers in Australia after their number peaked in Queensland in 1971 (Table 16.6). Lack of data for New South Wales precluded any assessment of the

Table 16.6. The number of mechanical cane harvesters in Queensland and New South Wales, 1971–1995.

Year	QLD	NSW	Year	QLD	NSW
1970	1 478	*	1983	1 709	#
1971	1 899	*	1984	1 706	#
1972	1 852	*	1985	1 724	#
1973	1 794	*	1986	1 676	#
1974	1 728	*	1987	1 593	#
1975	1 719	#	1988	1 502	37
1976	1 775	#	1989	1 452	36
1977	1 791	#	1990	1 350	34
1978	1 780	#	1991	1 355	36
1979	1 760	#	1992	1 287	36
1980	1 728	#	1993	1 266	27
1981	1 717	#	1994	1 238	27
1982	1 699	#	1995	1 200	27

* Mechanical harvesting just being introduced so numbers were probably less than thirty.

Data not available.

Source: Based upon figures supplied by the Australian Sugar Milling Council and P. Connell and B. Borrell, *Costs and Regulation of Cane Harvesting Practices* (Canberra, 1987), p. 12.

overall trend in that state, but it was probable that the number of cane harvesters declined throughout the mid-1980s, after peaking sometime c. 1980, following the introduction of mechanical harvesting in New South Wales during the mid-1970s. New South Wales figures became available for the late 1980s and suggest that the number of cane harvesters had stabilised around 35–37, until a rationalisation in the early 1990s. Most of this decline in cane harvester numbers occurred as a result of structural adjustments in the harvesting sector caused by amalgamations in harvesting group arrangements. Consequently, the amount harvested per group in each Australian sugar-producing district had gradually increased. All districts followed a similar trend, which was characterised by rapid growth over the periods 1970–1975 and 1985–1995, except the Southern Queensland region over the latter period (Table 16.7). Depressed sugar prices, rising equipment costs and the need to achieve economies of scale contributed to the growth in these periods, as individual contractors either exited the industry or sought to harvest larger amounts of cane. The average amount of cane harvested in Central and Southern Queensland remained considerably smaller than in the Lower Burdekin, North Queensland and New South Wales regions, a consequence of the persistence of numerous cane harvesters in these areas (see discussion later in chapter). The large increase in the amount harvested per

Table 16.7. The average amount harvested (tonnes) per cane harvester group in Australian sugar-producing regions, 1970–1995.

Year	NSW	Southern QLD	Central QLD	Lower Burdekin	North QLD	AUST.
1970	*	6 522	6 137	12 582	11 873	9 278.5
1975	18 535	9 353	9 882	23 149	14 780	15 139.8
1980	31 465	10 649	9 814	25 176	15 742	16 652.2
1985	31 338	12 112	8 923	26 808	16 679	19 172.0
1990	37 359	13 895	12 774	35 827	23 566	24 684.2
1995	65 454	19 478	20 820	49 896	39 589	39 047.4

Source: Based upon figures in Table 3.1 in L. Brennan, 'A systems analysis of cane harvesting and transport in the Mackay sugar industry'. PhD thesis, University of Queensland, 1998, p. 33.

group in Central Queensland between 1990 and 1995 was largely attributable to the rapid growth of group size in the Proserpine Mill supply area.[125]

In Australia by the mid-1970s, a new mechanical cane harvester with ancillary equipment such as trailers, haul-out tractors and maintenance equipment cost from $25 000 to $40 000, depending on the harvester type and its capacity. Therefore, substantial tonnages of sugar cane needed to be harvested per machine or it was very costly to the owner if the machine was idle or not allotted appropriate tonnages to harvest. Churchward and Belcher suggested in 1972 that a harvest of 10 000 to 15 000 tonnes per machine per season was required in Australia to cover costs, including depreciation on the smaller machines, whilst the more elaborate high-output machines should have been allotted 15 000 to 20 000 tonnes per season to cover costs. They concluded that there was 'considerable waste of capital and capacity' in the harvesting sector, an observation supported by the figures in Table 16.7. Moreover, the capacity of new cane harvester models operating on a single shift increased from about 20 000 tonnes per season in 1974 to 45 000 tonnes in 1982.[126] Therefore, it was not surprising that the Sugar Industry Working Party of 1985, chaired by Russell Savage, concluded that the costs for harvesting sugar cane and transporting it to Australian sugar mills could be reduced by larger harvester groups. Five years later, another Sugar Industry Working Party, chaired by E. Fitzpatrick, recommended that harvest groups of 50 000 tonnes or more in size were needed to be cut on a two shift basis in order to give savings on rollingstock to Australia's sugar mill-owners and economies of scale to the cane harvesters.[127] In 1995, only the New South Wales and Lower Burdekin cane harvester groups had reached this figure. In Southern and Central Queensland, cane harvester groups were less than half the recommended size (Table 16.7). Consequently, despite the trend towards fewer cane harvesters and larger harvester groups, the capacity of many mechanical cane harvesters in Australia had never been fully utilised between 1975 and 1995.

Various reasons have been suggested to explain the persistence of small harvest groups in the Australian cane harvesting sector between 1975 and 1995. A study by the QCGC in 1984 found that many canegrowers in Queensland continued harvesting with their own small cane harvesters because they wished to remain independent of contractors or other canegrowers. Some were forced to be independent when the farm was located in an isolated area where no contractor was available. Other canegrowers sought to maintain the capacity to harvest small tonnages if intermittent wet weather occurred during the harvest season. Peter Connell and Brent Borrell, economists with the Australian Bureau of Agricultural and Resource Economics, suggested that some Australian canegrowers persisted with cutting their own cane because they would otherwise consider themselves under-employed, the switch to minimum tillage reducing the amount of time needed on cultivation tasks. Connell and Borrell also suggested that generous taxation advantages resulting from accelerated depreciation and investment allowances on farm machinery, together with access to low cost capital before deregulation of Australia's financial and foreign exchange markets during the late 1980s, meant that the annual cost of capital to Australian canegrowers was less than the cost to the economy as a whole – basically Australian canegrowers could afford to own or have a part interest in a cane harvester because it was cheaper, relative to that of other inputs. In the mid-1990s, the Boston Consulting Group concluded that local board awards still determined the arrangements for harvesting, thereby contributing to the persistence of so many small harvest groups. Equity tended to be maintained between harvest groups because of the existence of an Equity Committee, consisting of representatives of mill-owners and canegrowers.[128]

After twenty years of amalgamations, independent contractor and grower-contractor had emerged as the most common categories of cane harvester ownership in Australia by 1995 (Table 16.8). 'Rising equipment costs, together with the potential to achieve economies of scale, have encouraged this type of harvesting', wrote Lisa Brennan. For grower-contractors, contracting provided an economic incentive to enlarge their seasonal business and so become large enough to justify the investment in harvesting equipment – an Austoft 7000 cost $280 000 in the early 1990s.[129] Grower-contractors were most popular in North Queensland where just over 60% of the crop was removed by this form of ownership category. Independent contractors were used most widely throughout the Lower Burdekin, where they harvested almost three-quarters of the sugar cane in 1995. In contrast, a third of the cane harvest groups in Central Queensland were still canegrowers cutting their own cane, although collectively they harvested only a small proportion (13.6%) of the region's crop (Table 16.8). This arrangement in Central Queensland duplicated pre-mechanical harvesting group structures when canegrowers relied on one or two canecutters to harvest their cane, instead of the larger gangs used in more northern sugar-producing districts. Central Queensland canegrowers were more involved in the loading and harvesting of their cane.[130]

A different cane harvester ownership pattern had emerged in New South Wales by 1995. At the outset of mechanical harvesting in New South Wales, canegrowers formed cooperatives to operate the machines. 'From the Queensland experience we could see that contract harvesting would get dearer and that contractors would be reluctant to invest in new machines', claimed Vince

Table 16.8. Categories of mechanical cane harvester ownership in Australia, 1995 season.

	Grower cut own cane		Grower & contractor		Coopera-tive		Cooper-ative & contractor		Indepen-dent con-tractor	
District	%*	%**	%*	%**	%*	%**	%*	%**	%*	%**
North QLD	12.5	5.6	52.1	60.4	3.1	2.4	8.3	2.4	24	29.1
Lower Burdekin	2.9	0.4	25.7	25.6	0	0	0	0	71.4	74
Central QLD	33.5	13.6	35.7	32.5	9.7	11.7	1.5	11.7	19.5	30.5
Southern QLD	#	#	#	#	#	#	#	#	#	#
NSW	0	0	0	0	100	100	0	0	0	0

* % of total groups ** % of total crop # Data not available

Source: L. Brennan *et al.*, 'Improving sugarcane harvesting procedures', in B. Keating and J. Wilson (eds.), *Intensive Sugarcane Production. Meeting the Challenges Beyond 2000* (Wallingford, UK, 1997), p. 329.

Castle in 1993. Castle, a Harwood Mill supplier who had been Chairman of the Palmers Island Harvesting Cooperative since 1978, claimed that the local cane-growers felt that they had enough mechanical expertise to operate the machines successfully. Moreover, these cooperatives were willingly merged to achieve economies of scale. Three harvesting cooperatives working in the Harwood Mill supply area, for example, amalgamated their operations in 1984, forming a single harvesting unit that aimed to cut about 200 000 tonnes of members' cane. By 1993, the expanded Bushgrove/Lawrence Canegrowers' Cooperative Society Ltd. operated four Toft 6500 full track cane harvesters, twelve transporters and four mobile workshops to provide service to the units in the field. It supplied a quarter of all the cane processed at Harwood Mill.[131] Moreover, individually-owned cane harvesters in New South Wales had vanished by 1995, with all the sugar cane being harvested by cooperatively-operated machines (Table 16.8).

As the harvester sector rationalised the number of operational machines each year, the mill-owners faced rising road transport costs, a consequence of increased oil prices in the mid-1970s. This trend prompted another expansion in the length of mill tramlines during the 1980s. Several examples will illustrate this response. In the Invicta Mill supply area, a 46 km tramline extension was completed between Clare and Dalbeg in early 1980, eliminating the costly method of hauling sugar cane by road from the farms at Dalbeg and Millaroo to the former terminus of the tramline at Clare. The project was an initiative of the canegrowers who formed the Upper Burdekin Cooperative Association Ltd. The borrowed funds for the project were to be repaid by levies on the canegrowers and by renting the line to the mill which maintained and operated the facility. In contrast, the Proserpine Cooperative Sugar Milling Association Ltd. funded the building of a 30 km tramline extension to the Elaroo locality during 1981 and 1982. In dollar terms it cost Proserpine Mill 31% more to shift 115 000 tonnes of sugar cane from the Elaroo area by road than it did to shift 835 000 tonnes by tramline from all other localities during the 1980 season.[132]

As new areas were cultivated with sugar cane during the late 1980s and early 1990s, the mill-owners had to decide if the harvested cane would be transported by rail or road. The Mossman Central Mill Co. Ltd. concluded that the only feasible way to get the harvested cane from the new assignments on the Atherton Tableland was to move it down the range by road to an assembly point where the cane bins could be loaded onto the mill's tramlines. Road haulage was also used to move the harvested cane from the Murray Upper locality to Mourilyan Mill, the Bundaberg Sugar Co. using four company-owned prime moversfitted with tipping cages.[133] In contrast, cane grown in the Murray Upper locality destined for Tully Mill was moved over a 10.4 km tramline built as part of the Murray Valley Infrastructure/Riversdale Water Management Project, one of the projects funded by the Sugar Infrastructure Package of 1993.

CSR, however, was faced with a different transport problem at its Plane Creek Mill. Existing canegrowers south of Koumala continued relying on the Queensland government railway to deliver their harvested sugar cane to Plane Creek Mill. A successful application under the Sugar Infrastructure Package of 1993 enabled $21 million to be spent on building the 44 km Plane Creek Southern Cane Railway Extension during 1994 and 1995. Its completion meant the end of transporting sugar cane on government railways in Queensland, a practice that had commenced almost 110 years earlier.[134] Moreover, it reinforced the importance of the narrow gauge tramlines in 1995 as the main cane transport method in twenty-three of the twenty-eight Australian sugar mill supply areas; road transport was used most extensively in only five mill supply areas, namely Maryborough and Rocky Point and the three New South Wales sugar mills.

In retrospect, the harvesting and transport of sugar cane in Australia between 1915 and 1995 occurred over two distinct phases. The period 1915–1950 was characterised by manual cutting and loading of sugar cane: attempts to harvest sugar cane during the pre-World War II years by machines failed. Harvested sugar cane was mostly moved to the sugar mills in rail trucks with stake ends for whole stalk cane hauled over mill-owned narrow gauge tramlines. In some mill supply areas, the sugar cane was also moved to the factory on government railways, by motor lorries or punts (mostly in New South Wales). Arrangements for the harvesting and transport of sugar cane altered after 1950. Persistent post-World War II labour shortages led to the eventual mechanisation of the loading and then harvesting of sugar cane between 1950 and 1975. This profound technological change radically altered many sugar industry practices that had persisted in some instances for seventy years – provision of cane barracks, annual sign-ons, shed design and the shape of fields. The transport of harvested sugar cane to the sugar mills also altered. Punting ended, the amount of sugar cane moved by government railways fell substantially and the rail trucks with stake ends vanished, their place taken by box-like cane bins or canetainers. Change did not cease when the harvesting cycle had finally been mechanised, but persisted during the 1980s and early 1990s, as the number of cane harvesters were rationalised as their capacity grew, green cane harvesting made a comeback and the last of the sugar cane transported over a government railway in Queensland was finally shifted via a mill-owned narrow gauge tramline. All these radical technological innovations, however, were not occurring in isolation. At the sugar mills, as will be seen in the next chapter, the installation of new machinery – often labour-saving machinery – and process technologies were also transforming this sector of the Australian sugar industry.

Notes

1 R. Shlomowitz, 'Team work and incentives: the origins and development of the butty gang system in Queensland's sugar industry, 1891–1913', *J. of Comparative Economics*, Vol. 3 (1979), pp. 47–48; C. Morton, 'Dry cleaned cane is not clean enough', *North Australian Monthly*, September 1958, p. 49.

2 D. Conomos, *The Greeks in Queensland. A History from 1859–1945* (Brisbane, 2002), p. 334; D. Menghetti, 'Canecutters', in B. Gammage and P. Spearritt (eds.), *Australians 1938* (Sydney, 1987), p. 274; B. Balanzategui, *Gentlemen of the Flashing Blade*. Studies in North Queensland History No. 15 (Townsville, 1990), pp. 48–50.

3 Conomos, *The Greeks in Queensland*, pp. 334–335.

4 Easterby, *The Queensland Sugar Industry*, p. 43; *ASJ*, 9 September 1915, p. 462, 4 November 1915, p. 553 & 10 December 1937, p. 539; B. Scott, *Tough in the Old Days* (Adelaide, 1979), p. 113.

5 Balanzategui, *Gentlemen of the Flashing Blade*, pp. 26–28; D. Formiatti, 'Sugar Cane Barracks of Far North Queensland'. B. Architecture thesis, Uni. of Qld., 1984, pp. 44–66.

6 Menghetti, 'Canecutters', pp. 273–278; Balanzategui, *Gentlemen of the Flashing Blade*, pp. 58–85.

7 Report of the Piddington Royal Commission, *CPP*, Vol. 4 (1920–21), p. 916.

8 Shogren, 'The politics and administration of the Queensland sugar industry to 1930', pp. 272–273.

9 *ASJ*, 7 September 1916, p. 469.

10 Easterby, *The Queensland Sugar Industry*, p. 53; D. Murphy, *T.J. Ryan. A Political Biography* (Brisbane, 1975), pp. 179, 202 & 232–233; Connolly, *John Drysdale and the Burdekin*, p. 155.

11 Connolly, *John Drysdale and the Burdekin*, p. 161; *BC*, 31 May 1917, p. 6; Murphy, *T.J. Ryan*, pp. 253–254; Shogren, 'The politics and administration of the Queensland sugar industry to 1930', p. 292.

12 Report of the Piddington Royal Commission, *CPP*, Vol. 4 (1920–21), pp. 916–917; Conomos, The Greeks in Queensland, pp. 331 & 334.

13 K. Saunders, 'The workers' paradox: indentured labour in the Queensland sugar industry to 1920', in K. Saunders (ed.), *Indentured Labour in the British Empire* (Canberra,

1984), pp. 244–245; Condong Mill Cane Insp., AR 1919, p. 3; Condong Mill Cane Insp., AR 1920, p. 6; Condong Mill Cane Insp., AR 1921, p. 4, all in CSRR, Z/109, Box 462.

14 Douglass, *From Italy to Ingham*, pp. 147–148; *ASJ*, 9 April 1925, p. 58 & 8 May 1925, p. 66.

15 Douglass, *From Italy to Ingham*, pp. 148–155; Hudson, *Sweet Success*, pp. 78–104; K. Kennedy, 'The South Johnstone strike, 1927', in D. Murphy (ed.), *The Big Strikes. Queensland 1889–1965* (Brisbane, 1983), pp. 174–185.

16 *ASJ*, 5 June 1930, p. 147 & 3 July 1930, p. 251–252; L. Henderson, 'Italians in the Hinchinbrook Shire, 1921–1939'. B.A. Hons thesis, James Cook Uni., 1978, pp. 251–253.

17 Douglass, *From Italy to Ingham*, pp. 174–186; Hambledon Mill Cane Insp., AR 1931, p. 10, CSRR, Z/438, Box 38; Henderson, 'Italians in the Hinchinbrook Shire, 1921–1939', pp. 33–34.

18 Condong Mill Cane Insp., AR 1931, p. 8, CSRR, Z/109, Box 462; B. Bourke, 'Cane harvesting and transport', *Proc. QSSCT*, 1931, p. 41; *QAJ*, Vol. 30, no. 9 (September 1928), p. 198; Ev. A. Morris, Mang., Mt Bauple Cooperative Sugar Milling Assoc., in 'Royal Commission on Sugar Peaks and Cognate Matters Evidence, 1939', p. 193, CSRR, Z/109, Box 48.

19 B. Kerr and K. Blyth, *They're All Half Crazy. 100 Years of Mechanical Cane Harvesting* (Brisbane, 1993), pp. 22–23; Burrows and Morton, *The Canecutters*, pp. 59–60.

20 Kerr and Blyth, *They're All Half Crazy*, pp. 30–33; Burrows and Morton, *The Canecutters*, pp. 64–69; Smith, *Sweet Beginnings*, pp. 376–377.

21 C. Young, 'The development of the Fairymead cane harvester', *Proc. ISSCT*, 1950, pp. 304–305.

22 *ASJ*, 6 August 1931, pp. 265 & 281; Kerr and Blyth, *They're All Half Crazy*, pp. 25–27; G. Burrows and R. Shlomowitz, 'The lag in the mechanization of the sugarcane harvest: some comparative perspectives', *Agricultural History*, Vol. 66, no. 3 (Summer 1992), p. 69.

23 *ASJ*, 6 March 1913, p.829; Morton, *By Strong Arms*, p. 95; B. Bourke, 'Cane harvesting and transport', *Proc. QSSCT*, 1931, p. 42.

24 Mulgrave Central Mill Co., AR 1926, Mitchell Library.

25 CSR, Cane Report 1915, p. 24, CSRR, Z/109, Box 378; C. Morton, *By Strong Arms*, p. 84; *QAJ*, Vol. 28, no. 6 (1 December 1927), p. 573; R. Harman, 'Manufacture of raw sugar – Australian practice', *Proc. ISSCT*, 1935, p. 45; BCSM, ARs, 1920 to 1928, in *QPP*, various volumes.

26 CSR, Cane Report 1915, p. 24, CSRR, Z/109, Box 378; CSR, Cane Report 1919, p. 29; CSR, Cane Report 1924, p. 29; CSR, Cane Report 1926, p. 21; CSR, Cane Report 1928, p. 32; all in CSRR, Z/109, Box 386.

27 D. Menghetti, 'The Weil's disease strike, 1935', in D. Murphy (ed.), *The Big Strikes. Queensland 1889–1965* (Brisbane, 1983), pp. 203–206; B. Penrose, 'Medical experts and occupational illness: Weil's disease in North Queensland, 1933–1936', *Labour History*, No. 75 (November 1998), p. 129.

28 *ASJ*, 4 September 1934, p. 325 & 6 November 1934, pp. 432–433 & 443; Penrose, 'Medical experts and occupational illness: Weil's disease in North Queensland, 1933–1936', pp. 130–131.

29 Penrose, 'Medical experts and occupational illness: Weil's disease in North Queensland, 1933–1936', p. 133; Menghetti, 'The Weil's Disease Strike, 1935', pp. 207–214; CSR, Cane Report 1935, p. 42, CSRR, Z/109, Box 387; *ASJ*, 10 October 1935, p. 389 & 395–397.

30 Queensland Director-General for Health & Medical Services (hereafter QDHMS), AR 1936–37, *QPP*, Vol. 2 (1937), p. 1063; QDHMS, AR 1937–38, *QPP*, Vol. 2 (1937), p. 939; QDHMS, AR 1938–39, *QPP*, Vol. 2 (1937), pp. 1166–1167; *ASJ*, 10 November 1936, pp. 459.

31 Douglass, *From Italy to Ingham*, pp. 242–245.

32 Douglass, *From Italy to Ingham*, pp. 245–246; *ASJ*, 13 October 1942, pp. 248–249.

33 *ASJ*, 14 September 1942, p. 226; Macknade Mill Cane Insp., AR 1942, pp. 14–15, CSRR, Z/109, Box 472; BSES, AR 1942–43, p. 3; A. Mansfield *et al.*, *Report of the Royal Commission Appointed to Investigate Certain Aspects of the Sugar Industry* (Brisbane, 1943), p. 3.

34 BSES, AR 1942–43, p. 4; A. Mansfield *et al.*, *Report of the Royal Commission Appointed to Investigate Certain Aspects of the Sugar Industry*, p. 3.

35 Kerr and Blyth, *They're All Half Crazy*, p. 44; Mansfield *et al.*, *Report of the Royal Commission Appointed to Investigate Certain Aspects of the Sugar Industry*, pp. 11–12; for details about the Toft loader see *ASJ*, 10 August 1940, p. 293 & 12 September 1940, pp. 327–328.

36 Mansfield *et al.*, *Report of the Royal Commission Appointed to Investigate Certain Aspects of the Sugar Industry*, pp. 11–12.

37 *ASJ*, 12 August 1943, p. 163, 15 June 1945, p. 109, 15 September 1945, p. 257, 15 October 1945, p. 309 & 15 November 1945, p. 349; Hudson, *Sweet Success*, p. 154; Macknade Mill Cane Insp., AR 1943, p. 7, CSRR, Z/109, Box 472; 'Manpower Report', in Tariff Board Report 1944, Attachment B, p. 102, CSRR, Z/109, Box 52.

38 *ASJ*, 13 October 1942, p. 247; 'Manpower Report' in Tariff Board Report 1944, Attachment B, pp. 105–106, CSRR, Z/109, Box 52; Manning, *In their own hands*, p. 279; BSES, AR 1943–44, p. 3.

39 Manning, *In their own hands*, pp. 279–280; Balanzategui, *Gentlemen of the Flashing Blade*, p. 68; see CSR, Cane Reports 1939 to 1945, in CSRR, Z/109, Box 388.

40 Hudson, *Sweet Success*, p. 155; BSES, AR 1945–46, p. 3; BSES, AR 1946–47, p. 8; BSES, AR 1947–48, p. 9.

41 Mulgrave Central Mill Co. Ltd., AR 1947; Macknade Mill Cane Insp., AR 1946, p. 6 & Macknade Mill Cane Insp., AR 1947, p. 7, both in CSRR, Z/109, Box 472; Victoria Mill Cane Insp., AR 1947, p. 13, CSRR, Z/109, Box 474; Hambledon Mill Cane Insp., AR 1947, p. 8, CSRR, Z/438, Box 38.

42 Balazategui, *Gentleman of the Flashing Blade*, p. xiii. Douglass, *From Italy to Ingham*, pp. 261–264; Burrows and Morton, *The Canecutters*, pp. 110; BSES, AR 1947–48, p. 9.

43 Burrows and Morton, *The Canecutters*, pp. 111–112; *ASJ*, 15 September 1948, p. 37; BSES, AR 1948–49, p. 10; Balazategui, *Gentleman of the Flashing Blade* , p. 21.

44 Balazategui, *Gentleman of the Flashing Blade*, pp. xiv; Burrows and Morton, *The Canecutters*, p. 120; for the performance of the gangs of displaced persons see *ASJ*, 15 June 1948, pp. 171–172 & 11 July 1948, pp. 233–234.

45 Douglass, *From Italy to Ingham*, pp. 264 & 277–279; Burrows and Morton, *The Canecutters*, pp. 140–141; *ASYB*, 1956, p. 123.

46 Burrows and Morton, *The Canecutters*, pp. 141–143; Douglass, *From Italy to Ingham*, pp. 283–284.

47 Burrows and Morton, *The Canecutters*, pp. 143–144; Douglass, *From Italy to Ingham*, pp. 286–288.

48 Douglass, *From Italy to Ingham*, pp. 288–290; Victoria Mill Cane Insp., AR 1960, p. 11; CSRR, Z/109, Box 475.

49 Kerr and Blyth, *They're All Half Crazy*, pp. 36 & 40– 41; *ASJ*, 13 July 1944, pp. 141–142 & 14 December 1944, pp. 306–307.

50 Hambledon Mill Cane Insp., AR 1945, p. 3, CSRR, Z/438, Box 38; Macknade Mill Cane Insp., AR 1945, p. 15, CSRR, Z/109, Box 472; Victoria Mill Cane Insp., AR, 1945, pp. 16–17, CSRR, Z/109, Box 474; Kerr and Blyth, *They're All Half Crazy*, p. 42; *ASJ*, 15 September 1945, p. 263 & 15 October 1945, pp. 302 & 306.

51 Kerr and Blyth, *They're All Half Crazy*, pp. 34–37 & 52–53; *ASJ*, September 1964, p. 415; S. Toft, *Report of Queensland Canegrowers' Council Delegation to USA, Cuba and Hawaii, Nov. 1945–Jan. 1946* (Brisbane, 1946), p. 49.

52 Victoria Mill Cane Insp., AR 1947, p. 16 & Victoria Mill Cane Insp., AR 1948, p. 18, both in CSRR, Z/109, Box 474; Harwood Mill Cane Insp., AR 1946, p. 7 & Harwood Mill Cane Insp., AR 1947, p. 8, both in CSRR, Z/109, Box 470; Smith, *Sweet Beginnings*, p. 354.

53 Burrows and Morton, *The Canecutters*, p. 146; *ASJ*, 26 June 1961, pp. 205–206.

54 President's AR 1952–53, p. 2, Richmond River Branch of the NSWCGA, Wardell; Clarence River Branch NSWCGA, AR 1952–53, in Minute Book, Clarence River Branch NSWCGA, 1951–1965, Maclean; *The Producer's Review*, 15 February 1955, p. 85; *ASYB*, 1956, p. 109.

55 *ASJ*, 26 June 1961, p. 206; Burrows and Morton, *The Canecutters*, p. 146.

56 *ASJ*, 26 June 1961, p. 206; Burrows and Morton, *The Canecutters*, pp. 145 & 150; *The Producers' Review*, 15 October 1956, p. 46 & 15 September 1957, p. 40.

57 *ASJ*, 26 June 1961, pp. 208–210; Goondi Mill Cane Insp., AR 1960, p. 11, CSRR, Z/109, Box 464; Burrows and Morton, *The Canecutters*, p. 147.

58 Numbers of Toft or front-end loaders in the Broadwater Mill and Condong Mill supply areas were listed yearly in the Cane Inspectors' ARs, CSRR, Z/109, Boxes 460–462 & 470–471; see also CSR, AR Cane Season 1967 Part 1, p. 35, CSRR, Z/109, Box 390.

59 President's AR, 1956–57, pp. 2 & 4, President's AR, 1960–61, p. 4, President's AR,1962–63, p. 5, Richmond River Branch of the NSWCGA, Wardell; Burrows and Morton, *The Canecutters*, p. 149.

60 Burrows and Morton, *The Canecutters*, pp. 145–146; Victoria Mill Cane Insp., AR 1954, pp. 14–15, CSRR, Z/109, Box 475; C. Story, 'Mechanical loading in the Mackay district, 1955', *Proc. QSSCT*, 1956, pp. 34–36.

61 CSR, AR 1948–49, p. 2, CSRR, 142/3530; Hambledon Mill Cane Insp., AR 1957, p. 1, CSRR, Z/109, Box 468; NSWCGA, Minute Book, 1950–1970, entry for meeting of 23–24 April 1956, Wardell; *ASJ*, 19 August 1957, p. 355.

62 CSR Ltd., *Clean cane and the work of cane inspectors* (Sydney, 1959), pp. 14–16; CSR, Raw Sugar Manufacture Report 1957, pp. 115–119, CSRR, Z/109, Box 382.

63 *The Producers' Review*, April 1971, pp. 56–60.

64 Kerr and Blyth, *They're All Half Crazy*, pp. 38–39; *ASJ*, 15 August 1949, p. 313; *ASYB*, 1956, pp. 125–126.

65 Kerr and Blyth, *They're All Half Crazy*, pp. 47–51; BSES, AR 1948–49, p. 10; Kerr, *Pioneer Pageant*, p. 202; *ASYB*, 1956, pp. 127–128; King, Mungomery and Hughes, *Manual of Cane-Growing*, 1953 edition, pp. 194–195.

66 Kerr and Blyth, *They're All Half Crazy*, pp. 40–45; *ASJ*, 15 November 1948, p. 505.

67 Kerr and Blyth, *They're All Half Crazy*, p. 60.

68 BSES, AR 1950–51, p. 11; Hungerford, 'The Bundaberg district sugar industry 1945 to 1995', p. 44; C. Young, 'The development of the Fairymead cane harvester', *Proc. ISSCT*, 1950, p. 306.

69 *ASJ*, 15 October 1948, p. 426; Kerr and Blyth, *They're All Half Crazy*, p. 55; Manning, *In their own hands*, p.290; C. Story, 'Mechanical harvesting and co-operative ownership', *Proc. QSSCT*, 1955, pp. 141–145.

70 Kerr and Blyth, *They're All Half Crazy*, p. 48.

71 Kerr and Blyth, *They're All Half Crazy*, pp. 57–58, 62–64; Hungerford, 'The Bundaberg district sugar industry 1945 to 1995', p. 46; Burrows and Morton, *The Canecutters*, pp. 174–176; *ASYB*, 1975, pp. 193–194.

72 J.K. Gaunt, 'Development of the Massey-Ferguson system of mechanically harvesting sugar cane', *Proc. QSSCT*, 1964, pp. 27–36; Burrows and Morton, *The Canecutters*, pp. 152–169 & 172; *ASJ*, November 1958, pp. 669–671 & October 1974, p. 345.

73 Hungerford, 'The Bundaberg district sugar industry 1945 to 1995', pp. 45 & 47; Burrows and Morton, *The Canecutters*, p. 172; *ASYB*, 1972, p. 161.

74 Kerr, *Southern Sugar Saga*, p. 133.

75 Kerr and Blyth, *They're All Half Crazy*, p. 110; Burrows and Morton, *The Canecutters*, p. 224.

76 President of Richmond River Branch NSWCGA, AR 1965–66, Wardell; Smith, *Sweet Beginnings*, p. 377–378; CSR, AR Cane Season 1965. Part 1, p. 34 & CSR, AR Cane Season 1966. Part 1, pp. 34–35, both in CSRR, Z/109, Box 390.

77 CSR, AR 1968–69, p. 16; Minutes of a Special Meeting of NSWCGA, 20 March 1969, in NSWCGA Minute Book, 1950–1970, Wardell; President of Richmond River Branch NSWCGA, AR, 1976–77, Wardell; Smith, *Sweet Beginnings*, pp. 376–383; Burrows and Morton, *The Canecutters*, p. 202.

78 *ASYB*, 1972, p. 165; Burrows and Morton, *The Canecutters*, pp. 180 & 224; E. Churchward and R. Belcher, 'Some economic aspects of mechanical harvesting in Queensland', *Proc. QSSCT*, 1972, p. 32.

79 King, Mungomery and Hughes, *Manual of Cane-Growing*, 1965 revised edition, p. 178.

80 B. Egan and C. Rehbein, 'Bacterial deterioration of mechanically harvested cut-up sugar cane during storage over weekends', *Proc. QSSCT*, 1963, pp. 11–23; B. Egan, 'Investigations into the sour storage rot problem', *Proc. QSSCT*, 1964, pp. 15–25; B. Egan, 'Post-harvest deterioration losses in sugar cane in Queensland', *Proc. ISSCT*, 1968, pp. 1729–1735; King, Mungomery and Hughes, *Manual of Cane-Growing*, 1965 revised edition, p. 178; *ASJ*, February 1968, p. 639.

81 See and Crouch, 'Mechanisation of sugar cane harvesting', p. 43.

82 Department of Labour and National Service, *Men and Machines in Sugar Cane Harvesting*, pp. 23–24; *ASYB*, 1971, pp. 19–20.

83 J. Buzacott, 'New look implement sheds', *CGQB*, Vol. 25, no. 2 (1 October 1961), p. 65; J. Usher, 'A home-made stone picker', *CGQB*, Vol. 36, no. 4 (1April 1973),

pp. 140–141; *ASYB*, 1963, p. 127; Goondi Mill Cane Insp., AR 1973, p. 35, CSRR, Z/109, Box 478.

84 Department of Labour and National Service, *Men and Machines in Sugar Cane Harvesting*, pp. 31–34 & 44; *ASYB*, 1972, p. 125.

85 Hudson, *By the Banyan*, p. 145; for a collection of stories about the fate of some North Queensland cane barracks see E. Navarre, *The Cane Barracks Story* (Cairns, 2007).

86 For a full account of the union's struggles see Burrows and Morton, *The Canecutters*, pp. 132–134 & Hungerford, 'The Bundaberg District Sugar Industry 1945–1985', pp. 47–49.

87 Kerr, *Only Room For One*, p. 105; Kerr, *A Crystal Century*, p. 118; W. Scott, *Tough in the old days* (Adelaide, 1979), pp. 135–136.

88 Kerr, *Sugar at Maryborough*, p. 185; CSR, *Newsletter* No. 64, April 1961, p. 4, CSRR, Z/109, Box 371.

89 Harwood Mill Cane Insp., AR 1938, p. 3, CSRR, Z/109, Box 470; G. Willis, *Harvesting and Transport of Sugar Cane in Australia* (Townsville, 1972), p. 47; Burrows and Morton, *The Canecutters*, pp. 190–191.

90 *ASJ*, 15 January 1954, p. 653.

91 *ASJ*, 15 January 1954, pp. 653 & 657.

92 *ASYB*, 1962, p. 121; King, Mungomery and Hughes, *Manual of Cane-Growing*, 1953 edition, pp. 198–199; Kerr, *Top Mill in the Valley*, p. 94.

93 Kerr, *Only Room For One*, p. 111; J. Dixon, 'Sugar milling by CSR', in A. Lowndes (ed.), *South Pacific Enterprise. The Colonial Sugar Refining Company Limited* (Sydney, 1956), p. 144.

94 *ASJ*, September 1961, pp. 481–483; Manning, *In their own hands*, pp. 286–289; E. H. Atherton, *Plane Creek Central Mill Company Limited 1937–1974* (Sarina?, 1980), p. 8; Armstrong, 'The sugar tramways of northern New South Wales', p. 128.

95 Kerr, *Pioneer Pageant*, p. 197; Kerr, *Southern Sugar Saga*, p. 129.

96 *ASJ*, January 1958, p. 757 & January 1964, p. 718.

97 *ASJ*, January 1964, pp. 717–719 & January 1965, pp. 744–745; Kerr, *Southern Sugar Saga*, p. 129; Kerr, *Top Mill in the Valley*, p. 63.

98 AR BSES 1954–55, p. 82; *ASYB,* 1962, p. 141; Kerr, *Only Room For One*, p. 110; Kerr, *Northern Outpost,* p. 135 & 151; I. Crellin, 'Australian sugar tramways. The challenge of the 1980s', *Light Railways*, No. 66 (October 1979), p. 12; W. Chapman, 'Development in cane transport at Victoria Mill', *The Producers' Review*, April 1968, pp. 14–15.

99 C. Murry and R. Pierce, 'Trends in cane transport from harvester to factory', *Proc. ASSCT*, 1980, pp. 27–28; Crellin, 'Australian sugar tramways. The challenge of the 1980s', p. 12; J. Allen, 'Advances in mill machinery in 70's', *ASJ*, October 1971, p. 391.

100 Murry and Pierce, 'Trends in cane transport from harvester to factory', pp. 27–28; Chapman, 'Development in cane transport at Victoria Mill', p. 15; Rolleston, *The Defiance*, p. 98; Kerr, *A Crystal Century*, pp. 129–130; P. Connell and B. Borrell, *Costs and Regulation of Cane Harvesting Practices* (Canberra, 1987), p. 14.

101 *ASJ*, July 1964, pp. 283 & 285 & November 1964, p. 619; Burrows and Morton, *The Canecutters*, p. 215; Kerr, *A Century of Sugar*, p. 171.

102 *ASJ*, July 1965, pp. 339–345.

103 *ASJ*, September 1965, pp. 497–513; Burrows and Morton, *The Canecutters*, pp. 216–217.

104 *ASJ*, September 1966, p. 383 & December 1966, pp. 593–595; *QPD*, Vol. 244, 1966–67, p. 2145; *The Producers' Review*, February 1967, pp. 6–8 & April 1968, p. 3.

105 Burrows and Morton, *The Canecutters*, p. 220; *The Producers' Review*, March 1968, p. 104.

106 Kerr, *Southern Sugar Saga*, p. 135; *ASYB*, 1962, p. 45; Fairymead Sugar Co. Ltd., AR 1968, p. 4.

107 CSR, AR 1968–69, p. 16; *The Producers' Review*, March 1969, p. 120; July 1969, p.iii & 7 November 1969, p. 5.

108 *The Producers' Review*, June 1971, p. 31; Burrows and Morton, *The Canecutters*, pp. 221–224; CSR, *Newsletter* No. 135, November 1973, p. 12, CSRR, Z/109, Box 372.

109 Kerr, *Northern Outpost*, p. 152–153; *ASYB*, 1971, pp. 16–17; *ASYB*, 1972, p. 165.

110 *ASYB*, 1972, pp. 31 & 165–166; T. Fuelling, 'Tipper bin development and performance', *Proc. QSSCT*, 1977, p. 165; T. Fuelling, 'New developments in cane transport', *Proc. QSSCT*, 1978, pp. 235–239.

111 *ASYB*, 1971, p. 25; *ASYB*, 1976, p. 179.

112 *ASYB*, 1976, pp. 179–180; D. Steele, 'Multi-Lift – NSW Success Story', *ASJ*, May 1983, p. 137; Smith, *Sweet Beginnings*, pp. 366–370.

113 Reid, *The Long View*, p. 97; *ASYB*, 1970, p. 111; CSR, *Newsletter* No. 135, November 1973, p. 15, CSRR, Z/109, Box 372.

114 Reid, *The Long View*, p. 97; Crellin, 'Australian sugar tramways. The challenge of the 1980s', pp. 15 & 26; R. Vickers and L. Schafer, 'The economic significance of the Tully cane transport schedule', *Proc. QSSCT*, 1969, pp. 219–220; Willis, *Harvesting and Transport of Cane Sugar Cane in Australia*, p. 74.

115 Kerr, *A Crystal Century*, pp. 117, 123 & 135; Kerr, *Top Mill in the Valley*, p. 95; *The Producers' Review*, September 1967, p. 9; Hudson, *By the Banyan*, pp. 245 & 249.

116 Crellin, 'Australian sugar tramways. The challenge of the 1980s', pp. 18–19 & 21; W. Chapman, 'Development in cane transport at Victoria Mill', *The Producers' Review*, April 1968, p. 13; Kerr, *A Crystal Century*, pp. 119 & 136.

117 Hudson, *By the Banyan*, p. 244.

118 Willis, *Harvesting and Transport of Cane Sugar Cane in Australia*, pp. 30 & 46.

119 *ASJ*, 27 September 1962, pp. 433 & 435 & October 1969, p. 369; D. Braddock, 'Cane transport and receival at Maryborough Mill', *Proc. ASSCT*, 1982, p. 321; Willis, pp. 80–81.

120 Queensland Commissioner for Railways, AR 1959–60, 1970–71 & 1980–81; Willis, *Harvesting and Transport of Cane Sugar Cane in Australia*, p. 41.

121 Willis, *Harvesting and Transport of Cane Sugar Cane in Australia*, p. 102; *ASJ*, September 1958, p. 488; Harwood Mill Cane Insp., AR 1949, p. 6 & Harwood Mill Cane Insp., AR 1956, p. 4, both in CSRR, Z/109, Box 470; Clarence River Branch NSWCGA, AR 1957–58, Maclean; Smith, *Sweet Beginnings*, p. 358.

122 CSR, *Newsletter* No. 105, April 1968, p. 3, CSRR, Z/109, Box 372; CSR, AR 1966–67, p. 15; D. Steele, 'Multi-Lift – NSW Success Story', *ASJ*, May 1983, p. 137; Burrows and Morton, *The Canecutters*, p. 202.

123 AC, November 1983, pp. 37–38; P. Makepeace, 'Weight transfer reduces field damage', *CGQB*, Vol. 45, no. 1 (July 1981), pp. 20–21; Kerr and Blyth, *They're All Half Crazy*, pp. 108 & 119; *ASYB*, 1986, p. 132.

124 L. Brennan *et al.*, 'Improving sugarcane harvesting procedures', in B. Keating and J. Wilson (eds.), *Intensive Sugarcane Production. Meeting the Challenges Beyond 2000* (Wallingford, UK, 1997), p. 329; Kerr and Blyth, *They're All Half Crazy*, pp. 108.

125 L. Brennan, 'A systems analysis of cane harvesting and transport in the Mackay sugar industry'. PhD thesis, Uni. of Qld., 1998, pp. 33–35.

126 E. Churchward and R. Belcher, 'Some economic aspects of mechanical cane harvesting in Queensland', *Proc. QSSCT*, 1972, p. 35; P. Connell and B. Borrell, *Costs and Regulation of Cane Harvesting Practices* (Canberra, 1987), p. 12.

127 A. Petersen *et al.*, 'Capacity and trends in mechanical harvesting machines in Australia', *Proc. ASSCT*, 1984, p. 120; R. Savage *et al.*, *Sugar Industry Working Party Report* (Brisbane, 1985), p. 31; E. Fitzpatrick *et al.*, *Report of the Sugar Industry Working Party* (Brisbane, 1990), p. 73.

128 Petersen *et al.*, 'Capacity and trends in mechanical harvesting machines in Australia', p. 119; Connell and Borrell, *Costs and Regulation of Cane Harvesting Practices*, p. 13; Boston Consulting Group, *Report to the Sugar Industry Review Working Party. Analysis of Issues and Identification of Possible Options* (Sydney, 1996), p. 151; Brennan, 'A systems analysis of cane harvesting and transport in the Mackay sugar industry', p. 242.

129 Brennan, 'A systems analysis of cane harvesting and transport in the Mackay sugar industry', pp. 34–35; Kerr and Blyth, *They're All Half Crazy*, p. 109.

130 Brennan, 'A systems analysis of cane harvesting and transport in the Mackay sugar industry', p. 244–245.

131 *ASYB*, 1985, p. 182; *AC* 26 July 1993, p. 24.

132 C. Guesdon, 'The Clare-Dalbeg tramway project', *Proc. ASSCT*, 1981, pp. 95–96; *ASYB*, 1981, p. 169; *ASJ*, October 1981, p. 389; Kerr, *A Crystal Century*, p. 146.

133 Kerr, *Northern Outpost*, pp. 172–173; *Bundaberg Sugar News*, Summer 1995, p. 19.

134 *AC*, 18 September 1995, p. 10; Hudson, *By the Banyan*, p. 244; Bill Kerr, *Southern Cane Railway* (Brisbane, 1995), pp. 2 & 14.

Chapter 17
Regulating and Modernising the Milling Sector

In August 1915, Premier Thomas Ryan (Labor) introduced a Regulation of Sugar Cane Prices Bill into the Queensland Parliament. Ryan claimed the proposed legislation provided for the 'appointment of independent tribunals – impartial competent tribunals – whose duty it will be to fix the price of sugar cane'. As mentioned in Chapter 10, the Central Board was the centrepiece of the new cane price-fixing system. Local sugar cane prices boards (hereafter local boards) on which canegrowers and mill-owners had equal representation could also be formed in those districts where twenty or more canegrowers desired to have such a board established to determine local cane prices each year, the conditions associated with the delivery of harvested cane to the sugar mills, the awarding of any penalties for diseased, frosted or burnt sugar cane and the compilation of lists of approved cane varieties. This information became known as a local or mill award and would be published annually in the *Queensland Government Gazette*. Any local board could delegate its power to make an award to the Central Board or where local boards did not exist the Central Board could create a local award or when local boards could not reach an agreement, the Central Board acted as an appeal board. Mill-owners or any ten canegrowers could also appeal to the Central Board against the provisions in specific local awards. In addition, the proposed legislation permitted the Central Board to appoint independent check chemists (later called cane testers) at each sugar mill. These officers were appointed for the protection of the canegrower and it was proposed that their duties would include the supervision of the weighing, sampling and analysis of the sugar cane. Funds to pay for the operation of the boards and the hiring of check chemists were to be raised from a levy on the amount of sugar cane harvested by each canegrower.[1]

Essentially, the newly elected Ryan government sought to ensure that the state's canegrowers were given 'an adequate and fair price for their cane'. However, the proposed legislation would radically alter the relationship between mill-owners and canegrowers. The latter would be unable to send their cane to different sugar mills each year in districts where competition existed for cane between mill-owners (e.g. Mackay; Bundaberg) and they would be required to exercise more care with their cultivation and grow only approved cane varieties which had good sucrose content. William Lennon, the Queensland Minister for Agriculture and Stock, thought such conditions were perfectly justifiable, as the State government had spent a considerable sum on the BSES and its officers had been advising canegrowers on the best cane varieties to grow in particular districts.[2] Mill-owners, if they had not already done so, would have to introduce chemical control at their sugar mills and accept the presence of government-appointed check chemists upon their premises. Moreover, for the system to work effectively, mill-owners and canegrowers in Queensland were required to maintain and reveal details about their operating costs and the value of their land, buildings and machinery. Such revelations would not be welcomed by some mill-owners, especially CSR which had resisted the 1912 Royal Commission's attempts to ascertain details about the firm's profits (see Chapter 9).

Upon examining the bill, the ASPA initially expressed no opposition to its introduction, although it thought the legislation 'will certainly not ensure satisfaction to all who may be brought under its operations'. CSR vehemently opposed the proposed legislation and the possible interference in its operations. Even before the bill had received the Queensland Governor's approval, Edward W. Knox had asked CSR's Sydney solicitors for legal opinions in September 1915 on the validity of the legislation and whether the firm could be represented by a solicitor or counsel at local board hearings. The Leader of the Queensland Opposition, Edward Macartney (Toowong), opposed the bill on the grounds that it 'went a long distance along the road to nationalisation of our industries'. Other members of the Queensland Legislative Assembly had concerns about the drafting of the legislation. William Murphy (Burke) claimed the bill was flawed as it contained no provision to 'compel' mill-owners to operate their sugar mill and crush the harvested cane at the price fixed by a local board or the Central Board. Lt-Col. Colin Rankin (Burrum), who had unsuccessfully promoted cane price-fixing legislation in 1913 (see Chapter 8), thought that the inclusion of a Central Board was fatal as all its time would be devoted to hearing appeals from local boards and that the finalisation of awards would be delayed.[3]

Despite the concerns raised by some members of the Legislative Assembly, only minor amendments to the Cane Prices Bill were made by the Legislative Council and the bill was assented to by the Governor of Queensland on 6 October 1915. Twenty-nine local boards were constituted in December 1915, with the lands of the canegrowers who supplied cane to each sugar mill during 1915 being assigned to a particular sugar mill.[4]

In January 1916, the local boards held their first meetings to make retrospective awards for the 1915 crushing season. Six local boards agreed upon a mill award, but twenty-one local boards delegated the decision to the Central Board. In half the cases, the Central Board simply ratified the current arrangements between the mill-owners and canegrowers, allowing a variety of payment systems to continue in existence (e.g. flat rate per ton of cane; payment by cane variety; sliding rate for ccs or individual analysis). Increases of between 1s and 2s per ton of cane were granted in other mill awards. Four of these mill awards containing increased prices for cane applied to CSR-owned sugar mills, so the potential for conflict between CSR and the Queensland government escalated. Half a dozen Southern Queensland sugar mills (e.g. Alberton; Rocky Point; Junction; Carbrook) were initially excluded from the operation of the legislation as they did not have at least twenty suppliers who could request the formation of a local board.[5]

Three months after the gazettal of the 1915 mill awards, the local boards began considering the awards for the 1916 crushing season. By June 1916, thirty-six mill awards had been gazetted, with twenty local boards reaching an agreement.[6] An analysis of these first mill awards for 1915 and 1916 revealed many similarities that subsequently became the general conditions which regulated the milling system in Queensland. Mill-owners were required to accept all sugar cane containing over 7% ccs. They had to provide free of charge sufficient portable rail track for the removal of the sugar cane from the field to the mill's tramline, and railage haulage and river transport allowances were to be paid by the mill-owner when the cane was transported to a sugar mill by other than the mill's tramline. Each canegrower was allotted a daily delivery tonnage. Badly topped, dirty, burnt, frosted and gumming disease-affected cane attracted penalties. Canegrowers were expected to cultivate approved cane varieties, not burn their cane unless permitted by the mill-owner and supply correctly topped and clean cane. The earliest awards also emphasised the importance of cultivating cane varieties with high sucrose content and contained lists of disapproved cane varieties such as Dupont, Cheribon, Otamite, Malabar and Tanna. This feature was especially emphasised in the mill awards for the Mackay district, where the local boards noted that there was an emphasis on growing certain cane varieties notoriously deficient in sugar, although they yielded heavy tonnages of cane per hectare. Mackay canegrowers were restricted to having only 40% of their assignments devoted to low sucrose content varieties such as Cheribon, Otamite and Malabar.[7]

Challenging, amending and implementing the Queensland cane price-fixing legislation, 1916–1925

Immediately following the introduction of Queensland's cane price-fixing legislation, Edward W. Knox advised all CSR's Queensland sugar mill managers not to provide any local boards with information on production costs, as the firm sought to avoid giving 'facilities to the Local Boards to formulate claims'. In February 1916, CSR unsuccessfully sought the Central Board's permission to operate a common cane supply agreement for its six Queensland sugar mills covering the crops of 1915 and 1916. [8] Undeterred by this setback, CSR then tried to make the system unworkable when the local boards for its sugar mills commenced their deliberations in early 1916. Edward W. Knox instructed all CSR's officers to claim at local board hearings that the proceedings were illegal and the matter should be considered by the Central Board, where the millers' representative was John Smith, then CSR's Inspector of Mills. CSR's officers were advised to disclose nothing about costs or profits and to reply that 'accounts are made up in Sydney and not communicated to them'. CSR's strategy in operation can be gleaned from the deliberations of the Goondi Local Board, which met at the beginning of April 1916. Sir Edward Mitchell KC appeared on CSR's behalf. For nineteen days he argued before the local board that it had no authority to fix the prices for sugar cane. Eventually, the local board fixed a price acceptable to CSR and its cane-growers, but as the historian Denis Murphy noted, 'it was an unsatisfactory prospect to have a senior counsel appearing in a tribunal established to allow grower and miller to sit down together and decide on a mutually agreeable price'.[9]

As the local boards held their deliberations, CSR also took out three writs against the Ryan government in Queensland – one seeking an injunction restraining the Central Board from authorising and directing its check chemists to enter the company's sugar mills and using its laboratory apparatus, and two referring to Goondi and Victoria Mills, where it was claimed that the awards were invalid as the local boards had not been constituted for 1916. CSR also claimed that payment by ccs as outlined in the local awards was not possible as three-quarters of its chemists had enlisted.[10] CSR's challenge to the legality of check chemists entering its premises was resolved on 15 September 1916, when the Supreme Court of Queensland found in favour of CSR in respect to its objections; two months later in respect of the Goondi and Victoria writs, the Supreme Court of Queensland found in the Queensland government's favour, arguing that once a local board had been created by gazettal it existed until the Order-in-Council was rescinded or amended. CSR appealed to the High Court of Australia against this judgment, but was again rebuffed with the High Court upholding the lower court's findings.[11]

Parliamentary opposition to the cane price-fixing legislation culminated in December 1916 when the Queensland Legislative Council refused to permit proposed amendments to the cane price-fixing legislation.[12] Undeterred, the Ryan government allowed the existing legislation to function, and thirty-five mill awards were gazetted between April and July 1917, covering the majority of the state's sugar mills. CSR again proved troublesome, with the Central Board refusing to recongnise the agreements made between CSR and the suppliers to its Hambledon and Homebush Mills, instead of a local board award. In their place, the Central Board made its own award for both sugar mills, arguing that they could be enforced under the legislation by penalty for any breach.[13]

The Ryan government, heartened by the smooth operation of the cane price-fixing legislation during early 1917, went on the offensive in the Queensland Parliament. In August 1917, William Lennon, the Queensland Minister for Agriculture and Stock, reintroduced the same bill to amend the *Regulation of Sugar Cane Prices Act* of 1915 that had been lost in 1916, although a new clause had been added giving the Queensland government the authority to take over a sugar mill which was not paying the award price for cane as determined by the local boards or Central Board. Predictably, the new bill met with strong opposition from the Queensland Legislative Council, CSR and the ASPA. Eventually, after considerable negotiations with the Legislative Council, the legislation was passed in November 1917 with numerous amendments. The most significant changes included: banning members of the legal profession from appearing before local boards; permitting the Central Board to change the assignment of particular lands from one sugar mill to another; compelling mill-owners to crush frosted or damaged cane, even if a canegrower's land was not assigned to that sugar mill; the establishment of a firm start date for each sugar mill; the provision of a proper and sufficient laboratory at each sugar mill for the testing and analyses of sugar cane juice; check chemists could, under certain circumstances, inspect any farm or sugar mill or any books or records of these relating to the manufacture of sugar or growing of sugar cane; check chemists could use a sugar mill's laboratory; mill-owners and canegrowers could not contract out of a local board; and the Queensland government could force mill-owners to pay the award for cane or have their sugar mills taken over by the state.[14]

The earliest operation of the cane price-fixing legislation was also hindered by some canegrowers who purportedly were the beneficiaries of the act, although in some instances this hindrance was not intentional. Under the requirements of the legislation, Queensland canegrowers were supposed to supply the Central Board in early 1916 with details about their assets, liabilities, cultivation costs and income from the sale of sugar cane in 1915. The Central Board needed this information so that it fixed the price of cane at an appropriate level to cover the canegrowers' production costs. Failure to comply could result in a fine. However, the response to the Central Board's request was acutely embarrassing. Of the 987 canegrowers in the Mackay district, for example, only 25 supplied the required information. 'Farmers find keeping accounts and returns irksome', claimed John Broadhurst, an Isis district canegrower who appeared before the Queensland Legislative Council's Select Committee investigation into amendments to the cane price-fixing legislation in December 1916. Few of the state's canegrowers kept comprehensive records. The Central Board declined to fine those canegrowers who had not responded to its request for financial details.[15]

Further attempts by the Central Board during 1917 failed to secure information on the production costs of the state's canegrowers, with less than 5% of Queensland's canegrowers making financial returns and most returns yielded little useful information. The ASPA did not assist the process, advising its canegrower members that the returns were not compulsory and 'you need not send them in unless you like'. In the midst of the deliberations over the mill awards in early 1917, John Smith, the mill-owners' representative on the Central Board, suggested that the state's canegrowers were deliberately with holding the details – he was probably correct for the Piddington Royal Commission concluded that some canegrowers feared any details they supplied would be used to their detriment. Eventually, the Central Board sent its Secretary to various sugar-producing districts in early 1919 and he managed to assemble some data on farmer production costs for 1918. The Piddington Royal Commission also investigated the matter in 1919, creating a generalised estimate of the production costs incurred by 'an average efficient farmer', using evidence and statements from the mill-owners who cultivated cane. Its conclusion, much to the annoyance of some unionists and canegrowers, was that Queensland canegrowers had their production costs covered by the payment of 'very satisfactory prices per ton of cane' and that CSR paid 'comparatively high prices' for its cane.[16]

From 1918 onwards, the cane price-fixing legislation in Queensland operated without any major rancour between mill-owners and canegrowers, although minor disputes over the interpretation of clauses in the legislation and whether or not mill-owners breached the legislation were being heard in different Police and Magistrates Courts throughout Queensland.[17] CSR even ceased its opposition to the legislation, the firm devising a scheme to circumvent the Queensland cane price-fixing legislation. Taking advantage of the free trade among the states' provision of Section 92 of the Australian constitution, CSR convinced suppliers to its Childers Mill to sell their cane to a Sydney-based intermediary, who then sold it to CSR. To secure this agreement, CSR abandoned payment by ccs and reverted to a scheme where cane payment depended on the number of tonnes of cane required to make one tonne of sugar and, apart from penalties for burnt and dirty cane, all cane received the same payment. This arrangement, accepted by the Central Board on the grounds that 90% of the Childers Mill suppliers agreed to the scheme, was then adopted by all CSR's sugar mills during 1918. Such a system applied to CSR's Queensland mills until at least the outbreak of World War II.[18] However, following continued complaints by some Queensland canegrowers, the cane price-fixing legislation was revised again in 1921, the major change being that the Central Board could recognise contracts or agreements between mill-owners and canegrowers as long as they were for no more than three years' duration and that they had been signed by 85% of a mill's canegrowers who supplied at least 60% of the sugar mill's cane.

Francis Maxwell, an engineer and sugar technologist, compared the different cane payment systems throughout the world's major sugar-producing countries in the mid-1920s. He claimed that Queensland's cane price fixing legislation made the state 'unique amongst cane sugar countries'.[19] Moreover, the regulated fixing of cane prices remained peculiar to Queensland, as it never spread to New South Wales, where CSR was the only miller. Instead, prices for cane in New South Wales continued to be determined by CSR and the terms and conditions that governed the sale of cane between miller and supplier were outlined in one-year agreements during the late 1910s and then five-year agreements after 1920. During the late 1910s, CSR negotiated these agreements with different farmer groups, but after the formation of the NSW Cane Growers' Association in 1922, this organisation advised if the clauses in the agreements were acceptable and if the farmers should sign the agreements. An analysis of some surviving agreements preserved in CSR's archival records for the period 1915–1955 revealed, however, that they contained many terms and conditions which applied in Queensland – canegrowers would grow only approved cane varieties, cane containing less than 7% ccs would not be processed, the price per tonne of cane was calculated according to ccs and prices would vary according to whether the cane was considered 'sound' or 'damaged'.[20]

Initially, local boards fixed the price for sugar cane in their particular mill supply areas, taking into consideration the estimated ccs of the harvested cane, the crushing capacity and efficiency of a particular sugar mill, the selling price of raw and refined sugar, and the labour conditions under which the sugar cane was grown, harvested and delivered to a mill. However, the Central Board changed the policy when dealing with appeals from the Mackay district in 1919. It imposed uniformity in the cane prices throughout the entire district, claiming that the difference in price led to applications by suppliers of low paying sugar mills for re-assignment to better paying sugar mills. This policy was re-iterated in 1920, when the Central Board refused the Australian Estates and Mortgage Co. Ltd. permission to pay its Kalamia Mill suppliers less than the price being received for the cane delivered to the neighbouring sugar mills in the Lower Burdekin district. By 1924, the Central Board had implemented a uniform price across the entire state. In 1925, there was an increase of 10d per ton for all sugar cane in areas north of Bundaberg, but in 1928 and 1931, by respective increases of 6d and 4d in the Bundaberg mill supply areas, the price was brought to the same level as in the north. Uniform statewide pricing then ceased until it was re-introduced in 1936, and applied to all mill supply areas, except for the Maryborough Sugar Factory and Rocky Point Mill, which did not accept the statewide price for cane until 1958 and 1976, respectively. In 1980, this situation prompted The Hon. Mr. Justice Ronald Matthews, President of the Industrial Court of Queensland, to comment that the legislation's intent had all but vanished, with the local boards no longer considering the issue of cane prices, leaving the matter entirely to the Central Board.[21]

The earliest Central Boards also had to divide the estimated profits of the industry on a fair and equitable basis to both mill-owners and canegrowers. Reaching a decision was hindered by a lack of information on production costs (see earlier discussion), but the Central Board, probably somewhat arbitrarily, adopted a ratio of 2:1 – two thirds to the canegrowers and a third to the mill-owners. This division was confirmed in September 1918 by the United Cane Growers' Association, the organisation asserting that mill-owners now paid canegrowers 64–75% of the gross value of a ton of sugar, whereas previously the level fluctuated between 47 and 55%.[22] Better figures on production costs did become available, but The Hon. Mr. Justice Ronald Matthews claimed in 1980 that with the lapse of time, the ratio of 2:1 acquired an 'aura of sanctity'. Yet a closer examination of the Queensland canegrowers' share of proceeds shows that since 1960 their share has fallen from around 68% to 62% in 1990, a consequence of markedly improved mill efficiencies.[23]

During the initial operation of the cane price-fixing legislation in Queensland, mill-owners paid their canegrowers under various arrangements. By 1920, CSR had adopted the 'sugar in bag' scheme, which paid farmers for their sugar cane on the basis of how many tonnes of cane were required to make one tonne of 94 n.t. sugar from the total crop at each sugar mill. Many other mill-owners purchased sugar cane using ccs, known as payment by chemical analysis or individual analysis. A third arrangement was the relative percentage scheme, adopted by the suppliers of Babinda, Mossman and Mulgrave Mills in 1917, 1918 and 1919, respectively. The scheme had been worked out in 1915 by Claude O'Brien, chemist and technical member of the Central Board during the late 1910s, and Dr John Reed, the former manager of CSR's Hambledon Mill.[24] Under the relative percentage scheme, the crushing season was divided into weekly periods. Cane having ccs above the average for the week (i.e. all cane crushed) is paid for at a proportionately higher rate and cane falling below the standard receives a proportionately lower rate. The values used to make the calculations were an approximate average of the ccs values set out in the Central Board's scale of cane prices. The scheme established a community of interest, encouraging all canegrowers to cultivate high sucrose-yielding cane varieties and ensured that canegrowers who had their cane harvested early and late in the crushing season were not penalised.

Before the advent of computers, the introduction of the relative percentage scheme at a sugar mill created much extra clerical work, so it is not surprising that only two other Queensland sugar mills – South Johnstone and Mourilyan – had adopted modifications of the relative percentage scheme between 1925 and 1940. [25] CSR's Queensland sugar mills introduced the scheme during World War II; their New South Wales sugar mills followed in the mid-1950s. During early 1950, Queensland canegrowers in the remaining mill supply areas again considered the relative percentage scheme, and several more sugar mills changed to it during the crushing season, namely Maryborough, Proserpine, Marian, Farleigh and Millaquin. By 1975, twenty-nine of the thirty-one mill supply areas in Australia used some form of the relative percentage scheme.[26]

Mill closures and upgrades, 1915–1929

The additional cost imposed upon Queensland's mill-owners by the Dickson Award, and the increased prices some of the state's mill-owners paid for their cane under the *Regulation of Sugar Cane Prices Act* of 1915, quickly had repercussions. At Bundaberg, the owners of Miara and Goodwood Mills concluded that they were unable to work the factories profitably under the new conditions and decided to shut them after the 1917 crushing season. Twelve months later, another two smaller sugar mills in the Bundaberg district – Waterloo and Baffle Creek – also closed, while a third – Invicta – was dismantled and re-erected in the Haughton River district during 1919. Nerang Mill was shut by the Comptroller of the Queensland Bureau of Central Sugar Mills at the end of the 1918 crushing season, after he concluded that the sugar mill with its limited cane supply would never operate profitably under the new cane price-fixing legislation, although the Queensland government was forced to write off £35 000.[27]

Three other firms were engaged in mill closures during the early 1920s. At Mackay, CSR closed its Homebush Mill after the 1921 crushing season. Years of conflict following the introduction of the cane price-fixing legislation by what CSR termed a 'hostile minority of the canegrowers' wanting a local board, and this lack of loyalty by its suppliers, probably influenced CSR's decision. Moreover, CSR wished to upgrade its other four northern sugar mills and could reduce this expense by transferring the various parts of the plant from a closed sugar mill, especially as the cost of new machinery had risen sharply after World War I.[28] Three years later, another Mackay sugar mill closed when the Australian Estates and Mortgage Co. Ltd. shut its Palms Mill, which had never been fully repaired after being badly damaged during the 1918 cyclone that devastated Mackay. The firm's Mackay sugar interests were merged with the Pleystowe Central Mill Co. Ltd. to form Amalgamated Sugar Mills Ltd. Another mill-owner with a factory struggling to make money was the Queensland National Bank. Its Doolbi Mill in the Isis district was short of cane – there were too many sugar mills in this drought prone locality. In 1924, the Queensland National Bank sold Doolbi to the Isis Central Sugar Mill Co. Ltd. CSR was very annoyed by this arrangement, as the firm had hoped to secure Doolbi's cane for its Childers Mill. The final outcome of this deal eventually occurred in 1932, when CSR closed its Childers Mill, the firm giving up trying to operate a factory that was persistently short of cane.[29]

In 1925, forty sugar mills operated in Australia. Six of these canegrower owned sugar mills were still controlled by the Queensland Bureau of Central Sugar Mills, its officers attempting to ensure that they worked efficiently so the companies met their repayments to the Queensland Treasury. The Central Board, however, noted in 1926 that the government-controlled central sugar mills were struggling to be profitable, on account of their ill-selected location or inadequate capacity. The Central Board refused to fix a lower price for sugar cane at these central sugar mills, as this solution would depart from the Board's policy of district uniformity, so it was prepared for some of these factories to cease operations. Consequently, the Queensland Treasurer, at the end of the 1926 crushing season, notified the suppliers of South Johnstone, North Eton, Gin Gin and Mt Bauple Mills of the government's intention to cease managing their affairs during 1927. These central sugar mills were returned to grower ownership under *The South Johnstone, Gin Gin, North Eton and Mount Bauple Sugar Works Act* of 1927, thereby effectively ending the existence of the Queensland Bureau of Central Sugar Mills. Suppliers of Tully and Proserpine Mills finally obtained control over their mills in 1931, although loan repayments from the central milling companies continued during the 1930s and 1940s. Mt Bauple Mill, however, closed at the conclusion of the 1950 crushing season, the growers' cooperative giving up the struggle to modernise the factory and repay its government debt after years of drought, severe frosts, a limited supply of cane and poor management. In 1959, the Tully Cooperative Sugar Milling Association Ltd. was the last central milling company to clear its debt to the Queensland Treasury.[30]

During the introduction of the cane price-fixing legislation, the Central Board assumed a minimum standard for mill efficiency, so canegrowers were assured that a reasonable amount of sugar in their cane had been recovered during the manufacturing process. As a means of comparing the efficiency of one Queensland sugar mill with another, the Central Board adopted a measurement known as the coefficient of work. The Central Board took a coefficient of work of 90% as the standard when considering the claims of the mill-owners in relation to cane prices. No allowance for any failure to reach that standard was permitted by the Central Board, but mill-ownerswho had improved the efficiency of their milling plant were allowed to retain all the profits accruing through having obtained greater efficiency than demanded by the Central Board.[31]

This incentive had prompted a few Queensland sugar mill-owners to continue upgrades commenced before 1915 or undertake new upgrades. The Australian Sugar Co. Pty. Ltd., for instance, had installed a new vacuum pan at Mourilyan Mill in 1918 and in June 1919 was awaiting the arrival of two new vacuum pans from England in order to complete the extensive overhaul of the factory which had commenced in 1914. Moreton Mill's triple effets were upgraded to a quadruple set in 1917. The Queensland Bureau of Central Sugar Mills authorised the installation of a new vacuum pan at Proserpine Mill in 1918 and a new vacuum pan and centrifugals at North Eton Mill in early 1919.[32] In addition, the Queensland government acquired a license in

1917 to use the newly patented Petree-Thomas process in the state-controlled central sugar mills, and by 1919 it had been adopted in at least nine Queensland sugar mills, namely Babinda, Kalamia, Proserpine, Millaquin, Fairymead, Bingera, Gin Gin, Doolbi and Racecourse Mills.[33] Developed by Cuthbert Petree, a CSR-employee, and Nehemiah Thomas, a Sydney-based chemist, this new procedure recovered an additional one and a half tonnes of sugar each week and eliminated the old-fashioned plate and frame filter presses (see Plate 17.1), thereby saving on the cost of the filter cloth, as well as dispensing with the eight to ten labourers who worked at this mill station. Instead of filtering the muds from the clarifiers through the plate and frame filter presses, they were mixed with the juice from the second set of crushing rollers and passed through a secondary clarifier. The muds created during this process (i.e. secondary muds) were then mixed with the maceration water and sprayed onto the bagasse as it went through the crushing rollers, the idea being that it would be filtered out on the bagasse and pass harmlessly to the boiler furnaces. [34]

Overall, however, most Australian mill-owners had been reluctant to improve their sugar mills substantially during the late 1910s. Indeed, Walter Adams, the general manager of Bundaberg Foundry Ltd., reported a decline in the demand for new sugar machinery and maintenance orders since the

Plate 17.1. Plate and frame filter presses at Moreton Mill, c. 1945.

Source: Image reproduction courtesy of Sunshine Coast Libraries, Picture Ref. No. M734213.

introduction of the cane price-fixing legislation. This approach had certainly been adopted by CSR, the firm introducing a policy of minimal expenditure on its Queensland sugar mills after 1915 until the uncertainty surrounding the operation of the cane price-fixing legislation had been resolved.[35] Yet the generally smooth operation of the cane price-fixing legislation after 1917 and the substantial increase in the Commonwealth government's guaranteed price for a ton of raw sugar from £21 to £30 6s 8d in 1920 (see Chapter 18), restored the mill-owners' confidence in the industry's future. They embarked upon substantial mill improvements during the 1920s.

Some machinery for these upgrades throughout the early 1920s came from the sugar mills that had closed during the late 1910s. Alberton Mill's owner secured Waterloo Mill's triple effets after the latter had ceased operations. In 1920, William Heck purchased much of the machinery lying idle at the closed Nerang Mill, and used it in the extensive upgrade to his Rocky Point Mill. After closing the Palms Mill, the Australian Estates and Mortgage Co. Ltd sent much of the machinery to Kalamia Mill where it formed the centre piece of that factory's major refurbishment.[36] Other mill-owners sourced their new machinery from Australian manufacturers. Most of the new machinery installed during the extensive upgrade to Mulgrave Mill in 1922 came from Walkers Ltd. or the Bundaberg Foundry Ltd. These two firms also supplied much of the new machinery that contributed to the substantial expansion in the capacity of Isis Mill during early 1922, and built all the machinery for Tully Mill, which was erected during 1925. One of their main local rivals, Evans, Deakin & Co. of Brisbane, constructed the machinery used to improve Bingera Mill in 1924.

English or Scottish sugar machinery makers, however, were still favoured by a few Australian mill-owners, despite the imported machinery attracting a tariff. During 1925 and 1926, George Fletcher & Co. of Derby built a new set of crushing rollers for the Plane Creek Mill, and three new sets of crushing rollers and the new steel building for the crushing house at Marian Mill. Duncan Stewart & Co. of Glasgow, one of George Fletcher & Co.'s major rivals in Great Britain, constructed the three new sets of crushing rollers and a new shredder that formed the core of the substantial £110 000 upgrade to Moreton Mill during 1926.[37]

The technical standards of some Queensland sugar mills during the late 1910s left much to be desired. Australian sugar mills generally had levellers (i.e. rotating knives), a Krajewski crusher or a National shredder. The latter piece of machinery broke the stalks of cane into smaller pieces as they passed through two revolving metal rollers, the top one of which revolved at a different speed to the bottom one. Very few factories had both levelling knives and crushers or shredders. Three sets of crushing rollers were normal in Australian sugar mills, although Mossman, Pleystowe and Palms Mills in 1919 still only had two sets of crushing rollers or double crushing. Juice clarification in the Maryborough Sugar Factory and Mt Bauple Mill continued occurring in open clarifiers; conical, closed continuous subsiders had been installed in CSR's sugar mills twenty years earlier. Chemical control had only been instigated at the Maryborough Sugar Factory for the first time in 1918, and that had only happened because of the introduction of the cane price-fixing legislation.[38]

Non-CSR Queensland sugar mills during the late 1910s had an average coefficient of work of 90% (Table 17.1). Harry Easterby, the general superintendent of the BSES, suggested at the Piddington Royal Commission in 1919 that a sugar mill with a coefficient of work of around 92–95% was efficient; 90–92% was a fairly efficient; less than 90% was inefficient. Easterby presented data gathered for twenty-three non-CSR Queensland sugar mills in 1917 and 1918 – in each year half were considered inefficient, and only a quarter were assessed to be efficient. After mill inspections and hearing the evidence of mill-owners or mill managers, the Piddington Royal Commissioners concurred with Easterby's assessment, and concluded that some Australian sugar mills displayed a lack of balance between the crushing capacity of the milling station and the evaporating plant which reduced economical sugar recovery. Moreover, the

Table 17.1. Selected crushing statistics for Queensland sugar mills, 1919–1932.

Year	Av. weekly amount of cane crushed per sugar mill (tonnes)	Approx. av. coefficient of work (%)	Year	Av. weekly amount of cane crushed per sugar mill (tonnes)	Approx. av. coefficient of work (%)
1919	3 078	90	1926	4 183	96
1920	2 864	90	1927	4 395	96
1921	3 005	92	1928	4 936	97.5
1922	3 321	93	1929	5 428	97.5
1923	3 478	93	1930	5 640	97.0
1924	3 589	94	1931	5 939	96.5
1925	3 769	95	1932	6 629	97.0

Note: These figures most likely did not include data for CSR mills.

Source: Based upon figures in J. MacGibbon, 'Some statistics of the Queensland sugar industry, 1915–1932', *Proc. QSSCT,* 1934, p. 125.

Commissioners remarked upon the presence of dirty milling machinery and the absence of cleanliness throughout many Australian sugar mills, features which they concluded hindered efficient sugar production.[39]

Easterby acknowledged the BSES needed to supply more technical assistance and advice to the state's sugar mill engineers. To improve the BSES's capacity to offer technical advice on milling, Norman Bennett was appointed the BSES travelling scholar who took up the study of sugar mill technology. He left Queensland in 1924 and visited Java, Scotland, Louisiana, Cuba, Hawaii and the Philippines.[40] In Bennett's absence, the performance of Queensland's sugar mills steadily improved during the 1920s – the average coefficient of work had reached 95% by 1925 and the average amount of cane handled weekly had risen from around 3 000 to 3 700 tonnes (Table 17.1). By the mid-1920s, the average amount of cane crushed yearly by an Australian sugar mill had increased to around 100 000 tonnes, or double the figure processed in the late 1910s (Figure 17.1).

The improved technical performance of Queensland's sugar mills during the early 1920s resulted from a combination of practices. In some instances, the mill upgrades aimed to bring the very backward technology used in the sugar mill up to long-established standards. Alberton Mill, for instance, received its first triple effet in 1924. CSR had pioneered the use of this technology in its Australian sugar mills forty years earlier. Some Australian mill-owners installed new machinery or process technology that had not been used universally before 1915. The Racecourse Central Sugar Co. Ltd. and the Mossman Central Mill Co. Ltd. fall into this category. During early 1922, each company funded the conversion of existing triple effets into quadruple effets at their respective mills.[41] Other mill-owners expanded the overall capabilities of their factories, especially in light of the upsurge in the area under sugar cane in many districts during the 1920s. Two examples will illustrate this trend. Mulgrave Mill was upgraded by the installation of new juice heaters, centrifugals, a vacuum pan and a set of quadruple effets. Its three existing 1.37 m crushing rollers were also replaced with new 1.52 m crushing rollers and engines. All this machinery, except the crushing rollers, was converted to being electrically driven. Moreover, two additional new boilers were installed to supply the extra power needed to operate the larger crushing rollers. The factory's annual capacity of 10 000 tonnes of sugar rose to 15 000 tonnes. The Australian Estates and Mortgage Co. Ltd. also enlarged the capacity of its Kalamia Mill substantially during 1925 and 1926, using some of the machinery from the closed Palms Mill. Just over £100 000 was spent on duplicating the milling train, installing a new set of triple effets and rebuilding the factory's boiler

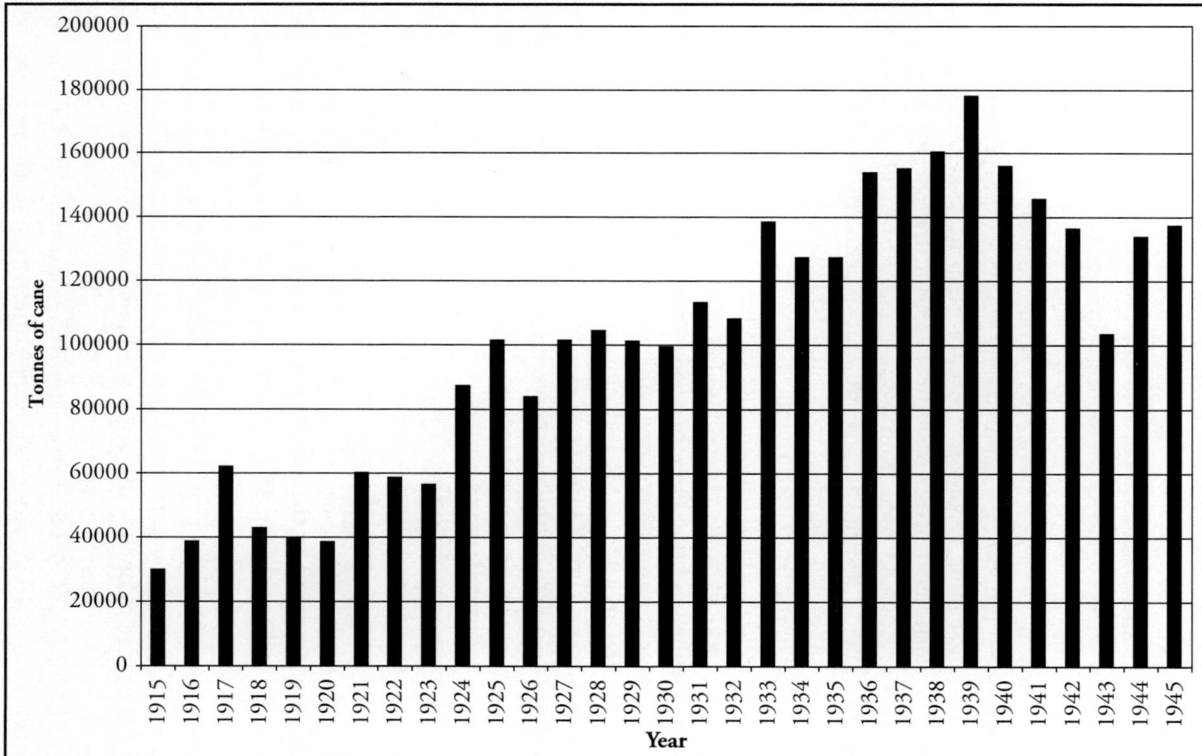

Figure 17.1. The average tonnage of cane crushed yearly by Australian sugar mills, 1915–1945

Source: Calculated from figures in the *Year Book of Australia*.

station, chimney, sugar store and spray pond. Consequently, Kalamia Mill became only the second Australian sugar mill to have a double crushing train, although this arrangement was short-lived, being abandoned in 1932.[42]

Technological innovations, imported from overseas or developed locally, were installed in Australian sugar mills during the period 1915 to 1930. Automatic feed regulators – new equipment for controlling the level of liquors in effets – had been developed in 1911 by Grosvenor Francis, an engineer and manager of Doolbi Mill, near Childers. This equipment was adopted in several Australian sugar mills during the late 1910s, including CSR's New South Wales sugar mills and their Childers Mill in Queensland.[43] Difficulties associated with wet bagasse – created through maceration – not feeding easily into the crushing rollers and the poor natural feeding properties of cane varieties like Badila were partially overcome by the development of horizontal pusher feeders. Consisting of rectangular steel bars driven by cranks and rods, the device was placed before a set of crushing rollers and thrust the bagasse into the feed opening of the crushing rollers. Their origins are unclear, but the device had been developed by the late 1910s – Doolbi and Moreton Mills were equipped with one by 1919 and CSR had installed them in their factories by 1925.[44] Better cane preparation before it entered the crushing rollers also attracted the attention of CSR's engineers who experimented with different arrangements for levellers during the late 1910s. The results from these trials led to more knives being fitted at the head of the cane carrier, their speeds being increased from 75 r.p.m. to 400–500 r.p.m. and a reduction in the clearance between the knives and cane, thereby increasing the amount of disintegration. Further CSR research into cane preparation led to the introduction of bent tip knives at Goondi Mill in 1926. Subsequently, they became standard equipment at all CSR sugar mills. Calandria pans which contained a large number of vertical tubes instead of the usual spiral coils to heat the juice were adopted at Racecourse and Farleigh Mills in 1927 and 1929, respectively. These units promised a greater heating surface in a smaller space than traditional coil pans.[45]

The BSES Division of Mill Technology and Queensland Society of Sugar Cane Technologists, 1928–1945

Upon Bennett's return from overseas in early 1928, the Division of Sugar Mill Technology was formally constituted within the BSES. Bennett began visiting non-CSR mills in Queensland, gathering information about their technical performance. On 29 May 1928, Bennett attended a meeting at Mackay where representatives of the Queensland mill-owners, excluding CSR, had assembled to consider the formation of a 'technological department'. Bennett spoke about the milling procedures he had observed when overseas and suggested the formation of a Technological Association. The state's sugar mill chemists and engineers could belong to this organisation which would meet annually to discuss the problems they had encountered during the year. John Inverarity, the Pleystowe Mill Manager (later Manager at Kalmia Mill) advised the gathering on the findings from his recent visit to Hawaii. He had witnessed that the island's forty mill-owners used a uniform system of technical control so that every week standarised milling results were collated by the Technological Department of the Experiment Station of the Hawaiian Sugar Planters' Association and circularised to each sugar mill. Results from each sugar mill could, therefore, be easily compared. In addition, advice on milling problems could be sought from the Technological Department. The representatives of the Queensland mill-owners agreed that the technical and economic control of Queensland's sugar mills needed to be improved, although the general consensus was that it should not be left to the BSES to develop this scheme, but that the mill-owners needed to take responsibility for its implementation.[46]

Inverarity relayed the proposal to form some 'technological department' to the ASPA. The organisation's executive considered the proposal at a special meeting in July 1928 and resolved initially to fund a report upon the state of sugar milling in Queensland. William McAllep, the Chief Sugar Technologist with the Experiment Station of the Hawaiian Sugar Planters' Association, was asked to complete this report. During the latter months of 1928, he visited eighteen of the larger Queensland sugar mills, mostly in the Bundaberg, Mackay and Lower Burdekin districts, although no CSR mills were included on his itinerary. McAllep delivered his report in December 1928 and it contained a scathing assessment of milling practices in Queensland. McAllep considered that the high extraction rates were obtained from the dubious practice of maceration, rather than pressure. Machinery was not utilised to its fullest. Between a quarter and half of the steam came from extra fuel sources (i.e. coal and/or wood), whereas Hawaiian sugar mills were self-sufficient in fuel, relying only on bagasse. Analytical methods needed to be made uniform and chemical control could be improved, as too much time was spent analysing juice samples for cane payments. McAllep recommended the establishment of an independent technical bureau supported by the sugar mills it served.[47]

Even though the ASPA had commissioned the McAllep report, the organisation was reluctant to fund the establishment of a separate technological department under its auspices. Instead, the BSES in March 1929 convened a meeting of mill managers, engineers and chemists at Mackay to discuss the milling sector's concerns. This gathering resolved to constitute the Queensland Society of Sugar Cane Technologists (its name was changed to the Australian Society of Sugar Cane Technologists in 1979). The new organisation did not enjoy immediate popularity and it was regarded with considerable suspicion by some Queensland mill-owners. Yet the annual meetings of the state's mill chemists and engineers at different sugar-producing centres in Queensland grew in popularity, and the scope of the Society's activities was widened to include an agricultural section during its yearly meetings and the publication of technical papers on particular milling topics in its annual proceedings. The other important event of 1929 was the inauguration of the Mutual Control Scheme, an arrangement that provided for the interchange of factory operating data between non-CSR sugar mills. The BSES facilitated the operation of this scheme by establishing its metrology service which standarised laboratory apparatus and techniques. By 1930, the Mutual Control Scheme operated on a fortnightly basis, and was maintained until activities were temporarily abandoned during World War II.[48]

The BSES Division of Mill Technology was enlarged in 1930 following the appointment of E.R. (Roly) Behne and experimental work commenced at a new laboratory built at Mackay. The Division's earliest work then received a setback following the resignation of Bennett in 1932. Research activities were then transferred back to Brisbane. Over the next decade, Behne was joined by a succession of engineers and chemists, the most noted being Norman Smith, who pioneered the concept of fractional liming during clarification tests in 1936 and Ir. J. Eigenhus, who developed the BSES's pH meter (see discussion later in the chapter). None, however, stayed with the organisation and Behne again found himself as the only member of the division in 1942, a position he occupied until 1945. Despite the instability in staff numbers, the BSES Division of Mill Technology during the 1930s provided mill-owners with a regular metrology service and the latest technical information. The compilation of the *Laboratory Manual for Queensland Sugar Mills* was commenced in 1933, the first edition being published the following year. The volume was well received and adopted as the standard reference work on analytical procedures in Queensland sugar mills. A second edition of the *Laboratory Manual* was published in 1939.[49]

Experimental work by the BSES Division of Mill Technology staff during the 1930s led to several major innovations in milling procedures and the development of new equipment, some of which was adopted immediately during the latter part of the decade. Norman Smith and his colleagues conducted clarification tests between 1935 and 1937. Smith observed that the P.O.J. cane varieties after cold liming – adding the lime to the cold juice before heating – produced very turbid juice and the subsidisation process or the period when the muds settle out of the limed juice in the subsiders took much longer compared to the juice from other cane varieties (e.g. Badila, M. 1900, E.K. 28). Investigations revealed that the juice from the P.O.J. cane varieties contained large quantities of starch which hindered effective clarification. To overcome this difficulty, Smith developed the fractional liming with double heating method in 1936, and although there is no certainty that this discovery was made prior to that of the sugar chemist J.G. Davies in Trinidad, it was developed independently. The process involved liming the cold juice to a pH of 6.2 to 6.4. The limed juice was then heated to boiling, re-limed to a pH of 7.6 to 8.2, and again heated until boiling.[50] Eigenhuis conducted experiments into the design of a convenient and inexpensive pH meter for the continuous measurement of pH under factory conditions. While the eventual design was based upon that devised by the sugar technologists in Java, several important improvements such as the use of mains electricity instead of batteries and enclosing the instrument in a neat aluminum case rendered the instrument more robust and better suited to routine mill use. Behne's investigations into prepatory devices found that National shredders performed very little useful work and could be replaced with high powered revolving knives or the Searby shredder.[51]

Recovering more sugar in depressed and troubled times, 1930–1945

In 1929, the average price per ton of sugar received by Australian mill-owners was £20. For the next five years, the average price per ton of sugar in Australia dropped a little each year, so that in 1934 it stood at £15. Average prices per ton of sugar then remained around £15 for the remainder of the 1930s. Consequently, Australian mill-owners faced a major cost-price squeeze. To cope with the reduced financial returns, they could adopt the following strategies: expand the amount of cane crushed and produce more sugar to cover their costs; recover more sugar from their existing crops for the same wages outlay; or reduce costs by introducing labour- and/or fuel-saving strategies.

Most Australian mill-owners, as will be shown shortly, adopted various strategies to remain economically viable during the 1930s. Yet planning the responses to this major cost-price squeeze was complicated as it occurred against the backdrop of the Peak Year Scheme which theoretically restricted annual sugar output. Strict adherence to the Peak Year Scheme, however, had broken down by the mid-1930s (see Chapter 10).

Some Australian mill-owners tackled the cost-price squeeze by improving the extractive capabilities of their factories, thereby reducing the amount of juice lost in the fibre that eventually becomes bagasse. These improvements could be achieved by a variety of mechanical changes. Better preparation of the cane before it entered the crushing rollers assisted extraction. In 1930, most Australian sugar mills were still equipped with National shredders. Bingera and Tully Mills were the exception, containing Searby shredders which consisted of revolving metal hammers that pulverised the cane in preparation for entry to the crushing rollers. Pioneered in Hawaii during 1916, Searby shredders obtained a much higher juice cell rupture than other shredders. They were installed in the place of National shredders in several CSR-owned mills during the 1930s, namely Hambledon (1934), Condong (1936) and Macknade (1939).[52] Other mill-owners decided that additional juice could be recovered by subjecting the bagasse to extra pressure, instead of increasing the overall amount of cane crushed. This strategy involved expanding the milling train. Consequently, a fourth set of crushing rollers was added to several Queensland sugar mills, including Cattle Creek (1931), Mossman (1935) and Isis (1936), while a fifth set of crushing rollers had been installed at a few of the state's sugar mills by 1939, namely Victoria (1935), Kalamia (1935) and South Johnstone (1938).[53]

Higher crushing rates could be achieved through the installation of pusher feeders, which enabled more cane to be processed without lengthening the crushing season. Horizontal pusher feeders, as mentioned earlier, had probably been developed during the 1910s. By the mid-1920s, they were joined by the vertical pusher feeders – rectangular steel bars that pressed the bagasse against the feed roller, thereby extracting additional liquid, particularly at the second, third or fourth sets of crushing rollers which handled very wet bagasse after the use of heavy maceration. They had been installed at several Queensland sugar mills by 1936, namely Farleigh (1930), Racecourse (1932), Proserpine (1933), Tully (1934), South Johnstone (1934), Mulgrave (1934), Pioneer (1934), Kalamia (1934), Inkerman (1934) and Maryborough (1936). Many of these sets of vertical pusher feeders were made at Cameron's Foundry, Mackay, the firm specialising in the construction of such devices. Together, both types of pusher feeders were known as double pushers. They generally increased the crushing rate by between 10 and 15%, but had a somewhat uneven feeding effect due to their intermittent operation. Therefore, double pushers had not eliminated the problem of poor feeding entirely.[54]

CSR had been experimenting with ways to improve cane preparation and extraction since the late 1910s. In 1928, the firm's engineer at Lautoka Mill (Fiji) developed what became known as a Pender leveller. This equipment consisted of a series of circular discs fitted to a shaft. At their outer edges they supported a number of blades, with the cutting edges running parallel to the shaft. By 1935, Pender levellers in place of revolving knives had been installed at Harwood, Victoria and Macknade Mills, but they never became popular and were used only to a limited extent in other Australian sugar mills. In the mid-1930s, CSR's engineering department also developed the continuous pressure feeder. This machinery was designed to replace pusher feeders and consisted of a pair of heavy, grooved rollers, usually installed ahead of the final set of crushing rollers, the bagasse being forced through a tapering chute into the mouth of the crushing rollers. The feeder rollers extracted a considerable portion of the juice from the fibre. The feed to the mill, besides being under constant pressure, was much drier and so possessed better feeding qualities, thereby increasing the capacity and efficiency of the crushing rollers. The action of the feeder was continuous and uniform, unlike the intermittent action of pushers, so this feature contributed to an improvement in crushing rates or in a lower moisture content of the bagasse, or both. The equipment, therefore, allowed larger amounts of cane to be crushed more effectively, even under heavy imbibition or maceration. However, continuous pressure feeders were very costly to install, and despite their advantages, they had only been placed in three CSR sugar mills by the outbreak of World War II, namely Hambledon (1937), Condong (1938) and Goondi (1938).[55]

Australian sugar mill-owners during the 1930s also sought to reduce fuel costs by economising on steam usage throughout the manufacturing process. This goal could be achieved by installing devices to aid in the combustion of fuel. In 1930, Millaquin Mill became the first Queensland sugar mill to be equipped with instruments that measured the efficiency of the combustion in the furnaces and operated an automatic boiler feed-water regulator. A year later, the installation of a draught fan at Cattle Creek Mill improved the draught at the boiler station, thereby cutting firewood consumption in half. A similar arrangement was added to Racecourse Mill in 1932. By 1938, many Australian sugar mills had been equipped with air preheaters or devices that used the heat recovered from the boiler flue gases to heat the air before combustion occurred in the boilers with the primary objective of increasing the thermal efficiency of the process.[56] Entirely new and more modern boilers that provided extra steam more efficiently were installed in several Australian sugar mills, including Bingera (1932 & 1938), Isis (1934), Mossman (1935) and Pleystowe (1939). Despite these advances, few Australian sugar mills on the eve of World War II could rely only on bagasse as fuel, so considerable quantities of wood and coal continued being burnt each crushing season.[57]

More economical steam utilisation also reduced fuel costs and could be achieved by a combination of various strategies. One approach favoured by many Australian mill-owners during the 1930s involved the installation of additional effets which improved a mill's overall steam economy, although it was rarely practical to go beyond five or six evaporators. Some of the last triple effets in Queensland's sugar mills were converted to quadruple effets during the 1930s and a trend had commenced towards using quintuple effets (Table 17. 2). Quadruple effets were converted to quintuple effets in a handful of Australian sugar mills, namely Mossman (by 1935), Victoria (by 1935), Hambledon (by 1935) and Harwood (1939) – the latter three sugar mills were operated by CSR. Steam or vapour bleeding – the use of vapours taken from the evaporators for juice heating and vacuum pan operation – also contributed

Table 17.2. The number of different types of effets in selected Queensland sugar mills, 1932–1940.

Year	Triple	Quadruple	Quintuple
1932	2	22	0
1933	2	22	0
1934	2	21	1
1935	2	20	2
1936	2	19	3
1937	2	16	6
1938	1	16	7
1939	0	14	10
1940	0	14	10

Note: For this entire period, thirty-three sugar mills operated in Queensland. Most likely the above figures excluded CSR sugar mills and the smaller sugar mills in Southern Queensland (e.g. Rocky Point; Eagleby).

Source: E. Behene, 'Progress in milling in Queensland factories', *Proc. QSSCT,* 1942, p. 84.

to the economical use of steam. In 1932, only one Queensland sugar mill used steam bleeding; this had risen to twenty by 1940.[58] Improved steam economy could also be achieved by the wider use of electricity to drive auxiliary machinery. New electricity-generating plants and/or electric motors that drove smaller mill units such as the centrifugals were installed in several Queensland sugar mills during the 1930s, namely Moreton (1930), Farleigh (1930), Marian (1932), Racecourse (1934), Isis (1935), Cattle Creek (1935), Proserpine (1936), Tully (by 1938) and South Johnstone (1939). Yet no Australian sugar mill had adopted the electrical drive for its milling train.[59]

Recovering more sugar from the subsider muds also attracted the attention of some Australian sugar milling companies during the 1930s. The Petree-Thomas process had not proven entirely successful. The presence of the mud had a detrimental effect on the feeding of the cane fibre through the final set of crushing rollers, and laboratory experiments had shown that the sucrose lost in the bagasse was increased by the presence of the mud. To overcome the deficiencies of the Petree-Thomas process, John Killer, the remarkably practical chief engineer at Millaquin Mill, developed the Killer system of clarification and maceration (or dilution and decantation) in 1930. Primary subsider muds were diluted with water and again subsided in settling tanks, the clear juice from this second subsidence being used for maceration, and the remaining settlings going into the mill drains.[60] The Killer process quickly gained favour, having been adopted in at least 40% of Australian sugar mills by 1935, although no CSR sugar mill had installed the arrangement, the firm opting to persist with plate and frame filters and double pressing (Table 17.3). Mill-derived pollution, however, increased as the muddy solution flowed through the mill drains into watercourses, although in some instances the polluted waters were allowed to settle in what became known as mud ponds (e.g. Pioneer Mill).

The widespread popularity of the Killer system was short-lived, although the process could be found in some Australian sugar mills in the early 1950s. Instead, Australian sugar mill-owners turned to the continuous rotary vacuum filters. These devices had become well established in the chemical industry by the mid-1920s, but had not been applied to the filtration of sugar mill muds until c. 1930. The equipment consisted of a hollow drum rotating about a horizontal axis and partially submerged in the muds to be filtered. The muds were sucked onto the mesh of the drum and sprinkled with water to elute more sucrose. The filtrate containing most of the mud's residual sugar was returned to the clarifier; the residue known as filter cake or filter press mud could be discarded, although in Australia it was sent to mud hoppers

Table 17.3. Type of filtration system used in selected Queensland sugar mills, 1935.

Type of system	Number	Name of sugar mills
Killer	13	Bingera; Millaquin; Farleigh; Racecourse; Pleystowe; Cattle Creek; Pioneer; Kalamia; Inkerman; Tully; Mourilyan; South Johnstone; Mossman
Plate and frame	8	Plane Creek; North Eton; Victoria*; Macknade*; Goondi*; Babinda; Mulgrave; Hambledon*
Petree-Thomas	1	Marian
Rotary vacuum	2	Fairymead; South Johnstone

* CSR-owned sugar mill

Source: Based upon details in 'Mill machinery lists for 23 Queensland mills', in *Handbook for 5th Congress of ISSCT,* Brisbane, Aug–Sept 1935, David North Collection MS 115, Adolph Basser Library, Australian Academy of Science.

conveniently located outside the factory and became a source of fertiliser. Rotary vacuum filters had two main advantages over other filtration methods – firstly, more sugar was recovered from the subsider muds, and secondly, less mud made its way back to the subsiders in re-circulated mixed juices.[61] By 1935, continuous rotary vacuum filters were found in at least two Australian sugar mills, namely Fairymead and South Johnstone. Over the next few years, several Australian sugar mills, including those operated by CSR, were upgraded by either the imported Oliver-Campbell brand of rotary vacuum filter or the Wizard rotary vacuum filter which was manufactured in Australia. By 1940, rotary vacuum filters were considered to be 'standard equipment' in Australian sugar mills. Initially, however, these earliest rotary vacuum filters suffered from poor filtering efficiency, but their operation was greatly improved by mixing the mud with sufficiently fine bagacillo (i.e. the fraction of fine fibre particles from bagasse) to make it more adhesive.[62]

Improved juice extraction increased the volume of syrup that had to be processed at many Australian sugar mills during the 1930s. Consequently, the crystallisaton station needed to be expanded to match the crushing capabilities of the milling trains. During the early 1930s, the number of vacuum pans was increased at several Australian sugar mills, including Cattle Creek (1932), Racecourse (1932), Maryborough (1932), Proserpine (1936) and Macknade (1939). New and enlarged vacuum pans were also installed at a handful of Australian sugar mills, namely Hambledon (1934), Pleystowe (1934), Broadwater (1938) and Fairymead (1939). Queensland's first Webre calandria pan became operational at Tully Mill in October 1935. This unit contained screw propellers or impellers that introduced mechanical circulation into the pan in an attempt to overcome the defects of natural circulation in a vacuum pan. Its performance lived up to expectations – faster boiling, especially with low-grade massecuites.[63]

The operation of these vacuum pans also came under greater scientific control following the identification of the conductivity-viscosity-supersaturation relationship in Java during the late 1920s and early 1930s. Conductivity is approximately inversely proportional to the viscosity of a solution, so the conductivity of a solution of electrolytes decreases with the viscosity, due to the effect of viscosity in retarding the movement of ions in the solution. By using electrodes of a robust type suitable for installation in a vacuum pan, together with simple instruments to record the conductivity of the boiling juice, the entire pan operations could be placed under complete scientific and automatic control, resulting in the production of better quality massecuite. During 1934 and 1935, pan control based on conductivity was installed in at least six Queensland sugar mills, namely Bingera, Millaquin, Pioneer, Kalamia, Tully and Mourilyan. Three years later, E. (Roly) Behne and G. Jenkins of the BSES claimed that practically every vacuum pan in the state had been equipped with a conductivity set. Boiling massecuite automatically under complete control and to a predetermined schedule was now being undertaken in several Queensland sugar mills.[64]

External forces also prompted the production of better quality massecuite at Australian sugar mills during the mid-1930s. Prior to 1933, Australian sugar boilers aimed to produce a massecuite which would centrifugal easily with a resultant sugar of required polarisation and net titre. Refiners, however, desired a raw sugar that would be easy to filter (i.e. known as its filterability) and of a colour that could be easily decolourised in the char filters. In 1933, Australian sugar mills produced raw sugar with irregular and fine grains, and poor filterability, traits not favoured by either the Australian refiners or the overseas sugar refineries which increasingly processed the bulk of Australia's raw sugar.[65] The adoption of the conductivity method assisted in producing the type of sugar required by the refiners. Moreover, changes in the pan stage of Australian sugar mills had a beneficial effect on the centrifugal station. Improvements in the quality of the massecuite meant it fugalled more freely. Steam or water washing was reduced to a minor amount or eliminated completely, thus reducing the demand for steam and the costs of operating the centrifugal station. This trend also resulted in a smaller return of sugar to the pan stage as molasses, effectively increasing the yield of sugar per tonne of massecuite.[66]

Clarification also came under greater scrutiny in Australian sugar mills during the 1930s. At the beginning of the decade, the amount of lime added to the juice so that it reached the correct pH level was controlled by colorimetric determination in the sugar mill (i.e. use of test papers and colour charts), checked with the quinhydrone electrode in the laboratory. Automatic control had been adopted in at least Tully, Kalamia and Babinda Mills by 1935, using the imported Leeds and Northrup recording potentiometer with Micromax controlling mechanism and antimony electrode. These devices did not become popular due to their high cost. Instead, the BSES pH meter was preferred, with sixteen Queensland sugar mills having adopted the equipment by 1937. This greater emphasis on more accurate pH determination of the juice during clarification occurred as a result of the introduction of fractional liming. First used at Millaquin and Bingera Mills in 1936, fractional liming was then adopted at other Queensland sugar mills that crushed an increasing proportion of the P.O.J. cane varieties. [67]

By the outbreak of World War II, the average annual crushing per sugar mill in Australia stood at around 180 000 tonnes, almost 65 000 tonnes more than 1931 (Figure 17.1). This additional cane had been handled by quite a substantial increase in the average hourly crushing rate of non-CSR mills from just under 47 tonnes per hour in 1933 to nearly 63 tonnes per hour in 1939 (Figure 17.2). Yet the average figures hide some significant variations in the performance of the country's sugar mills. At one end of the spectrum were the sugar mills that crushed record amounts of sugar cane each year. During the 1930s, these mills included South Johnstone (1930), Mulgrave (1931, 1935, 1936), Tully (1932, 1933, 1938), Inkerman (1934) and Victoria (1937, 1939). These sugar mills consistently processed between 230 000 and 310 000 tonnes of cane annually. In contrast, the Queensland sugar mills that had not brought their equipment up to modern standards were discussed by Dr H. (Bill) Kerr during his evidence to the Royal Commission into Sugar Peaks in 1939. Kerr remarked that the present low sugar prices had precluded machinery upgrades at these sugar mills, and the owners were often only able to carry out normal maintenance work by levying their suppliers. Rocky Point Mill was one of these factories. It handled 10 000–15 000 tonnes of cane annually, but this low throughput meant the sugar mill only worked two shifts daily in 1939. Frank Heck, the mill's owner, claimed this arrangement was less economical than the continuous running which characterised many of the larger Australian sugar mills.[68]

Mobilisation for war did not immediately curtail some of the planned mill overhauls and upgrades. At Pleystowe Mill, further factory electrification

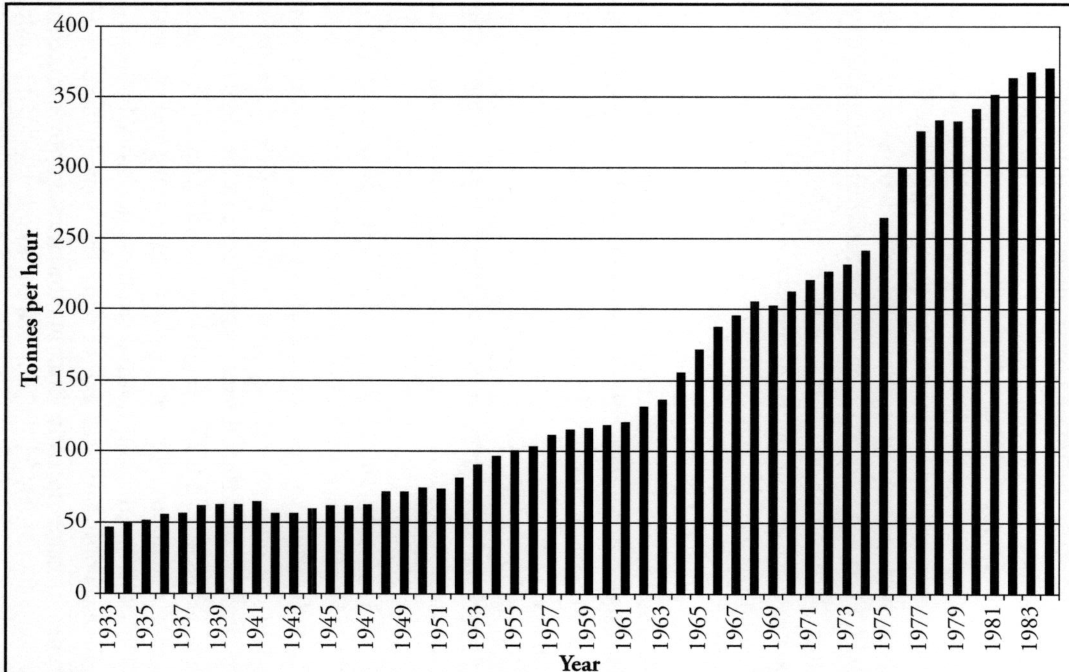

Note: From 1933 to 1955, the sources do not mention if the performances of CSR sugar mills were included in the figures. From 1956 to 1975, the figures definitely exclude CSR sugar mills. From 1976 onwards, figures include the performance of CSR sugar mills.

Figure 17.2. The average crushing rate (tonnes per hour) at Queensland sugar mills, 1933–1984.

Source: Compiled from figures in the *Year Book of Australia*, *ASYB* and the BSES, ARs.

commenced in 1940, when a number of existing steam driven units were converted to electric motor drive and a new electricity generating plant was installed to provide for immediate requirements and to leave a reasonable margin for future expansion. A cane washing unit or cane cleaner was installed at Fairymead Mill in 1940, following the Fairymead Sugar Co. Ltd. introducing mechanical harvesting on its plantation (see Chapter 16 for full details). High pressure sprays and electrically-driven stripping rollers in the cane cleaner removed the trash and dirt from the mechanically harvested cane. Maintenance costs for this unit, however, were quite high until the initial operating problems were resolved. At the Maryborough Sugar Factory, mechanical cane unloaders were erected for the 1941 crushing season, saving two men per shift at a time when labour was scarce.[69] However, capital investment in the Australian sugar mills was curtailed substantially and annual mechanical overhauls were restricted between 1942 and 1945, given the shortages of labour and spare parts.

As the war progressed, Australian sugar mills were placed on emergency footing. They became covered by protective shields of camouflage paint and camouflage nets. The floors of sugar storage areas were kept clear of flammable material to reduce the risk of fire from incendiary bombs. Volunteer guard duty from May until the end of the crushing season was undertaken by mill staff or retired soldiers to prevent any acts of sabotage which could restrict or prevent crushing. Additions to sugar storage facilities at each Australian sugar mill enabled them to cope with the slower transport of bagged sugar due to a shortage of ships. Following the outbreak of hostilities with Japan in December 1941, more precautions were taken to protect vital sections of the sugar mills against air raid damage. Air raid shelters were constructed for mill employees. A small army group moved through the North Queensland sugar mills, inspecting and painting white crosses on equipment at varied spots throughout the factories. Questioned by mill managers, they admitted that the sites marked were for the placement of demolition charges so that the machinery would not fall into the hands of the invading Japanese. Crushing in the sugar mills during 1942 and 1943 occurred under 'blackout' conditions, with considerable expenditure incurred to comply with blackout standards, whilst at the same time providing lighting for employees to efficiently operate equipment and machinery.[70]

Shortages of mill workers became a major concern during 1942, even though the Australian sugar industry had become a 'protected industry', and exemptions from war service were given to electricians, engineers and tradesmen holding key positions essential to keeping the sugar mills operating. By then, women had gained employment in Australian sugar mills as laboratory assistants or weighbridge clerks, and when qualified, as check chemists or cane testers. However, there were not enough mill workers and canecutters available during the 1942 crushing season to maintain continuous crushing. Many Queensland sugar mills worked for only two shifts or on occasions were forced intermittedly to close down for several days. At the end of 1942, the state's sugar mills had crushed at 80% of their pre-war capacity.[71]

Following its acceptance of most recommendations from the 1942–1943 Mansfield Royal Commission into how the Queensland sugar industry could cope with the 1943 crushing season, the Queensland government (Cooper; Labor) restricted each sugar mill to an average of 75% of its previously best seasonal weekly crushing rate for the 1943 season. Consequently, many Queensland sugar mills north of Mackay again worked only two shifts daily; the remainder of the state's sugar mill's worked three shifts daily but with reduced hourly crushing rates. Most Australian sugar mills, however, ceased operations at times to avoid exceeding the weekly tonnages stipulated by the government regulations. Crushing capacity fell to 63.9% of normal production. Some easing of this restriction occurred during 1944 and 1945, and several Australian sugar mills resumed three shifts (e.g. Victoria and Macknade).[72]

Post war mill overhauls and upgrades, 1946 to 1969

During the late 1940s, all Australian mill-owners embarked upon a renewal process, in some instances outlaying large sums of capital to acquire new or second hand machinery. The upgrade to Babinda Mill cost £475 000; at Mulgrave Mill, £400 000 was spent on new machinery. More modest outlays of £40 000 and £60 000 improved the machinery at Cattle Creek Mill and Isis Mill, respectively.[73] Some of this machinery was again sourced from overseas, although local machinery manufacturers such as Walkers Ltd. (Maryborough) and Cameron's Foundry (Mackay) were kept busy supplying smaller pieces of equipment. Yet because so many Australian sugar mills were engaged in upgrades, other mill-owners were able to secure second-hand machinery for their factories. Two examples will illustrate this trend. W. H. Heck and Sons Pty. Ltd. improved Rocky Point Mill in 1946 by purchasing a surplus continuous subsider and juice heater from the Maryborough Sugar

Factory. One improvement at Cattle Creek Mill in 1948 included the addition of a set of second-hand crushing rollers obtained from Plane Creek Mill.[74]

All these upgrades, however, occurred against a backdrop of post-war shortages in materials and delays in delivering machinery, rail and shipping freight increases of about 40% and increases in the cost of inputs such as coal, wrapping materials, container board, paper and jute sacks. The introduction of the forty-hour week in 1948 also reduced the capacity of Australian sugar mills to handle the larger crops being harvested without commencing the season unduly early or crushing after Christmas. Macknade and Victoria Mills, for example, were forced to commence crushing in May 1948 in order to process all the cane grown in their mill supply areas.[75]

The capital expenditure at the Australian sugar mills during the late 1940s was directed to a range of priorities. In some instances, the sugar mill's overall capacity to handle larger crops was improved by the duplication of existing machinery or upgrades to the capacity of existing machinery. New boilers to provide the extra steam needed for enlarged crushing trains were installed at several Queensland sugar mills, namely Racecourse, Mulgrave, Mossman and Isis. A second Wizard rotary vacuum filter was installed at Isis Mill in 1948.[76] Improved extraction was sought by alterations to the milling trains. At Cattle Creek Mill, for example, new sets of crushing rollers were installed in 1948 and 1950, replacing the smaller 1.37 m wide crushing rollers with crushing rollers having a width of 1.5m. Even larger sets of crushing rollers, ranging from 1.83 m to 2.1 m in width, were installed in several sugar mills, including Racecourse (1946), Fairymead (1948), Kalamia (1949) and Babinda (by 1950). CSR installed a Searby shredder at its Victoria and Goondi Mills in 1946 and 1949, respectively.[77] Other Australian sugar mill engineers brought the machinery in their factories up to standards that had existed in some Australian sugar mills before World War II. The triple effet at Mt Bauple Mill, for example, was replaced in 1949 with a quadruple effet purchased from Racecourse Mill. The Killer process at Racecourse Mill was abandoned after the installation of a Wizard rotary vacuum filter in 1946. The process of electrification continued in some Australian sugar mills, namely Racecourse (1946) and Isis (1947).[78]

As the immediate post-war upgrades occurred in Australia's sugar mills, the ASPA expressed dissatisfaction with the BSES's performance, especially in the area of sugar mill technology. The organisation began promoting the idea in 1948 of a more intensive research effort directed towards the problems associated with sugar milling. A conference of mill-owners in January 1949 supported the ASPA's proposal to form a Sugar Research Institute (SRI), its operations being based in Mackay. Initially with twenty-six members, the SRI grew to incorporate every Australian sugar mill when CSR joined as a member in March 1976.[79] Details about its operation are documented thoroughly in Jane Reid's history of the organisation and will not be repeated in this account, although some of its main achievements include the development of cane transport scheduling programs, a SRI clarifier that drastically reduced sugar losses and lime consumption, the treatment of sugar mill effluent and air pollution control at sugar mills (see discussion later in chapter). Running parallel to these investigations was research by the BSES Division of Mill Technology. Its inquiries, suspended entirely during World War II, had resumed under the direction of Don Nicklin, who guided the section until 1969; his place was then taken by Phillip Atherton. Topics investigated by the BSES mill technologists during the 1950s and 1960s included the benefits of different clarifying agents such as magnesium oxide and soda ash, by-product utilisation, deterioration of chopped cane, juice scale control of sugar mills and the performance of different types of centrifugals. In addition, the BSES Division of Mill Technology continued offering advice to mill-owners on a wide range of topics, including electricity generation and distribution, boiler testing and evaluation, effet capacity testing and sugar quality problems, and the Metrology Section continued serving the industry by standardising factory analytical equipment. Revisions of the *Laboratory Manual for Queensland Sugar Mills* were published in 1954 and 1961.[80]

Machinery details on thirty of the thirty-five Australian sugar mills were published for the first time in the 1952 edition of the *Australian Sugar Year Book*. From this information, it was possible to conclude that 80% of the country's sugar mills were equipped with four sets of crushing rollers. Nearly half of the sugar mills now relied solely on either bent or straight revolving knives as their preparatory device, the National shredder having fallen out of favour. Despite its advantages, the Searby shredder remained a feature in only seven factories, mostly CSR-owned sugar mills. The continuous pressure feeder had been installed in few Australian sugar mills. Juice clarification in conical subsiders only occurred in half the Australian sugar mills, while four Australian sugar mills were equipped with a combination of conical subsiders and multi-tray clarifiers. The remaining ten Australian sugar mills relied solely on multi-tray clarifiers, mostly a Bach subsider, which was characterised by the injection of steam into the mud compartment at each revolution of a stirrer, in order to vent any gas which accumulated in the juice as it was being clarified. The new types of multi-tray clarifiers had some advantages, including their compactness and vapour sealing and were considered by some engineers to be an improvement on the conical subsiders. Half

Plate 17.2. Installing new effets at Moreton Mill, 1952.

Source: Image reproduction courtesy of Sunshine Coast Libraries, Picture Ref. No. M754209.

the Australian sugar mills were now equipped with one Wizard or Oliver Campbell rotary vacuum filter and the clarified juice was boiled in either quadruple or quintuple effets. Most Australian sugar mills had three to six vacuum pans, which were overwhelmingly the coil variety. Some vacuum pans were the Halpin coil pan, designed by J. Halpin, the Chief Engineer at Pleystowe Mill. In this unit, the heating surface consisted of a flattened coil section with the tubes in a concentric circular array. A few Australian sugar mills operated calandria pans, including the Webre pan.[81] Rocky Point, Qunaba, Pioneer and Moreton Mills and the Maryborough Sugar Factory remained the country's least technologically advanced sugar mills. Each still only had two or three sets of crushing rollers, and employed plate and frame presses or the Killer process.

Following the Central Board's approval of increased mill peaks during the early 1950s, major upgrades occurred in many Australian sugar mills. Considerable sums of capital were invested in these upgrades. In February 1953, the Australian Sugar Co. Pty. Ltd. announced that the firm intended to spend £350 000 over the next three years on improvements at its Mourilyan

Mill. A similar amount was spent on modernising Cattle Creek Mill. CSR devoted £750 000 towards modernising its Goondi Mill, but the single largest upgrade occurred at Victoria Mill, where CSR spent £1.3 million on duplicating the milling train and associated plant, so that the factory could handle 21 000 tonnes of cane weekly. Even the owners of Rocky Point Mill embarked upon a modernisation program, following advice from the BSES in 1954 on ways to improve the sugar mill's capabilities. Overall, the country's mill-owners outlaid £23 million on improvements to their sugar mills and transport networks between 1950 and 1955.[82]

The bigger crops and growing demand for Australian sugar, however, led to another round of mill upgrades between 1958 and 1968. Large amounts of capital were again devoted to these mill improvements: Pioneer Sugar Mills Ltd.'s major extensions and improvements in mill plant during the early 1960s cost an estimated £250 000; the Haughton Sugar Co. Ltd. spent £490 000 on additional plant and equipment between 1964 and 1968; nearly £170 000 was outlaid to purchase new equipment for the Maryborough Sugar Factory during the 1960s; and the Moreton Central Sugar Mill Co. Ltd. undertook an expansion program that cost £150 000. In New South Wales, CSR embarked upon a major modernisation program at its factories, after threats to close the sugar mills prompted an upsurge in cane cultivation during the late 1950s (see Chapter 10). By 1961, CSR had spent £1.1 million on major upgrades at its Condong, Harwood and Broadwater Mills.[83]

Improvements to the milling train were the focus of many upgrades during the 1950s and 1960s. Mills without shredders or the old-fashioned National shredders acquired Searby shredders, so that they had become almost universal by 1969 (Table 17.4). The long intermediate carriers and large maceration baths were abandoned. Under this arrangement, known as the 'Nobel' maceration system, the crushed fibre was washed with a countercurrent of hot juice and water, and could take twenty to thirty minutes to pass through the milling train. However, such a slow process could not be maintained when so much additional sugar cane had to be crushed.[84] Extra sets of crushing rollers were added to the milling trains in most Queensland sugar mills over this two decade period. In 1951, seven Queensland sugar mills had less than four sets of crushing rollers (Table 17. 4). During the 1950s, a fourth set of crushing rollers was added to some of these sugar mills, namely Pioneer (1952), Qunaba (1954), Moreton (1956) and Invicta (1958). These sugar mills now had a crushing train similar to that reached in many Australian sugar mills during the 1930s. Other milling companies copied the innovation which first appeared in Australia during the late 1930s – a fifth set of crushing rollers.

Table 17.4. Statistics on milling train arrangements in Queensland's sugar mills, 1951 and 1969.

	1951	1969
No. of sets of crushing rollers	122+	158
No. of mills with two or three sets of crushing rollers	7	1
No. of mills with four sets of crushing rollers	21	10/1*
No. of mills with five sets of crushing rollers	3	10/1**
No. of mills with six or seven sets of crushing rollers	0	9
No. of mills with a shredder and a Searby shredder (in brackets)	17 (7)	26 (21)

Notes: Details about the milling train for one Queensland sugar mill were not provided. It was assumed to have at least three sets of crushing rollers. * 'B' Train at Victoria Mill. ** 'A' train at Victoria Mill.

Source: Based upon machinery details in the *ASYB,* 1952 and 1970.

Such upgrades occurred at several sugar mills, including Bingera (1953), Fairymead (1958), Gin Gin (1963) and Mossman (1964). Some mill-owners, however, did not stop at five sets of crushing rollers as they sought to handle the larger crops being produced by their suppliers. During the 1960s, a sixth set of crushing rollers was installed at several Queensland sugar mills, namely Mossman (1965), Racecourse (1966), Bingera (1967) and Marian (1967).[85] By 1969, mill engineers at Pioneer and Inkerman Mills had gone even further, installing a seventh set of crushing rollers. Consequently, there were 158 sets of crushing rollers in Queensland by 1969, 36 more than in 1951 (Table 17.4).

The exception to this trend during the 1960s was the Fairymead Sugar Co. Ltd, the firm deciding that improved extraction could be obtained by diffusion, not extra sets of crushing rollers. In 1965, a Burnett diffuser was erected between the second and third sets of crushing rollers at Fairymead Mill, the crushed fibre being washed by a countercurrent of hot water and juice before being subjected to extra pressure in the third, fourth and fifth sets of crushing rollers. At the time of its erection, it was the world's largest single-unit diffuser.[86]

The engines powering the crushing rollers also changed during the 1950s. In 1951, a new set of crushing rollers with 2.1 m wide rollers installed at South Johnstone Mill was driven by a steam turbine engine, the first to be used in an Australian sugar mill. Imported from Mirrlees Watson Co. Ltd., Scotland, the success of this innovation then prompted other Australian mill-owners, albeit slowly initially, to replace the traditional reciprocating or piston engines that powered the crushing rollers. By 1965, the BSES reported that when new sets of crushing rollers were installed or it became necessary to increase the speed and power of existing sets of crushing rollers, the choice of a turbine had become 'universal'. During the 1965 season, twenty-eight turbines ranging from 600–1 000 hp replaced existing reciprocating engines, so that half of all sets of crushing rollers in Australian sugar mills were powered by turbine engines.[87]

Major changes also occurred in the feeding of the shredded pieces of cane through the crushing rollers. During 1953, automatic carrier speed control equipment was installed and operated for the first time at Proserpine, Tully and Mourilyan Mills. This equipment enabled the carrier to be slowed or shut down quickly if blockages arose, a distinct advantage when trying to handle larger volumes of cane.[88] Many Australian sugar mills continued relying on pusher feeders to maintain an adequate feeding of the crushing rollers, as continuous pressure feeders remained expensive. However, experiments in 1954 at Fairymead and Pleystowe Mills with an apron feeder or conveyor, another type of continuous feeding device, proved successful. This equipment was arranged so that the head of the conveyor pressed the cane fibre against the top roller. Apron feeders were then adopted as a cheaper alternative to continuous pressure feeders, so that Australian sugar mills during the mid-1950s were equipped with combinations of pushers, continuous pressure feeders or apron feeders.[89]

H. (Bert) Donnelly, Chief Engineer at Pioneer Mill in the late 1950s, recognised that pushers were at their limit at around 100 tonnes per hour, but that the crushing rollers needed to be worked at a higher rate to handle the larger crops. Moreover, overseas sugar mills operated effectively without pushers, relying on gravity feeding alone. With assistance from Dr H. (Bill) Kerr of the SRI, he conceived the idea of improving roller feed by eliminating the pusher and allowing the fibre to fall through a long, near vertical chute, the pressure at the bottom of such a chute being sufficient to give a substantial degree of compaction to the feed and to act as an effective feeding device (Plate 17.3). Successful trials of a Donnelly chute or the high, gravity feed, closed chute occurred at Pioneer Mill during the 1956 and 1957 crushing seasons. Knowledge of their success spread quickly – Mossman Mill had installed Donnelly's chutes for the 1958 crushing season – and they rapidly replaced the horizontal and vertical pushers at Australian sugar mills during the late 1950s and early 1960s. Their uptake was rapid because they delivered a better performance than pushers, cost considerably less than pressure feeders, and contained no moving parts that could break. By 1966, the BSES reported that only six sets of pushers were now used in

Plate 17.3. A Donnelly chute on No. 2 set of crushing rollers, Pioneer Mill, 1958.

Source: *ASJ*, June 1958, p. 233. Image reproduction courtesy of the Australian Cane Farmers Association Limited.

Queensland's sugar mills. The late John Kerr, the author of several sugar mill histories, has claimed that the Donnelly chute was perhaps the single most successful invention in sugar milling technology during the last half century, having been widely adopted throughout the world to achieve higher crushing rates. [90]

Donnelly chutes improved feeding in the milling train, but Australian research into the topic continued during the 1960s. In 1961, Don Shann, a SRI researcher conducting measurements at Fairymead Mill, concluded that it was possible to develop an effective, yet inexpensive light-duty feeder. A light feeder or an SRI feeder consisted of two, light cast iron rollers grooved similarly to the mill rollers and driven from the top and feed roller shafts by roller chains and sprockets. These rollers compressed the feed slightly without expressing juice, and a closed chute presented the compressed mat of fibre to the crushing rollers. Some feeding force was available as in the continuous pressure feeder, but the device, being of light construction and not designed to express juice, was much cheaper and simpler than the continuous pressure feeder – they could be readily fabricated in sugar mill workshops and even installed during a weekend. Within three years of its development, sixty such feeders had been installed in Queensland sugar mills. They were often used in conjunction with Donnelly chutes which fed the compressed mat of fibre into the crushing rollers. In addition, CSR had also showed a renewed interest in this field, testing a toothed pressure feeder with spiky rollers during 1963. The device was patented by CSR in 1966 and it soon appeared in other Australian sugar mills. By 1970, at least half the intermediate crushing rollers in Australian sugar mills had been equipped with toothed pressure feeders.[91]

Upgrades also occurred at other Australian sugar mill stations during the 1950s and 1960s. The clarification capabilities of Australian sugar mills were enhanced, following complaints from Canadian refiners about the filterability of Australian raw sugar in 1954, and the need to match the increased crushing rates. CSR's research during the mid-1950s into the clarification process revealed that the turbidity (or clarity) and phosphate content of the clarified juice affected filterability in the refinery. Inadequate clarifier capability and poor flocculant (i.e. chemical precipitate) formation contributed to these features. To remedy these deficiencies, Australian sugar mills during the late 1950s and 1960s were equipped with extra clarifiers and the old-fashioned conical subsiders were replaced with the multi-tray Dorr clarifier or Bach subsiders, which both gave greater capacity for the same floor space.[92] By 1969, conical subsiders were only found in six Queensland sugar mills –Tully, North Eton, Cattle Creek, Maryborough, Rocky Point and Moreton. Synthetic polyelectrolytes or flocculants such as Dow E.T. 81, Lytron X-886, Sedipur TF2 and Separan AP273 improved clarification, and by the mid-1960s were used at all CSR mills and all the Lower Burdekin, Bundaberg and Mackay sugar mills.[93]

Muds from the clarifiers ceased to be treated in the out-dated plate and frame presses or by the Killer process, as all Queensland sugar mills were equipped with rotary vacuum filters during the 1950s. Additional sets of these filters were added to many Queensland sugar mills so that the total number per mill had risen to two or three by 1969 – Tully and

Victoria Mills even had four rotary vacuum filters by 1969. Included in these upgrades were new types of filters. Eimcobelt filters had become popular in the Mackay district sugar mills by 1969. This device used a cloth filter medium that passed around a series of rollers after it left the rotating drum. The cloth was then washed thoroughly before returning to the drum for the next cycle. In the late 1960s, CSR had also commenced experimenting with a Feinc string discharge filter at Victoria Mill. This type of filter also used a synthetic cloth as a filter medium, around which passed a series of strings, set close together; the strings lifted the filter cake from the cloth when they left the main drum, discharging it when they passed over a small roller before returning to the filter drum. These new devices assisted in the recovery of more sugar from the muds.[94] Overall, seventy-three rotary vacuum filters were found in Queensland sugar mills in 1969, compared to twenty in 1951, and the treatment area stood at 2 918 m^2, almost five times the area that existed in 1951 (Table 17.5).

Extra juice could only be handled efficiently by improvements in the sugar boiling station, so new, enlarged and extra effet vessels were installed at all Australian sugar mills during the 1950s and 1960s.[95] A few Queensland sugar mills, namely Babinda, Mourilyan, Victoria, Macknade and Inkerman, were even equipped with two sets of quadruple or quintuple effets. Consequently, the total area of evaporating equipment in Queensland sugar mills doubled between 1951 and 1969 (Table 17.5). Some evaporating vessels in these sets of effets were also fitted with entrainment arrestors – equipment placed at the top of the vessel that forced a change in the direction of the vapour leaving the evaporator. This promoted the coalescence of small liquid droplets to form larger ones which fall back into the boiling liquid, thereby avoiding the loss of sucrose. These devices were various shapes and consisted of different materials, including a wood maze separator comprised of grid layers of timber strips, layers of metal wire mesh known as the Yorkmesh demister, and an umbrella-shaped separator made from metal.[96] In the crystallisation stage, an additional eighty vacuum pans were added to Queensland sugar mills between 1951 and 1969, leading to two and a half times more overall volume in these pans. The calandria type of vacuum pan became more popular during the early 1960s, so that approximately equal numbers of coil and calandria pans were used by 1969 (Table 17.5).

Continuous centrifugals to handle low-grade massecuite (i.e. a mixture of molasses and sugar crystals of variable shape) had also made their appearance in several Australian sugar mills during the early 1960s, including Condong (1962), Farleigh (1962), Plane Creek (1962), Gin Gin (1963) and Broadwater (1965). By 1968, 28% of the low-grade centrifugals at member sugar mills in

Table 17.5. Statistics on the clarification, evaporation and crystallisation stations of Queensland's sugar mills in 1951 and 1969.

	1951	1969
No. of conical subsiders	65	15
No. of tray subsiders	18	88
No. of rotary mud filters	20	73
Total treatment area of mud filters (m^2)	604.1	2 918
No. of mills with a single set of triple effets	1	0
No. of mills with a single set of quadruple effets	9	14
No. of mills with a single set of quintuple effets	19	10
No. of mills with more than one set of effets	0	7
Total area of evaporating equipment (m^2)	85 220	171 153
No. of vacuum pans	146	227
No. of coil pans	84	78
No. of calandria pans	22	83
Total volume of pans (m^3)	2 776	6 588

Notes: (1) Descriptions of the machinery in the New South Wales sugar mills were not listed in the 1952 edition of the *ASYB*, but do appear in the 1970 edition. Therefore, to make a proper comparison, the details on the three New South Wales sugar mills have been omitted. (2) Details on the number of mud filters and treatment area for some sugar mills were not included in the machinery descriptions. Hence, the figures are under-estimates. (3) Details about vacuum pans did not always distinguish their type. Hence, the figures on the number of coil and calandria vacuum pans are underestimates.

Source: Based upon details in the *ASYB*, 1952 and 1970.

the Queensland Mutual Control Scheme were this type of machine; a decade later they had virtually replaced their batch counterparts. This new equipment separated sugar from molasses in a continuous process. Massecuite flowed into the centrifugal in a continuous stream and the molasses and sugar were discharged at separate outlets. Continuous centrifugals had several advantages over the batch type, including lower initial purchase cost, reduced maintenance costs, less power use and approximately twice the throughput of their batch counterparts. Yet these devices had one major problem – they caused excessive damage to sugar crystals which was unacceptable in the production of A and B grade sugars (i.e. high quality standard shipment sugar) because the Queensland Sugar Board required specific crystal shapes and sizes.[97]

Extra steam was needed to drive these additional sets of enlarged crushing rollers, bigger clarifiers and larger vacuum pans. Therefore, boiler upgrades became necessary at many Australian sugar mills between 1954 and 1966.[98] The amount of steam generated by these boilers also increased – thirteen new boilers became operational during 1965 and of these only a third had a capacity of less than 45 350 kg of steam per hour, and the largest could produce 63 500 kg of steam per hour. This amount, however, was eclipsed in 1966 by the commissioning of a new 136 100 kg of steam per hour boiler at Pleystowe Mill. Baggase continued to be the main fuel used in these boilers, but the limited bagasse storage facilities at Australian sugar mills changed during the mid-1950s. A separate bagasse storage facility was commissioned at Pleystowe Mill during 1953 – a first in Queensland. Rubber conveyor belts for the storage and return of bagasse were the only connections between the boiler station and its storage facility. With this arrangement, bagasse handling became a part-time job for only one worker, so it was a major labour-saving development that soon became a feature at other Australian sugar mills during boiler station upgrades.[99]

Electrification of the auxiliary power drives in Australian sugar mills continued during the 1950s. A 2 000 kW turbo-alternator was installed at Inkerman Mill in 1953 – the largest low voltage generating unit in Australia at the time. By 1956, twenty-seven Queensland sugar mills (87%) had been equipped with turbo-alternators. Gin Gin Mill, one of the last Queensland sugar mills to use electricity for lighting only, was finally equipped with an electricity generating plant in 1959, having previously relied on steam only.[100] All new generating plants complied with the standard voltage and frequency used by the regional electricity boards in the sugar-producing districts, facilitating the use of regional electricity board power during the slack season and weekends when the sugar mills were not operational, and permitting the export of surplus electricity to regional electricity boards during the crushing season. Further modifications to the electrification of the Australian sugar mills occurred during the period 1964–1967, when the number of high voltage schemes increased from two to twelve, following the decision to change the system of generation and distribution from 240 or 415 to 3 300 or 11 000 volts. The power was generated at high voltage and distributed with high voltage cable to strategic points around the sugar mill where the step-down transformers broke down the voltage to the required level. Eight factories decided upon 3 300 volts; the remainder 11 000 volts. Fairymead Mill and the three Lower Burdekin sugar mills were amongst those factories equipped with transformers.[101]

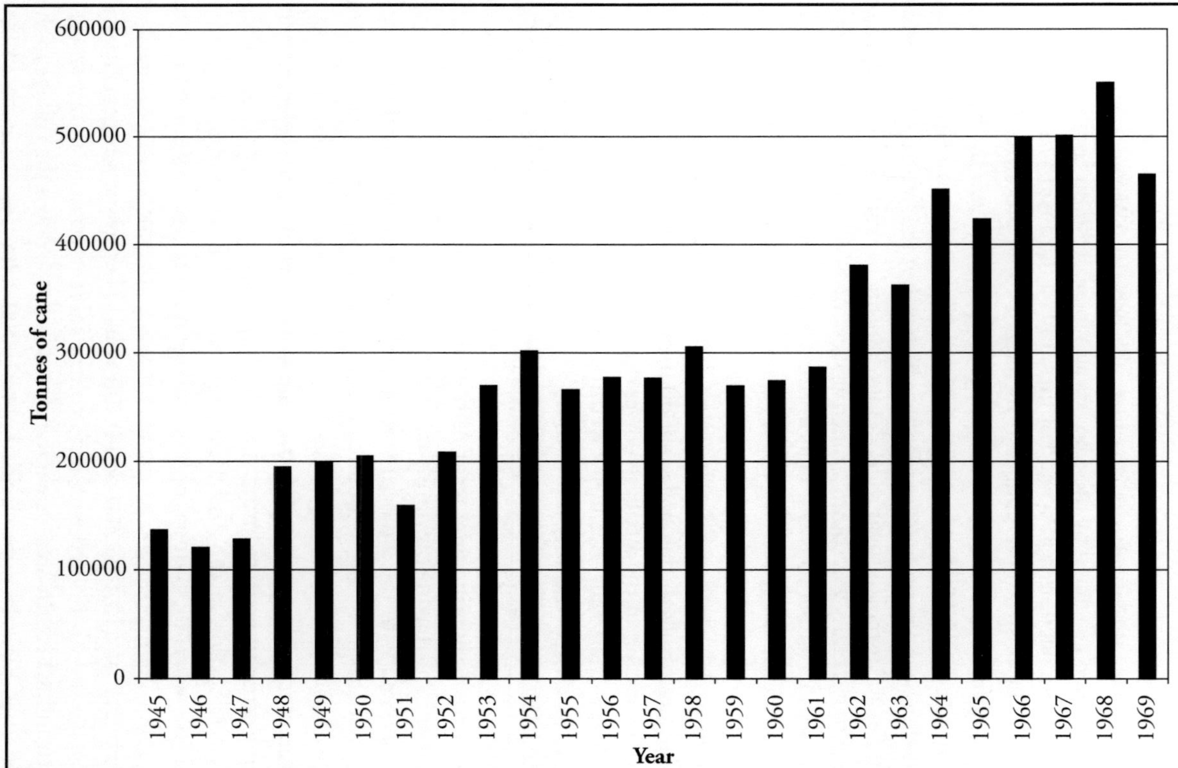

Figure 17.3. The average tonnage of cane crushed yearly at Australian sugar mills, 1945–1969.

Source: Compiled from figures in the *Year Book of Australia*, various issues.

Before 1951, Tully Mill was the only Australian sugar mill with an average crushing rate that had exceeded 100 tonnes of cane per hour. By the mid-1950s, many Australian sugar mills had exceeded that level – South Johnstone Mill had even reached 150 tonnes of cane per hour, the highest rate for an Australian sugar mill with a single crushing train. Moreover, the hourly crushing rate continued to climb in Australian sugar mills. By 1965, Inkerman Mill handled 305 tonnes of cane per hour.[102] Overall, the average hourly crushing rates for the non-CSR Queensland sugar mills trebled during this period, increasing from around 60 tonnes per hour in 1945 to just over 200 tonnes per hour in 1970 (Figure 17.2). As the Australian sugar mills processed more cane per hour, they began breaking records for the overall amount of cane processed during a week and a season. In 1954, Inkerman Mill crushed 521 323 tonnes of cane, the first occasion any Australian sugar mill with a single crushing train handled over half a million tonnes of cane. In addition, seven Australian sugar mills with single crushing trains – South Johnstone, Macknade, Proserpine, Farleigh, Pleystowe, Marian and Plane Creek – crushed greater than 18 000 tonnes of cane weekly for the first time. Consequently, the average amount of cane crushed yearly per Australian sugar mill rose from around 200 000 tonnes to approximately 500 000 tonnes (Figure 17.3). Rocky Point Mill, Australia's smallest sugar mill had even been transformed, managing to work three shifts during the crushing season since 1959.[103]

A growing concern about the environment

Sugar production in Australia was traditionally a polluting industrial activity. Large quantities of polluted water could be generated at a sugar mill, a result of tainting with the following contaminants: scale from cleaning evaporating equipment and juice heaters; cleaning agents such as caustic soda; grease, oil, bagasse and bagacillo; syrup spillages at the centrifugals; sugar lost at weekend cleaning of evaporating vessels; and filter mud. These untreated waste waters were usually discharged into drains that emptied into the nearest natural watercourse, where the mill-owner hoped that the diluting effect of the receiving water would rid them of the problem. Fly-ash from the sugar mill's boilers and from some larger particles of unburnt bagasse polluted the air. Stack emissions from mill chimneys often resulted in nearby settlements being covered in fly-ash during the crushing season, while the residents of those

Australian towns established close to sugar mills (e.g. Nambour; Tully; Mossman) were subjected to increased levels of noise, a result of the operation of machinery and the movement of locomotives and rail trucks. Moreover, the manufacture of sugar produces three by-products: bagasse; molasses and filter mud. Each could become a major pollutant if their final disposal was not handled correctly.

Watercourses throughout coastal eastern Australia quickly became 'rivers of refuse' following the establishment of sugar mills during the late nineteenth century. Residents of Waterford, near Beenleigh, for example, complained in 1882 that all the 'refuse' from Waterford Mill was being discharged into the Logan River, creating a 'bad stench' and ruining the drinking water. In 1894, Sandy Creek, downstream of CSR's Homebush Mill, was described as a 'seething mass of corruption with thousands of fish floating dead upon its surface or stretched out on its banks'.[104] A few Australian sugar mill-owners, however, tried reducing their environmental impact. By 1912, the untreated waste waters from Homebush and Farleigh Mills, for example, were discharged into storage ponds that had been created by building a dam across a watercourse near the sugar mill. These waste waters were then discharged only during periods of heavy flow in nearby rivers. Molasses was burnt in the boiler furnaces or converted into a stock food supplement at some Australian sugar mills during the 1900s (e.g. Fairymead; Childers). Australian distilleries also used some molasses to manufacture rum. These activities, however, hardly impacted upon the volume of the product being run into the watercourses. Municipal councils in Queensland threatened sugar mill-owners with fines if they did not cease polluting waterways, and although some fines were issued, the discharge of unwanted molasses and untreated waste waters into creeks and streams near sugar mills continued during the early part of the twentieth century, mostly unabated. [105]

A small proportion of the unwanted molasses in Queensland ceased making its way into the watercourses throughout Australia's sugar-producing districts during the late 1920s. Part of the molasses commenced being sold as a fertiliser (see Chapter 11), while some of the by-product was converted into fuel after 1925, following the formation of the Australian National Power Alcohol Co., a mill-owner supported organisation which operated a power alcohol plant at Sarina. The burning of molasses as a fuel in the boiler furnaces continued at a few Australian sugar mills, but mostly the practice had been abandoned by 1930 because it produced large amounts of clinker in the furnaces. Thus, unwanted molasses and the untreated waste waters from Australian sugar mills continued being discharged into nearby watercourses. Moreover, often this waste water was tainted heavily with muddy residues during the period when the Killer process operated at some Queensland sugar mills. Norman Bennett hinted that public complaints about such practices were either not being made or ignored, for he mentioned at a meeting of Queensland sugar technologists in 1930 that the state's sugar mill-owners had 'not been severely dealt with by the health authorities', but the day was coming when, like Natal, the state's sugar mills would have to treat all their waste waters.[106]

Complaints about fish kills in North Queensland sugar-producing districts during the late 1940s and early 1950s eventually prompted a few mill-owners to try and reduce the pollution of nearby watercourses. Five shallow settling ponds were built near Pleystowe Mill in 1959 to receive the factory's effluent. In 1966, waste water from Moreton Mill commenced being pumped along a pipeline to settling ponds built on farms owned by the Moreton Central Sugar Mill Co. Ltd. Untreated waste waters from Proserpine Mill started being discharged into effluent ponds with aerators during 1969.[107] Other mill-owners had constructed cooling ponds, which allowed torricellian waters – spent cooling waters used at the effets – to cool before they were released back into watercourses (see Plate 7.4). Yet these efforts were overshadowed by the lack of attention to the problem by other sugar mill-owners. In 1960, the Federation of Chambers of Commerce of Far North Queensland complained to the Queensland Treasurer that waste waters from local sugar mills had killed countless fish in the Mossman and Mulgrave Rivers and Babinda Creek. An investigation by one of the Queensland Department of Local Government's

Plate 17.4. The cooling pond at Qunaba Mill, c. 1965.

Source: *ASYB*, 1965, p. 415. Image reproduction courtesy of Rural Press Queensland Pty. Ltd.

sewerage engineers in 1965 found that three sugar mills in the Mackay district – Racecourse, Farleigh and Marian – ran their untreated waste waters into drains that emptied into various local creeks. The 4 km drain running from Racecourse Mill to Bakers Creek had now been nicknamed 'Stink Gully' by the local residents because of its foul odours. The Wildlife Preservation Society of Queensland and the Johnstone Shire Council complained about the fish kills in the North and South Johnstone Rivers in 1966 – effluent from the three local sugar mills were blamed. In the same year, the Queensland government received a report that the Haughton River, downstream of the Invicta Mill for several kilometres, was polluted with thousands of dead fish. Effluent from the local sugar mill was again blamed for killing the fish.[108]

The increasing number of complaints about polluted watercourses near sugar mills in Queensland prompted the Queensland Department of Harbours and Marine to threaten the Racecourse Cooperative Sugar Milling Association in November 1965 with prosecution under the *Fisheries Act*. Effluent from Racecourse Mill was killing fish in Bakers Creek. This threat was never implemented. Almost two years later on 26 October 1967, Frederick Graham (Labor; Mackay) delivered a speech to the Queensland Parliament in which he described Bakers Creek as being no more than a 'stagnant rotten pool', and the Queensland government had done nothing about the matter for years as it was afraid of offending the Directors of the Racecourse Cooperative Sugar Milling Association.[109] With the issue now attracting the attention of politicians, the Directors of the Racecourse Cooperative Sugar Milling Association authorised the building of 1.5 m deep settling ponds during 1968. Here the mill's waste water underwent natural biological and chemical processes until the biochemical oxygen demand (BOD) was low enough to allow its discharge into natural waterways. In the same year, Gibson & Howes Ltd. starting impounding the effluent from Bingera Mill behind a specially constructed dam, using the water for irrigation on its plantation; a similar arrangement was implemented at Isis Mill. During 1969, CSR commenced a program of enhanced mill housekeeping, installing drip trays and grease catchers on all gearings and drives in mill machinery and removing spilt sugar and bagasse from the mill floors before they were hosed.[110] Yet these efforts were not widespread. Results from water quality surveys undertaken by the Queensland Department of Local Government officers in 1970 showed that almost half of Queensland's sugar mills undertook no remedial measures to reduce the effect of water pollution. Consequently, the level of pollution in Mossman and South Johnstone

Rivers and Babinda and Sandy Creeks, for instance, continued to be assessed as either severe or moderate.[111]

In response to the growing public concern about water pollution near Australian sugar mills, the SRI had commenced investigations in 1965 into ways to reduce the problem. These investigations, which increased markedly after 1967, discovered that high intensity surface aeration by some mechanical means would improve the BOD_5 (i.e. average BOD over 5-days) removal and reduce obnoxious odours of waste waters. This work culminated in the building of an aerator in the effluent pond at Moreton Mill during 1968. Surveys at Queensland's sugar mills also showed that the average sugar loss at an individual factory was equivalent to approximately 0.5% of the total sugar production – in 1971 this amounted to a loss of 13 200 tonnes of sugar in Queensland. This loss came from spillages of juice, massecuite and raw sugar and vapour escapes in the effets (i.e. entrainment). As these studies were being undertaken, the SRI advised its member sugar mills to implement interim measures such as good housekeeping inside the sugar mills, intelligent regulation of discharges and the provision of emergency storage for waste waters.[112]

During the late 1960s, CSR had also commenced investigations into ways to reduce the amount of pollutants being discharged with waste waters at its Australian sugar mills. Trials involving a pilot liquid effluent treatment system at Victoria Mill were conducted between 1969 and 1971. Factory waste waters, including spray pond overflows, were pumped through a treatment plant that removed fibrous matter and grease before being filtered through two ponds – one operating under anaerobic conditions and the other under aerobic conditions – and two clarifying stations that removed sludge. These trials showed that the BOD_5 of mill waste water could be readily degraded so that the final effluent contained on average 26 mg/litre BOD_5, which was low enough for its discharge into natural waterways. By way of comparison, tertiary treated sewage had a BOD of around 20 mg/litre.[113]

Following the gazettal of the *Clean Waters Act* of 1971 in Queensland, the state's sugar mill-owners were required to consider water pollution more seriously. The legislation, although passed by the Queensland Parliament in December 1971, did not come into force until 30 June 1973. Sugar mill-owners were given six months to obtain a license to discharge waste waters into watercourses or drains and to provide data about the flow, temperature, pH and suspended solids and BOD levels in mill effluent. Regulations associated with the *Clean Waters Act* in Queensland set the desirable minimum level for dissolved oxygen (DO) concentration at 5 mg/litre (although 3 mg/litre was the minimum for a coarse fish population) and BOD at 20 mg/litre. Waste waters returned to streams from Queensland sugar mills were also not to exceed the river temperature by specified amounts. At Macknade Mill on the Herbert River, for example, the temperature difference permitted was 10°C, while for Goondi Mill on the North Johnstone River this difference could be 17°C. Provision was made for progressive compliance over four years. The Water Quality Council of Queensland (WQCQ) was established to monitor compliance with the legislation, although initially the Queensland sugar industry had no representative on this statutory authority. This omission was rectified during 1975 when Francis (Joe) McAvoy, Chairman of the QCGC, became the representative of Queensland's primary industries on the Council. Dr John Allen, the SRI Director, also joined as a member in 1979.[114]

Some Queensland sugar mill-owners monitoring these legislative developments had commenced trying to reduce the amount of effluent at their factories during the early 1970s. CSR measured the volume and source of waste water at each of its Queensland sugar mills. A fully operational waste water treatment plant was finally commissioned at Victoria Mill in 1973; in the same year a liquid effluent treatment plant came into operation at Farleigh Mill. By mid-1973, surveys at most Queensland sugar mills had identified the quantities of waste water and pollutants being produced in the factories. In many instances the pollution was associated with a loss of sugar during the manufacturing process – spillages of molasses and syrup and sucrose in condenser water. At some Queensland sugar mills, clean torricellian waters had commenced being segregated from water polluted by spillages or cleaning agents, so that small volumes of heavily contaminated water could be dealt with more effectively. Leaks from pumps and valves throughout the sugar mills had also been reduced.[115]

Other sugar mill-owners in Queensland found the new regulations difficult to meet, and a Queensland sugar industry representation was made to the WQCQ during early 1975, seeking less stringent standards than those found in the licenses just issued in January 1975. After considering this submission, the WQCQ announced in June 1975 that Queensland sugar mills would be subject to a 'General Interim Standard', which required discharge water to contain 20 mg/litre BOD and 30 mg/litre suspended solids, but the restriction on the temperature difference between intake water from streams and waste water would be removed. Segregation of torricellian and process waters was to continue and the Queensland sugar mills were given another four years to comply. A condition of this retreat from the earlier, harsher standards was that the Queensland sugar industry commission an independent consultant to study the effects of sugar mill effluent on water quality and aquatic life in watercourses.[116]

During the late 1970s, the issue of waste waters polluting streams continued to be addressed at many Queensland sugar mills. Effluent treatment plants or ponds were built at several sugar mills, namely Macknade, Goondi, Tully, South Johnstone, Cattle Creek and Farleigh. A cooling tower was constructed at Mourilyan Mill in 1978 to reduce the temperature of waste water being returned to the South Johnstone River; a similar cooling facility was erected at Goondi Mill so that the water returned to the North Johnstone River had been sufficiently cooled. Large earth dams were built near Millaquin and Qunaba Mills, impounding effluent that was later used for irrigation.[117] Yet in 1979 the results from the stream survey financed by the Queensland sugar industry showed that the South Mossman, South Johnstone and Mulgrave Rivers and Babinda, Reliance and Amhurst Creeks were still moderately to severely polluted by sugar mill effluent.

The WQCQ responded to this report by announcing that some Queensland sugar mills had been too slow in undertaking the necessary work towards complying with water quality regulations and that their priorities needed to be 'reconsidered', but no legal action was taken against offending sugar mills.[118] Moreover, a WQCQ report in 1982 concluded that no Queensland sugar mill was fully compliant with the licence standard of BOD_5 for torricellian waters. However, a few Queensland sugar mill-owners interviewed during the report's compilation complained that their water intake had BOD values of 50–150 mg/litre, so it was unfair to expect them to discharge effluent that was cleaner than their available water.[119]

Meeting their waste water licence requirements remained a matter of concern for some Queensland sugar mill-owners during the late 1980s and early 1990s. CSR reported that its licence conditions had been met at Victoria Mill until 1988, with the final effluent being suitable for discharge into watercourses under normal conditions. However, as factory crushing rates increased, leading to higher volumes of water use, compliance with the licence conditions became the exception rather than the rule. During 1993, CSR was forced to upgrade its activated sludge treatment plant. Artificial wetlands were also constructed near Victoria Mill during November 1993. They were used to process partially treated waste waters from the sugar mill. Together these two changes improved the factory's compliance with its licence conditions. At South Johnstone Mill, accidental spillages of oil during July and October 1990 and boiling molasses in September 1992 led to the South Johnstone River being badly polluted. In June 1993, the South Johnstone Cooperative

Milling Association took steps to ensure that similar incidents would not occur again. The organisation outlaid $650 000 on three major projects for effluent monitoring and early detection and prevention of spills, including a system to divert potential polluting material away from the river.[120]

Air pollution had been taken seriously by only a few Queensland sugar mill-owners before 1965. During 1956, the Millaquin Sugar Co. Ltd. installed fly-ash arrestors on two of its seven boilers at Millaquin Mill, possibly attempting to reduce the fly-ash nuisance in the city of Bundaberg. In 1959, complaints from the residents of Nambour about soot, led to a fly-ash arrestor being installed on the chimney at Moreton Mill. Residents of Maryborough in 1960 organised a petition against the bagasse and smoke nuisance from the chimneys at the Maryborough Sugar Factory. This petition led to the Maryborough Sugar Factory Prop. Ltd. dismantling two smaller smokestacks in 1961, and replacing them with one, large 27 m high chimney fitted with fly-ash arrestors.[121]

All Queensland sugar mill-owners, however, had to consider air pollution following the introduction of a *Clean Air Act* which was passed by the Queensland Parliament in late 1963, although it did not come into operation in regional areas (i.e. those localities outside Brisbane and Ipswich) until 26 September 1970. The clean air legislation required two distinct parameters be controlled: particulate emission from boiler stacks (fly-ash) and smoke emission. Industries appearing on a schedule of air polluters would be required to obtain a license which could prescribe both the quantity and quality of dischargeable pollution. Scheduled industries, which included sugar mills, were given until September 1977 to meet the required standards. The process would be monitored by the Air Pollution Council of Queensland (APCQ). Before the release of the Clean Air Regulations of 1968, the APCQ met with the Air Pollution Committee of the Queensland Sugar Industry during 1967. At the conclusion of these meetings, the APCQ agreed that the amount of knowledge needed to formulate a sensible emission standard for the state's sugar industry had not been fully gathered and that the SRI should continue its investigations into the matter. Any emission standards for the sugar industry proclaimed under the proposed Clean Air Regulations would be an interim measure.[122]

Using the data gained from the SRI and CSR's investigations into reducing fly-ash emissions at sugar mills, the Air Pollution Committee of the Queensland Sugar Industry wrote to the APCQ on 4 December 1970, proposing that the Clean Air Regulations of 1968 be amended to allow Queensland sugar mills an allowable emission standard of 1.15 grains per cubic metre of effluent gas. Subsequent meetings in early 1971 with Dr Alan Gilpin, the Director of the APCQ, reached a compromise about emission levels at Queensland's sugar mills, with the Queensland sugar industry accepting an emission standard of 0.8 grains per cubic metre of waste gas. Subsequently, the APCQ accepted this recommendation and the Queensland Clean Air Act Regulations of 1968 were amended on 1 June 1972 to incorporate an emission standard that applied only to sugar mills. New boilers installed in sugar mills, however, had to meet an emission standard of 0.69 grains per cubic metre of waste gas. As a result of these interactions, Dr John Allen of the SRI was invited to join the APCQ, so the Queensland sugar industry now had a representative on this organisation. The Clean Air Act Regulations were reviewed in October 1976, but the special arrangements for sugar mills were not changed.[123]

To meet the new air quality standards, Queensland's sugar mills were equipped with fly-ash collection systems during the mid-1970s, including wet arresters. Fly-ash dewatering plants were erected as part of the air pollution control measures at CSR's North Queensland sugar mills. Sluicing water washed the boiler gases as they flowed through the scrubber vessels. This process created a watery sludge which was later separated in the fly-ash dewatering plant, the fly-ash and boiler ash being mixed together for final disposal.[124] In addition, the APCQ imposed stringent conditions on the numerous new boilers built during the 1970s as part of the industry expansion (see discussion later in the chapter). They had to be constructed with appropriate fly-ash collection systems and the associated chimney stacks were to be of a height considered suitable for dispersing pollutants from the boilers. Low sulfur (0.5–2.0% S) fuel oil could only be burnt in these boilers.[125]

As a result of these new arrangements, ash quantities recovered by the collectors greatly increased during the 1970s. By 1977, Graham Sawyer and Rod Cullen of the SRI estimated that the total quantity of wet ash that needed to be handled at Queensland's sugar mills annually would exceed one million tonnes. Disposing of this recovered ash created a new potential pollution problem. The WQCQ had already foreseen this possibility. Any increase in polluting discharges were not permitted when Queensland's sugar millowners renewed their water discharge licences in 1976. This change aimed to prohibit the discharge of ash-laden water from ash sluicing circuits or fly-ash dewatering plants.[126] Various methods were used to overcome this problem. At Bingera, Racecourse and Mulgrave Mills, the fly-ash was mixed with the rotary filter muds and used as fertiliser by the mills' suppliers. The ash-laden waters from Fairymead, Millaquin and Qunaba Mills were sent to settling ponds where the ash accumulated during the crushing season. It was removed during the slack season and used primarily as land fill or by farmers as a soil conditioner. CSR developed a combined mill waste disposal system at its Queensland sugar mills during the mid-1970s, mixing the fly-ash with the rotary filter mud and sludge from the effluent treatment plants, producing a mixture that was sold to canegrowers for use as fertiliser.[127]

In New South Wales, CSR had to address environmental concerns, following the introduction of the *Clean Air Act* of 1961 and the *Clean Waters Act* of 1970. CSR had commenced better mill housekeeping by 1970, trying to contain grease and oil spillages and sugar losses, so that they did not pollute waste waters. In 1971, drain modifications were completed specifically at Broadwater Mill to reduce the amount of pollutants entering the mill's waste waters. By 1976, all the stacks at CSR's three New South Wales sugar mills had been fitted with dry fly-ash arrestors, enabling their chimney emissions to be within the upper limit of 0. 46 grains per cubic metre of waste gas fixed by the *Clean Air Act*.[128] CSR's efforts to reduce the environmental impacts associated with sugar production were maintained after the NSW Sugar Milling Coop. assumed control of the three sugar mills in New South Wales. The new boiler installed at Harwood Mill in 1984 was equipped with fly-ash arrestors, so that smoke and particulate emissions were below the regulation limits. Noise reduction measures were installed during 1988–1989 and by 1994, all mill mud and fly-ash was being recycled as top dressing on farms, and all waste waters from the milling process were being recycled in the milling process, so any nutrients ultimately joined the ash and mud disposal systems. Wet scrubbers were installed during the Broadwater Mill boiler upgrade during 1994-1995, thereby reducing particulate emissions even further. Yet equipment failure at Condong Mill in August 1994 resulted in the escape of boiler feed water with a low level of sugar into the storm water drains leading into the Tweed River. Following this incident the firm developed an Environmental Management Plan which identified and eliminated potential environmental hazards at its factories.[129]

Towards crushing 38 million tonnes of sugar cane annually

'The mill began to take on a new look', wrote the canegrower Kenneth Manning, as he recalled the changes to Farleigh Mill during the 1970s in his history of the factory.[130] His words aptly described the wide scale upgrades occurring at all Australian sugar mills during the 1970s and early 1980s. CSR alone was estimated to have spent $43 million on its four Queensland sugar mills during the mid-1970s; Australian Estates Co. Ltd. outlaid $13.7 million to increase the capacity of its three sugar mills; the NSW Sugar Milling Coop., over the period 1979–1982, invested $25.09 million in new machinery and buildings for the three sugar mills it purchased from CSR.[131] The stimulus for such widespread overhauls to mill buildings and mill equipment was a succession of decisions by the Central Board which increased mill peaks by just over 1 million tonnes between 1970 and 1984 (see Chapter 10). Australian sugar mill-owners were required to handle larger volumes of cane annually, without unduly extending the crushing season, and with limited overtime harvesting.

Improved hourly crushing rates were essential to deal with the expected larger crops. Previously, when Australian sugar mill-owners were faced with this challenge, they added extra sets of crushing rollers to the milling train. A few Australian sugar mill-owners adopted this approach during the 1970s and early 1980s. The NSW Sugar Milling Coop., for example, funded the conversion of the milling train in each of its sugar mills from three to four sets of crushing rollers.[132] A similar upgrade occurred at Invicta Mill, where an entirely new building was erected to house an expanded milling train that went from four to five sets of crushing rollers. Overall, however, most Australian sugar mill-owners during the 1970s and early 1980s did not enlarge their milling trains, with the majority of the country's sugar mills in 1985 having either four or five sets of crushing rollers, similar to crushing arrangements in 1969 (Table 17. 6).

Two other strategies were adopted during the 1970s to improve the hourly crushing rates. The first involved duplicating the milling train and occurred only at Inkerman Mill. By 1970, this sugar mill already had seven sets of crushing rollers so realistically it was not possible to add any more units to the milling train. Therefore, Pioneer Sugar Mills Ltd. decided in 1974 that a second milling train, equipped with a diffuser, would handle the larger crops (see Plate 17.5). Inkerman then became the second Australian sugar mill to operate with a diffuser.[133] Other Australian sugar mill-owners, however, concluded that improved crushing rates could be achieved by the addition of newer sets of crushing rollers with wider rollers. At Mulgrave Mill, for instance, the older 1.8 m wide crushing rollers were replaced with new 2.13 m wide crushing rollers. Similar upgrades occurred at Farleigh Mill. By 1974, the mill's five sets of crushing rollers were all 2.13 m wide, turbine driven and controlled from a central console equipped with a closed circuit television. Farleigh Mill was further upgraded in 1983 by the installation of 2.75 m wide crushing rollers, the world's largest set of crushing rollers at that time. To power these larger sets of crushing rollers, forty reciprocating

Table 17.6. Statistics on milling train arrangements at Australian sugar mills, 1969, 1985 and 1995.

	1969	1985	1995
No. of operational sugar mills	34	33	28
No. of mills with two or three sets of crushing rollers	4	1	2
No. of mills with four sets of crushing rollers	11	15	10
No. of mills with five sets of crushing rollers	11	13	15
No. of mills with six or more sets of crushing rollers	9	5	3
No. of double train mills	2	3	5
No. of mills with a diffuser	1	2	3
No. of mills with shredders only	0	19	17
No. of mills with both shredders and knives	26	12	6
No. of mills with other combinations of shredders, knives and levelers	8	2	4

Notes: (1) Details about Mourilyan Mill's crushing train for 1995 were not included in the machinery descriptions. (2) In 1985 and 1995, Inkerman and Invicta Mills had two sets of crushing rollers as part of the second milling train that was associated with their diffuser unit. (3) Details about CSR's Ord River Mill were not included.

Source: Based upon mill machinery details in the *ASYB*, 1970, 1986 and 1996.

Plate 17.5. The diffuser at Inkerman Mill, 1977.

Source: BSES, AR 1976–77, p. 17. Image reproduction courtesy of BSES Limited.

engines with historical links to the past were replaced during the 1970s with more powerful turbines.[134]

Additional cane arriving at Australian sugar mills also strained the capacity of their unloading, carrier and cane preparation facilities. In several instances, these stations were upgraded to avoid and overcome bottlenecks occurring at the beginning of the milling trains and to cope with the changeover to mechanically harvested cane. An example of such an upgrade occurred at Marian Mill, where a new cane receiving station and cane carrier were built to handle the expected larger crops being delivered to the crushing rollers. Similar improvements occurred at Moreton Mill. The cane carrier was rebuilt entirely and automated using electric motors from Sweden. Rotary tipplers or side tipplers were installed at most Australian sugar mills during the 1970s (see Plate 17.6). They enabled faster emptying of either single or two cane bins laden with chopped cane.[135] The revolving knives on the carrier were removed in other Australian sugar mills, because they now handled chopped cane entirely. By 1985, 60% of Australian sugar mills relied on shredders only (Table 17.6). This change permitted the application of more power to the shredders, thereby encouraging the installation of larger and more powerful shredders of advanced design known as heavy duty shredders. Developed in conjunction with Walkers Ltd. during the

Plate 17.6. A rotary tippler in operation at Fairymead Mill, c. 1974.

Source: *ASYB*, 1974, p. 349. Image reproduction courtesy of Rural Press Queensland Pty. Ltd.

late 1960s, they replaced the Searby shredders. Shredder drives not already driven by turbines were converted to turbines.[136]

Juice clarification was performed overwhelmingly in Dorr, RapiDorr, ATV or Bach multi-tray units during the 1960s, but Ted Whayman and David Hale, two SRI researchers, were skeptical of their efficiency, especially as throughput was not improving, even with the use of flocculants. Discarding existing designs, Whayman developed his own clarifier in which he removed the internal multi-trays, replacing them with an arrangement whereby the juice entered at the top, flowed downwards before being withdrawn at many points. The unit was about one-fifth the volume of the usual plant for the same juice capacity, and settling time fell from two hours to thirty-five minutes. Its design meant that Australian sugar mills could improve their existing clarifiers by removing the trays and performing some minor redesign work, instead of purchasing entirely new units.

The first SRI clarifier was installed at Gin Gin Mill in 1969 and its superior performance quickly attracted the attention of Australia's sugar mill engineers – eight more of the units were installed at other Australian sugar mills during 1970 and 1971. By 1985, just over half the subsiders in Australian sugar mills were SRI clarifiers, although the units were not favoured by CSR, its sugar mills continuing to be mostly equipped with Dorr or ATV multi-tray units. Moreover, with its faster throughput, the SRI clarifier reduced the need for several clarifiers per mill, so the overall number of clarifiers in Australian sugar mills declined during the 1970s and early 1980s. In 1995, the SRI clarifier was still the commonest subsider found in Australian sugar mills (Table 17.7). Moreover, it had also been installed in overseas sugar mills, the unit being marketed under license by the UK sugar mill machinery manufacturer, Fletcher Stewart Ltd.[137]

To effectively boil the greater amounts of juice being produced during the 1970s and early 1980s, the total evaporating surface of the effets in all Australian sugar mills was increased by almost 130 000 m^2 between 1969 and 1985. This increase was achieved by a combination of converting quadruple effets to quintuple effets and/or installing larger quadruple or quintuple effets in the sugar mills (Table 17.7). The heating surface ranged from 15 000 to 18 000 m^2 in some of the newer evaporator units in several sugar mills, namely Tully, Mulgrave, Victoria, Inkerman and Fairymead. Many of the last vessels in the effets had been fitted with CSR's patented vertical louvre separator, developed in the early 1970s. This new entrainment device consisted of two horizontal rows of corrugated plates one on top of each other that were fitted at the top of the evaporator. The corrugated shape of the plates caused larger

Table 17.7. Statistics on the clarification, boiling and crystallisation stations at Australian sugar mills, 1969, 1985 & 1995.

	1969	1985	1995
Total no. of subsiders & no. of SRI subsiders (in brackets)	109 (1)	53 (28)	47 (30)
No. of rotary vacuum filters	80	85	77
Total treatment area of rotary vacuum filters (m^2)	3109	7713	7935
No. of mills with a single set of quadruple effets	10	11	3
No. of mills with a single set of quintuple effets	14	16	17
No. of mills with a single set of sextuple effets	0	0	1
No. of mills with more than one set of effets	10	5	6
Total area of evaporating equipment (m^2)	180 855	322 074	359 221
No. of vacuum pans	243	257	215
No. of coil pans	89	46	24
No. of calandria pans	88	143	141
No. of continuous pans	0	4	11
Total volume of pans (m^3)	6 588	13 829	12 466

Notes: (1) Details on subsiders and mud filters were not given for Mourilyan Mill. (2) Descriptions of the type of vacuum pans and their area in some instances were not given. Therefore, the figures are under-estimates. (3) Details about CSR's Ord River sugar mill were not included.

Source: Based upon mill machinery details in the *ASYB*, 1970, 1986 and 1996.

droplets to form, thereby recovering more of the escaping vapour.[138] New and enlarged vacuum pans with greater heating surfaces and up to 164 m^3 in capacity were being added to Australian sugar mills. These pans had been fully designed and fabricated in Australia and often exceeded in size the largest pans found overseas. By 1985, the total number of vacuum pans in Australian sugar mills had increased slightly compared to the level reached in 1969, but the volume of these units had doubled (Table 17.7). Moreover, increasingly these new vacuum pans were calandria pans, including those with mechanical stirrers.[139]

In 1970, crystallisation remained a batch process in Australian sugar mills, but if it could be converted to a continuous process, then the overall effect at this sugar mill station would be greater efficiencies, higher

Plate 17.7. Australia's first continuous vacuum pan for cane sugar production at Mossman Mill, 1976. Evans Deakin Industries Ltd. undertook the design, manufacture, installation and commissioning of the unit.

Source: *ASYB*, 1981, p. 84. Image reproduction courtesy of Rural Press Queensland Pty. Ltd.

throughput and reduced labour costs. Continuous boiling had first been attempted unsuccessfully at Kalamia Mill during the early 1940s. The SRI revisited the idea of continuous boiling during the late 1960s and early 1970s, concentrating initially on handling low-grade massecuite, but the engineers had to design a vacuum pan that produced sugar comparable in crystal size and suitability for export to that produced in batch units. A SRI-designed continuous low-grade vacuum pan was trialed at Rocky Point Mill during 1974 and 1975, and produced sugar suitable for export. The first continuous low-grade vacuum pan using the SRI design was installed at Mossman Mill in 1976 (see Plate 17.7). By 1984, this equipment incorporating SRI-design elements and/or overseas features had been added to three other Australian sugar mills, namely Tully, Racecourse and Mourilyan.[140]

Higher crushing rates led to increased demands for steam. To meet these requirements, new and enlarged boilers and powerhouses and accompanying high stacks were installed at many Australian sugar mills, including Harwood, South Johnstone, Pleystowe, Macknade, Invicta, Isis, North Eton, Bingera, Millaquin, Mourilyan, Inkerman and Farleigh, to list a few. Some of these new boiler stations were massive. At Babinda Mill, for example, the new boiler station erected during 1974 was equal in height to a ten storey building. Most of these new boilers generated between 110 and 180 tonnes of steam/hour, although the newly installed units at Babinda, Farleigh, South Johnstone and Tully Mills were capable of producing between 200 and 220 tonnes of steam/hour. Bagasse was the normal fuel used in these boilers, but several were fitted with auxiliary oil firing and/or were suitable for coal firing in case of a bagasse shortage. New, circular bagasse storage tanks were also erected at several Australian sugar mills equipped with new boilers in order to keep them fully supplied with fuel (e.g. Harwood; Pioneer; Victoria; Macknade; Marian; North Eton).[141]

During the 1970s, the thirty Queensland sugar mills used around 36 000 tonnes of furnace oil annually as auxiliary fuel for their bagasse-fired boilers. It was used at the start of the season, during extended wet weather periods, if milling problems produced wet bagasse that could not be used in the boilers or when insufficient bagasse was stored in the factory's bagasse storage bins. Reducing this reliance on furnace oil became a priority throughout the late 1970s, following the oil price increases of the mid-1970s and a major Australian shortage of furnace oil in 1979. Millaquin Mill switched to using a mixture of coal and bagasse, the coal being supplied from a nearby colliery. Sawdust secured from a nearby sawmill became the alternate fuel at the Maryborough Sugar Factory and Racecourse Mill.

Not all Australian sugar mills could access sawdust, but they all produced bagasse. However, boiler stations in most Australian sugar mills burnt all the bagasse generated during the crushing season, so as to avoid the cost and problems associated with the disposal of this waste material. Yet SRI investigations showed that sugar mill boilers could still operate efficiently, even though burning less bagasse. Thus, the cheapest way of using less oil was to burn less bagasse, store more bagasse and to reclaim it when the sugar mill needed fuel for the boiler furnaces. All Australian sugar mills had some form of 'on-line' bagasse storage bins to provide sufficient bagasse for start-ups, finishes and stoppages, but 'off-line' storage was often insufficient and poorly protected from the weather. Reclaiming this bagasse also presented problems, as the tractors/loaders were slow and inefficient.[142]

Trials were conducted by the SRI and CSR into ways of preserving bagasse more effectively. Investigations into pelleting bagasse at Millaquin Mill using a commercially available pelleting machine concluded that this method was too slow to be effective. The researchers also created one tonne bales of bagasse and developed special equipment for debaling the bagasse at the boiler station. The stored bales, however, were easily soaked during rain and usually ended up with a higher than acceptable moisture content. During 1980, many Australian sugar mills experimented with storing the bagasse in the open, either under no covers or placed under various types of covers, including tarpaulins and fiberglass reinforced plastics. CSR's efforts in developing the tarpaulin cover loose storage technique for bagasse was recognised in 1981, when the firm received the National Energy Management Award from the Department of National Development and Energy. By the early 1990s, covering piles of bagasse with nylon reinforced PVC covers provided the best storage at a cheap price. Reclaiming this bagasse meant that furnace oil consumption by the Australian sugar industry during the early 1990s had fallen to around 6 500 tonnes or 220 tonnes per factory. About a third of the country's sugar mills also used wood chips, waste wood and sawdust from timber yards or sawmills.[143]

More mill automation occurred during the 1970s and early 1980s through the introduction of computers. The Mossman Central Mill Co. Ltd. led the way, relying on a computer in Brisbane in 1967 to calculate all payments for sugar cane delivered by suppliers. The firm went further in 1971, purchasing an IBM 1800 process control computer. An automatic weighbridge was installed, and following parallel testing of the automated and manual systems of cane payment calculations during 1972, the clerical work and manual recording were eliminated. Each cane bin arriving at Mossman Mill was equipped with coded metallic squares attached to it, enabling sensing equipment to record the bin number and weight automatically. This information was stored in the computer's memory, later to be linked with details from the automatic juice sampling machine for calculating payments to individual farmers. Further advances led to the development of programs which sensed information from instruments throughout the sugar mill, leading to the automatic process control of the milling train and clarification station.[144] CSR was also an early adopter of computerised process control, having installed computers at Victoria Mill in 1972. By 1976, process control had been extended to Macknade Mill and all CSR's sugar mills used computers for data processing and storage. The computers in CSR's four North Queensland sugar mills were linked together by an on-line commercial data processing network, which was managed by the computer services department at Victoria Mill. By 1976, other Australian sugar mills, namely Marian, Fairymead, Mourilyan and Tully had been equipped with computers, which assisted in the calculation of wages and cane payments, control of sampling and sample tracking and the control of juice flow, pH levels and boiling of massecuite.[145]

All these engineering developments meant that the average crushing rate at Queensland's sugar mills rose from just over 200 tonnes per hour in 1970 to 370 tonnes per hour by 1984 (Figure 17.2). Consequently, the average amount of cane being processed at each Australian sugar mill by 1984 had risen to nearly 800 000 tonnes, compared to 518 000 tonnes in 1970 (Figure 17.4). Yet this trend occurred against a backdrop of growing mill-owner concern about their future cane supply, following suitable cane growing lands being acquired for competing uses such as urban development, rural subdivision and horticulture. By 1980, studies had been commissioned into the suitability of land for cane growing in the localities surrounding Moreton and Rocky Point Mills. During the 1980s, sugar caneland suitability studies were completed for several other districts, namely Proserpine (1981), Maryborough (1983), Mossman (1983), Mackay (1985), Isis (1985), Cairns (1985), Herbert River (1987) and Sarina (1987). The findings from these studies were used to identify unassigned land that could be developed for cane growing during future expansions or to replace land that had been converted to other uses.[146] At least two milling companies took a more proactive approach to preserving their cane supply. During the early 1970s, the Moreton Central Sugar Mill Co. Ltd. acquired several farms throughout the Maroochy district; this policy was continued by Howard Smith Industries Pty. Ltd. after its acquisition of Moreton Mill in 1976. In the mid-1980s, the Maryborough Sugar Factory Ltd. commenced acquiring farms throughout the Maryborough district, having secured seven by 1993. They were equipped with travelling irrigators and leased to canegrowers. An additional 330 ha was acquired by the firm in 1994 at Goomboorian, near Gympie, the land being developed into a new cane growing area.[147]

The 1970s also witnessed major changes to the ownership of Australia's sugar mills, after a long period of stability (Table 17.8). The Fairymead Sugar Co. Ltd, Gibson & Howes Ltd. and the Millaquin Sugar Co. Ltd. merged their milling interests between 1972 and 1974, creating the Bundaberg Sugar Co. (hereafter Bundaberg Sugar). In 1975, this new firm controlled four sugar mills – Fairymead, Bingera, Millaquin and Qunaba. As ownership of the Bundaberg sugar mills became concentrated in the hands of one company, CSR commenced a series of manoeuvers that radically altered its investment in sugar milling. The firm acquired the Australian Estates Co. Ltd. in June 1975,

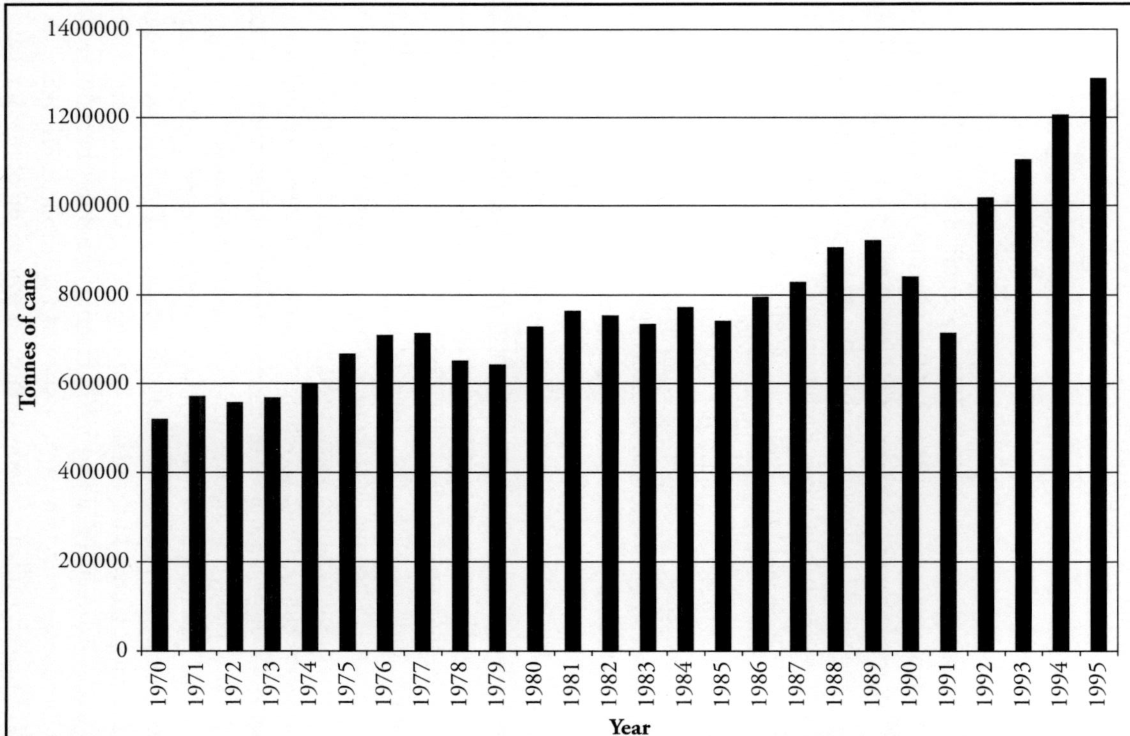

Note: The 1995 figure does not include the 80 000 tonnes of cane crushed at CSR's Ord River Mill during its first season.

Figure 17.4 The average tonnage of cane crushed yearly at Australian sugar mills, 1970–1995.

Source: Based upon figures in the *Year Book of Australia* and *ASYB*, various issues.

Table 17.8. The number of Australian sugar mills classified according to ownership, 1925–1995.

Category of ownership	1925	1945	1965	1975	1985	1995
CSR	8	7	7	10	7	8
Bundaberg Sugar Co.	*	*	*	4	4	6
Other public companies	10	10	10	5	6	1
Canegrower-owned companies	17	17	16	13	15	13
Private companies or individuals	5	1	1	1	1	1
TOTAL	40	35	34	33	33	29

* Company did not exist

Source: Based primarily on ownership information in the *ASYB*, various issues.

giving it control over Pleystowe, Invicta and Kalamia Mills, the latter two factories being located in the Lower Burdekin district. CSR could now operate in the Lower Burdekin district, where yields were high and assured because of irrigation and land was available for expansion, whereas two of CSR's northern mills – Goondi and Hambledon – were landlocked with little prospect of larger crops.[148] At the time of this takeover, CSR was also in negotiation with its New South Wales suppliers about the sale of the New South Wales sugar mills, which the firm had considered for some time to be unprofitable. This deal was eventually finalised in July 1978, when the newly formed NSW Sugar Milling Coop. raised enough funds to purchase Condong, Broadwater and Harwood Mills from CSR.[149]

As raw sugar prices declined in the early 1980s, sugar milling profitability was reduced substantially – Russell Savage's committee of inquiry into the Queensland sugar industry estimated that a Queensland sugar mill's operating surplus over the period 1982–1985 was insufficient to cover interest repayments and depreciation. In 1982, emergency funding totaling $10 million was provided to Tully, Babinda, South Johnstone, Proserpine, Marian and Cattle Creek Mills by the Queensland government (Bjelke-Petersen; Coalition of National and Liberal Parties), enabling the factories to operate. The NSW Sugar Milling Coop. also suffered major financial stress in the mid-1980s. To pay for ongoing operations during 1983, the firm invoked a levy of thirty cents per tonne of cane on its members, deferred loan repayments and closed its three sugar mills for ten weeks in early 1985 with all staff accepting ten weeks leave without pay.[150] Other Australian sugar mill-owners sought new sources of income to carry them over during this difficult period. The Mossman Central Mill Co. Ltd. ventured into tourism and mariculture; the Proserpine Coop. Sugar Milling Association Ltd. purchased a tourist resort; the North Eton Coop. Sugar Milling Association Ltd. established a subsidiary company that contracted with local canegrowers to supply cut flowers for the domestic and export markets. Organised mill tours commenced at Tully Mill in the hope that the public's fascination with sugar mills would pay dividends.[151]

The falling profitability of sugar production, however, prompted another wave of takeovers and sugar mill closures in Australia during the late 1980s and early 1990s. The first company to act was CSR. The firm began acquiring shares in Pioneer Sugar Mills Pty. Ltd., the owner of Pioneer and Inkerman Mills on the Lower Burdekin and Plane Creek Mill near Sarina. After a successful takeover of Pioneer Sugar Mills Pty. Ltd. in 1987, CSR controlled ten Queensland sugar mills, but the firm sought to concentrate its milling interests in the Herbert and Lower Burdekin districts. Subsequently, it closed Goondi Mill in early 1987 under controversial circumstances following political interference from the Queensland government (Bjelke-Petersen; National Party). Assignments were reallocated to Mourilyan and Babinda Mills. Hambledon Mill's closure at the end of the 1991 crushing season was less controversial, with its assignments being transferred to Mulgrave Mill.[152] In 1988, CSR also sold Pleystowe Mill to a newly formed Mackay Sugar Cooperative Association Ltd. (hereafter Mackay Sugar), which emerged following a merger of the five Mackay canegrower-owned milling companies in late 1987. However, in return for financial assistance from the Commonwealth and Queensland governments to enable the purchase of Pleystowe Mill from CSR, Mackay Sugar had to consider mill rationalisation. Pleystowe had been identified for closure during the negotiations in the lead up to the creation of Mackay Sugar, but the sugar mill was spared closure (although it was eventually shut after the 2008 crushing season). Instead, the independent Mill Closure Committee established by Mackay Sugar announced in mid-1988 that two sugar mills would close: North Eton following the 1988 season and Cattle Creek after the 1991 crushing. To compensate, Marian Mill's capacity was to be increased so that it could handle the extra cane.[153]

Bundaberg Sugar had not remained inactive during these developments. At the end of 1985, Bundaberg Sugar closed Qunaba Mill, which was landlocked and unable to secure any additional cane supplies, thereby making it the least efficient of the firm's sugar mills. Mourilyan and Moreton Mills were also added to Bundaberg Sugar's assets in 1988, after Howard Smith Industries Pty Ltd., their owner, decided to concentrate on its core businesses of coal mining, maritime activities and heavy engineering following financial losses in the previous year – the unprofitable sugar companies were not considered central to the firm's interests.[154] Bundaberg Sugar also purchased the financially struggling Babinda Mill in 1989, much to the disquiet of the grower-owned companies operating Mulgrave, Tully and South Johnstone Mills. With the purchase of Babinda Mill, Bundaberg Sugar now processed around 16% of the sugar cane produced in Australia. Yet the newly enlarged and profitable Bundaberg Sugar – the firm had surplus cash estimated at $40 million and no net debt at 30 April 1991 – became the target of an unwanted takeover in June 1991, becoming part of the British-based Tate and Lyle International, the world's largest sweetener manufacturer.[155]

By 1995, the ownership of Australian sugar mills had become concentrated in fewer companies, compared to thirty years earlier (Table 17.8). CSR still owned the largest number of sugar mills, followed closely by Bundaberg Sugar, with both firms achieving their dominance by absorbing canegrower-owned companies or public companies such as Pioneer Sugar Mills Pty. Ltd. or Australian Estates Co. Ltd. Despite some public concern that canegrower-owned milling companies were under threat in Australia, they still controlled thirteen sugar mills in 1995, although two of them – South Johnstone and Tully – had just rejected a hostile takeover bid in 1994 from Bundaberg Sugar. The demise of the canegrower-owned milling companies in Far North Queensland during the 1980s, however, had prompted the emergence of a greater sense of cooperation between the surviving canegrower-owned sugar mills. Tully, South Johnstone, Mulgrave and Mossman Mills, the four canegrower-owned factories in Far North Queensland, eventually formed Sugar North Ltd. in 1993. This new organisation shared the resources of a Human Resources Manager, undertook a number of operational activities utilising jointly owned equipment and identified operational cost savings – gaining the economies of scale and resource sharing achievable by belonging to a group.[156]

The interests of the sugar milling companies since the 1910s had been promoted by three voluntary organisations: the ASPA formed in 1907; the Cooperative Sugar Millers' Association which had operated since 1980; and the Proprietary Sugar Miller's Association Pty. Ltd., established on 4 April 1927. Yet the sugar mill ownership and legislative changes affecting the Australian sugar industry during the late 1980s led to members of each organisation agreeing that their interests needed to be represented in a unified manner. The Cooperative Sugar Millers' Association and Proprietary Sugar Miller's

Association Pty. Ltd. decided to wind up those companies in 1987, while the ASPA miller members agreed to withdraw from their association. Together they formed the Australian Sugar Milling Council Pty. Ltd. and focused upon industrial relations and milling policy. All Australian mill-owners became members of the new organisation. The canegrower members left in the ASPA voted to continue as a voluntary and independent association, although it was renamed the Australian Cane Farmers Association Ltd.[157]

One issue considered immediately by the newly formed Australian Sugar Milling Council Pty. Ltd. concerned crushing arrangements during an expansionary phase. Debt-ridden milling companies had curtailed capital expenditure on mill overhauls and upgrades during the mid-1980s.[158] A lack of expanded mill capacities, however, became all too apparent during the late 1980s, when Australian sugar mill-owners were required to process the larger crops that were being harvested annually due to the expansion in assignments. The harvesting season of June to November could have been increased, but this solution was not ideal as cane is often not ready for harvesting before June (i.e. reduced ccs) and delays from wet weather could occur during December. Continuous crushing, instead of weekend breaks for sugar mill maintenance, however, was a possible solution, enabling sugar mill capacity to be increased by as much as 35%. It was first mooted as a way to handle larger crops in July 1986 at a hearing before the Queensland Industrial Commission, but never eventuated during the 1986 crushing season. The arrangement was finally adopted the following year, when Mourilyan Mill became the first Australian sugar mill to introduce continuous crushing, despite objections from its suppliers. Its use spread to New South Wales in 1988.[159] The concept caused considerable division throughout the sugar-producing communities in Queensland, but other milling companies, assisted by the Australian Sugar Milling Council Pty. Ltd. and the QCGA, eventually negotiated agreements with their local Cane Growers' Executives, unions and employees to develop satisfactory roster arrangements permitting the use of continuous crushing. During 1994, twenty of the twenty-eight Australian sugar mills crushed on a seven-day basis.[160]

Miller-canegrower relationships were slightly deregulated in Queensland during the early 1990s. Following the abolition of the Central Board in 1991, its role as a regulator of the amount of sugar produced in Queensland was transferred to the Queensland Sugar Corporation (QSC). Local boards assumed a greater role – as actually envisaged by the original 1915 legislation – being required by the legislation to specify each year a base price or base prices for sugar cane of different quality. The existing cane price formula was to be regarded as a 'bottom line' in such negotiations, and the amount of ccs continued to be one aspect of quality which had to be taken into account when determining a base price for sugar cane. Local boards were also required to oversee any special payments or incentives that were outside the formulated price for cane. An example of one such arrangement occurred in CSR's mill supply areas. Instead of incurring capital expenditure on mill upgrades, the firm extended the crushing season to maximise yearly throughput, but paid a 'ccs underwriting allowance' to compensate canegrowers for the lower sugar content of cane delivered during the extended period. Hearing appeals from the local boards and dispute resolution were transferred to the Sugar Industry Tribunal, consisting of three members who either had a legal background or some sugar industry experience. In contrast, unregulated arrangements between the miller and canegrowers persisted in New South Wales, where the NSW Sugar Milling Coop., like CSR previously, continued to enter into five-year contracts (known as a Memorandum of Understanding) with its canegrowers. The price they received for their sugar cane was largely determined by a formula similar to that used by the local boards in Queensland.[161]

Under the *Sugar Industry Act* of 1991, the QSC had an obligation by July 1993 to report on the rules and procedures about the calculation and distribution to canegrowers of monies due to them from the sale of sugar. After examining alternative approaches to cane payment, the QSC suggested that its preferred option was to maintain the existing formula, retaining the existing base level of efficiency for canegrowers and mill-owners, but introducing an arrangement known as the 'Cooperative formula', whereby mill-owners and canegrowers shared the gain from improved efficiencies. After a year of negotiations, a Productivity/Cane Payment Package was adopted by Queensland's sugar producers in June 1994. Under this 1994 Package, mill-owners paid canegrowers an additional 25 cents per tonne of cane in 1994 and 1995, and canegrowers and mill-owners agreed to work together at the local level to identify and implement further productivity/cost reduction measures worth another 50 cents per tonne of cane.[162]

Investment enabling Australian sugar mill capacity to catch up with production increases finally occurred during the early 1990s. Mackay Sugar initiated this new wave of sugar mill upgrades by making improvements during 1990 to the milling train at Farleigh Mill, the vacuum pans at Pleystowe Mill and the milling train and centrifugals at Racecourse. The organisation spent a further $30.5 million during 1991, duplicating the milling train at Marian Mill, thereby enabling a crushing rate of 670 tonnes per hour. An upgrade costing nearly $3 million was carried out at South Johnstone Mill in early 1991.

The work increased the mill's crushing rate from 440 to 460 tonnes per hour. Just over $4 million was spent at Proserpine Mill during 1991 and 1992 on building a new bagasse storage bin, bagasse handling equipment and wet dust collection equipment to meet air emissions standards. A further $12 million was spent at Proserpine Mill throughout 1993 and 1994 on the erection of a new cooling tower and a 200 tonnes of steam/hour bagasse-fired boiler.[163] A similar sized boiler was commissioned by the Isis Central Sugar Mill Co. Ltd. in May 1994. The project reached a conclusion in early 1996.

CSR's projects during the early 1990s, however, overshadowed the amount of capital investment by other Australian sugar milling companies. During 1993 and 1994, the firm spent $75 million upgrading its two Herbert River sugar mills. The improvements permitted Victoria Mill to increase its average crushing capacity from 830 tonnes of cane per hour in 1993 to 870 tonnes per hour in 1994. Booming cane expansion in the Lower Burdekin district also led to a $148 million upgrade at CSR's Invicta Mill. Large-scale factory rebuilding commenced in 1994, and was finally completed in 1997. These improvements aimed to increase the mill's crushing rate from 400 tonnes per hour in 1993 to 1 000 tonnes per hour in 1997. The centre piece of this development was a 55 m high and 11 m wide diffuser – the world's largest at the time – that could extract 1.0% more sugar out of cane than the traditional Australian milling train of multiple sets of crushing rollers. A new 320 tonne/hour bagasse-fired boiler – the world's largest bagasse-fired boiler at the time – provided the steam to drive all of this new machinery. CSR also invested another $40 million in its new high-tech Ord River Mill which operated for the first time in 1995.[164]

New innovations in sugar milling technology continued to be adopted at some Australian sugar mills during the late 1980s and early 1990s. In 1988, a SRI continuous high-grade vacuum pan was installed at the Maryborough Sugar Factory. Constructed by Walkers Ltd., it was the first installation of a continuous vacuum pan for high-grade boiling in Australia, and followed the successful implementation over a number of years of continuous boiling technology for low-grade massecuite production. Earlier use of continuous vacuum pans for high-grade boiling had been delayed by perceived difficulties in processing high purity massecuites to yield crystals of export sugar quality and concerns about excessive formation of crusted material on the internal surfaces of the continuous pan disrupting production. By 1995, continuous high-grade vacuum pans had also been installed in Tully, Inkerman, Invicta and Victoria Mills, the latter three being CSR-owned factories.[165] Continuous centrifugals developed to handle all grades of massecuite had been installed at five Queensland sugar mills by 1995, including Kalamia and Inkerman Mills, two CSR-owned mills. Known as the NQEA Superfugal, this new machine permitted the production of high-grade sugars by overcoming the problem of sugar crystal breakage.[166] In 1994, South Johnstone Mill established another first for an Australian sugar mill, when work began on an innovative cascade aeration system to cool water being discharged into the river. Instead of processing its heated water through cooling towers, it flowed over 150 m of artificial rapids which reduced the temperature of the water flowing back into the South Johnstone River from 45°C to 39°C. CSR's Ord River Mill also contained a number of technological innovations, including a CSR-designed vertical continuous vacuum pan, two continuous high-grade batch centrifugals and two continuous low-grade centrifugals. Such a high level of automation and fully continuous operation meant that the sugar mill required only thirty people to operate it, whereas a conventional sugar mill of that size relied upon 100 workers.[167]

By 1995, twenty-nine sugar mills operated in Australia: twenty-five in Queensland, three in New South Wales, and one in Western Australia. Included amongst these factories were CSR's Victoria Mill, the largest sugar mill in Australia, and one of the largest sugar mills in the world.[168] Overall, the majority of Australian sugar mills, excluding the Ord River Mill, were equipped with either four of five sets of crushing rollers, bigger milling trains of six or seven sets of crushing rollers having gone out of favour during the 1980s. Five Australian sugar mills were now equipped with double trains – diffusers operated in three of these factories (Table 17.6). Juice was clarified in mostly SRI or ATV clarifiers and the muds treated by rotary vacuum filters. Clarified juice was boiled in a single set of quintuple effets in 60% of Australian sugar mills; this task was performed in multiple effet sets in 20% of Australian sugar mills. Harwood Mill was unique, having a set of sextuple effets that had been installed to reduce the steam and fuel imbalance arising following the construction of a sugar refinery at Harwood in 1989 (see Chapter 18 for a full discussion).[169] The total area of evaporating equipment had reached a record high, despite the sugar mill closures between 1985 and 1995. Crystallisation occurred overwhelmingly in calandria vacuum pans, although continuous vacuum pans were gaining popularity (Table 17.7). Overall, the average amount of sugar cane handled annually by each sugar mill had risen sharply from around 900 000 tonnes in the late 1980s to 1.28 million tonnes in 1995 (Figure 17.4). Together, Australian sugar mills in 1995 had the capacity to handle almost 38 million tonnes of cane, a record amount.

Notes

1 *QPD*, Vol. 120 (1915), p. 339.
2 *QPD*, Vol. 120 (1915), p. 418.
3 *ASJ*, 9 September 1915, p. 401; CSR, Half Yearly Report, 30 September 1915, CSRR, P 10/75; E.W. Knox to Messrs Minter, Simpson & Co., Solicitors, Sydney, 2 September 1915, pp. 12–123, CSRR, 142/1137; *QPD*, Vol. 120 (1915), pp. 377 & 378.
4 *QPD*, Vol. 120 (1915–1916), pp. 377–378 & 460; *QGG*, Vol. 105, pp. 1971–1972; Shogren, 'The politics and administration of the Queensland sugar industry to 1930', pp. 259–263.
5 *ASJ*, 6 January 1916, p. 695; see the awards published in *QGG*, Vol. 106 (1916), pp. 1326, 1489–1490, 2368 & 2374 & Vol. 107 (1916), pp. 131–135 & 439–445.
6 Based upon an analysis of the awards published in *QGG*, Vol. 106 (1916), pp. 1609, 1610, 1634, 1852 & 2330 & Vol. 107 (1916), pp. 75–78, 131–135 & 439–445; see also Alcorn and Dunn, *Moreton Sugar Mill*, p. 75.
7 *QGG*, Vol. 107 (1916), pp. 439–445.
8 E.W. Knox to Macknade Mill Mang., 4 November 1915 & E.W. Knox to Goondi Mill Mang., 18 November 1915, CSRR, 142/1061, pp. 114–117 & 138; *ASJ*, 30 September 1915, p. 473 & 9 March 1916, pp. 838–840 & 850; *BC*, 9 February 1916, p. 8.
9 E.W. Knox, 'General Instructions as to Procedure Before Local Boards, 23 March 1916', 8 pp., CSRR, 142/3544; *BC*, 11 April 1916, p. 8; D. Murphy, *T.J. Ryan. A Political Biography* (Brisbane, 1975), p. 169.
10 *ASJ*, 7 September 1916, pp. 467–468.
11 Shogren, 'The politics and administration of the Queensland sugar industry to 1930', p. 274; *ASJ*, 10 October 1916, p. 533–534 & 9 November 1916, pp. 608–612; Qld. State Reports 1916, no. 238 & Qld. State Reports 1917, no. 1; Appeals by CSR to High Court of Australia re: Central Sugar Cane Prices Awards, A10044, 1916/5, 1916/6 & 1916/8, National Archives of Australia.
12 *ASJ*, 8 June 1916, p. 156; Murphy, *T.J. Ryan*, p. 212; for additional details see 'Report and Minutes of Evidence of the Select Committee on the Regulation of Sugar Cane Prices Act Amendment Bill', *QPP,* Vol. 2 (1916–17), pp. 1263–1351.
13 *QGG*, Vol. 108 (1917), p. 1870.
14 *ASJ*, 9 August 1917, p. 294, 8 November 1917, pp. 516–517 & 6 December 1917, pp. 588–589; Murphy, *T.J. Ryan*, pp. 295–296 & 343–344; *QPD*, Vol. 128 (1917), pp. 3464–3465; compare 6 George V No. 5 & 8 George V No. 18, *Queensland Statutes.*
15 Report of the Piddington Royal Commission, *CPP*, Vol. 4 (1920–21), p. 921; Ev. J. Broadhurst, Q. 119, in 'Report and Minutes of Evidence of the Select Committee on the Regulation of Sugar Cane Prices Act Amendment Bill', *QPP,* Vol. 2 (1916–17), p. 1281.
16 *QGG*, Vol. 109 (1917), p. 416; G. Pritchard, ASPA, to J. Clayton, Sec., Childers Mill Cane Growers' Assoc., 13 March 1918, OM BG/2/1, Envelope 12, & G. Pritchard, ASPA, to W. Compton, Proserpine, 28 December 1916, OM BG/2/2, Envelope 3, JOL; Report of the Piddington Royal Commission, *CPP*, Vol. 4 (1920–21), pp. 922–923.
17 For some of these disputes at Mackay see Manning, *In Their Own Hands*, pp. 217–221.
18 Kerr, *Only Room For One*, pp. 91–92; Kerr, *A Century of Sugar*, pp. 114–116; CSR, Half Yearly Report, 31 March 1918, CSRR, P 10/75; see also letters of agreement between

CSR, its agents and canegrowers between 1918 and 1921 in CSRR, 142/3545; CSR, *Newsletter* No. 3, August 1945, pp. 6–7, CSRR, Z/109, Box 371.

19 Ev. A. Dean (p. 366), in 'Piddington Royal Commission Evidence', CSRR, Z/109, Box 45; cf. 8 George V, No. 18 & 12 George V, No. 8, Queensland Statutes; *ASJ*, 8 April 1921, p. 4; F. Maxwell, *Economic Aspects of Cane Sugar Production* (London, 1927), p. 65.

20 Ev. T. McMahon, Pres. of the Clarence River Farmers' and Settlers' Association (p. 89), in 'Piddington Royal Commission Evidence', CSRR, Z/109, Box 45; Smith, *Sweet Beginnings*, pp. 286–287; for examples of these agreements see CSRR, Z/109, Box 370.

21 R. Matthews, 'An attempt to understand why the price of sugar cane has all but ceased to be regulated by *The Regulation of Sugar Cane Prices Acts* 1915–1966', *Discussions of ASSCT*, 1980, pp. 16–17.

22 Matthews, 'An attempt to understand why the price of sugar cane has all but ceased to be regulated by *The Regulation of Sugar Cane Prices Acts* 1915–1966', p. 18; *The Producers' Review*, 10 September 1918, p. 35.

23 Matthews, 'An attempt to understand why the price of sugar cane has all but ceased to be regulated by *The Regulation of Sugar Cane Prices Acts* 1915–1966', p. 18; Industry Commission, *The Australian Sugar Industry. Report No. 19* (Canberra, 1992), pp. 193–194.

24 *ASJ*, 4 July 1924, pp. 235–236; *QGG*, Vol. 110 (1918), p. 1339 (Babinda Mill award); for an example of CSR's payment scheme see *QGG*, Vol. 118 (1922), p. 835 (Childers Mill Award); Mulgrave Central Mill, *50 years in retrospect 1895–1945: a souvenir of the golden jubilee of the Mulgrave Central Mill* (Gordonvale, 1947), pp. 86–87.

25 *ASJ*, 4 July 1924, pp. 235–236; C. Waddell, 'Principles of a relative percentage payment scheme', *Proc. ISSCT*, 1950, p. 613; Anon, *The Mourilyan Story*, p. 8.

26 Goondi Mill Cane Insp., AR 1945, p. 17, CSRR, Z/109, Box 465; *The Producers' Review*, 15 April 1950, p. 40A & 15 August 1950, p. 40 A; Kerr, *Crystal Century*, p. 115; for New South Mills see CSR, Agricultural Circulars 1953–1956 , CSRR, Z/109, Box 370; *ASJ*, 15 March 1950, p. 761 & 15 May 1950, p. 131; A. Saranin, 'Sugar cane payment systems', *Sugar Technology Reviews*, Vol. 3 (1975), p. 222.

27 Kerr, *Southern Sugar Saga*, pp. 84–86; R. Longhurst, *Nerang Shire. A History to 1949* (Nerang, 1994), p. 153; Griggs, 'Plantation to small farm', pp. 327–330.

28 CSR, Half Yearly Report, 31 March 1921, CSRR, P 10/75; Kerr, *A Century of Sugar*, pp. 116–117.

29 Kerr, *Pioneer Pageant*, pp. 165–166; Kerr, *Only Room For One*, pp. 83–87 & 94–97.

30 BCSM, AR 1926–27, *QPP*, Vol. 2 (1927), p. 607; QAG, AR 1928–29, *QPP*, Vol. 1 (1929), p. 48; QAG, AR 1958–59, *QPP*, Vol. 1 (1959–60), p. 143; Hudson, *By The Banyan*, pp. 73–74; Kerr, *A Crystal Century*, p. 91; Dignan, *The Story of Kolan*, pp. 102–104; *ASJ*, 15 June 1945, pp. 113–114; Kerr, *Sugar at Maryborough*, pp. 195–198.

31 Easterby, *The Queensland Sugar Industry*, p. 144; *The Producers' Review*, 10 October 1919, p. 37.

32 Ev. A. Thorp, North Eton Mill (p. 425); Ev. R. Sloan, Mourilyan Mill (p. 323); Ev. S. Mason, Proserpine Mill (p. 390); & Ev. G. Greathead, Moreton Mill (p. 510); all in 'Piddington Royal Commission Evidence', CSRR, Z/109, Box 45.

33 Indenture between C. Petree and the Qld. Treasurer re: improvements in cane sugar manufacture, 1917, Item 1101498, QSA; Alcorn and Dunn, *Moreton Sugar Mill*,

p. 76; Kerr, *Southern Sugar Saga*, p. 89; Ev. F. Harris, Babinda Mill (p. 312); Ev. A. Cruickshank, Kalamia Mill (p. 366); Ev. S. Manson, Proserpine Mill (p. 390); Ev. H. Young, Fairymead Mill (p. 467); Ev. W. Gibson, Bingera Mill (p. 482); Ev. M. Stapleton, Gin Gin Mill (p. 495); Ev. G. Griffin, Doolbi Mill (p. 496); & Ev. J. McDonald, Racecourse Mill (p. 417); all in 'Piddington Royal Commission Evidence', CSRR, Z/109, Box 45.

34 G. Jenkins, *Introduction to Cane Sugar Technology* (Amsterdam, 1966), pp. 174–175 & 211–212; J. Davies, *The Principles of Cane Sugar Manufacture* (London, 1938), pp. 38–39.

35 CSR, Half Yearly Report, 30 September 1916, CSRR, P 10/75; Ev. W. Adams, 30 July 1919, in 'Piddington Royal Commission Evidence', p. 489, CSRR, Z/109, Box 45.

36 *ASJ*, 9 December 1921, p. 529 & 5 September 1924, p. 402; Kerr, *Pioneer Pageant*, pp. 165.

37 *ASJ*, 10 March 1922, p. 717, 9 June 1922, p. 169, 6 June 1924, p. 193 & 3 December 1925, p. 564; H. Easterby, 'Cane Crop Prospects', *QAJ*, Vol. 25, no. 5 (1 May 1926), p. 417; H. Easterby, 'Cane Crop Prospects', *QAJ*, Vol. 26, no 2 (1 August 1926), p. 114; Easterby, *The Queensland Sugar Industry*, p. 121.

38 Ev. R. Clarke, Pleystowe Mill (p. 401); Ev. W. McCready, Palms Mill (p. 419); Ev. G. Muntz, Mossman Mill (p. 237); Ev. T. Braddock, Maryborough Sugar Factory (p. 508); & Ev. J. Paddle, Mt Bauple Mill (p. 506); all in 'Piddington Royal Commission Evidence', CSRR, Z/109, Box 45.

39 Report of the Piddington Royal Commission, *CPP*, Vol. 4 (1920–21), p. 933.

40 Easterby, *The Queensland Sugar Industry*, p. 224; for Bennett's biographical details see *ASJ*, April 1974, p. 69.

41 Anon, 'The Logan Cane Crop', *QAJ*, Vol. 22, no. 3 (September 1924), p. 154; *ASJ*, 10 March 1922, p. 717.

42 *ASJ*, 10 March 1922, p. 725; BSES AR 1921–22, *QPP*, Vol. 2 (1923), p. 277; Griggs, 'Plantation to small farm', pp. 381–382; Kerr, *Black Snow and Liquid Gold*, p. 193; Anon, *Kalamia Sugar Mill. The First 100 Years* (Ayr?, 1984), p. 5; Mulgrave Central Mill, *50 years in retrospect 1895–1945: a souvenir of the golden jubilee of the Mulgrave Central Mill*, p. 80.

43 Kerr, *Only Room For One*, p. 81; *ASJ*, 6 July 1916, p. 302.

44 Ev. G. Francis, Doolbi Mill (p. 496) & Ev. G. Greathead, Moreton Mill (p. 511); both in 'Piddington Royal Commission Evidence', p. 496, CSRR, Z/109, Box 45; Jenkins, *Introduction to Cane Sugar Technology*, pp. 39–40; J. Clayton, 'Some features of milling in Queensland', *Proc. ISSCT*, 1953, p. 631.

45 E. Marriott, 'Milling practice', *Proc. ISSCT*, 1935, p. 545; Easterby, *The Queensland Sugar Industry*, p. 120; C. Anderson, 'Technology Changes in the Australian Sugar Industry'. M. Eng. thesis, Uni. of Qld., 1981, p. 51.

46 Minutes of Manufacturer's Conference to form a Technological Department, 29 May 1928 & Minutes of a Special Meeting of ASPA Executive 16 July 1928, in AER, 165/194, NBAC.

47 W. McAllep, 'Report on Queensland sugar factories, 15 December 1928', AER, 165/194, NBAC.

48 Easterby, *The Queensland Sugar Industry*, pp. 224–226; BSES, AR 1928–29, *QPP*, Vol. 2 (1929), p. 981.

49 J. Clayton, 'The work of the Division of Mill Technology', in BSES, *Fifty Years of Scientific Progress* (Brisbane, 1950), pp. 73–75.

50 Clayton, 'The work of the Division of Mill Technology', p. 75; N. Smith, 'Notes on clarification', *Proc. QSSCT*, 1936, pp. 45–49; J. Davidson, 'Notes on the clarification of the juices from P.O.J. canes', *Proc. QSSCT*, 1937, pp. 237–242; Jenkins, *Introduction to Cane Sugar Technology*, pp. 174–175 & 266.

51 E. Behne, 'Present practice in raw sugar manufacture in Queensland', *Proc. ISSCT*, 1938, p. 1056; E. Behne, *Cane Preparation. Part 11: The National Shredder (continued), High Speed Knives and the Searby Shredder*. BSES Technical Communication, 1941, p. 26.

52 Mill machinery details for 23 Queensland mills are to be found in *Handbook for 5th Congress of ISSCT*, Brisbane, Aug–Sept 1935, David North Collection MS 115, Adolph Basser Library, Australian Academy of Science (hereafter Congress Handbook 1935); CSR, Mill Engineering and Fuel Notes Summary, 1934–1945, p. 5, CSRR, Z/109, Box 537; Hugot, *Handbook of Cane Sugar Engineering*, pp. 57–59; Anderson, 'Technology Changes in the Australian Sugar Industry', p. 35.

53 *ASJ*, 9 February 1935, p. 91; Hudson, *Sweet Success*, p. 208; Kerr, *Only Room for One*, p. 105; Kerr, *Top Mill in the Valley*, p. 37; Congress Handbook 1935, pp. 44 & 49.

54 E. Powell, *Elements of Raw Sugar Manufacture* (Sydney, 1955), p. 46; *ASJ*, 8 October 1931, p. 381 & 8 May 1934, p. 99; Kerr, *A Century of Sugar*, p. 141; Kerr, *Sugar at Maryborough*, p. 184; Kerr, *A Crystal Century*, p. 97; C. Murry and D. Shann, 'Crushing mill feeding. A review of Queensland practice', *Sugar Technology Reviews*, Vol. 1 (1969), p. 45.

55 Marriott, 'Milling practice', p. 545; E. Behne and G. Jenkins, 'Present practice in raw sugar manufacture in Queensland', *Proc. ISSCT*, 1938, p. 1049; J. Dixon, 'Sugar milling by CSR', in A. G. Lowndes (ed.), *South Pacific Enterprise* (Sydney, 1956), p. 141; Hugot, *Handbook of Cane Sugar Engineering*, pp. 79–80; CSR, Mill Engineering and Fuel Notes Summary, 1934–1945, p. 6, CSRR, Z/109, Box 537.

56 *ASJ*, 7 August 1930, p. 325; Kerr, *Top Mill in the Valley*, p. 37; Kerr, *Southern Sugar Saga*, p. 102; Kerr, *A Century of Sugar*, pp. 141–142; Behne and Jenkins, 'Present practice in raw sugar manufacture in Queensland', p. 1052; BSES, AR 1932–33, *QPP*, Vol. 2 (1934), p. 825.

57 Kerr, *Southern Sugar Saga*, p. 102; Kerr, *Only Room for One*, p. 105; A. Crees, 'New boiler installation at Mossman Mill', *Proc. ISSCT*, 1935, pp. 129–130; Extract from Notes of GSC's visit to Queensland July–September 1938, AER, 165/197, p. 311 (red), NBAC; W. Stoward, 'Boiler and furnace installation at Bingera', *Proc. QSSCT*, 1939, p. 95; E. Behne & G. Jenkins, 'Present practice in raw sugar manufacture in Queensland', *Proc. ISSCT*, 1938, p. 1052; *ASYB*, 1940–41, p. 229.

58 E. Behne, 'Progress in milling in Queensland factories', *ASYB*, 1942–43, p. 163; Congress Handbook, 1935, pp. 32, 36, 37, 41, 49, 61 & 63.

59 *ASYB*, 1940–41, pp. 229 & 255; *ASJ*, 7 August 1930, p. 322, 8 October 1931, p. 381, 4 August 1932, p. 221, 8 May 1934, p. 99, 9 February 1935, p. 95 & 12 April 1938, p. 99; Minutes of 14th AGM of Amalgamated Sugar Mills Ltd., 18 May 1939, AER, 165/197, p. 360 (red), NBAC; Kerr, *Only Room For One*, p. 105; Kerr, *Crystal Century*, p. 97; A. Sheppeard, 'The relationship between regional electric authorities and sugar factories', *Proc. QSSCT*, 1946, p. 77; Hudson, *Sweet Success*, p. 112.

60 D. McBryde, *Notes on Raw Sugar Manufacture*. BSES Division of Mill Technology Bulletin No. 1 (1935), p. 12; Kerr, *Southern Sugar Saga*, p. 102; G. Harman, 'Manufacture

of raw sugar – Australian practice', *Proc. ISSCT*, 1935, p. 46; Jenkins, *Introduction to Cane Sugar Technology*, p. 212; J. Hayden, 'Economic treatment of subsider muds', *Proc. QSSCT*, 1931, p. 65.

61 Reid, *The Long View*, p. 54; Hugot, *Handbook of Cane Sugar Engineering*, pp. 338–342; Mackay Institute of Mill Engineers, 'Advantages and disadvantages of a rotary mud filtration plant', *Proc. QSSCT*, 1939, p. 27; Jenkins, *Introduction to Cane Technology*, pp. 212–213.

62 CSR, Mill Engineering and Fuel Notes Summary, 1934–1945, p. 7, CSRR, Z/109, Box 537; *ASYB*, 1940–41, p. 231; *ASJ*, 11 March 1937, p. 691.

63 Kerr, *Sugar at Maryborough*, pp. 182; Kerr, *A Century of Sugar*, pp. 142; Kerr, *Top Mill in the Valley*, p. 37; Kerr, *Crystal Century*, p. 97; *ASJ*, 13 July 1939, p. 251; CSR, Mill Engineering and Fuel Notes Summary, 1934–1945, pp. 8–9, CSRR, Z/109, Box 537; Anderson, 'Technology Changes in the Australian Sugar Industry', p. 51; Jenkins, *Introduction to Cane Sugar Technology*, p. 293.

64 Jenkins, *Introduction to Cane Sugar Technology*, pp. 320–322; Behne and Jenkins, 'Present practice in raw sugar manufacture in Queensland', p. 1060; *ASYB*, 1940–41, p. 233; Congress Handbook, 1935, pp. 30, 31, 44, 52 & 55; S. Best, 'Electric conductivity control in the boiling of low grade massecuite', *Proc. QSSCT*, 1935, pp. 89–90.

65 V. Thorp, 'The refining quality of sugars in the Mackay district. Report covering research work for 1933 and 1934 seasons', *Proc. ISSCT*, 1935, p. 759 & 761; BSES, *Laboratory Manual for Queensland Sugar Mills* (Brisbane, 1934), p. 86; D. McBryde, 'Pan boiling by conductivity control in Queensland', *Proc. ISSCT*, 1935, pp. 306–307.

66 V. Thorp, 'The refining quality of sugars in the Mackay district', p. 763; *ASYB*, 1940–41, p. 237.

67 BSES, AR 1933–34, *QPP*, Vol. 2 (1935), p. 795; *ASYB*, 1940–41, p. 231; Behne and Jenkins, 'Present practice in raw sugar manufacture in Queensland', pp. 1056–1058; Congress Handbook 1935, pp. 45 & 58.

68 *ASYB*, 1956, p. 45; Ev. F.W. Heck (p. 48) and Ev. H.W. Kerr (p. 1006), in 'Royal Commission on Sugar Peaks and Cognate Matters Evidence, 1939', CSRR, Z/109, Box 48.

69 *ASJ*, 9 November 1940, p. 427; D. Nicklin, 'The Pleystowe electrification', *Proc. QSSCT*, 1941, p. 127; Kerr, *Sugar at Maryborough*, p. 201; R. Beale and L. Pashley, 'The Fairymead cane cleaner', *Proc. QSSCT*, 1958, p. 93.

70 Anon, *Kalamia Sugar Mill. The First 100 Years*, pp. 6–7; Kerr, *Top Mill in the Valley*, p. 39; G. Burrows and C. Morton, *The Canecutters* (Melbourne, 1986), p. 102; Morton, *By Strong Arms*, p. 124; BSES, AR 1940–1941, p. 4; BSES, AR 1942–1943, p. 4.

71 *ASJ*, 13 August 1942, pp. 181, 182, 221 & 222; BSES, AR 1942–1943, pp. 3 & 20; Kerr, *Crystal Century*, p. 104.

72 CSR, Raw Sugar Manufacture Report 1942, p. 1, CSRR, Z/109, Box 380; CSR, Raw Sugar Manufacture Report 1944, p. 2 & CSR, Raw Sugar Manufacture Report 1945, p. 7, both in CSRR, Z/109, Box 381; BSES, AR 1943– 44, pp. 3 & 25.

73 Kerr, *Only Room For One*, p. 108; Ev. N. Burtt, Mang. Babinda Central Mill, p. 1178 & Ev. H. Whitaker, Mang., Mulgrave Central Mill, p. 1206, in 'Mansfield Royal Commission Evidence, Vol. 5, 1950', CSRR, Z/109, Box 56.

74 For sources of machinery see Kerr, *Southern Sugar Saga*, p. 118; Kerr, *A Century of Sugar*, p. 152; & Kerr, *Northern Outpost*, p. 129; for purchases of second hand machinery see Kerr, *Sugar at Maryborough*, p. 205 & Kerr, *Top Mill in the Valley*, p. 42.

75 Reid, *The Long View*, p. 35; *ASYB*, 1949, pp. 119 & 156; CSR, Mill Engineering and Fuel Summary Notes, 1947–48, p. 20, CSRR, Z/109, Box 537; for complaints about the poor material supply position see Mulgrave Central Mill, AR 1950, ML.

76 Kerr, *Northern Outpost*, p. 129; Kerr, *Only Room for One*, p. 108; Mulgrave Central Mill, AR 1949, ML; *ASJ*, 15 September 1947, p. 309.

77 Kerr, *Top Mill in the Valley*, p. 42; Kerr, *Southern Sugar Saga*, p. 119; Kerr, *A Century of Sugar*, p. 152; CSR, Chemical Report 1946, p. 10 & CSR, Chemical Report 1949, p. 11, both in CSRR, Z/109, Box 381; Ev. N. Burtt, Mang. Babinda Central Mill, p. 1178, in 'Mansfield Royal Commission Evidence, Vol. 5, 1950', CSRR, Z/109, Box 56; *ASJ*, 15 January 1949, p. 591.

78 Kerr, *Sugar at Maryborough*, p. 197; Kerr, *A Century of Sugar*, p. 152; Kerr, *Only Room for One*, p. 108.

79 Reid, *The Long View*, pp. 17–26; *ASYB*, 1977, p. 75.

80 BSES, *75 Years of Scientific Progress. Bureau of Sugar Experiment Stations 1900–1975* (Brisbane, 1975), pp. 32–35.

81 J. Clayton, 'Some features of milling in Queensland', *Proc. ISSCT*, 1953, p. 630; *ASYB*, 1941, p. 231; Jenkins, *Introduction to Cane Sugar Technology*, pp. 289; C. Venton, 'Pans and pan boiling in Queensland', *Proc. ISSCT*, 1953, pp. 778–779.

82 *ASYB*, 1952, p. 297; *ASYB*, 1962, p. 141; CSR, *Newsletter*, No. 35, March 1954, p. 22, CSRR, Z/109, Box 371; Colin Wadell, 'Technical report on Rocky Point Sugar Factory, 21 December 1954', PRV 14721/1/108, Box 45: Rocky Point Mill, QSA; Minute Book, Australian Sugar Co. Pty. Ltd., entry for 16 February 1953, MSS 3565/32, ML; *ASJ*, 15 July 1957, p. 227.

83 *ASYB*, 1962, p. 105; Haughton Sugar Co. Ltd., AR 1963–64, p. 2; Kerr, *Sugar at Maryborough*, p. 222; Alcorn and Dunn, *Moreton Sugar Mill*, p. 158; *The Producers' Review*, 15 February 1963, p. 3.

84 E. Jensen, 'Diffusion-together with its effect on process and recovery', *Proc. QSSCT*, 1968, pp. 229–230.

85 *ASJ*, June 1967, p. 197, Kerr, *Southern Sugar Saga*, pp. 120, 121, 130, 131& 137; Kerr, *A Century of Sugar*, p. 173; *ASJ*, September 1958, p. 517; Kerr, *Northern Outpost*, p. 148; Alcorn and Dunn, *Moreton Sugar Mill*, p. 141.

86 Anon, 'The Fairymead Diffuser', *ASYB*, 1967, p. 87; D. Foster and D. Shann, 'Experiences with cane diffusion in Queensland', *Proc. ISSCT*, 1971, pp. 1267–1280; Jensen, 'Diffusion-together with its effect on process and recovery', pp. 229–234.

87 BSES, AR 1952–53, p. 49; BSES, AR 1967–68, p. 82; *ASYB*, 1967, p. 85.

88 Hudson, *Sweet Success*, p. 208; *ASJ*, 15 October 1953, p. 465.

89 Jenkins, *Introduction to Cane Sugar Technology*, pp. 41 & 43–44; BSES, AR 1954–55, p. 81; Anderson, 'Technology change in the Australian sugar industry', pp. 40–41; for the installation of apron feeders at Racecourse and Plane Creek Mills see *ASJ*, July 1957, p. 265.

90 Reid, *The Long View*, p. 63; Anderson, 'Technology change in the Australian sugar industry', p. 41; Kerr, *Black Snow and Liquid Gold*, p. 249; *ASJ*, June 1958, pp. 233–234; W. Harper, 'Recent milling train improvements at Mossman', *Proc. QSSCT*, 1961, pp. 89–90; *ASYB*, 1967, p. 85.

91 Anderson, 'Technology change in the Australian sugar industry', pp. 41–43; Reid, *The Long View*, p. 82; Murry and Shann, 'Crushing mill feeding. A review of Queensland

practice', pp. 46–47; N. Farmer, 'The toothed roller continuous pressure feeder', *Proc. ISSCT*, 1977, pp. 2241–2249; D. Shann, 'The development of the S.R.I. light duty feeder', *Proc. ISST*, 1977, pp. 2347–2344.

92 For complaints about Australian sugar and CSR's research into clarification see CSR, Raw Sugar Manufacturing Report 1954, p. 124, CSRR, Z/109, Box 382 & Jenkins, *Introduction to Cane Sugar Technology*, pp. 200 & 404–405; for upgrades to the clarification stations see Kerr, *Southern Sugar Saga*, p. 130; CSR, Raw Sugar Manufacturing Report 1955, p. 29, CSRR, Z/109, Box 382; CSR, Raw Sugar Manufacturing Report 1960, p. 91; & CSR, Raw Sugar Manufacturing Report 1963, p. 13, both in CSRR, Z/109, Box 383.

93 Reid, *The Long View*, pp. 54–55 & 89; CSR, Raw Sugar Manufacture Report 1959, p. 33, CSRR, Z/109, Box 382; Kerr, *A Century of Sugar*, p. 173; *ASYB*, 1969, mill section: lists of machinery.

94 Jenkins, *Introduction to Cane Sugar Technology*, p. 219; *ASYB*, 1970, pp. 246 & 279–311.

95 BSES, AR 1967–68, p. 82; Kerr, *Only Room for One*, p. 116; *ASJ*, June 1964, p. 249; Kerr, *Sugar at Maryborough*, pp. 216 & 222; Haughton Sugar Co. Ltd., AR 1967–68, p. 2 & Haughton Sugar Co. Ltd., AR 1968–69, p. 2; Kerr, *A Century of Sugar*, p. 173; Alcorn & Dunn, *Moreton Sugar Mill*, p. 159; Kerr, *Top Mill in the Valley*, pp. 49–50; *ASJ*, 15 July 1957, p. 231; *The Producers' Review*, 15 April 1963, p. 61.

96 Jenkins, *Introduction to Cane Sugar Technology*, p. 230; R. Beale, 'Entrainment prevention at Gin Gin', *Proc. QSSCT*, 1959, pp. 205–207; A. Claire and T. Mulvena, 'Entrainment prevention at Racecourse Mill', *Proc. QSSCT*, 1967, pp. 107–111.

97 *The Producers' Review*, 15 February 1963, p. 76; CSR, *Newsletter* No. 77, June 1963, CSRR, Z/109, Box 372; L. Kirby, 'Continuous centrifugals', *Proc. QSSCT*, 1969, pp. 77–78; Kerr, *Southern Sugar Saga*, p. 130; CSR, Raw Sugar Manufacture Report 1965, p. 4, CSRR, Z/109, Box 383; Reid, *The Long View*, p. 93; B. Sockhill, 'Low grade fugals – continuous or batch', *Proc. QSSCT*, 1970, p. 299; R. Swindells and L. Kirby, 'Continuous high grade massecuite separation', *Proc. ASSCT*, 1981, p. 49.

98 *The Producers' Review*, 15 February 1963, pp. 3 & 76, 15 April 1963, p. 61 & 15 June 1963, p. 94; CSR, Raw Sugar Manufacture Report 1952, p. 10, CSR, Raw Sugar Manufacture Report 1954, pp. 9 & 11, CSR, Raw Sugar Manufacture Report 1955, p. 13, all in CSRR, Z/109, Box 382; *ASJ*, June 1964, p. 249; Kerr, *Sugar at Maryborough*, p. 215; Kerr, *A Century of Sugar*, p. 173; Alcorn and Dunn, *Moreton Sugar Mill*, p. 158; Kerr, *Southern Sugar Saga*, p. 130.

99 BSES, AR 1953–54, p. 82; BSES, AR 1964–65, p. 76; BSES, AR 1965–66, p. 75.

100 BSES, AR 1949–50, p. 48; BSES, AR 1953–54, p. 81; BSES, AR 1955–56, p. 93; Kerr, *Southern Sugar Saga*, p. 130; Kerr, *Sugar at Maryborough*, pp. 216; Minute Book, Australian Sugar Co. Pty. Ltd., entry for 16 February 1953, MSS 3565/32, ML.

101 BSES, AR 1953–54, p. 80; BSES, AR 1954–55, p. 83; BSES, AR 1962–63, p. 82; BSES, AR 1964–65, p. 79; BSES, AR 1967–68, p. 82; *ASJ*, 30 September 1963, p. 471; Kerr, *A Crystal Century*, p. 106.

102 BSES, AR 1954–55, p. 81; Hudson, *Sweet Success*, p. 209; *ASYB*, 1967, p. 85.

103 BSES, AR 1954–55, p. 4; *ASYB*, 1959, p. 121; Langford and Thomis, *Rocky Point and the Heck family*, p. 62.

104 Kerr, *Pioneer Pageant*, p. 171; *Logan Witness*, 26 August 1882, p. 3; for other examples of sugar mills polluting streams see Kerr, *Sugar at Maryborough*, p. 48; H. Gregory, *Making Maroochy: a history of the land, the people and the Shire* (Brisbane, 1991), p. 47; & J. Brunnich, Qld. Government Agricultural Chemist, 'Report on Pollution of Tinana Creek water by sewage from Mt Bauple Central Mill, 6 May 1898', AGS/N94, QSA.

105 Kerr, *Pioneer Pageant*, p. 171; Kerr, *Only Room for One*, p. 68; W. Farquhar, CSR Insp. of Mills, to E.W. Knox, 19 May 1908, CSRR, 142/2896, p.8; Homebush Mill Mang. to E.W. Knox, 25 March 1912, CSRR, 142/257, Letter No. 846.

106 Kerr, *Pioneer Pageant*, p. 172; P. Phillips, *Sarina Shire in Retrospect* (Sarina?, c. 1988), pp. 185–188; BSES, AR 1928–29, *QPP*, Vol. 2 (1929), p. 983; J. Hayden, 'Economic treatment of subsider muds', *Proc. QSSCT*, 1931, p. 66.

107 D. Dunstan, 'Effects of sugar mill effluent on fish life', *Fisheries Newsletter* (Fisheries Division of the Council for Scientific and Industrial Research), Vol. 14, no. 5 (May 1955), pp. 15 & 23; Hudson, *By the Banyan*, p. 199; *ASJ*, 30 July 1962, p. 279; J. Bond and E. McNeil, 'Treatment of sugar mill waste by shallow ponding', *Proc. QSSCT*, 1976, p. 317; Alcorn and Dunn, *Moreton Sugar Mill*, p. 163; Kerr, *A Crystal Century*, p. 138.

108 For these complaints see 'Correspondence re: destruction of fish by pollution of rivers and streams by sugar mill wastes, 1960–1969', in Qld. Treasury Dept., SRS6232/7/1216, Item 957275, QSA.

109 A. Peel, Director, Qld. Department of Harbours & Marine, to Mang., Racecourse Cooperative Sugar Milling Assoc., 11 November 1965, in Qld. Treasury Dept., SRS6232/7/1216, Item 957275, QSA; *QPD*, Vol. 246, 1967–68, pp. 1179–1180.

110 Kerr, *A Century of Sugar*, pp. 180–181; *ASJ*, November 1968, pp. 434–435; *Bundaberg News Mail*, 5 October 1968 & 21 November 1968; CSR, Raw Sugar Manufacturing Report 1969, p. 122, CSRR, Z/109, Box 384.

111 'Notes on Pollution Problems, Queensland Sugar Mills: Results of Surveys, 1970', PRV 1472/1/42, Box 19, File 3P/8/2- Pollution: General (Item 896497), QSA.

112 Reid, *The Long View*, p. 101; D. Bevan, 'The disposal of sugar mill effluents', *Sugar y Azucar*, June 1973, pp. 38–39.

113 R. Bathgate, J. Keniry and A. Strong, 'Treatment of sugar mill waste waters', *Proc. ISSCT*, 1977, pp. 2509–2515; CSR, Raw Sugar Manufacturing Report 1971, pp. 147–148, CSRR, Z/109, Box 384.

114 WQCQ, AR 1972–73, p. 1; WQCQ, AR 1975–76, p. 1; WQCQ, AR 1978–79, p. 1; CSR, Raw Sugar Manufacturing Report 1974, p. 112, CSRR, Z/109, Box 385.

115 CSR, Raw Sugar Manufacturing Report 1970, pp. 92–93, CSRR, Z/109, Box 384; CSR, Raw Sugar Manufacturing Report 1973, pp. 113–114, CSRR, Z/109, Box 385; *ASJ*, June 1973, p. 170; K. McNeil, J. Bond and W. Gampe, 'Effluent treatment at Farleigh Mill', *Proc. QSSCT*, 1974, pp. 235–236; J. Sestero and J. Logan, 'Mill housekeeping', *Proc. QSSCT*, 1972, pp. 165–168.

116 WQCQ, AR 1974–75, p. 4; CSR, Raw Sugar Manufacturing Report, 1975, p. 84, CSRR, Z/109, Box 385.

117 Bathgate, Keniry and Strong, 'Treatment of sugar mill wastes', p. 2509; Hudson, *By the Banyan*, pp. 201–202; Hudson, *Sweet Success*, p. 235; *Bundaberg Sugar News*, March 1977, p. 5; C. Henderson and K. Peatey, 'Effluent control at Cattle Creek', *Proc. ASSCT*,

1981, pp. 153–157; Anon, *The Mourilyan Story*, p. 9; CSR, Raw Sugar Manufacturing Report 1974, p. 112, CSRR, Z/109, Box 385.

118 WQCQ, AR 1978–79, p. 5; *ASYB*, 1978, pp. 67–68; R. Pearson and L. Penridge, *An ecological survey of selected rivers in Queensland: final report* (Townsville, 1979), p. xviii.

119 L. Haworth, *Progress Made by Queensland Sugar Mills Towards Meeting Water Quality Licence Conditions* (Brisbane, 1982), various pages, in Coordinator General's Office, Queensland, SRS 1158/1/8736, Item 577216, QSA.

120 J. Joyce and B. Ferris, 'Evaluation of an upgrade to CSR Victoria Mill's activated sludge plant', *Proc. ASSCT*, 1994, pp. 229–233; W. Dawson, G. Likacs and D. Sockhill, 'Establishment of constructed wetlands at Victoria Mill', in *Proc. National Conference on Wetlands for Water Quality Control* (Townsville, 1995), pp. 61–69; Hudson, *Sweet Success*, pp. 244–248.

121 BSES, AR 1956–57, p. 94; *ASJ*, July 1959, p. 351; Kerr, *Sugar at Maryborough*, p. 218.

122 APCQ, AR 1970–71, pp. 1–3.

123 APCQ, AR 1970–71, p. 3; APCQ, AR 1972–73, p. 14; APCQ, AR 1976–77, p. 2; *ASJ*, June 1972, p. 151; Reid, *The Long View*, p. 99.

124 For examples of Queensland sugar mills being equipped with fly-ash collectors see Hudson, *Sweet Success*, p. 235 & Kerr, *A Century of Sugar*, p. 181; for CSR's fly-ash dewatering plant see CSR, *Newsletter* No. 159, January 1980, pp. 21–22, CSRR, Z/109, Box 376.

125 APCQ, AR 1974–75, p. 3; APCQ, AR 1975–76, p. 4; APCQ, AR 1976–77, p. 2; APCQ, AR 1978–79, pp. 4–5.

126 G. Sawyer and R. Cullen, 'A review of ash handling systems in Queensland sugar mills', *Proc. ISSCT*, Vol. 3, 1977, p. 263; Haworth, *Progress Made by Queensland Sugar Mills Towards Meeting Water Quality Licence Conditions*, p. 5.

127 *Bundaberg Sugar News*, March 1977, p. 5; R. Jones and R. Dyne, 'Boiler ash separation by ponding', *Proc. QSSCT*, 1977, pp. 335–339; N. Maclean, 'Agronomic aspects of fly ash disposal by incorporation with filter mud in the Herbert River district', *Proc. QSSCT*, 1976, p. 83; J. Watt and A. Morton, 'Combined mill waste disposal', *Proc. ASSCT, 1979*, pp. 167–171.

128 CSR, Raw Sugar Manufacturing Report 1972, pp. 90–91 & CSR, Raw Sugar Manufacturing Report 1974, pp. 112, both in CSRR, Z/109, Box 385.

129 Briody, *Sunshine Sugar*, p. 118; NSW Sugar Milling Coop., AR 1988–89, p. 10; NSW Sugar Milling Coop., AR 1993–94, p. 17; NSW Sugar Milling Coop., AR 1994–95, p. 14; NSW State Pollution Control Commission, AR 1983–84, p. 53.

130 Manning, *In their Own Hands*, p. 302.

131 CSR, *Newsletter* No. 142, September 1975, p. 10, CSRR, Z/109, Box 372; Sir Charles Johnston, Chairman, Australian Estates, to All Shareholders, 25 February 1975, AER, 165/218; Briody, *Sunshine Sugar*, p. 35.

132 *ASJ*, June 1981, p. 192.

133 P. Kelly and R. Porter, 'Operation of the Inkerman diffuser', *Proc. QSSCT*, 1978, pp. 19–25; *ASYB*, 1977, pp. 313–314.

134 Mulgrave Central Mill, AR 1976–77; Manning, *In Their Own Hands*, pp. 302–303; J. Allen, 'Advances in mill machinery in 70's', *ASJ*, October 1981, p. 391.

135 Alcorn and Dunn, *Moreton Sugar Mill*, p. 173; *ASJ*, March 1976, p. 617; for the installation of rotary bin tipplers see Rolleston, *The Defiance*, p. 106; Mulgrave Central Mill, AR 1973; *ASJ*, July 1973, p. 214; & CSR, Raw Sugar Manufacture Report 1974, p. 30, CSRR, Z/109, Box 385.

136 Rolleston, *The Defiance*, p. 106; Anderson, 'Technology change in the Australian sugar industry', pp. 35–37; J. Allen, 'Advances in mill machinery in 70's', *ASJ*, October 1981, p. 391; P. Rein, *Cane Sugar Engineering* (Berlin, 2007), p. 87.

137 Reid, *The Long View*, pp. 101–103; BSES, AR 1970–71, p.57; *ASYB*, 1986, see machinery details for CSR mills; Rein, *Cane Sugar Engineering*, p. 231.

138 Haworth, *Progress Made by Queensland Sugar Mills Towards Meeting Water Quality Licence Conditions*, p. 26; C. Henderson *et al.*, 'The vertical louvre entrainment separator', *Proc. ASSCT*, 1981, pp. 145–151; P. Wright, 'Entrainment separation devices for raw sugar factories', *Proc. ISSCT*, Vol. 1 (1989), p. 169.

139 *ASYB*, 1972, p. 331; *ASYB*, 1973, p. 27; *ASJ*, June 1976, p. 159; *ASYB*, 1979, p. 288; C. Anderson, 'Technology changes in the Australian sugar industry', pp. 52–58; Rolleston, *The Defiance*, p. 108, Anon, *The Mourilyan Mill Story*, p. 9; . Allen, 'Advances in mill machinery in 70's', *ASJ*, October 1981, p. 395.

140 Reid, *The Long View*, pp. 137–140; *ASYB*, 1977, p. 84; Hudson, *By the Banyan*, p. 329; Anon, *The Mourilyan Story*, p. 9; R. Broadfoot and P. Wright, 'The continuous low grade pan at Mossman', *Proc. QSSCT*, 1978, pp. 71–177.

141 Howard Smith Ltd., AR 1974, p. 6; *ASJ*, May 1973, pp. 128 & 160, September 1973, p. 319, December 1973, p. 466, March 1976, p. 619 & May 1981, p. 140; Kerr, *Southern Sugar Saga*, pp. 150 & 152; *The Producers' Review*, February 1971, p. 60; Rolleston, *The Defiance*, pp. 106–107; *ASYB*, 1981, p. 58; J. Allen, 'Advances in mill machinery in 70's', *ASJ*, October 1981, p. 391; NSW Sugar Milling Coop., AR 1980–81, p. 11.

142 *Bundaberg Sugar News*, Summer 1979, p. 6; Reid, *The Long View*, pp. 158–159; Kerr, *A Century of Sugar*, p. 192; *ASYB*, 1983, p. 167; R. Cullen, C. Murry and C. Anderson, *Final Report. Elimination of Use of Fuel Oil for Steam Generation in the Sugar Industry* (Canberra, 1982), p. 6; R. Jones and D. Maddison, 'Coal as a fuel in baggase fired boilers', *Proc. ASSCT*, 1980, pp. 133–140.

143 Reid, *The Long View*, p. 159; *AC*, 29 August 1988, p. 12; Cullen, Murry and Anderson, *Final Report. Elimination of Use of Fuel Oil for Steam Generation in the Sugar Industry*, pp. 1–4; V. Matson, 'Energy conservation and management in Australian raw sugar factories', *Proc. ISSCT*, Vol. 1, 1995, pp. LXXVI–LXXVII.

144 Kerr, *Northern Outpost*, p. 157; R. Allaway, 'Three years of computer process control at Mossman', *Proc. QSSCT*, 1974, pp. 83–90; *ASYB*, 1968, p. 1.

145 *ASYB*, 1977, pp. 325 & 361; Hudson, *By the Banyan*, p. 268; Anon, *The Mourilyan Mill Story*, p. 8; Kerr, *Only Room for One*, p. 128; CSR, *Newsletter* No. 145, July 1976, pp. 23–27, CSRR, Z/109, Box 372; Kerr, *A Crystal Century*, p. 142.

146 M. Capelin, 'Planning strategies to resolve land-use conflict between the sugar industry and urban related development in Australia'. M. Urban & Regional Planning thesis, Uni. of Qld., 1989, pp. 123–124.

147 Alcorn & Dunn, *Moreton Sugar Mill*, pp. 173 & 185; *AC*, 2 November 1992, p. 2, 19 April 1993, p. 5 & 28 November 1994, p. 5.

148 Bundaberg Sugar Co., AR 1975–76; Kerr, *Southern Sugar Saga*, pp. 130–131 & 148–152; CSR, *Newsletter* No. 140, March 1975, p. 30, CSRR, Z/109, Box 372; Kerr, *Black Snow and Liquid Gold*, p. 247.

149 NSWCGA, Minute Book, vol. 3, 1970–1989, entry for meeting 7 February 1974; Briody, *Sunshine Sugar*, pp. 11–24; Smith, *Sweet Beginnings*, pp. 392–393.

150 R. Savage *et al.*, *Sugar Industry Working Party Report* (Brisbane, 1985), pp. 82–83; Briody, *Sunshine Sugar*, pp. 58–59; *ASYB*, 1986, p. 128; *ASYB*, 1983, p. 135.

151 Kerr, *Southern Sugar Saga*, pp. 162–167; Hudson, *By the Banyan*, pp. 281–282; *ASYB*, 1985, p. 163; *ASYB*, 1988, p. 136; Rolleston, *The Defiance*, p. 105; *AC*, June 1986, p. 17.

152 Kerr, *Black Snow and Liquid Gold*, p. 249; Hudson, *Sweet Success*, pp. 215–228; *AC* January 1987, p. 9; *ASYB*, 1988, p. 121; Morton, *By Strong Arms*, pp. 171–173.

153 *ASYB*, 1987, pp. 107–108; Kerr, *A Century of Sugar*, pp. 205–214; Kerr, *Top Mill in the Valley*, pp. 111–112; *AC*, May 1988, p. 11.

154 Bundaberg Sugar Co., AR 1986–87; *ASYB*, 1986, p. 127; *ASYB* 1989, p. 47; Alcorn and Dunn, *Moreton Sugar Mill*, p. 199.

155 Hudson, *Sweet Success*, pp. 215–216 & 228–230; *AC*, 24 October 1988, p. 7, 8 May 1989, p. 1, 31 July 1989, p. 8 & 4 December 1989, p. 5; Bundaberg Sugar Co., *Why you should reject the inadequate takeover offer from Tate & Lyle* (Brisbane, 1991), pp. 2–4.

156 Hudson, *Sweet Success*, pp. 232–233; Hudson, *By the Banyan*, pp. 293–310; *ASYB*, 1994, pp. 11–12; *ASYB*, 1995, p. 13; *ASYB*, 1997, p. 198.

157 ASMC, AR 1988, p. 5.

158 For the impact and extent of the reduction in mill expenditure see M. McGovern, *The economic impact of the sugar industry on the Herbert River and Mackay regional economies* (Armidale, 1988).

159 *AC*, July 1986, pp. 14–15 & 10 October 1988, p. 19; *ASYB*, 1989, p. 105; Briody, *Sunshine Sugar*, p. 81.

160 ASMC, AR 1989, p. 21; ASMC, AR 1994, p. 4; *AC*, 26 February 1990, p. 6; *ASYB*, 1992, p. 39.

161 Industry Commission, *The Australian Sugar Industry. Report No. 19* (Canberra, 1992), pp. 109 & 193.

162 ASMC, AR 1995, p. 7.

163 Mackay Sugar, AR 1990, p. 7; Kerr, *Top Mill in the Valley*, p. 112; Kerr, *A Crystal Century*, p. 174; *ASYB*, 1985, p. 102; *ASYB*, 1992, p. 38.

164 *AC*, 22 March 1993, p. 3 & 18 October 1993, p. 6; *ASYB*, 1995, pp. 22 & 128; *ASYB*, 1996, p. 18.

165 R. Broadfoot, K. Miller and L. Davies, 'Commissioning trials on the SRI continuous high grade pan at Maryborough Factory', *Proc. ASSCT*, 1989, pp. 152–153; Reid, *The Long View*, p. 143; *ASYB*, 1996, pp. 118, 120 & 125.

166 C. Greig, E. White and C. Murry, 'Assessing the performance of continuous fugals', *Proc. ASSCT*, 1992, pp. 287–296; L. Kirby *et al.*, 'Performance of a new design of continuous centrifugal', *Proc. ISSCT*, 1995, pp. CLI–CLVII; Rein, *Cane Sugar Engineering*, pp. 450– 452.

167 Hudson, *Sweet Success*, p. 243; *ASYB*, 1996, pp. 18 & 20.

168 *ASYB*, 1996, p. 117.

169 R. Dephoff, 'Sextuple evaporation at Harwood Mill', *Proc. ASSCT*, 1992, pp. 230–237.

CHAPTER 18
The Refining, Marketing and Pricing of Sugar

On 6 April 1915, E.W. Knox wrote to the Hon. William Holman, Premier of New South Wales, advising him of a serious shortfall in the amount of locally-produced sugar that would be available for the Australian market later in the year, a result of a severe summer drought in Queensland.[1] As news of this expected shortage became more widely known, some federal parliamentarians in early May 1915 questioned the genuineness of the shortage which they believed had been manufactured by CSR for its own benefit. They called for the Commonwealth government to take control of the commodity 'out of the hands of the sugar kings'. Yet the shortage was real. By the third week of May, country storekeepers were running out of sugar and could not get answers from either the Millaquin Sugar Co. Ltd. or CSR about future supplies. By then, William M. (Billy) Hughes, the Attorney-General, had commenced an investigation into a request by five of the Australian state governments for the Commonwealth to participate in an arrangement, supported by all states except Queensland, whereby CSR would import 16 300 tonnes of sugar from Java to meet the expected shortage and that this sugar would be sold at £25 per ton, but the import duty of £6 would be suspended for a period.

The findings from Hughes' inquiry, released to the press on 31 May 1915, concluded that the sugar shortage for the 1915 season would be 16 300 tonnes, but an additional 80 000 to 100 000 tonnes might be needed by April 1916 to meet Australian demand. Hughes accepted CSR's warning of serious shortages ahead, but was concerned that the firm had waited for so long to do anything about it. Moreover, Hughes concluded that CSR was partially responsible for the shortage. Despite the Commonwealth's embargo on sugar exports, introduced in September 1914 to stop Australian sugar mill-owners from exporting sugar to take advantage of the increase in sugar prices brought about by the commencement of World War 1, CSR had exported 3 050 tonnes of Fiji-produced sugar lying in bond in Australia to Canada on 20 April 1915.[2]

In the face of a looming sugar shortage, the Commonwealth government allowed CSR to import sugar from Java, thereby meeting the 16 300 tonnes shortfall in domestic production. However, Hughes and CSR were deadlocked over the price to be received for this sugar. Some Queensland sugar mill-owners, in the face of such uncertainty, had refused to sign agreements with their suppliers and the start of the yearly crushing season was in jeopardy. Thomas Ryan, the Premier of Queensland since 22 May 1915, made resolving this issue one of his first priorities. He travelled to Melbourne on 18 June 1915 for discussions with both parties to the disagreement. After a week of negotiations, an agreement was reached whereby the Queensland government would, as the agent for the Commonwealth government, acquire the entire output of raw sugar at a fixed price, compulsorily in its own state, and for the relatively small amount produced in New South Wales by arrangement with CSR. This raw sugar would be sold to the Commonwealth government at cost. The latter agreed to supply raw and refined sugar to domestic consumers at the lowest possible price, impose an embargo on imports and exports of sugar and took responsibility for purchasing sufficient quantities of sugar from abroad, to make up any shortfall in production for domestic requirements. The official justification for such interventionist measures was twofold: firstly, to maintain adequate supplies of sugar to all Australian consumers during the war by preventing Australian sugar mill-owners and refiners exporting their sugar to capitalise on higher prices abroad; and secondly, to ensure that Australian consumers were not forced to pay exorbitant prices for sugar during the wartime period.[3]

On 30 June 1915, the Ryan government in Queensland issued an emergency proclamation acquiring all raw sugar in Queensland, beginning 'single desk' selling of the product in Australia. Its actions were validated in August 1915 when the Queensland Parliament passed the *Sugar Acquisition Act* of 1915. In the meantime, the Commonwealth government as the receiver of the country's raw sugar had to decide upon its refining. Two choices were possible: establish a Commonwealth refinery as had been suggested during the early 1910s (see Chapter 9) or use the two existing refiners. In the absence of records or even oral tradition about the sequence of events in this matter during June 1915, it is difficult to speculate on what course of action was considered by the Commonwealth government (Fisher; Labor) or CSR's motivations during this crucial period. The outcome, however, was clear. On 6 July 1915, the Commonwealth government and CSR reached an agreement whereby CSR would refine and distribute all the sugar acquired by the Commonwealth at a fixed fee per tonne, except for 25 400 tonnes that would be retained for the Millaquin Sugar Co. Ltd. The Commonwealth acquired the sugar that had already arrived from Java and agreed to pay CSR the net cost of any future imports to make up the shortfall of 16 300 tonnes. Negotiations were being held with the state governments to reach an agreement whereby refined sugar would be retailed throughout Australia at 3d per lb. The existing refiners' discounts allowed to wholesale merchants would be extended to all manufacturers, but no other discounts would be allowed. The Commonwealth government implemented its side of the agreement when the Sugar Purchase Bill was introduced into the Commonwealth Parliament on 31 August 1915. It was rushed through both Houses of Parliament and received the Royal assent on 6 September 1915. The Commonwealth government had regained control over the Australian sugar industry after relinquishing its authority only three years earlier.[4] Hughes, who became Prime Minister in October 1915, was reputed to have said, 'Now I have CSR in the shafts, all I have to do is hold the reins'.[5]

Marketing and refining under Commonwealth control, 1915–1922

Owing to the production of Australian sugar being insufficient to meet the increase in Australia's demand for sugar during the late 1910s and early 1920s, the Commonwealth government imported varying amounts of sugar, reaching a maximum amount of 118 000 tonnes in 1920–1921. Most of this sugar came from Java or Fiji, but smaller consignments were sourced from Taiwan, Cuba and Peru. To finance the purchase of sugar imports, the Commonwealth Parliament passed legislation authorising the Treasurer to borrow money from the Commonwealth Bank of Australia. Initially, CSR had purchased and paid for the foreign sugar required by the Commonwealth government, but the financing of foreign sugar imports from July 1920 onwards was handled by the Commonwealth Bank of Australia, which claimed it could negotiate foreign drafts at a more favourable exchange rate.[6]

Between 1915 and 1922, CSR processed around 200 000 tonnes of raw sugar annually or the bulk of the raw sugar melted in Australia during this period. CSR met this increased demand for refined sugar in Australia without any difficulties, despite some disruptions to its refined sugar output because of a fire in the refined sugar store at Pyrmont Refinery in October 1918. Its only rival, the Millaquin Sugar Co. Ltd., however, struggled during this period. Under Commonwealth government control, the Millaquin Sugar Co. Ltd. was allowed to melt a maximum of 25 400 tonnes of raw sugar each year. This amount was processed easily at the Millaquin Refinery during the late 1910s, but the refinery in the early 1920s handled only half its allocation. William Hart, Chairman of

Directors of the Millaquin Sugar Co. Ltd., told shareholders in September 1922 that the firm was hindered in its operations, as it could only refine for six months of the year. Since 1917, the crushing rate required to handle the expanding crop by the associated Millaquin crushing or juice mill had turned the refinery into a raw sugar mill for the crushing season. Substantial capital investment was needed to make the crushing mill into a complete raw sugar mill separate from the Millaquin Refinery, as the firm could not crush cane and refine at the same time. Consequently, there was 'practically no profit' in refining, and the Directors were not inclined to invest in this project presently.[7]

Under Commonwealth control, the wholesale price of refined sugar had been increased slightly in early 1916, but since then had remained unchanged (Table 18.1). In order to prevent high domestic sugar prices from placing exporters of manufactured goods containing sugar at a competitive disadvantage in overseas markets, the Commonwealth government (Hughes;

Table 18.1. Wholesale and retail prices of refined sugar in Australia, July 1915–August 1941.

Date of determination	Price to miller and canegrower per ton of raw sugar (£/s/d)	Wholesale price per ton of 1A grade refined sugar* (£/s/d)	Retail price per lb (d)
19 July 1915 to 15 Jan. 1916	18/0/0	25/10/0	3
16 Jan. 1916 to 30 June 1917	18/0/0	29/5/0	3½
1 July 1917 to 24 March 1920	21/0/0	29/5/0	3½
25 March 1920 to 30 June 1920	21/0/0	49/0/0	6
1 July 1920 to 31 October 1922	30/6/8	49/0/0	6
1 Nov. 1922 to 30 June 1923	30/6/8	42/0/0	5
1 July 1923 to 21 Oct. 1923	27/0/0	42/0/0	5
22 Oct. 1923 to 31 Aug. 1925	26/0/0	37/11/4	4½
1 Sept. 1925 to 31 Aug. 1931	26/10/0**	37/6/8	4½
1 Sept. 1931 to 4 Jan. 1933	22/0/0	37/6/8	4½
5 Jan. 1933 to 31 Aug. 1936	22/0/0	33/4/0	4
1 Sept. 1936 to 31 Aug. 1941	22/0/0	33/4/0	4

* Sugar traditionally used in the home ** This amount is an average for the period, due to fluctuations in the returns from the exports of raw sugar.

Source: *Year Book of Australia*, 1946–47, p. 942.

Nationalist) had introduced towards the end of 1917 an export sugar rebate that applied to sugar used in the manufacture of jams, jellies and canned fruits. CSR thought this interference in the pricing of sugar was intolerable and announced in April 1918 that it would be 'a grievous mistake' to pursue the current arrangements after the conclusion of the war. Prime Minister Hughes, however, was not so certain. Even though the Commonwealth government removed some wartime enactments associated with the *War Precautions Act* of 1914 after the end of World War I, its wartime marketing control over wheat, wool and sugar were not relinquished, but extended under the *Commercial Activities Act* of 1919. Arrangements for sugar after that date would be determined when the Commonwealth government received a report from a Royal Commission set up to investigate what controls should exist over the Australian sugar industry after late 1920. The three appointees to this Royal Commission were Albert Piddington and Nicholas Lockyer, who had been Chief Commissioner and Commissioner on the Interstate Commission, respectively, and Stephen Mills, a barrister.

The Piddington Royal Commission took a considerable time to carry out its investigations, obtaining three extensions of the period in which to complete its report. Proceedings were conducted on a 'free and easy plan – a go-as-you-please style – with everybody interrogating', observed wrote Harry Easterby, the BSES Director during the 1920s. A report from the Piddington Royal Commission was eventually delivered on 27 February 1920, although not publically released until 18 March 1920. The Commission expressed concern about the charges paid for the services of the two Australian sugar refiners in connection with the refining of the sugar and recommended that a thorough investigation be made into the actual refining costs before any new agreements were reached with the two refiners. Control over the country's sugar industry should continue to be exercised by an organisation, possibly called the Commonwealth Sugar Commission, and that the price of sugar should be increased from £21 to £22 per ton, especially since wages to Queensland sugar industry workers had been increased under the McCawley Award of 1919.[8]

By early 1920, Queensland's sugar mill-owners and canegrowers were concerned that the continuation of the price of £21 per ton of raw sugar or £29 5s per ton of refined sugar was unreasonable in view of their costs and prevailing market conditions – overseas suppliers were receiving around £57 per ton of raw sugar, brought about by a shortage of production caused by the wartime disruption to the European beet sugar-producing areas. In addition, Queensland's sugar producers had become alarmed at the interstate campaign by southern Australian fruit growers and jam manufacturers to abolish the embargo on sugar imports and to allow sugar into Australia free of all duty or excise.[9] A conference of Australian sugar industry leaders, chaired by William Gillies, the Queensland Minister for Agriculture and Stock, met in Brisbane on 17 February 1920. They decided to ask Prime Minister Hughes to increase the price of raw sugar to £30 6s 8d per ton (or £49 per ton of refined sugar) and to seek a new agreement covering three years. A delegation of Australian sugar industry representatives met Hughes for the first time on 27 February 1920. After negotiations lasting almost a month, Hughes agreed to a price increase and a three-year agreement on the following conditions: the total price increase of £9 6s 9d should be split so that £4 went to the Australian sugar mill-owners and the rest to the canegrowers; a council of representatives from the sugar industry would meet in February each year to decide if the price of sugar needed to be increased; wages to sugar industry workers were not to be altered during the crushing season; and the wages paid to sugar industry workers in 1920 were not to be increased during the term of the agreement, except to meet any increased cost of living.[10] Subsequently, the retail price of sugar rose sharply to 6d per lb (Table 18.1).

Hughes' action ignored the recommendations of the Piddington Royal Commission, but ensured no disruptions would occur to the crushing seasons in the immediate future. The Sugar Agreement of 1920 was also a partial solution to how much control the Commonwealth government should exercise over the Australian sugar industry, for it provided another three years to consider the extent of sugar industry regulation. Domestic consumers, fruit growers and jam manufacturers in southern Australia were strongly opposed to Hughes' policy, and they commenced calling for the abolition of the customs duty on imported sugar. To counteract these anti-sugar industry sentiments, the ASPA published a series of small pamphlets during 1921 and 1922. These publications highlighted the value of the sugar industry to the country, its role in peopling the tropical regions in Australia and arguing for a 'stable protective policy'. Canegrowers, the ASPA and the Sugar Committee of the Queensland Council of Agriculture, formed in 1922, also began pressuring the Commonwealth government to increase the duty on imported foreign sugars and to extend the Sugar Agreement beyond 30 June 1923.[11]

By early 1922, Hughes' Nationalist government was resisting moves to renew the Sugar Agreement, favouring relinquishing control over industries regulated during World War I, although the Commonwealth was not adverse to increasing protection for the Australian sugar industry. The matter of increasing the duty on imported sugar was referred to the Tariff Board (later the Industries Assistance Commission). Its report was presented to the Commonwealth Parliament in September 1922. The Tariff Board noted that a duty of £6 per ton on imported sugar had existed since 1901, but this protection was inoperable because of the Commonwealth's embargo on sugar imports and exports, but it would apply again after the expiry of the Sugar Agreement on 30 June 1923. If Commonwealth control was relinquished over the Australian sugar industry, the Tariff Board recommended a duty of £9 6s 8d per ton on imported raw cane sugar and £11 6s 8d on imported beet sugar. Hughes' government accepted the recommendations of the Tariff Board, but opposition to further support for the Australian sugar industry in the Commonwealth Parliament resulted in the Custom Tariff (Sugar Duties) Bill being amended so that both imported cane and beet sugar were subjected to a duty of £9 6s 8d per ton under the legislation eventually passed by the Commonwealth Parliament.[12]

As Hughes' Nationalist government considered the level of protection for the Australian sugar industry, some sections of the community in southern Australia were expressing considerable dissatisfaction about the domestic price for refined sugar, and the continued ban on sugar imports. In response to appeals from the public for cheaper sugar prices and lobbying by the Queensland sugar mill-owners who wanted higher prices for their raw sugar, a Commonwealth Parliamentary Joint Committee of Public Accounts began an inquiry in June 1922 into the supply of sugar products to the Australian public. The parliamentarians were particularly interested in determining the lowest retail price at which cane sugar could be sold in Australia under the existing arrangements without loss to the Commonwealth government. However, the Public Accounts Committee, which reported in October 1922, could not agree on a final report. The compilers of the Minority Report complained about a lack of detailed information on production costs, but concluded that £26 would be a sufficient price for a ton of raw sugar, and that refined sugar should retail at 4½d per lb. This amount could be further reduced with 'more efficient organisation and methods of production and a careful revision of all charges'. The clear message from the parliamentarians who compiled the Minority Report was that the Commonwealth's involvement in the sugar business had seen a financial loss to the Australian taxpayer, led to inflated prices for the consumer, been poorly supervised, and that CSR appeared to have done handsomely from the arrangements.[13]

In December 1922, Hughes' Nationalist government was defeated at federal elections. However, the opposition Australian Labor Party did not assume office as the Nationalist Party formed a coalition with the Country

Party led by Earle Page, although the Country Party made Hughes' resignation the price for joining the coalition. Stanley Bruce assumed the leadership of the Nationalist Party and became the new Australian Prime Minister. Representatives of the Australian sugar industry finally met Prime Minister Bruce on 2 March 1923 and were advised that the Commonwealth government wished to relinquish control over the Australian sugar industry. Bruce would not consider renewing the Sugar Agreement of 1920, but he accepted some of the industry's suggestions. Briefly summarised, the following conditions were conceded by the Commonwealth: the industry should form a pool free of Commonwealth control to purchase raw sugar at £27; negotiations should be held with CSR about the firm continuing to refine and distribute sugar in Australia; an embargo against imported 'black' grown sugar would be maintained until 30 June 1925, after which the industry would be protected by a customs duty; the price of raw sugar for the 1924 and 1925 seasons was to be determined by a tribunal, but it could not exceed £27 per ton; and the export sugar rebate would be maintained. Representatives of the Australian sugar industry, although concerned about the drop in the sugar price, agreed to the Commonwealth's conditions, which were formalised in the second Sugar Agreement. This arrangement was to last initially for two years.[14]

First export markets, 1923–1945

To implement the Sugar Agreement of 1923, the Queensland government (Theodore; Labor) used its powers under the *Sugar Acquisition Act* of 1915 to establish the Queensland Sugar Board (hereafter Sugar Board), which was to undertake all investigations, negotiations and recommendations regarding the acquisition of raw sugar in Queensland and its delivery to the place of refinement, payment of the proclaimed price to the sugar mill-owners and the acquisition of the sugar produced in New South Wales. The initial three members of the Sugar Board were appointed in July 1923 and consisted of William Short (Chairman), George Pritchard (Secretary of the ASPA) and Thomas Powell (representative of the canegrowers). On 11 August 1923, another member, Alexander Innes, representing the mill-owners, was added to the Board. The Sugar Board, like the Commonwealth government in 1915, had to decide upon a refiner for the country's raw sugar, how to distribute refined sugar products throughout Australia, and what arrangements needed to be implemented for the export of surplus raw sugar. CSR and the

Millaquin Sugar Co. Ltd. were appointed again the as the domestic marketing and refining agents for the Sugar Board, and CSR was assigned the task of arranging the transport of the sugar, the collection of proceeds and the development of the Australian sugar industry's marketing strategies. For their services, the refining companies continued receiving payments to reimburse them for the actual cost of refining and distributing refined sugar products throughout the country (i.e. toll refining).[15] This arrangement was renewed each year and lasted until deregulation of the Australian sugar industry in 1989.

As the Sugar Board and CSR arranged for the first substantial amount of raw sugar exports from Australia during 1924, Prime Minister Bruce changed his mind about the future protective arrangements for the Australian sugar industry. He announced in December 1924 that the Commonwealth government was prepared to extend the embargo on the importation of sugar into Australia for another three years, subject to the industry agreeing to certain concessions. In return for this concession, the Australian sugar industry agreed in early 1925 to the wholesale price of refined sugar being reduced from £37 11s 4d per ton to £37 6s 8d per ton in September 1925, but accepted that the retail price for sugar in the capital cities should remain at 4½ d per lb (Table 18.1). All refined sugar used by exporters of approved manufactured products (e.g. canned fruit; jams; confectionary; cordials) attracted a rebate of £6 5s 1d per ton and the operation of the export rebate system was to be administered by the Export Sugar Committee, composed of a chairman nominated by the Commonwealth government and one nominee of each the Sugar Board and the manufacturers of exported products containing sugar. The Queensland government, on behalf of the Australian sugar industry, was to cover all financial losses associated with the export of surplus sugar. These prices and conditions were renewed for another three years in 1928 and were to expire in August 1931.[16]

Annual demand for sugar in Australia had stabilised around 55 kg per capita during the 1920s, the highest apparent average annual sugar consumption per capita globally. Even though Australians were not consuming extra sugar per capita, the country's population growth during the 1920s meant that the sugar refiners had to supply more refined and raw sugar domestically. To meet this demand, meltings of raw sugar at CSR's Pyrmont and Yarraville Refineries slowly increased during the 1920s (Figure 18.1). During 1923, the Millaquin Sugar Co. Ltd. also invested £60 000 in new machinery (i.e. triple effets, vacuum pans; crystallisers) to convert the Millaquin crushing mill into a factory that could manufacture raw sugar, thereby allowing the Millaquin Refinery to refine throughout the entire year.[17]

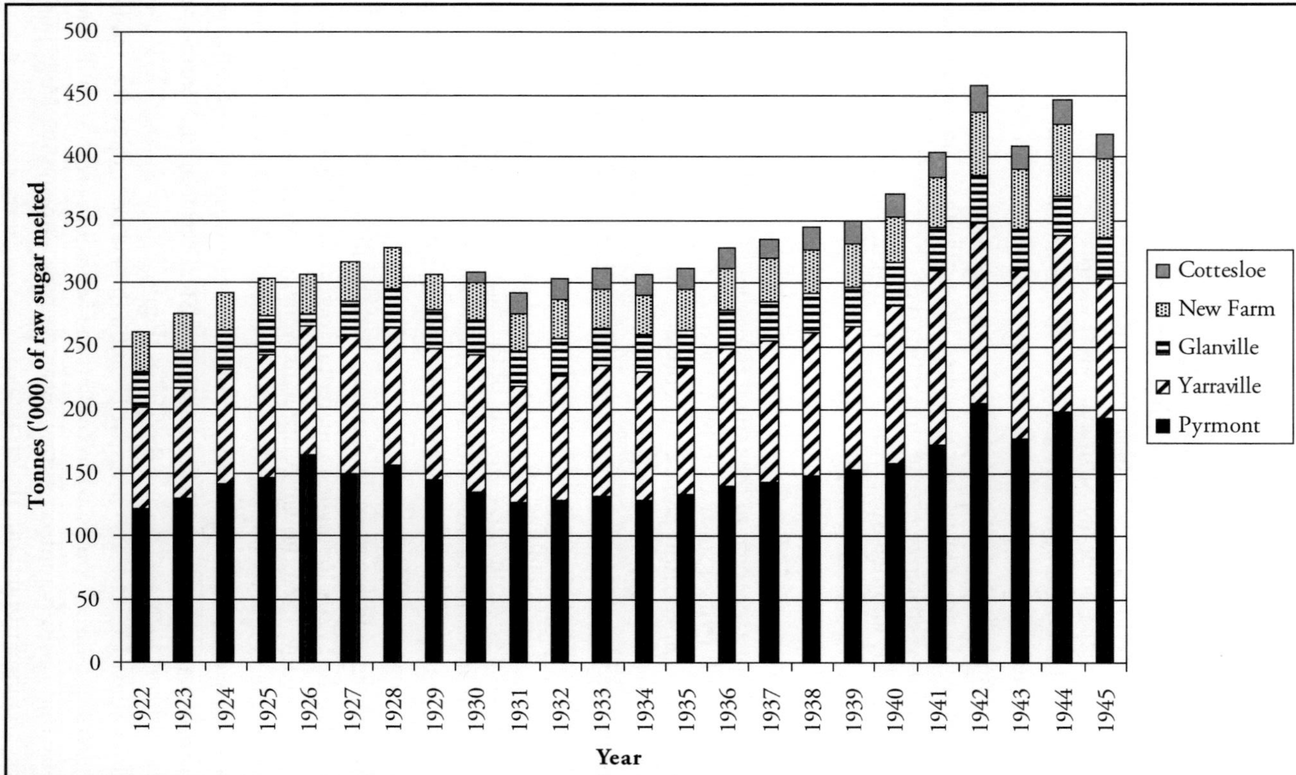

Figure 18.1. Tonnes ('000) of raw sugar melted at CSR's refineries, 1922–1945.

Source: Based upon figures in CSR's Annual Refinery Reports, in CSRR, Z/109, Boxes 396–398

Australia's sugar refining capacity, however, was severely tested during the mid-1920s. On 24 February 1926, CSR's Glanville Refinery in Adelaide was almost completely destroyed by fire. Reconstruction took nearly eight months, with demand being met by overtime work at CSR's other Australian refineries, especially Pyrmont. Just over a year later on 28 April 1927, a fire destroyed a large part of the Millaquin Refinery and 2 600 tonnes of refined sugar. The decision to rebuild was made within three weeks of the fire, with the Millaquin Sugar Co. Ltd. being assured by William Forgan Smith, the Acting Queensland Premier, that the Queensland government supported the continued existence of the refinery. The reconstructed refinery commenced operations on 21 August 1928, but still only operated for a limited period each year in 1930, refining around 12 700 tonnes of raw sugar annually.[18]

To handle the additional meltings required to meet the Australian demand for refined sugar products during the 1920s, CSR engaged in modest refinery upgrades. In 1924, the raw sugar store at its New Farm Refinery was extended. As space was limited on both the southern and northern ends of the existing store, the most feasible solution involved extending the store beyond the boundaries of the existing site. This move required the realignment of both the railway line and a nearby street. The Queensland government (Theodore; Labor) agreed to this change, after CSR surrendered a portion of the housing allotments it had acquired around the refinery. At Yarraville Refinery, new boilers fired with little hand labour were installed in 1926, duplicating the earlier success of these vessels at Pyrmont Refinery. An entirely new power house was built at Pyrmont in 1923, and extended in 1928; other additions during the 1920s included extensions to the boiler station and melting house. Electricity began driving the refinery's centrifugals and calandria vacuum pans replaced the less efficient vacuum pans equipped with coils.[19] However, CSR decided in 1927 that the cost of expanding the mechanical capacities of its existing refineries to meet the Australian population growth was equal to building a new refinery, given the recent increases in the cost of materials and labour. Therefore, the firm decided to build an entirely new refinery in Western Australia.[20]

CSR's new Western Australian refinery was located between Perth and Fremantle, at Cottesloe in the Mosman Park district. The site was located on a high bluff overlooking the Swan River and its selection was not popular amongst Perth residents who complained to CSR that the buildings would not 'add to the beauty of the very picturesque river bank which they will occupy'. E.W. Knox, now CSR's Chairman of Directors, was not too worried about the protests, saying that the firm would do what it could to minimise the impact of the refinery buildings on its surrounds. Subsequently, CSR provided for the planting of extensive lawns on the slopes and shade trees to hide the refinery buildings. At the time of the protests, Knox's main concern was the rising costs of building the new refinery. In 1928, the estimated cost of a modest refinery capable of turning out 400 tonnes of white sugar weekly was £190 000. No provision was made for making hessian bags – these were to be supplied from the other refineries – and there was to be no separate melthouse, but a combination of melting and filtration in a two-storey lean-to against the raw sugar store, which would also store the bags of refined sugar. Eighteen months later, the construction costs had escalated to £240 000, a result of substantial increases in the prices of locally-sourced materials which proved to be much higher than Sydney or Melbourne and the 'go slow' methods of the workers trying to save their jobs, fearing unemployment as the Great Depression commenced.[21] Cottesloe Refinery eventually began operations in May 1930, supplying the Western Australia market with refined sugar manufactured from melting between 15 000 and 17 000 tonnes of raw sugar annually (Figure 18.1).

As CSR sought to ensure its refineries could meet the Australian demand for refined sugar products during the late 1920s and 1930s, the Sugar Board had to sell increasing amounts of raw sugar that was surplus to domestic requirements. Brokerage was entrusted to the famous London-based sugar brokers Czarnikow Ltd., who arranged for the sale of the majority of the raw sugar to English refiners, although just how much could not be calculated with certainty due to the fragmentary nature of the Sugar Board records from this period: export statistics had to be assembled from other sources and in some instances estimated (Table 18.2). The requirements of the British market, however, influenced the production activities at Australia's sugar mills. Prior to 1928, some Australian sugar mills made high polarisation sugar over 99°S (i.e. measurement of sucrose content in sugar, with 100°S being the maximum or 100% sucrose), but the British Preferential Tariff on sugar from the Dominions was varied in 1928 so that raw sugar polarising 99ºS or over would in future be penalised £1 1s 4d per ton compared with raw sugar polarising under 99ºS. Consequently, Australian sugar mill-owners were forced to produce lower strength sugars.[22]

Australian sugar exports to Canada also grew between 1925 and 1935 (Table 18.2). By the mid-1930s, Canadian sugar refiners absorbed around 10–12% of Australia's sugar exports. Moreover, by then the Sugar Board had realised the need to find some new markets for the country's growing

Table 18.2. Exports of Australian raw sugar by destination, 1924/1925–1939/1940.

Year	Great Britain	Canada	Year	Great Britain	Canada
1924–25	*	12 517	1932–33	142 681	48 174
1925–26	*	15 017	1933–34	287 883	24 782
1926–27	*	33 986	1934–35	311 344	49 285
1927–28	14 0716 #	12 506	1935–36	321 490	43 159
1928–29	183 761 #	5 214	1936–37	350 520	65 024
1929–30	203 000 #	*	1937–38	365 760	69 088
1930–31	*	*	1938–39	398 272	50 800
1931–32	275 742	16 514	1939–40	*	*

* Data not available # Estimate

Source: Based upon figures in the *ASJ*, 8 May 1930, p. 81, H. Prinsen Geerligs and R. Prinsen Geerligs, *Cane Sugar Production 1912–1937* (London, 1938), p. 156 & Queensland Agent-General in London, ARs 1936–1939, in *QPP*, various volumes.

volume of raw sugar exports. During early 1939, the Sugar Board asked the Commonwealth Government's Trade Commissioner in Washington to inquire if some of Australia's surplus raw sugar could be exported to the USA. Investigations suggested that the USA would be an unlikely market as its tariff legislation constituted a barrier to the importation of Australian raw sugar.[23]

The production of lower polarising raw sugar in Australia from 1928 onwards forced changes to the refining process used at CSR's refineries. A third of the raw sugar arriving at CSR's refineries in 1927 was of such high quality that it could be dissolved immediately in hot water (i.e. direct melting) and sent to filtration. From 1928 onwards, all raw sugar arriving at CSR's refineries contained more impurities. This change meant the additional regular use of affination. By 1931, CSR complained that the raw sugar received at its refineries contained far too many hard, syrupy lumps and had very poor filterability, a trait not favoured by any refiner. Raw sugar from the Bundaberg district, especially the output of Fairymead Mill, had the worst filterability. In 1934, CSR's Research Department began surveying the quality of the raw sugar being received at its refineries, identifying the poorly performing Australian sugar mills and offering advice on how to improve the quality of raw sugar. Despite some improvements in the quality of raw sugar arriving at CSR's refineries during the mid-1930s, the firm's research chemists still complained about the irregular grain size of the sugar being processed in 1938, with the worst raw sugar coming from Isis, Fairymead and Gin Gin Mills.[24]

CSR may have been trying to deliver high-quality refined sugar to Australian consumers during the late 1920s, but they were complaining about the embargo on sugar imports inflating the retail price for white sugar. Henry League of Melbourne published a pamphlet in which he claimed that Australian consumers paid too much for sugar and that the Australian sugar industry had 'fully exploited the sympathy of people of the Southern States for the alleged hardships which have to be endured by those engaged in developing northern areas'. The Australian Sugar Consumers' Association, an organisation consisting of housewives and manufacturers of sweets, jams, cordials and canned fruit products, also published a pamphlet which refuted the Australian sugar industry's claim that it was responsible for peopling northern Australia, as the cultivation of sugar cane only occupied patches of coastal country separated by long distances. CSR was again accused of prospering because of its virtual monopoly over the supply of a commodity that enjoyed complete protection in the home market. To counteract these arguments, the QCGC commissioned an album of photographs showing the Australian sugar industry in its various production phases and examples of factories which used sugar in southern Australia; this pictorial survey then toured the country with sugar industry representatives who promoted the benefits of protection for the Australian sugar industry.[25] An organisation known as the Queensland Sugar Industry Defence Committee also emerged in Melbourne. It published a small booklet in 1930 refuting the claims against the sugar industry, and authorised some racist newspaper and journal advertisements, which claimed the 'colour menace' of Asia was a very real danger to Australia, and that the only bulwark was the sugar industry.[26]

To decide the future Australian sugar price, the Commonwealth government (Scullin; Labor) in August 1930 appointed another Royal Commission into the Australian sugar industry, headed by the Hon. John Gunn, the Commonwealth Director of Development. The Gunn Royal Commission heard evidence from various Australian manufacturers who claimed that the high domestic price of sugar resulted in higher prices for sweets, cordials, jams and canned fruits, thereby affecting the sales of these products. The majority of commissioners, comprising the representatives of the mill-owners and canegrowers (William Short, Sugar Board, and Frederick Curlewis, Cairns), a Commonwealth government representative (Albert Townsend, accountant of the Department of Trade and Customs), a sugar employees' representative (Clarence Fallon, AWU) and a fruit-growers' representative (William Young, Victoria), recommended the same sugar price for another three years and the continuance of the embargo on sugar imports for another five years. This embargo was subject to a definite contribution of £315 000 a year to Australia's fruit-growers. These funds were in lieu of the concessionary price paid for refined sugar by either the manufacturers engaged in processing all fruit products for home consumption (i.e. domestic sugar rebate), or the manufacturers of jams, jellies and canned fruit for export (i.e. export sugar rebate). The former arrangement had been introduced specifically to encourage the Australian fruit-growing industry. Three commissioners, comprising the chairman, and the representatives of manufacturers and domestic consumers (Eliza Morgan, Western Australia, and Francis Dutton, Sydney) also recommended the embargo on sugar imports for another five years, but proposed a reduction of £2 6s 8d in the price of raw sugar for two years and action to reduce overproduction in Queensland.

The Commonwealth government accepted the majority report, and the Sugar Agreement was renewed virtually as recommended. The Queensland government, on behalf of the Australian sugar industry, was to contribute £315 000 annually to the Fruit Industry Sugar Concession Committee (FISCC), a new organisation that would oversee the allocation of the funds to Australia's fruit-growers. Provisions were inserted for the variation of the price in abnormal circumstances, but the former price for a ton of raw and refined sugar was guaranteed for another three years.[27]

Immediately, an indignant propaganda campaign against the agreement commenced in southern Australia, provoked by the continuance of the Great Depression and by what seemed to be the maintenance of the same income to a favoured section of the community by the prohibition on competition and price-fixing. A June 1931 meeting in Hobart, attended by five of the six Tasmanian senators, for instance, called for the Tasmanian government to test the validity of the embargo in the High Court of Australia.[28] The Scullin government refused to reverse its decision and mounting anger over the way it was handling the depression led to its defeat at the general elections on 19 December 1931. The new Commonwealth government consisted of the United Australia Party led by ex-Labor Joseph Lyons in coalition with the Country Party led by Earle Page. Within six weeks of its election, Prime Minister Lyons had announced a conference to renew the terms of the Sugar Agreement, so that the Australian sugar industry could share in the general reduction in income suffered by practically every section of the Australian community during the previous two years. The ASPA and QCGC expressed astonishment at such a suggestion, and went on the offensive, claiming that Australians and the Commonwealth government had forgotten about the benefits accruing from the export of 300 000 tonnes of sugar in their zeal to achieve lower retail prices for consumers. Advertisements were again inserted in Australian newspapers claiming that the sugar industry was 'responsible for peopling the empty north'.[29]

The conference of Australian sugar industry interests convened by Prime Minister Lyons met in Brisbane on 27 July 1932. A year later, Senator Thomas Crawford, the President of the ASPA, remarked in his annual report that the Australian sugar industry representatives during this meeting came to realise that the embargo on sugar imports itself was 'resting on a very slender basis politically', so they agreed to a series of tentative proposals, including a price reduction for sugar. During late July 1932, Prime Minister Lyons continued to be reminded that sections of the Australian public opposed the special treatment afforded to the country's sugar industry. A series of anti-embargo meetings were held in South Australia and Western Australia. Lyons also received a telegram from the Australian Sugar Consumers' Association. This organisation strongly objected to the embargo on sugar imports, claiming that it was an 'unwarranted and oppressive burden upon the people of Australia and should be immediately removed'.

The matter was finally resolved at another meeting between Prime Minister Lyons and representatives of the Australian sugar industry on 3 September 1932. The Australian sugar industry agreed to a reduction in the retail price of sugar in all the metropolitan cities of the Commonwealth after 1 January 1933. This price reduction from 4½d to 4d was equivalent to an effective drop in the price of refined sugar from £37 6s 8d to £33 4s per ton (Table 18.1). In return, a new Sugar Agreement preserving the embargo on sugar imports and reducing the Australian sugar industry's assistance to the Australian fruit industry to £200 000 would operate until 31 August 1936. This arrangement pleased few. Federal parliamentarians such as Walter Nairn (Perth; United Australia Party) complained that sugar producers received an 11% cut in the price of their product, but other agricultural products were down 40% – sugar was still the 'darling of politics'. The Federal Housewives' Conference opposed the embargo, claiming that the Australian sugar industry should not be pampered. Subsequent meetings of canegrowers in the major Australian sugar-producing districts condemned the deal that their representatives had made on their behalf. Such vilification, however, was probably unwarranted for they had saved the embargo on imports of cheaper produced sugar.[30]

During the early 1930s, the Australian sugar industry had failed to convince sections of the community that it deserved special treatment, particularly protection against cheap imports. To counteract negative publicity about the Australian sugar industry, produced mainly by the Australian Sugar Consumers' Association, the ASPA spearheaded a campaign between 1932 and 1938 to improve the industry's image in southern Australia. Pamphlets and small booklets were published contradicting the claims made by the Australian sugar industry's opponents. Instead of focusing upon the 'defence of the nation' approach used previously, these new publications highlighted the importance of the Australian sugar industry when it came to creating employment and export earnings, and assisting the fresh fruit industry and manufacturers of jams, cordials and other canned fruit products in Australia via the domestic and export sugar rebates. Sections in these publications defended CSR, refuting claims that the company had made enormous profits at the expense of the consumer, and that the country was fortunate to have such a firm with the resources to conduct such a huge and risky business.[31]

In the midst of this propaganda blitz by the Australian sugar industry, Prime Minister Lyons, previously a critic of the sugar industry, had a volte-face about its value to the country. He noted that exports of raw sugar were worth £2.5 million and this figure would increase when world sugar prices returned to normal. In September 1934, he announced: 'We cannot let the sugar industry slip. We will not let it slip'. Negotiations then commenced with the Queensland government (Forgan Smith; Labor) about the renewal of the Sugar Agreement. Eventually, Prime Minister Lyons announced in February 1935 that the Commonwealth and Queensland governments had agreed upon the terms and conditions of the new Sugar Agreement. Anti-Sugar Agreement propaganda commenced being published immediately, but the protests against the embargo on sugar imports were less organised and vocal than the 1932–1933 campaign. Opposition to the legislation ratifying the Sugar Agreement during its consideration by the Commonwealth Parliament throughout October–December 1935 was limited to complaints about the excessive profits made by CSR and grumblings about the price of sugar still being excessive. An attempt to have the terms of the proposed Sugar Agreement investigated by a parliamentary inquiry failed, and the legislation was passed by the Commonwealth Parliament in December 1935.

The *Sugar Agreement Act* of 1935 extended the Sugar Agreement for another five years from 1 September 1936, continued the embargo on sugar imports, left the price of refined sugar unchanged at £33 4s per ton and maintained the retail sugar price at 4d per lb in Australian capital cities. Revenue from the Australian sugar industry to the FISSC, however, was increased from £200 000 to £216 000. Even though the Australian sugar industry failed to achieve a sugar price rise, mill-owners and canegrowers were generally satisfied with the arrangements negotiated by W. Forgan Smith, the Premier of Queensland, in view of the industry's pre-occupation with avoiding overproduction (see Chapter 10).[32]

Even though some Australian consumers complained about the high retail price of sugar during the 1930s, it did not deter them from purchasing the product. Annual melts at CSR's sugar refineries dipped slightly during the Great Depression (Figure 18.1), but the Australian fondness for sweet things returned once economic conditions improved. Average apparent annual sugar consumption per capita in Australia during the late 1930s exceeded 60 kg, the highest on record. CSR, however, had plenty of refining capacity to meet this upsurge in demand. The only major upgrade of significance at CSR's refineries during the mid-1930s was the replacement of the phosphatation-Taylor bag filters using bone char paste at the New Farm Refinery with the carbonation process and diatomite pressure filtration in order to improve the clarification of the sugar liquor. Taylor bag filters and bone char paste were also abandoned at Millaquin Refinery in 1939, their place being taken by diatomite filtration. This process consisted of melting the raw sugar, adding the diatomaceous earth to the liquid solution, and then filtering under pressure through two Sweetland filters. The clear filtered liquor was then passed over bone char.[33]

Wartime difficulties

The plentiful supply of sugar for Australian consumers was threatened by the commencement of World War II, but the Commonwealth government was not forced to introduce sugar rationing for consumers until August 1942. Australians regarded the sugar ration of only half a kilogram each week a serious hardship, despite it being double the amount provided to English, American and Canadian consumers during the wartime emergency.[34] Even though the Australian public suffered sugar rationing, CSR's refineries melted record amounts of sugar during the early 1940s (Figure 18.1). This upsurge was a direct consequence of three factors. Firstly, the armed forces demanded additional sugar products, especially golden syrup. Secondly, abnormal buying by householders stockpiling refined sugar products commenced during late 1941 and continued throughout early 1942. Thirdly, the spread of hostilities throughout South-East Asia prompted the Commonwealth government in early 1942 to announce a build up of six months reserve of essential commodities – sugar was included in the list of such commodities.

To meet this demand, CSR eliminated the manufacture of all 'fancy' products for civilian use, including icing sugar, tablet sugar and brewers' crystals, and concentrated on the production of 1A grade refined sugar, the sugar of uniform crystals traditionally used in the home.

Extensive alterations to plant, equipment and packaging routines were made at the Australian sugar refineries to ensure the production targets were met, particularly as the refined products supplied to the armed services had to be packed differently. Instead of using hessian or jute bags, the refined sugar products destined for the troops were placed in specially designed containers to withstand rough handling and exposure to tropical conditions for long periods without cover. Tablet sugar was made only for the armed forces, being the most suitable form in which sugar could be conveniently packed in the ration packs for the servicemen.[35]

Prices for refined sugar products during the war years continued to be controlled by the Commonwealth government. The Sugar Agreement of 1936 expired in August 1941, but had been renewed under the Sugar Agreement of 1940, thereby extending pricing arrangements from September 1941 for another five years. The Sugar Agreement of 1940 continued the past obligations on the sugar industry. Negotiations between the Commonwealth and Queensland governments to reach this agreement were not conducted this time against a background of a consumer campaign to remove the embargo on sugar imports. Yet the debate over the legislation to ratify the agreement in the Commonwealth Parliament during May 1940 was marked by an unsuccessful attempt by Senator Stanley Amour (New South Wales; Labor) to delay its passage by referring the matter to a Royal Commission into the profits being made by CSR from its refining arrangements with the Sugar Board. Senator Amour again attacked CSR as being 'one of the greatest monopolies in Australia', but his claims were not supported by the other senators who mostly stated that the matter had been investigated by the Gunn Royal Commission in 1930 and there was no need to re-open the matter.[36]

Sugar Agreements were rarely altered during their period of existence, the most noted exception being Prime Minister Lyons' changes to the Sugar Agreement of 1930. Yet on 19 May 1943, a deputation representing Australia's canegrowers met with Prime Minister John Curtin. They claimed their profits had fallen below a reasonable level, a consequence of the decline in raw sugar production and increased operating costs since the outbreak of the war. The deputation asked that the price for a ton of raw sugar should rise by £4 13s 6d. Curtin was ambivalent about a price increase, preferring to assist the Australian sugar industry by a subsidy, but he wanted the matter examined by a Tariff Board inquiry before making any decision.

The Tariff Board hardly moved quickly upon the canegrowers' request, for it did not hold public inquiries into the Australian sugar industry's production costs until May–June 1944. Evidence was gathered from representatives of the country's sugar mill-owners and canegrowers and the Tariff Board members visited only Marian and Pleystowe Mills. On the basis of this testimony and brief foray into Queensland, the Tariff Board concluded that there was no evidence of abnormal numbers of canegrowers being forced from the Australian sugar industry by economic stress. No Australian sugar mills were in danger of ceasing operations, although if the Tariff Board had inquired a little more carefully it may have discovered that Mt Bauple Mill was in a dire financial situation, and subsequently closed in 1950. The Tariff Board recommended that the 'industry is able to carry on for the present'. Consequently, Prime Minister Curtin declined to modify the Sugar Agreement of 1940. The QCGC displayed its annoyance with Curtin by describing the entire process – perhaps correctly – as entirely 'perfunctory'.[37] Moreover, when 'Ned' Hanlon, the Acting Queensland Premier, announced the terms of the Sugar Agreement of 1946 on 4 October 1945, they contained no provisions for an increase to the retail price of sugar.

Bulk handling and new overseas markets, 1945–1979

The Allied war victory did not bring about an immediate end to sugar rationing in Australia. Nevertheless, some restrictions on sugar use were reduced during 1945 and 1946, with consumers being permitted to obtain 1 kg of raw sugar instead of their refined sugar ration. Sugar rationing finally ceased on 2 July 1947. Australians, however, had lost none of their desire for sweet things during the wartime period of sugar deprivation. Average annual apparent sugar consumption per capita reached 56.2 kg during the period 1946–1950, slightly down on pre-war levels. Yet CSR struggled to meet the immediate post-World War II demand for refined sugar products in Australia. Indeed, W. Rothe, CSR's general manager, advised the firm's shareholders that 'on 3rd July 1947 when sugar rationing was lifted, the stock position was more precarious than at any stage during or after the war'.[38]

A combination of factors affected the performance of CSR's refineries, which previously had always managed to keep the Australian market fully supplied with refined sugar products. A lack of ships, labour disputes on the wharves and the difficult attitude of the waterside workers delayed the movement of raw sugar – the daily rate of loading at North Queensland ports in 1947 had fallen to less than half that of 1939. Consequently, CSR's sugar refineries increasingly suffered raw sugar shortages or complete exhaustion of the stock. On several occasions the refineries had to reduce production or close down temporarily. The firm also suffered from a shortage of workers at its Adelaide, Sydney and Melbourne refineries, especially from 1947 to 1949, hindering maximum output and causing a sharp drop in annual meltings (Figure 18.2). By 1948, CSR's refineries manufactured refined sugar products at about 80% of their mechanical capacities. Moreover, Pyrmont Refinery shut down entirely for seven weeks during November–December 1948, a consequence of running out of coal because of the miners' strike on the New South Wales coalfields. Raw sugar had to be sold in considerable quantities in Victoria and South Australia to supplement refined sugar deliveries.[39] Some of this production shortfall at CSR's refineries during the late 1940s was met by the Millaquin Sugar Co. Ltd. operating its Millaquin Refinery longer than the usual two runs of ten weeks each year. The quantities of raw sugar melted at the Millaquin Refinery ranged from 26 800 tonnes in 1947 to 41 200 tonnes in 1951, compared to 16 000 tonnes during the mid-1930s.[40]

The delays in shipping raw sugar to the Australian refineries and refined sugar shortages prompted calls for the erection of new sugar refineries in Australia. Sir Robert Cosgrove, Premier of Tasmania, for instance, called upon CSR in March 1949 to build a refinery in Tasmania, although his request was ignored by CSR. A month later the annual conference of the QCGC recommended the erection of another sugar refinery in Queensland. The Sugar Board was not impressed with this discussion, advising the QCGC that the capacity of the state's two refineries were more than sufficient to meet Queensland's requirements for refined sugar products. By 1951, CSR had also become annoyed with the constant calls for new refineries and advised the QCGC that the shortages of refined sugar products were not caused by a lack of mechanical capacity at its Australian sugar refineries. Moreover, it would take at least five years or more to plan, erect and commence production at any new sugar refinery. The Hanlon (Labor) government in Queensland also advised the QCGC that it did not favour a proposal for a new sugar refinery, preferring to see an upgrade at Millaquin

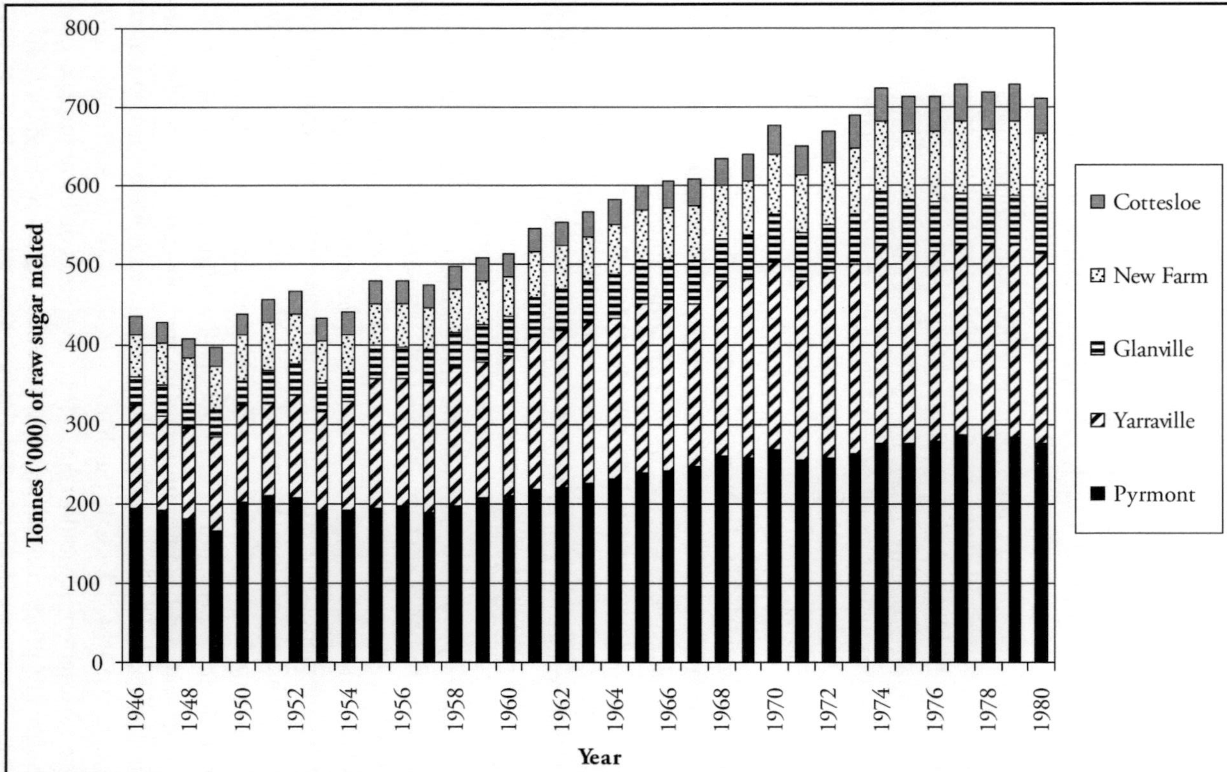

Figure 18.2. Tonnes ('000) of raw sugar melted at CSR's refineries, 1946–1980.

Source: Based upon figures in CSR's Annual Refinery Reports, in CSRR, Z/109, Boxes 398–402

Refinery if additional refining capacity was eventually needed.[41] Even when sugar shortages were overcome by the introduction of bulk handling in the mid-1950s, the issue of erecting a sugar refinery in North Queensland was re-visited by the ASPA's annual conferences in 1954 and 1960 and the QCGC's annual conference in 1970.[42]

CSR was convinced that the post-World War II difficulties in supplying refined sugar would not be overcome by building more refineries, but by addressing the issue of the performance of Queensland's waterside workers. In the mid-1940s, the loading and unloading of bagged sugar was still extremely labour-intensive, time consuming, costly and had changed little since 1915, except for the introduction at some ports (e.g. Bundaberg; Cairns) of mechanical conveyor loading of sacks in place of slings (see Plate 18.1). Edward Ritchie Knox, CSR's Chairman of Directors, claimed in 1947 that the real reason for the shipping delays was 'the absence of the will to work' amongst Queensland's waterside workers, and that the mechanical loading gear in existence at the ports was being deliberately operated ineffectively or restrictions placed on its use.

Plate 18.1. Mechanical conveyor loading of bagged sugar at the Port of Cairns, 1939. This mechanical handling equipment was installed in 1924.

Source: Image No. 108815, John Oxley Library. Image reproduction courtesy of the State Library of Queensland, Brisbane.

CSR's concerns about the performance of Queensland's waterside workers had seen the company investigate the system of mechanically loading and unloading raw sugar used at Swedish and American sugar refineries before World War II, but its inquiries found no great economic advantages to the practice during pre-war conditions and cost levels. However, the economic situation after World War II had changed. If sugar could be transported in loose bulk it would avoid the delays associated with the reliance on Australian waterside workers and eliminate the cost of jute bags which had become considerably more expensive after World War II. Consequently, Edward R. Knox announced in June 1947 that CSR intended spending £350 000 on building a bulk storage bin for 40 640 tonnes of raw sugar at Pyrmont Refinery. The raw sugar would be emptied from the sacks immediately after discharge from the vessels, instead of the full sacks being stacked and subsequently emptied at the affination station. This arrangement would enable experience and data to be accumulated on the bulk handling of sugar, with a view to its wider adoption by the sugar industry.[43]

Instead of proceeding with the scheme immediately, CSR hesitated and conducted further study into the matter. The firm concluded that it would be more advantageous to concentrate capital expenditure on a comprehensive plan based on handling in loose bulk from the sugar mills to the refineries. Mechanised handling and storage would then apply throughout the entire system for transporting raw sugar to both domestic and overseas refineries. CSR planned to make the first major step towards the bulk handling of raw sugar at its two Herbert River district sugar mills, shipping through the Port of Lucinda. The firm would provide the entire capital and own all facilities at both ends. At this point, the Sugar Board and other Queensland sugar mill-owners began expressing concern about CSR's proposals and how such a system could be adopted rapidly and equitably by the entire Australian sugar industry. The Sugar Board and leaders of the sugar industry associations in Queensland then began planning the introduction of a comprehensive mechanised bulk sugar transport system in Australia. In the meantime, CSR's refinery operations continued to be hampered by slow deliveries of raw sugar.[44]

To test the suitability of shipping Australian raw sugar in bulk, the Sugar Board arranged for an experimental cargo to be loaded into the ship S.S. *Artemisia* at the Port of Mackay in late 1951. The raw sugar supplied by all the Mackay sugar mills was transported to the wharf in bags and 'bled' (or poured) from the bags into the ship's hold. This first shipment took three weeks to load, slightly longer than stacking bags in the hold. The ship left for London on 8 November 1951 and arrived six weeks later in the

United Kingdom. The sugar reached London in excellent condition and flowed freely into the barges when unloaded from the S.S. *Artemisia.* The success of this experiment prompted the Sugar Board to arrange for further trial shipments from the Ports of Cairns, Townsville and Mackay to the United Kingdom during 1952. Opposition to the departure from normal bag loading by the waterside workers led to considerable industrial trouble at the Ports of Townsville and Cairns. Yet the new arrangements were continued during 1953, so that almost one-third of Australia's sugar exports were shipped in this manner. This arrangement also alleviated the bag shortage, as the same sacks could be used repeatedly.[45]

In September 1951, the Sugar Board had also commissioned independent reports into the bulk handling of sugar by Mr. E. Plomley (CSR) and Mr. F. Chapman, the Chief Technologist with Tate and Lyle, the main English refining company. After considering these reports and the results from the trial shipments of sugar in bulk, the Sugar Board advised the Gair (Labor) government in Queensland that the conversion of some of the state's ports to handle bulk sugar should commence at the Ports of Mackay and Lucinda, and that the warehouse type of storage building used in Hawaii was the most suitable and cheapest type for use in Queensland. The Premier of Queensland, Vince Gair, announced on 5 August 1952 that the Queensland government had approved the building of bulk sugar handling facilities at Mackay and Lucinda, and that Queensland Railways had been preparing for the wider use of bulk sugar handling by experimenting with different rail wagon designs for the transport of bulk sugar. As planning commenced to build the first bulk handling facilities at Mackay and Lucinda, CSR had to decide about the type of unloading facilities to be erected at its refineries. Two of its engineers were sent to the USA, including Hawaii, and Europe in 1953 to study bulk sugar loading and how other commodities were transported in loose form. After considering the overseas arrangements, CSR decided that it would not copy them but install an unloading system akin to and developed from the iron ore unloading gantries at Port Kembla.[46]

To introduce its bulk handling arrangements, two relatively small bulk storage and loading installations commenced operating at Harwood and Broadwater Mills in 1954. Front-end loaders transferred the raw sugar housed in a storage shed onto a portable hopper which fed a conveyor belt taking the sugar directly to a small coastal freighter's hold. The raw sugar was unloaded in bulk by a grab on the coal unloading crane at Pyrmont Refinery. The following year, a large-scale mechanised bulk unloading and storage installation costing £800 000 was completed at Pyrmont Refinery. Unloading was carried out by special shore-based grabs and cranes. They emptied the ships, except for about 100 tonnes of sugar, without the employment of any workers on the ships. During 1955, temporary equipment was also provided for unloading sugar at Yarraville Refinery.[47] The results from these trials convinced CSR that their strategy was correct. Casual workers or carters were employed during the process of unloading, instead of waterside workers. Moreover, work commenced immediately when the ship arrived and continued on a 24-hour, seven-day-a week basis until discharging had been completed.

CSR then extended the system of bulk handling to Glanville (1959), Cottesloe (1961) and New Farm (1961). Unloading arrangements at the latter two refineries, however, were slightly different to that developed at Pyrmont. New Farm Refinery, with its up-river location, did not have deep water or the wharf and crane facilities to accommodate ships big enough to justify the expense of erecting a wharf-side bulk unloading facility. The refinery had been receiving about 40% of its raw sugar in bulk by road from Condong Mill since 1956, so CSR decided to redesign the sugar receiving station at New Farm to accept sugar by train coming from Maryborough and Moreton Mills and road transport from Rocky Point and Condong Mills. Sugar fell or was dropped by grab into a receiving hopper which fed conveyors to an enlarged bulk sugar storage facility. Similarly, raw sugar destined for Cottesloe Refinery with its up-river location, was unloaded at Fremantle and taken by road to the refinery, where it was tipped into the new receiving station. A conveyor system then delivered it to the new bulk sugar store.[48]

General planning continued for a more accelerated development of bulk sugar handling facilities in Queensland during 1954 and 1955. Sections covering the erection of bulk sugar installations at Queensland ports were included in the *Queensland Harbours Act* of 1955. A harbour board, with the approval of the Queensland government, could proceed with the construction of bulk sugar loading facilities, and when the terminal was completed, the Sugar Board assumed responsibility for the control, management, maintenance and operation of the bulk sugar handling facilities. The Sugar Board, however, was obliged to reimburse the harbour board concerned over a fixed period of years for the capital expenditure involved in the terminal's construction. The harbour board was allowed interest on its original investment, at a rate approved by the Queensland Treasurer. The *Harbours Act* also allowed the Sugar Board to delegate authority to local committees to run the terminals on its behalf. Mackay, Bundaberg, Cairns and Townsville Harbour Boards assumed responsibility for the construction and financing of the bulk sugar terminals at their respective ports. At Lucinda and Mourilyan Harbours where no harbour boards existed, the Queensland Treasury became the constructing authority.[49]

By 1955, the Mackay Harbour Board's scheme, sponsored by the Sugar Board, was underway, with 1957 adopted as the target date for completion. Construction ran to schedule so that the first raw sugar began arriving at the Mackay Bulk Sugar Terminal on 20 June 1957. Six weeks later the first ship docked at the Mackay Bulk Sugar Terminal to begin loading sugar from the 1957 crop.[50] Over the next seven years, another five bulk sugar terminals became operational in Australia (Table 18.3). The largest terminals were at Mackay and Townsville, where the single storage sheds had a capacity of 152 400 tonnes, with equipment for loading ships at a maximum rate of 760 tonnes per hour. Major engineering work and dredging widened and deepened the entrance of Mourilyan Harbour so that larger vessels could reach the Mourilyan Bulk Sugar Terminal, while

Table 18.3. Initial and subsequent capacity at Australian bulk sugar terminals, 1957–1980.

Bulk sugar terminal	Year of first operation	Initial capacity (tonnes)	Subsequent upgrades until 1980	Storage capacity in 1980 (tonnes)
(a) Cairns	1964	101 600	2nd shed (1972)	234 000
(b) Mourilyan	1960	142 240	Shed extension (1972)	175 000
(c) Lucinda	1958	76 200	2nd shed (by 1970); 3rd shed (1979)	231 000
(d) Townsville	1959	142 240	2nd shed (1965)	285 000
(e) Mackay	1957	152 400	2nd shed (1963); 3rd shed (1963); 4th shed (1980)	667 000
(f) Bundaberg	1958	50 803	No 1 shed extension (1963); 2nd shed (1965); No. 1 shed extension (1977)	316 000

Source: (a) *ASJ*, December 1972, p. 449 (b) ASJ, December 1972, p. 449 (c) QSB, AR 1979–80, p. 4 (d) *ASYB*, 1964, p. 45 (e) *ASYB*, 1964, p. 45; G. Lewis, *A History of the Ports of Queensland* (Brisbane, 1973) p. 262 (f) G. Lewis, *A History of the Ports of Queensland* (Brisbane, 1973) p. 262. Tonnage figures from E. Driscoll, 'Recent developments in the Queensland sugar industry', *Geography*, Vol. 51 (1966), p. 148 & B. Hoyle, 'Ports and hinterlands in an agricultural economy: the case of the Australian sugar industry', *Geography*, Vol. 69, no. 4 (October 1984), p. 314.

an entirely new location with access to deeper water towards the mouth of the Burnett River was selected for the site of the Bundaberg Bulk Sugar Terminal.

Bagged sugar continued to be available from Australia during the late 1950s and 1960s, despite the conversion to bulk sugar handling during this period. The Sugar Board recognised that not all Australia's raw sugar customers, notably Japan, were equipped to accept raw sugar in bulk. In Southern Queensland, the Maryborough Sugar Factory and Gin Gin Mill produced bagged sugar for shipping through the Port of Urangan, while Hambledon and Mulgrave Mills exported bagged sugar out of the Port of Cairns until the end of the 1962 and 1963 crushing seasons, respectively. From 1964 onwards, all Australian raw sugar exports were handled from the sugar mills to the ports in bulk. The Sugar Board, however, was conscious that some small overseas customers still preferred to receive bagged sugar. Consequently, a small bagging station was maintained at the Port of Townsville. A drier climate compared to the other northern sugar ports made the Port of Townsville more suitable for packing bagged sugar and exporting it in good condition, and the port received many overseas ships that loaded diverse cargoes.[51]

To complement the development of the bulk sugar terminals, the state's sugar mills installed new plant for handling sugar in bulk. Raw sugar, instead of being bagged at the sugar mills, went by conveyor belt to large single or double storage hoppers or sugar bins (Plate 18.2). Bulk boxes on trucks, semi-trailers or rail wagons were drawn underneath the sugar bins and received the raw sugar, before being dispatched to the bulk sugar terminal.[52] On arrival, the bulk boxes were weighed at the receiving station and then tipped sideways, the contents falling through a grille onto a conveyor leading to massive storage sheds of around 300 m in length and 50 m high, with a floor space in excess of one hectare. When a ship needed to be loaded, a conveyor belt carried the raw sugar to a weighing tower and then to a mobile gantry loader above the sugar wharf. Freighters which required three weeks in port to load 10 000 tonnes of bagged sugar were usually loaded in two days using bulk handling methods.

As a result of these developments, the number of labourers were reduced at the sugar mills and loading ports. At CSR's sugar mills in the early 1950s, for instance, eleven men per week were employed filling, sewing, wheeling and stacking bags of sugar; with bulk, the job was done by three men. Loading of the stacked bagged sugar at the sugar mill required casual gangs of approximately twelve men; with bulk handling, the task was handled by five men. At the loading ports, hundreds of men employed on a casual basis

Plate 18.2. The sugar bin at Pioneer Mill, c. 1965 (right foreground). Sugar was loaded into bulk boxes on rail wagons (foreground) for dispatching to the Townsville Bulk Sugar Terminal.

Source: *ASYB*, 1965, p. 326. Image reproduction courtesy of Rural Press Queensland Pty. Ltd.

were replaced by small teams of trained, permanently employed men using machines. Figures assembled by the historian Ross Fitzgerald showed that the number of wharf labourers employed daily at the Port of Townsville had been reduced from an average of 549.2 in 1955 to 183.8 by 1962. A similar trend was observed at the Port of Cairns, where daily wharf employment fell from 517.9 to 99 over the same period. In other North Queensland ports, the reduction in labourers was as high as 90%.[53]

Arrangements for the movement of Queensland's raw sugar to the sugar refineries and export markets during the early 1950s revolved around the main export ports of Cairns, Lucinda, Townsville, Bowen, Mackay, Urangan and to a lesser extent, Bundaberg (Figure 18.3). Much of the bagged sugar reached these ports, except Cairns, by either rail or road transport. Bagged sugar from Mossman, Goondi, Mourilyan and South Johnstone Mills reached the Port of Cairns by small coastal sugar lighters, the harbour facilities at Mourilyan and Port Douglas being unsuitable to receive larger, ocean going vessels.[54] Raw sugar from Millaquin and Qunaba Mills supplied Millaquin Refinery; part of the bagged sugar from the sugar mills in the Bundaberg and Maryborough districts was shipped through the Fairymead wharf or the Bundaberg wharf, both on the Burnett River, or increasingly out of the Port of Urangan, due to difficulties in bringing shipping up the Burnett River, which suffered from flooding and shallow

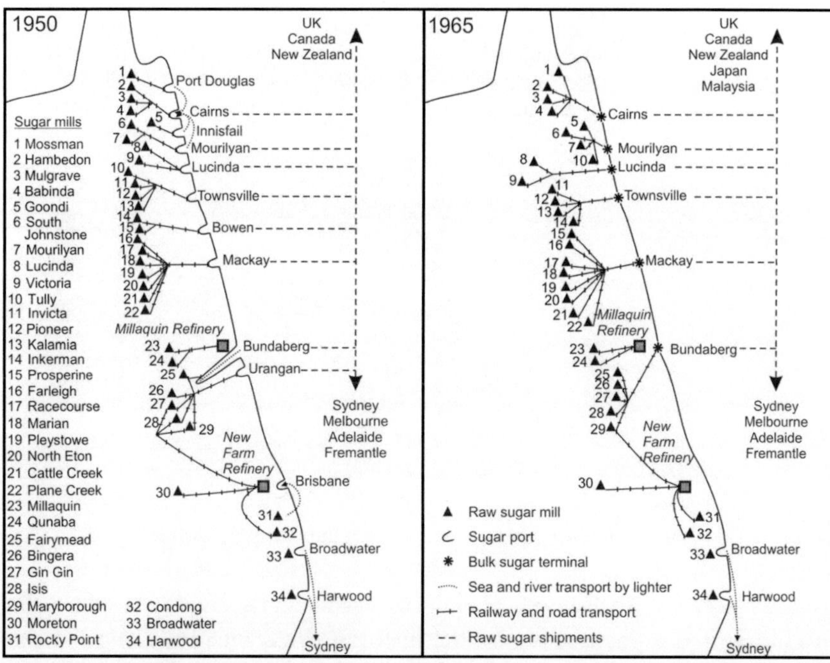

Figure 18.3. Schematic diagram showing the transport arrangements for raw sugar in Australia, 1950 and 1965.

Source: Based upon details in G. Lewis, *A History of the Ports of Queensland. A Study in Economic Nationalism* (Brisbane, 1973), passim.

depths. In addition, some of this bagged sugar was railed to the New Farm Refinery in Brisbane. In addition, Rocky Point Mill, south of Brisbane, sent its raw sugar by coastal lighters to New Farm Refinery.[55] These arrangements had persisted since the late 1910s.

As the bulk sugar terminals became operational, the transport of Queensland's raw sugar began altering during the late 1950s and early 1960s, causing a decline in trade at several Queensland ports. The first to suffer were the Ports of Bowen and Port Douglas. Until 1957, the former port handled the shipment of about 100 000 to 120 000 tonnes of bagged sugar annually from Inkerman and Proserpine Mills. However, Bowen ceased to be a sugar port when the sugar from Inkerman and Proserpine Mills was sent to the Townsville and Mackay Bulk Sugar Terminals, respectively, after they became operational. Port Douglas was also closed as a sugar port at the end of 1957. Lightering of bagged sugar to the Port of Cairns ceased, forcing Mossman Mill to send bagged sugar on pallets to Cairns using its own fleet of seven large freighter trailers, before they were converted to bulk sugar handling.[56] After 1962, the Port of Urangan suffered a similar fate, when all the sugar exported from the Bundaberg and Maryborough districts was either railed or road-hauled in bulk to the Bundaberg Bulk Sugar Terminal or sent by rail to CSR's New Farm Refinery. The Port of Cairns had its sugar shipments reduced by more than a half in the early 1960s. It continued receiving raw sugar from the four nearest mills (i.e. Mossman, Hambledon, Mulgrave and Babinda), but the cessation of lighter movements meant that sugar from Goondi, Mourilyan and South Johnstone Mills was exported through the new Mourilyan Bulk Sugar Terminal. In 1959, Tully Mill, which had previously sent its sugar by rail to Townsville, switched to exporting its sugar initially through the Lucinda Bulk Sugar Terminal and then out of the Port of Mourilyan once its bulk handling facilities became operational.[57]

The newly established bulk sugar handling system and orderly deliveries of raw sugar to customers was severely tested during 1963, following the almost total destruction of the bulk sugar storage shed at Townsville by fire on 9 May 1963. Concentration of vessels at Townsville enabled the progress of reconstruction to continue and still cope with deliveries from the local sugar mills, but created problems at other Queensland ports where stocks accumulated, filling the bulk sugar terminals to capacity. Towards the end of the crushing season the second sugar storage shed at the Port of Mackay became partially available, easing the congestion considerably.[58]

The enlargement of storage facilities at the Port of Mackay had been undertaken in anticipation of the need for additional storage space given the likelihood of further expansion in the Australian sugar industry. Subsequently, second sheds were built at the Ports of Bundaberg, Lucinda and Townsville during the mid-1960s (see Plate 18.3). Continued growth in Australia's raw sugar output during the early 1970s prompted further extensions to existing storage sheds or the erection of new sheds (Table 18.3). The biggest development occurred at Lucinda, where the storage capacity was not only increased in 1979, but was accompanied by the construction of a new wharf, 5.8 km off-shore in deep water, and a cyclone resistant, high speed enclosed conveyor belt connecting the wharf with the sugar storage sheds. This upgrade enabled Lucinda to be converted from a domestic to an export shipping terminal, and all Australian domestic sugar needs were then met from the southernmost terminal at the Port of Bundaberg, thereby saving the industry considerable freight charges. At the Port of Mackay, the commissioning of the third and fourth storage sheds ensured that the port had the world's largest storage complex for raw sugar in 1980, and lifted the total storage capacity of the state's bulk sugar terminals to almost 2 million tonnes.[59]

Plate 18.3. An aerial view of the Lucinda Bulk Sugar Terminal, 1969.

Source: Queensland Department of Harbours and Marine, AR 1968–69, p. 10. Image reproduction courtesy of the Queensland Department of Transport and Main Roads.

The Sugar Board embarked upon the conversion to bulk handling against a background of having to secure new markets for Australia's burgeoning raw sugar output. Reconstructing the destination of Australia's sugar exports between 1950 and 1979 was hindered by a lack of consistent data, with most figures being presented over a financial year, before a switch to using calendar years as the period of measurement. Despite the difficulties with the data, it is clear that Great Britain remained Australia's most important customer between 1954 and 1961. However, under the Sugar Board's arrangements with the British Ministry of Food, some of this sugar was reserved for Canada and New Zealand, which together took about a third of Australia's annual raw sugar output. British Malaya and Hong Kong purchased small amounts of sugar during the mid-1950s, but Japan was the major new customer for this period (Figure 18.4). Australian sugar was successfully marketed in Japan for the first time in 1954. Subsequently, CSR and the Sugar Board put considerable effort into building up relationships with Japanese sugar buyers. The signing of an Australia-Japan Trade Agreement, which became effective on 6 July 1957, also assisted in the development of this relationship, for it provided an assurance that import controls would not be operated in a manner discriminatory to Australian sugar. By 1961, Japan was Australia's third most important customer for sugar. Moreover, the Sugar Board had arranged with several Queensland sugar mills in the Cairns and Lower Burdekin districts to manufacture a special low polarising sugar for the Japanese market. Known as the JA brand of sugar, it had to have a polarisation of not more than 97.80°S, not less than 97.00°S, and as near as practical to 97.60°S.[60]

Japanese sugar refiners, however, were not always impressed with the quality of Australia's raw sugar, particularly its filterability. This complaint had also been made regularly by CSR's production chemists at its Australian refineries.[61] Acutely conscious of the need to preserve the burgeoning Japanese market, the Sugar Board decided in 1960 to remove the cause of dissatisfaction to the Japanese refiners by entering into a special arrangement with the three Cairns district sugar mills, providing them with a financial incentive for improved filterability. Special steps taken in 1960 at the three Cairns district sugar mills producing sugar for Japan included the employment of one or two additional chemists to take extra measurements and additional staff to supervise the factory processes. These arrangements improved the quality of the JA sugar, with the Sugar Board announcing in mid-1961 that the 'current Japanese attitudes to Australian raw sugar are in

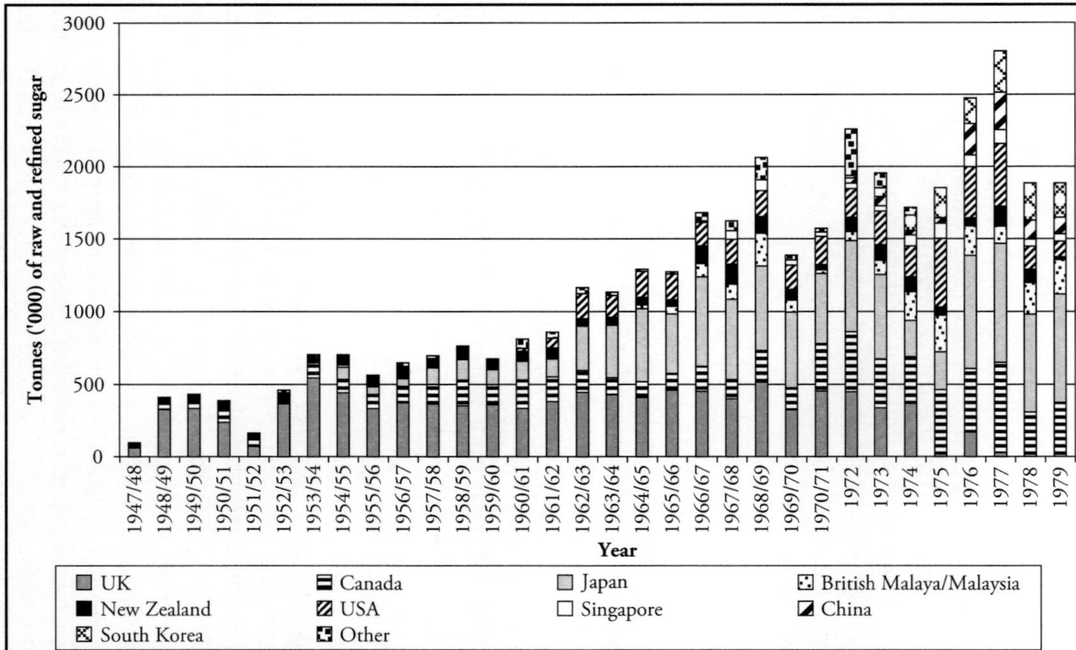

Notes: (1) The figures for Malaysia from 1961 to 1963 were a combined total for Malaysia and Singapore.
(2) Most exports were raw sugar, with less than 50 000 tonnes of exports each year being refined sugar.
(3) The 'Other' category included the island nations of Oceania, Papua New Guinea and Ceylon.

Figure 18.4. Exports ('000 tonnes) of Australian raw and refined sugar by destination, 1947/1948–1979.

Source: Based upon figures in the following publications: Commonwealth Bureau of Census and Statistics, *Overseas Trade Bulletins,* 1949/50–1970/71; *ASYB*, 1978, p. 59; and Proprietary Sugar Millers' Association, ARs, 1979 & 1980.

sharp contrast to their criticisms and objections up to the end of the 1959 season'. This rapid turnaround in the attitudes of the Japanese sugar refiners prompted the Sugar Board to announce the introduction of bonuses on good filtering Brand 1 sugar (i.e. raw sugar destined for any market, except Japan, with a polarisation between 98°S and 99°S), ranging from 3d to a maximum of 3s 6d per ton with effect from the beginning of the 1961 season, and to institute penalties of the same amount on poor filtering Brand 1 sugar from the beginning of the 1962 season. Corresponding penalties on poor filtering JA sugar were also introduced from the commencement of the 1962 season. The Sugar Board continued stressing the need for improved raw sugar quality, announcing in 1965 that all Australian sugar mills had to be fitted with vibrating screens for lump removal during the 1966 crushing season.[62]

The destination of Australia's raw sugar exports changed during the 1960s. Japan overtook Canada as Australia's second most important market for raw sugar. Moreover, the Sugar Board arranged for Japan to be supplied with cane invert, which was manufactured at Pleystowe and Racecourse Mills between 1967 and 1972. Invert is a honey coloured liquid of similar consistency to molasses, although made up of glucose and fructose. The Japanese distillers used the product during the manufacture of a Japanese whisky and alcohol for sake.[63] The USA also emerged as a major new customer in the early 1960s, and Malaysia and Singapore continued to be important purchasers of Australia's raw sugar (Figure 18.4).

Countries importing Australian raw sugar, however, changed further during the 1970s. From 1972 onwards, China and South Korea sourced some raw sugar from Australia. The opening up of these two new markets helped compensate for the loss of Great Britain as a major customer, following that country's accession to the EEC in 1973. Great Britain purchased its last Australian raw sugar in 1977. Most purchasers of Australia's raw sugar had been locked into long-term contracts for periods of three to six years by the Sugar Board, so the Australian sugar industry had certainty in its overseas markets, although the importers realised that they were dealing with a reliable supplier that delivered a high quality product. Yet on occasions during the 1970s, the Sugar Board had enough raw sugar so that it could meet demands from occasional customers such as Sweden and Finland (1971–1973), Tunisia (1972) and the USSR (1972–1974). This growing volume of sugar exports ensured that Australia became the world's third largest sugar exporter by 1979, ranked behind only Cuba and the EEC.[64]

Marketing arrangements, refinery upgrades and new products, 1950–1979

Marketing arrangements for sugar products in Australia between 1950 and 1979 continued to be determined by the Commonwealth government through a succession of mostly five-year Sugar Agreements with the Queensland government (Table 18.4). The terms and conditions of these agreements were identical to pre-World War II agreements, except the clause in which the Queensland government agreed to accept responsibility for any loss arising from the export of surplus sugar was removed under the Sugar Agreement of 1975.[65] The Australian sugar industry continued assisting the Australian manufactured fruits industry by maintaining its contributions (i.e. domestic and export sugar rebates) to the FISCC. This arrangement was considered by the Sugar Enquiry Committee of 1960. Its members concluded that there was no reason why the Australian sugar industry should continue providing a domestic rebate to the processed fruits industry and recommended relieving the sugar industry of this responsibility. This suggestion was not accepted by the Commonwealth government (Menzies; Coalition of Liberal and Country Parties) and the existing arrangement remained in place until the Sugar Agreement of 1975 suspended payments to the FISCC. These payments were only to be resumed when the FISCC's accumulated funds of $1 million fell to $200 000, and thereafter the funds provided by the Australian sugar industry to the FISSC would only be to the extent needed for payment of the domestic sugar rebate and the other operating costs of the Committee.[66]

Under the Sugar Agreements, successive Commonwealth governments continued fixing the refined sugar price in Australia. On several occasions over this three decade period, the Commonwealth agreed to increase the price of a ton of raw sugar, in some instances quite substantially. The first of these major price increases occurred in December 1947, when a delegation of Australian sugar industry representatives convinced Prime Minister Ben Chifley that the rising costs of wages, coal, hessian bags and shipping freights warranted an increase in the price of sugar. The increase granted by the Commonwealth government restored the price of a ton of 1A grade refined sugar to the level that had existed in 1932, thereby forcing up the retail sugar price to 4½d per lb in each Australian state capital city. A further increase of £12 per ton was permitted under the Sugar Agreement of 1951, but within twelve months the Australian sugar industry had asked Prime Minister Robert Menzies for another price increase, again claiming rising production costs were not being covered by the price they received for a ton

Table 18.4. Australian Sugar Agreements and their impact upon the wholesale and retail prices of refined sugar, 1946–1979.

Sugar Agreement of:	Period of Agreement	Wholesale price per ton (£ s d) or tonne ($) of 1A grade refined sugar	Equivalent retail price in capital cities (per lb or kg)
(a) 1946	1.9. 46–31.8.1951; amended Dec. 1947	£33 4s (Sept. 1946 to Dec. 1947) £37 6s 8d (Dec 1947 to Nov. 1949) £41 9s 4d (Nov. 1949 to July 1951)	4 d 4½ d 5 d
(b) 1951	7. 7. 1951–31.8. 1956	£53 6s 8d (July 1951 to March 1952) £65 12s 10d (March 1952 to Oct. 1952) £73 16s 11d (Oct. 1952 to May 1956) £82 1s (May 1956 to Sept. 1956)	6½ d 8 d 9 d 10 d
(c) 1956	1.9. 1956–31.8.61; extended to 31.5.62	£82 1s (Sept. 1956 to May 1960) £90 5s 2d (May 1960 to May 1962)	10 d 11 d
(d) 1962	1.9.1961–31.8. 1967; extended four times to 30.6.1969	£90 5s 2d or $180.52 (May 1962 to June 1967) $206.72 (June 1967 to June 1969)	9.2 cents 10.5 cents
(e) 1969	1.7. 1969–30.6 1974; extended to 31.1.1975	$206.72 (July 1969 to 31.1. 75)	10. 5 cents per lb. /23 cents kg
(f) 1975	1.2.1975–30.6.1979; extended to 30.9.79	$219.50 (Jan.–Dec. 1975); $240 (August 1976 to June 1977); $266.80 (June 1977 to June 78); $297.44 (June 1978 to Sept. 1979)	24 cents kg; 26 cents kg; 29 cents kg; 33 cents kg.

Sources: (a) *ASYB*, 1946–47, pp. 81–82; *ASYB,* 1947–48, p. 162–163 (b) *Year Book of Australia,* 1955, p. 778 (c) *Year Book of Australia,* 1958, p. 893 (d) *Year Book of Australia,* 1966, pp. 914 & 917 (e) *Queensland Year Book,* 1971/72, p. 332 (f) IAC, *The Sugar Industry. Report No. 209* (Canberra, 1979), p.30; *Queensland Year Book,* 1976, p. 376; *Queensland Year Book*, 1977, p. 360; *Queensland Year Book,* 1978, p. 285; *Queensland Year Book,* 1979, p. 338.

of sugar. In March 1952, Menzies acceded to a substantial interim price increase of £12 and referred the matter of additional price increases to a committee of inquiry. The 1952 Sugar Committee of Inquiry was split on the matter of price, making three separate recommendations for a further increase in the retail sugar price (i.e. an increase of ½d, ¾d and 1d per lb of sugar). The Menzies government accepted the maximum increase, leading to the price of refined sugar rising by another £8 per ton (or 1d/lb.) in October 1952.[67]

The price for a ton of refined sugar in Australia rose substantially in 1956 and 1960, again in response to the industry's rising production costs, but price rises during the early 1960s were much smaller. Eventually these price rises ceased, and the refined sugar price in Australia remained unaltered between June 1967 and January 1975. After this lengthy period of price stability, the Sugar Agreement of 1975 granted an immediate 7.9% increase in the price of refined sugar in Australia. Moreover, an important new provision of the Sugar Agreement of 1975 was a review of the sugar prices during the term of the agreement. Reviews occurred three times during the term of the Sugar Agreement of 1975, lifting the domestic price of refined sugar from $219 to $297 per tonne, causing the retail price for a kilogram of sugar to rise from 24 to 33 cents (Table 18.4).

During the 1950s, the average annual apparent sugar consumption per capita in Australia dropped slightly compared to the late 1940s, but still remained around 52 kg per capita. Post-war immigrants to Australia did not have such sweet tastes, thereby lowering per capita consumption rates. In the 1950s, Poles consumed 50% of the Australian per capita consumption, Yugoslavs and Italians 30% and Greeks 20%.[68] Nevertheless, according to figures collected during the Second World Food Survey conducted by the Food and Agriculture Organisation in 1953, Australians still had the highest average annual sugar consumption per head of any country in the world. Moreover, annual apparent sugar consumption in Australia continued to average around 52 kg per capita during the 1960s and 1970s. This persistent high sugar use, together with a growing Australian population, meant that the overall domestic demand for refined sugar products continued growing in Australia during the period 1950 to 1979.[69]

CSR anticipated some post-war increase in the demand for sugar in Australia, and had commenced planning for refinery upgrades, even before the cessation of hostilities in 1945. The implementation of these upgrades, however, had been retarded by delays in securing machinery from overseas. Pyrmont Refinery was a major recipient of this funding, receiving two new high-pressure boilers and generating plant for electricity, a wharf upgrade and new mechanical coal unloaders and concrete silos for coal. Its bone char kilns from 1955 onwards were also converted from coke to oil firing. Some older bone char kilns during 1958 were replaced with a new direct-fired, multiple-hearth bone char kiln which was controlled automatically. This new equipment reactivated char that had been saturated with impurities and had an output that was equal to six to nine old type kilns. By the mid-1950s, Pyrmont was the largest industrial establishment operated by CSR and the hub of their refining and research operations.[70]

Improvements were also completed at CSR's other sugar refineries. At Yarraville Refinery, the output of refined sugar was raised from 3 050 tonnes to 5 100 tonnes weekly by the addition of a new calandria vacuum pan and new electric semi-automatic centrifugals. Its bone char kilns were converted to oil firing, thereby putting char revivification under better control. Improved clarification was the focus of upgrades at Cottesloe during 1956, following the installation of a phosphatation-flotation plant in 1956. Under this new arrangement, the raw sugar liquor was mixed with phosphoric acid, causing the formation of a floating precipitate of calcium phosphate and impurities which were removed by a revolving scraper, leaving behind clear liquor. A phosphatation-flotation plant, new boiler station and new coal-handling facilities were installed during upgrades at Glanville Refinery in the mid-1950s. These improvements increased Glanville Refinery's processing capacity by 50%, enabling it to manufacture over 1 200 tonnes of refined sugar weekly.[71]

Before the refinery upgrades had commenced, CSR's refining costs had escalated during the late 1940s, a consequence of a combination of factors. The cost of hessian bags had risen almost 50% during 1948, a result of the interruption to their supply caused by the partitioning of British India. By 1951, CSR could not even secure hessian bags, and was forced to purchase Japanese calico as a substitute. Shortages of workers at the refineries contributed to an increase in overtime wages. Inferior coal with high ash content caused the average production of refined sugar per tonne of coal to fall from five to four tonnes, yet the price of coal had risen during the late 1940s.[72] Some of these increased costs were covered when the Sugar Board agreed to increased refining charges, but CSR also installed mechanical coal loading equipment which reduced the number of workers needed at each refinery's wharf. By the mid-1950s, the other machinery changes at the refineries had enabled higher melting rates, improved efficiencies and reduced maintenance. In addition, coal and jute shortages had disappeared.[73]

CSR reduced its refining costs even further during the late 1950s by achieving economies in steam usage. In 1952, an average of 30 tonnes of coal was required to refine 100 tonnes of raw sugar; the amount had fallen to 16.2 tonnes by 1961. This reduction had been achieved by better automatic control keeping the densities of liquors and syrups at the highest practicable level so that the minimum amount of steam was required during the processing, improved process scheduling which reduced steam blow-off periods (i.e. when steam cannot be used), and increased attention to heat conservation in pipes and vessels.[74]

During the 1950s, bone char usage at CSR's refineries also came under greater scrutiny. Tests at Pyrmont Refinery showed that large amounts of bone char were used per 100 melts of raw sugar, and that the product was probably being wasted or used inefficiently. Moreover, the discharging of 'supposedly' exhausted char entailed the manual emptying of the cisterns and shoveling the char onto conveyor belts for disposal or revivification, particularly arduous and messy tasks. During 1958, J. Kenny, a member of CSR's Research Division, found that the use of very finely divided bone char improved the rate of decolourisation of the clear filtered liquid by a factor of about 100 compared to normal pieces of bone char. Furthermore, his experiments showed a reduction of about 50% in bone char usage for the same sugar production. This discovery prompted immediate action, the finer bone char being adopted at CSR's refineries, leading to a marked reduction in its use per 100 melts of sugar. Further research in this area resulted in Glanville Refinery abandoning the use of bone char entirely in 1974, its place taken by an ion-excluding resin.[75]

More modest upgrades occurred at CSR's sugar refineries during the 1960s and 1970s, with the focus being upon adopting technological advances in refining and bulk handling. At New Farm Refinery, a new refined sugar warehouse was completed during 1961. During the late 1970s, further extensions occurred to both the raw and refined sugar storage facilities, including the erection of new bulk storage bins and liquid storage tanks. In the late 1960s, CSR decided that the phosphatation-flotation process used at Glanville and Cottesloe Refineries was not delivering the expected results and it was replaced with carbonation, thereby standardising this phase of the refining process at CSR's sugar refineries. During the late 1960s, mechanical stirrers were added to the vacuum pans in each CSR sugar refinery, contributing to an increase in the volume which could be boiled in each cycle and a 5–10% increase in the yield of crystals, resulting in fewer boilings to achieve the same amount of sugar. Manually operated and semi-automatic centrifugals were replaced with fully automatic continuous or batch recycling centrifugals. Some boilers at Cottesloe and Yarraville Refineries were converted from coal to oil firing, thereby increasing boiler efficiency and evaporating capacity in the refineries, but reducing the costs of maintenance and steam generation. Productivity gains were achieved by the installation of automatic control at all the carbonation plants and affination and fugal stations.[76] These changes were sufficient to allow a steady increase in the amount of sugar melted at CSR's sugar refineries during the 1960s and 1970s (see Figure 18.2).

Following the introduction of clean air and clean water legislation in all Australian states during the 1960s and 1970s, CSR had to reduce the emissions and dirty water discharges at each of its sugar refineries. At Pyrmont Refinery, smoke emissions from the bone char retorts exceeded permissible limits. Instead of trying to rectify this matter, CSR decided in July 1975 to close the bone char-making plant, after concluding that the cost of manufacturing bone char at Pyrmont Refinery exceeded the purchase price from commercial producers of the product. Bone char was then imported from overseas manufacturers located in Brazil, Portugal and Japan. Smoke and dust from the boilers at Pyrmont Refinery were reduced after the commissioning of a $1.2 million fabric filter installation in 1977. This installation received the inaugural State Pollution Award of New South Wales. Bone char kilns and boilers at Cottesloe Refinery were converted to natural gas firing by 1977, eliminating the risk of smoke emission. Airborne lime dust at Pyrmont and Cottesloe Refineries were reduced by enclosing their lime receiver hoppers. By 1977, char drain water and boiler discharges at the refineries were being treated to ensure that suspended solids and BOD_5 levels conformed to state government regulations. Floor maintenance had been completed to eliminate seepage of contaminants.[77]

At the conclusion of World War II, CSR's main refinery products were various grades of refined sugar (i.e. 1A or 1XD), caster sugar, icing sugar, golden syrup, treacle and a small quantity of molasses which was sold to the company's distilleries. However, a new product known as liquid sugar – a solution consisting of about two-thirds sugar and one third water – commenced being manufactured at Pyrmont Refinery during 1949, and supplied by road tanker to two large Sydney breweries. Liquid sugar is a much less stable commodity than granulated sugar, requiring special equipment (i.e. storage vats and pipelines) and care in all stages of its use to avoid contamination. Moreover, liquid sugar was not suitable for all manufacturers, especially the jam makers. Some manufacturers interested in the product had to convert their plant which was designed for handling granulated sugar, so there were delays in its uptake by CSR's customers. By 1957, a quarter of the sugar required by manufacturers in Sydney was being delivered in the liquid form by road tankers and the product had just commenced being supplied in Melbourne. New Farm Refinery began manufacturing liquid sugar in 1960.[78] Two years later, CSR released superfine liquid sugar, a special grade of liquid sugar for soft drink manufacturers. By 1978, 14% of CSR's sugar sales were liquid sugar.

Further product developments occurred in 1968 when CSR Sugar Cubes commenced being produced at Pyrmont Refinery. Each cube contained one level teaspoon of sugar and the new product was developed specifically for caterers, replacing the old-style rectangular tablets of refined sugar known as 'loaf sugar'. In 1978, special caster sugar became available following requests from manufacturers such as Cottee's General Foods, after the firm had commenced making powder-based drinks that required particularly fine-grained sugar.[79]

Traditionally, most refined sugar was delivered to Australian manufacturers and wholesalers in 70 lb (32 kg) hessian bags and 140 lb (63.5 kg) jute sacks. The development of liquid sugar partly eliminated this arrangement, but CSR during the late 1950s also commenced examining the practicability of supplying granulated sugar by using bulk sugar tankers. CSR first delivered refined sugar in bulk on 9 October 1958 to Weston's biscuit factory in western Sydney. By 1960, manufacturers could secure refined sugar in bulk from New Farm, Pyrmont and Yarraville Refineries.[80]

During the 1950s, CSR also investigated providing sugar in smaller paper packages for retail sale, following the growing demand from the grocery trade – particularly self service stores – for grocery items to be pre-packed. After a lengthy trial period, CSR established its own retail packaging plant at Pyrmont Refinery and commenced supplying standard granulated white sugar (1A brand) in 5 lb pre-packed retail packages to the New South Wales market in July 1958. By late 1962, these pre-packed packages were available in all Australian states, although they were now being issued in 2 and 4 lb packages, following a survey of the buying habits of Australian housewives. High speed parceling machines were installed at the refineries to improve the capacity of the retail packaging plant. Production of new packs of soft brown sugar in clear polythene 1lb and 2 lb packets commenced in 1962; this type of packaging was extended in 1972 when CSR released for the first time pre-packed 1 kg packets of raw sugar. Granulated refined sugar, however, was still packed in bags and sacks during the 1950s and 1960s, but these were phased out by early 1970. In their place, customers received their sugar in 30 kg multiwall paper bags. CSR noted that even though the hessian and jute bags had served the industry well, they were no longer considered a satisfactory package for a 'pure food'. Automatic packing and palletising of these bags had been introduced by 1977.[81]

The Millaquin Sugar Co. Ltd. remained CSR's only rival in the supply of refined sugar products to the Australian market between 1950 and 1979. For most of this period, the firm's general manager was Sir Douglas Forbes, who held the post as a consequence of his general managership of the Queensland National Bank. Forbes and the Millaquin Board of Directors were very cautious in their outlook throughout the 1950s and 1960s.[82] Any major expansion plans at Millaquin Refinery were shunned during the post-World War II period, the company continuing to melt mostly the raw sugar from its own two sugar mills. In 1955, this amount had declined to 24 000 tonnes, after the increased activity of the late 1940s, but over the next two decades it slowly rose, in line with the expansion of output at its sugar mills. By 1980, Millaquin Refinery turned out around 35 000 tonnes of refined sugar products, which supplied about a third of Queensland's requirements.[83] However, some small changes were made in the refinery. In 1959, the diatomite-Sweetland filter process for removing impurities from melted raw sugar was replaced with the phosphatation-flotation process. This change reduced bone char use from 37 tons to 24 tons per 100 tons of raw sugar melted and produced refined sugar of a better quality. Two Hesser packing machines for pre-packaged sugar were installed in 1961 as part of the retailing changeover to pre-packaged groceries which amounted to 40% of the refinery's output by 1968.[84]

Declining domestic consumption, deregulation and its aftermath, 1980–1995

Concern about the levels of sugar consumption in Australia became more widespread during the late 1960s and 1970s, as increasing numbers of academics and health professionals gained publicity by claiming that sugar was responsible for a host of human ailments. In 1980, a study commissioned by the Sugar Board found that a significant proportion of the Australian population believed tooth decay and weight problems were caused by sugar. Not unexpectedly, the average annual apparent consumption of sugar over the five-year period 1980/1981 to 1984/1985 stood at just under 50 kg per capita or approximately a 7% decline compared to the previous five-year period. The Australian sugar industry, especially the refiners, became alarmed at this trend. Such a drop in sugar consumption was estimated to be equivalent to a market loss of 50 000 tonnes of sugar annually, worth around $25 million a year.[85] Moreover, this decline occurred at a time of moderate Australian population growth fuelled by a sharp rise in net migration, making this trend particularly worrying for the Australian sugar industry.

The Sugar Board claimed that the main factors behind this sudden decline in apparent sugar consumption per capita were the increased domestic sugar price (see Table 18.4), a growing switch to artificial sweeteners (e.g. saccharin; cyclamate; aspartame) due to the aggressive advertising of such products, and the public's increased concern about nutrition. In 1981, the Sugar Board, assisted by CSR and the Millaquin Sugar Co. Ltd., commenced a decade long Australian campaign with the following three main goals: changing the negative attitudes held about sugar by large sections of the Australian public, politicians, health professionals and the media; improving the nutrition education for the Australian public; and slowing and reversing the per capita decline in Australian sugar consumption. Funds totaling $4.5 million a year were dedicated to this program over a five-year period, with three-quarters of the expenditure being directed towards media advertising. CSR dedicated another $1.5 million a year towards sales promotion at the retail level.[86]

This campaign began in February 1981 and ended in 1990. It involved a variety of activities. The Sugar Board distributed a nutrition booklet to schools, dentists and medical practitioners throughout Australia. Two nutritionists were employed by the Australian sugar industry to establish a comprehensive and freely available nutrition data base. As information about the latest research on sugars and human nutrition was gathered, it was also published in small booklets titled *Sugar and nutrition abstracts*. These booklets were initially issued every two months, but later became quarterly publications distributed to every Australian doctor, dentist, pharmacist and nutritionist between September 1983 and June 1991. A booklet on diabetes management for dieticians was also sponsored by the Australian sugar industry. Television and magazine advertising informed the Australian public about the role of sugar in a nutritionally balanced and varied diet, and reassured the public that they were not eating more sugar now than fifty years ago. Five, sixty second television advertisements were produced on the theme of 'Sugar – A Part of Life'. They dealt specifically with sugar's role in a balanced diet, sugar being a natural product and that the production of sugar was an important and major Australian manufacturing industry (see Plate 18.4). In order to encourage the public to associate sugar with healthy activity, CSR had commenced supporting New South Wales Little Athletics during the late 1970s. Funding for athletics events at all levels then commenced in 1984. As a consequence of this sponsorship, advertisements promoting sugar as 'an energy food' or 'a natural part of life' appeared at athletics carnivals throughout Australia. In addition, the Australian sugar industry sponsored 'Athletics Towards 2000', a development program initiated by the Australian Athletics Union for Junior Development and Get Fit Australia programs.[87]

Plate 18.4. One of the 1985 television advertisements promoting sugar in Australia.

Source: *AC*, June 1985, p. 43. Image reproduction courtesy of CANEGROWERS, Brisbane, and the Queensland Sugar Limited, Brisbane.

Eighteen months into the Australian sugar industry's campaign, follow-up market research study found that Australian sugar consumers now viewed sugar more positively, with only 20% of respondents believing that eating sugar should be avoided as far as possible.[88] By the conclusion of the Australian sugar industry's $20 million awareness campaign in 1990, two-thirds of Australian consumers thought 'too much fuss' was made about sugar and that at least seven out of ten Australians thought a little sugar was good for you and sugar had a place in a balanced diet. Despite this 'supposed' changed attitude of Australian consumers towards sugar, the Australian sugar industry had not slowed or reversed the decline in Australian apparent sugar consumption per capita. The 'Natural Part of Life' campaign initially lifted overall domestic sugar sales during the mid-1980s, but these increased sales did not come from domestic consumers buying more sugar in the supermarket, but from the non-alcoholic beverage, preserved foods, dairy foods and confectionary sectors purchasing more sugar. Australian annual apparent sugar consumption per capita did rise slightly from 48.1 kg in 1984–1985 to 49.9 kg in 1985–1986 and 1986–1987. However, the gains were short-lived, with annual apparent sugar consumption per capita falling again in the late 1980s and early 1990s. [89]

Declining apparent sugar consumption per capita in Australia during the 1980s did not deter CSR from modernising its Australian sugar refineries to improve efficiencies. Over the period 1980 to 1982, $10.4 million was spent on energy-saving modifications and the replacement of one of the two bone char decolourisation plants at Pyrmont Refinery in 1981 with a granulated carbon decolourisation plant using carbon made from anthracite. This alteration replaced the bone char method to decolourise sugar liquor and had the advantage of requiring less factory space, labour and maintenance and the granulated carbon removed more impurities from the liquor than bone char. These changes contributed to better energy use, especially as the bone char kilns were eliminated, with CSR's sugar refineries reducing energy consumption by about 4% between 1980 and 1982. Granulated carbon plants had been erected at Yarraville and New Farm Refineries by 1984 and the remaining bone char plant at Pyrmont Refinery was replaced by the granulated carbon system in 1986. Computer-based control systems affecting the affination, carbonation, filtration and decolourisation stations were installed at Pyrmont Refinery during 1985.[90] In contrast, Bundaberg Sugar continued to slowly increase the mechanical capacity at Millaquin Refinery, turning out enough refined sugar products to supply 5% of the domestic market (or approximately 40 000 tonnes by 1989). However, the refinery was becoming increasingly out-dated in its technology, lacking automation at some refinery stations and bone char was still being used to decolourise sugar liquor in 1989.[91]

Prices for refined sugar products continued to be controlled by the Commonwealth government during the 1980s. However, the Sugar Agreement of 1979 had changed arrangements for the supply of refined sugar a little, for it allowed the adoption of ex-refinery pricing which gave customers a choice of delivery options. Customers could choose their own carrier to pick up their sugar directly from the refineries, a previously unpermitted practice. During 1980, approximately 20% of the domestic sales of refined sugar were picked up from the sugar refineries by customers. In addition, a new formula was introduced to determine the price of sugar on 1 July each year, after taking into consideration the change in the Consumer Price Index over the preceding twelve months. These conditions were maintained under the final Sugar Agreement which came into force on 1 July 1984, although prices were to be fixed bi-annually, on 1 January and 1 July each year. In addition, a small relaxation of the embargo on sugar imports was permitted to allow for the importation of small consignments of packaged sugar products and the domestic sugar rebate was discontinued. Under these new arrangements, the price of refined sugar rose from $353 per tonne in January 1980 to $629.19 per tonne in June 1989.[92]

During the negotiations to reach the Sugar Agreement of 1984, the Commonwealth government (Hawke; Labor) signaled to the Australian sugar industry that it must embrace some form of deregulation. Despite this warning, the Australian sugar industry was taken by surprise in May 1988 with the announcement of the following new administrative arrangements: the embargo on sugar imports would be removed on 1 July 1989; the administered price arrangements for refined sugar in Australia would cease on 30 June 1989; and the Australian sugar industry would be protected by a tariff, which would decrease in size through time. These changes, however, did not affect the Queensland government's compulsory acquisition of all the raw sugar manufactured in the state. Yet the traditional relationship between Australia's two refiners and the Sugar Board for the refining and marketing of sugar products in Australia ended. From 1 July 1989, Australian refiners could purchase raw sugar to suit their own requirements (i.e. they were not acting on behalf of the Sugar Board) and they assumed responsibility for their own marketing arrangements. Sales of raw sugar were made on a commercial basis related to the equivalent price at which sugar could be imported (i.e. import parity pricing), including the tariff. The Sugar Board, however, retained CSR to provide agency services in relation to export marketing, initially for a period of three years.[93]

Immediately following the announcement of deregulation in May 1988, the NSW Sugar Milling Coop. made another attempt at severing its ties with the Queensland sugar industry, the organisation having grown increasingly discontented with the sugar marketing arrangements imposed by the Sugar Board during the 1980s. A failed attempt had been made during the early 1980s to acquire a partner to market direct consumption raw sugar that the NSW Sugar Milling Coop. contemplated manufacturing at its New South Wales sugar mills. The NSW Sugar Milling Coop. had commenced negotiations with a proposed distributing company, Green's General Foods Pty. Ltd., but CSR scuttled the deal by acquiring 50% of the shares in that company. The second attempt by the NSW Sugar Milling Coop. to end its relationship with the Sugar Board was successful and had far-reaching ramifications for the domestic marketing of refined sugar products in Australia. In October 1988, the NSW Sugar Milling Coop. signed an agreement to build a jointly owned sugar refinery in New South Wales with the Manildra Group, a New South Wales based company that distributed flour, glucose and starch. The Manildra Group would market the refined and direct consumption raw sugars under the brand *Sunshine Sugar*. A site directly beside Harwood Mill was selected as the location for the new sugar refinery. Harwood Mill's boiler had the capacity to supply both the steam and electricity for a refinery. Moreover, the site was close to the nearby Port of Yamba which could receive large ships, thereby enabling the transport of sugar in bulk, and the railway at Grafton, so sugar could be trucked to Grafton, stored and railed to southern centres.[94]

Tate and Lyle Process Technology (Tate and Lyle Industries Ltd.) provided the design and process technology for the new sugar refinery, the first to be built in Australia in almost seventy years. Initially, the sugar refinery was constructed with the capacity to produce 250 000 tonnes of refined sugar each year, the design emphasising the need for minimum staffing levels and computerised process control. Despite extensive building activities commencing near Harwood Mill in late 1988, the Sugar Board had no idea that the NSW Sugar Milling Coop. was planning such a radical decision. Inquiries about the nature of extensions occurring at Harwood Mill received a standard reply: the NSW Sugar Milling Coop. was erecting 'a new sugar storage shed'. Eventually, Greg Messiter, the General Manager of the NSW Sugar Milling Coop., advised the Sugar Board on 20 February 1989 that New South Wales would market sugar independently of the Sugar Board and its agents from 1 July 1989. This announcement sent 'major shock waves through the Queensland cane growing industry', reported the *Australian Sugar Year Book*. Queensland's Primary Industries Minister, Neville Harper, criticised the decision stating, 'it would cause instability in the Australian industry at a time when unity was needed due to the impending lifting of the import embargo'. With the new refinery buildings underway, the Sugar Board could do little to reverse the decision by the NSW Sugar Milling Coop.[95]

The new refinery commissioned by Manildra Harwood Sugars made its first batch of refined sugar at the end of September 1989. Despite some initial commissioning problems, the refinery was melting 250 tonnes of raw sugar daily by November 1989; this figure had risen to 600 tonnes per day by February 1990, or close to the factory's full capacity. From the outset, Manildra Harwood Sugars produced refined sugar products with a customer rather than production focus. The new firm promised and delivered refined sugar in whatever quantity asked for by the customer, on a 24-hour, seven-day-a-week basis. This approach redefined customer service in the marketing of sugar domestically in Australia. Major domestic sugar users liked the new service and Manildra Harwood Sugars quickly obtained 25% of the domestic market during its first two years in operation, holding this market share throughout the rest of the 1990s.[96] The only major downside during its early operations was the need to purchase 40 000 tonnes of raw

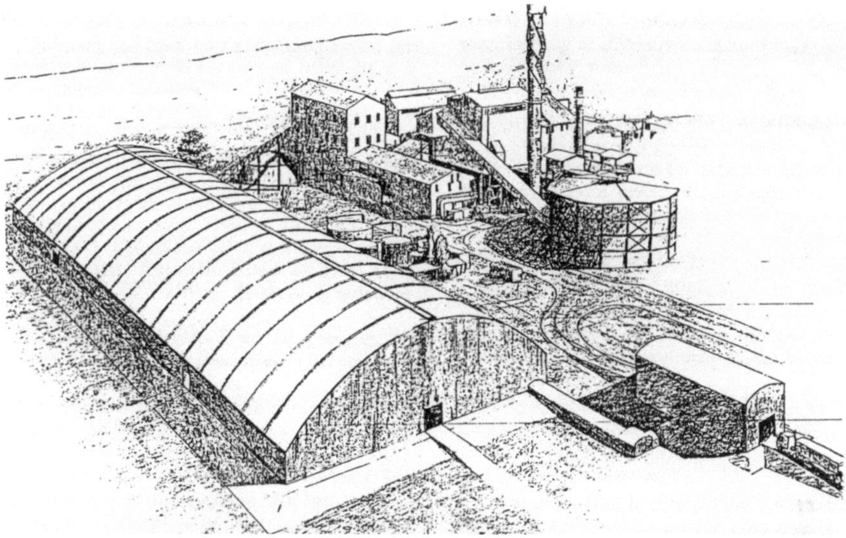

Plate 18.5. A sketch of the Manildra Harwood Sugars Refinery and Harwood Mill complex showing the sugar shed, foreground, the refinery, centre, and sugar mill, background, c. 1990.

Source: *ASYB,* 1990, p. 16. Image reproduction courtesy of Rural Press Queensland Pty. Ltd.

sugar from the Sugar Board during 1990. This shortfall in local raw sugar production was overcome by the mid-1990s, and the refinery manufactured 5 000 tonnes of refined sugar weekly by 1995. Moreover, the new firm had diversified its range of products to include one, two and three kilogram packages that were distributed to retail outlets.[97]

The Manildra Harwood Sugars Refinery sliced about $200 million off CSR's revenue during its first year in operation. Because of this loss, CSR closed Glanville Refinery in April 1991. Yarraville Refinery commenced supplying the South Australian market, as it did a century earlier. Productivity improvements were achieved through additional automation and more efficient work practices at the refineries, thereby keeping the increase in refining costs below the rate of inflation. However, these changes were not enough to correct the imbalance between sales and production capacity, so Pyrmont Refinery was shut in September 1992 and the site was sold for inner city residential development. Capacity was further increased at Yarraville and New Farm Refineries, enabling them to produce enough refined sugar products to serve the demands of consumers in eastern Australia. Part of these upgrades included the introduction of robot palletisers that dramatically increased packaging efficiency, while Yarraville Refinery received a new $8 million carbon decolourisation plant that economically produced liquid sugar for bulk delivery to brewers and food manufacturers. At New Farm Refinery, a change from five to seven days per week continuous operation commenced in August 1992.[98]

During these changes, CSR also investigated the possibility of reducing refining costs by refining sugar in a raw mill. Modifications costing $2.4 million were made to Kalamia Mill, enabling a new refining process to be trialed in March and April 1992, but this new process technology was not implemented after these initial investigations. CSR also considered the possibility of producing higher grade, semi-refined sugar known as Very Low Colour (VLC) sugar at its sugar mills, reducing the amount of further processing needed at the remaining refineries. By 1995, however, these semi-refined sugars had only been made towards the end of the crushing season at Plane Creek Mill.[99]

Unlike CSR, deregulation prompted Bundaberg Sugar to announce in May 1990 that it intended to expand operations by supplying the domestic market with more sugar products, lifting its market share above 5%. Bundaberg Sugar had acquired Hardings Manufactures Pty. Ltd., a small independent retail packager and wholesaler of dry goods (including sugar, salt, icing sugar and dried fruit) in 1987, so the firm now had a distribution agent. The markets of Southern Queensland were located close to Moreton Mill at Nambour, so the factory was selected to produce, using its current raw sugar as feedstock, food grade raw sugar for industrial use at prices less than refined sugar. To achieve this change, Moreton Mill was equipped with a vibrating screen fitted with magnets to remove any foreign matter and a new building to house the bagging plant. In addition, about $12 million was also spent over 1991 and 1992 to almost quadruple the capacity of Millaquin Refinery to approximately 150 000 tonnes.[100] This increase involved replacing the bone char cisterns with a granular activated carbon system and the installation of a new bagging line, two automatic palletisers, extra storage bins, an upgraded process control system and a new vacuum pan. These upgrades greatly enhanced the refinery's capacity for the removal of colour and impurities, thereby contributing to the production of a better quality refined sugar. They were combined with the release in 1992 of a new range of products (e.g. tubs of sugar; single serve sachets) packaged in vibrantly coloured packaging (i.e. yellow, purple and green). By late 1994, Bundaberg Sugar had captured about 15% of the domestic market in refined sugar products, including the lucrative contract to supply liquid sugar to a new Coca-Cola Amatil bottling plant in Brisbane.[101]

As the expansion at Millaquin commenced, two new firms entered the refined sugar business. In August 1991, the Golden Circle Cannery in Brisbane decided to erect a small liquid sugar plant to refine raw sugar for use in the cannery. Golden Circle, however, had no intention of competing with the other Australian sugar refiners on the domestic market. In contrast, Mackay Sugar announced in July 1992 that the company intended to construct a new sugar refinery capable of producing 350 000 tonnes of refined sugar per annum on the Racecourse Mill site, a 45 000 tonnes white sugar silo at the Port of Mackay, and a bulk refined sugar receiving and distribution depot at Sydney. These new facilities were part of a $100 million joint venture with ED & F Man Holdings Ltd., a UK-based commodity group that had considerable experience in sugar refining and the marketing of refined sugar.[102] The Mackay Refined Sugars Pty. Ltd. joint venture agreements were signed on 15 September 1992 and work commenced on building the $51 million new refinery.

Racecourse Refinery incorporated the latest technological advances in refining, including decolourisation over granular activated carbon and labour-saving technologies. It was designed for continuous operation, the factory retrieving raw sugar from the Mackay Bulk Sugar Terminal during the non-crushing season. Following the successful commissioning of the refinery in June 1994, the factory then melted at close to full capacity during its first eighteen months in operation. Twenty-seven full-time employees operated the Racecourse Refinery, compared to around 200 employees at a CSR refinery, thereby giving it a significant cost advantage over CSR. However, a lower than budgeted share of the Australian market resulted in a first year loss for Mackay Refined Sugars, despite its reduced labour costs and the product meeting market expectations.[103]

During late 1994, the three other Australian sugar refiners commenced price discounting after the packaged sugar from Mackay Refined Sugars commenced being released onto the domestic market. CSR in particular slashed its refined sugar prices to the point of loss in an effort to lock Mackay Refined Sugars out of the domestic market. Following media speculation about these practices, the Australian Trades Practices Commission announced an inquiry in November 1994 into possible unfair competition in the refined sugar market. At the end of April 1995, the Commission closed its investigations, taking no action against CSR. After the inquiry had been announced, ED & F Man Holdings Ltd. and Mackay Refined Sugars commenced legal proceedings against CSR in the Federal Court of Australia in Brisbane, seeking $60 million in damages. The litigants alleged that CSR sold substantial quantities of refined sugar below the cost of production, provided substantial customer rebates on previously agreed contracts, and falsely suggested that Mackay Refined Sugars was unable to meet its contracts. The Federal Court of Australia scheduled the case to be heard in April 1997, but subsequent negotiations between both parties led to the abandonment of the court case and a merger of refining interests to create Sugar Australia Pty. Ltd. in March 1998.[104]

Deregulation and then escalating competition in the domestic refined sugar market prompted a renewed interest in the export of refined sugar. The Sugar Industry Task Force set up in 1992 considered the matter, but was not particularly enthusiastic about the proposal in their final report, reminding Simon Crean, the Commonwealth's Minister for Primary Industries and Energy, that the cost of producing refined sugar was not always 'necessarily recoverable in the market place on a sustained basis'.[105] CSR, however, was pragmatic enough to realise that the expanded refining capacity in Australia had created the potential to supply the lucrative white sugar export market. Hence it commenced negotiations with Mackay Refined Sugars about the possibility of a new refined sugar production, marketing and distribution venture in Australia. In August 1993, the firms announced the formation of a joint venture in which CSR would have a 50% interest in the new company, while Mackay Sugar and ED & F Man Holdings Ltd. each would have a 25% holding in the venture. The new arrangement would use CSR's existing refineries and the refinery being constructed by Mackay Refined Sugars, the refined bulk sugar storage facilities of both firms in Mackay, Sydney and Brisbane, and a carrier ship that had an onboard bagging plant allowing bulk loading in Australia and discharge of either bulk or bagged sugar at unloading.[106]

After considering the proposed merger, the Australian Trade Practices Commission rejected this joint sugar venture in December 1993 because it considered competition in the refined sugar market would be lessened substantially, and that the proposed arrangement protected the position of the dominant refiner CSR, allowing it to close New Farm Refinery and gain access to more efficient capacity at Racecourse Refinery. This outcome meant that Mackay Refined Sugar's new refinery increased refining capacity in Australia to 1.3 million tonnes, considerably beyond the annual domestic demand of around 850 000 tonnes. With the domestic sugar market trading at below export parity due to the price war between refiners, the only option was to export the refined sugar. During late 1994 and 1995, refined sugar produced by Mackay Refined Sugars was exported to Indonesia, China, Hong Kong, the Philippines, Sri Lanka and India. A similar export drive into New Zealand was successful, with the firm capturing some 40% of the industrial market, although returns were well below cost because of the sustained price war involving the New Zealand Sugar Co., a CSR subsidiary. Bundaberg Sugar also commenced exporting to Singapore, the Philippines, Papua New Guinea and New Zealand. Consequently, during 1994/1995, Australian refineries exported more than 160 000 tonnes of refined sugar, compared to around 40 000 tonnes in 1990/1991.[107]

Throughout the 1980s and early 1990s, the Sugar Board (later the Queensland Sugar Corporation) arranged for the continued export of raw and refined sugar from the bulk sugar terminals, although exports of refined sugar products became the responsibility of the individual refiners after deregulation. To handle the growing volume of exported sugar, the Sugar Board authorised a range of upgrades at the bulk sugar facilities during the 1980s and early 1990s. These improvements included the construction of new rail loop receiving systems at Mackay and Townsville (1981–1982), additional storage for 70 000 tonnes at Mackay (1983) and new road receival facilities at Mackay (1991). A greater degree of automation in terminal operations and the implementation of 24-hour (three shifts) receiving of sugar commenced in the mid-1980s. At the same time, the progressive introduction of B-double road truck haulage receival facilities at the Cairns, Mourilyan, Mackay and Bundaberg Bulk Sugar Terminals increased transport efficiency and resulted in significant cost savings for the sugar industry.[108]

Increased sugar production in Southern Queensland outstripped the storage capacity of the New Farm Refinery, forcing Moreton Mill to send its sugar output by road to Bundaberg during 1982 and 1983. This costly exercise resulted in the Sugar Board deciding to build a new bulk sugar terminal towards the mouth of the Brisbane River. Capable of storing 120 000 tonnes of raw sugar received from the New South Wales and Southern Queensland sugar mills, the new sugar terminal received its first sugar on 20 June 1985. The new deep water facilities which lifted Queensland's bulk sugar storage capacity to 2.09 million tonnes were designed to accept larger vessels than those using the Clarence River in northern New South Wales. However, the Brisbane Bulk Sugar Terminal became a 'white elephant' following deregulation in 1989, with the NSW Sugar Milling Coop. preferring to export its sugar out of the Port of Yamba, near Grafton.[109]

During the 1980s, Australia's main customers for its sugar exports were Japan, Canada, China, Malaysia, Singapore and South Korea (see Figure 18.5). These export destinations were again secured by long-term contracts. An important new customer after 1981 was the USSR. Following the break up of the Soviet Union, Russia continued to purchase Queensland sugar,

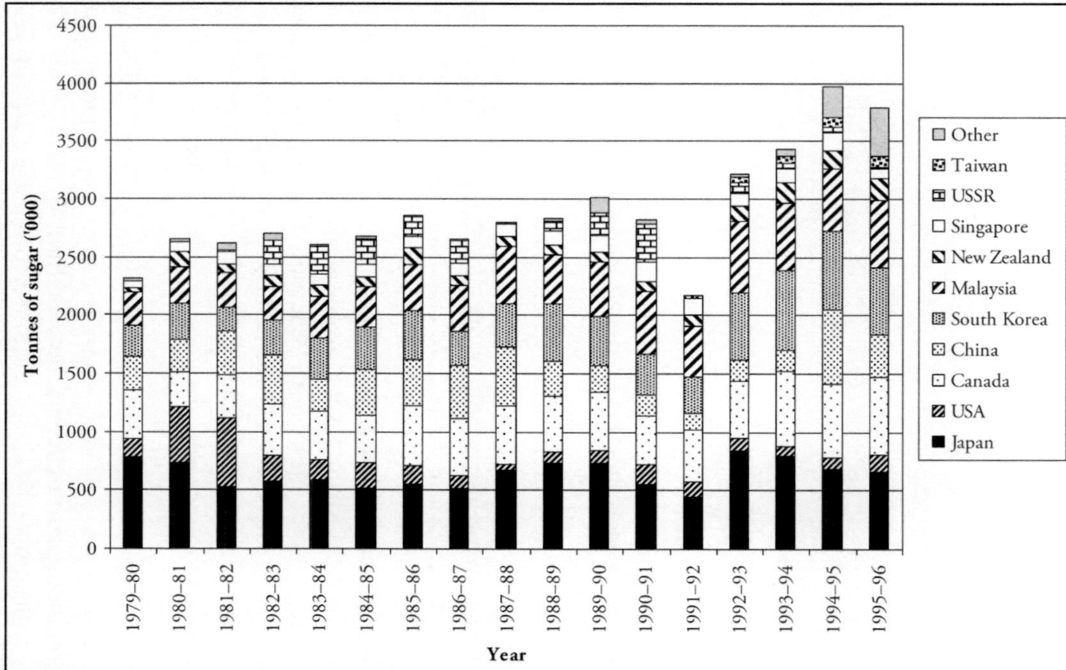

Notes: (1) Figures to 1990–91 are for exports of raw and refined sugar; from 1991–92 onwards the figures are for exports of raw sugar only. (2) Figures for the USSR after 1990–91 are for Russia only. (3) The 'Other' category includes exports to island countries of Oceania, Papua New Guinea, Dubai, Egypt, Croatia and Slovenia.

Figure 18.5. Exports ('000 tonnes) of Australian raw and refined sugar by destination, 1979/1980–1995/1996.

Source: Based upon figures in the *ASYB*, various issues.

although not on the same scale as the USSR. Exports to the USA, however, fell following the introduction of increasingly restrictive import quotas which reduced Australia's access to the American market. Refined sugar exports also fell after 1983, following the decision of Papua New Guinea to cease the import of sugar after the establishment of its domestic sugar industry. Yet these losses were compensated by the emergence of new destinations for Queensland's raw sugar, a consequence of the Queensland Sugar Corporation actively targeting potential new markets, particularly former communist countries in Europe and Middle Eastern countries. The first sale of sugar to Taiwan was made in 1991. Raw sugar was sold directly to a processor in Bulgaria for the first time in 1993. Direct sales of raw sugar were made for the first time to Egypt, Dubai and Vietnam in 1994 and the Philippines in 1995. Moreover, following severe economic dislocation in Cuba and dry weather conditions in Thailand, Queensland emerged as the largest global exporter of raw sugar during 1994 and 1995.[110]

In retrospect, the discussion at the beginning of this chapter showed that the Commonwealth government in 1915 commenced regulating the refining, marketing and pricing of sugar in Australia. This intervention which commenced as a wartime measure lasted for another seventy-four years, albeit with some modifications. CSR's dominance over the supply of refined sugar products in Australia remained unchallenged during this period. Such a regulated system was partially deregulated in 1989, weakening CSR's hold over the Australian refined sugar market. Yet 'single desk' selling persisted, as the Queensland Sugar Corporation continued being the sole purchaser of raw sugar in Queensland. This arrangement remained the last vestige of an elaborate regulated system for the marketing and refining of sugar in Australia, protecting Australian sugar mill-owners and refiners against sugar imports, but possibly denying Australian consumers access to cheaper sugar for much of the twentieth century.

Notes

1 CSR, Half Yearly Report, 31 March 1915, p. 1, CSRR, P10/75; *ASJ*, 4 February 1915, pp. 759 & 760; Shogren, 'The politics and administration of the Queensland sugar industry to 1930', p. 246.

2 *Commonwealth Parliamentary Debates*, Vol. 76 (1914–1915), pp. 3203–3204, 3207 & 3343; Shogren, 'The politics and administration of the Queensland sugar industry to 1930', pp. 247–249; Murphy, *T.J. Ryan*, p. 112.

3 Shogren, 'The politics and administration of the Queensland sugar industry to 1930', p. 250; Murphy, *T.J. Ryan*, p. 113.

4 Shogren, 'The politics and administration of the Queensland sugar industry to 1930', p. 252; Memo. of Agreement between the Commonwealth of Aust. and CSR, *CPP*, Vol. 4 (1914–1917), pp. 1067–1068; Memo. of Agreement between the Commonwealth of Aust. and the Millaquin Sugar Co. Ltd., 15 July 1915, A3280, P3850, National Archives of Australia, Canberra.

5 H. Herbert, 'The Australian sugar industry', in A.G. Lowndes (ed.), *South Pacific Enterprise. The Colonial Sugar Refining Company* (Sydney, 1956), p. 112.

6 Minority Report from the Joint Committee of Public Accounts Upon Sugar, October 1922, pp. 5–6, & Commonwealth of Aust., Sugar Control Accounts, 1915–1922, pp. 20–29, both in TR 1809/1, Box 2, QSA.

7 CSR, Half Yearly Report, 31 March 1919, p. 2, CSRR, P10/75; *ASJ*, 8 September 1922, p. 365; Commonwealth of Aust., Sugar Control Accounts, 1915–1920, p. 33, TR18091/1, Box 2, QSA.

8 CSR, Half Yearly Report, 31 March 1918, p. 2, CSRR, P 10/75; Easterby, *The Queensland Sugar Industry*, p. 59; Shogren, 'The politics and administration of the Queensland sugar industry to 1930', pp. 296–298; Report of the Piddington Royal Commission, *CPP*, Vol. 4 (1920–21), p. 953.

9 For details about anti-embargo meetings see *ASJ*, 10 July 1919, pp. 201–203 & 7 November 1919, pp. 473–475 & 481–482.

10 Shogren, 'The politics and administration of the Queensland sugar industry to 1930', pp. 299–230.

11 ASPA, *The Australian Sugar Industry* (Brisbane?, 1921); ASPA, *Queensland Sugar Producers Have Been Maligned* (Brisbane?, 1922); ASPA, *White Australia's Great Sugar Industry Only Can Keep Tropical Australia White* (Brisbane?, 1922).

12 Shogren, 'The politics and administration of the Queensland sugar industry to 1930', pp. 315–316.

13 Majority and Minority Reports from the Joint Committee of Public Accounts Upon Sugar, October 1922, various pages, in TR 1809/1, Box 2, QSA.

14 The Fruit Industry Sugar Concession Committee, *The Fruit Industry Sugar Concession Committee from 1931 to 1963* (Canberra, 1964), p. 19; Easterby, *The Queensland Sugar Industry*, p. 62.

15 *QGG*, Vol. CXXI (1923), pp. 15 & 464.

16 Easterby, *The Queensland Sugar Industry*, p. 66; The Fruit Industry Sugar Concession Committee, *The Fruit Industry Sugar Concession Committee from 1931 to 1963*, p. 20.

17 C. Robertson, *World Sugar Production and Consumption: An Economic and Geographical Survey* (London, 1934), pp. 107 & 111; Kerr, *Southern Sugar Saga*, p. 95.

18 CSR Ltd., *Glanville Refinery from Beginning to End: 100 Years of Sugar Refining in South Australia* (Adelaide, 1991), p. 3; Kerr, *Southern Sugar Saga*, pp. 95–96.

19 CSR, AR 1925–26, p. 1, CSRR, 142/3529; Allom, Lovell and Marquis-Kyle, *New Farm Sugar Refinery Conservation Study* (Brisbane, 1990), p. 13; C. Davis, 'The Pyrmont Sugar Refinery with particular reference to char'. PhD thesis, Uni. of Sydney, 1984, p. 192.

20 CSR, Half Yearly Report, 30 September 1927, p. 2, CSRR, 142/3529.

21 *ASJ*, 5 June 1930, pp. 187–188; CSR, Half Yearly Report, 31 March 1929, p. 2 & CSR, Half Yearly Report, 31 March 1930, p. 2, both in CSRR, 142/3529; Memo: cost of

Cottesloe, 13 June 1930, CSRR, Z/303, Box 49, File E 2.0. Folder 21, Un-numbered document; CSR, *History of Cottesloe Refinery*, 12 pp. pamphlet, date unknown, CSRR, Z/109, Box 369.

22 Qld. Agent-General, London, AR 1928, *QPP*, Vol. 2 (1929), pp. 993–994; Qld. Agent-General, London, AR 1934, *QPP*, Vol. 2 (1934), p. 809; BSES, AR 1928–29, *QPP*, Vol. 2 (1929), p. 983; P. Goldfinch, CSR Acting Gen. Mang., to Chairman of Directors, Amalgamated Sugar Mills Ltd., Melbourne, 19 May 1928, AER, 165/194, pp. 398–399, NBAC.

23 Commonwealth of Australia (Department of Commerce), *Report on the Market for Sugar in the United States* (Canberra, 1939), p. 11.

24 Based upon CSR's Annual Refinery Reports, 1927–1938, in Z/109, Boxcs 396 & 397.

25 H. League, *The sugar embargo: is Queensland a parasitic state?* (Melbourne, c. 1930), pp. 5–7; Sugar Consumers' Association (Australia), *Sugar embargo: the story of a greedy industry* (Melbourne, c. 1930), pp. 7 & 8; for a copy of this album see Queensland Cane Growers' Council, 'The Queensland Sugar Industry and those it benefits from Mossman to Perth, 1930', National Library of Australia.

26 The Queensland Sugar Defence Committee, *The Truth About the Queensland Sugar Industry* (Melbourne, 1930).

27 *ASJ*, 6 October 1932, pp. 349–350; The Fruit Industry Sugar Concession Committee, *The Fruit Industry Sugar Concession Committee from 1931 to 1963*, pp. 2–3.

28 *ASJ*, 2 July 1931, p. 234, 5 November 1931, p. 446–447 & 3 December 1931, pp. 463 & 508–509.

29 *ASJ*, 4 February 1932, p. 579; for an example of one of these advertisements see *Labour Daily*, 22 April 1932.

30 *ASJ*, 4 August 1932, pp. 223–228 & 267–269, 8 September 1932, pp. 274–274a & 279–291 & 6 October 1932, pp. 335–337; Telegram from Australian Sugar Consumers' Assoc. to Hon. J. Lyons, 30 July 1932, in MS 1186, Papers of the Citizen's League of South Australia, National Library of Australia.

31 ASPA and Queensland Cane Growers' Council, *The Australian Sugar Industry* (Brisbane, 1932); ASPA, *"Reciprocity" is the new catch–word in trade: interdependence should dissipate jealousies* (Brisbane, 1934?); Sugar Growers of Australia, *The Australian Sugar Industry* (Brisbane, 1933 & 1934); ASPA and Queensland Cane Growers' Council, *Queensland Sugar Industry and Interstate Relationships* (Brisbane, 1934); The Sugar Industry Organisations, *The Australian Cane Sugar Industry* (Brisbane, 1938).

32 *ASJ*, 4 September 1934, pp. 300–301, 9 November 1935, pp. 413–421, 12 December 1935, pp. 469–471 & 11 January 1936, pp. 529–531 & 561–564; for an example of anti-embargo material see E. Craigie, *The Case Against the Embargo* (Melbourne, 1935).

33 P. Saranin, 'A flocculation-flotation process for clarification of sugar refinery melt using phosphoric acid and lime'. M. Eng. thesis, Uni. of Qld., 1963, p. 6; *ASYB*, 1969, p. 191; CSR, 'Refinery Production Data, 1935–1980', Table 26, CSRR, Z/109, Box 402; P. Griggs, 'A Natural Part of Life: The Australian sugar industry's campaign to reverse declining Australian sugar consumption', *J. of Australian Studies*, No. 87 (2006), p. 144.

34 M. McKernam, *All In. Fighting the War at Home*. (1983; reprinted, Sydney, 1995), p 158; R. Hutchinson, *Food for the People of Australia* (Sydney, 1958), p 17.

35 CSR, Half Yearly Report, 31 March 1942, pp. 1–2, CSRR, 142/3530; Ev. J. Dixon, CSR, in 'Sugar Inquiry Committee Evidence, 1952', p. 673, CSRR, Z/109, Box 57; CSR, *Newsletter* No. 25, March 1951, p. 24, CSRR, Z/109, Box 371.

36 *ASJ*, 11 June 1940, pp. 151–152 & 159–161.

37 *ASYB*, 1945–46, p. 32; Tariff Board's Report: Subsidy for the Raw Sugar Producing Industry, 30 August 1944, pp. 1–3 & 26–29, in CSRR, Z/109, Box 52.

38 W. Rothe, CSR Gen. Mang., Memo to All Shareholders re: sugar supply, 7 November 1947, pp. 1–3, & CSR, AR 1942–43, p. 2, in CSR 142/3530.

39 BSES, AR 1946–47, p. 9; W. Rothe, CSR Gen. Mang., Memo to All Shareholders re: sugar supply, 7 November 1947, p. 1; W. Rothe, CSR, Memo to All Shareholders re: sugar supply, 4 November 1948, pp. 1–2; W. Rothe, CSR, Memo To All Shareholders, re: sugar supply, 3 November 1949, p. 1; CSR, AR 1946–47, p. 2; CSR, AR 1949–50, p. 1, all in CSRR 142/3530; CSR, *Newsletter* No. 16, November 1948, pp. 126–127, CSRR, Z/109, Box 371; *ASYB*, 1949, p. 119.

40 Ev. G. Goding, Sec. AWU, Bundaberg, in 'Sugar Inquiry Committee Evidence, 1952', p. 206, CSRR, Z/109, Box 55; A. Jones, Qld. Minister for Labour & Industry, to R. Muir, Sec., QCGC, 28 February 1952, in File 5R/1/3, No. 2, PRV 14721/1/75, File: Sugar Refineries, Box 37, QSA; W. Forgan Smith, Chairman QSB, to Hon. E. Walsh, Queensland Treasurer, 5 June 1952, in SRS 1043/1/487, Box 155, QSA.

41 For a large collection of newspaper articles and letters about this issue see Files 5R/1/3, No. 1 & No. 2, in PRV14721–1–75, Box 37, File: Sugar Refineries, QSA.

42 *ASJ*, 15 April 1954, pp. 23–25 & 14 April 1960, p. 37; *The Producers' Review*, 15 March 1970, p. 107.

43 BSES, AR 1946–47, p. 9; CSR, AR 1946–47, pp. 2–3, CSRR, 142/3530; *ASYB*, 1953, p. 165; CSR Ltd., *Bulk Handling of Raw Sugar* (Sydney, 1959), p. 4.

44 CSR Ltd., *Bulk Handling of Raw Sugar*, pp. 4–5; *ASYB*, 1953, p. 165.

45 K. Mulherin, 'Bulk handling in the Australian sugar industry', *Quarterly Review of Agricultural Economics*, Vol. 10, no. 3 (July 1958), pp. 113; Kerr, *Pioneer Pageant*, p. 198; *ASYB*, 1953, p. 19; *ASYB*, 1954, p. 131; CSR Ltd., *Bulk Handling of Raw Sugar*, p. 5.

46 CSR Ltd., *Bulk Handling of Raw Sugar*, pp. 5–6; *ASJ*, 15 December 1952, pp. 639–643; *ASYB*, 1953, p. 122; *ASYB*, 1953, pp. 119–121; *ASYB*, 1955, p. 27.

47 CSR Ltd., *Bulk Handling of Raw Sugar*, p. 6; D. Hull, 'Queensland sugar ports: labour and technological change', *J. of Australian Political Economy*, No. 6 (1979), pp. 61–62; *ASYB*, 1956, p. 139.

48 *ASJ*, 15 May 1959, p. 176 & 25 July 1961, pp. 297–298; *ASYB*, 1962, p. 47; CSR, *Newsletter* No. 70, April 1962, p. 3, in CSRR, Z/109, Box 371.

49 Mulherin, 'Bulk handling in the Australian sugar industry', p. 114; CSR Ltd., *Bulk Handling of Raw Sugar*, p. 7.

50 *ASJ*, 15 July 1957, p. 271; Kerr, *Pioneer Pageant*, p. 198.

51 *ASJ*, 25 October 1961, p. 503, 28 February 1962, p. 796 & April 1963, p. 13; Mulherin, 'Bulk handling in the Australian sugar industry', p. 114.

52 Mackay Institute of Sugar Mill Engineers, 'The bulk handling of sugar in the Mackay district mills', *Proc. QSSCT*, 1958, pp. 13–19; Kerr, *Only Room for One*, p. 116; Alcorn and Dunn, *Moreton Sugar Mill*, p. 151; *ASJ*, 15 August 1958, pp. 401 & 428.

53 E. Chapman, 'Queensland ports and bulk shipment of Australian raw sugar', *Geography*, Vol. 47 (1962), p. 31; R. Fitzgerald, *From 1915 to the Early 1980s. A History of Queensland* (Brisbane, 1984), pp. 200–201; CSR, Raw Sugar Manufacture Report 1954, p. 136, CSRR, Z/109, Box 382.

54 *ASJ*, 25 July 1961, p. 340, P. Ryle, *The Sugar Lighters of Cairns and District* (Cairns, 2001) & C. Lack, 'From mosquito fleet to bulk handling in the sugar industry', *J. Royal Historical Society of Queensland*, Vol. 6, no. 3 (1960–61), pp. 625–626.

55 For the problems of shipping in the Burnett River see J. Kerr, *Bundaberg: The Persistent Port* (Bundaberg, 1996), pp. 52–61.

56 Kerr, *Northern Outpost*, p. 132–133; *ASJ*, July 1958, p. 361.

57 Chapman, 'Queensland ports and bulk shipment of Australian raw sugar', pp. 311–312; Kerr, *Sugar at Maryborough*, p. 219; *ASJ*, 29 September 1961, p. 449; *ASYB*, 1960, p. 3; Lack, 'From mosquito fleet to bulk handling in the sugar industry', pp. 625–626 & 632; A. Donnelly, *The Port of Townsville* (Townsville, 1959), p. 314.

58 *ASJ*, March 1964, p. 885.

59 *ASYB*, 1977, p. 67; *ASYB*, 1978, pp. 213–214; B. Hoyle, 'Ports and hinterlands in an agricultural economy: the case of the Australian sugar industry', *Geography*, Vol. 69, no. 4 (October 1984), p. 314.

60 *ASYB*, 1964, p. 101; CSR, 'Evidence to 1961 Sugar Enquiry Committee. Statement No. 5: Developments in export raw sugar', pp. 3 & 11, CSRR, Z/109, Box 59.

61 CSR, Annual Refinery Report 1947, p. 10 & CSR, Annual Refinery Report 1950, p. 9, in CSRR, Z/109, Box 398.

62 *ASYB*, 1962, p. 97; *ASYB*, 1963, p. 103; *ASJ*, 26 June 1961, p. 275; CSR, Raw Sugar Manufacture Report 1965, pp. 81–82, CSRR, Z/109, Box 383.

63 Kerr, *A Century of Sugar*, p. 175; *ASYB*, 1968, p. 93.

64 IAC, *The Sugar Industry. Report No. 209* (Canberra, 1979), p. 19.

65 *ASJ*, March 1975, p. 571.

66 Australia. Sugar Enquiry Committee, *Summary of Conclusions and Recommendations from Report of the Sugar Enquiry Committee of 1960* (Canberra, 1961), pp. 7 & 10; *ASJ*, March 1975, p. 571.

67 *ASYB*, 1947–48, pp. 162–163; ASYB, 1953, pp. 40–41; D. Stalley 'The Sugar Industry', in A. Hunter (ed.), *The Economics of Australian Industry* (Melbourne, 1963), p. 380.

68 A. McMichael, 'Changes in gastrointestinal cancer risks in migrants to Australia: clues to dietary influences', *Food Technology in Australia*, Vol. 35, No. 2 (February 1983), p 89.

69 R. Hutchinson, *Food for the People of Australia* (Sydney, 1958), p. 80; Griggs, 'A Natural Part of Life', p. 144.

70 W. Rothe, CSR Gen. Mang., Memo to All Shareholders re: sugar supply, 18 November 1946, pp. 2–3, CSRR, 142/3530; Ev. J. Dixon, CSR, in 'Sugar Industry Inquiry Evidence 1952', pp. 1009–1010, in CSRR, Z/109, Box 58; Davis, 'The Pyrmont Sugar Refinery with particular reference to Char', p. 202.

71 CSR, AR 1951–52, p. 2, CSRR, P10/75; CSR, *Newsletter* No. 36, June 1954, p. 118, CSR, *Newsletter* No. 47, December 1957, pp. 8–9 & CSR, *Newsletter* No. 62, December 1960, pp. 32–33, all in CSRR, Z/109, Box 371; *ASYB*, 1953, p. 152; CSR, *Raw and Refined*. Issue No. 59, August 1958, last page, CSRR, Z/454, Box 13.

72 CSR, AR 1947–48, p. 2, CSRR, 142/3530; *ASYB*, 1949, p. 119; Ev. J. Dixon, CSR, in 'Sugar Industry Inquiry Evidence, 1952', pp. 745–751, CSRR, Z/109, Box 57.

73 Based upon correspondence and sugar refining agreements in SRS 1043/1/487, Box 155, QSA.

74 CSR, *Newsletter* No. 67, October 1961, pp. 3–6, CSRR, Z/109, Box 371; CSR, 'Evidence to Sugar Inquiry Committee 1960: Statement No. 2: The cost of sugar refining and administration', p. 4, CSRR, Z/109, Box 59.

75 Davis, 'The Pyrmont Sugar Refinery with particular reference to Char', pp. 228 & 356; J. Kenny, 'The use of fine bone char in a continuous counter-current process for the decolorization of sugar liquors'. PhD thesis, Uni. of Sydney, pp. xiii to xv; 'Proportion of Char used per 100 melts' in Annual Refinery Process Tables, 1961–1980, CSRR, Z/109, Boxes 398–402.

76 CSR Ltd., *1893–1993: spirit of New Farm: celebrating a hundred years at CSR's New Farm sugar refinery* (Brisbane, 1993), p. 8; CSR, *Newsletter* No. 84, August 1964, p. 22, CSR, *Newsletter* No. 90, August 1965, pp. 14–15 & CSR, *Newsletter* No. 111, April 1969, pp. 21–22, all in CSRR, Z/109, Box 372; CSR, 'Refinery Production Data, 1935–1980', Table 26, CSRR, Z/109, Box 402; CSR, Report on Refinery Operations 1976–77, p. 1, CSRR, Z/109, Box 402.

77 Davis, 'The Pyrmont Sugar Refinery with particular reference to Char', p. 356; CSR, Report on Refinery Operations 1976–77, p. 1 & CSR, Report on Refinery Operations 1977–78, pp.1, 8–10 & 39, both in CSRR, Z/109, Box 402.

78 CSR, AR 1949–50, p. 1, CSRR, 142/3530; CSR, *Newsletter* No. 21, March 1950, pp. 15–16 & CSR, *Newsletter* No. 47, December 1957, p. 14, in CSRR, Z/109, Box 371; CSR, Annual Refinery Report 1950, p. 91, CSRR Z/109, Box 398 & CSR, Refinery Process Tables, 1960, CSRR, Z/109, Box 399.

79 Michels Warren Pty. Ltd., *The Sugar House. CSR Pyrmont Refinery 1878–1992* (Sydney?, 1992), pp. 17 & 25; CSR, *Newsletter* No. 107, August 1967, p. 3, CSRR, Z/109, Box 372; CSR, AR 1969–70, p. 23; CSR, *Raw and Refined* No. 3, Nov–Dec. 1978, CSRR, Z/454, Box 12; IAC, *The Sugar Industry. Report No. 209* (Canberra, 1979), p. 31.

80 CSR, *Newsletter* No. 59, June 1960, p. 14–15, CSRR, Z/109, Box 371.

81 *ASYB*, 1963, p. 137; CSR, *Raw and Refined.* Issue No. 59 (August 1958), CSRR. Z/454, Box 13; CSR, *Newsletter* No. 69, February 1962, p. 11, CSRR, Z/109, Box 371; CSR, *Newsletter* No. 104, February 1968, p. 3 & CSR, *Newsletter* No. 149, p. 30, both in CSRR, Z/109, Box 372; ASJ, December 1972, p. 440.

82 Lance Jones & Co., *The Australian Sugar Industry* (Brisbane?, 1975), p. 61; David Watson, 'Forbes, Sir Douglas Stuart Mcgregor Gordon (1890–1973)', *ADB*, Vol. 14 (1996), pp 197–198.

83 IAC, *The Sugar Industry Report No. 209* (Canberra, 1979), p. 23; QSB, AR 1981–82, p. 9; J. Dixon, 'Sugar marketing operations', in A. Lowndes (ed.), *South Pacific Enterprise. The Colonial Sugar Refining Company Limited* (Sydney, 1956), p. 151; J. Donaldson, Chairman, QSB, to Under-Secretary, Premier's Dept., 19 June 1956, SRS 1043/1/487, Box 155, QSA.

84 Kerr, *Southern Sugar Saga*, p. 129; Millaquin Sugar Co. Ltd., Chairman's Address, Annual Shareholders' Meeting 14 September 1961, ML; Saranin, 'A flocculation-flotation process for clarification of sugar refinery melt using phosphoric acid and lime', pp. 6, 16 & 213; *ASYB*, 1969, pp. 189–190.

85 QSB, AR 1979–1980, p 7; *ASYB*, 1984, p 132; Griggs, 'A Natural Part of Life', pp. 144 & 146–147.

86 *ASYB*, 1982, p. 233; *AC*, June 1985, p 42.

87 Griggs, 'A Natural Part of Life', pp. 141–154.

88 *AC,* June 1985, p. 43.

89 Australian Sugar Industry Associations, *Living Directions Australian Sugar Industry* (North Sydney, 1990), p. 18; QSB, AR 1987–1988, p. 4; Griggs, 'A Natural Part of Life', p. 144.

90 CSR, AR 1982–1983, p. 17; Allom, Lovell & Marquis-Kyle, *New Farm Sugar Refinery Conservation Study*, p. 15; CSR, *Raw and Refined* No. 14, Spring 1980, p. 1, CSR, *Raw and Refined* No. 21, Spring 1985, p. 1 & CSR, *Raw and Refined* No. 22, Autumn 1986, p. 1, all in CSRR, Z/454, Box 13.

91 For details about the need for Millaquin Refinery upgrades see P. Manson. and R. Swindells., 'Recent developments at Bundaberg Refinery', in The Institution of Engineers, Australia, *Conference papers: development in the Wide Bay Burnett/Central Engineering Conference* (Brisbane, 1993), p. 7.

92 QSB, AR 1979–1980, p. 6; QSB, AR 1983–1984, p. 8; QSB, AR 1988–1989, p. 14; *Queensland Year Book*, 1986, p. 373–374.

93 CSR, AR 1989–1990, p. 24.

94 Smith, *Sweet Beginnings*, p. 396; Briody, *Sunshine Sugar*, pp. 82–84.

95 Briody, *Sunshine Sugar*, pp. 85–89; *ASYB*, 1990, p. 12.

96 Briody, *Sunshine Sugar*, pp. 91–94.

97 *AC*, 6 March 1995, p. 5; Minutes of the Annual General meeting of the Richmond River Branch of the NSWCGA, 6 April 1995, in Minute Book, Richmond River Branch of the NSWCGA, August 1993–March 1996, Wardell; Industry Commission, *The Australian Sugar Industry*. Report No. 19, March 1992 (Canberra, 1992), p. 110.

98 N. Austin, 'Sugar made sweeter', *The Bulletin*, 11 September 1990, p. 52; CSR, AR 1990–91, p. 18; CSR, AR 1991–92, p. 18; CSR, AR, 1992–93, p. 18; CSR Ltd., *1893–1993: spirit of New Farm*, pp. 9–10.

99 *ASYB*, 1993, p. 13; CSR, AR 1994–1995, p. 18; *AC*, 11 December 1995, p. 6.

100 *AC*, 12 September 1988, p. 1; Bundaberg Sugar, AR 1989–1990, p. 8; *ASYB*, 1993, p. 28; Alcorn and Dunn, *Moreton Sugar Mill*, pp. 201–202.

101 Austin, 'Sugar made sweeter', p. 82; *Bundaberg Sugar News,* Summer 1991, p. 19 & Winter 1995, p. 21; P. Manson and R. Swindells., 'Recent developments at Bundaberg Refinery', in The Institution of Engineers, Australia, *Conference papers: development in the Wide Bay Burnett/Central Engineering Conference* (Brisbane, 1993), pp. 2–19.

102 Industry Commission, *The Australian Sugar Industry*. Report No. 19, March 1992 (Canberra, 1992), p. 132; Mackay Sugar, AR, 1992, p. 4.

103 *AC*, 4 April 1994, p. 18; Mackay Sugar, AR 1995, p. 14; Mackay Sugar, AR 1996, p. 7; Austin, 'Sugar made sweeter', p. 82.

104 M. Massey, 'Commercial reality hits the sugar club', *Business Review Weekly*, 26 September 1994, p. 83; CSR, AR 1994–95, p. 18; Mackay Sugar, AR 1995, p. 14; *ASYB*, 1996, p. 30; M. West, 'Mackay serves $60 m claim against CSR', *Sydney Morning Herald*, 5 December 1994, p. 21.

105 *ASYB*, 1991, p. 13; ABARE, *The Australian Sugar Industry in the 1990s*, pp. 15–16; Fitzpatrick *et al.*, *Report of the Sugar Industry Working Party* (Brisbane, 1990), pp. 61–62; Sugar Industry Task Force (Chairman: Brian Courtice), *Sugar Industry Task Force Report* (Canberra, 1993), p. 34.

106 *AC*, 23 August 1993, p. 6.

107 *AC*, 29 November 1993, p. 12. Mackay Sugar, AR 1995, p. 14; Boston Consulting Group, *Report to the Sugar Industry Review Working Party, Analysis of Issues and Identification of possible options* (Sydney, 1996), p. 21; CANEGROWERS, *Cane Growers' Information Handbook 1995–1996* (Brisbane, 1996), p. 37.

108 QSB, AR 1981–1982, p. 6; QSB, AR 1982–1983, p. 4; QSB, AR 1984–1985, p. 10; QSB, AR 1986–1987, p. 6; QSB, AR 1987–19888, p. 6; QSC, AR 1991–1992, p. 23, QSC, *Sugar Notes* (Brisbane, 1997), p. 27.

109 *ASYB*, 1983, pp. 168–169; Proprietary Sugar Millers' Association, AR 1983, p. 14; QSB, AR 1985–1986, p. 9; Alcorn and Dunn, *Moreton Sugar Mill*, p. 196.

110 CSR, AR 1983–1984, p. 11; ASMC, AR 1994, p. 6; QSC, AR 1992–1993, p. 9; QSB, AR 1986–1987, p.6; QSC, AR 1993–1994, p. 15; *ASYB*, 1996, p. 52.

Conclusion

In 1995, sugar had been manufactured continuously in Australia since 1864. From a global perspective, sugar production in Australia has occurred over a relatively short period of time; not even 150 years. Commercial sugar production has existed in Brazil, Cuba and parts of the West Indies for at least 300 years. Yet during this short time span, the Australian sugar industry experienced dynamic growth rates, expanding raw sugar production from a mere 3 588 tonnes in 1870 to 114 291 tonnes in 1900. By 1924, Australian sugar producers had satisfied domestic demand for sugar, which stood at around 430 000 tonnes annually. Subsequently, the Australian growth rates for sugar output have been a little less dynamic, but nonetheless substantial. Between 1924 and 1954, the national sugar output trebled from 432 585 tonnes to almost 1.3 million tonnes, even though the Queensland government attempted to slow production for much of the period before 1939. By 1974, the output of sugar had reached 2.8 million tonnes, twice the 1954 level. In 1995, Australia produced almost 5 million tonnes of sugar, a record amount. This steadily growing output – the amount produced regulated between 1945 and 1990 – ensured that Australia emerged as a reliable exporter of raw sugar, eventually becoming one of the world's top four exporters of raw cane sugar during the 1980s and early 1990s.

The location of sugar-producing districts in Australia has been remarkably stable. After a period of pioneering experimentation, the cultivation of sugar cane since 1900 has been overwhelmingly confined to the coastal districts between Mossman and Mackay in North Queensland and between Bundaberg in Southern Queensland and Grafton in northern New South Wales. Expansion in output came about through more intensive use of land and eventually the cultivation of more sloping and waterlogged land in these districts or more marginal land agriculturally.

This stability resulted from a combination of four main factors. Firstly, investors in Australian sugar mills by 1900 recognised that the most suitable localities for sugar production were in Queensland north of the Tropic of Capricorn. Secondly, the Commonwealth government which relieved South Australia of its control over the Northern Territory in 1911 never attempted to promote the cultivation of sugar cane in this region, despite parts of the area being suitable for the cultivation of sugar cane, especially if assisted by irrigation. Thirdly, successive Western Australian governments also showed little interest in encouraging the cultivation of sugar cane in the state's tropical regions until the 1980s. Fourthly, the Queensland government tried to maintain a tight control over sugar production after 1925 via its agencies the Central Sugar Cane Prices Board and the Queensland Sugar Board, thereby seeking to avoid overproduction, a condition which had persistently plagued the international sugar market. When any expansion in output was sanctioned by the Queensland government, equity issues were important, with existing canegrowers mostly receiving preference in the allocation of increased assigned areas. This process ensured that cane growing remained concentrated in existing localities. The persistence of this pattern for eighty years finally ended in the late 1980s and early 1990s. A lack of suitable land for any further expansion in many North Queensland mill supply areas prompted the Queensland government to approve the development of new sugar-producing localities on the Atherton Tableland in North Queensland, and the Western Australian government authorised the establishment of a sugar industry in the Ord River Irrigation Area.

The districts where Australian sugar production eventually emerged, however, were not entirely ideal for the cultivation of sugar cane. Cold winters and frost make the climate marginal for growing sugar cane in northern New South Wales. CSR was forced to accept a two- year growing cycle and the firm's cane variety improvement program placed considerable emphasis on identifying frost-tolerant cane varieties. Regular frosts also troubled the canegrowers in the Bundaberg, Isis and Maryborough districts, requiring the adoption of more frost-tolerant cane varieties. Variable amounts of annual rainfall that was marginal to the successful cultivation of sugar cane affected canegrowers in the Lower Burdekin, Proserpine, Mackay, Bundaberg, Isis and Maryborough districts. The adoption of irrigation overcame this deficiency. Poor drainage led to waterlogged paddocks, which contributed to reduced yields and outbreaks of chlorotic streak and leaf scald diseases in northern New South Wales and North Queensland. This limitation was eventually overcome by the building of an extensive network of drainage channels during the post-World War II period. Together, these modifications ensured that the districts with less than ideal physical environments became consistent producers of sugar, thereby contributing to the substantial expansion in Australia's sugar output, especially after World War II.

During its development, the Australian sugar industry was a laggard in some areas. Breeding new cane varieties at sugar experiment stations, for example, was slow to emerge in Australia. An Australian-bred new hybrid cane variety – H.Q. 426 – suited to Australian conditions was not developed by CSR until 1906, while the BSES's cane variety breeding program only commenced in the mid-1920s, and yielded few useful commercial cane varieties before 1940. In contrast, the first specially-bred hybrid Barbadian and Demeraran cane varieties were cultivated during the 1890s; some were even imported into Australia throughout the late 1890s. For the Australian sugar industry's first twenty years, its pioneering mill-owners relied very much on imported developments in mill technology such as double and treble crushing and triple effets. Sutton's Atmospheric Evaporating Pan, the first real Australian sugar milling innovation, was not installed in some Australian sugar mills until the mid-1880s. A few additional Australian milling innovations such as Henry Braby's bagasse-drying contrivance, CSR's continuous conical subsider and Smith Williams David's mechanical cane unloaders were developed between 1885 and 1915, but Australian sugar mill-owners, even CSR, continued relying heavily on new imported milling technology such as hydraulic pressure attachments, revolving knives, shredders and Yaryan evaporators. Eventually, however, the Australian sugar milling sector after 1915 became more innovative, developing new process technologies and machinery such as vertical and horizontal pusher feeders, the Petree-Thomas process, CSR's continuous pressure feeder, Donnelly chutes, fractional liming with double heating and the SRI clarifier.

On the other hand, the Australian sugar industry was an innovator or near pioneer in a number of fields. The debate over the ability or otherwise of 'whites' to permanently settle the tropics and manage hard physical labour in the tropics was finally put to rest by the Australian sugar industry. By 1915, a production system comprising central sugar mills supplied by small family farms owned by Europeans relying on 'white' field and mill workers, mostly of Anglo-Celtic descent, characterised the Australian sugar industry. This system was unique amongst the world's cane sugar industries and showed that Europeans could live, work and thrive in the tropics. Such reliance on more expensive European labour, however, prompted Australian canegrowers to develop and adopt labour-saving cultivation machinery such as mechanical cane planters, mechanical weeders and rotary hoes, making them world leaders in this field until at least World War II. The Australian sugar industry was not entirely the pioneer in the adoption of mechanical loading and harvesting of sugar cane. However, it can be counted amongst the first wave of sugar-producing countries to widely mechanise this phase during the growing of crops of sugar cane. In addition, although the Australian sugar industry did not pioneer the bulk handling of raw sugar, it was amongst those initial countries to mechanise the transport of raw sugar during the mid-1950s, and has subsequently developed the largest and most advanced bulk sugar storage and handling facilities in the world.

Some important themes in the history of Australia as a nation have also been associated with the development of the Australian sugar industry. Before 1900, canegrowers in Queensland, especially in the northern sugar-producing districts, relied heavily on cheap, indentured labour, mostly Pacific Islanders. The implementation of the notorious White Australia policy purged the Australian sugar industry of its non-European workers and canegrowers. Protectionism versus free trade, another of the national debates in Australia since Federation, also involved the Australian sugar industry. In return for the Australian sugar industry agreeing to shed its reliance on coloured labour, successive Commonwealth governments protected its 'white' canegrowers from imports of cheaper overseas-produced sugar where the sugar mill-owners had access to cheap labour. However, protection for the Australian sugar industry was a contentious policy during the 1920s and 1930s. Over these two decades, Australian consumers of refined sugar imports argued unsuccessfully for the removal of the embargo on sugar imports so they could access cheaper sugar.

Peopling the new Australian state, especially its northern tropical regions, and defence against potential overseas threats, particularly from Asia, were a major concern of Commonwealth governments between 1901 and 1960. A thriving and expanding Australian sugar industry was seen by some politicians and sections of the Australian community as a vehicle for achieving this goal, without too much cost to the nation. Representatives of the Australian sugar industry also used this supposed 'role' as a justification for the continued protection and regulation of the industry, particularly during the 1920s and 1930s. However, deregulation and opening Australia's economy to international competition had emerged as a new force in Australian political thinking during the 1970s and 1980s, and the Australian sugar industry had to adjust to these new realities as it was partially deregulated, albeit reluctantly. By then the old shibboleths used by representatives of the Australian sugar industry no longer held much sway in Canberra.

In this book, CSR has assumed a prominent role. As an author, it has been easy in many instances to provide CSR's perspective about a range of issues and to outline the firm's strategies because so much of the organisation's earliest records have been preserved. Yet CSR's prominent role in this book has not come about due to the organisation's good record keeping practices. The firm was a major driving force in the development of an Australian sugar industry and the significant part it played cannot be overlooked. CSR has been responsible for about a quarter of the country's raw sugar output for much of the period since 1920. Under the guidance of Edward Knox and his son, Edward W. Knox, CSR led the way in the application of science to agricultural and milling problems in the Australian sugar industry until at least World War II. CSR bred new cane varieties, identified and controlled outbreaks of cane diseases, tried to control pests such as the beetle borer and cane grubs, promoted the benefits of good drainage and plant nutrition, and developed new milling procedures and machinery (e.g. conical subsiders). Despite the firm's attempts to deny its competitors knowledge of these advances, other Australian sugar milling companies often followed CSR's lead, recognising that the organisation was usually the country's pioneer in technological advances in sugar production. In addition, the firm had a virtual monopoly over the refining sector between 1850 and 1880 and 1900 and 1989, supplying the bulk of the refined sugar products in Australia. This near monopoly, however, led to CSR being reviled by parliamentarians, canegrowers, workers, unions, and domestic consumers. Such vilification was in this author's opinion not always justified. CSR's adoption of clever business practices, quality control measures (i.e. mill and plantation inspectorate) and industrial and agricultural research ensured that the firm produced a high quality product which was readily purchased by Australian consumers.

The environmental consequences of sugar production in eastern Australia have been examined in this book. Large-scale deforestation in many coastal districts of eastern Australia has been associated with the growing of sugar cane. Beyond the initial clearing of the native vegetation to create arable land, the production of sugar placed other demands on the native forests. The vegetation was cleared to provide timber for buildings and railway sleepers, to supply the firewood for the boiler furnaces in the sugar mills until at least 1950, and to protect sugar cane fields from infestation with cane grubs. Attempting to protect crops of sugar cane was responsible for the introduction into Queensland of two wildlife species that have now become notorious pests: the Indian myna bird (*Acridotheres tristis*) and the cane toad (*Bufo marinus*). Poor cultivation methods, an emphasis on greater output that forced the use of sloping and waterlogged land and the assignment system which made it difficult to rest eroding land or change where sugar cane was cultivated led to a soil erosion problem in parts of the sugar-producing lands in eastern Australia between 1930 and 1985. A promotion of better drainage led to the creation of many new channels where none previously existed, the straightening and deepening of existing watercourses, the loss of freshwater wetlands and the emergence of an acid sulphate soil threat, especially in Southern Queensland and northern New South Wales.

Finally, this book has highlighted that few individuals or successive governments of varying political persuasions showed much concern about the environmental damage the growing of sugar cane did to the coastal lands of eastern Australia during the late nineteenth and much of the twentieth centuries. Instead both governments and the organisations representing the Australian sugar industry focused upon the continued expansion of sugar cane growing and increased productivity through the breeding of new cane varieties, better water management, disease reduction, pest control and soil enhancement through the use of fertiliser and trace elements. Changing public opinion in the 1960s and 1970s, however, forced a reassessment of what sugar production had done to the soils, wildlife, drainage networks and native forests of eastern Australia and the in-shore reefs of the Great Barrier Reef. From the late 1980s onwards, the Australian sugar industry became more concerned about the environment, and commenced, albeit reluctantly initially, to reduce the impact cane growing had on the landscape. Moreover, during the late 1990s and early 2000s, some Australian canegrowers have sought to repair the environmental damage of their predecessors through tree planting schemes, especially replanting riparian vegetation, and establishing sediment traps and artificial wetlands to stop eroded materials reaching watercourses. Yet they have a long way to go before the environment resembles even a fraction of what the pre-European landscape looked like when those initial canegrowers commenced cutting down the forests on their newly selected blocks of land with the hope that the nearby sugar mill being erected could eventually produce a half decent product that would fill the sugar bowls on Australian kitchen tables.

Select Bibliography

I. Official Records, Manuscript

1. National Archives of Australia

A3280/P3850: Sugar Agreements between the Commonwealth and Millaquin Sugar Co. Ltd., 1915–1922

A10044/1916/5–8: Appeals by Colonial Sugar Refining Co. to High Court of Australia in the matter of Central Sugar Cane Prices Board Awards for Victoria, Goondi and Hambledon Mills, 1916

A1608/B23/2/3: Sugar Industry in Wartime – General Representations, 1942–1943

2. Public Record Office, Victoria

Defunct Trading Companies (Series 932/PO):

Alstonsville Sugar Co. Ltd., No. 808
Bloomfield River Sugar Co. Ltd., No. 741
Burdekin Delta Sugar Co. Ltd., No. 582
Crystal Brook Sugar Co. Ltd., No. 678
Hamleigh Sugar Co. Ltd., No. 575
Macknade Sugar Co. Ltd., No. 714
Mackay Sugar Co. Ltd., No. 572
Melbourne Mackay Sugar Co. Ltd., No. 753
Melbourne Mackay Queensland Sugar Co. Ltd., No. 581
Oakenden Sugar Co. Ltd., No. 737
Pleystowe Sugar Co. Ltd., No. 671
Pyramid Sugar Co. Ltd., No. 784

3. Queensland State Archives

a. Agriculture and Stock

AGS/1: Reports and correspondence relating to visit to British New Guinea, 1893 & 1895

AGS/5–7: Register of certificates of exemption granted to aliens under the *Sugar Cultivation Act* of 1913, 1913–1923
AGS/N92: Correspondence relating to establishment of Queensland Bureau of Sugar Experiment Stations, 1899–1903
AGS/N94: Correspondence concerning introduction and propagation of sugarcane and eradication of cane grubs, 1895–1898
AGS/N96: Correspondence relating to the *Sugar Cultivation Act* of 1913

b. Company's Office

A/21276, 3/1: Pimpama Sugar Co. Ltd.
A/21276, 4/1: Maryborough Sugar Co. Ltd.
A/21349, 92/2: Brisbane Refining Co. Ltd.
A/21277, 8/1: Bowen Sugar Co. Ltd.
A/21280, 35/1: Pioneer Floating Sugar Mill Co. Ltd.
A/21306, 57/1: Doughby Creek Sugar Co. Ltd.
A/21332, 207/1: Buderim Mt Sugar Co. Ltd.
A/21333, 216/1: Pandora Sugar Co. Ltd.
A/21355, 238/2: Urangan Sugar Co. Ltd.
A/21358, 178/2: Yeppoon Sugar Co. Ltd.
A/21502, 110/5: Yatala Sugar and Saccheurea Factory Co.

c. Co-ordinator General's Office

SRS 1158/1/8736: Water Quality Council of Queensland report on progress made by Queensland sugar mills towards meeting water quality conditions, 1982 (Item 577216)

d. Lands

LAN/213–217: Sugar selection files: blocks leased under the 66th Clause of the *Crown Lands Alienation Act* of 1868
LAN/AK 12: Correspondence concerning land to be opened for sugar cane development, 1894–1899
LAN/AK 97: Applications from selectors for sugar cane plants, 1881–1883

e. Premier's and Cabinet

SRS 1043/1/487, Box 155: Sugar Refining Agreements, 1938–1958 (Item 538260)

f. Queensland Cane Growers' Council

PRV 14721–1–74, Box 37: Marketing and Export of Sugar, 1941–1946
PRV 14721–1–75, Box 37: Sugar Refineries, 1946–1958
PRV 14721–1–76, Box 37: Refining, managing and selling charges, 1952–1968
PRV 14721–1–11: Overproduction General, 1941–1961
PRV 14721–1–42, Box 19, File 3P/8/2: Pollution: General (Item 896497)
RSI 14844–1–11: Commonwealth government sugar control accounts, 1915–1922 (Item 334649)
RSI 14844–1–12: Report of the Joint Committee of Public Accounts on Sugar, 1922 (Item 334650)

g. Treasury

A/20286–A/20288: Correspondence concerning Moreton, Mt Bauple and Mossman Central Mills, 1895–1910
A/20292: Correspondence concerning Nerang, North Eton and Plane Creek Central Mills, 1895–1910
A/20293: Correspondence concerning Pleystowe, Proserpine and Racecourse Central Mills, 1895–1910
A/26105: Indenture between Queensland Treasurer and Cuthbert Petree re: improvements in cane sugar manufacture, 1917 (Item 1101498)
SRS6232/7/1216: Correspondence about destruction of fish by pollution of rivers and streams by sugar mill wastes, 1960–1969 (Item 957275)

4. State Records Authority of New South Wales

a. Colonial Secretary

In-letters from Port Macquarie, 1822–24 (Series 4/1815)

b. Public Works Department

17/1254, File D39/1552: Moball and Crabbes Creek Drainage Trust, 1921–1939
17/1254, File D40/1995: Chinderah and Murwillumbah Drainage Trust, 1938–1940
17/1256, File D49/553: Mooball and Crabbes Swamp Drainage Union, 1921–1949

17/1265, File D56/376: Duranbah Swamp Drainage Union, 1956–1967
17/1265, File D59/394: Harwood Island No. 1 & No. 2 Basins Drainage Union, 1959–69
17/1266, File D60/410: Condong Drainage Union, 1941–1961
17/1276, File 19/276: Cudgen Drainage Union, 1909–1919
17/1279, File 23/539: Chinderah Drainage Trust 1909–1923
17/1281, File 25/158: Keith Hall Drainage Union, 1912–1925

II. Official Material, Printed

1. Commonwealth Government

a. Parliamentary Papers

Vol iv 1906. Report and evidence of the Royal Commission on Customs and Excise Tariffs
Vol iii 1912. Report of the Royal Commission on the Sugar Industry of Australia
Vol iv 1913. Evidence of the Royal Commission on the Sugar Industry of Australia
Vol iv 1920–21. Report of the Piddington Royal Commission on the Sugar Industry
Vol iii 1929–31. Report of the Sugar Industry Committee on Subjects on which the Committee reached unanimous conclusions (Gunn Inquiry)
Vol iii 1929–31. Minority Report of the Sugar Industry Committee (Gunn Inquiry)
Vol iii 1929–31. Majority Report of the Sugar Industry Committee (Gunn Inquiry)

b. Industries Assistance Commission (later Industry Commission)

Report No. 209 (1979): The Sugar Industry
Report No. 314 (1983): The Sugar Industry: interim report, 18 February 1983
Report No. 332 (1983): The Sugar Industry, 11 November 1983
Report No. 19 (1992): The Australian Sugar Industry

2. Queensland Government

a. *Votes and Proceedings of the Queensland Parliament,* 1860–1900; *Queensland Parliamentary Papers,* 1900–1950

i. Sections referred to annually

Reports of the Agent–General for Queensland, 1925–1950
Reports of the Auditor–General, selected years, 1888–1950
Reports of the Department of Agriculture & Stock, 1888 to 1930
Reports of the Department of Public Lands, 1870–1915 (Extracts from Annual Reports of Land Commissioners for various districts)
Reports of the Director of the Brisbane Botanic Gardens, 1863–1880
Reports of the Director of the Bureau of Sugar Experiment Stations, 1901–1935
Reports Upon Government Central Sugar Mills, 1905 to 1930

ii Royal Commissions and Inquiries

Report on the Establishment of Central Sugar Mills, Vol. 1, 1885
Report on the Central Sugar Mill System at Mackay, Vol. 3, 1888
Royal Commission to Enquire into the General Condition of the Sugar Industry, Vol 4, 1889
Report of the Inquiry into Central Sugar Mill System, Vol 3, 1890
Report by Dr Maxwell upon Central Sugar Mills, Vol. 1, 1904
Royal Commission into Number of Pacific Islanders to be Deported from Queensland at the end of 1906, Vol. 2, 1906
Royal Commission into the Advisability of Establishing more Central Mills, Vol. 2, 1911–1912
Report of the Board of Inquiry into the Sugar Industry, Vol. 2, 1916–1917
Report and Minutes of Evidence of the Royal Commission Appointed to Inquire into the Most Suitable Locations for Sugar Mills Which May be Erected in the Near Future, Vol. 2, 1923

b. Published Annual Reports

Air Pollution Council of Queensland, 1970–1987
Bureau of Sugar Experiment Stations, 1936–1995
Commissioner for Railways, 1945–1980
Irrigation and Water Supply Commission, 1950–1978

Queensland Sugar Board, 1978–1995
Ripple Creek Drainage Board, 1986–1990
Rural Lands Protection Boards, 1986–1995
Smithfield Drainage Board, 1987–1990
Water Quality Council of Queensland, 1972 –1987
Water Supply Commission, 1979–1992

c. Statistical registers

Statistical Register of Queensland, 1859–1868; later *Statistics of the Colony of Queensland,* 1869–1900 and *Statistics of the State of Queensland,* 1901–1945

Queensland Government Statistical Bulletins, various issues, 1935–1950

3. New South Wales Government

i. *Votes and Proceedings*

Report of the Parliamentary Inquiry into the Railway from Grafton to Tweed, Vol. 6, 1890

Petition from the canegrowers of Lismore and district concerning establishment of central mills, Vol.4, 1895

4. Victoria

i. *Victorian Parliamentary Papers*

Report and minutes of evidence of the Royal Commission on the Tariff, Vol. 4, 1883

Report and minutes of evidence of the Board of Inquiry into the Effect of the Fiscal System of Victoria upon Industry and Production, Vol. 2, 1894 & Vol. 2, 1895–1896

III. Manuscript and Business Records

1. Australian Capital Territory

a. Adolph Basser Library, Australian Academy of Science
David North Collection, MS 115.

b. National Library of Australia

Papers of the Australian Sugar Consumers' Association, 1932 (part of MS 1186, Papers of the Citizens' League of South Australia)

c. Noel Butlin Archives Centre, Australian National University

i. Australian Estates and Mortgage Co. Ltd. (and later companies).

158/189–191: Manager's Annual Reports, Kalamia Mill, 1929–1931
165/136–144: Mail Letterbooks, 1895–1922
165/193–197: Correspondence relating to the formation of Amalgamated Sugar Co. Ltd. and subsequent operation of Pleystowe Mill and Palms plantation, 1907–1940
165/218: Takeover by CSR Ltd.
165/305: Correspondence, annual reports and maps relating to Seaforth, Kalamia and Palms sugar plantations, 1897–1905

ii. Colonial Sugar Refining Company (CSR):

The CSR records are very extensive. Detailed references to the files used are fully provided in the footnotes. The main series of records consulted are listed below:

Deposit 142 – Head Office Sydney: Inward and Outward Correspondence

142/40–70	Melbourne In, 1892–1894
142/235–238	Melbourne Out, 1892–1894
142/815–827	Broadwater Mill Out, 1890–1900
142/872–885	Broadwater Mill In, 1902–1927
142/844–856	Broadwater Mill Out, 1918–1936
142/909	Hambledon Nursery, In 1904–1907
142/1010–1014	Hambledon Mill Out, 1899–1903
142/1254–1258	Homebush Mill In, 1904–1916
142/1404–1413	Childers Mill In, 1903–1916
142/1418–1436	Childers Mill Out, 1894–1915
142/1219–1223	Condong Mill In, 1919–1924
142/1171–1178	Condong Mill Out, 1915–1928
142/1230–1235	Homebush Mill Out, 1888–1895
142/1264–1271	Brisbane Branch In, 1903–1907
142/1325–1342	Brisbane Branch Out, 1892–1908

142/1418–1436	Childers Mill Out, 1894–1915
142/1455–1461	Goondi Mill Out, 1888–1897
142/1546–1568	Victoria Mill Out, 1883–1915
142/1708–1727	Harwood Mill Out, 1889–1915
142/1770–1778	Harwood Mill In, 1904–1915
142/1787	Harwood Mill In, 1893–1895
142/2321–2332	Superintendent of Clarence River Mills Out, 1870–1882
142/2339–2341	Queensland Out, 1881–1884
142/2345–2356	Parbury, Lamb & Co. (Brisbane agents) In, 1881–1892
142/2705–2711	Sugar Imports Out, 1863–1878
142/2764–2765	Australasian Sugar Co.,1847–1857
142/2853	Robey & Co.,1853–1855
142/2854–2897	Inspector of Mills Out, 1881–1910
142/2940	G. Kottman, Inspecting Chemist In, 1885–1888
142/3171	Mackay Out (W.H. Paxton, agent), 1880–1883
142/3356–3357	Parbury, Lamb & Co. (Brisbane agents) In,1876–1881
142/3505–3519	Agricultural and chemical circulars to millers, contractors and mill managers, 1873–1915
142/3527–3531	Annual Reports, 1887–1957
142/3593–3594	Price Lists at Millaquin & CSR Refineries

Deposit Z/109: Head Office Library

Boxes 175–177	Half Yearly Statements, 1855–1888
Boxes 181–183	Inspecting Chemist Letterbooks, 1884–1907
Box 195	Annual Mill Reports and Chemical summaries, various mills, 1902–03
Box 239–243	Reports of David North, CSR's plant pathologist, 1910–1958
Boxes 311–317	Raw Sugar Purchases, 1890–1915
Box 318	Mill Inspector Letterbook, 1897–1900
Box 370	Agricultural Circulars, 1953–1973
Boxes 371–373	Monthly Newsletters, 1945–1982
Box 374	Annual Reports of CSR's New South Wales Mills, 1870–1882
Box 375–77	Chemical reports on CSR's mills, 1884–1914
Boxes 380–386	Annual Reports on Raw Sugar Manufacture, 1934–1983
Boxes 387–391	Annual Reports on Cane Crops and Technical Fieldwork, 1916–1978

Boxes 396–403	Annual Reports on Refineries
Boxes 460–478	Annual Cane Inspector's reports, various mills, 1909–1974
Box 495	Records of land ownership and leases
Box 496	Experiments on subsoiling and mole drainage, Broadwater Experiment Station
Box 520	Report on Pioneer and Inkerman Mills, re: possible purchase, 1930–1931
Box 545	C. Venton, 'Report to New South Wales Canegrowers on the Feasibility of their purchase of NSW mills from CSR Ltd., September 1974'.

Deposit Z/303: Head Office Archives

Box 1	CSR Co. history
Box 3	Extracts from Board Minutes, 1855–1887
Boxes 13–15	Sugar refineries
Box 18	Scientific cultivation methods
Box 19	Manufacturing experiments
Box 21	Early milling technology
Box 28	Sugar growing Norfolk Island
Box 28A	New Guinea cane collecting expeditions, 1893–1977
Boxes 33–37	NSW sugar industry-history
Box 38	Keith Hall estate
Boxes 40–46	Qld sugar industry-history
Boxes 47–48	Victorian Sugar Co., 1857–1887
Box 49	Glanville and Cottesloe Refineries, 1880–1980
Box NS 239	D. North's reports on pests, diseases and cane varieties at CSR's mills, 1910–1915

Deposit N/126: Goondi Mill

Technical Field Officers' Letters, 1932–1968 (Boxes 129–132)

Deposit N/141: Macknade Mill

Cane Inspector's Annual Reports, 1907–1917 (Boxes 1–2).
Cane Inspector's Correspondence, OUT, 1907–1912 (Box5).

Deposit Z/395: Technical and Field Department, Macknade Mill

Nursery correspondence to Head Office, 1918–1978 (Boxes 1–10)
Nursery Correspondence, Hambledon Mill, to Head Office, 1918–1979 (Boxes 18–21)
Macknade Nursery Reports, 1909–1914 (Boxes 43–44)
Pathology Reports, 1955–1962 (Box 57)
Head Office Letters and Circulars, 1897–1907 (Boxes 75–78).

2. New South Wales

a. Mitchell Library (Sydney)

Mss 3565/32: Australian Sugar Co. Pty. Ltd., Minute Books, Meetings of Directors, 1914–1960 (Mourilyan Mill)
Mss 98/20: Edward Knox's Business Correspondence, 1884–1889
Millaquin Sugar Co. Ltd., Annual Reports, 1964–1974
Millaquin Sugar Co. Ltd., Annual address of Chairman of Board of Directors, 1961–64 & 1971–74
Mulgrave Central Mill Co., Annual Reports, 1924–1977

b. Richmond River Historical Society (Lismore)

William Clement (of Alpha Sugar Mill) papers, Boxes 148–154 & 691.
Extracts from the Diaries of George Towner, 1856–1920
Charles Jarrett, The late William Smith, 4 pp. typescript, no date.
W. Meston, Notes about the Meston family, the sugar industry and about Woodburn after 1876, 8 pp. typescript, no date.

c. Richmond River Branch of the New South Wales Cane Growers' Association (Wardell).

Annual Reports of the Richmond River Branch of the New South Wales Cane Growers' Association, 1948 to 1995/1996
Annual Reports of the New South Wales Sugar Milling Co-operative Ltd., 1978/1979–1995/1996
Minute Books, Meetings of Directors, New South Wales Cane Growers' Association, 1927–2000
Minute Books, Meetings of Directors, Richmond River Branch of the New South Wales Cane Growers' Association, 1990–1996

d. Clarence River Branch of the New South Wales Cane Growers Association (Maclean)

Annual Reports of the Clarence River Branch of the New South Wales Cane Growers' Association, 1972/1973–1995/1996

Minute Books, Meetings of Directors, Clarence River Branch of the New South Wales Cane Growers' Association, 1925–1965

e. Tweed River Branch of the New South Wales Cane Growers' Association (Murwillumbah)

Annual Reports of the Tweed River Branch of the New South Wales Cane Growers' Association, 1964/1965–1995/1996

Minute Books, Meetings of Directors, Tweed River Branch of the New South Wales Cane Growers' Association, 1924–1996

3. Queensland

a. Cattle Creek Sugar Co. Ltd.

Cane delivery journal, 1906–1915

Register of crop liens and mortgages, 1905–1941

Minute Books, Meetings of Directors, 2 vols., 1905–1920

b. Innisfail Historical Society

Mourilyan Syndicate Ltd., Minute Books, Meetings of Directors, 2 vols., 1907–1914

c. James Cook University Library: Special Collection Archives

i. Charles Young papers (founder of Kalamia Mill)

CY/LB/1: Charles Young's Letterbook of correspondence to Union Mortgage and Agency Co., Melb., 1897–1899

CY/LB/2: Private Letterbook of Charles Young, 1893–1900

ii. Pioneer Sugar Mills (Pty.) Ltd. Records

PMR/1AL/1–PMR/1AL/7: Ledger journals, 1899–1930

PMR/[CT/1] Boxes 7–9: Correspondence and transactions, 1883–193

PMR/JA/1–JA/5: Pioneer Estate Account journals, 1894–1911

PMR/LB/25–LB/48: Outward correspondence, 1897–1928

PMR/[M/P] Box, Folio 1: Maps and plans, 1888–1905

iii. John Oxley Library

Arthur Neame, Diary, 1870–1897 (edited by Sidney May), 164 pp. typescript

iv. Kalamia Mill, via Ayr

Ledger books, 1905–1909
Account journal, 1897–1904
Manager's Letterbook, 1897–1907
Manufacturer's crushing book, 1902–1908
Letterbooks, 2 vols., 1897–1923

v. Marian Central Mill, via Mackay

Minute Books, Meetings of Directors, 3 vols, 1898–1913

vi. Mossman Central Mill Co. Ltd., Mossman

Minute Books, Meetings of Shareholders, 3 vols., 1893–192
Minute Books, Meetings of Directors, 3 vols., 1897–1905
Financial Ledgers, 2 vols., 1898–1910
Outward Correspondence Letterbooks, 10 vols., 1895–1915.

vii. North Eton Cooperative Sugar Milling Association Ltd.

Minute Books, Meetings of Directors, 2 vols, 1897–1907

viii. Proserpine Cooperative Sugar Milling Association Ltd.

Cane credits register, 1905–1930

ix. Racecourse Central Sugar Co. Ltd.

Annual reports, 1894–1910
Annual Balance Sheets, 1901–1910.
Minute Books, Meetings of Directors, 6 vols., 1887–1906

4. Victoria

a. ANZ Group Archive, Melbourne

i. Bank of Australasia

A/8/60–A/8/74: Inward letters to Bank Secretary, London, from Superintendent, Melbourne, 1886–1893

A/30/13: Inward letters to Manager, Melbourne, from Bank Superintendent, 1885–1892
A/268/4: Inward letters to Bank Superintendent from local managers and branch inspectors, 1887–1891
A/300/1: Ashburton and Pyramid sugar plantations, 1888–1893

ii. Union Bank of Australia Ltd.

U/60/1–U/60/15: Inward letters to Standing Committee, London, from General Manager, Melbourne, 1895–1907
U/103/35–U/103/72: Inward letters, London, from Melbourne Office, 1889–1907
U/219: Half yearly reports of the New South Wales Division, 1898–1901 (Rous plantation)
U/220: Half yearly reports of the Queensland Division, 1899–1907 (Mourilyan and Rockholme plantations)

b. Melbourne University Archives, Melbourne

i. Australian Estates and Mortgage Co. Ltd.

London Office Books (outward correspondence), Vols. 1 (1896)–Vol. 28 (1927) (Kalamia, Seaforth and Palms Mills)
London Correspondence: General Manager's private and confidential letters to the Chairman of Board of Directors, London, Vol 1 (1907)–Vol. 5 (1917).

ii. Gibbs, Bright & Co, sugar brokers

MS 11044: F. Keating's Letter books (outward correspondence), Vol. 1, 1884–1887
MS 1110A: Private Letters of Vicary Gibbs, 1883–1884.

iii Swallow & Ariell

F. T. Derham Letterbooks (outward correspondence), Vol. 1, 1885–1890 (Hambledon plantation)

c. National Australia Bank Group Archives, Melbourne

i. Queensland National Bank

A/QNB/301–311: Minute Books, Meetings of Directors, 1873–1906
A/QNB/652: General Manager's Letterbook, 1881–1892

BR/QNB/156/121& 122: Sydney Branch, Inward Letters from General Manager, Brisbane, 1884–1898
C/QNB/326/115: Half Yearly Balance Books, Bundaberg Branch, 1883–1894
Half Yearly Balance Books, Beenleigh Branch, 1873–1887
Half Yearly Balance Books, Maryborough Branch, 1880–1888

ii. Commercial Banking Company of Sydney

Minute Books, Meetings of Directors, Minute Books, No. 8(1872)–No. 13 (1889); Maclean (NSW) Branch, Special Letterbook No. 1 (outward reports to Head Office), 1893–1894; Maryborough (Qld) Branch, Outward Letterbook (correspondence to Head Office, Sydney), Vols. 2 (1879)–Vol. 6 (1898)

5. Scotland

a. Business Records Centre, Glasgow University, Glasgow

i. A. & W. Smith & Co. Ltd.

UGD 118/1, 1/2/3–1/2/9 Order Books No. 6 (1866–1869)–No. 12 (1883–1886)
UGD 118/1, 1/2/26–1/2/39 Sub-Order books No. 26 (1872–1874)-No. 39 (1908–1909)

ii. Mirrlees Watson & Co.

UGD 118/2, 2/1/1–2/1/21 Job Books No. 20 (1899–1900)-No. 21 (1915–16)
UGD 118/2, 2/4/1–2/4/17 Order Books 1860–1898
UGD 118/2, 2/35 Boiler Order Book
UGD 118/2, 2/36 Engine Order Book
UGD 118/2, 2/37 Mills and Krajewski Crusher Order Book
UGD 118/2, 2/7/1– 2/7/7 Drawing Book No. 1 (1848–1878)–No. 7 (1910–1915)

iii. W. & A. McOnie

UGD 118/4, 4/1/1–4/1/4 Order Books 1861–1869
UGD 118/4, 4/2/1–4/2/65 Drawing Order Book No. 1 (1872)-No. 65 (1914–1916)

IV. Serials

a. Newspapers

These were examined in detail for the years indicated:

Australian Town and Country Journal, 1870–1885
Australian Tropiculturist, 1895–1898
Brisbane Courier, 1862–1885
Bundaberg Mail, 1892–1900
Bundaberg Star, 1875–1879
Cairns Post, 1884–1886, 1897–1915
Capricornian (Rockhampton), 1880–1895
Clarence and Richmond River Examiner, 1875–1900
Clarence River Advocate (Maclean), 1898–1905
Geraldton Advocate and Johnstone River Guardian, 1895–1896
Illustrated Sydney News, 1868–1873
Logan Witness, 1878–1893
Mackay Mercury, 1870–1885
Mackay Standard, 1878–1879; 1884–1895
Manning River News, 1867–1871
Maryborough Chronicle, 1880–1895
Nambour Chronicle, 1904–1908
North Queensland Herald, 1893–1915
Northern Star, 1877–1900
Northern Territory Times, 1880–1885
Port Denison Times, 1866–1870; 1880–1884
Queenslander, 1866–1900
Richmond River Express, 1878–1900
Rockhampton Morning Bulletin, 1880–1884
Sydney Gazette, 1801–1811, 1827–28
The Mirror (Geraldton), 1896
Tweed and Brunswick Advocate, 1888–1896
Tweed Herald and Brunswick Chronicle,1898–1901
The Weekly Herald and Queensland Agricultural, Mining and Pastoral Advertiser, 1864–1868

b. Journals and Proceedings

Agricultural Gazette of New South Wales, 1890–1940
Australian Canegrower, 1979–1996
Australian Cordial Maker, 1905–1907
Australian Sugar Journal, 1909–1970
Australian Brewers' Journal, 1882–1893
Cane Grower and Tropical Cultivator, 1915
International Sugar Journal, 1899–1915
Proceedings of the Australian Society of Sugar Cane Technologists, 1979–2000
Proceedings of the International Society of Sugar Cane Technologists, 1924–1986
Proceedings of the Queensland Society of Sugar Cane Technologists, 1929–1978
The Cane Growers' Quarterly Bulletin, 1933–1982
The Producers' Review, 1954–1975
The Queensland Agricultural Journal, 1897–1940
The Australian Sugar Planter, 1883
The Journal of Commerce (Melbourne), 1885–1895
The Planter and Farmer (Brisbane), 1882–1891
The Sugar Cane (Manchester), 1875–76; 1880–1897
The Sugar Journal and Tropical Cultivator (Mackay), 1892–1905

c. Other annual publications

Australian Sugar Year Book, 1941–1996
Bundaberg Sugar News, 1975–1995
Port Denison Times Almanac, 1868
Pugh's Almanac, 1865–1885
Sands and McDougall's Melbourne, suburban and country directory, 1870–1895
Slater's Queensland Almanac, 1868–1885
Willmett's Northern Queensland Almanac and Directory, 1876–1879; 1883; 1888

d. Bureau of Sugar Experiment Stations (Queensland) Publications

BSES Bulletin, No. 1 (1983)–No. 58 (1997)
Division of Entomology, *Bulletins*, No. 1 (1914)–No. 22 (1935)
Division of Pathology, *Bulletins* No. 1 (1923)–No. 4 (1933)

e. Other annual reports

Australian Sugar Milling Council Pty. Ltd., 1988–1996
Bundaberg Sugar Co. Ltd., 1974–1995
CSR Ltd., 1970–1999
Haughton Sugar Co. Ltd., 1967–1971
Howard Smith Ltd., 1974–1978
Mackay Sugar Cooperative Association Ltd., 1989–1996
Pioneer Sugar Mills Ltd., 1969–1986
Proprietary Sugar Millers' Association, 1974–1986

V. Books, Articles and Pamphlets

a. Works published prior to 1915

Agricultural Society of Southern Queensland. *Cultivation of Sugar Cane, Manufacture of Sugar; Diseases of Cane, Mixed Farming and Labor Question*. Brisbane: George Slater & Co., 1875.

Anon. *The Sugar Industry, Bundaberg District, Currajong Creek and North Kolan*. Bundaberg: Reprint of a Series of articles in the *Bundaberg Mail*, 1893.

Bell, Frederick. *Handbook of Practical Directions for Sugar Cane Planting, Sugar Making and the Distillation of Rum*. Sydney: J.J. Moore, 1866.

Black, Maurice H. *The Sugar Industry of Queensland: to the agricultural community of Great Britain*. London: Lake & Sison, 1894.

————. *The Sugar Industry*. London: n. pub., 1897.

Campbell, Walter. *Extracts from Reports on Certain Agricultural Districts of New South Wales*. Sydney: New South Wales Government Printer, 1888.

Clarke, Percy. *The 'New Chum' in Australia, or the Scenery, Life and Manners of Australians in Town and Country*, 2nd ed. London: J.S. Virtue, 1886.

Cobb, Nathan. *Plant Diseases and Their Remedies: Diseases of the Sugar-Cane*. Sydney: New South Wales Government Printer, 1893.

Craig, G. "The Sugar Industry", *Queensland Review*, Aug. 1885, pp. 54–64.

Colonial Sugar Refining Company. *Notes on Cane Cultivation in Queensland*. Sydney: Colonial Sugar Refining Company, 1914.

Daintree, Richard. *Analysis of Queensland Sugar and Wheat Soils*. London?: Queensland Government?, 1874.

de Keating, Thomy. Sugar *Cultivation*. Brisbane: *Brisbane Courier*, 1865.

Easterby, Harry. *The Cultivation of Cane Upon Old Lands*. Queensland Bureau of Sugar Experiment Stations Bulletin: General Series, No. 1, 1913.

Fletcher, Price. "The Sugar Industry of Queensland", in *Queensland: Its Resources and Institutions*. Ed. Price Fletcher, Brisbane: Queensland Government Printer, 1886, n.pag.

Giraud, A.A. *The Sugar Cane: Its Culture and Manufacture in Queensland.* Brisbane: Watson, Ferguson and Co., 1883.

Goodwin, Andrew. *The Sugar Industry of the Lower Burnett in 1882.* Bundaberg: n.pub., 1882.

Hinchcliffe, John. *The Sugar Cane in Queensland: its natural history, botanical structure etc.* Brisbane: George Slater, 1868.

Liversidge, Archibald. *Diseases in the Sugar Cane, Queensland.* Sydney: n. pub., 1876.

Lyne, Charles. *The Industries of New South Wales.* Sydney: New South Wales Government Printer, 1882.

Knox, Edward W. "On an Application of Chemical Control to a Manufacturing Business", *Proceedings of the Second Meeting of the Australasian Association for the Advancement of Science*, Melbourne, 1890, pp. 372–379.

Koebele, A. "Sugar-Cane Insects in New South Wales", *Insect Life*, 4 (1891–92): 385–389.

Mackay, Angus. (ed.) *The Sugar Cane in Australia. A Series of Essays Upon Cultivation and Manufacture.* Brisbane: George Slater & Co., 1870.

———. *The Semi-Tropical Agriculturalist and Colonists' Guide.* Brisbane: George Slater & Co., 1875.

———. *The Sugar Cane in Australia: practical details concerning cane, sorghum and beet cultivation and sugar making, with latest improvements in machinery.* Sydney: Town & Country Journal, 1883.

———. *The Australian Agriculturist.* 3rd edition. Sydney: George Robertson & Co., 1897.

Maxwell, Walter. *Report Upon An Investigation Into the Condition of the Sugar Industry in Queensland.* Brisbane: Government Printer, 1900.

Munro, Aeneas. *The Sugar Fields of Mackay. Mackay*: Hodges and Chataway, 1895.

Queensland Department of Agriculture and Stock. *Sugar Planting in Queensland.* Brisbane: Government Printer, 1893.

Queensland Government Intelligence and Tourist Bureau. *Queensland Sugar Industry.* Brisbane: Government Printer, 1913.

Roth, Henry L. *A Report on the Sugar Industry in Queensland.* Brisbane: Gordon and Gotch, 1880.

———. *The Sugar Industry in Queensland: The labour difficulties it has to contend with and an exposure of some of the fallacies which have been stated concerning it.* Mackay: n.p., 1883.

Shaw, Flora. *Letters from Queensland. London*: Macmillan & Co., 1893.

Strachan, J. W. "Sugar Cultivation in Queensland", *Transactions of the Philosophical Society of Queensland, 1859–1872* (5pp. reprint from *Queensland Daily Guardian,* 30 March 1867).

Tryon, Henry. *Gumming of Cane.* Brisbane: Department of Agriculture and Stock, 1895.

Articles and Books Published Since 1915

Anon. *Kalamia Sugar Mill: The First One Hundred Years 1884–1984.* Ayr: n.pub, 1984.

Anon. *The Mourilyan Story. 100th Anniversary of the Operations of Mourilyan Sugar Mill.* Innisfail?: Howard Smith Industries, 1984.

Alcorn, Berenis and Dunn, Robin. *Moreton Sugar Mill. Sweet Heart of Nambour.* Nambour: B. Alcorn and R. Dunn, 1997.

Allen, John. "Australian Sugar Technology – A Leader in the Development of Process Technologies", *Chemeca 83. Proceedings of the Eleventh Australian Conference on Chemical Engineering*, September 1983, pp. 1–7.

Allom, Lovell and Marquis-Kyle. *New Farm Sugar Refinery Conservation Study*. Brisbane: Allom, Lovell and Marquis-Kyle, 1990.

Anderson, Ben. *The History of Irrigation in the Bundaberg Region.* Brisbane: Strand Publishing, 1978.

Armstrong, J. and Verhoeven, G. *The Innisfail Tramway: the story of the history and development of the Geraldton Shire tramway and the Mourilyan tramway of North Queensland.* Brisbane: Australian Railway Historical Society, Queensland Division, 1973.

Armstrong, J. B. "Aspects of Japanese Immigration to Queensland before 1900", *Queensland Heritage,* 2, 9 (1973): 3–9.

Armstrong, J. "The Sugar Strike, 1911", in *The Big Strikes, Queensland 1889–1965.* Ed. Denis Murphy. Brisbane: University of Queensland Press, 1983, pp. 100–116.

Armstrong, J. "The Sugar Tramways of Northern New South Wales", *Bulletin Australian Railway Historical Society,* 27 (1976): 118–132.

————. "Millaquin Sugar Mill Tramways", *Bulletin Australian Railway Historical Society,* 36 (1985): 147–160.

Atherton, E.H. *Plane Creek Central Mill Company Limited, 1937–1974.* Mackay?: Pioneer Sugar Mills Ltd., 1980.

Australia. Sugar Inquiry Committee. *Report on the Australian Sugar Industry.* Canberra: Commonwealth of Australia, 1952.

Australia. Sugar Enquiry Committee. *Summary of Conclusions and Recommendations from Report of the Sugar Enquiry Committee of 1960.* Canberra: Commonwealth of Australia, 1961.

Australian Bureau of Agricultural and Resource Economics (ABARE). *The Australian Sugar Industry in the 1990s. Submission 91.5 to the Industry Commission.* Canberra: ABARE, 1991.

Australian Sugar Producers' Association Ltd. and Queensland Cane Growers' Council. *The Australian Sugar Industry.* Brisbane: Australian Sugar Producers' Association Ltd. and Queensland Cane Growers' Council, 1932.

————. *"Reciprocity" is the new catch-word in trade: inter-dependence should dissipate jealousies.* Brisbane: Australian Sugar Producers' Association Ltd. and Queensland Cane Growers' Council, c. 1934.

————. *Queensland Sugar Industry and Interstate Relationships.* Brisbane: Australian Sugar Producers' Association Ltd. and Queensland Cane Growers' Council, c. 1934.

Balazategui, Bianka V. *Gentlemen of the Flashing Blade.* Studies in North Queensland History No. 15. Townsville: Department of History & Politics, James Cook University of North Queensland, 1990.

Barnes, Arthur. *The Sugar Cane.* New York: John Wiley & Sons, 1974.

Birrell, W.K. *The Manning Valley.* Brisbane: Jacaranda, 1987.

Bindon, G. and Miller, D.P. "Sweetness and Light: Industrial Research in the Colonial Sugar Refining Company, 1855–1900", in *Australian Science in the Making.* Ed. Rod Home. Surry Hills, NSW: Cambridge University Press in association with the Australian Academy of Science, 1988, pp. 170–194.

Birch, A. "The Implementation of the White Australia Policy in the Queensland Sugar Industry 1901–1912", *Australian Journal of Politics and History*, 11 (1965): 198–210.

————. "The Origins of the Colonial Sugar Refining Company 1841–55", *Business Archives and History*, 5, 1 (1965): 21–31.

————. "The Organisation and Economies of Pacific Island Labour in the Australian Sugar Industry 1863–1906", *Business Archives and History*, 6 (1966): 53–76.

Blackburn, Frank. *Sugar-cane.* London: Longman, 1984.

Blume, Helmut. *Geography of Sugar Cane: environmental, structural and economical aspects of cane sugar production.* Berlin: Verlag Dr. Albert Bartens, 1985.

Boileau, Joanna. *Caldera to the Sea: a History of the Tweed Valley.* Murwillumbah: Tweed Shire Council, 2006.

Bolton, Geoffrey. *A Thousand Miles Away: A History of North Queensland to 1920.* Canberra: Australian National University Press, 1972.

Boston Consulting Group, *Main Report to the Sugar Industry Review Working Party. Analysis of Issues and Identification of Possible Options.* Sydney: Boston Consulting Group, 1996.

Briody, Patrick. *Sunshine Sugar. The History of the New South Wales Sugar Milling Cooperative Limited.* Brisbane: New South Wales Sugar Milling Cooperative Limited, 2003.

Browning, J. "Rous Sugar Mill Tramways", *Light Railways*, 22, No. 88 (1985): 3–9.

Bundaberg Sugar Company, *Why you should reject the inadequate takeover offer from Tate & Lyle.* Brisbane: Bundaberg Sugar Company Ltd., 1991.

Burrows, Geoff and Morton, Clive. *The Canecutters.* Melbourne: Melbourne University Press, 1986.

Burrows, Geoff and Shlomowitz, Ralph. "The lag in mechanisation of the sugar cane harvest: some comparative perspectives", *Agricultural History*, 66 (1992): 61–75.

Bureau of Sugar Experiment Stations, Queensland. *75 Years of Scientific Progress. Bureau of Sugar Experiment Stations 1900–1975.* Brisbane: Bureau of Sugar Experiment Stations, 1975.

Calcino, D. *Australian Sugarcane Nutrition Manual.* Brisbane: Bureau of Sugar Experiment Stations, 1995.

Capelin, Mike. *Moreton Mill Area – a sugar caneland suitability study.* Division of Land Utilisation Technical Bulletin No. 37. Brisbane: Queensland Department of Primary Industries, 1979.

Chenoweth & Associates. *Nature Conservation Issues in Sugar Cane Areas.* Brisbane: Queensland Department of Environment and Heritage, 1995.

Clark, John M. "Developments in the Queensland Sugar Industry", *Quarterly Review of Agricultural Economics,* 12, 1 (January 1959): 24–31.

Collinson, James. "The Origin and Growth of the Sugar Industry in the Cairns District", *Journal of the Royal Historical Society of Queensland,* 3 (1945): 260–4.

Commonwealth of Australia (Department of Commerce), *Report on the Market for Sugar in the United States.* Canberra?: Department of Commerce, 1939.

Connell, Peter and Borrell, Brent., *Costs and Regulation of Cane Harvesting Practices.* Canberra: Australian Government Publishing Service, 1987.

Connolly, Roy. *John Drysdale and the Burdekin.* Sydney: Ure Smith Pty. Ltd., 1964.

Conomos, Denis. *The Greeks in Queensland. A History from 1859–1945.* Brisbane: Copyright Publishing Co. Pty. Ltd., 2002.

Corkhill, Alan. *Queensland and Germany. Ethnic, Socio-Cultural Political and Trade Relations 1838–1991.* Melbourne: Academia Press, 1992.

Corris, Peter. "Pacific Island Labour Migrants in Queensland", *Journal of Pacific History,* 5 (1970): 43–64.

————. "White Australia in Action: the Repatriation of Pacific Islanders from Queensland", *Historical Studies,* 15 (1972): 237–50.

Corris, Peter. Passage, *Port and Plantation. A History of Solomon Islands Labour Migration 1870–1914.* Melbourne: Melbourne University Press, 1973.

Courtenay, P. Phillip. *Plantation Agriculture.* rev. ed., Boulder, Colorado: Westview Press, 1980.

Crellin, Ian. "Australian Sugar Tramways: the Challenge of the 1980s", *Light Railways,* No. 66 (October 1979): 3–32.

CSR Ltd. *1893–1993: spirit of New Farm: celebrating a hundred years at CSR's New Farm sugar refinery.* Brisbane: CSR Ltd., 1993.

————. *Bulk handling of Raw Sugar: An Outline of an Important Advance in the Australian Sugar Industry and an Explanation of the role of the CSR Co. in the complex teamwork which produced this advance.* Sydney: CSR Ltd., 1959.

————. *Clean Cane and the Work of Cane Inspectors.* Sydney: CSR Ltd., 1959.

————. *Glanville Refinery from Beginning to End: 100 Years of Sugar Refining in South Australia.* Adelaide: CSR Ltd., 1991.

————. *The Preparation of Land For Sugar Cane.* Sydney: CSR Ltd., 1958.

————. *Victoria Mill, 1953.* Sydney: CSR Ltd., 1953. 32 pp. booklet issued to mark the reconstruction of Victoria Mill.

CSR Ltd. Sugar Division. *Ord Sugar industry Feasibility Study. A Report to the Government of Western Australia.* Sydney: CSR Ltd., 1976.

Daley, Ben and Griggs, Peter. "Mining the Reefs and Cays: Coral, Guano and Rock Phosphate Extraction in the Great Barrier Reef, Australia, 1844–1940", *Environment and History,* 12, 4 (November 2006): 395–433.

Daley, Louise. *Men and a River. Richmond River District, 1828–1895.* Melbourne: Melbourne University Press, 1966.

De Lepervanche, Marie. *Indians in a White Australia.* Sydney: George Allen & Unwin, 1984.

Department of Labour and National Service, Australia. *Men and Machines in Sugar Cane Harvesting.* Employment and Technology Series No. 7. Melbourne: Department of Department of Labour and National Service, 1970.

Dignan, Don. *The Story of Kolan.* Brisbane: Smith and Paterson, 1964.

————. "Chiaffredo Venerano Fraire, 1852–1931", in *The Queensland Experience. The Life and Work of Fourteen remarkable Immigrants.* Ed. Maximillian Brandle. Brisbane: Phoenix Publications, 1991, pp. 49–75.

Donovan, Peter. *A Land Full of Possibilities. A History of South Australia's Northern Territory.* Brisbane: University of Queensland Press, 1981.

Douglass, William. *From Italy to Ingham. Italians in North Queensland.* Brisbane: University of Queensland Press, 1995.

Driscoll, E. "Recent Developments in the Queensland Sugar Industry", *Geography,* 51 (1966): 147–149.

Easterby, Harry. *Summary of Some Experiments conducted with Sugar Cane and of Soil and other Chemical Analyses from 1902 to 1923.* Queensland Bureau of Sugar Experiment Stations Bulletins: General Series, Bulletin No. 4, 1924.

————. *The Queensland Sugar Industry: an historical review.* Brisbane: Queensland Government Printer, c. 1933.

Edwards, Geoffrey. "Two Government Failures? A Tale of Sugar and Wool", *Review of Marketing and Agricultural Economics,* 61, 2 (August 1993): 97–112.

England, M. "William Dart, Sugar Planter of St. Lucia", *Journal of the Royal Historical Society of Queensland,* 19, 2 (2005): 628– 643.

Farnfield, Jean. "Cotton and the Search for an Agricultural Staple in Early Queensland", *Queensland Heritage,* 2, 4 (1971): 20–25.

Farquhar, Ian. *Howard Smith shipping: enterprise and diversity 1854–2001.* Melbourne: Nautical Association of Australia, 2002.

Fitzpatrick, E., Watson, D. and Soper, F. *Report of the Sugar Industry Working Party, 5 June 1990.* Brisbane: Queensland Department of Primary Industries, 1990.

Foley, Patrick, Fyffe, Ronald and Gritching, Wolfgang. *Sugar People Take a Caning: a Report on the Impact of the Decline of the Sugar Industry on families in the Mulgrave Shire, North Queensland.* Townsville: Department of Behavioural Sciences, James Cook University of North Queensland, 1986.

Food and Agriculture Organisation. *The World Sugar Economy in Figures 1880–1959.* Rome: Food and Agriculture Organisation, 1960?

Galassi, Fernando. *Sotto La Croce del Sud (Under the Southern Cross). The Jumma Immigrants of 1891.* Studies in North Queensland History No. 16, Townsville: Department of History and Politics, James Cook University of North Queensland, 1991.

Galassi, Fernando. "From the Land of the Rising Sun to the Sugar-Growing Districts of North Queensland; Japanese Immigration 1892–1901", *Journal of the Royal Historical Society of Queensland,* 17 (2001): 427–432.

Gallagher, J. *Summary of Sugar Cane Research, 1950–1985.* Perth: Western Australia Department of Agriculture, 1986.

Galloway, Jock. *The Sugar Cane Industry. An Historical Geography from its Origins to 1914.* Cambridge: Cambridge University Press, 1989.

Gibbs, Harry. *Report of the Committee of Inquiry into Matters Concerning the Expansion of the Australian Sugar Industry.* Brisbane: Queensland Government Printer, 1963.

Gistitin, Carol. *Quite a Colony: South Sea Islanders in Central Queensland,* 1887–1993. Brisbane: Aebis Publishing, 1995.

Graves, Adrian. "The Abolition of the Queensland Labour Trade: Politics or Profit", in *Essays in the Political Economy of Australian Capitalism.* Eds. Edward Wheelwright and Ken Buckley. Sydney: Australian and New Zealand Book Company, 1975, Vol. 4, pp. 41–57.

————. "Crisis and Change in the Queensland Sugar Industry, 1862 – 1906", in *Crisis and Change in the International Sugar Economy 1860–1914.* Eds. Bill Albert and Adrian Graves. Norwich: ISC Press, 1984, pp. 261–279.

————. "Crisis and Change in the Australian Sugar Industry, 1914–1939", in *The World Sugar Economy in War and Depression 1914 – 1940.* Eds. Bill Albert and Adrian Graves. London: Routledge, 1988, pp. 142–156.

————. *Cane and Labour. The Political Economy of the Queensland Sugar Industry.* Edinburgh: Edinburgh University Press 1993.

Griggs, Peter. "Rust Disease Outbreaks and their Impact on the Queensland Sugar Industry, 1870–1880", *Agricultural History,* 69, 3 (1995): 413–437.

————. "The Origins and Early Development of the Small Cane Farming System in Queensland, 1870–1915", *Journal of Historical Geography,* 23, 1 (1997): 46–61.

————. "Alien Agriculturalists: non-European Small Farmers in the Australian Sugar Industry, 1880–1920", in *White and Deadly: Sugar and Colonialism.* Eds Pal Ahluwalia, Bill Ashcroft and Roger Knight. New York: Nova Science Publishers Inc., 1999, pp. 135–156.

————. "Sugar Demand and Consumption in Colonial Australia", in *Food, Power and Community. Essays in the History of Food and Drink.* Ed. Robert Dare. Adelaide: Wakefield Press, 1999, pp. 74–90.

————. "The Decline of Competition: The Emergence of a Duopoloy in the Australian Sugar Refining Sector 1841–1915", *Australian Geographer,* 32, 3 (2001): 359–376.

————. "Sugar Plantations in Queensland 1864–1912: Origins, Characteristics, Distribution and Decline", *Agricultural History,* 74, 3 (2000): 587–608.

————. "Improving Agricultural Practices. Science and the Australian Sugar Canegrower, 1864–1915", *Agricultural History* 78, 1 (2004): 1–33.

————. "Entomology in the Service of the State: Queensland Scientists and the Campaign against the Cane Grub, 1895–1950", *Historical Records of Australian Science,* 16, 1 (2005): 1–29.

————. "Soil Erosion, Scientists and the Development of Conservation Tillage Techniques in the Queensland Sugar Industry, 1935–1995", *Environment and History,* 12, 3 (2006): 233–268.

————. "A Natural Part of Life: The Australian Sugar Industry's Campaign to Reverse Declining Australian Sugar Consumption", *Journal of Australian Studies,* No. 87 (2006): 141–154.

————. "Defeating Cane Diseases: Plant Pathologists and the Development of Disease Control Strategies in the Australian Sugar Industry, 1920–1950", *Historical Records of Australian Science,* 18, 1 (2007): 43–73.

————. "Deforestation and Sugar Cane Growing in Eastern Australia", *Environment and History,* 13, 3 (2007): 255–283.

Gregory, Helen. *Making Maroochy: a history of the land, the people and the Shire.* Brisbane: Boolarong with the Maroochy Shire, 1991.

Gutteridge Haskins and Davey Pty. Ltd. *Environmental Audit of the Queensland Canegrowing Industry.* Report to CANEGROWERS. Brisbane: Gutteridge Haskins and Davey Pty. Ltd., 1996.

Higman, Barry. "Sugar Plantations and Yeoman Farming in New South Wales", *Annals of the Association of American Geographers,* 58 (1968): 697–719.

Hillock, Ian M. *Broken Dreams and Broken Promise. The Cane Conspiracy. Plantation Agriculture in the Northern Territory 1878–1889.* Darwin: Northern Territory University Press, 2000.

Hogarth, Mac and Allsopp, Peter (eds.). *Manual of Cane Growing.* Brisbane: Queensland Bureau of Sugar Experiment Stations, 2000.

Holz, G. *Rocky Point – a sugar caneland suitability study.* Division of Land Utilisation Technical Bulletin No. 38. Brisbane: Queensland Department of Primary Industries, 1979.

Hoyle, Brian. *The Spatial Structure of the Australian Sugar Industry.* Department of Geography Discussion Paper No. 22, Southampton: University of Southampton, 1983.

Hudson, Alan. *Sweet Success: A Story of South Johnstone Mill.* Brisbane: Christopher Beck Books, 1995.

————. *By the Banyan: Tully Sugar, the first 75 years.* Brisbane: Christopher Beck Books, 2000.

Hughes, Cecil G., "The Production and Testing of Sugar-cane Seedlings", *Queensland Journal of Agricultural Science,* 6, 1 (March 1949): 1–60.

Hungerford, Linda. "A Short History of Sugar Cane Farming in the Bundaberg Region", *Central Queensland Journal of Regional Development,* 3, 1 (June 1994): 48–52.

Hungerford, Linda. "Australian Sugar in the Global Economy: Recent Trends, Emerging Problems", in *Globalisation and Agri-Food Restructuring. Perspectives from the Australasia Region.* Eds. David Burch, Roy Dickinson and Geoffrey Lawrence. Aldershot, UK: Avebury, 1996, pp. 127–136.

Hunt, Doug. "Exclusivism and Unionism: Europeans in the Queensland Sugar Industry 1900–1910", in *Who are our enemies?: Racism and the Australian Working Class.* Eds. Ann Curthoys and Andrew Markus. Sydney: Hale and Iremonger, 1978, pp. 80–95.

Isis Land Use Committee. *Report on a Land Use Study of the Isis District.* Brisbane: Queensland Department of Primary Industries, 1971.

Jeans, Denis. *An Historical Geography of New South Wales to 1901.* Sydney: Reed Education, 1972.

Jones, Dorothy. *Cardwell Shire Story.* Brisbane: Cardwell Shire Council, 1961.

————. *Hurricane Lamps and Blue Umbrellas: a History of the Shire of Johnstone to 1973.* Cairns: Bolton Printers, 1973.

————. *Trinity Phoenix: a History of Cairns and District.* Cairns: Cairns and District Centenary Committee, 1976.

Johnson, Andrew, McDonald, Geoffrey, Shrubsole, Dan and Walker, Dan. "Natural Resource Use and Management in the Australian Sugar Industry: Current Practice and Opportunities for Improved Policy, Planning and Management", *Australian Journal of Environmental Management,* 5 (1998): 97–108.

Jowett, Des and Weickhardt, Ian. *Return to Rosstown. Railways, Land Sales and Sugar Beet Ventures in Caulfield.* Mordiallac, Victoria: Rosstown Historical Research Group, 1978.

Kane, R. 'Drainage Development – Fairymead Plantation', in *Conference on Agricultural Engineering.* Ed. Institution of Engineers, Australia. National Committee on Agricultural Engineering. Canberra: Institution of Engineers, Australia, 1984, pp. 4–8.

Keating, Brian and Wilson, J. (eds.). *Intensive Sugarcane Production: Meeting the Challenges Beyond 2000.* Wallingford, UK: CAB International, 1997.

Kennedy, Kett. "The South Johnstone Strike" in *The Big Strikes. Queensland 1889–1965.* Ed. Denis Murphy. Brisbane: University of Queensland Press, 1983, pp. 174–185.

Kerr, Bill. *Southern Cane Railway.* Brisbane: CANEGROWERS Plane Creek Area, 1995.

Kerr, Bill and Blyth, Ken. *They're All Half Crazy. 100 Years of Mechanical Harvesting.* Brisbane: Canegrowers, 1993.

Kerr, H. William and Bell, Arthur. *The Queensland Agricultural and Pastoral Handbook.* Vol. 4: Sugar Cane and Its Culture. Brisbane: Queensland Government, 1939.

————. *The Queensland Cane Growers' Handbook.* Brisbane: Bureau of Sugar Experiment Stations, 1939.

Kerr, John. *Northern Outpost.* Mossman: Mossman Central Mill Co. Ltd., 1979.

————. *Pioneer Pageant: A History of the Pioneer Shire.* Mackay: Pioneer Shire Council 1980.

————. *Southern Sugar Saga: A History of the Sugar Industry in the Bundaberg District.* Bundaberg: Bundaberg Sugar Co. Ltd., 1983.

————. *Sugar at Maryborough: 120 years of Challenge.* Maryborough: The Maryborough Sugar Factory Ltd., 1987.

————. *A Century of Sugar. Racecourse Sugar Mill, Mackay.* Mackay: Mackay Sugar Co-operative Association Ltd., 1988.

————. *Top Mill in the Valley. Cattle Creek Sugar Mill, Finch Hatton, 1906–1990.* Brisbane: Boolarong Publications and the Mackay Sugar Co-operative Association Ltd., 1991.

————. *Only Room For One. A History of Sugar in the Isis District.* Childers: Isis Central Sugar Mill Co. Ltd., 1996.

————. *A Crystal Century: The History of Proserpine's Sugar Mill, Its Farms and Its People.* Proserpine: Proserpine Co-operative Sugar Milling Association Ltd., 1997.

King, Norman. "Soils Investigations in the Sugar-Cane Producing Areas of Bundaberg", *The Queensland Journal of Agricultural Science,* 6, 3 (September 1949): 1–176.

King, Norman, Mungomery, Reginald and Hughes, Cecil G. (eds.). *Manual of Cane-Growing.* Melbourne: Angus and Robertson, 1953.

King, Norman, Mungomery, Reginald and Hughes, Cecil G. (eds.). *Manual of Cane-Growing.* rev. ed. Melbourne: Angus and Robertson, 1965.

Koivukangas, Olavi. *Sea, Gold and Sugarcane: attraction versus distance, Finns in Australia,* 1851–1947. Turku: Institute of Migration, 1986.

Lack, Clem. "From Mosquito Fleet to Bulk handling in the Sugar Industry", *Journal of the Royal Historical Society of Queensland,* 6, 3 (1960–61), pp. 617–634.

Landells, J.H. *A History of Plane Creek Mill, Sarina.* Mackay: n.pub., 1937.

Langford, Muriel and Thomis, Malcolm. *Rocky Point and the Heck Family: 100 years of sugar milling in South East Queensland.* Brisbane.: n.p., 1979.

League, Henry. *The Sugar Embargo: Is Queensland a Parasitic State?* Melbourne: Henry League, c. 1930.

Le Plan Community Research and Planning and Morton Consulting Services Pty. Ltd. *The New South Wales Sugar Industry. Planning for the Future.* New South Wales: New South Wales Sugar Milling Cooperative Limited, 1991.

Lewis, Glen. *A History of the Ports of Queensland. A Study in Economic Nationalism.* Brisbane: University of Queensland Press, 1973.

Longhurst, Robert. *Nerang Shire. A History to 1949.* Nerang: Albert Shire Council, 1994.

Lowndes, A. (ed.). *South Pacific Enterprise: The Colonial Sugar Refining Co. Ltd.* Sydney: Angus and Robertson, 1956.

Manning, Kenneth. *In their Own Hands: A North Queensland Sugar Story.* Mackay: Farleigh Co-operative Sugar Milling Association Ltd., 1983.

Mansfield, Alan. *Report of the Royal Commission Appointed to Investigate Certain Aspects of the Sugar Industry.* Brisbane: Queensland Government Printer, 1943.

Mansfield, Alan. *et al. Report of the Royal Commission Appointed to Prepare an Orderly Plan for the Development of the Sugar Industry.* Brisbane: Queensland Government Printer, 1951.

Manson, P. and Swindells, R. "Recent developments at Bundaberg Refinery", in *Conference papers: development in the Wide Bay Burnett/Central Engineering Conference.* Ed. The Institution of Engineers, Australia. Brisbane: The Institute of Engineers, Australia, 1993, pp. 2–19.

Maxwell, Francis. *Economic Aspects of Sugar Cane Production.* London: Norman Rodger, 1927.

May, Cathie. *Topsawyers: The Chinese in Cairns 1870 to 1920.* Studies in North Queensland History, No. 6. Townsville: History Department, James Cook University of North Queensland, 1984.

McDonald, Lorna. *Rockhampton: A History of City and District.* Brisbane: University of Queensland Press, 1981.

McDougall, W. "An Investigation of the Rat Pest Problem in Queensland cane fields: 1. Economic Aspects", *Queensland Journal of Agricultural Science,* 1, 2 (June 1944): 31–47.

McGuire, Peter. *Irrigation of Sugarcane.* Brisbane: Queensland Government Printer, 1991.

Mercer, Patricia and Moore, Clive. "Australia's Pacific Islanders 1906–1972", *Journal of Pacific History,* 13 (1978): 90–101.

Menghetti, Diane. "The Weil's Disease Strike, 1935", in *The Big Strikes. Queensland 1889–1935.* Ed. Denis Murphy. Brisbane: University of Queensland Press, 1983, pp. 202–216.

Mercer, Patricia. *White Australia Defied: Pacific Islander Settlement in North Queensland.* Studies in North Queensland History No. 21. Townsville: History and Politics Department, James Cook University of North Queensland, 1995.

Moore, Clive. *Kanaka. A History of Melanesian Mackay.* Port Moresby: Institute of Papua New Guinea Studies and the University of Papua New Guinea, 1985.

Morton, Clive. *By Strong Arms.* Gordonvale: The Mulgrave Central Mill Co. Ltd., 1995.

Moynagh, Michael. "The Negotiation of the Commonwealth Sugar Agreement 1949–1951", *Journal of Commonwealth and Comparative Politics,* 15, 2 (July 1977): 170–190.

Mulgrave Central Mill. *50 years in retrospect 1895–1945: a souvenir of the golden jubilee of the Mulgrave Central Mill.* Gordonvale: Mulgrave Central Mill Co. Ltd., 1947.

Munro, D. "Gilbert & Ellice Islanders on Queensland cane fields 1894–1899", *Journal of the Royal Historical Society of Queensland,* 14, 11 (1992): 449–465.

New South Wales Sugar Industry Study Group. New *South Wales Sugar Industry Study Group Report to the Minister for Agriculture and Fisheries.* Sydney: Department of Agriculture, North Coast Region, 1985.

North, David. *The Control of Cane Diseases.* Agricultural Technical Report No. 6. Sydney: CSR Co. Ltd., 1923.

North, David. *Leaf Scald, a bacterial disease of Sugar Cane.* Agricultural Technical Report No. 8. Sydney: CSR Co. Ltd., 1926.

North, David. *The Gumming Disease of the Sugar Cane: its Dissemination and Control.* Agricultural Technical Report No. 10. Sydney: CSR Co. Ltd., 1935.

Parsons, T.G. "Melbourne Money and Queensland Sugar. The Experience of Swallow and Ariell in the 1880s", *Victorian Historical Magazine,* 44 (1973): 32–39.

Pioneer Sugar Mills. *Pioneer Sugar Mills (Pty) Limited, 1884–1958.* Brandon, Qld: Pioneer Sugar Mills (Pty) Ltd., 1958.

Pioneer Sugar Mills. *Pioneer Sugar Mill – 100 years.* Brisbane: Pioneer Sugar Mills (Pty) Ltd., 1984.

Penrose, B. "Medical Experts and Occupational Illness: Weil's Disease in North Queensland, 1933–1936", *Labour History,* 75 (1998): 125–143.

Powell, E.G. *Elements of Raw Sugar Manufacture.* Sydney: CSR Ltd., 1955.

Price, C. with Baker, E. "Origins of Pacific Island Labourers in Queensland, 1863–1940: A Research Note", *Journal of Pacific History,* 11, 2 (1976): 106–121.

Prinsen Geerligs, Hendrik and Prinsen Geerligs, R. *Cane Sugar Production,* 1912–1937. London: Norman Rodger, 1938.

Queensland Sugar Corporation. *The Production and Marketing of Queensland Raw Sugar: Response to the Industry Commission's Draft Report on the Australian Sugar Industry.* Brisbane: Queensland Sugar Corporation, 1991.

Queensland Sugar Defence Committee. *The Truth About the Queensland Sugar Industry.* Melbourne: The Committee, 1930.

Reid, Jane. *The Long View. The History of the Sugar Research Institute 1949–1999.* Mackay: Sugar Research Institute, 1999.

Rein, Peter. *Cane Sugar Engineering.* Berlin: Verlag Dr. Albert Bartens KG, 2007.

Reynolds, Yvonne. "Sweet Surrender: Sugar Production at St Helena Penal Establishment", in *Brisbane: Moreton Bay Matters.* Ed. Murray Johnson. Brisbane History Group Papers No. 19. Brisbane: Brisbane History Group, 2002, pp. 19–26.

Ricaud, C., Egan, Brian, Gillaspie, A. and Hughes, C. Graham (eds.). *Diseases of sugar cane: major diseases.* Amsterdam: Elsevier, 1989.

Robertson, L., Allsopp, P., Chandler, K. and Mullins, R. "Integrated Management of Canegrubs in Australia: Current Situation and Future Research Directions", *Australian Journal of Agricultural Research,* 46 (1995): 1–16.

Robinson, Guy. "Deregulation and Restructuring in the Australian Sugar Industry", *Australian Geographical Studies,* 33, 2 (1995): 212–227.

Roderick, Colin. "T.A. Scott and his work at Port Macquarie", *Journal of the Royal Australian Historical Society,* 44 (1958): 23–30.

Rolleston, Frank. *The Defiance. The Story of North Eton Co-operative Sugar Milling Association Ltd, 1888–1987.* Mackay: North Eton Co-operative Sugar Milling Association, 1987.

Ryle, Peter. *The Sugar Lighters of Cairns and district.* Cairns: Peter Ryle, 2001.

Saranin, Alexander. "Sugar Cane Payment Systems", *Sugar Technology Reviews,* 3 (1975): 155–238.

Saunders, Kay. "Troublesome Servants: The Strategies of Resistance Employed by Melanesian labourers on Plantations in Colonial Queensland", *Journal of Pacific History,* 14 (1979): 168–83.

———. *Workers in Bondage: The Origins and Bases of Unfree Labour in Queensland 1824–1916.* Brisbane: University of Queensland Press, 1982.

———. "The Workers Paradox: Indentured Labour in the Queensland Sugar Industry to 1920", in *Indentured Labour in the British Empire.* Ed. Kay Saunders. Canberra: Croom Helm, 1984, pp. 213–259.

Savage, Russel *et al. Sugar Industry Working Party Report.* Brisbane: Sugar Industry Working Party, 1985.

See, J. and Crouch, H. "Mechanisation of Sugar Cane Harvesting". *Australian Department of Labour and National Service Personnel Bulletin,* 1963: 40–48.

Shlomowitz, Ralph. "Team Work and Incentives: The Origins and Development of the Butty Gang System in Queensland's Sugar Industry, 1891–1931", *Journal of Comparative Economics,* 3 (1979): 41–53.

———. "The Search for Institutional Equilibrium in Queensland's Sugar Industry, 1884–1913", *Australian Economic History Review,* 19 (1979): 91–122.

———. "Markets for Indentured and Time-Expired Melanesian Labour in Queensland, 1863–1906", *Journal of Pacific History,* 16 (1981): 70–91.

———. "The Profitability of Indentured Melanesian Labour in Queensland", *Australian Economic History Review,* 22 (1982): 49–67.

———. "Melanesian Labour and the Development of the Queensland Sugar Industry, 1863–1906", *Research in Economic History,* 7 (1982): 327–361.

————. "Time-Expired Melanesian Labour in Queensland: An Investigation of Job Turnover 1884–1906", *Pacific Studies,* 8, 2 (1985): 25–44.

Senate Standing Committee on Industry, Science and Technology. *Assistance for the Sugar Industry: Report.* Canberra: Australian Government Publishing Service, 1989.

Skerman, Percy., Fisher, A. and Lloyd, Peter. *Guiding Queensland Agriculture 1887–1987.* Brisbane: Queensland Department of Primary Industries, 1988.

Smith, Graham. *Sweet Beginnings: a history of sugar cane growing on the Richmond River of New South Wales.* Coraki, NSW: G. Smith, 1991.

Southgate, John. *The Commonwealth Sugar Agreement 1951–1974.* London: C.Czarnikow Ltd., 1984.

Spenceley, Geoff. 'Mrs Beet and Mr Cube: The Resolution of the International Sugar Crisis from an Australian Perspective' in *Competing for the Sugar Bowl.* Eds. Roger Munting and Tamas Szmrecsányi. St Katharinen, Germany: Scripta Mercaturae Verlag, 2000, pp. 105–126.

Stalley, D. 'The Sugar Industry' in *The Economics of Australian Industry.* Ed. Alex Hunter. Melbourne: Melbourne University Press, 1963.

Stephens, C. *The Nature and Incidence of Soil Erosion on the Sugar Cane Fields of Queensland.* Divisional Report No. 20 of 1945, Division of Soils, Council for Scientific and Industrial Research, 1945.

Sugar Consumers' Association. *Sugar Embargo: the Story of a Greedy Industry.* Melbourne: Sugar Consumers' Association, c. 1930.

Sugar Industry Review Working Party (Queensland). *Sugar-winning globally. Report of the Sugar Industry Review Working Party.* 2 vols. Brisbane: Sugar Industry Review Working Party, 1996.

Sugar Industry Taskforce. *Report of the Sugar Industry Taskforce.* Canberra: Department of Primary Industries and Energy, 1993.

Toft, Stan. *Report of Queensland Canegrowers' Council Delegation to USA, Cuba and Hawaii, November 1945–January 1946.* Brisbane: Queensland Canegrowers' Council, 1946.

Veitch, Robert. *The Cane Beetle Borer in Australia.* CSR Agricultural Circular No. 3. Sydney: CSR Co. Ltd., 1917.

Willis, Geoffrey. *The Harvesting and Transport of Sugar Cane in Australia.* Department of Geography Monograph Series, No. 3. Townsville: James Cook University of North Queensland, 1972.

Wilson, T., Wissemann, A. and Dwyer, G. *Report on the Soil Conservation Practices and Related Attitudes of Innisfail Canegrowers.* Project Report No. Q082015. Brisbane: Queensland Department of Primary Industries, 1982.

Wood, Andrew. "Management of crop residues following green cane harvesting of sugarcane in North Queensland", *Soil and Tillage Research,* 20, 1 (1991): 69–85.

Wood, Charles. "The Queensland sugar industry as depicted in the Whish and Davidson diaries", *Journal of the Royal Historical Society of Queensland,* 7, 3 (1964–65): 563–83.

Wood, Charles. *Sugar Country: A Short History of the Raw Sugar Industry of Australia 1864–1964.* Brisbane: Queensland Cane Growers' Council, 1965.

Unpublished Theses

Adamson, Shirley. The Queensland Sugar Industry 1860–1917. B.A. Hons., University of Queensland, 1953.

Anderson, Clyde. Technology Changes in the Australian Sugar Industry. M. Eng. University of Queensland, 1981.

Bardsley, Kenneth. The Maroochy Sugar Region: A Geographical Analysis. B.A. Hons., University of Queensland, 1959.

Balanzategui, Bianka. V. The Material Aspects of the Tropical North Queensland Sugar Cane Industry, 1872 to 1955. M.A. Material Culture Unit, James Cook University of North Queensland, 1994.

Brennan, Lisa. A Systems Analysis of Cane Harvesting and Transport in the Mackay Sugar Industry. PhD., University of Queensland, 1998.

Capelin, Mike. Planning Strategies to Resolve Land-use Conflicts between the Sugar Industry and Urban-related Development in Australia. M. of Urban and Regional Planning, University of Queensland, 1989.

Cavanagh, Jo-Anne. Organochlorine Insecticide Usage in the Sugar Industry of the Herbert and Burdekin River Regions: chemical, biological and risk assessments. PhD., James Cook University, 2000.

Davies, R.E. The Clarence River and the City of Grafton: Economic and Social Development. M.A., University of Sydney, 1953.

Davis, Charles. The Pyrmont Sugar Refinery with particular reference to Char. PhD., University of Sydney, 1984.

Formiatti, Dennis. Sugar Cane Barracks of Far North Queensland. B. Architecture, University of Queensland, 1984.

Gaylard, M.J. Economic Development in the Maroochy District until 1915. B.A. Hons., University of Queensland, 1967.

Griggs, Peter. Plantation to Small Farm: A Historical Geography of the Lower Burdekin Sugar Industry. PhD., University of Queensland, 1990.

Hayes, Lincoln. Pacific Islanders on Queensland Plantations: archaeological landscapes of power and survival in the 19th century. PhD., James Cook University, 2000.

Henderson, Lyn. Italians in the Hinchinbrook Shire, 1921–1939. B.A. Hons, James Cook University of North Queensland, 1978.

Higman, Barry. Yeomen and Plantation Landscapes in the Sugar Industry of New South Wales. B.A. Hons, University of Sydney, 1967.

Hungerford, Linda. Sugarcane farming in the Bundaberg district 1945 to 1985: major issues, problems and highlights from the growers' perspective. M.A. University of Queensland, 1991.

Kenny, J.W. The use of fine bone char in a continuous counter-current process for the decolorization of sugar liquors. PhD., University of Sydney, 1958.

Mason, Karen. Dr Walter Maxwell and the Queensland Sugar Industry 1899–1908. B.A. Hons, University of Queensland, 1987.

Mercer, Patricia. The Survival of a Pacific Islander Population in North Queensland 1900–1940. PhD., Australian National University, 1981.

Moore, Clive. Kanaka Maratta: A History of Melanesian Mackay. PhD., James Cook University of North Queensland., 1981.

Myers, David. Land-use Conflicts between the Sugar Industry and Urban Development and Rural Residential Development: a Case Study of Mulgrave Shire. B. Regional and Town Planning, University of Queensland, 1991.

Prove, Brian. A Study of the Hydrological and Erosional Processes Under Sugar Cane Culture on the Wet Tropical Coast of North Eastern Australia. PhD., James Cook University of North Queensland, 1991.

Saranin, Alexander. A Flocculation-flotation Process for Clarification of Sugar Refinery Melt using Phosphoric Acid and Lime. M. Eng., University of Queensland, 1963.

Shogren, Diana. The Politics and Administration of the Queensland Sugar Industry to 1930. PhD., University of Queensland, 1980.

Tanzer, John. An Investigation of New Settlement in the Sugar Industry as a Result of Post-War Expansions. A Case Study in the Herbert River District, North Queensland. B. Economics thesis, James Cook University of North Queensland, 1979.

Wegner, Janice. Hinchinbrook: The Hinchinbrook Shire Council, 1879–1979, M.A., James Cook University of North Queensland, 1984.

Index